Spon's Mechanical and Electrical Services Price Book

2012

Spon's Mechanical and Electrical Services Price Book

Edited by

Davis Langdon, An AECOM Company

Engineering Services

2012

Forty-third edition

Spon Press
an imprint of Taylor & Francis
LONDON AND NEW YORK

First edition 1968
Forty-third edition published 2012
by Spon Press
2 Park Square, Milton Park, Abingdon, Oxon, OX14 4RN

Simultaneously published in the USA and Canada
by Spon Press
711 Third Avenue, New York, NY 10017

Spon Press is an imprint of the Taylor & Francis Group, an informa business

British Library Cataloguing in Publication Data
A catalogue record for this book is available from the British Library

ISBN 13: 978-0-415-68066-0 (hardback)
ISBN 13: 978-0-203-15707-7 (ebook)

ISSN: 0305-4543

Typeset in Arial by Spon Press

MIX
Paper from
responsible sources
FSC
www.fsc.org   FSC® C004839

Printed and bound in Great Britain by
TJ International Ltd, Padstow, Cornwall

Contents

Preface

The Forty-third Edition of *SPON'S Mechanical and Electrical Services Price Book* continues to cover the widest range and depth of services, reflecting the many alternative systems and products that are commonly used in the industry as well as current industry trends.

The UK construction industry is being pulled in different directions. Outlook for many contractors is uncertain as the advent of a new financial year heralds the real start of the government's public sector cutbacks. The private sector is making more positive sounds but there is still a lag in bringing projects to market. At the same time there have been soaring input costs as global uncertainties and improving demand send commodity prices upwards.

These conflicting signals are resulting in contractors submitting prices that are even more difficult to predict than ever.

- Davis Langdon tender index shows that tender prices have fallen approximately 2.5% over the last 12 months
- Building costs pulled in different directions with labour rates frozen and materials prices rising. Building costs have been rising steadily since 2009 – an increase of 3.5% over the last year and projected to rise between 5% and 4.5% over the next 2 years
- Consumer price inflation is at twice government target and forecast to go higher before heading back towards target in 2012
- Manufacturing input costs at very high levels but output prices unable to match input cost increases
- Steel price increases push construction prices towards much higher levels of inflation
- More wage agreements frozen although builders await possible new settlement
- Construction earnings still falling but redundancies declining.

In the short term conditions will continue to be tough and it looks increasingly likely that a two speed building services contracting industry will emerge. Forecasts are that London and the South East will experience a small upturn towards the end of this year and into the next, but the remainder of the UK will lag behind.

The continuing strengths of commodity prices worldwide means that there is limited scope for further falls in prices in the year to come. With labour rates holding, and market competition remaining aggressive, pressures still remain on Contractors margins.

Before referring to prices or other information in the book, readers are advised to study the `Directions' which precede each section of the Material Costs/Measured Work Prices. As before, no allowance has been made in any of the sections for Value Added Tax.

The order of the book reflects the order of the estimating process, from broad outline costs through to detailed unit rate items.

The approximate estimating section has been thoroughly reviewed to provide up to date key data in terms of square metre rates, all-in-rates for key elements and selected specialist activities and elemental analyses on a comprehensive range of building types.

The prime purpose of the Materials Costs/Measured Work Prices part is to provide industry average prices for mechanical and electrical services, giving a reasonably accurate indication of their likely cost. Supplementary information is included which will enable readers to make adjustments to suit their own requirements. It cannot be emphasized too strongly that it is not intended that these prices should be used in the preparation of an actual tender without adjustment for the circumstances of the particular project in terms of productivity, locality, project size and current market conditions. Adjustments should be made to standard rates for time, location, local conditions, site constraints and any other factor likely to affect the costs of a specific scheme. Readers are referred to the build up of the gang rates, where allowances are included for supervision, labour related insurances, and where the percentage allowances for overhead, profit and preliminaries are defined.

Readers are reminded of the service available on the Spon's website detailing significant changes to the published information: www.pricebooks.co.uk/updates

As with previous editions the Editors invite the views of readers, critical or otherwise, which might usefully be considered when preparing future editions of this work.

Whilst every effort is made to ensure the accuracy of the information given in this publication, neither the Editors nor Publishers in any way accept liability for loss of any kind resulting from the use made by any person of such information.

In conclusion, the Editors record their appreciation of the indispensable assistance received from the many individuals and organizations in compiling this book.

DAVIS LANGDON, An AECOM Company
Engineering Services
MidCity Place
71 High Holborn
London WC1V 6QS

Telephone: 0207 061 7000
Facsimile: 0207 061 7061

Special Acknowledgements

comunica

Comunica plc
The Hallmarks
146 Field End Road
Eastcote
Pinner
Middlesex
HA5 1RJ
Tel: 020 8429 9696
Fax: 020 8429 4982
Email: enquiries@comunicaplc.co.uk
www.comunicaplc.co.uk

HOTCHKISS

Hampden Park Industrial Estate
Eastbourne
East Sussex
BN22 9AX
Tel : 01323 501234
Fax : 01323 508752
E-Mail : info@Hotchkiss.co.uk
www.Hotchkiss.co.uk

Abbey

Abbey Thermal Insulation Ltd
23-24 Riverside House
Lower Southend Road
Wickford
Essex
SS11 8BB
Tel: 01268 572116
Fax: 01268 572117
E-mail: general@abbeythermal.com

MITIE

The Counting House
1st Floor
53 Tooley Street
London
SE1 2QN
Tel: 020 7022 8400
Fax: 020 7022 8401
E-mail: info@mitie.co.uk

T.Clarke
BUILDING SERVICES GROUP

Stanhope House
116-118 Walworth Road
London
SE17 1JY
Tel: 020 7358 5000
E-mail: info@tclarke.co.uk
www.tclarke.co.uk

TROX®TECHNIK
The art of handling air

TROX UK LTD
Caxton Way
Thetford
Norfolk
IP24 3SQ
Grilles, Air Filters, FCU's, VAV's
Tel: 01842 754545
Fax: 01842 763051
www.troxuk.co.uk

SPON'S PRICEBOOKS 2012

Acknowledgements

The editors wish to record their appreciation of the assistance given by many individuals and organizations in the compilation of this edition.

Manufacturers, Distributors and Sub-Contractors who have contributed this year include:-

A C Plastics Industries Ltd
Armstrong Road
Daneshill East
Basingstoke RG24 8NU
Tel: 01256 329334
Fax: 01256 817862
www.acplastiques.com
GRP Water Storage Tanks

Actionair
Joseph Wilson Ind. Estate
South Street
Whitstable
Kent CT5 3DU
Tel: (01227) 276100
Fax: (01227) 264262
Email: sales@actionair.co.uk
www.actionair.co.uk
Dampers

Alfa Laval Limited
Unit 1, 6 Wellheads Road
Farburn Industrial Estate
Dyce
Aberdeen AB21 7HG
Tel: 01224 424300
Fax: 01224 725213
www.alfalaval.com
Heat Exchangers

Aquilar Limited
Dial Post Court
Horsham Road
Rusper
West Sussex RH12 4QX
Tel: 08707 940310
Fax: 08707 940320
www.aquilar.co.uk
Leak Detection

Balmoral Tanks
Balmoral Park
Loirston
Aberdeen AB12 3GY
Tel: 01224 859000
Fax: 01224 859123
www.balmoral-group.com
GRP Water Storage Tanks

Biddle Air Systems Ltd
St. Mary's Road, Nuneaton
Warwickshire CV11 5AU
Tel: +44 (0) 24 7638 4233
Fax: +44 (0) 24 7637 3621
Email: info@biddle-air.co.uk
Air Curtains

Braithwaite Engineers Ltd
Neptune Works
Uskway
Newport
South Wales NP9 2UY
Tel: 01633 262141
Fax: 01633 250631
www.braithwaite.co.uk
Sectional Steel Water Storage Tanks

Broadcrown Limited
Alliance Works
Airfield Industrial Estate
Hixon
Staffs ST18 0PF
Tel: 01889 272200
Fax: 01889 272220
www.broadcrown.co.uk
Generators

Caradon Stelrad Ideal Boilers
PO Box 103
National Avenue
Kingston-upon-Hall
North Humberside HU5 4JN
Tel: 08708 400030
Fax: 08708 400059
www.rycroft.com
Boilers/Heating Products

Carrier Air Conditioning
United Technologies House
Guildford Road
Leatherhead
Surrey KT22 9UT
Tel: 0870 6001100
Fax: 01372 220221
www.carrier.uk.com
Chilled Water Plant

Chloride Power Protection
Unit C, George Curl Way
Southampton SO18 2RY
Tel: 023 8061 0311
Fax: 023 8061 0852
www.chloridepower.com
Static UPS Systems

Cooper Lighting and Security
London Project Office
Suite 8, King Harold Court
Sun Street
Waltham Abbey
Essex EN9 1ER
Tel: 01302 303303
Fax: 01392 367155
www.cooper-ls.com
Emergency Lighting and Luminaires

Danfoss Flowmetering Ltd
Magflo House
Ebley Road
Stonehouse
Glos GL10 2LU
Tel: 01453 828891
Fax: 01453 853860
www.danfoss-randall.co.uk
Energy Meters

Dewey Waters Limited
Cox's Green
Wrington
Bristol BS40 5QS
Tel: 01934 862601
Fax: 01934 862604
www.deweywaters.co.uk
Tanks

Dunham-Bush Limited
8 Downley Road
Havant
Hampshire PO9 2JD
Tel: 02392 477700
Fax: 02392 450396
www.dunham-bush.com
Convectors and Heaters

EMS Radio Fire & Security Systems Limited
Technology House
Sea Street
Herne Bay
Kent CT6 8JZ
Tel: 01227 369570
Fax: 01227 369679
www.emsgroup.co.uk
Security

Engineering Appliances Ltd
Unit 11
Sunbury Cross Ind Est
Brooklands Close
Sunbury On Thames TW16 7DX
Tel: 01932 788888
Fax: 01932 761263
Email: info@engineering-appliances.co.uk
www.engineeringappliances.com
Expansion Joints, Air and Dirt Separators

Enviroplas Limited
Unit 2
Shepard Cross Street
Lancashire
Bolton
BL1 3DE
Tel: 01204 844744
Fax: 01204 841500
Email: sales@enviroplas.co.uk
Plastic Ductwork

FCS Ductwork Limited
3rd Floor
Thomas Telford House
1 Heron Quay
Canary Wharf
London E14 5JD
Tel: (020) 7987 7692
Fax: (020) 7537 5627
www.fcsgroup.co.uk
Fire Rated Ductwork

FKI Hawker Siddeley
Falcon Works
P O Box 7713
Meadow Lane
Loughborough
Leicestershire LE11 1ZF
Tel: 01495 331024
Fax: 01495 331019
www.fkiswitchgear.com
HV Supply, Cables and HV Switchgear and Transformers

Flakt Woods Limited
Axial Way
Colchester CO4 5ZD
Tel: 01206 222555
Fax: 01206 222777
Fans

Hall Fire Protection Limited
186 Moorside Road
Swinton
Manchester M27 9HA
Tel: 0161 793 4822
Fax: 0161 794 4950
www.hallfire.co.uk
Fire Protection Equipment

Halton
5 Waterside Business Park
Witham
Essex CM8 3YQ
Tel: 01376 503040
Fax: 01376 503060
www.haltongroup.com
Chilled Beams

Hattersley, Newman, Hender Ltd
Burscough Road
Ormskirk
Lancashire L39 2XG
Tel: 01695 577199
Fax: 01695 578775
Email: uksales@hattersley-valves.co.uk
www.hattersley.com
Valves

Hitec Power Protection Limited
Unit B21a
Holly Farm Business Park
Honiley
Kenilworth
Warwickshire CV8 1NP
Tel: 01926 484535
Fax: 01926 484336
www.hitecups.co.uk
Uninterruptible Power Supply (Rotary/Diesel)

Honeywell CS Limited
Honeywell House
Anchor Boulevard
Crossways Business Park
Dartford
Kent DA2 6QH
Tel: 01322 484800
Fax: 01322 484898
www.honeywell.com
Control Components

Hoval Limited
Northgate
Newark
Notts NG24 1JN
Tel: 01636 672711
Fax: 01636 673532
www.hoval.co.uk
Boilers

HRS Hevac Ltd
10-12 Caxton Way
Watford Business Park
Watford
Herts WD18 8JY
Tel: 01923 232335
Fax: 01923 230266
www.hrshevac.co.uk
Heat Exchangers

Hudevad
Bridge House
Bridge Street
Walton on Thames
Tel: 01932 247835
Fax: 01932 247694
www.hudevad.co.uk
Radiators

Hydrotec (UK) Limited
Hydrotec House
5 Mannor Courtyard
Hughenden Avenue
High Wycombe HP13 5RE
Tel: 01494 796040
Fax: 01494 796049
www.hydrotec.com
Chemical Treatment

IAC
IEC House
Moorside Road
Winchester
Hampshire SO23 7US
Tel: 01962 873000
Fax: 01962 873102
www.industrialacoustics.com
Attenuators

IC Service & Maintenance Ltd
Unit K3 Temple Court
Knights Place
Knight Road
Strood
Kent ME2 2LT
Tel: 01634 290300
Fax: 01634 290700
www.icservice.biz
Fire Detection & Alarm

Ideal Boilers
P O Box 103
National Avenue
Kingston Upon Hull
East Yorkshire HU5 4JN
Tel: 01482 492251
Fax: 01482 448858
www.idealboilers.com
Boilers

Kampmann
Benson Environmental Limited
47 Central Avenue
West Molesey
Surrey KT8 2QZ
Tel: (020) 8783 0033
Fax: (020) 8783 0140
www.diffusionenv.com
Trench Heating

Kiddie Fire Protection Services
Enterprise House
Jasmine Grove
London SE20 8JW
Tel: (020) 8659 7235
Fax: (020) 8659 7237
www.kfp.co.uk
Fire Protection Equipment

Metcraft Ltd
Harwood Industrial Estate
Littlehampton
West Sussex BN17 7BB
Tel: 01903 714226
Fax: 01903 723206
www.metcraft.co.uk
Oil Storage Tanks

Mitsubishi Electric Europe B.V.
Unit 8, Electra Park
Bidder Street
Canning Town
London E16 4ES
Tel (switchboard): 0207 511 5664
www.mitsubishi-lifts.co.uk
Lifts and Escalators

Osma Underfloor Heating
18 Apple Lane
Sowton Trade City
Exeter
Devon EX2 5GL
Tel: 01392 444122
Fax: 01392 444135
www.osmaufh.co.uk
Underfloor Heating

Pullen Pumps Limited
158 Beddington Lane
Croydon CR9 4PT
Tel: (020) 8684 9521
Fax: (020) 8689 8892
www.pullenpumps.co.uk
Pumps, Booster Sets

Reliance Hi-tech
Boundary House
Cricketfield Road
Uxbridge
Middlesex UB8 1QG
Tel: 01895 205000
Fax: 01895 205100
www.reliancesecurity.co.uk
Access Control and Security Detection and Alarm

Rycroft
Duncombe Road
Bradford BD8 9TB
Tel: 01274 490911
Fax: 01274 498580
www.rycroft.com
Storage Cylinders

Saint-Gobain PAM UK Ltd
Lows Lane
Stanton-By-Dale
Ilkeston
Derbyshire DE7 4RU
Tel: 0115930 5000
Fax: 0115932 9513
www.saint-gobain-pam.co.uk
Cast Iron Pipework

Sanber Ltd
3 Newnham Street
Astley Bridge
Bolton
BL1 8QA
Tel: 01204 596 015
Fax: 01204 598 751
Email: jane.holland@sanberlabservices.co.uk
www.sanberlabservices.co.uk
Plastic Ductwork

Schneider Electric Limited
120 New Cavendish Street
London W1W 6XX
Tel: 0870 608 8608
www.schneider-electric.com

SF Limited
Pottington Business Park
Barnstaple
Devon EX31 1LZ
Tel: 01271 326633
Fax: 01271 334303
Flues

Simmtronic Limited
Waterside
Charlton Mead Lane
Hoddesdon
Hertfordshire EN11 0QR
Tel: 01992 456869
Fax: 01992 445132
www.simmtronic.com
Lighting Controls

Socomec Limited
Knowl Piece
Wilbury Way
Hitchin
Hertfordshire SG4 0TY
Tel: 01462 440033
Fax: 01462 431143
www.socomec.com
Automatic Transfer Switches

Spirax-Sarco Ltd
Charlton House
Cheltenham
Gloucestershire GL53 8ER
Tel: 01242 521361
Fax: 01242 573342
www.spiraxsarco.com
Traps and Valves

Tyco Limited
Unit 6 West Point Enterprize Park
Clarence Avenue
Trafford Park
Manchester M17 1QS
Tel: 0161 875 0400
Fax: 0161 875 0491
www.tyco.com
Fire Protection

Utile Engineering Company Ltd
Irthlingborough
Northants NN9 5UG
Tel: 01933 650216
Fax: 01933 652738
www.utileengineering.com
Gas Boosters

Woods of Colchester
Tufnell Way
Colchester
Essex CO4 5AR
Tel: 01206 544122
Fax: 01206 574434
Air Distribution, Fans, Anti-vibration mountings

Davis Langdon,
An AECOM Company

Program, Cost, Consultancy
www.davislangdon.com
www.aecom.com

A leading light in our industry

SPON'S PRICEBOOKS 2012

Spon's Architects' and Builders' Price Book 2012

DAVIS LANGDON

The most detailed, professionally relevant source of UK construction price information currently available anywhere.

Major and minor works are now combined. New items include Omnidec concrete upper floors, Slimdek & Ribdeck upper floors, insulated breathable membranes, Bradstone EnviroMasonry products, ASSA Abloy ironmongery doorsets, and the Aquacell stormwater drainage attenuation system. More has been given for cavity closures and insulation, suspended ceilings, flooring Altro products, cladding, resins, Velfac windows, and Decra roof tiles. And it now presents cost models for a community centre, a car showroom, and a primary school extension.

Hbk & electronic package*:
880pp approx.: 978-0-415-68063-9: **£145**

electronic package only:
978-0-203-15719-0: **£145**
(inc. sales tax where appropriate)

Spon's Mechanical and Electrical Services Price Book 2012

DAVIS LANGDON ENGINEERING SERVICES

Still the only comprehensive and up to date annual services engineering price book available for the UK. This year's delivers a market update of labour rates and daywork rates, material costs and prices for measured works, and all-in-rates and elemental rates in the Approximate Estimating section. Engineering features that have been revised in line with changes to legislation and regs on Part L, CO_2 targets and renewables. Feed-In Tariffs have been overhauled and a new feature has been added for infrastructure.

Hbk & electronic package*:
800pp approx.: 978-0-415-68066-0: **£145**

electronic package only:
978-0-203-15707-7: **£145**
(inc. sales tax where appropriate)

Spon's External Works and Landscape Price Book 2012

DAVIS LANGDON

The only comprehensive source of information for detailed external works and landscape costs.

This year provides more approximate estimate composite costs items than before, and plenty of new and overhauled measured works items: swimming pools with accessories, Instant hedges, landscape design, garden design, living walls, mechanical ground clearance, demolitions, and timber decking.

Hbk & electronic package*:
600pp approx.: 978-0-415-68065-3: **£115**

electronic package only:
978-0-203-15709-1: **£115**
(inc. sales tax where appropriate)

Spon's Civil Engineering and Highway Works Price Book 2012

DAVIS LANGDON

In line with today's climate assumptions on overheads and profits and on preliminaries have been kept low and labour rates have been adjusted. Some materials such as steel products, structural sections and reinforcement show significant rises in price

Hbk & electronic package*:
752pp approx.: 978-0-415-68064-6: **£155**

electronic package only:
978-0-203-15716-9: **£155**
(inc. sales tax where appropriate)

---

*Receive our ebook free when you order any hard copy Spon 2012 Price Book, with free estimating software to help you produce tender documents, customise data, perform word searches and simple calculations. Or buy just the ebook, with free estimating software.

Visit **www.pricebooks.co.uk**

To Order: Tel: +44 (0) 1235 400524 **Fax:** +44 (0) 1235 400525

Post: Taylor and Francis Customer Services,
Bookpoint Ltd, 200 Milton Park, Abingdon, Oxon, OX14 4SB, UK

Email: book.orders@tandf.co.uk

A complete listing of all our books is on:
www.sponpress.com

Spon Press
an imprint of Taylor & Francis

Spon's Asia-Pacific Construction Costs Handbook

4th Edition

Edited by **Davis Langdon & Seah**

Spon's Asia Pacific Construction Costs Handbook includes construction cost data for 20 countries. This new edition has been extended to include Pakistan and Cambodia. Australia, UK and America are also included, to facilitate comparison with construction costs elsewhere. Information is presented for each country in the same way, as follows:

• key data on the main economic and construction indicators.

• an outline of the national construction industry, covering structure, tendering and contract procedures, materials cost data, regulations and standards

• labour and materials cost data

• Measured rates for a range of standard construction work items

• Approximate estimating costs per unit area for a range of building types

• price index data and exchange rate movements against £ sterling, $US and Japanese Yen

The book also includes a Comparative Data section to facilitate country-to-country comparisons. Figures from the national sections are grouped in tables according to national indicators, construction output, input costs and costs per square metre for factories, offices, warehouses, hospitals, schools, theatres, sports halls, hotels and housing.

This unique handbook will be an essential reference for all construction professionals involved in work outside their own country and for all developers or multinational companies assessing comparative development costs.

April 2010: 234x156: 480pp
Hb: 978-0-415-46565-6: **£120.00**

To Order: Tel: +44 (0) 1235 400524 **Fax:** +44 (0) 1235 400525
or Post: Taylor and Francis Customer Services,
Bookpoint Ltd, Unit T1, 200 Milton Park, Abingdon, Oxon, OX14 4TA UK
Email: book.orders@tandf.co.uk

For a complete listing of all our titles visit:
www.tandf.co.uk

Taylor & Francis
Taylor & Francis Group

PART 1

Engineering Features

This section on Engineering Features, deals with current issues and/or technical advancements within the industry. These shall be complimented by cost models and/or itemized prices for items that form part of such.

The intention is that the book shall develop to provide more than just a schedule of prices to assist the user in the preparation and evaluation of costs.

Understanding the CDM 2007 Regulations

2nd Edition

O. Griffiths et al.

Almost 3000 lives have been lost in the UK construction industry over the last twenty-five years, in addition to those seriously injured or made ill. The need to reduce this rate has required tight controls to be introduced throughout the planning and management of construction projects in the UK. The Construction (Design & Management) Regulations 2007 outline the responsibilities and liabilities for the various professionals and agents involved.

Straightforward and practical, *Understanding the CDM 2007 Regulations* demonstrates the rationale behind the regs, covers the duties of the five core duty holders (client, CDM coordinator, designer, principal contractor and contractor), explains the importance of the hazard management process on every project and also sets out the consequences of failing to successfully plan, design and manage for safety.

Any client, architect, engineer, CDM co-ordinator, project manager, construction professional, or student will find this a simple but thorough and dependable guide and should value the management toolkit and the numerous practical examples of best practice and guidance on how to use the Approved Code of Practice appropriately. This book shows how to unleash the potential of the regulations and add real value to the industry.

December 2010: 352pp
Pb: 978-0-415-55653-8: **£29.99**

To Order: Tel: +44 (0) 1235 400524 **Fax:** +44 (0) 1235 400525
or Post: Taylor and Francis Customer Services,
Bookpoint Ltd, Unit T1, 200 Milton Park, Abingdon, Oxon, OX14 4TA UK
Email: book.orders@tandf.co.uk

For a complete listing of all our titles visit:
www.tandf.co.uk

Renewable Energy Options

This article focuses on building-integrated options rather than large-scale utility solutions such as wind farms, which are addressed separately, and provides an analysis of where they may be best installed.

The legislative background, imperatives and incentives

In recognition of likely causes and effects of global climate change, the Kyoto protocol was signed by the UK and other nations in 1992, with a commitment to reduce the emission of greenhouse gases relative to 1990 as the base year.

The first phase of European Union Emission Trading Scheme (EU ETS) covers the power sector and high-energy users such as oil refineries, metal processing, mineral and paper pulp industries. From 1 January 2005, all such companies in all 25 EU member states must limit their CO_2 emissions to allocated levels in line with Kyoto. The EU ETS principle is that participating organizations can:

- Meet the targets by reducing their own emissions, or
- Exceed the targets and sell or bank their excess emission allowances, or
- Fail to meet the targets and buy emission allowances from other participants.

These targets can only be met by either energy efficiency measures or making using more renewable energy instead of that derived from fossil fuels.

In the UK, the Utilities Act (2000) requires power suppliers to provide some electricity from renewables, starting at 3% in 2003 and rising to15% by 2015. In a similar way to EU ETS, generating companies receive and can trade Renewables Obligation Certificates (ROCs) for the qualifying electricity that they generate. For small renewable generators, Renewable Energy Guarantee of Origin Certificates (REGOs) has been introduced, in units of 1 kWh.

The initial focus is on Carbon Dioxide (CO_2), and the goals set by the UK government are:

- 20% emission reduction by 2010 (and 10% of UK electricity from renewable sources)
- 60% emission reduction by 2050
- Real progress towards the 60% by 2020 (and 20% of UK electricity from renewable sources).

Four years after the introduction of ROCS, it was estimated that less than 3% of UK electricity was being generated from renewable sources. A 'step change' in policy was required, and the Office of the Deputy Prime Minister (ODPM) published 'Planning Policy Statement 22 (PPS 22): Renewable Energy' in order to promote renewable energy through the UK's regional and local planning authorities.

Local Planning & Building Regulations

More than 100 local authorities have already embraced PPS 22 and its Companion Guide published in December 2004, by adopting pro-renewables planning policies. Others are expected to follow; the typical requirement is for 10% of a site's electricity or heat to be derived from renewable sources, but at least one authority has already 'raised the bar' to 15%. The London Plan states that 'The Mayor will and boroughs should require major developments to show how the development would generate a proportion of the site's electricity or heat from renewables'.

Energy Performance Certificates (EPC)

The EU Directive on the Energy Performance of Buildings (EPBD) requires that Energy Performance Certificates must be prominently placed on all buildings open to the public and commercial buildings built, sold or let from January 2006, thus enabling prospective purchasers and tenants to be more aware of a building's energy performance.

Assessing the carbon emissions associated with the operation of buildings is now an important part of the overall early design process for planning approval. Methods are set out in:

- Part L of the UK Building Regulations: (Conservation of fuel & power)
- BREEAM – (Building Research Establishment Environmental Assessment Method)
- Standard Assessment Procedure (SAP) for Energy Rating.

On-site renewable energy sources are taken into account, but developers are not allowed to rely on any 'green tariff' as part of an assessment.

Technology options and applications

- **Wind Generators** – In a suitable location, wind energy can be an effective source of renewable power generation. Even without grant aid, an installed cost range of £2500 to £5000 per kW of generator capacity has been established over the past few years. The most common arrangement is a machine with three blades on a horizontal axis; all mounted on a tower or, increasingly for small generators in inner city areas, on top of a building. Average site wind speeds of 4 m/s can produce useful amounts of energy from a small generator up to say 3 kW, but larger generators require at least 7 m/s. A small increase in average site wind speed will produce a large increase in the output power. There is a need for inverters, synchronizing equipment and metering for a grid connection.

  Third party provision through an Energy Service Company (ESCo) can be successful for larger installations co-located with or close to the host building, especially in industrial settings where there may be less aesthetic or noise issues than inner city office or residential. The ESCo provides funding, installs and operates the plant and the client signs up for the renewable electrical energy at a fixed price for a period of time.

- **Building Integrated Photovoltaics (BIPV)** – Photovoltaic materials, commonly known as solar cells, generate direct current electrical power when exposed to light. Solar cells are constructed from semi-conducting materials that absorb solar radiation; electrons are displaced within the material, thus starting a flow of current through an external connected circuit. Conversion efficiency of solar energy to electrical power is improving with advances in technology and ranges from 7% to 18% under laboratory conditions. In practice, however, allowing for typical UK weather conditions, an installation of at least 7 m² of the latest high-efficiency hybrid modules is needed to produce 1000 watts peak (1 kWp), yielding perhaps 800 kWh in a year. Installed costs range from £300 to £450/m² for roof covering, and from £850 to £1300/m² for laminated glass.

- **Ground Source Heat Pumps** – The ground temperature remains substantially constant throughout the year and heat can be extracted by circulating a fluid (normally water) through a system of pipes and into a heat exchanger. An electrically driven heat pump is then used to raise the fluid temperature via the compression cycle, and hot water is delivered to the building load as if from a normal boiler (albeit at a somewhat lower temperature than a normal boiler).

  Most ground heat systems consist of a cluster of pipes inserted into vertical holes typically 50 to 100 metres deep depending on space and ground type. Costs for the drilling operation vary according to location, site access and ground conditions. A geological investigation may be needed minimize the risk of failure and to improve cost certainty.

  Such systems can achieve a Coefficient of Performance (COP = heat output /electrical energy input) of between 3 and 4, achieving good savings of energy compared with conventional fossil fuels. Installed costs are in the range £800 to £1200/kW depending on system size and complexity.

- **Borehole Cooling** – The constant ground temperature is well below ambient air temperature during the summer, so 'coolth' can be extracted and used to replace or, more likely commercially, to supplement conventional building cooling systems. Such borehole systems may be either 'open' – discharging ground water to river or sewer after passing it through a heat exchanger, or 'closed' – circulating a fluid (often water) through a heat exchanger and vertical pipes extending below the water table.

  Ground source heating and cooling systems are only 'partial renewable energy' because they rely on electrical power, mainly for pumping. Considerable carbon savings can be justifiably claimed, however, by avoiding the use of fossil fuel for heating and electrical power to drive conventional chillers. Indicative system costs are from £200 to £250/kW.

- **Solar Water Heating** – Simple flat-plate water-based collector panels have been used successfully on South-facing roofs over many years in the UK – especially by DIY enthusiasts prepared to devise their own simple control systems. The basic principle is to collect heat from the sun and circulate it to pre-heat space heating or domestic hot water, in either a separate tank or a twin coil hot water cylinder. Purpose-designed, evacuated tube collectors have been developed to increase performance and a typical $4\,m^2$ installed residential system has a cost range from £2500 to £4000 depending on pipe runs and complexity. Such a system could produce approximately 2000 kWh saving in energy use per year. Commercial systems are simply larger and slightly more complex but should achieve similar performance; low-density residential, retail and leisure developments with washrooms and showers may be suitable applications having adequate demand for hot water.

- **Biomass Boilers** – Wood chips or pellets derived from waste or farmed coppices or forests are available commercially and are considered carbon neutral, having absorbed carbon dioxide during growth. With a suitable fuel storage hopper and automatic screw drive and controls, biomass boilers can replace conventional boilers with little technical or aesthetic impact. They do, however, depend on a viable source of fuel, and there is a requirement for ash removal/disposal as well as periodic de-coking. In individual dwellings, space may be a problem because a biomass boiler does not integrate readily into a typical modern kitchen. Biomass boilers are available in a wide range of domestic and commercial sizes. For a large installation, they are more likely to form part of a modular system rather than to displace conventional boilers entirely. There is a cost premium for the biomass storage and feed system, and the cost of the fuel is currently comparable with other solid fuels. As an addition to a conventional system, installed costs could range from £200 to £250/kW.

- **Biomass Combined Heat & Power (CHP)** – Conventional CHP installations consist of either an internal combustion engine or a gas turbine driving an alternator, with maximum recovery of heat, particularly from the exhaust system. For best efficiency, there needs to be a convenient and constant requirement for the output heat energy, and the generated electricity should also be utilized locally, with any excess exported to the grid.

  Unless a source of fuel is available from landfill gas, or from a local biomass digester, then an on-site biomass to gas conversion plant would be needed to fuel the CHP engine. Considering the cost implications for biomass storage and handling as described for boilers, it appears that biomass CHP will only be viable in specific circumstances, with installed system costs in the order of £2500 to £3000/kW (electrical).

Investment 'yield' table for various renewable technologies

It can be quite difficult to compare renewable options in terms of how much energy they might save on a particular project, and how that translates into CO_2, especially if more than one option appears to be feasible. The table illustrates the potential saving per £100,000 of renewable investment i.e. £100,000 is the notional 'extra over' cost of introducing a proportion of renewable energy into the particular building service. Grant aid has been ignored in the table.

The photovoltaic options indicated include no allowance for the displacement of conventional building fabric. In practice, the 'yield per £100,000' may be higher, e.g. if PV is fully-integrated.

Renewable technology	Candidate buildings	Prerequisites	Potential barriers	Annual saving per £100,000 of capital cost	
				kWh	Kg CO_2
Tower-mounted wind generators	A	F	Environmental impact. Site space for large turbines	100,000	43,000 c.f. elec.
Building-mounted 'micro wind'	B	G	Environmental impact. Roof space for small turbines.	40,000	17,200 c.f. elec.
*Photovoltaic roof or panels	B	H	Available roof space	12,500	5,375 c.f. elec.
*Photovoltaic rain screen or glass	C	H	None	9,000	3,870 c.f. elec.
Passive solar water heating	D	J	None	50,000	9,500 c.f. gas
Ground source heat pump	B	K	Site space for pipes	40,000	7,600 c.f. gas
Borehole cooling	D	K	Site space for pipes	12,000	5,160 c.f. elec.
Biomass boilers	B	L	Environmental impact, & maintenance	100,000	19,000 c.f. gas
Biomass CHP	E	M	Environmental impact, & maintenance	28,000 + 63,000	12,000 c.f. elec. + 12,000 c.f. gas

Key:

A	Industrial, distribution centres	G	Average site wind speed minimum 3.5 m/s
B	Most types of building	H	Roughly south-facing, un-shaded
C	Prestige offices or retail	J	Roughly south-facing, un-shaded – for hot water
D	Residential and commercial, hotels & leisure	K	Feasible ground conditions
E	Industrial, Hotel, leisure, hospital	L	Space and convenient source of fuel
F	Average site wind speed minimum 7 m/s	M	Space & convenient source of fuel – for summer heat

The following exclusions relate to all of the aforementioned indicative costs:

Inflation beyond second quarter 2005, maintenance charges, general builders work, main contractor's overheads, profit and attendance, main contract preliminaries, professional and prescribed fees, contingency and design reserve, grant aid, tax allowances, Value Added Tax. Price levels indicated are based on provincial locations.

Grey Water Recycling and Rainwater Harvesting

The potential for grey water recycling and rainwater harvesting for both domestic residential and for various types of commercial building, considering the circumstances in which the systems offer benefits, both as stand-alone installations and combined.

Water usage trends

Water usage in the UK has increased dramatically over last century or so and it is still accelerating. The current average per capita usage is estimated to be at least 150 litres per day, and the population is predicted to rise from 60 million now to 65 million by 2017 and to 75 million by 2031, with an attendant increase in loading on water supply and drainage infrastructures.

Even at the current levels of consumption, it is clear from recent experience that long, dry summers can expose the drier regions of the UK to water shortages and restrictions. The predicted effects of climate change include reduced summer rainfall, more extreme weather patterns, and an increase in the frequency of exceptionally warm dry summers. This is likely to result in a corresponding increase in demand to satisfy more irrigation of gardens, parks, additional usage of sports facilities and other open spaces, together with additional needs for agriculture. The net effect, therefore, at least in the drier regions of the UK, is for increased demand coincident with a reduction in water resource, thereby increasing the risk of shortages.

Water applications and re-use opportunities

Average domestic water utilization can be summarized as follows, as a percentage of total usage (Source: Three Valleys Water):

- Wash hand basin 8%
- Toilet 35%
- Dishwasher 4%
- Washing machine 12%
- Shower 5%
- Kitchen sink 15%
- Bath 15%
- External use 6%.

Water for drinking and cooking makes up less than 20% of the total, and more than a third of the total is used for toilet flushing. The demand for garden watering, although still relatively small, is increasing year by year and coincides with summer shortages, thereby exacerbating the problem.

In many types of building it is feasible to collect rainwater from the roof area and to store it, after suitable filtration, in order to meet the demand for toilet flushing, cleaning, washing machines and outdoor use – thereby saving in many cases a third of the water demand. The other, often complementary recycling approach, is to collect and disinfect 'grey water' – the waste water from baths, showers and washbasins. Hotels, leisure centres, care homes and apartment blocks generate large volumes of waste water and therefore present a greater opportunity for recycling. With intelligent design, even offices can make worthwhile water savings by recycling grey water, not necessarily to flush all of the toilets in the building but perhaps just those in one or two primary cores, with the grey water plant and distribution pipework dimensioned accordingly.

Intuitively, rainwater harvesting and grey water recycling seem like 'the right things to do' and rainwater harvesting is already common practice in many counties in Northern continental Europe. In Germany, for example, some 60,000 to 80,000 systems are being installed every year – compared with perhaps 2,000 systems in the UK. Grey water recycling systems, which are less widespread than rainwater harvesting, have been developed over the last 20 years and the 'state of the art' is to use biological and UV (ultra-violet) disinfection rather than chemicals, and to reduce the associated energy use through advanced technology membrane micro filters.

In assessing the environmental credentials of new developments, the sustainable benefits are recognized by the Building Research Establishment Environmental Assessment Method (BREEAM) whereby additional points can be gained for efficient systems – those designed to achieve enough water savings to satisfy at least 50% of the relevant demand. Furthermore, rainwater harvesting systems from several manufacturers are included in the 'Energy Technology Product List' and thereby qualify for Enhanced Capital Allowances (ECAs). Claims are allowed not only for the equipment, but also to directly associated project costs including:

- Transportation – the cost of getting equipment to the site
- Installation – cranage (to lift heavy equipment into place), project management costs and labour, plus any necessary modifications to the site or existing equipment
- Professional Fees – if they are directly related to the acquisition and installation of the equipment.

Rainwater Harvesting

A typical domestic rainwater harvesting system can be installed at reasonable cost if properly designed and installed at the same time as building the house. The collection tank can either be buried or installed in a basement area. Rainwater enters the drainage system through sealed gullies and passes through a pre-filter to remove leaves and other debris before passing into the collection tank. A submersible pump, under the control of the monitoring and sensing panel, delivers recycled rainwater on demand. The non-potable distribution pipework to the washing machine, cleaner's tap, outside tap and toilets etc could be either a boosted system or configured for a header tank in the loft, with mains supply back-up, monitors and sensors located there instead of at the control panel.

Calculating the collection tank size brings into play the concept of system efficiency – relating the water volume saved to the annual demand. In favourable conditions – ample rainfall and large roof collection area – it would be possible in theory to achieve almost 100%. In practice, systems commonly achieve 50 to 70% efficiency, with enough storage to meet demand for typically one week, though this is subject to several variables. As well as reducing the demand for drinking quality mains supply water, rainwater harvesting tanks act as an effective storm water attenuator, thereby reducing the drainage burden and the risk of local flooding which is a benefit to the wider community. Many urban buildings are located where conditions are unfavourable for rainwater harvesting – low rainfall and small roof collection area. In these circumstances it may still be worth considering water savings through grey water recycling.

Grey Water Recycling

In a grey water recycling system, waste water from baths, showers and washbasins is collected by conventional fittings and pipework, to enter a pre-treatment sedimentation tank which removes the larger dirt particles. This is followed by the aerobic treatment tank in which cleaning bacteria ensure that all bio-degradable substances are broken down. The water then passes onto a third tank, where an ultra-filtration membrane removes all particles larger than 0.00005 mm, (this includes viruses and bacteria) effectively disinfecting the recycled grey water. The clean water is then stored in the fourth tank from where it is pumped on demand under the control of monitors and sensors in the control panel. Recycled grey water may then be used for toilet and urinal flushing, for laundry and general cleaning, and for outdoor use such as vehicle washing and garden irrigation – a substantial water saving for premises such as hotels. If the tank becomes depleted, the distribution is switched automatically to the mains water back-up supply. If there is insufficient plant room space, then the tanks may be buried but with adequate arrangements for maintenance access. Overflow soakaways are recommended where feasible but are not an inherent part of the rainwater harvesting and grey water recycling systems.

Combined rainwater/Grey water systems

Rainwater can be integrated into a grey water scheme with very little added complication other than increased tank size. In situations where adequate rainwater can be readily collected and diverted to pre-treatment, then heavy demands such as garden irrigation can be met more easily than with grey water alone. An additional benefit is that they reduce the risk of flooding by keeping collected storm water on site instead of passing it immediately into the drains.

Indicative system cost and payback considerations

Rainwater Harvesting Scenario: Office building in Leeds having a roof area of 2000 m² and accommodating 435 people over 3 floors. Local annual rainfall is 875 mm and the application is for toilet and urinal flushing. An underground collection tank of 25,000 litres has been specified to give 4.5 storage days.

Rainwater harvesting cost breakdown	Cost £
System tanks and filters and controls	18,000
Mains water back-up and distribution pump arrangement	4,000
Non-potable distribution pipework	1,000
Connections to drainage	1,000
Civil works and tank installation (assumption normal ground conditions)	7,000
System installation & commissioning	2,000
TOTAL COST	33,000

Rainwater harvesting payback considerations

Annual water saving: 1100 m³ @ average cost £2.00/m³ = £2200
Annual Maintenance and system energy cost = £700

Indicative payback period = 33,000/1,500 = 22 years

Grey water recycling scenario: Urban leisure hotel building with 200 bedrooms offers little opportunity for rainwater harvesting but has a grey water demand of up to 12000 litres per day for toilet and urinal flushing, plus a laundry. The grey water recycling plant is located in a basement plant room.

Grey water recycling cost breakdown	Cost £
System tanks and controls	34,000
Mains water back-up and distribution pump arrangement	4,000
Non-potable distribution pipework	4,000
Grey water waste collection pipework	4,500
Connections to drainage	1,000
System installation & commissioning	5,000
TOTAL COST	52,500

Grey water recycling payback considerations

Annual water saving: 4260 m³ @ average cost £2.00/m³ = £8,520
Annual Maintenance and system energy cost = £2,000

Indicative payback period = 52,500/6,520 = 8 years

Exclusions

- Site organization and management costs other than specialist contractor's allowances
- Contingency/design reserve
- Main contractor's overhead and profit or management fee
- Professional fees
- Tax allowances
- Value Added Tax
- Inflation beyond third quarter 2007.

Conclusions

There is a justified and growing interest in saving and recycling water by way of both grey water recycling and rainwater harvesting. The financial incentive at today's water cost is not great for small or inefficient systems but water costs are predicted to rise and demand to increase – not least as a result of population growth.

The payback periods for the above scenarios are not intended to compare to potential payback periods of rainwater harvesting against grey water recycling but rather to illustrate the importance of choosing 'horses for courses'. The office building with its relatively small roof area and limited demand for toilet flushing results in a fairly inefficient system. Burying the collection tank also adds a cost so that payback exceeds 20 years. Payback periods of less than 10 years are feasible for buildings with large roofs and a large demand for toilet flushing or for other uses. Therefore sports stadia, exhibition halls, supermarkets, schools and similar structures are likely to be suitable.

The hotel scenario is good application for grey water recycling. Many hotels and residential developments will generate more than enough grey water to meet the demand for toilet flushing etc. and in these circumstances large quantities of water can be saved and recycled with attractive payback periods.

Ground Water Cooling

The use of ground water cooling systems, considering the technical and cost implications of this renewable energy technology.

The application of ground water cooling systems is quickly becoming an established technology in the UK with numerous installations having been completed for a wide range of building types, both new build and existing (refurbished).

Buildings in the UK are significant users of energy, accounting for 60% of UK carbon emissions in relation to their construction and occupation. The drivers for considering renewable technologies such as groundwater cooling are well documented and can briefly be summarized as follows:

- Government set targets – The Energy White Paper, published in 2003, setting a target of producing 10% of UK electricity from renewable sources by 2010 and the aspiration of doubling this by 2020.
- The proposed revision to the Building Regulations Part L 2006, in raising the overall energy efficiency of non domestic buildings, through the reduction in carbon emissions, by 27%.
- Local Government policy for sustainable development. In the case of London, major new developments (i.e. City of London schemes over 30,000 m^2) are required to demonstrate how they will generate a proportion of the site's delivered energy requirements from on-site renewable sources where feasible. The GLA's expectation is that, overall, large developments will contribute 10% of their energy requirement using renewables, although the actual requirement will vary from site to site. Local authorities are also likely to set lower targets for buildings which fall below the GLA's renewables threshold.
- Company policies of building developers and end users to minimize detrimental impact to the environment.

The ground as a heat source/sink

The thermal capacity of the ground can provide an efficient means of tempering the internal climate of buildings. Whereas the annual swing in mean air temperature in the UK is around 20 K, the temperature of the ground is far more stable. At the modest depth of 2 m, the swing in temperature reduces to 8 K, while at a depth of 50 m the temperature of the ground is stable at 11–13°C. This stability and ambient temperature therefore makes groundwater a useful source of renewable energy for heating and cooling systems in buildings.

Furthermore, former industrial cities like Nottingham, Birmingham, Liverpool and London have a particular problem with rising ground water as they no longer need to abstract water from below ground for use in manufacturing. The use of groundwater for cooling is therefore encouraged by the Environment Agency in areas with rising groundwater as a means of combating this problem.

System types

Ground water cooling systems may be defined as either open or closed loop.

Open loop systems

Open loop systems generally involve the direct abstraction and use of ground water, typically from aquifers (porous water bearing rock). Water is abstracted via one or more boreholes and passed through a heat exchanger and is returned via a separate borehole or boreholes, discharged to foul water drainage or released into a suitable available source such as a river. Typical ground water supply temperatures are in the range 6–10°C and typical re-injection temperatures 12–18°C (subject to the requirements of the abstraction licence).

Open loop systems fed by groundwater at 8°C, can typically cool water to 12°C on the secondary side of the heat exchanger to serve conventional cooling systems.

Open loop systems are thermally efficient but overtime can suffer from blockages caused by silt, and corrosion due to dissolved salts. As a result, additional cost may be incurred in having to provide filtration or water treatment, before the water can be used in the building.

Abstraction licence and discharge consent needs to be obtained for each installation, and this together with the maintenance and durability issues can significantly affect whole life operating costs, making this system less attractive.

Closed loop systems

Closed loop systems do not rely on the direct abstraction of water, but instead comprise a continuous pipework loop buried in the ground. Water circulates in the pipework and provides the means of heat transfer with the ground. Since ground water is not being directly used, closed loop systems therefore suffer fewer of the operational problems of open loop systems, being designed to be virtually maintenance free, but do not contribute to the control of groundwater levels.

There are two types of closed loop system:

Vertical Boreholes –   Vertical loops are inserted as U tubes into pre drilled boreholes, typically less than 150 mm in diameter. These are backfilled with a high conductivity grout to seal the bore, prevent any cross contamination and to ensure good thermal conductivity between the pipe wall and surrounding ground. Vertical boreholes have the highest performance and means of heat rejection, but also have the highest cost due to associated drilling and excavation requirements.

                     As an alternative to having a separate borehole housing the pipe loop, it can also be integrated with the piling, where the loop is encased within the structural piles. This obviously saves on the costs of drilling and excavation since these would be carried out as part of the piling installation. The feasibility of this option would depend on marrying up the piling layout with the load requirement, and hence the number of loops, for the building.

Horizontal Loops –   These are single (or pairs) of pipes laid in 2 m deep trenches, which are backfilled with fine aggregate. These obviously require a greater physical area than vertical loops but are cheaper to install. As they are located closer to the surface where ground temperatures are less stable, efficiency is lower compared to open systems. Alternatively, coiled pipework can also be used where excavation is more straightforward and a large amount of land is available. Although performance may be reduced with this system as the pipe overlaps itself, it does represent a cost effective way of maximizing the length of pipe installed and hence overall system capacity.

The case for heat pumps

Instead of using the groundwater source directly in the building, referred to as passive cooling, when coupled to a reverse cycle heat pump, substantially increased cooling loads can be achieved.

Heat is extracted from the building and transferred by the heat pump into the water circulating through the loop. As it circulates, it gives up heat to the cooler earth, with the cooler water returning to the heat pump to pick up more heat. In heating mode the cycle is reversed, with the heat being extracted from the earth and being delivered to the HVAC system.

The use of heat pumps provides greater flexibility for heating and cooling applications within the building than passive systems. Ground source heat pumps are inherently more efficient than air source heat pumps, their energy requirement is therefore lower and their associated CO_2 emissions are also reduced, so they are well suited for connection to a groundwater source.

Closed loop systems can typically achieve outputs of 50 W/m (of bore length), although this will vary with geology and borehole construction. When coupled to a reverse cycle heat pump, 1 m of vertical borehole will typically deliver 140 kWh of useful heating and 110 kWh of cooling per annum, although this will depend on hours run and length of heating and cooling seasons.

Key factors affecting cost

- The cost is obviously dependent on the type of system used. Deciding on what system is best suited to a particular project is dependent on the peak cooling and heating loads of the building and its likely load profile. This in turn determines the performance required from the ground loop, in terms of area of coverage in the case of the horizontal looped system, and in the case of vertical boreholes, the depth and number or bores. The cost of the system is therefore a function of the building load.

- In the case of vertical boreholes, drilling costs are significant factor, as specific ground conditions can be variable, and there are potential problems in drilling through sand layers, pebble beds, gravels and clay, which may mean additional costs through having to drill additional holes or the provision of sleeving etc. The costs of excavation obviously make the vertical borehole solution significantly more expensive than the equivalent horizontal loop.

- The thermal efficiency of the building is also a factor. The higher load associated with a thermally inefficient building obviously results in the requirement for a greater number of boreholes or greater area of horizontal loop coverage, however in the case of boreholes the associated cost differential between a thermally inefficient building and a thermally efficient one is substantially greater than the equivalent increase in the cost of conventional plant. Reducing the energy consumption of the building is cheaper than producing the energy from renewables and the use of renewable energy only becomes cost effective, and indeed should only be considered, when a building is energy efficient.

- With open loop systems, the principal risk in terms of operation is that the user is not in control of the quantity or quality of the water being taken out of the ground, this being dependent on the local ground conditions. Reduced performance due to blockage (silting etc.) may lead to the system not delivering the design duties whilst bacteriological contamination may lead to the expensive water treatment or the system being taken temporarily out of operation. In order to mitigate the above risk, it may be decided to provide additional means of heat rejection and heating by mechanical means as a back up to the borehole system, in the event of operational problems. This obviously carries a significant cost. If this additional plant were not provided, then there are space savings to be had over conventional systems due to the absence of heating, heat rejection and possibly refrigeration plant.

- Open loop systems may lend themselves particularly well to certain applications increasing their cost effectiveness, i.e. in the case of a leisure centre, the removal of heat from the air-conditioned parts of the centre and the supply of fresh water to the swimming pool.

- In terms of the requirements for abstraction and disposal of the water for open loop systems, there are risks associated with the future availability and cost of the necessary licenses; particularly in areas of high forecast energy consumption, such as the South East of England, which needs to be borne in mind when selecting a suitable system.

Whilst open loop systems would suit certain applications or end user clients, for commercial buildings the risks associated with this system tend to mean that closed loop applications are the system of choice. When coupled to a reversible heat pump, the borehole acts simply as a heat sink or heat source so the problems associated with open loop systems do not arise.

Typical costs

Table 1 gives details of the typical borehole cost to an existing site in Central London, using one 140 m deep borehole working on the open loop principle, providing heat rejection for the 600 kW of cooling provided to the building. The borehole passes through rubble, river gravel terraces, clay and finally chalk, and is lined above the chalk level to prevent the hole collapsing. The breakdown includes all costs associated with the provision of a working borehole up to the well head, including the manhole chamber and manhole. The costs of any plant or equipment from the well head are not included.

Heat is drawn out of the cooling circuit and the water is discharged into the Thames at an elevated temperature. In this instance, although the boreholes are more expensive than the dry air cooler alternative, the operating cost is significantly reduced as the system can operate at around three times the efficiency of conventional dry air coolers, so the payback period is a reasonable one. Additionally, the borehole system does not generate any noise, does not require rooftop space and does not require as much maintenance.

This is representative of a typical cost of providing a borehole for an open loop scheme within the London basin. There are obviously economies of scale to be had in drilling more than one well at the same time, with two wells saving approximately 10% of the comparative cost of two separate wells and four wells typically saving 15%.

Table 2 provides a summary of the typical range of costs that could expected for the different types of system based on current prices.

Table 1: Breakdown of the Cost of a Typical Open Loop Borehole System

Description	Cost £

<u>General Items</u>

• Mobilization, Insurances, demobilization on completion	20,000
• Fencing around working area for the duration of drilling and testing	2,000
• Modifications to existing LV panel and installation of new power supplies for borehole installation	14,000

<u>Trial Hole</u>

• Allowance for breakout access to nearest walkway (Existing borehole on site used for trial purposes, hence no drilling costs included)	3,000

<u>Construct Borehole</u>

• Drilling, using temporary casing where required, permanent casing and grouting	31,000

<u>Borehole Cap and Chamber</u>

• Cap borehole with PN16 flange, construct manhole chamber in roadway, rising main, header pipework, valves, flow meter	12,000
• Permanent pump	13,000

<u>Samples</u>

• Water samples	300

<u>Acidization</u>

• Mobilization, set up and removal of equipment for acidization of borehole, carry out acidization	11,500

<u>Development and Test Pumping</u>

• Mobilize pumping equipment and materials and remove on completion of testing	4,000
• Calibration test, pre-test monitoring, step testing	3,500
• Constant rate testing and monitoring	19,000
• Waste removal and disposal	3,000

<u>Reinstatement</u>

• Reinstatement and Making Good	1,500

Total	**137,800**

Table 2: Summary of the Range of Costs for Different Systems

System	Range Small–4 kWth	Medium–50 kWth	Large–400 kWth	Notes
Heat pump (per unit)	£ 3,500–4,500	£ 30,000–40,000	£ 140,000–170,000	
Slinky pipe (per installation) including excavation	£ 3,000–4,000	£ 40,000–50,000	£ 360,000–390,000[1]	[1] Based on 90 nr 50 m lengths
Vertical, closed (per installation) using structural piles	N/A	£ 40,000–60,000	Not available	Based on 50 nr piles. Includes borehole cap and header pipework but excludes connection to pump room and heat pumps
Vertical, closed (per installation) including excavation	£ 2,000–3,000	£ 60,000–80,000	£ 360,000–390,000	Includes borehole cap and header pipework but excludes connection to pump room and heat pumps
Vertical, open (per installation) including excavation	£ 2,000–3,000	£ 45,000–65,000	£ 330,000–360,000	Excludes connection to pump room and heat exchangers

Fuel Cells

The application of fuel cell technology within buildings

Fuel cells are electrochemical devices that convert the chemical energy in fuel into electrical energy directly, without combustion, with high electrical efficiency and low pollutant emissions. They represent a new type of power generation technology that offers modularity, efficient operation across a wide range of load conditions, and opportunities for integration into co-generation systems. With the publication of the energy white paper in February this year, the Government confirmed its commitment to the development of fuel cells as a key technology in the UK's future energy system, as the move is made away from a carbon based economy.

There are currently very few fuel cells available commercially, and those that are available are not financially viable. Demand has therefore been limited to niche applications, where the end user is willing to pay the premium for what they consider to be the associated key benefits. Indeed, the UK currently has only one fuel cell in regular commercial operation. However, fuel cell technology has made significant progress in recent years, with prices predicted to approach those of the principal competition in the near future.

Fuel cell technology

A fuel cell is composed of an anode (a negative electrode that repels electrons), an electrolyte membrane in the centre, and a cathode (a positive electrode that attracts electrons). As hydrogen flows into the cell on the anode side, a platinum coating on the anode facilitates the separation of the hydrogen gas into electrons and protons. The electrolyte membrane only allows the protons to pass through to the cathode side of the fuel cell. The electrons cannot pass through this membrane and flow through an external circuit to form an electric current.

As oxygen flows into the fuel cell cathode, another platinum coating helps the oxygen, protons, and electrons combine to produce pure water and heat.

The voltage from a single cell is about 0.7 volts, just enough for a light bulb. However by stacking the cells, higher outputs are achieved, with the number of cells in the stack determining the total voltage, and the surface area of each cell determining the total current. Multiplying the two together yields the total electrical power generated.

In a fuel cell the conversion process from chemical energy to electricity is direct. In contrast, conventional energy conversion processes first transform chemical energy to heat through combustion and then convert heat to electricity through some form of power cycle (e.g. gas turbine or internal combustion engine) together with a generator. The fuel cell is therefore not limited by the Carnot efficiency limits of an internal combustion engine in converting fuel to power, resulting in efficiencies 2 to 3 times greater.

Fuel cell systems

In addition to the fuel cell itself, the system comprises the following sub-systems:

- A fuel processor – This allows the cell to operate with available hydrocarbon fuels, by cleaning the fuel and converting (or reforming) it as required
- A power conditioner – This regulates the dc electricity output of the cell to meet the application, and to power the fuel cell auxiliary systems
- An air management system – This delivers air at the required temperature, pressure and humidity to the fuel stack and fuel processor
- A thermal management system – This heats or cools the various process streams entering and leaving the fuel cell and fuel processor, as required
- A water management system – Pure water is required for fuel processing in all fuel cell systems, and for dehumidification in the PEMFC.

The overall electrical conversion efficiency of a fuel cell system (defined as the electrical power out divided by the chemical energy into the system, taking into account the individual efficiencies of the sub-systems) ranges from 35–55%. Taking into account the thermal energy available from the system, the overall or cogeneration efficiency is 75–90%.

Also, unlike most conventional generating systems (which operate most efficiently near full load, and then suffer declining efficiency as load decreases), fuel cell systems can maintain high efficiency at loads as low as 20% of full load.

Fuel cell systems also offer the following potential benefits:

- At operating temperature, they respond quickly to load changes, the limiting factor usually being the response time of the auxiliary systems
- They are modular and can be built in a wide range of outputs. This also allows them to be located close to the point of electricity use, facilitating cogeneration systems
- Noise levels are comparable with residential or light commercial air conditioning systems
- Commercially available systems are designed to operate unattended and manufactured as packaged units
- Since the fuel cell stack has no moving parts, other than the replacement of the stack at 3–5 year intervals there is little on-site maintenance. The maintenance requirements are well established for the auxiliary system plant
- Fuel cell stacks fuelled by hydrogen produce only water, therefore the fuel processor is the primary source of emissions, and these are significantly lower than emissions from conventional combustion systems
- Since fuel cell technology generates 50% more electricity than the conventional equivalent without directly burning any fuel, CO_2 emissions are significantly reduced in the production of the source fuel
- Potentially zero carbon emissions when using hydrogen produced from renewable energy sources
- The facilitation of embedded generation, where electricity is generated close to the point of use, minimizing transmission losses
- The fast response times of fuel cells offer potential for use in UPS systems, replacing batteries and standby generators.

Types of fuel cell

There are four main types of fuel cell technology that are applicable for building systems, classed in terms of the electrolyte they use. The chemical reactions involved in each cell are very different.

Phosphoric Acid Fuel Cells (PAFCs) are the dominant current technology for large stationary applications and have been available commercially for some time. The only working fuel cell installation in the UK, in Woking, uses a PAFC, rated at 200 kW. There is less potential for PAFC unit cost reduction than for some other fuel cell systems, and this technology may be superseded in time by the other technologies.

The Solid Oxide Fuel Cell (SOFC) offers significant flexibility due to its large power range and wide fuel compatibility. SOFCs represent one of the most promising technologies for stationary applications. There are difficulties when operating at high temperatures with the stability of the materials, however, significant further development and cost reduction is anticipated with this type.

The relative complexity of Molten Carbonate Fuel Cells (MCFCs) has tended to limit developments to large scale stationary applications, although the technology is still very much in the development stages.

The quick start-up times and size range make Proton Exchange Membrane Fuel Cells (PEMFCs) suitable for small to medium sized stationary applications. They have a high power density and can vary output quickly, making them well suited for transport applications as well as UPS systems. The development efforts in the transport sector suggest there will continue to be substantial cost reductions over both the short and long term.

All four technologies remain the subject of extensive research and development programmes to reduce initial costs and improve reliability through improvements in materials, optimization of operating conditions and advances in manufacturing. It is expected that all types will be commercially available in limited markets by 2006 and with mass market availability by 2010.

The market for fuel cells

The stationary applications market for fuel cells can be sectaries as follows:

- Distributed generation/CHP – For large scale applications, there are no drivers specifically advantageous to fuel cells, with economics (and specifically initial cost) therefore being the main consideration. So, until cost competitive and thoroughly proven and reliable fuel cells are available, their use is likely to be limited to niche applications such as environmentally sensitive areas from 2005. Wider commercialization is likely closer to 2010, with high temperature cells (MCFCs and SOFCs) being most suitable, although PEMFCs may preferable in specific areas, i.e. where hydrogen is available.

- Domestic and small scale CHP – The drivers for the use of fuel cells in this emerging market are better value for customers than separate gas and electricity purchase, reduction in domestic CO_2 emissions, and potential reduction in electricity transmissioncosts. However, the barriers of resistance to distributed generation, high capital costs and competition from Stirling engines needs to be overcome. Commercialization depends on cost reduction, and successful demonstration which is expected to begin in the next 2–3 years, leading to wider commercialization before 2010. Systems based on SOFCs and PEMFCs are being developed for this application.

- Small generator sets and remote power – The drivers for the use of fuel cells are high reliability, low noise and low refuelling frequencies, which cannot be met by existing technologies. Since cost is often not the primary consideration, fuel cells will find early markets in this sector. Existing PEMFC systems are close to meeting the requirements in terms of cost, size and performance. Small SOFCs have potential in this market, but require further development.

Cost comparison

Table 1 provides an indication of the capital and operating costs of the different fuel cell types, together with comparative figures for the existing technologies. The projected figures for the fuel cell technologies are based on economies typically achieved through mass manufacture.

The projected costs for 2007 show the fuel cell technologies still being significantly more expensive than the existing technologies. To extend fuel cell application beyond niche markets, their cost needs to reduce significantly. The successful and wide-spread commercial application of fuel cells is dependent on the projected cost reductions indicated, with electricity generated from fuel cells being competitive with current centralized and distributed power generation.

It has been estimated that if the cost reductions are met, fuel cells could achieve up to 50% penetration of the global distributed energy market by 2020.

Typical current project costs

Table 2 gives an indication of the typical cost breakdown to be expected for the installation of a fuel cell system in the UK. This is based on information from the manufacturer and from economic evaluation/feasibility studies, since with only one working system installed to date in the UK there is no available accurate cost data.

The unit is a standard commercially available PAFC, complete with fuel processor, fuel stack and power conditioning system. The parameters are as follows:

- Rating – 200 kW/235 kVA, 400 V, 3ph
- Power generating efficiency – 40%
- Heat output – 204 kW, 60°C hot water
- Fuel, consumption – Natural gas, 54 m³/hr
- External location
- Size – 5.5 m × 3 m × 3 m (h)
- Weight – 20 tons
- Noise level at full load – 62 dBA at 10 m.

The above illustrates the fact that the costs to supply, install and set to work a modestly sized fuel cell unit are prohibitively high, compared with incumbent generator technologies. Despite being the only commercially available unit, only 220 units have been sold worldwide, and so the full benefits of volume manufacture have not been realized, and a proportion of the costs associated with developing the unit is included within the unit cost.

Conclusions

Despite significant growth in recent years, fuel cells are still at a relatively early stage of commercial development, with prohibitively high capital costs preventing them from competing with the incumbent technology in the market place. However, costs are forecast to reduce significantly over the next five years as the technology moves from niche applications, and into mass production.

However, in order for these projected cost reductions to be achieved, customers need to be convinced that the end product is not only cost competitive but also thoroughly proven, and Government support represents a key part in achieving this.

The Governments of Canada, USA, Japan and Germany have all been active in supporting development of the fuel cell sector through integrated strategies, however the UK has been slow in this respect, and support has to date been small in comparison. It is clear that without Government intervention, fuel cell applications may struggle to reach the cost and performance requirements of the emerging fuel cell market.

Table 1: Cost Comparison for Stationary Generation Equipment

Period	Fuel cell type			
	PEMFC	PAFC	MCFC	SOFC
Capital costs (£/kW)	£	£	£	£
2004	2,600–6,500	2,000–3,400	2,000–5,000	5,000–10,000
2006	1,500	1,900	1,850	2,300
2010	700	1,500	950	1,000
2015 (Projected)	500	1,300	700	750
2020 (Projected)	300	1,100	550	550
Operating costs (£/kW/h)				
2004	0.04–0.06	0.08	0.04–0.08	0.05
Maintenance costs (£/kW/h)				
2006	0.003–0.01	0.003–0.01	0.003–0.01	0.003–0.01

Period	Conventional systems			
	Internal combustion – generator	Micro turbine – generator	Gas turbine	Steam turbine
Capital costs (£/kW)	£	£	£	£
2004	–	–	–	–
2006	200–820	450–820	450–570	500–630
2015 (Projected)	–	–	–	–
2020 (Projected)	–	–	–	–
Operating costs (£/kW/h)				
2004	–	–	–	–
Maintenance costs (£/kW/h)				
2006	0.005–0.01	0.002–0.01	0.002–0.06	0.003

Table 2: Breakdown of Typical Project Costs

Item	£
Fuel Cell Stack	225,000
Fuel Processor	85,000
Plant	113,600
Labour	113,400
Total	**567,000**
Delivery to site (Outside USA)	20,000

Installation costs (site and application dependent)

Standard (Generation Only)	50,000
Non Standard	80,000
CHP	100,000–200,000
Spares	10% of Capital Cost
Maintenance	16,000–26,000 pa
Total (£/kW)	**3,549–4,415**

Fuel stack

Fuel Stack Replacement	150,000 every 4 years
Fuel Stack Refurbishment	120,000 every 4 years

Note: Excludes Incoming Gas Supply, BWIC, Main Contractor's Overheads and Profit, Attendance, Preliminaries and Professional Fees etc, VAT.

Biomass Energy

The potential for biomass energy systems, with regards to the adequacy of the fuel supply and the viability of various system types at different scales.

Biomass heating and combined heat and power (CHP) systems have become a major component of the low-carbon strategy for many projects, as they can provide a large renewable energy component at a relatively low initial cost. Work by the Carbon Trust has demonstrated that both large and small biomass systems were viable even before recent increases in gas and fuel oil prices, so it is no surprise that recent research by South Bank University into the renewables strategies to large London projects has found that 25% feature biomass or biofuel systems.

These proposals are not without risk, however. Although the technology is well established, few schemes are in operation in the UK and the long-term success depends more on the effectiveness of the local supply chain than the quality of the design and installation.

How the biomass market works

Biomass is defined as living or recently dead biological material that can be used as an energy source. Biomass is generally used to provide heat, generate electricity or drive CHP engines. The biomass family includes biofuels, which are being specified in city centre schemes, but which provide lower energy outputs and could transfer farmland away from food production.

In the UK, much of the focus in biomass development is on the better utilization of waste materials such as timber and the use of set-aside land for low-intensity energy crops such as willow, rather than expansion of the biofuels sector. There are a variety of drivers behind the development of a biomass strategy. In addition to carbon neutrality, another policy goal is the promotion of the UK's energy security through the development of independent energy sources. A third objective is to address energy poverty, particularly for off-grid energy users, who are most vulnerable to the effects of high long-term costs of fuel oil and bottled gas.

Biomass' position in the zero-carbon hierarchy is a little ambiguous in that its production, transport and combustion all produce carbon emissions, albeit most is offset during a plant's growth cycle. The key to neutrality is that the growing and combustion cycles need to occur over a short period, so that combustion emissions are genuinely offset. Biomass strategy is also concerned about minimizing waste and use of landfill, and the ash produced by combustion can be used as a fertilizer.

Dramatic increases in fossil fuel prices have swung considerations decisively in favour of technologies such as biomass. Research by the Carbon Trust has demonstrated that, with oil at $50 (£25) a barrel, rates of return of more than 10% could be achieved with both small and large heating installations. CHP and electricity-only schemes have more complex viability issues linked to renewable incentives, but with oil currently trading at over ($100) £50 a barrel and a plentiful supply of source material, it is argued that biomass input prices will not rise and so the sector should become increasingly competitive.

The main sources of biomass in the UK include:

- Forestry crops, including the waste products of tree surgery industry
- Industrial waste, particularly timber, paper and card: timber pallets account for 30% of this waste stream by weight
- Woody energy crops, particularly those grown through 'short rotation' methods such as willow coppicing
- Wastes and residues taken from food, agriculture and manufacturing.

Biomass is an emerging UK energy sector. Most suppliers are small and there remains a high level of commercial risk associated with finding appropriate, reliable sources of biomass. This is particularly the case for larger-scale schemes such as those proposed for Greater London, which will have sourced biomass either from multiple UK suppliers or from overseas. Many have adopted biofuels as an alternative.

The UK's only large-scale biomass CHP in Slough has a throughput of 180,000 tonnes of biomass per year requiring the total production of more than 20 individual suppliers – not a recipe for easy management or product consistency. However, Carbon Trust research has identified significant potential capacity in waste wood (5 million to 6 million tonnes) and short-rotation coppicing, which could create the conditions for wider adoption of small and large-scale biomass.

Biomass technologies

A wide range of technologies have been developed for processing various forms of biomass, including anaerobic digesters and gasifiers. However, the main biomass technology is solid fuel combustion, as a heat source, CHP unit or energy source for electricity generation.

Solid fuel units use either wood chippings or wood pellets. Wood chippings are largely unprocessed and need few material inputs, other than seasoning, chipping and transport. Wood pellets are formed from compressed sawdust. As a result they have a lower moisture content than wood chippings and consistent dimensions, so are easier to handle but are about twice as expensive.

Solid fuel burners operate in the same way as other fossil fuel-based heat sources, with the following key differences:

- Biomass heat output can be controlled but not instantaneously, so systems cannot respond to rapid load changes. Solutions to provide more flexibility include provision of peak capacity from gas-fired systems, or the use of thermal stores that capture excess heat energy during off-peak periods, enabling extended operation of the biomass system itself
- Heat output cannot be throttled back by as much as gas-fired systems, so for heat-only installations it may be necessary to have an alternative summer system for water heating, such as a solar collector
- Biomass feedstock is bulky and needs a mechanized feed system as well as extensive storage
- Biomass systems are large, and the combustion unit, feed hopper and fuel store take up substantial floor area. A large unit with an output of 500 kW has a footprint of 7.5 m × 2 m
- Biomass systems need maintenance related to fuel deliveries, combustion efficiency, ash removal, adding to the lifetime cost
- Fuel stores need to be physically isolated from the boiler and the rest of the building in order to minimize fire risk. The fuel store needs to be sized to provide for at least 100 hours of operation, which is approximately 100 m² for a 500 kW boiler. The space taken up by storage and delivery access may compromise other aspects of site planning
- Fire-protection measures include anti-blowback arrangements on conveyors and fire dampers, together with the specification of elements such as flues for higher operating temperatures
- Collocation of the fuel source and burner at ground level require larger, free-standing flues.

As a result of these issues, which drive up initial costs, affect development efficiency and add to management overheads, take-up of biomass has initially been mostly at the small-scale, heat-only end of the market, based on locally sourced feedstock. In such systems the initial cost premium of the biomass boiler can be offset against long-term savings in fuel costs.

Sourcing biomass

Compared with solar or wind power installations, the initial costs of biomass systems are low, the technology is well established and energy output is dependable. As a result, the real challenge for successful operation of a biomass system is associated with the reliable sourcing of feedstock.

Heat-only systems themselves cost between £150 and £750 per kW (excluding costs of storage), depending on scale and technology adopted. This compares with a typical cost of £50 to £300 per kW for a gas-fired boiler – which does not require further investment in fuel or thermal storage bunkers.

As a high proportion of lifetime cost is associated with the operation of a system, availability of good-quality, locally sourced feedstock is essential for long-term viability – particularly in areas where incentivization through policies like the Merton rule is driving up demand.

Research funded by BioRegional in connection with medium-scale biomass systems in the South-east shows that considerable feedstock is already in the system but far more is required to respond to emerging requirements.

The researchers estimate that the existing biomass resource within 25 km of London totals 330,000 tonnes a year, sourced from waste wood, energy crops and forestry (tree-surgery) byproducts. However, they calculate that a new 2,000-unit low-energy residential system would require 35,000 tonnes a year. This means that London's total bio-mass would have the potential to support the equivalent heating load of just 20,000 homes.

Fortunately, the scale of the UK's untapped resource is considerable, with 5 million to 6 million tonnes of waste wood going to landfill annually, and 680,000ha of set-aside land that could be used for energy crops without affecting agricultural output.

Based on these figures, it is estimated that 15% of the UK's building-related energy load could be supported without recourse to imported material. However, the supply chain is fragmented in terms of producers, processors and distributors – presenting potential biomass users with a range of complexities that gas users simply do not need to worry about. These include:

- Ensuring quality. Guaranteeing biomass quality is important for the assurance of performance and reliability. Variation in moisture content affects combustion, while inconsistent woodchip size or differences in sawdust content can result in malfunction. The presence of contaminants in waste wood causes problems too. High-profile schemes including the 180,000-tonne generator in Slough have had to shutdown because of variations in fuel quality. Use of pellets reduces to risk, but they are more expensive and require more energy for processing and transport
- Functioning markets. The scale of trade in biomass compares unfavourably with gas or oil, in that there are no standard contracts, fixed-price deals or opportunities for hedging which enable major users to manage their energy cost risk
- Security of supply. The potential for competing uses could lead to price inflation. Biofuels carry the greatest such risk, but many biomass streams have alternative uses. Lack of capacity in the marketplace is another security issue, with no mechanism to encourage strategic stockpiling for improved response to crop failure or fluctuations in demand
- Installation and maintenance infrastructure. The different technologies used in biomass systems creates maintenance requirements not yet met by a readily available pool of skilled system engineers.

Optimum uses of biomass technology

Biomass is a high-grade, locally available source of energy that can be used at a range of scales to support domestic and commercial use. Following increases in fossil fuel prices, one of the main barriers to adoption is, now, the capability of the supply chain.

The Carbon Trust's biomass sector review, completed before the large energy price rises in 2007, drew the following key conclusions about the most effective application of the technologies:

- Returns on CHP and electricity-generating systems depend heavily on government incentives such as renewable obligations certificates. Under the present arrangements, large CHP systems provide the best returns
- Heat-only systems are very responsive to changes in fuel prices, with systems at all scales providing returns in excess of 10% when oil prices are at $50 (£25) a barrel
- Small-scale heat-only plants produce the best returns, because the cost of the displaced fuel (typically fuel oil) is more expensive
- Small-scale electricity and large-scale heat-only installations produce very poor returns
- There is little difference in the impact of fuel type in the returns generated by projects.

The study also concluded that heat installations at all scales had the greater potential for carbon saving, based on a finite supply of biomass. This is because heat-generating processes have the greatest efficiency and, in the case of small-scale systems in isolated, off-grid dwellings, displace fuels such as oil that have the greatest carbon intensity. Ninety percent of the UK's existing biomass resource of 5.6 m tonnes per annum could be used in displacing carbon-intensive off-grid heating, saving 2.5 m tonnes of carbon emissions.

Small-scale systems are well established in Europe and the existing local supply chain suits the demand pattern. In addition, since the target market is in rural, off-grid locations, affected dwellings are less likely to suffer space constraints related to storage. As fuel costs continue to rise, the benefits of avoiding fuel poverty, combined with the effective reduction of carbon emissions from existing buildings, mean smaller systems are likely to offer the best mid-term use of the existing biomass supply base, with large-scale systems being developed as the supply chain matures and expands.

Large-scale systems also offer the opportunity to generate significant returns, but the barriers that developers or operators face are significant, particularly if there is an electricity supply component, which requires a supply agreement. However, while developers are required to delivery renewable energy on site, biomass in the form of biofuels, has the great attraction of being able to provide a scale of renewable energy generation that other systems such as ground source heating or photovoltaics simply cannot compete with.

Whether biomass plant should be used on commercial schemes in urban locations is potentially a policy issue. Sizing of both CHP and heat-only systems should be determined by the heat load, which for city-centre schemes may not be that large – affecting the potential for the CHP component. The costs of a district heating element on these schemes may also be prohibitively high, and considerations of biomass transport and storage also make it harder to get city-centre schemes to stack up.

It may be a more appropriate policy to encourage industrial users or large scale regenerators to take first call on the expanding biomass resource, rather than commercial schemes. The launch of a 45 MW biomass power station in Scotland illustrates this trend. Data shows that 50% of the market potential for industrial applications of CHP could utilize 100% of the UK's available biomass resource. The issues that city-centre biomass schemes face in connection with storage, transport, emissions and supply chain management might be better addressed by industrial users or their energy suppliers in low-cost locations rather than by developers in prime city-centre sites.

Indicative costs

System	Indicative load kW	Capital cost (£/kWh)
Gas-fired boiler	50	85
	400	45
Biomass-fired boiler	50	500
	500	250
Biomass-fired CHP	1,000	450

Allowance for stand-alone boiler house and fuel store £30,000–60,000 for 50 kWh system indicative costs exclude flues and plant room installation

Fuel costs

Wood chip: 1.5 to 2.5p/kWh
Wood pellet: 3 to 4.5p/kWh
Fuel oil: 4.3 to 4.8p/kWh
Natural gas: 2.5 to 3.2p/kWh
Bottled LPG: 6.8 to 7p/kWh

Energy Efficiency in New and Existing Buildings

F. Mackenzie

This BRE Trust report considers the relative impact on UK CO2 savings targets of constructing new zero-carbon buildings compared to improving the energy efficiency of the existing stock. Carbon dioxide emissions from buildings accounted for around 40% of total UK CO2 emissions in 2006. To achieve the government's challenging target of reducing greenhouse gas emissions by 80% by 2050, improving the energy efficiency of buildings - both new and existing - will clearly be vital. This report uses existing data to explore the extent to which improving the energy efficiency of the existing building stock would be a more cost-effective route for achieving CO2 savings than constructing new buildings to the higher levels of energy performance required to meet low- and zero-carbon targets.

September 2010: 32pp
Pb: 978-1-84806-137-8: **£30.00**

To Order: Tel: +44 (0) 1235 400524 **Fax:** +44 (0) 1235 400525
or Post: Taylor and Francis Customer Services,
Bookpoint Ltd, Unit T1, 200 Milton Park, Abingdon, Oxon, OX14 4TA UK
Email: book.orders@tandf.co.uk

For a complete listing of all our titles visit:
www.tandf.co.uk

Enhanced Capital Allowances (ECA), The Energy Technology Criteria List (ETCL) and The Water Technology List (WTL)

Background

This bulletin is intended to raise awareness of the existence, location and relevance of the ETCL and the WTL which are often referred to for simplicity as 'The Technology Lists'. The lists comprise the technologies and products which qualify for the UK Government's Enhanced Capital Allowances (ECA) scheme. Also included in the lists are the energy-saving and water-saving performance criteria for each product or technology.

What are ECAs?

Enhanced Capital Allowances are a form of tax relief enabling a business to claim 100% first-year capital allowances on their spending on qualifying plant and machinery instead of the normal reducing balance writing down allowances of 10% for integral features and 20% for main pool plant. There are two building services-related schemes for ECAs:

- Energy-saving plant and machinery
- Water conservation plant and machinery.

Businesses can write off the whole of the capital cost of their investment in these technologies against their taxable profits of the period during which they make the investment. This can deliver a helpful cash flow boost and a shortened payback period via both the energy, or water, saved and the allowances claimed.

ECAs are only available to investors of qualifying plant and machinery and not developers. Developers are 'traders' and they incur revenue expenditure and not capital expenditure and so cannot claim capital allowances. However the purchaser of a newly constructed development for investment can still make use of ECAs provided they are sold unused and the purchaser is a taxpayer. This can be a valuable selling point for developers if highlighted in the marketing literature.

Interest in ECAs has been highlighted by the changes to the regime which have reduced the value of allowances for many qualifying items. The introduction of Energy Performance Certificates and the Carbon Reduction Commitment Energy Saving Scheme will also promote the types of technologies that could qualify for ECAs.

Who manages the Technology Lists and where are they?

The lists are part of the ECA scheme which was developed by the Treasury, HM Revenue & and the Department of Energy & Climate Change (DECC). The ETL is managed and maintained by the Carbon Trust. The WTL is managed by Business Link and the Department for Environment Food & Rural Affairs (DEFRA). Both lists can be found at www.eca.gov.uk

What technologies and products are included in the ETCL?

- Air-to-air energy recovery
- Automatic monitoring and targeting
- Boiler equipment
- Combined heat and power
- Compact heat exchangers
- Compressed air equipment
- Heat pumps
- Heating ventilation and air conditioning equipment
- Lighting (high efficiency lighting units, lighting controls and LEDs)
- Motors and drives
- Pipework insulation
- Radiant and warm air heaters
- Refrigeration equipment
- Solar thermal systems
- Thermal screens
- Uninterruptible power supplies.

And the WTL?

- Cleaning in place equipment
- Efficient showers
- Efficient taps
- Efficient toilets
- Efficient washing machines
- Flow controllers
- Leakage detection equipment
- Meters and monitoring equipment
- Rainwater harvesting equipment
- Small scale slurry and sludge dewatering equipment
- Vehicle wash water reclaim units
- Water efficient industrial cleaning equipment
- Water management equipment for mechanical seals
- Water re-use.

Eligibility criteria

It is not possible to summarize here the eligibility criteria here because each technology has different definitions and considerations. Most of the above technologies have specific products listed that are eligible, but others are defined by detailed performance criteria. The key technologies worthy of note which operate on a Performance criteria are Lighting, CHP, Pipework Insulation, Automatic Monitoring & Targeting equipment and Water Re-use. Qualification is generally extremely strict. To give a brief example; Lighting doesn't merely concentrate on the lamp type used but also has very strict requirements for the light fitting itself, having to meet high Light Output Ratios, colour rendering requirements and use high frequency control equipment. CHP and AMT equipment is also heavily guided.

Scope of ECA claims

For products on the Technology Lists, claims will be considered for the cost of the equipment itself, and other costs directly involved in installing it. These include:

- Transportation – the cost of getting equipment to the site.
- Installation – cranage (to lift heavy equipment into place), project management costs and labour, plus any necessary modifications to the site or existing equipment.
- Preliminary costs and on-costs – if they are directly related to the acquisition and installation of the equipment.
- Professional Fees – if they are directly related to the acquisition and installation of the equipment.

Making a claim for ECAs

If investing in eligible technology claimants should submit their ECA claims as part of their normal Income or Corporation Tax return. The best advice we can give clients on relevant projects is to highlight the fact that their investment may be eligible, and introduce Davis Langdon Banking Tax & Finance to the client as early as possible in the design process so that they can offer a view on the potential for worthwhile tax relief before, during and after the investment programme.

Follow-up

For further general advice on any aspect of this Bulletin, please contact Andy White of Davis Langdon Banking Tax & Finance.

Daylighting

P. Tregenza et al.

This authoritative and multi-disciplinary book provides architects, lighting specialists, and anyone else working daylight into design, with all the tools needed to incorporate this most fundamental element of architecture. It includes:

- an overview of current practice of daylighting in architecture and urban planning

- a review of recent research on daylighting and what this means to the practitioner

- a global vision of architectural lighting which is linked to the climates of the world and which integrates view, sunlight, diffuse skylight and electric lighting

- up-to-date tools for design in practice

- delivery of information in a variety of ways for interdisciplinary readers: graphics, mathematics, text, photographs and in-depth illustrations

- a clear structure: eleven chapters covering different aspects of lighting, a set of worksheets giving step-by-step examples of calculations and design procedures for use in practice, and a collection of algorithms and equations for reference by specialists and software designers.

This book should trigger creative thought. It recognizes that good lighting design needs both knowledge and imagination.

August 2010: 216x138: 352pp
Pb: 978-0-415-56650-6: **£34.99**

To Order: Tel: +44 (0) 1235 400524 **Fax:** +44 (0) 1235 400525
or Post: Taylor and Francis Customer Services,
Bookpoint Ltd, Unit T1, 200 Milton Park, Abingdon, Oxon, OX14 4TA UK
Email: book.orders@tandf.co.uk

For a complete listing of all our titles visit:
www.tandf.co.uk

LED Lighting

LED Lighting is it the 'best thing since sliced bread'?

Background

There has been much 'talk' that with the introduction of low energy LED luminaires they may in time replace the traditional fluorescent office light fitting. To date however none of the major suppliers have been able to produce a direct replacement LED style unit for general office illumination. There are claims from 'less mainstream' suppliers that they have been able to manufacture a luminaire but as the 'big boys' are taking an opposite stance then it must be assumed that these are false claims. That is not to say one will not be produced at some time in the future and the first to do so will have laid the proverbial 'Golden Egg'.

Currently available types of LED luminaries

The initial concept of LED lighting was developed for decorative external lighting, as a 'white' LED was problematical to produce (at that time) and 'white' was generally 'engineered' by having several differing coloured LEDs as the light source. As a result there is a myriad of different styles, types and colours available on the market for the specialist external lighting market.

Also available are street lighting luminaires, which, with the recent advent of 'white' LEDs, has had the side benefit of dramatically improving the resolution of CCTV coverage to areas such as car parks and building elevations.

Generally speaking the only luminaires available for internal usage are limited to down lighters either small 'twinklies' or the more traditional larger size perimeter down lighter.

Warning though!

It should be noted that a 'standard LED Luminaire' is not generally available in an emergency version and they cannot be converted and a secondary emergency luminaire will be required.

Finally the market place has various 'replacement tubes' available where on the face of it you simply install a LED 'tube' – this is not the case. The 'tube' is simply a linear strip of LED's within a tube, additionally the luminaire requires modification and the photometrics of the two light sources are not directly comparable. It should also be remembered that all diffusers are specially designed for a fluorescent tube and will probably adversely affect the light distribution and will be completely different to that of a normal tube. So caution, when dealing with these, if requested by a client.

Cost implications

Manufacturers and suppliers make great hay regarding the energy savings that can be made against more traditional light sources.

The manufacturers' cost calculations should be carefully examined as in some instances they base their calculations on several incorrect assumptions; firstly they may state 365 twelve hour days and they make no reference to additional emergency luminaires but, even after correcting these incorrect assumptions, the payback and subsequent energy cost savings are dramatic.

Payback

As the LED luminaires are currently more expensive than more traditional sources of illumination the payback period is of particular importance as actual energy savings will only be realised after the LED luminaires if you like have paid for itself.

Therefore based on the following material costs only as it is assumed the labour content will be similar:

Power cost (£)/kWh	0.11
Days in use/year	260
Hours in use/day	12
No. of luminaires	25

	Typical manufacturers	Compact flourescent down lighter	Halogen downlight
Input Power (Watts)	12	24	50
Lifetime (hours)	50,000	12,000	2,500
Replacement Lamp Cost including labour (£)	-	15	13
Annual Energy Cost per lamp (£)	4.12	8.24	17.16
Total Energy for 25 luminaires per year (£)	103	206	429
Luminaire Unit Cost (£)	120	60	12

Supply only cost of 25 luminaires	3,000	1500	300
Supply only cost of 4 emergency luminaries or allowance for battery packs	700	400	400
Lamps	N/A	125	75
Total cost of luminaries/emergencies/lamps	3,700	2,025	775

Extra over cost of LED luminaires and emergency luminaires compared to traditional luminaires is		1,675	2,925
Yearly cost of energy and yearly re-lamping allowance	103	331	754
The calculated yearly energy saving of using LED luminaires when compared to traditional luminaires is		228	651
Therefore time taken in years to 'pay' for the additional cost of the LED luminaires based on the energy and re-lamping costs of 'traditional' luminaires in years is		**7.35**	**4.49**

Conclusion

- LED luminaires do save energy.
- LED luminaires are generally twice the price of a 'traditional' downlight.
- LED luminaires can not be converted to emergency so 'additional' luminaires required.
- LED replacement lamps have completely different light emitting characteristics and cannot be directly compared and the luminaire required modification to accept the LED lamp.

Building Performance Simulation for Design and Operation

J. Hensen et al.

Effective building performance simulation can reduce the environmental impact of the built environment, improve indoor quality and productivity, and facilitate future innovation and technological progress in construction. It draws on many disciplines, including physics, mathematics, material science, biophysics and human behavioural, environmental and computational sciences. The discipline itself is continuously evolving and maturing, and improvements in model robustness and fidelity are constantly being made. This has sparked a new agenda focusing on the effectiveness of simulation in building life-cycle processes.

Building Performance Simulation for Design and Operation begins with an introduction to the concepts of performance indicators and targets, followed by a discussion on the role of building simulation in performance-based building design and operation. This sets the ground for in-depth discussion of performance prediction for energy demand, indoor environmental quality (including thermal, visual, indoor air quality and moisture phenomena), HVAC and renewable system performance, urban level modelling, building operational optimization and automation.

Produced in cooperation with the International Building Performance Simulation Association (IBPSA), and featuring contributions from 14 internationally recognised experts in this field, this book provides a unique and comprehensive overview of building performance simulation for the complete building life-cycle from conception to demolition. It is primarily intended for advanced students in building services engineering, and in architectural, environmental or mechanical engineering; and will be useful for building and systems designers and operators.

January 2011: 536pp
Hb: 978-0-415-47414-6: **£65.00**

To Order: Tel: +44 (0) 1235 400524 **Fax:** +44 (0) 1235 400525
or Post: Taylor and Francis Customer Services,
Bookpoint Ltd, Unit T1, 200 Milton Park, Abingdon, Oxon, OX14 4TA UK
Email: book.orders@tandf.co.uk

For a complete listing of all our titles visit:
www.tandf.co.uk

Getting the Connection

Details the process involved in the provision of a new electrical supply to a site.

The provision of an electricity connection has both a physical and contractual element. Physically, it involves the design, planning and construction of electrical infrastructure (cables, switchgear, civil works), whilst contractually, it requires legal agreements to be drawn up and agreed (construction, connection, adoption).

Planning

The Planning Stage, typically RIBA Stage C, is where the site's developer (the Developer) should be formulating their plans for the scheme and, in doing so, consult the local distribution network operator's (the Host)[1] long-term development statement, which will identify potential connection point opportunities. This information is normally readily available from the Host and should provide an early indication of whether the Host's network may need to be reinforced before the development can be connected.

As the Planning Stage progresses, the Developer should discuss its proposals with the Host. Relatively simple connections for single building supplies are straight forward, however schemes of a more complex nature may require some form of feasibility study to be carried out to assess connection options and provide indicative costs for the contestable[2] and non-contestable[3] work elements.

Design

The Design Stage, typically RIBA Stage D, is the point at which the Developer submits its formal connection request to the Host. It is important that this is completed in accordance with the specific procedures of the Host, since if it does not include all supporting information required by the Host's application process, there is likely to be a delay in the processing of a firm offer. The Host is expected to provide a documented application process to assist the applicant make a complete application.

The convention is to request a Section 16/16A connection[4] where the terms are standard and non-negotiable. The alternative is a Section 22 connection[2] offer, where the terms are fully negotiable. If the Section 22 route is chosen then caution is required, as disputes over Section 22 agreements cannot be referred to the regulator for determination after the connection agreement has been signed; a post contract dispute will have to be pursued as a civil action. It is always advised that a Section 16/16A offer is requested before considering the Section 22 option.

On receipt of the Host's firm offer[5], the Developer generally has 90 days to accept its terms and to undertake its own review of the offer to ensure it meets its requirements. If for any reason the Developer and Host are unable to reach agreement of the terms, it is recommended that the Developer seeks specialist advice. It should be noted that in extreme circumstances, it may be necessary to refer an issue to regulator for determination but, be warned, this can be a lengthy process, taking upto 16 weeks to conclude.

---

1.   A distribution network operator (DNO) is a company that is responsible for the design, construction operation and maintenance of a public electricity distribution network. The host DNO is the electricity distribution network to which the development site will directly connect.
2.   Contestable is work in providing the connection that can be carried out by an accredited independent party.
3.   Non contestable is work that can only be carried by the Host.
4.   The Electricity Act 1989.
5.   Unless the offer becomes interactive, i.e. another connection scheme is vying for network capacity. Where the connection request becomes interactive the Developer has 30 days to accept the offer. The Host will inform the Developer if the connection request is interactive.

Competition

One of the key decisions affecting the way in which the connection process proceeds is whether the Developer wishes to introduce competition into the procurement process by appointing a third party to design and construct the contestable connection works. In these circumstances, the Developer can requisition a non contestable quotation from the Host or ask the third party to do so on its behalf. The Host is obliged to provide information within standardized time frames against which its performance is monitored by the industry regulator (Ofgem).

Contestable works are those that may be carried out either by the Host or by an approved contractor, on the Developer's behalf. This contrasts with non contestable works, which can only be carried out by the Host. Broadly speaking, the Host will make the connection to its network for the new supply and undertake any upstream network reinforcement works. All works downstream from the point of connection to the Host network into the site and to each building are contestable works. Therefore, the extent of the contestable works can vary significantly depending on where the point of connection is designated by the Host. Alternatively, the Developer can request an independent licensed network operator (IDNO) to adopt assets constructed by the third party.

If the Developer decides to contract with a third party to construct the contestable works, it is the Developer's responsibility to ensure that the construction works meet the Host's network adoption requirements.

Network Reinforcement

Reinforcement works may be required to increase the capacity of the network to enable the connection to meet a site's projected demand. In terms of the capacity made available by reinforcement, the following possible scenarios arise:

- Where reinforcement works are necessary for sole use by the Developer, the Developer is charged the full cost of the works. In some circumstances the most economic method of reinforcement may introduce spare network capacity in excess of the Developer's requirements. In such circumstances the Developer can receive a rebate where this spare capacity is absorbed by subsequent developments. However, the possibility of a rebate is time constrained and the original development will only qualify for a rebate for up to 5 years after the connection is completed.
- Where reinforcement works are necessary but also cater for the Host's future network requirements, the Developer is charged for a proportion of the cost of the works in the form of a capital contribution. This is calculated using cost apportionment factors for security (the ratio of capacity requested to that which is made available) and fault level (the ratio of fault level contribution of connection to that which is made available).

The charges levied for reinforcement works are attributable to those reinforcement works undertaken at one voltage level above the connection voltage only. That is, if connection voltage is 400V then costs are chargeable for reinforcement works undertaken at the next highest voltage, which is usually 11kV, and not for works conducted at the next highest voltage, which is usually 33 kV[6]. These deeper network reinforcement costs are recovered as part of the system charges built into the electricity supply tariffs.

Costs

The cost of the Developer's connection depends on the nature and extent of the works to be undertaken. The distance between the site and the Host's network, the size of the customer demand in relation to available capacity (and hence the potential need for reinforcement), customer-specific timescales and to an extent market value of raw material and labour, are all significant factors that will affect the cost.

6.   Some DNO's use voltages of 6.6kV and 22kV

As part of the firm offer received from the Host, the Developer is provided with a charging statement, which includes the charges to be levied for the following items:

- Assessment and Design – to identify and design the most appropriate point on the existing network for the connection
- Design Approval – to ensure design of a connection meets the safety and operation requirements of the Host
- Non-contestable Works/Reinforcement – to include circuits and plant forming part of the connection that can be undertaken by the Host only, including land rights issues and consents
- Contestable Connection Works – to include circuits and plant forming part of the connection that can be undertaken by approved contractors or the Host
- Inspection of Works – to ensure that works are being constructed in accordance with the design requirements of the Host
- Commissioning of Works – to include circuit outages and testing that will ensure that connection is safe to be energized.

In some circumstances, in addition to the cost of the physical connection works there may be chargeable costs associated with operation, maintenance repair and replacement of the new or modified connection. These are known as O&M costs and are chargeable as capitalized up-front costs where the Developer requests a solution that is in excess of the minimum necessary to provide the connection. For example, where extra resilience is requested for a connection, over and above that which the Host is obligated to provide, then O&M costs can be levied but only for the extra resilience element of the connection, not the total cost of the connection.

Spon's First Stage Estimating Handbook

3rd Edition

By **Bryan Spain**

Have you ever had to provide accurate costs for a new supermarket or a pub "just an idea...a ballpark figure..." ?

The earlier a pricing decision has to be made, the more difficult it is to estimate the cost and the more likely the design and the specs are to change. And yet a rough-and-ready estimate is more likely to get set in stone.

Spon's First Stage Estimating Handbook is the only comprehensive and reliable source of first stage estimating costs. Covering the whole spectrum of building costs and a wide range of related M&E work and landscaping work, vital cost data is presented as:

- Costs per square metre
- Principal rates
- Elemental cost analyses
- Composite rates

Compact and clear, *Spon's First Stage Estimating Handbook* is ideal for those key early meetings with clients. And with additional sections on whole life costing and general information, this is an essential reference for all construction professionals and clients making early judgements on the viability of new projects.

January 2010: 216x138: 244pp
Pb: 978-0-415-54715-4: **£45.00**

To Order: Tel: +44 (0) 1235 400524 **Fax:** +44 (0) 1235 400525
or Post: Taylor and Francis Customer Services,
Bookpoint Ltd, Unit T1, 200 Milton Park, Abingdon, Oxon, OX14 4TA UK
Email: book.orders@tandf.co.uk

For a complete listing of all our titles visit:
www.tandf.co.uk

Facade Systems

The relationship between the building envelope and the provision of cooling in a building and how this affects office costs.

Attaining legislative and aspirational environmental performance criteria for any new project represents a key deliverable for the design team, which entails an intrinsic link between the services strategy and building envelope solutions. In this cost model, Building Services Cost Specialist Davis Langdon Engineering Services investigates the impact that air conditioning solutions can have upon the building envelope and the potential affect that this can have upon costs.

The building envelope solution must accommodate a wide variety of criteria in addition to environmental performance, some of which could be perceived to contradict others. Figure 1 provides a simple illustration of the interrelated factors that need to be addressed in the design of a building facade. As environmental performance of our buildings becomes more stringent to reduce harmful greenhouse gas emissions it is clear that architectural intent could be heavily influenced by the ultimate building envelope solution. It may well be that certain criteria override others on specific projects, and so drive the ultimate solution.

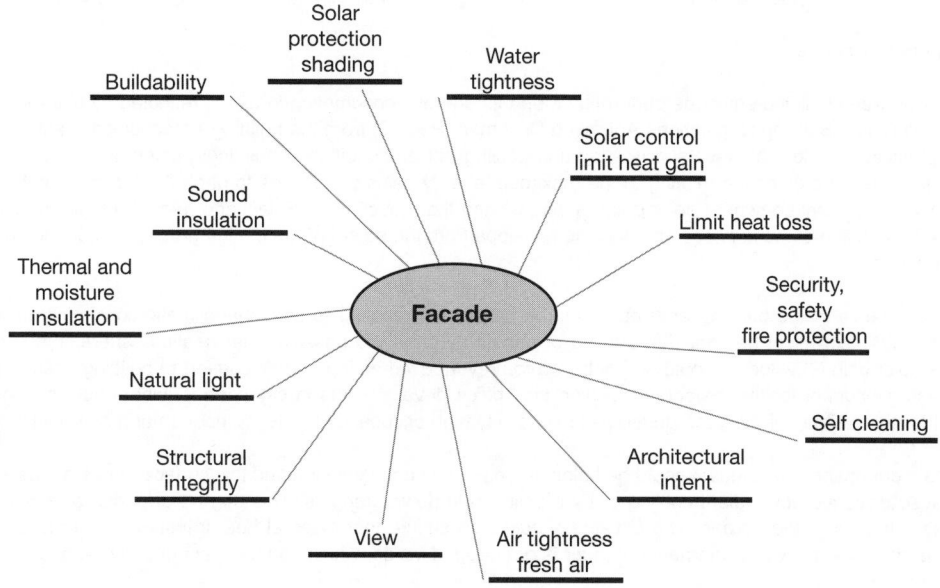

Figure 1 Functions of a facade

It is therefore reasonable to state that design of the building envelope, particularly in the case of modern commercial office buildings, upon which this article is based, entails a managed balance between its performance and that of the services, taking cognisance of the primary frame design, in order to achieve a solution that fulfils the brief and delivers high quality whilst being commensurate with project specific requirements, including budget constraints.

The building envelope provides an interface between the controlled internal environment of an office building and the variable external climate, thus ostensibly acting as a modifier of the direct effects of the climactic variables highlighted in Table 1, which are the primary parameters influencing thermal comfort and energy requirements within the building.

In naturally ventilated and/or mixed mode buildings, where site locations are conducive (noise and air pollution), the indoor temperatures follow the diurnal cyclic pattern of outdoor temperature and solar radiation. This relationship between the outdoor temperature, solar radiation and indoor temperature is determined at a building level by the orientation, building form, optical and thermo physical properties of the building envelope. Likewise in an air-conditioned building, the properties of the building envelope determine the amount of cooling required. Thus a climatically responsive design of the building envelope is key to improving thermal comfort conditions and reducing energy consumption.

When developing the design of new commercial office buildings, it should to be recognized that the balance between building envelope and services performance, whilst intrinsic, must dove tail with other key project specific criteria, including but not limited to:

- Architectural intent, to enable a high quality building solution, in terms of appearance and performance, ensuring whole life cost and performance efficiency
- Buildability, to ensure quality and build efficiency, thus keeping construction costs and programme to a minimum, taking cognisance of cleaning and long term maintenance of the building in use
- Procurement, ensuring that facade design solutions can be built by a wide range of appropriately skilled specialist contractors, to mitigate risk and ensure best value
- Cost, and the need to ensure that all criteria are aligned to maintain the commercial viability of the scheme.

Energy performance

The link between building envelope performance and its energy consumption in use is reflected by allowable CO_2 emissions (under Building Regulations Approved Document Part L2) from the building's mechanical plant. In order to keep within the allowable level the heating and cooling (for air-conditioned buildings) loads need to be limited. These limits provide guidance relating to the maximum energy gains and losses through the facade, notably heat loss (U values), management of solar gain (G values) and the reduction of air leakage rates. Thus a co-ordinated approach to design of the services and building envelope from the outset of the design process should be seen as a prerequisite.

The performance of the building envelope must be capable of reacting to seasonal climatic conditions. In winter months, solar gain might be considered to be of benefit by providing passive solar heating, whereas in summer months solar gain (G values) should be controlled to prevent overheating and discomfort to building users. This is particularly important for the majority of 'sealed box' office developments in city centres, where floor spaces are cooled for the majority of the year, driven by increasingly high occupancy levels, computers and artificial lighting.

Thermal performance (U values) should be balanced to prevent condensation and low surface temperatures, which can cause discomfort by radiant cooling or by creating cold down-draughts. This can be of particular concern for facades with large glazed areas. The U value should also ensure that external heat transference to the inside is reduced and conversely any internal heated air is not trapped inside, which could impact upon cooling loads.

Therefore, a key deliverable in the building envelope design is solar control, which can be achieved by a variety of solutions, ranging from cost effective high performance coatings on the glass, aided by potentially increased height insulated spandrel panels, possible further increase in solidity, again potentially cost effective, to solutions which attract higher cost premiums, notably external brise soleil (shading) or ventilated cavity curtain walling with motorized solar control blinds. All options have their own idiosyncratic pros and cons, in terms of capital and operating costs, plus associated risks.

The building envelope 'zone' depth can be affected by a range of issues, including excessive floor spans (dynamic wind pressure), introduction of external brise soleil (shading) and tolerance/vertical alignment of the system solution relative to slab edge. A large amount of commercial office building comprise curtain walling solutions, often unitized (prefabricated), which are set outbound of the primary structure and which thus maximize net floor area.

Adjustable external shading is considered to be more energy efficient compared to the relative inflexibility of fixed shading systems, however these solutions attract cost premiums from both an initial capital cost and potentially operating cost perspective, exacerbated by the increase in facade zone depth (reduction in net floor area).

Where external shading is deemed to be necessary for attaining calculated solar control, value engineered building envelope designs often entail fixed shading solutions. These comprise horizontal fins for south facing elevations to combat high angle sun and vertical fins for east and west orientation to accommodate lower angle sun paths. Clearly the effectiveness of fixed shading is reduced compared to adjustable or motorized blinds within a ventilated cavity system or possibly within double glazed units, and so it is common to utilize glass with high performance coating to supplement solar control performance of the fixed shading solution.

In addition to attracting a cost premium, the following issues also need to be considered when considering possible external shading solutions:

- Dictating architecture
- Potential reduced day light transmittance (exacerbated during inclement weather)
- Reduced accessibility to glass for cleaning and glass replacement
- Damage risk by cleaning cradles
- Necessity to clean shading frequently to prevent invalidation of warranty (for coated or anodized finishes).

Elevations orientated due north do not generally require solar control measures, which potentially enables more cost effective building envelope solutions to be adopted, thus assisting in reducing the building envelope global average cost, which enhances the commercial viability of a scheme.

The impact of different M&E solutions on the building envelope

The table used in the cost model below lists four common generic air conditioning solutions, each of which places more onerous performance requirements upon the building envelope solution.

Each of the systems has a maximum cooling output relative to the performance of the plant and equipment that comprise the system. Taking the internally generated heat gains out of the system capacity, such as those from lighting, smallpower and occupants, leaves the remainder to deal with the solar gain and so this dictates how hard the facade has to work in reducing this heat gain to a level that the air conditioning system can satisfactorily deal with. This could equally result in an expensive high performing glazed facade or alternatively one with smaller windows and a more heavyweight structure.

Whilst this may be considered an oversimplification, it does essentially describe the process.

Cost model

In order to evaluate the performance of different types of building envelope solutions pertinent to the variety of air conditioning systems in a commercial office building, target solar gains have been set relative to the different systems in terms of W/m of facade that each air conditioning system is capable of accommodating in terms of its output. These are summarized in Table 1 below.

The figures are derived from the typical available steady-state capacity of each air conditioning system being considered. Internal gains for occupants, lighting and small power are first subtracted based on BCO guideline figures and assuming no daylight control on lighting. An allowance for fabric/infiltration is also subtracted, leaving a net available system capacity for absorbing solar gain into the space. This solar gain capability is then translated into an allowable solar gain expressed as W/linear metre of facade, based on a typical perimeter office. This enables the figures to be independent of facade height or glazing area.

The G-values (which can be defined as the coefficient of permeability of total solar radiation energy) are then derived by taking a typical incident solar radiation value on a 2.6 m high facade panel and calculating the percentage of this energy that may be transmitted into the space to meet the solar gains per metre of facade set by the first process.

Table 1

Item	W/m of facade	G Value	Height	System
1	100	0.05	2.6	Passive chilled beams or displacement systems
2	200	0.10	2.6	Active chilled beams
3	300	0.15	2.6	Low capacity fan coil units or all air system
4	400	0.20	2.6	High capacity fan coil units but with poorer building energy performance

In terms of the cost model, Table 2 details the facade solutions that are proposed for each system that can satisfy the criteria outlined in Table 1, ranging in cost and specification. It assumes a typical 1.5 m module width unitized curtain walling solution for a storey height of 3.8 m and with a vision glass area of 1.5 m wide × 2.6 m high, in line with the Table 1 above. This assumes a 1.2 m high solid spandrel panel. The table demonstrates how the variety of G values are achievable with different building envelope solutions, based on a variation on theme from a base unitized 'chassis' solution, taking performance, architectural intent and cost into consideration.

The building envelope solutions adopted in the table are generic for the purpose of this article, but are representative of typical solutions for commercial office buildings. Clearly there are a variety of sub types, which could impact upon cost, architectural intent, buildability efficiency, net floor area, etc., all as covered above.

It can be seen that different facade solutions can satisfy the set performance criteria, ranging from the cost effective to the relatively high cost, depending on the criteria that are called for. Similarly, the same solutions may, in some cases, be appropriate for more than one air conditioning system.

Table 2

	Description	Cost/m² Facade	Comments
A	Unitized curtain wall, high performance (HP) coated glass sealed units with G value of 0.25, solid spandrel panels	£650/m²	Based on table above, this solution will not achieve compliance and will require additional solar control measures
A1	Unitized curtain wall, as described in A, but 50% solid panels in addition to spandrel panels	£750/m²	This option should be capable of achieving G values of 0.15 and 0.20, possibly with less dark HP glass, thus cost effective
A2	Unitized curtain wall, as described in A, but with external brise soleil (horizontal blades to south, vertical to east and west)	£800/m²	This option should be capable of achieving G values of 0.15 and 0.20, possibly with less dark HP glass, thus cost effective
A3	Unitized curtain wall, as described in A2 but with greater density of brise soleil for improved solar shading	£850/m²	This option should be capable of accommodating a G value of 0.10, including HP glass
B	Unitized externally ventilated cavity double wall facade, with cavity depth circa 100 mm. This solution does not require mechanically conditioned air within the cavity. Wider walk-in cavity facades attract significant cost premiums	£1100/m²	This option comprises motorized Venetian blinds within cavity, which requires electrical and BMS supply. For a 3.8 m storey height and no spandrels this solution should achieve a minimum G value of 0.10 with blinds closed

Conclusion

The above demonstrates how cost effective facade solutions could be possible for buildings, and recognizes the significant cost premium associated with upgrading to more complex solutions.

Achieving current Part L, and indeed the next iteration of Part L of the Building Regulations and producing energy efficient commercial buildings, may mean a shift away from the iconic highly glazed facade to more opaque, perhaps more heavyweight buildings, with greater attention to architectural detailing.

Developers and occupiers are increasingly recognizing the advantage of lower energy buildings with higher BREEAM ratings, of which the fabric plays a key part. This contributes to supporting corporate environmental policies, potentially attracting favourable funding terms, whilst facilitating lower cost in use.

Construction Contracts Questions and Answers

2nd Edition

By **David Chappell**

What they said about the first edition: "A fascinating concept, full of knowledgeable gems put in the most frank of styles… A book to sample when the time is right and to come back to when another time is right, maybe again and again." – *David A Simmonds,* Building Engineer *magazine*

• Is there a difference between inspecting and supervising?

• What does 'time-barred' mean?

• Is the contractor entitled to take possession of a section of the work even though it is the contractor's fault that possession is not practicable?

Construction law can be a minefield. Professionals need answers which are pithy and straightforward, as well as legally rigorous. The two hundred questions in the book are real questions, picked from the thousands of telephone enquiries David Chappell has received as a Specialist Adviser to the Royal Institute of British Architects. Although the enquiries were originally from architects, the answers to most of them are of interest to project managers, contractors, QSs, employers and others involved in construction.

The material is considerably updated from the first edition – weeded, extended and almost doubled in coverage. The questions range in content from extensions of time, liquidated damages and loss and/or expense to issues of warranties, bonds, novation, practical completion, defects, valuation, certificates and payment, architects' instructions, adjudication and fees. Brief footnotes and a table of cases will be retained for those who may wish to investigate further.

August 2010: 216x138: 352pp
Pb: 978-0-415-56650-6: **£34.99**

To Order: Tel: +44 (0) 1235 400524 **Fax:** +44 (0) 1235 400525
or Post: Taylor and Francis Customer Services,
Bookpoint Ltd, Unit T1, 200 Milton Park, Abingdon, Oxon, OX14 4TA UK
Email: book.orders@tandf.co.uk

For a complete listing of all our titles visit:
www.tandf.co.uk

Feed-In Tariffs (FITs)

To ensure that everyone plays their part towards achieving the Renewable Energy Strategy's (RES) target of 15% for renewable energy by 2020, the need to develop new methods of renewable energy in all sectors is apparent.

On 1st February 2010, the Department for Energy and Climate Change (DECC) announced their intention to implement tariff levels following the Energy Act 2008 introduction of Feed-in Tariffs (FITs). The FIT scheme is intended to encourage the adoption of small-scale low carbon technologies up to a maximum of 5 MW through tariff payments made on both generation and export of produced renewable energy.

Feed-in Tariffs (FITs) are effectively payments which are made for every kilowatt per hour of renewable electricity (kWhr) that is generated. Homeowners/businesses will have the opportunity to benefit financially from the scheme either by consuming it on-site, earning a generation tariff, or by selling spare capacity to the grid, generating an export tariff.

The Feed-in Tariffs Order 2010 came into effect on 1st April 2010. The schemes policy and tariff rates are set by the Government with the scheme being administered by energy suppliers and Ofgem.

Feed-in-Tariffs will be paid by electricity suppliers. Generators between 50 kW and 5 MW should apply to Ofgem for accreditation. Micro-generators can obtain accreditation via the Microgeneration Certification Scheme (MCS), before then applying to a supplier for a FIT agreement. All generation will be metered and FIT payments will be made in accordance with the Electricity Act 1989. The consumer will be responsible for the capital of plant; the necessary access/connections to the electricity and transmission systems and organization of the payment receipts.

FITs review

The first review of the Feed-in Tariffs scheme was scheduled to be carried out in 2012 however it has been brought forward by a year and the findings are due to be published by the end of 2011.

The review has been ordered to be carried out a year early for a number of reasons. The first is due to the high uptake of the Feed-in Tariff scheme; 21,000 installations have been registered as of February 2011. The other reason is the government's concern that the FITs scheme is being used for commercial gains. Findings suggest that large scale solar farms are being set up with the intention of making money from Feed-in Tariffs, this is soaking up money that was intended to help homes, communities and small businesses generate their own electricity.

FITs Amendment Order 2011

The Feed-in Tariffs Amendment Order 2010 was brought into affect 30th May 2011 and works alongside the first FIT Review. The amendment was carried out in light of the early experiences of implementing the FITs scheme.

The FITs Amendment Order 2011 introduces modifications and refinements to the 2010 Order. The aim of the Order is to provide greater clarity for the Authority, suppliers, generators and wider participants as to how the FITs scheme is intended to run.

There are three key areas of the 2010 Order that the FITs Amendment Order 2011 has identified as needing refinements:

- Providing transitional arrangements to enable micro-hydro schemes to obtain accreditation.
- Extend by one year the period in which eligible installations with a capacity of less than 50 kW can transfer from Renewable Obligations to FITs.
- Limiting the time period over which generators can potentially receive both grants and FITs, this is to reduce 'over-subsidy' and prevent FIT costs from rising unnecessarily.

Eligibility

Anyone who installs a qualifying renewable electricity system after 15th July 2009 is eligible to claim FITs. However, they will only be able to receive tariffs from April 2010.

The maximum declared net capacity for the renewable installations under the scheme is 5 MW and supports the following technologies:

- Solar Photovoltaic (PV)
- Wind
- Hydro
- Anaerobic Digestion
- MicroCHP (pilot programme with a 2 kW limit).

Installations applying for the scheme which have a capacity of 50 kW or less are required to use MCS eligible products installed by an MCS accredited installer. This requirement does not apply to anaerobic digestion installations or larger installations up to the scheme limit of 5 MW.

As of 1st April 2010, micro generators (<50 kW) in Anaerobic Digestion, Hydro, Solar PV and Wind will be ineligible for support under the Renewable Obligation (RO) due to FIT inclusion. Refurbished/renovated installations will be ineligible. Small generators that applied for accreditation on or after the 15th July 2009 and before 1st April 2010 will be eligible to transfer to FITs.

Generating stations whose electricity is sold under a Non-Fossil Fuel Obligation (NFFO) arrangement will be ineligible for FITs; however will remain eligible to receive RO support.

Any existing systems installed before 15th July 2009 will only qualify for the FIT scheme if they are under 50kW and registered with the RO.

Tariffs

The Feed-in Tariffs (FITs) provide three main financial benefits to the consumer:

- A 'Generation' tariff (up to 41.3p/kWhr depending on type and size) based on the total generation and the energy type
- An 'Export' tariff (3p/kWhr) for any surplus energy produced exported into the grid and due to producing a percentage of the energy used
- With the additional benefit of reducing electricity bills from your supplier for the remaining imported energy.

Renewable technology	Scale	Generation Tariff for new installations (years run from April to March)			Duration (years)
		Year 1: 2010–2011	Year 2: 2011–2012	Year 3: 2012–2013	
Anaerobic digestion	≤ 500 kW	11.5	12.1	12.1	20
Anaerobic digestion	> 500 kW	9.0	9.4	9.4	20
Hydro	< 15 kW	19.9	20.9	20.9	20
Hydro	> 15–100 kW	17.8	18.7	18.7	20
Hydro	> 100 kW–2 MW	11.0	11.5	11.5	20
Hydro	> 2 MW–5 MW	4.5	4.7	4.7	20
Micro-CHP	≤ 2 kW	10	10.5	10.5	10
Solar PV	≤ 4 kW (new build)	36.1	37.8	37.8	25
Solar PV	≤ 4 kW (retrofit)	41.3	43.3	43.3	25
Solar PV	> 4–10 kW	36.1	37.8	37.8	25
Solar PV	> 10–100 kW	31.4	32.9	32.9	25
Solar PV	> 100 kW–5 MW	29.3	30.7	30.7	25
Solar PV	Stand Alone System	29.3	30.7	30.7	25
Wind	≤ 1.5 kW	34.5	36.2	36.2	20
Wind	> 1.5–15 kW	26.7	28	28	20
Wind	> 15–100 kW	24.1	25.3	25.3	20
Wind	> 100–500 kW	18.8	19.7	19.7	20
Wind	> 500 kW–1.5 MW	9.4	9.9	9.9	20
Wind	> 1.5–5 MW	4.5	4.7	4.7	20
Existing generators transferred from RO		9.0	9.4	9.4	To 2027

Carbon Trading

What is carbon trading?

Carbon trading is currently the central pillar of the Kyoto Protocol and other international agreements aimed at slowing climate change.

Emissions' trading (also known as cap and trade) is a market-based approach used to control pollution by providing economic incentives for achieving reductions in the emissions of pollutants. Carbon emissions' trading is a form of emissions trading that specifically targets carbon dioxide.

A central authority (usually a governmental body) sets a limit or cap on the amount of a pollutant that can be emitted. The limit or cap is allocated or sold to firms in the form of emissions permits which represent the right to emit or discharge a specific volume of the specified pollutant. Firms are required to hold a number of permits (or credits) equivalent to their emission levels. The total number of permits cannot exceed the cap, limiting total emissions to that level. Firms that need to increase their emission permits must buy permits from those who require fewer permits. The transfer of permits is referred to as a trade. In effect, the buyer is paying a charge for polluting, while the seller is being rewarded for having reduced emissions. Thus, in theory, those who can reduce emissions most cheaply do so, achieving the pollution reduction at the lowest cost to society.

Carbon trading in the UK

New rules set out by the government will pitch some of the UK's largest organizations against each other in a drive to cut carbon dioxide (CO_2) emissions.

For the first time large non-energy intensive organizations, which account for about 10% of UK CO_2 emissions, will be legally bound to closely monitor and report their emissions from energy use in preparation for carbon trading.

The Government scheme, known as the CRC Energy Efficiency Scheme, will include businesses will be ranked according to reductions in energy use and improvements in energy efficiency alongside public sector organizations such as NHS trusts, local authorities and government departments.

Analysis for the Environment Agency suggests that the scheme could reduce CO_2 emissions by up to 11.6 million tonnes per year by 2020, the equivalent of taking four million cars off the road. It is also expected to save organizations money through reduced energy bills, benefiting the economy by at least £1billion by 2020.

During the introductory phase in 2012, allowances will be sold at a fixed price of £12 per tonne of CO_2. A further 15,000 organizations that use less than 6,000 MWh, but still have at least one half hourly electricity meter, will be obliged to register and declare their electricity use.

How the CRC Energy Efficiency Scheme works

The CRC Energy Efficiency Scheme will be phased in over three years. Once fully operational, CRC Participants (about 5,000 organizations) will be required to monitor their emissions and purchase allowances for each tonne of CO_2 they emit at the beginning of each reporting year.

After the three-year introductory phase, the total number of allowances will be capped, and these allowances will be auctioned, rather than sold at a fixed price. As a result, the cost of purchasing allowances should become higher making it financially more attractive for CRC Participants to reduce their CO_2 emissions by introducing energy saving measures. Participants that perform well will also be placed higher in the Performance League Table, which will be published annually by the Environment Agency. Being higher up the league table will have the added benefit of enhancing the organization's reputation.

The EU Emission Trading System (EU ETS)

The EU ETS is one of the policies introduced across the European Union (EU) to help it meet its greenhouse gas emissions reduction target under the Kyoto Protocol. The EU has to make an eight per cent reduction on 1990 levels by the first Kyoto Protocol commitment period (2008–2012).

Carbon floor

What is a carbon floor price and why is it needed?

Creating a carbon floor price in the UK essentially requires our industries to pay a top up if the market price for carbon falls below a certain level.

A carbon floor price is a regulatory/taxation policy that states that polluters must pay a minimum amount of money for the right to pollute. This is likely to take the form of a tax that requires those who qualify to make a payment to the Treasury. It is expected to replace the existing Climate Change Levy, which is a downstream tax on energy use rather than a direct upstream tax on greenhouse gas pollution.

Roughly half of Europe's emissions are covered by a European regulation that caps emissions (the EU Emissions Trading Scheme) requiring them to submit sufficient permits to cover their emissions. Permits, known as allowances, can be freely traded and the price someone is willing to pay to acquire them determines the price of pollution. At present because there are too many allowances available in the market compared to the demand prices are relatively low — at around €15 per tonne of carbon dioxide equivalent. This low price is not necessarily enough to dissuade polluters from continuing to emit and does not provide an attractive enough return for would-be investors in low carbon solutions. There is also the risk that it could fall even lower. This lack of price certainty is seen as a potentially important barrier to investment.

A carbon floor price is therefore primarily designed to attract low carbon investment into a country by making the price of pollution higher and increasing the rewards for low carbon projects. As explained below it is not in and of itself an environmental policy and in terms of value for money it must be assessed as an industrial policy.

The UK could decide to impose an immediately effective floor price, which takes the price of pollution all the way up to this projected level of cost. This would however have a very significant impact on the competitiveness of UK industry relative to competitors in the rest of Europe. It is much more likely that the floor price will initially be set either at today's market levels (around £13-14 per tonne) or slightly above, incorporating the cost of the Climate Change Levy which is currently equivalent to around £4–6 per tonne. A price escalator may also be built-in to steadily increase the floor over time.

Renewable Obligation Certificates (ROCs)

The Renewable Obligation (RO) is the main support scheme for renewable electricity projects in the UK, placing an obligation on licensed UK suppliers of electricity to increase their proportion of electricity production from renewable sources, or result in a penalty.

Since its introduction in 2002, it has succeeded in tripling the level of renewable electricity in the UK from 1.8% to 5.4% and is currently worth around £1 billion/year in support to the renewable electricity industry. The RO applies to all powered plant with a power capacity greater than 5MW. The target started at 3% and is presently at 11.4% rising incrementally to 15.4% by 2015. It is likely to be extended to 20% by 2020.

In April 2009, the introduction of banding under the Renewables Obligation Order 2009 meant different technologies receive different levels of support, providing a greater incentive to those that are further from the market.

The RO was extended from its current end date of 2027 to 2037, in April 2010, for new projects with a view to providing greater long-term certainty for investors and an increase in support for offshore wind projects.

The RO is administered by the Office of the Gas and Electricity Markets (Ofgem) and suppliers of electricity have to prove they have met this obligation, producing Renewable Obligation Certificates (ROCs) to renewable electricity generators at the end of each year.

A Renewable Obligation Certificate (ROC) is a green certificate that is issued by Ofgem to an accredited generator for eligible renewable electricity generated within the UK and supplied to customers in the UK by a licensed supplier. A ROC is issued for each megawatt hour (MWh) of eligible renewable output generated.

Failing to meet the obligation results in 'buy-out' fines being paid to Ofgem on the shortfall of every MWh sold that was not renewable. Ofgem then distributes the funds to all electricity supply companies possessing ROCs, the amount received being in proportion to the number of ROCs held (at the end of the year). If a supplier meets part or all of its RO, but other companies do not, the supplier who has ROCs will be rewarded with a share of the fines.

Previously, 1 ROC was issued for each megawatt hour (MWh) of eligible generation, regardless of technology. Since April 2009, the reforms introduced means that new generators joining the RO now receive different numbers of ROCs, depending on their costs and potential for large-scale deployment. For example, onshore wind continues to receive 1 ROC/MWh, whereas offshore wind and energy crops currently receive 2 ROCs/MWh.

Obligation periods are valid for a year, beginning on the 1 April to 31 March. Supply companies have until the 31 September following the period to submit sufficient ROCs to cover their obligation, or submit sufficient payment to Ofgem to cover their shortfall.

Buy-out price

Suppliers can meet all, or part of their obligations by making a buy-out payment. The buy-out price set by Ofgem for the compliance period of 2011–2012 is £38.69 per Renewables Obligation Certificate (ROC). The buy-out price sets the rate which suppliers must pay if they fail to meet their obligations under the scheme and is adjusted annually in accordance with the Retail Prices Index (RPI).

Buy-out fund redistribution

At the end of the year, the funds made to Ofgem are distributed to all the electricity suppliers possessing ROCs, with the amount received in proportion to the number of ROCs held. If a supplier meets all or part of its RO, it will be rewarded with a share of the buy-out fines.

Pricing

Due to ROCs having the potential to save the supplier from having to commit to a buy-out payment, it increases the price of the electricity. When the renewable generator sells the electricity to a supplier it is not uncommon for the ROC to be sold in addition ultimately forcing the cost of electricity upwards. Also, due to the fact that ROCs entitles suppliers to a share of the 'buy-out' fund at the end of year, increases its value. Electrical suppliers can benefit financially by participating in the RO system due to the renewable targets set by the Government likely to be under-fulfilled and the fact that the RO is not over-subscribed will result in the ROCs and their recycled values being worth more than the £38.69 per MWh.

e-ROC

The most efficient method of buying and selling Renewable Obligation Certificates (ROCs) is through the e-ROC on-line auctions. They offer renewable generators access to the whole supplier market in the UK, delivering high ROC prices for low fees. The average price of ROCs sold through the auctions 25 May 2011 was £51.24 and with the fees set at only 50p per ROC (subject to a minimum fee of £50), indicates a profitable return for those in participation. Auctions are operated by NFPAS, a subsidiary of the Non-Fossil Purchasing Agency Limited (NFPA) and are usually held four times a year. NFPAS runs regular e-ROC on-line auctions for the sale of Renewable Obligation Certificates (ROCs).

Eligibility

The reforms stated in Renewable Obligations Order 2009 introduced the concept of 'banding' for the Renewable Obligation Certificates (ROCs). The aim of ROC banding is to establish the number of ROCs per MWh that can be obtained according to the type of technology that is used to generate the renewable electricity.

There are 28 renewable technologies covered by ROCs Banding, resulting in an increasingly complex regulatory environment for technology providers, project developers and finance providers to navigate.

Band	Renewable technology	Level of banding (ROCs/MWh)
Established 1	Landfill Gas	0.25
Established 2	Sewage Gas Co-Firing of Non-Energy Crops (Regular) Biomass	0.5
Reference	Onshore Wind Hydro-Electric Co-Firing of Energy Crops Co-Firing of Biomass with CHP Energy from Waste with CHP Geo Pressure Pre-Banded Gasification Pre-Banded Pyrolysis Standard Gasification Standard Pyrolysis	1

Band	Renewable technology	Level of banding (ROCs/MWh)
Post-Demonstration	Offshore Wind Dedicated Regular Biomass Co-Firing of Energy Crops with CHP	1.5
Emerging	Wave Tidal Steam Advanced Gasification Advanced Pyrolysis Anaerobic Digestion Dedicated Energy Crops Dedicated Energy Crops with CHP Dedicated Regular Biomass with HP Solar Photovoltaic Geothermal Tidal Lagoons Tidal Barrages	2

Typical Engineering Details

In addition to the Engineering Features, Typical Engineering Details are included. These are indicative schematics to assist in the compilation of costing exercises. The user should note that these are only examples and cannot be construed to reflect the design for each and every situation. They are merely provided to assist the user with gaining an understanding of the Engineering concepts and elements making up such.

ELECTRICAL

- Urban Network Mainly Underground
- Urban Network Mainly Underground with Reinforcement
- Urban Network Mainly Underground with Substation Reinforcement
- Typical Simple 11 kV Network Connection For LV Intakes Up To 1000 kVA
- Typical 11 kV Network Connections For HV Intakes 1000 kVA To 6000 kVA
- Static UPS System – Simplified Single Line Schematic For a Single Module
- Typical Data Transmission (Structured Cabling)
- Typical Networked Lighting Control System
- Typical Standby Power System, Single Line Schematic
- Typical Fire Detection and Alarm Schematic
- Typical Block Diagram – Access Control System (ACS)
- Typical Block Diagram – Intruder Detection System (IDS)
- Typical Block Diagram – Digital CCTV.

MECHANICAL

- BMS Controls For Low Pressure Hot Water (LPHW)
- BMS Controls For Primary Chillers and Chilled Water
- Fan Coil Unit System
- Displacement System
- Chilled Ceiling System (Passive System)
- Chilled Beam System (Passive System)
- Variable Air Volume (VAV)
- Variable Refrigerant Volume System (VRV)
- Alternative All Air System (FGU)
- Reverse Cycle Heat Pump.

Urban Network Mainly Underground

Details: Connection to small housing development 10 houses, 60 m of LV cable from local 11/LV substation route in footpath and verge, 10 m of service cable to each plot in verge

Supply Capacity: 200 kVA

Connection Voltage: LV

Nr of Phases: 1Φ

Breakdown of Detailed Cost Information

	Labour	Plant	Materials	Overheads	Total
Cable	£233	£109	£ 1,153	£428	£1,923
Jointing	£1,567	£398	£555	£716	£3,236
Switchgear	-	-	-	-	-
Termination	£371	£94	£122	£167	£754
Transformer	-	-	-	-	-
Trench/Reinstate	£989	£596	£1,127	£1,102	£3,814
OHL LV	-	-	-	-	-
OHL HV	-	-	-	-	-
Other	-	-	-	-	-
Special/One-offs					-
Total Calculated Price	£3,160	£1,197	£2,957	£2,413	£9,727

Non-contestable Elements and Associated Charges ranges between 15–20%

Total Non-contestable Elements and Associated Charges   £1,459–£1,945

Grand Total Calculated Price excl. VAT   £11,186–£11,672

Urban Network Mainly Underground with Reinforcement

Details: Connection to small housing development 10 houses, 60 m of LV cable from local 11/LV substation route in footpath and verge, 10 m of service cable to each plot in verge. Scheme includes reinforcement of LV distribution board

Supply Capacity: 200 kVA

Connection Voltage: LV

Nr of Phases: 1Φ

Breakdown of Detailed Cost Information

	Labour	Plant	Materials	Overheads	Total
Cable	£233	£109	£1,153	£428	£1,923
Jointing	£1,567	£398	£555	£716	£3,236
Switchgear	£1,808	£574	£2,613	£1,425	£6,420
Termination	£371	£94	£122	£167	£754
Transformer	-	-	-	-	-
Trench/Reinstate	£989	£596	£1,127	£1,102	£3,814
OHL LV	-	-	-	-	-
OHL HV	-	-	-	-	-
Other	-	-	-	-	-
Special / One-offs					-
Total Calculated Price	£4,968	£1,771	£5,570	£3,838	£16,146

Non-contestable Elements and Associated Charges ranges between 15–20%

Total Non-contestable Elements and Associated Charges   £2,422–£3,229

Grand Total Calculated Price excl. VAT   £18,568–£19,375

Urban Network Mainly Underground with Substation Reinforcement

Details: Connection to small housing development 10 houses, 60 m of LV cable from local 11/LV substation route in footpath and verge, 10 m of service cable to each plot in verge. Scheme includes reinforcement of LV distribution board and new substation and 20 m of HV cable

Supply Capacity: 200 kVA

Connection Voltage: LV

Nr of Phases: 1Φ

Breakdown of Detailed Cost Information

	Labour	Plant	Materials	Overheads	Total
Cable	£276	£129	£1,599	£573	£2,576
Jointing	£1,866	£474	£783	£888	£4011
Switchgear	£5,537	£1,757	£15,488	£6,509	£29,291
Termination	£611	£155	£306	£305	£1,377
Transformer	£2,844	£902	£7,725	£3,277	£14,748
Trench/Reinstate	£1,536	£1,005	£1,926	£1,815	£6,282
OHL LV	-	-	-	-	-
OHL HV	-	-	-	-	-
Other	-	-	-	-	-
Special / One-offs					-
Total Calculated Price	£12,669	£4,422	£27,827	£13,367	£58,285

Non-contestable Elements and Associated Charges ranges between 15–20%

Total Non-contestable Elements and Associated Charges   £8,743–£11,657

Grand Total Calculated Price excl. VAT   £67,028–£69,942

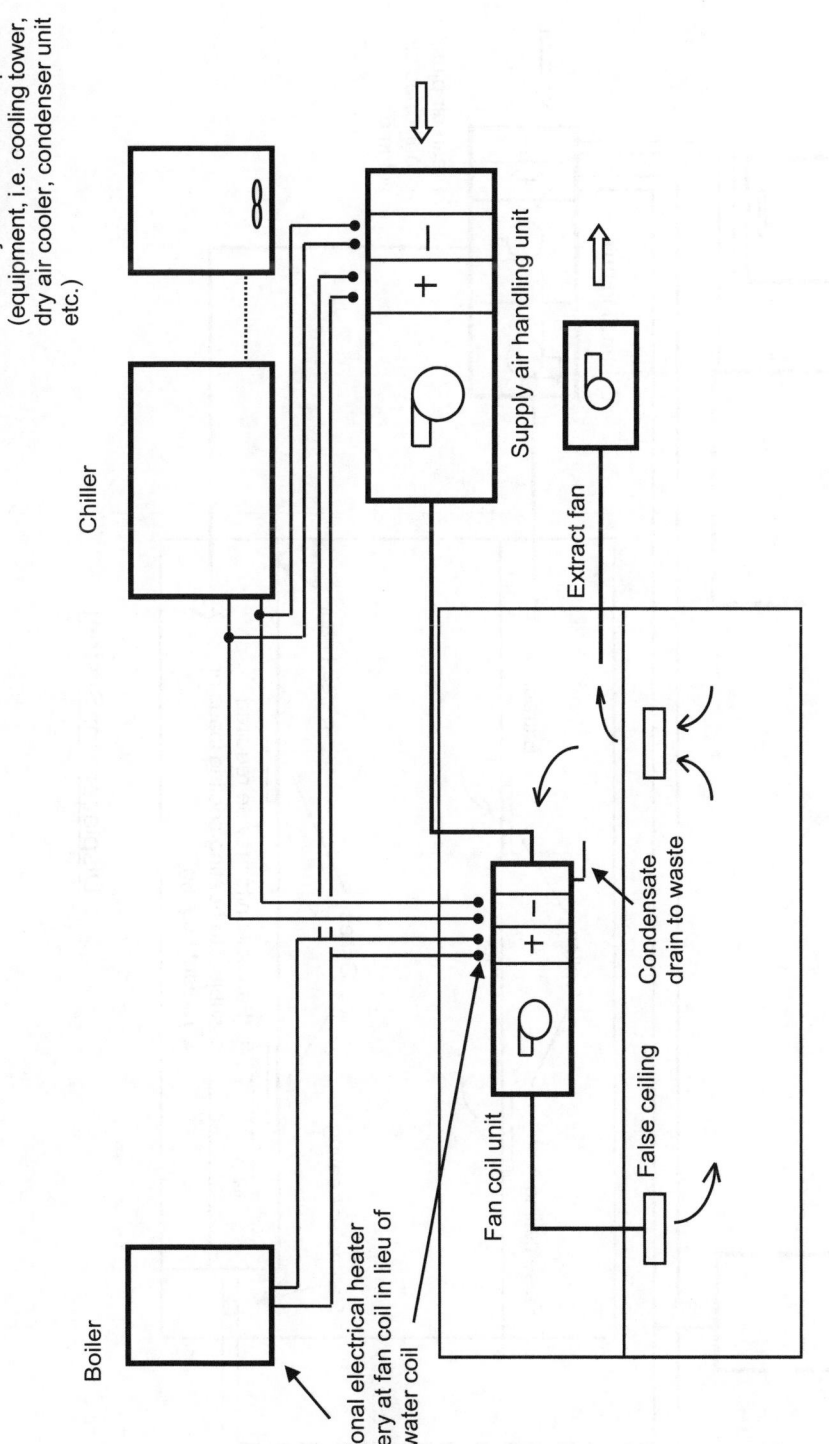

Heat rejection to atmosphere (equipment, i.e. cooling tower, dry air cooler, condenser unit etc.)

Chiller

Boiler

Optional electrical heater battery at fan coil in lieu of hot water coil

Supply air handling unit

Extract fan

Fan coil unit

Condensate drain to waste

False ceiling

Fan Coil Unit System

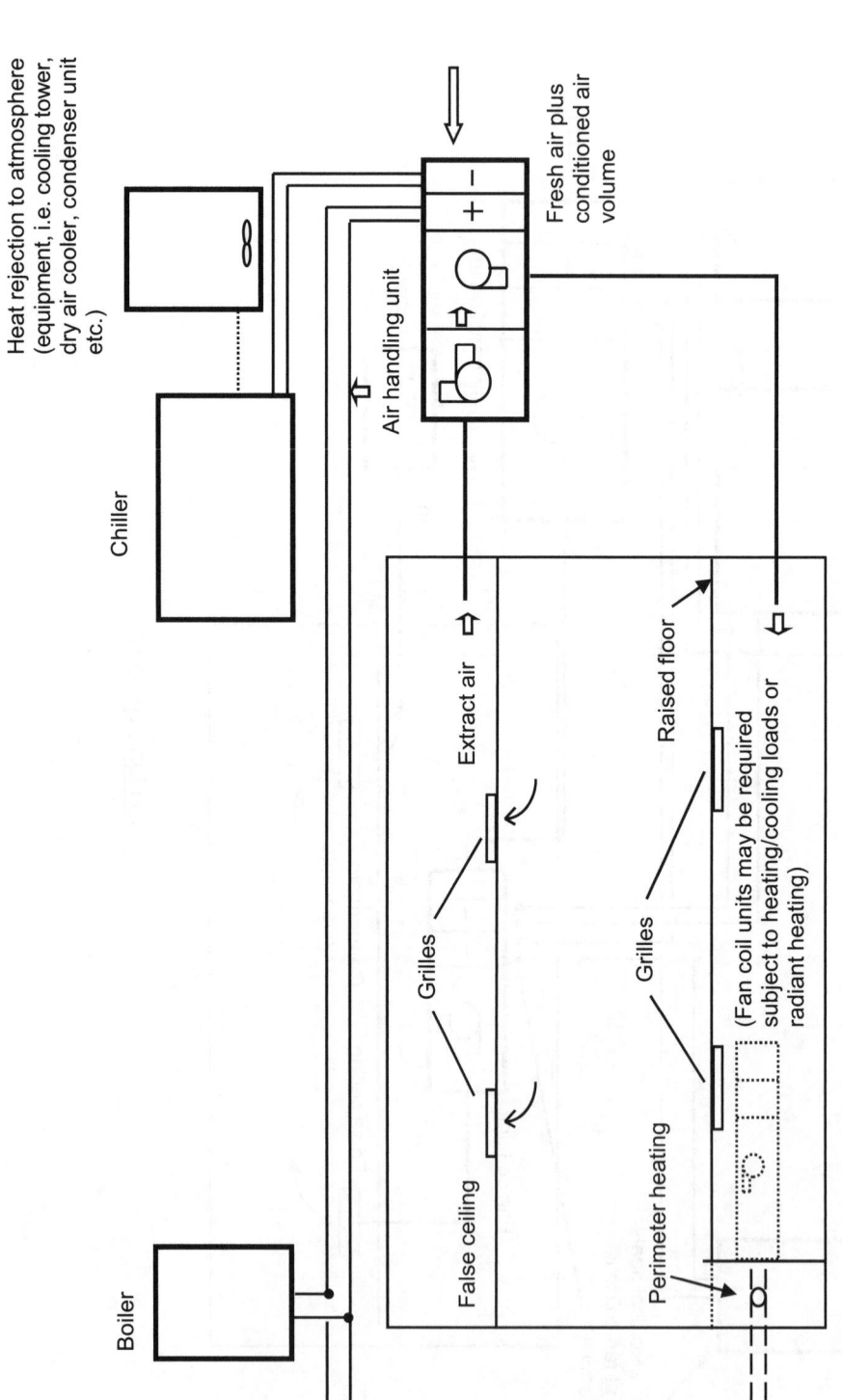

Heat rejection to atmosphere (equipment, i.e. cooling tower, dry air cooler, condenser unit etc.)

Chiller

Boiler

Air handling unit

Fresh air plus conditioned air volume

Extract air

Raised floor

Grilles

Grilles

False ceiling

Perimeter heating

(Fan coil units may be required subject to heating/cooling loads or radiant heating)

Displacement System

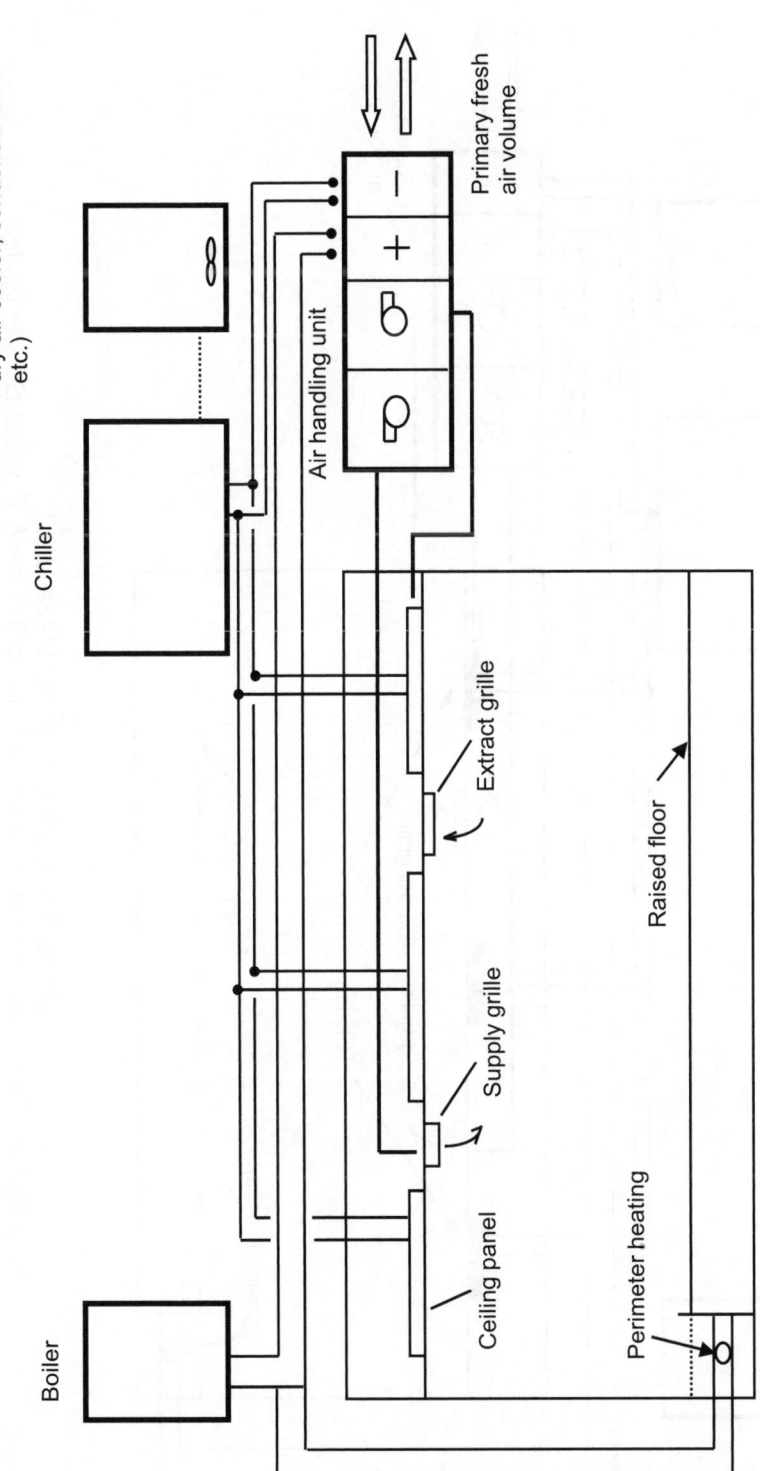

Heat rejection to atmosphere (equipment i.e. cooling tower, dry air cooler, condenser unit etc.)

Chiller

Boiler

Air handling unit

Primary fresh air volume

Extract grille

Supply grille

Ceiling panel

Perimeter heating

Raised floor

Chilled Ceiling System (Passive System)

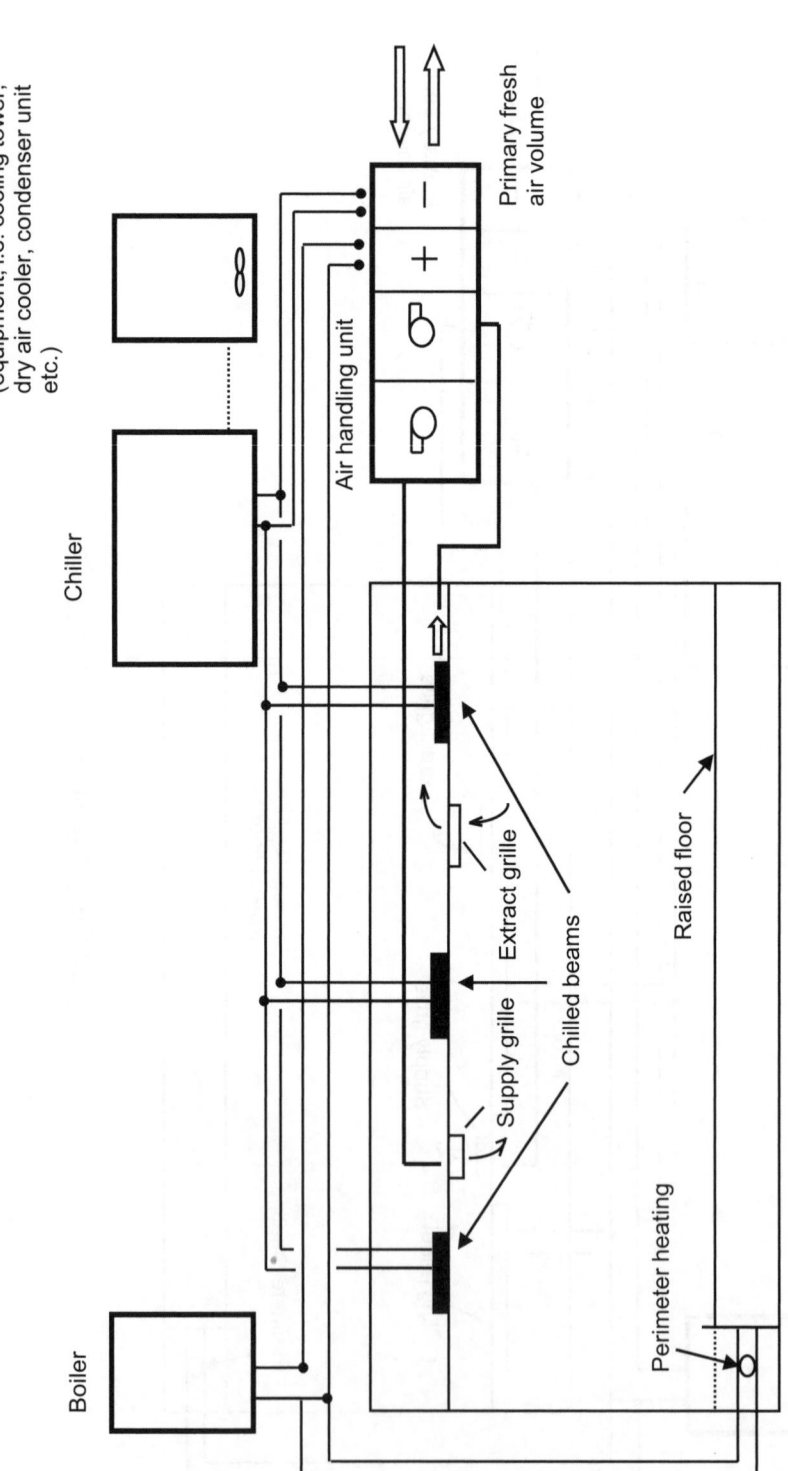

Heat rejection to atmosphere (equipment, i.e. cooling tower, dry air cooler, condenser unit etc.)

Chiller

Boiler

Air handling unit

Primary fresh air volume

Extract grille

Supply grille

Chilled beams

Raised floor

Perimeter heating

Chilled Beam System (Passive System)

Active Option Connects Air Supply Duct To Chilled Beams & Deletes Supply Grilles

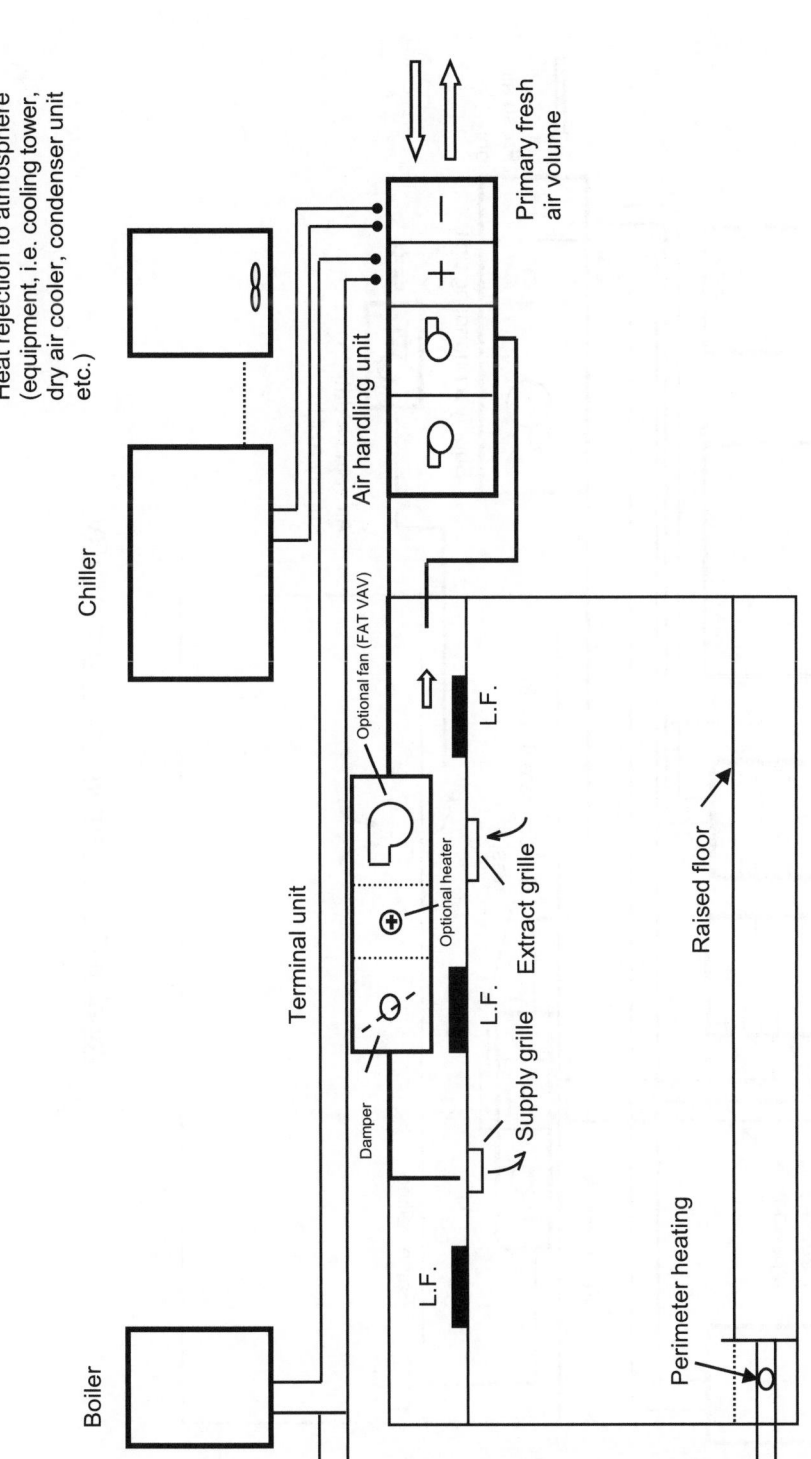

Heat rejection to atmosphere (equipment, i.e. cooling tower, dry air cooler, condenser unit etc.)

Chiller

Boiler

Air handling unit

Primary fresh air volume

Terminal unit

Optional fan (FAT VAV)

Optional heater

Damper

Supply grille

Extract grille

L.F.

L.F.

L.F.

Raised floor

Perimeter heating

Variable Air Volume (VAV)

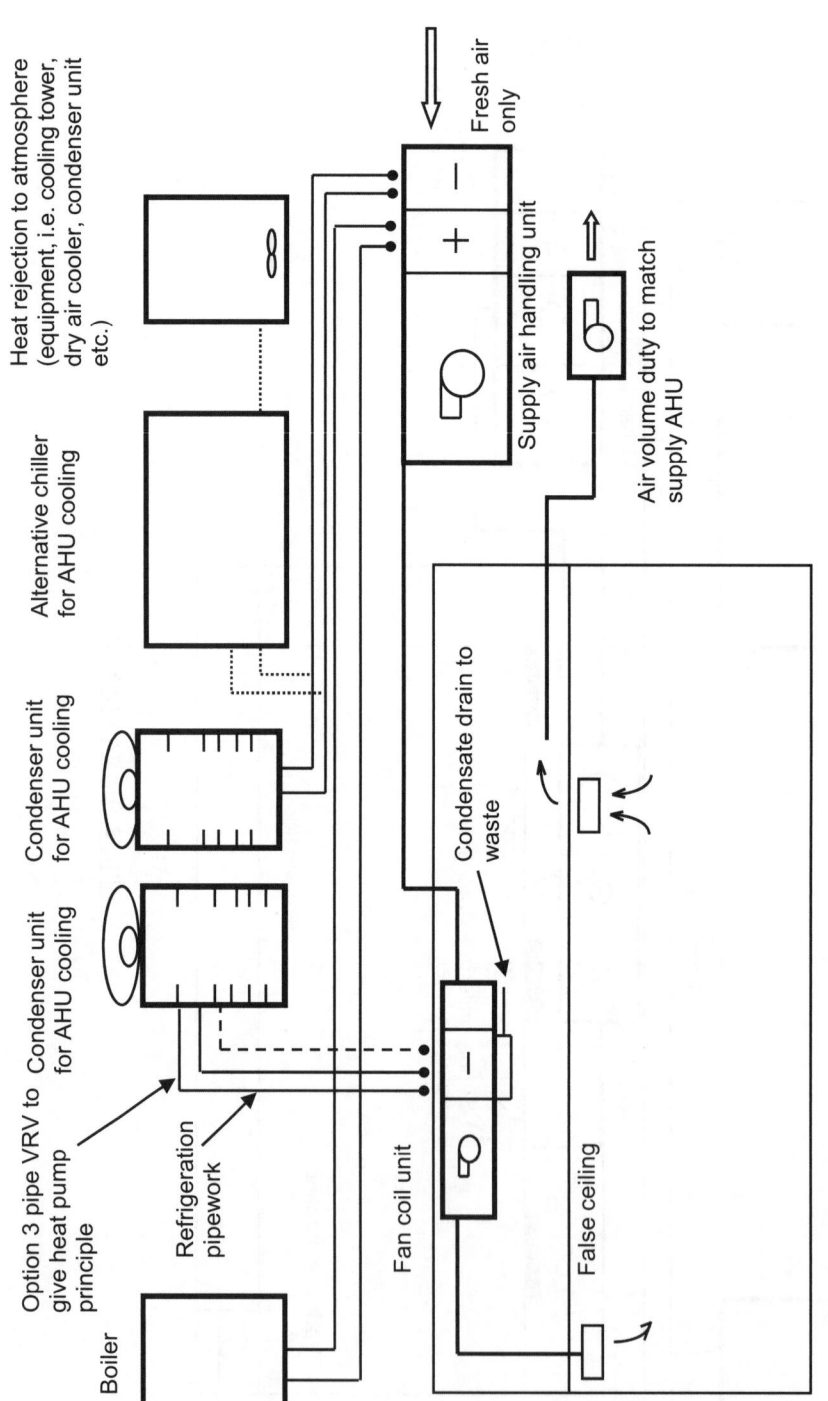

Heat rejection to atmosphere (equipment, i.e. cooling tower, dry air cooler, condenser unit etc.)

Alternative chiller for AHU cooling

Condenser unit for AHU cooling

Option 3 pipe VRV to Condenser unit give heat pump for AHU cooling principle

Refrigeration pipework

Boiler

Fresh air only

Supply air handling unit

Air volume duty to match supply AHU

Condensate drain to waste

Fan coil unit

False ceiling

Variable Refrigerant Volume System (VRV)

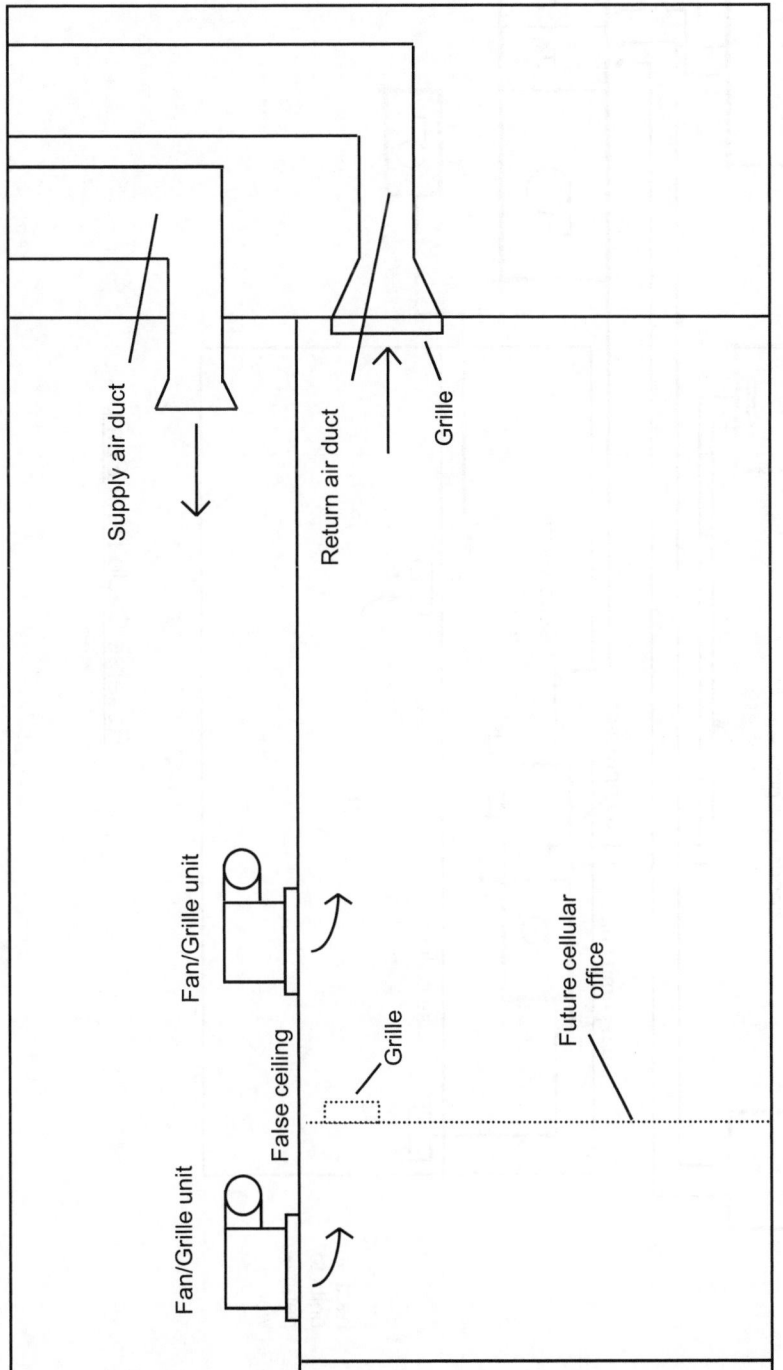

Supply air duct

Return air duct

Grille

Fan/Grille unit

Fan/Grille unit

False ceiling

Grille

Future cellular office

Alternative All Air System (FGU)

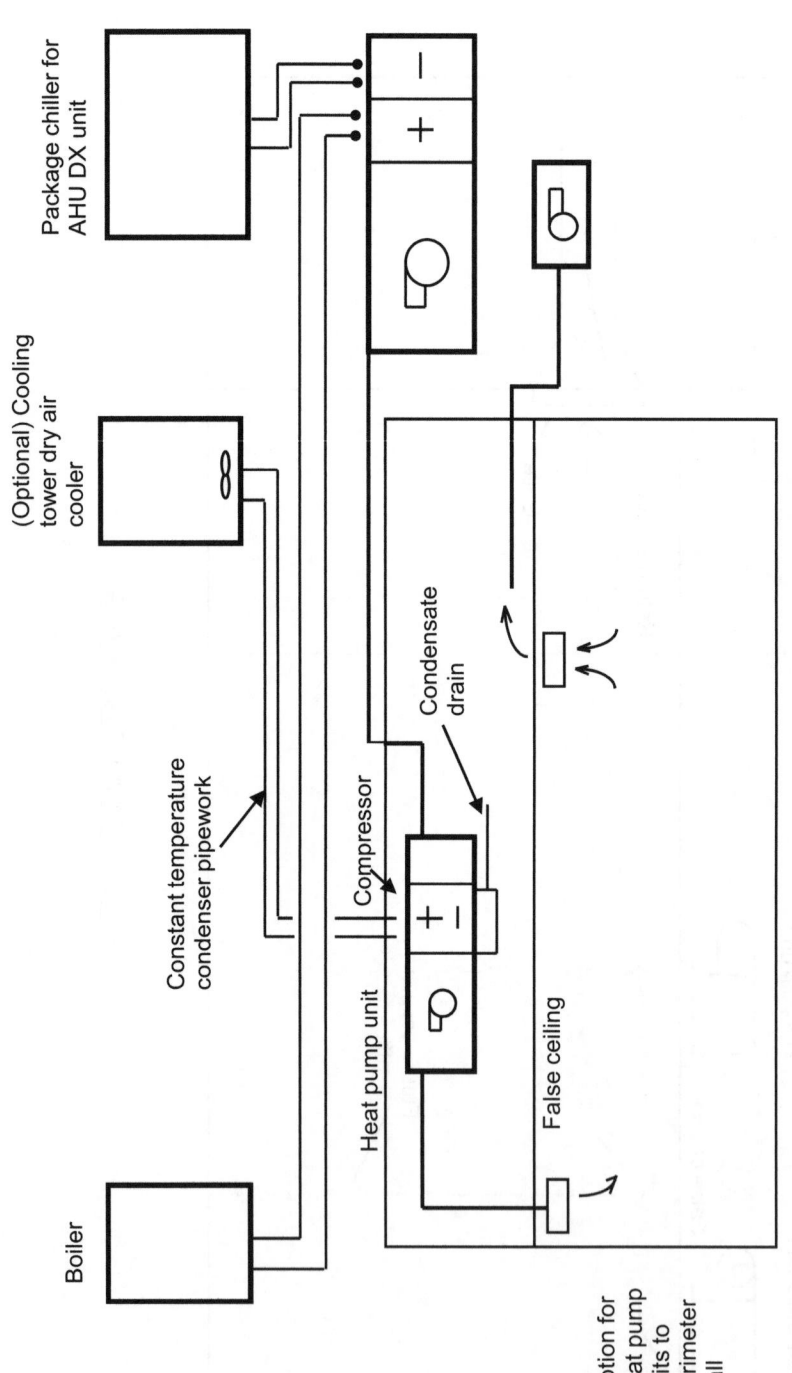

Package chiller for AHU DX unit

(Optional) Cooling tower dry air cooler

Constant temperature condenser pipework

Compressor

Condensate drain

Heat pump unit

Boiler

False ceiling

Option for heat pump units to perimeter wall

Reverse Cycle Heat Pump

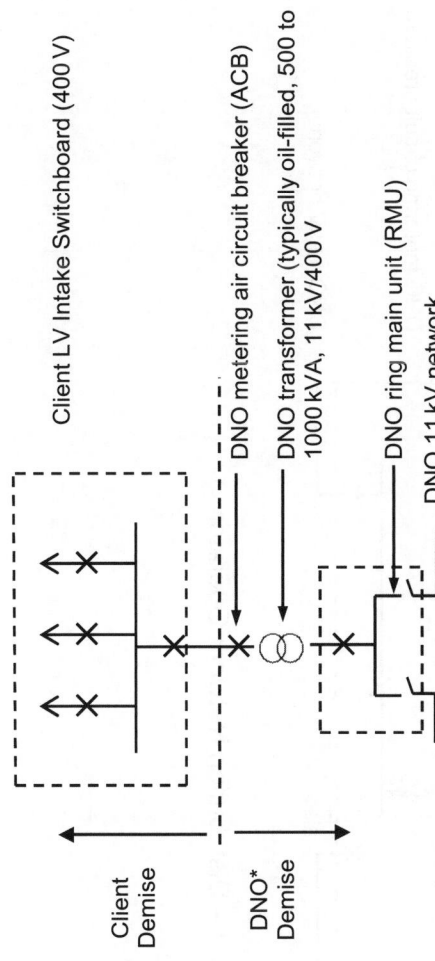

Client LV Intake Switchboard (400 V)

DNO metering air circuit breaker (ACB)

DNO transformer (typically oil-filled, 500 to 1000 kVA, 11 kV/400 V

DNO ring main unit (RMU)

DNO 11 kV network

Client Demise

DNO* Demise

Note: * DNO – Distribution Network Operator

Typical Simple 11 kV Network Connection For LV Intakes Up To 1000 kVA

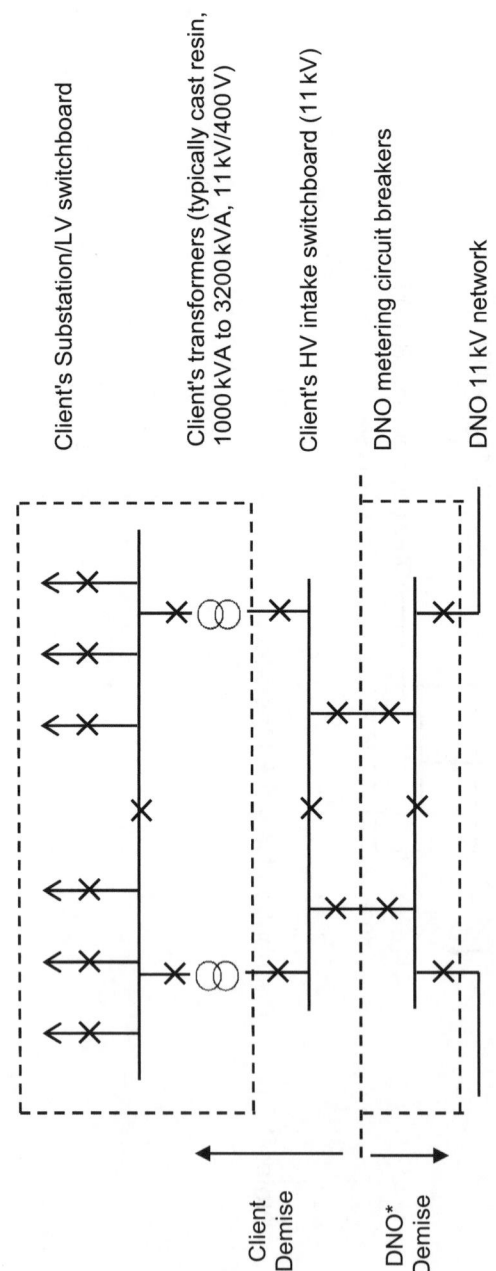

Client's Substation/LV switchboard

Client's transformers (typically cast resin, 1000 kVA to 3200 kVA, 11 kV/400 V)

Client's HV intake switchboard (11 kV)

DNO metering circuit breakers

DNO 11 kV network

Client Demise

DNO* Demise

Note: * DNO – Distribution Network Operator

Typical 11 kV Network Connections For HV Intakes 1000 kVA To 6000 kVA

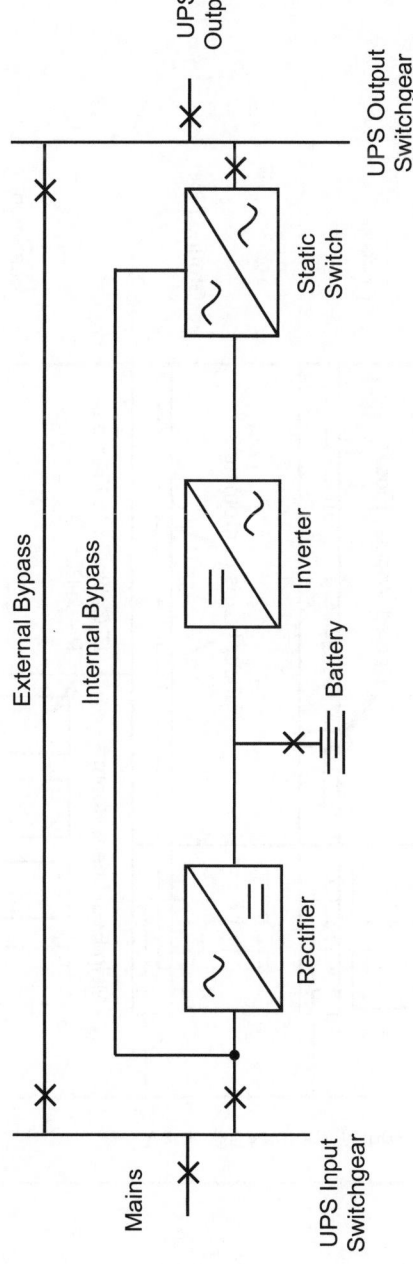

Static UPS System – Simplified Single Line Schematic For a Single Module

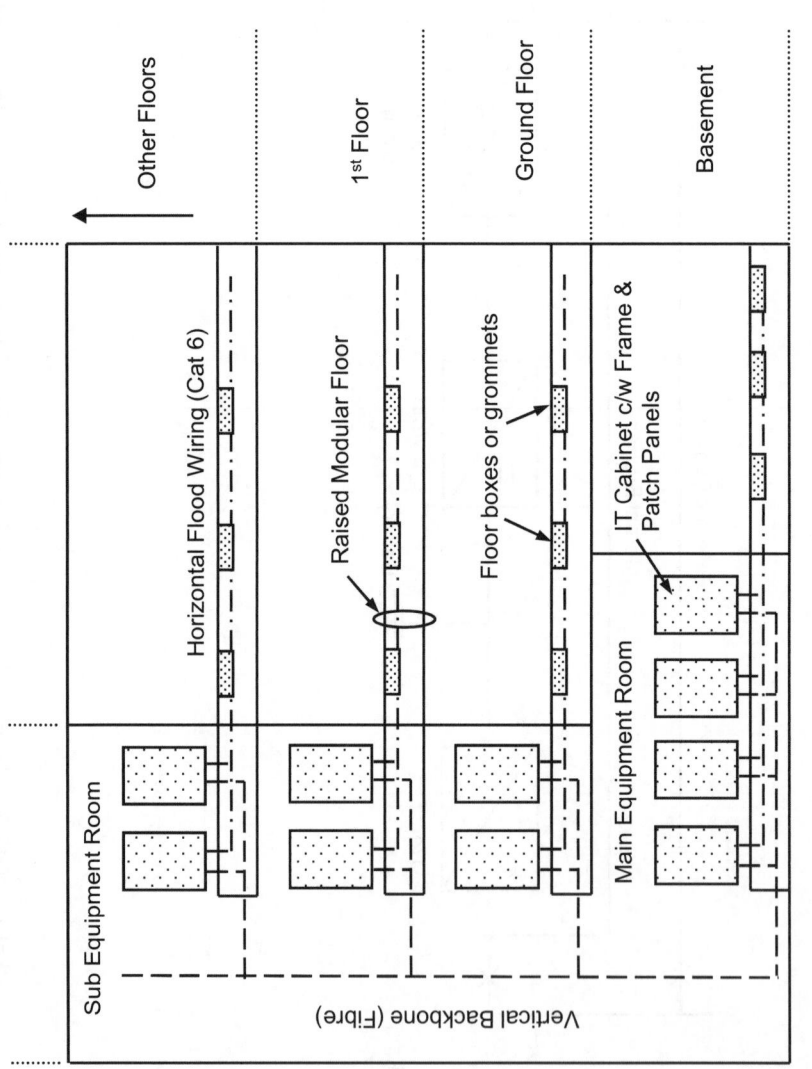

Typical Data Transmission (Structured Cabling)

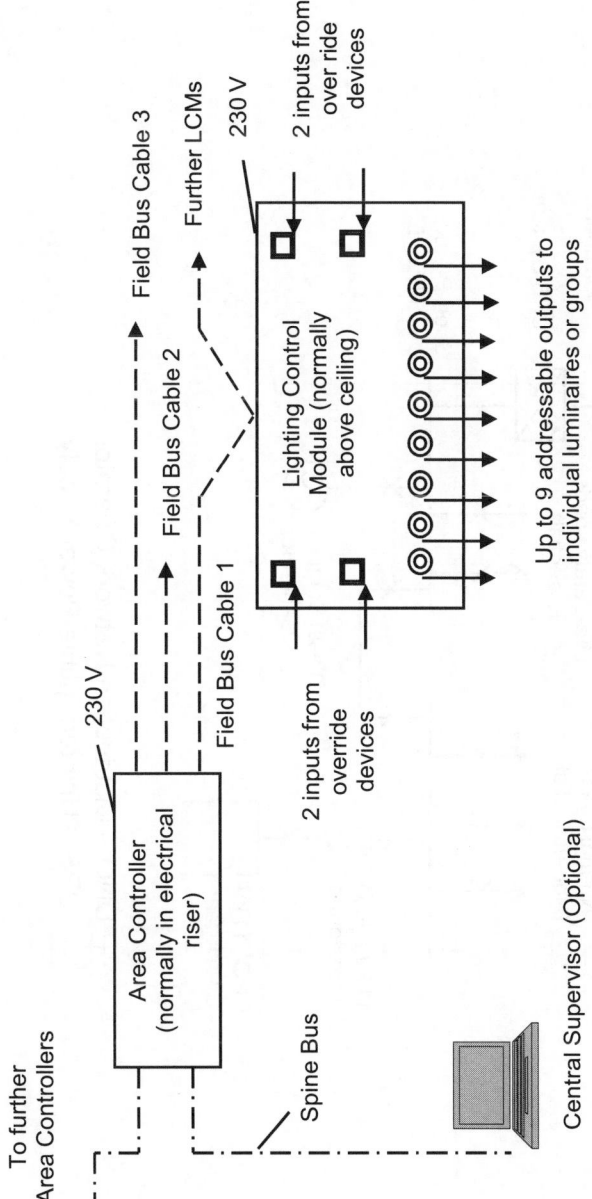

To further
Area Controllers

230 V

Area Controller
(normally in electrical
riser)

Spine Bus

Central Supervisor (Optional)

Field Bus Cable 1

Field Bus Cable 2

Field Bus Cable 3

Further LCMs

230 V

2 inputs from
over ride
devices

2 inputs from
override
devices

Lighting Control
Module (normally
above ceiling)

Up to 9 addressable outputs to
individual luminaires or groups

Typical Networked Lighting Control System

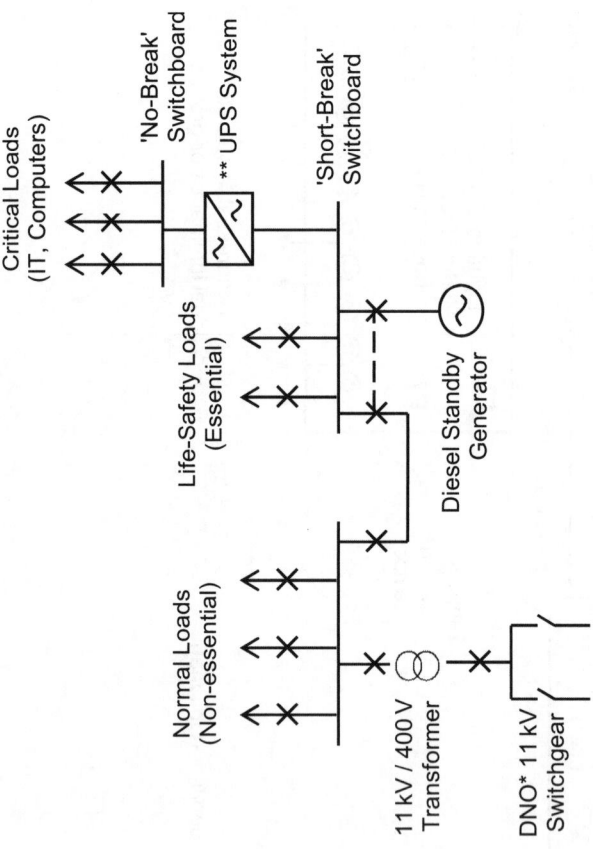

Critical Loads
(IT, Computers)

'No-Break' Switchboard

** UPS System

'Short-Break' Switchboard

Life-Safety Loads
(Essential)

Diesel Standby Generator

Normal Loads
(Non-essential)

11 kV / 400 V Transformer

DNO* 11 kV Switchgear

Note: * DNO – Distribution Network Operator
** UPS – Uninterruptible Power Supply

Typical Standby Power System, Single Line Schematic

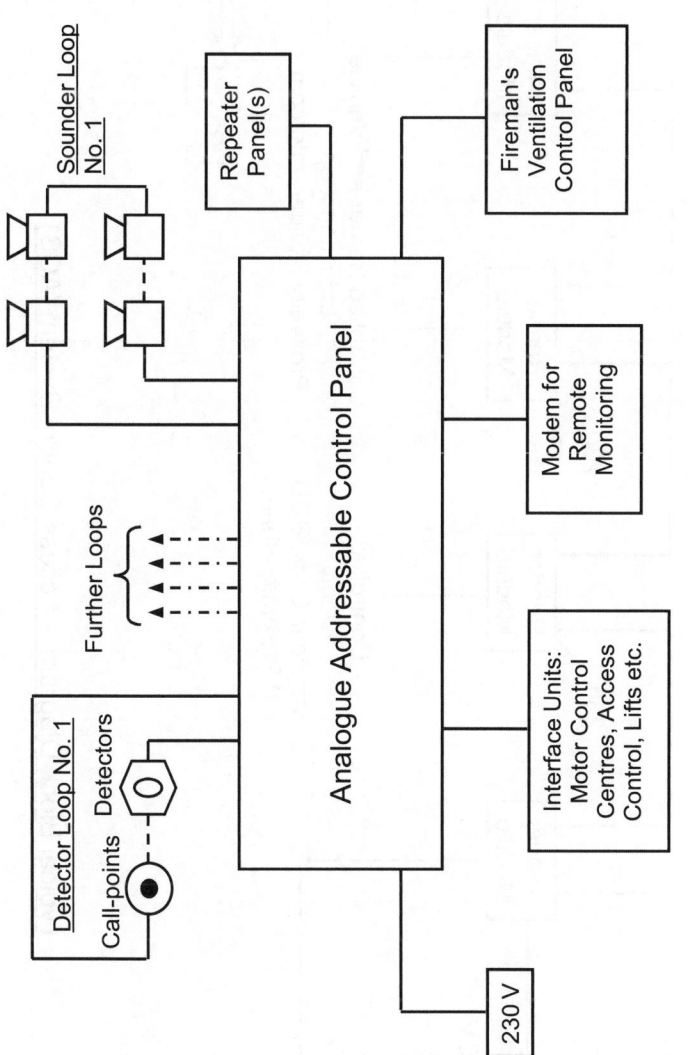

Typical Fire Detection and Alarm Schematic

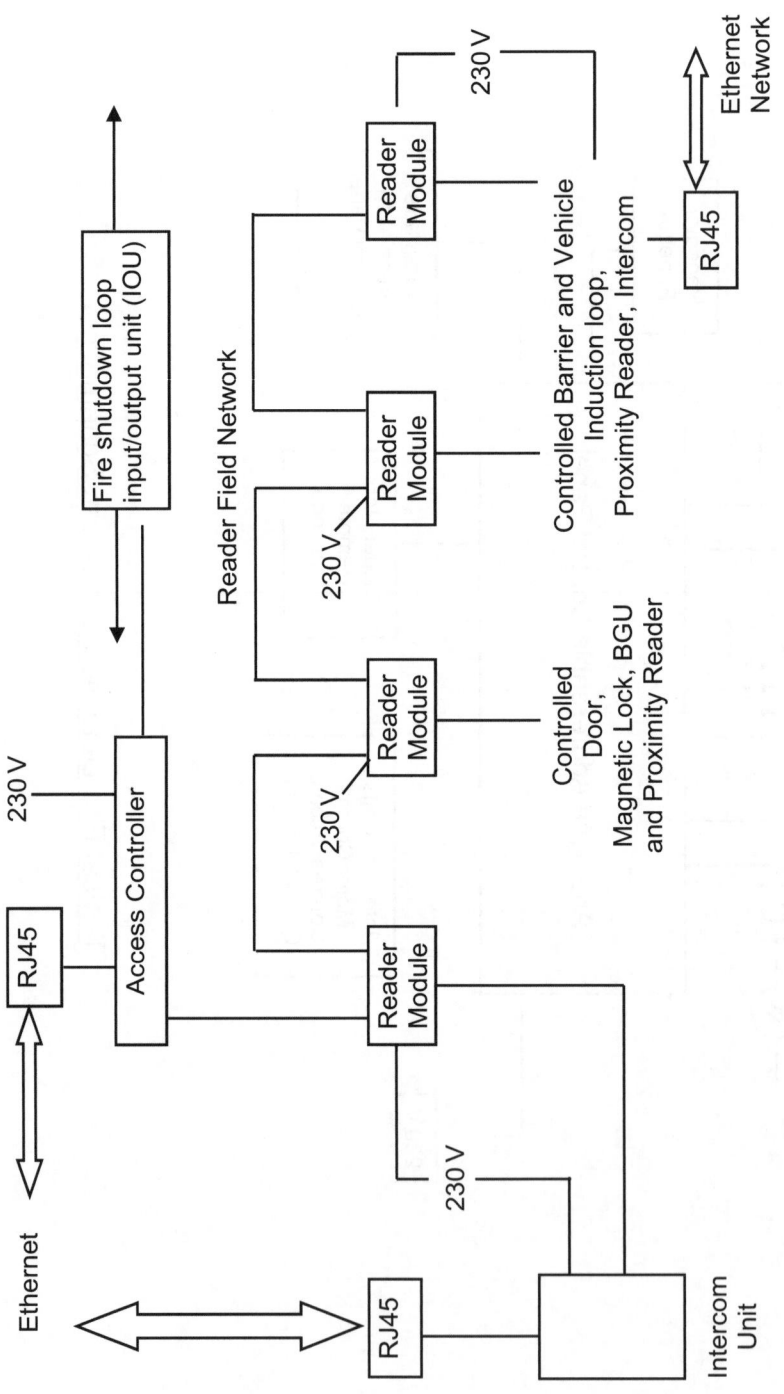

Typical Block Diagram – Access Control System (ACS)

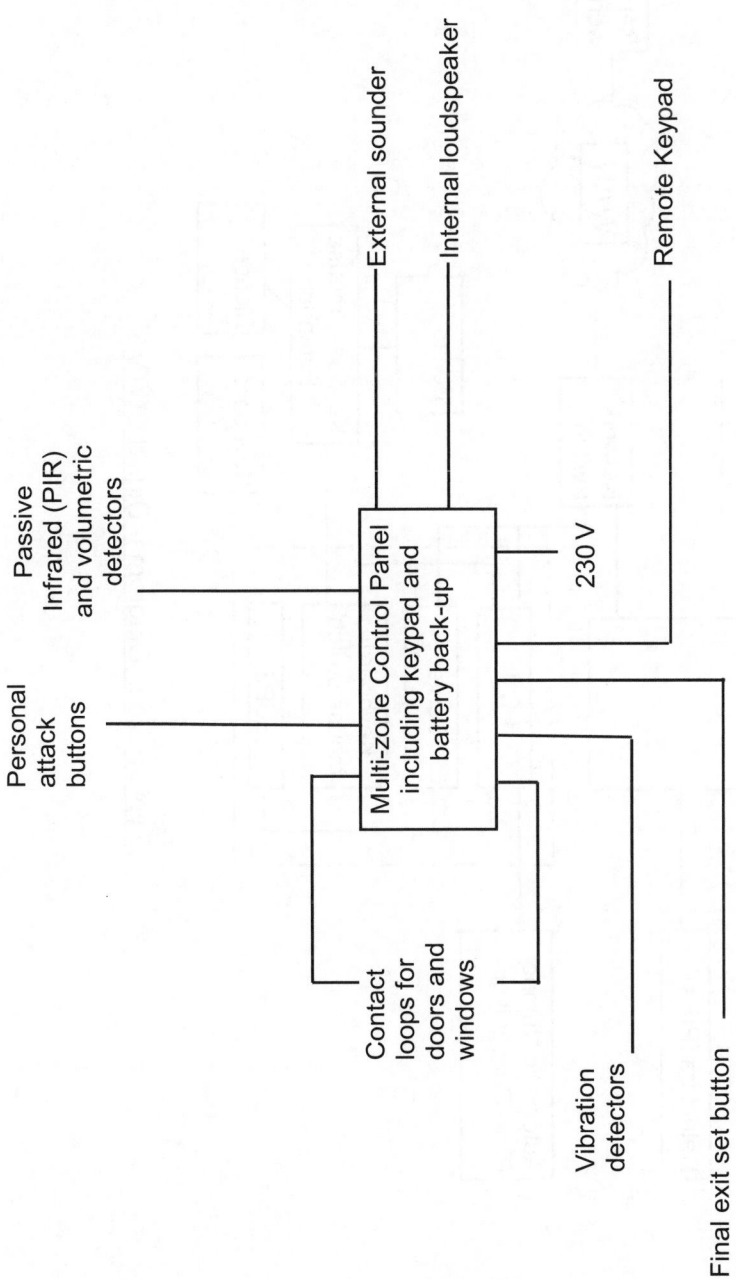

Typical block diagram – Intruder Detection System (IDS)

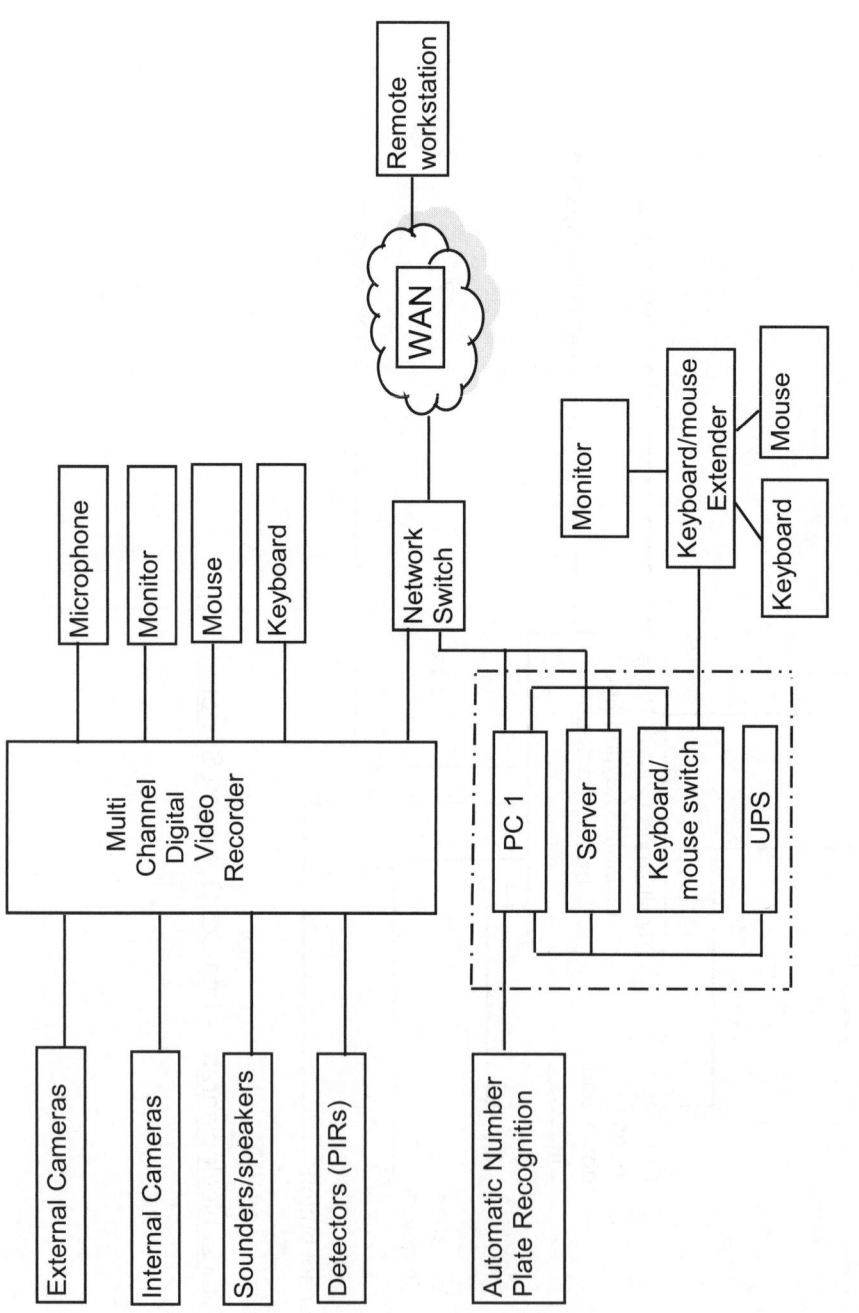

Typical Block Diagram – Digital CCTV

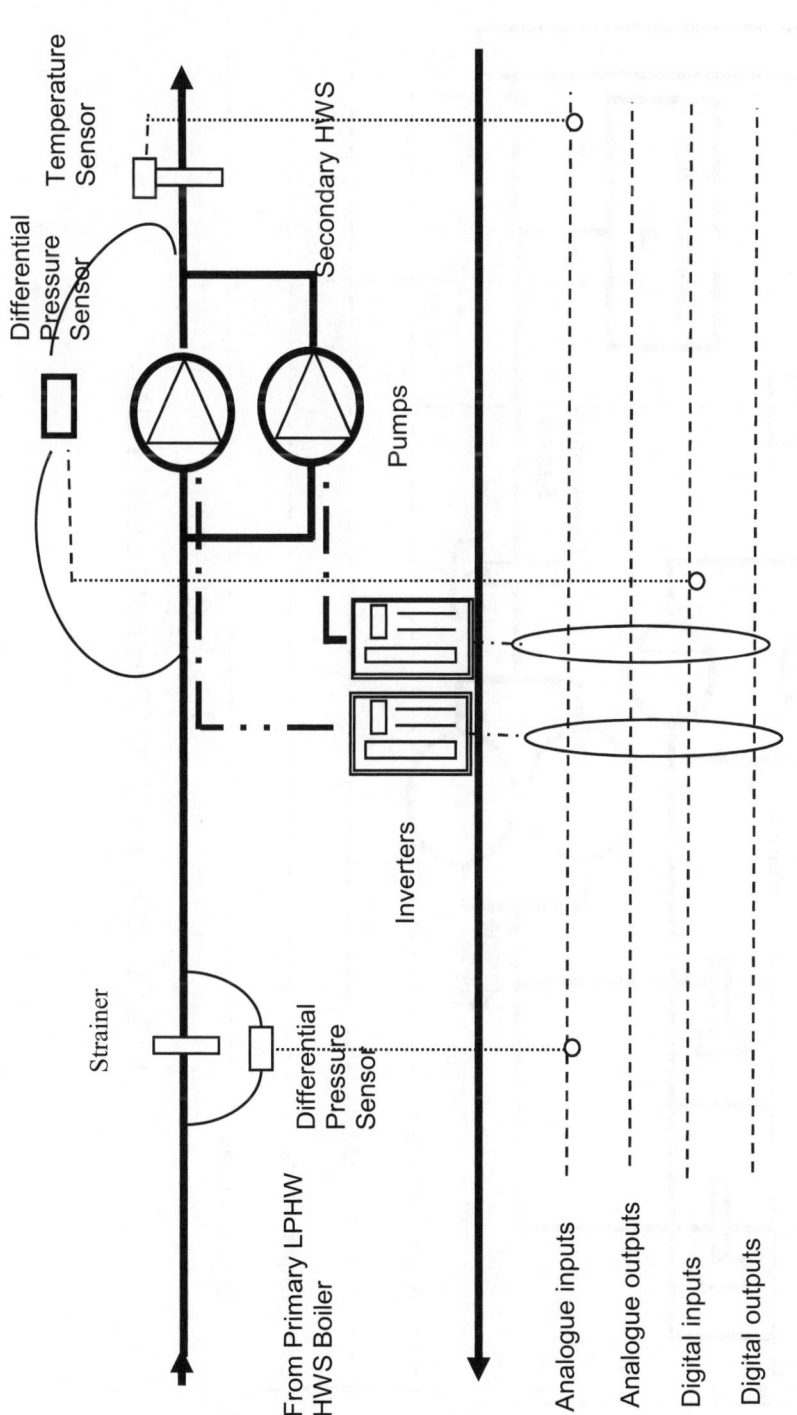

Temperature Sensor

Differential Pressure Sensor

Secondary HWS

Pumps

Differential Pressure Sensor

Strainer

Differential Pressure Sensor

Inverters

From Primary LPHW HWS Boiler

Analogue inputs

Analogue outputs

Digital inputs

Digital outputs

BMS Controls For Low Pressure Hot Water (LPHW)

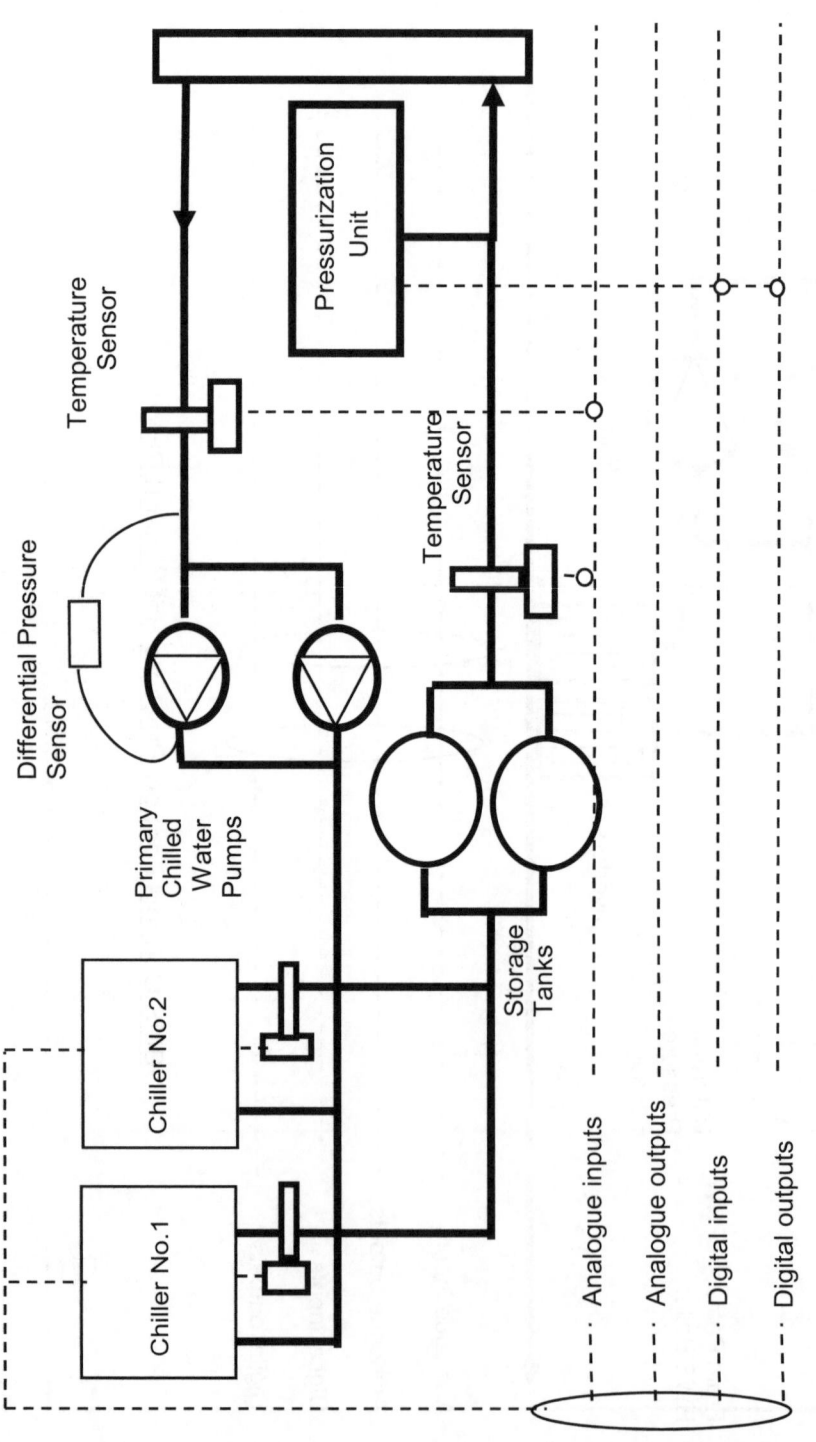

BMS Controls For Primary Chillers and Chilled Water

- - - - - - Analogue inputs
- - - - - - Analogue outputs
- - - - - - Digital inputs
- - - - - - Digital outputs

Understanding the CDM 2007 Regulations

2nd Edition

O. Griffiths et al.

Almost 3000 lives have been lost in the UK construction industry over the last twenty-five years, in addition to those seriously injured or made ill. The need to reduce this rate has required tight controls to be introduced throughout the planning and management of construction projects in the UK. The Construction (Design & Management) Regulations 2007 outline the responsibilities and liabilities for the various professionals and agents involved.

Straightforward and practical, *Understanding the CDM 2007 Regulations* demonstrates the rationale behind the regs, covers the duties of the five core duty holders (client, CDM coordinator, designer, principal contractor and contractor), explains the importance of the hazard management process on every project and also sets out the consequences of failing to successfully plan, design and manage for safety.

Any client, architect, engineer, CDM co-ordinator, project manager, construction professional, or student will find this a simple but thorough and dependable guide and should value the management toolkit and the numerous practical examples of best practice and guidance on how to use the Approved Code of Practice appropriately. This book shows how to unleash the potential of the regulations and add real value to the industry.

December 2010: 352pp
Pb: 978-0-415-55653-8: **£29.99**

To Order: Tel: +44 (0) 1235 400524 **Fax:** +44 (0) 1235 400525
or Post: Taylor and Francis Customer Services,
Bookpoint Ltd, Unit T1, 200 Milton Park, Abingdon, Oxon, OX14 4TA UK
Email: book.orders@tandf.co.uk

For a complete listing of all our titles visit:
www.tandf.co.uk

Energy Efficiency in New and Existing Buildings

F. Mackenzie

This BRE Trust report considers the relative impact on UK CO2 savings targets of constructing new zero-carbon buildings compared to improving the energy efficiency of the existing stock. Carbon dioxide emissions from buildings accounted for around 40% of total UK CO2 emissions in 2006. To achieve the government's challenging target of reducing greenhouse gas emissions by 80% by 2050, improving the energy efficiency of buildings - both new and existing - will clearly be vital. This report uses existing data to explore the extent to which improving the energy efficiency of the existing building stock would be a more cost-effective route for achieving CO2 savings than constructing new buildings to the higher levels of energy performance required to meet low- and zero-carbon targets.

September 2010: 32pp
Pb: 978-1-84806-137-8: **£30.00**

To Order: Tel: +44 (0) 1235 400524 **Fax:** +44 (0) 1235 400525
or Post: Taylor and Francis Customer Services,
Bookpoint Ltd, Unit T1, 200 Milton Park, Abingdon, Oxon, OX14 4TA UK
Email: book.orders@tandf.co.uk

For a complete listing of all our titles visit:
www.tandf.co.uk

Daylighting

P. Tregenza et al.

This authoritative and multi-disciplinary book provides architects, lighting specialists, and anyone else working daylight into design, with all the tools needed to incorporate this most fundamental element of architecture. It includes:

- an overview of current practice of daylighting in architecture and urban planning
- a review of recent research on daylighting and what this means to the practitioner
- a global vision of architectural lighting which is linked to the climates of the world and which integrates view, sunlight, diffuse skylight and electric lighting
- up-to-date tools for design in practice
- delivery of information in a variety of ways for interdisciplinary readers: graphics, mathematics, text, photographs and in-depth illustrations
- a clear structure: eleven chapters covering different aspects of lighting, a set of worksheets giving step-by-step examples of calculations and design procedures for use in practice, and a collection of algorithms and equations for reference by specialists and software designers.

This book should trigger creative thought. It recognizes that good lighting design needs both knowledge and imagination.

August 2010: 216x138: 352pp
Pb: 978-0-415-56650-6: **£34.99**

To Order: Tel: +44 (0) 1235 400524 **Fax:** +44 (0) 1235 400525
or Post: Taylor and Francis Customer Services,
Bookpoint Ltd, Unit T1, 200 Milton Park, Abingdon, Oxon, OX14 4TA UK
Email: book.orders@tandf.co.uk

For a complete listing of all our titles visit:
www.tandf.co.uk

Building Performance Simulation for Design and Operation

J. Hensen et al.

Effective building performance simulation can reduce the environmental impact of the built environment, improve indoor quality and productivity, and facilitate future innovation and technological progress in construction. It draws on many disciplines, including physics, mathematics, material science, biophysics and human behavioural, environmental and computational sciences. The discipline itself is continuously evolving and maturing, and improvements in model robustness and fidelity are constantly being made. This has sparked a new agenda focusing on the effectiveness of simulation in building life-cycle processes.

Building Performance Simulation for Design and Operation begins with an introduction to the concepts of performance indicators and targets, followed by a discussion on the role of building simulation in performance-based building design and operation. This sets the ground for in-depth discussion of performance prediction for energy demand, indoor environmental quality (including thermal, visual, indoor air quality and moisture phenomena), HVAC and renewable system performance, urban level modelling, building operational optimization and automation.

Produced in cooperation with the International Building Performance Simulation Association (IBPSA), and featuring contributions from 14 internationally recognised experts in this field, this book provides a unique and comprehensive overview of building performance simulation for the complete building life-cycle from conception to demolition. It is primarily intended for advanced students in building services engineering, and in architectural, environmental or mechanical engineering; and will be useful for building and systems designers and operators.

January 2011: 536pp
Hb: 978-0-415-47414-6: **£65.00**

To Order: Tel: +44 (0) 1235 400524 **Fax:** +44 (0) 1235 400525
or Post: Taylor and Francis Customer Services,
Bookpoint Ltd, Unit T1, 200 Milton Park, Abingdon, Oxon, OX14 4TA UK
Email: book.orders@tandf.co.uk

For a complete listing of all our titles visit:
www.tandf.co.uk

PART 2

Approximate Estimating

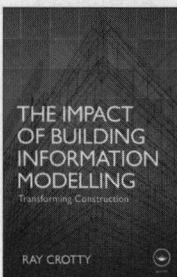

DIRECTIONS

The prices shown in this section of the book are average prices on a fixed price basis for typical buildings tendered during the second quarter of 2011. Unless otherwise noted, they exclude external services and professional fees.

The information in this section has been arranged to follow more closely the order in which estimates may be developed, in accordance with the RIBA stages of work:

a)   Cost Indices and Regional Variations – These provide information regarding the adjustments to be made to estimates taking into account current pricing levels for different locations in the UK.

b)   Feasibility Costs – These provide a range of data (based on a rate per square metre) for all-in engineering costs, excluding lifts, associated with a wide variety of building types. These would typically be used at work stage A/B (feasibility) of a project.

c)   Elemental Rates – The outline costs for offices have been developed further to provide rates for the alternative solutions for each of the services elements. These would typically be used at work stage C, outline proposal.

Where applicable, costs have been identified as Shell and Core and Fit Out to reflect projects where the choice of procurement has dictated that the project is divided into two distinctive contractual parts.

Such detail would typically be required at work stage D, detailed proposals.

d)   All-in-Rates – These are provided for a number of items and complete parts of a system i.e. boiler plant, ductwork, pipework, electrical switchgear and small power distribution, together with lifts and escalators. Refer to the relevant section for further guidance notes.

e)   Elemental Costs – These are provided for a diverse range of building types; offices, laboratory, shopping mall, airport terminal building, supermarket, performing arts centre, sports hall, luxury hotel, hospital and secondary school. Also included is a separate analysis of a building management system for an office block. In each case, a full analysis of engineering services costs is given to show the division between all elements and their relative costs to the total building area. A regional variation factor has been applied to bring these analyses to a common London base.

Prices should be applied to the total floor area of all storeys of the building under consideration. The area should be measured between the external walls without deduction for internal walls and staircases/lift shafts i.e. G.I.A (Gross Internal Area).

Although prices are reviewed in the light of recent tenders it has only been possible to provide a range of prices for each building type. This should serve to emphasize that these can only be average prices for typical requirements and that such prices can vary widely depending on variations in size, location, phasing, specification, site conditions, procurement route, programme, market conditions and net to gross area efficiencies. Rates per square metre should not therefore be used indiscriminately and each case needs to be assessed on its own merits.

The prices do not include for incidental builder's work nor for profit and attendance by a Main Contractor where the work is executed as a subcontract: they do however include for preliminaries, profit and overheads for the services contractor. Capital contributions to statutory authorities and public undertakings and the cost of work carried out by them have been excluded.

Where services works are procured indirectly, i.e. ductwork via a mechanical Subcontractor, the reader should make due allowance for the addition of a further level of profit etc.

COST INDICES

The following tables reflect the major changes in cost to contractors but do not necessarily reflect changes in tender levels. In addition to changes in labour and materials costs, tenders are affected by other factors such as the degree of competition in the particular industry, the area where the work is to be carried out, the availability of labour and the prevailing economic conditions. This has meant in recent years that, when there has been an abundance of work, tender levels have tended to increase at a greater rate than can be accounted for solely by increases in basic labour and material costs and, conversely, when there is a shortage of work this has tended to result in keener tenders. Allowances for these factors are impossible to assess on a general basis and can only be based on experience and knowledge of the particular circumstances.

In compiling the tables the cost of labour has been calculated on the basis of a notional gang as set out elsewhere in the book. The proportion of labour to materials has been assumed as follows:
Mechanical Services – 30:70, Electrical Services – 50:50, (1976 = 100)

Mechanical Services

Year	First quarter	Second quarter	Third quarter	Fourth quarter
1997	361	356	358	363
1998	365	363	368	373
1999	368	363	370	384
2000	384	386	388	400
2001	401	401	405	411
2002	411	411	410	442
2003	443	446	447	456
2004	458	464	467	482
2005	486	488	492	508
2006	513	522	527	533
2007	535	541	546	555
2008	555	560	567	584
2009	582	578	577	577
2010	576	583	585	P 592
2011	P 599	F 602	F 605	F 614
2012	F 619	F 622	F 625	F 636

Electrical Services

Year	First quarter	Second quarter	Third quarter	Fourth quarter
1997	411	411	411	411
1998	410	423	422	422
1999	433	432	431	446
2000	458	464	465	468
2001	485	484	484	487
2002	508	508	508	513
2003	530	533	533	541
2004	571	574	576	589
2005	607	608	607	615
2006	631	636	641	649
2007	665	666	668	676
2008	693	697	700	710
2009	720	721	724	737
2010	760	766	766	P 768
2011	P 777	F 781	F 783	F 792
2012	F 803	F 804	F 806	F 817

(P = Provisional)
(F = Forecast)

COST INDICES

Regional variations

Prices throughout this Book apply to work in the London area (see Directions at the beginning of the Mechanical Installations and Electrical Installations sections). However, prices for mechanical and electrical services installations will of course vary from region to region, largely as a result of differing labour costs but also depending on the degree of accessibility, urbanization and local market conditions.

The following table of regional factors is intended to provide readers with indicative adjustments that may be made to the prices in the Book for locations outside of London. The figures are of necessity averages for regions and further adjustments should be considered for city centre or very isolated locations, or other known local factors.

Greater London	1.03	Yorkshire & Humberside	0.91
South East	0.97	North West	0.92
South West	0.97	North East	0.90
East Midlands	0.89	Scotland	0.92
West Midlands	0.92	Wales	0.90
East Anglia	0.93	Northern Ireland	0.69

RIBA STAGE A FEASIBILITY COSTS

TYPICAL SQUARE METRE RATES FOR ENGINEERING SERVICES	
The following examples indicate the range of rates within each building type for engineering services, excluding lifts etc, utilities services and professional fees. Based on Gross Internal Area (GIA).	
Industrial Buildings	£/m² GIA
Factories	
Owner occupation: Includes for rainwater, soil/waste, LTHW heating via HL radiant heaters, BMS, LV installations, lighting, fire alarms, security, earthing	120
Owner occupation: Includes for rainwater, soil/waste, sprinklers, LTHW heating via HL radiant heaters, local air conditioning, BMS, HV/LV installations, lighting, fire alarms, security, earthing	180
Warehouses	
High bay for owner occupation: Includes for rainwater, soil/waste, LTHW heating via HL gas fired heaters, BMS, HV/LV installations, lighting, fire alarms, security, earthing	95
High bay for owner occupation: Includes for rainwater, soil/waste, sprinklers, LTHW heating via HL radiant heaters, local air conditioning, BMS, HV/LV installations, lighting, fire alarms, security, earthing	195
Distribution Centres	
High bay for letting: Includes for rainwater, soil/waste, LTHW heating via HL gas fired heaters, BMS, HV/LV installations, lighting, fire alarms, security, earthing	120
High bay for owner occupation: Includes for rainwater, soil/waste, sprinklers, LTHW heating via HL radiant heaters, local air conditioning, BMS, HV/LV installations, lighting, fire alarms, security	220
Office Buildings 5,000 m² to 15,000 m²	
Offices for letting	
Shell & Core and Cat A non air-conditioned; Includes for rainwater, soil/waste, cold water, hot water via local electrical heaters, LTHW heating via radiator heaters, toilet extract, controls, LV installations, lighting, small power (landlords), fire alarms, earthing, security & IT wireways	270
Shell & Core and Cat A non air-conditioned; Includes for rainwater, soil/waste, cold water, hot water, LTHW heating via perimeter heaters, toilet extract, controls, LV installations, lighting, small power (landlords), fire alarms, earthing, security wireways, IT wireways	285
Shell & Core and Cat A air-conditioned; Includes for rainwater, soil/waste, VRV 3 pipe heat pumps , toilet extract, BMS, LV installations, lighting, small power (landlords), fire alarms, earthing, security & IT wireways	420
Shell & Core and Cat A air-conditioned; Includes for rainwater, soil/waste, cold water, hot water via local electrical heaters, LTHW heating via perimeter heaters, 2 pipe , toilet extract, BMS, LV installations, lighting, small power (landlords), fire alarms, earthing, security & IT wireways	450
Offices for owner occupation	
Non air-conditioned; Includes for rainwater, soil/waste, cold water, hot water via local electrical heaters, dry risers, LTHW heating via radiator heaters, toilet extract, controls, LV installations, lighting, small power (landlords), fire alarms, earthing, security, IT wireways	300
Non air-conditioned; Includes for rainwater, soil/waste, cold water, hot water, dry risers, LTHW heating via perimeter heaters, toilet extract, BMS, LV installations, lighting, small power (landlords), fire alarms, earthing, security, IT wireways	325
Air-conditioned; Includes for rainwater, soil/waste, cold water, hot water via local electrical heaters, dry risers, LTHW heating via perimeter heating, 4 pipe air conditioning, toilet extract, kitchen extract, sprinkler protection, BMS, LV installations, life safety standby generators, lighting, small power, fire alarms, earthing, security, IT wireways small power, fire alarms L1/P1, earthing, security, and IT wireways	510

RIBA STAGE A FEASIBILITY COSTS

TYPICAL SQUARE METRE RATES FOR ENGINEERING SERVICES – cont'd	
Health and Welfare Facilities	£/m² GIA
District general hospitals	
Natural ventilation Includes for rainwater, soil/waste, cold water, hot water, dry risers, medical gases, LTHW heating via perimeter heating, toilet extract, kitchen extract, BMS, LV installations, standby generation, lighting, small power, fire alarms, earthing/lightning protection, nurse call systems, security, IT wireways	440
Natural ventilation Includes for rainwater, soil/waste, cold water, hot water, dry risers, LTHW heating via perimeter heaters, localized VAV air conditioning, kitchen/toilet extract, BMS, LV installations, standby generation, lighting, small power, fire alarms, earthing/lightning protection, nurse call systems, security, IT wireways	620
Private hospitals	
Air-conditioned; Includes for rainwater, soil/waste, cold water, hot water, dry risers, medical gases, LTHW heating, 2 pipe air conditioning, toilet extract, kitchen extract, BMS, LV installations, standby generation, lighting, small power, fire alarms, earthing, nurse call systems, nurse call system, security, IT wireways	700
Air-conditioned; Includes for rainwater, soil/waste, cold water, hot water, dry risers, medical gases, LTHW heating , 4 pipe air conditioning, kitchen/toilet extract, BMS, LV installations, standby generation, lighting, small power, fire alarms, earthing, nurse call system, security, IT wireways	750
Day care unit	
Natural ventilation; Includes for rainwater, soil/waste, cold water, hot water via local electrical heaters, medical gases, LTHW heating, toilet extract, kitchen extract, BMS, LV installations, lighting, small power, fire alarms , earthing, security, IT wireways	460
Comfort cooled; Includes for rainwater, soil/waste, cold water, hot water, medical gases, LTHW heating, DX air conditioning, kitchen/toilet extract, BMS, LV installations, lighting, small power, fire alarms, earthing, security, IT wireways	500
Entertainment and Recreation Buildings	
Non performing	
Natural ventilation; Includes for rainwater, soil/waste, cold water, central hot water, dry risers, LTHW heating, toilet extract, kitchen extract, controls, LV installations, lighting, small power, fire alarms, earthing, security, IT wireways	330
Comfort cooled; Includes for rainwater, soil/waste, cold water, hot water via local electrical heaters sprinklers/dry risers, LTHW heating , DX air conditioning, kitchen/toilet extract, BMS, LV installations, lighting, small power, fire alarms, earthing, security, IT wireways	525
Performing Arts (With Theatre)	
Natural ventilation; Includes for rainwater, soil/waste, cold water, central hot water, sprinklers/dry risers, LTHW heating, toilet extract, kitchen extract, controls, LV installations, lighting, small power, fire alarms, earthing, security, IT wireways	550
Comfort cooled; Includes for rainwater, soil/waste, cold water, central hot water, sprinklers/dry risers, LTHW heating, DX air conditioning, kitchen/toilet extract, BMS, LV installations, lighting including enhanced dimming/scene setting, small power, fire alarms, earthing, security, IT wireways including for production, audio and video recording, EPOS system	650
Sports halls	
Natural ventilation; Includes for rainwater, soil/waste, cold water, hot water gas fired heaters, LTHW heating, toilet extract, BMS, LV installations, lighting, small power, fire alarms, earthing, security	200
Comfort cooled; Includes for rainwater, soil/waste, cold water, hot water via LTHW heat exchangers, LTHW heating, air conditioning via AHU's, toilet extract, BMS, LV installations, lighting, small power, fire alarms, earthing, security	300

RIBA STAGE A FEASIBILITY COSTS

TYPICAL SQUARE METRE RATES FOR ENGINEERING SERVICES – cont'd	
Entertainment and Recreation Buildings – cont'd	£/m² GIA
Multi Purpose Leisure Centre	
Natural ventilation; Includes for rainwater, soil/waste, cold water, hot water gas fired heaters, LTHW heating, toilet extract, BMS, LV installations, lighting, small power, fire alarms, earthing, security, IT wireways	300
Comfort cooled; Includes for rainwater, soil/waste, cold water, hot water LTHW heat exchangers, LTHW heating, air conditioning via AHU, pool hall supply/extract, kitchen/toilet extract, BMS, LV installations, lighting, small power, fire alarms, earthing, security, IT wireways	400
Retail Buildings	
Open arcade	
Natural ventilation; Includes for rainwater, soil/waste, cold water, hot water, sprinklers/dry risers, LTHW heating, toilet extract, smoke extract, BMS, LV installations, life safety standby generators, lighting, small power, fire alarms, public address, earthing/lightning protection, security, IT wireways	320
Enclosed Shopping Mall	
Air-conditioned; Includes for rainwater, soil/waste, cold water, hot water, sprinklers/dry risers, LTHW heating, air conditioning via AHU's, toilet extract, smoke extract, BMS, LV installations, life safety standby generators, lighting, small power, fire alarms, public address, earthing/lightning protection, CCTV/security, IT wireways people counting systems	500
Department stores	
Air-conditioned; Includes for sanitaryware, soil/waste, cold water, hot water, sprinklers/dry risers, LTHW heating, air conditioning via AHU's, toilet extract, smoke extract, BMS, LV installations, life safety standby generators, lighting, small power, fire alarms, public address, earthing, lightning protection, CCTV/security, IT installation wireways.	375
Supermarkets	
Air-conditioned; Includes for rainwater, soil/waste, cold water, hot water, sprinklers, LTHW heating, air conditioning via AHU's, toilet extract, BMS, LV installations, lighting, small power, fire alarms, earthing, lightning protection, security, IT wireways, refrigeration	570
Educational Buildings	
Secondary schools (Academy)	
Natural ventilation; Includes for rainwater, soil/waste, cold water, central hot water, LTHW heating, toilet extract, BMS, LV installations, lighting, small power, fire alarms, earthing, security, IT wireways	320
Natural vent with comfort cooling to selected areas (BB93 compliant); Includes for rainwater, soil/waste, cold water, central hot water, LTHW heating, DX air conditioning, general supply/extract, toilet extract, BMS, LV installations, lighting, small power, fire alarms, earthing, security, IT wireways	390
Scientific Buildings	
Educational research	
Comfort cooled; Includes for rainwater, soil/waste, cold water, central hot water, dry risers, compressed air, medical gases, LTHW heating, 4 pipe air conditioning, toilet extract, fume, BMS, LV installations, lighting, small power, fire alarms, earthing, lightning protection, security, IT wireways	750
Air-conditioned; Includes for rainwater, soil/waste, laboratory waste, cold water, central hot water, specialist water, dry risers, compressed air, medical gases, steam, LTHW heating, VAV air conditioning, Comm's room cooling, toilet extract, fume extract, BMS, LV installations, UPS, standby generators, lighting, small power, fire alarms, earthing, lightning protection, security, IT wireways	1250

RIBA STAGE A FEASIBILITY COSTS

TYPICAL SQUARE METRE RATES FOR ENGINEERING SERVICES – cont'd	
Scientific Buildings – cont'd	£/m² GIA
Commercial research	
Air-conditioned; Includes for rainwater, soil/waste, laboratory waste, cold water, central hot water, specialist water, dry risers, compressed air, medical gases, steam, LTHW heating, VAV air conditioning, Comm room cooling, toilet extract, fume extract, BMS, LV installations, UPS, standby generators, lighting, small power, fire alarms, earthing, lightning protection, security, IT wireways	1250
Hotels	
1 to 3 Star; Includes for rainwater, soil/waste, cold water, hot water, dry risers, LTHW heating via radiators, toilet/bathroom extract, kitchen extract, BMS, LV installations, lighting, small power, fire alarms, earthing, lightning protection security, IT wireways	350 to 550
4 to 5 Star; Includes for rainwater, soil/waste, cold water, hot water, sprinklers, dry risers, 4 pipe air conditioning, kitchen extract, toilet/bathroom extract, BMS, LV installations, life safety standby generators, lighting, small power, fire alarms, earthing, security, IT wireways	750 to 850

RIBA STAGE C ELEMENTAL RATES

ELEMENTAL RATES FOR ALTERNATIVE ENGINEERING SERVICES SOLUTIONS

The following examples of building types indicate the range of rates for alternative design solutions for each of the engineering services elements based on Gross Internal Area for the Shell and Core and Net Internal Area for the Fit Out. Fit Out is assumed to be to Cat A standard.

Consideration should be made for the size of the building, which may affect the economies of scale for rates i.e. the larger the building the lower the rates

OFFICES	Shell & Core £/m² GIA	Fit Out £/m² NIA
MECHANICAL SERVICES		
Sanitaryware		
Building up to 3,000 m²	8 to 10	-
Building over 3,000 m² to 15,000 m² (low rise)	6 to 9	-
Disposal installation		
Building up to 3,000 m²	15 to 25	-
Building over 3,000 m² to 15,000 m²	15 to 20	-
Water installation		
Building up to 3,000 m²	15 to 25	-
Building over 3,000 m² to 15,000 m²	15 to 25	-
LPHW Heating Installation; including gas installations		
Building up to 3,000 m²	35 to 45	35 to 45
Building over 3,000 m² to 15,000 m²	40 to 45	25 to 35
Air Conditioning; including ventilation		
Comfort Cooling;		
2 pipe fan coil for building up to 3,000 m²	50 to 70	85 to 100
2 pipe fan coil for building over 3,000 m² to 15,000 m²	50 to 60	80 to 90
2 pipe variable refrigerant volume (VRV) for building up to 3,000 m²	40 to 50	60 to 70
Full air conditioning;		
4 Pipe fan coil for building up to 3,000 m²	80 to 100	130 to 150
4 Pipe fan coil for building over 3,000 m² to 15,000 m²	80 to 100	110 to 130
3 pipe variable refrigerant volume for building up to 3,000 m²	70 to 80	110 to 120
Ventilated (active) chilled beams for building over 3,000 m² to 15,000 m²	80 to 100	130 to 140
Chilled beam exposed services for building over 3,000 m² to 15,000 m²	80 to 100	200 to 215
Concealed passive chilled beams for building over 3,000m to 15,000 m²	80 to 100	110 to 130
Chilled ceiling for building over 3,000 m² to 15,000 m²	80 to 100	205 to 215
Chilled ceiling/perimeter beams for building over 3,000 m² to 15,000 m²	80 to 100	215 to 230
Displacement for building over 3,000 m² to 15,000 m²	80 to 90	70 to 90
Ventilation systems (excluding smoke extract)		
Building up to 3,000 m²	30 to 40	-
Building over 3,000 m² to 15,000 m²	25 to 35	-
Fire Protection over 3,000 m² to 15,000 m²		
Dry risers	3 to 5	-
Sprinkler installation	15 to 20	25 to 30
BMS Controls: including MCC panels and control cabling		
Full air conditioning	15 to 20	15 to 20

RIBA STAGE C ELEMENTAL RATES

ELEMENTAL RATES FOR ALTERNATIVE ENGINEERING SERVICES SOLUTIONS – cont'd		
OFFICES – cont'd	Shell & Core £/m² GIA	Fit Out £/m² NIA
ELECTRICAL SERVICES		
LV Installations		
Standby generators (life safety only)		
Buildings over 3,000 m² to 15,000 m²	15 to 20	-
LV distribution		
Buildings up to 3,000 m²	25 to 35	-
Buildings over 3,000 m² to 15,000 m²	30 to 40	-
Lighting Installations (including lighting controls and luminaries)		
Buildings up to 3,000 m²	15 to 20	40 to 60
Buildings over 3,000 m² to 15,000 m²	15 to 20	50 to 70
Small Power		
Buildings up to 3,000 m²	4 to 8	-
Buildings over 3,000 m² to 15,000 m²	4 to 6	-
Protective Installations		
Earthing		
Buildings up to 3,000 m²	2 to 3	1 to 2
Buildings over 3,000 m² to 15,000 m²	2 to 3	1 to 2
Lightning protection		
Buildings up to 3,000 m²	3 to 4	-
Buildings over 3,000 m² to 15,000 m²	2 to 3	-
Communication Installations		
Fire alarms (single stage)		
Buildings up to 3,000 m²	8 to 10	10 to 15
Buildings over 3,000 m² to 15,000 m²	10 to 12	10 to 15
Fire alarms (phased evacuation)		
Buildings over 3,000 m² to 15,000 m²	12 to 15	12 to 18
IT (wireways only)		
Buildings up to 3,000 m²	2 to 3	-
Buildings over 3,000 m² to 15,000 m²	2 to 3	-
Security		
Buildings up to 3,000 m²	5 to 8	-
Buildings over 3,000 m² to 15,000 m²	5 to 10	-
Electrical Installations for Mechanical Plant		
Buildings up to 3,000 m²	5 to 8	-
Buildings over 3,000 m² to 15,000 m²	5 to 10	5 to 10

RIBA STAGE C ELEMENTAL RATES

ELEMENTAL RATES FOR ALTERNATIVE ENGINEERING SERVICES SOLUTIONS – cont'd	
HOTELS	**£/m² GIA**
MECHANICAL SERVICES	
Sanitaryware	
2 to 3 Star	20 to 30
4 to 5 Star	20 to 50
Above ground disposal installation	
2 to 3 Star	20 to 35
4 to 5 Star	20 to 35
Water installation	
2 to 3 Star	28 to 38
4 to 5 Star	40 to 50
LPHW Heating Installation; including gas installations	
2 to 3 Star	30 to 45
4 to 5 Star	30 to 45
Air Conditioning; including ventilation	
2 to 3 Star – 4 pipe fan coil	200 to 240
4 to 5 Star – 4 pipe fan coil	200 to 240
2 to 3 Star – 3 pipe variable refrigerant volume	130 to 150
4 to 5 Star – 3 pipe variable refrigerant volume	130 to 150
Fire Protection	
2 to 3 Star – dry risers	8 to 12
4 to 5 Star – dry risers	8 to 12
2 to 3 Star – Sprinkler installation	20 to 25
4 to 5 Star – Sprinkler installation	20 to 30
BMS Controls: including MCC panels and control cabling	
2 to 3 Star	10 to 12
4 to 5 Star	20 to 35

RIBA STAGE C ELEMENTAL RATES

ELEMENTAL RATES FOR ALTERNATIVE ENGINEERING SERVICES SOLUTIONS – cont'd	
HOTELS – cont'd	£/m² GIA
ELECTRICAL SERVICES	
LV Installations	
Standby generators (life safety only)	
2 to 3 Star	10 to 20
4 to 5 Star	10 to 20
LV distribution	
2 to 3 Star	25 to 40
4 to 5 Star	35 to 50
Lighting Installations	
2 to 3 Star	10 to 35
4 to 5 Star	15 to 50
Small Power	
2 to 3 Star	5 to 10
4 to 5 Star	10 to 15
Protective Installations	
Earthing	
2 to 3 Star	1 to 2
4 to 5 Star	1 to 2
Lightning protection	
2 to 3 Star	1 to 2
4 to 5 Star	1 to 2
Communication Installations	
Fire alarms	
2 to 3 Star	10 to 20
4 to 5 Star	10 to 20
IT	
2 to 3 Star	10 to 15
4 to 5 Star	10 to 15
Security	
2 to 3 Star	15 to25
4 to 5 Star	15 to25
Electrical Installations for Mechanical Plant	
2 to 3 Star	5 to 8
4 to 5 Star	5 to 8

RIBA STAGE C ELEMENTAL RATES

ELEMENTAL RATES FOR ALTERNATIVE ENGINEERING SERVICES SOLUTIONS – cont'd		
RESIDENTIAL	Shell & core £/m² GIA	Fit out £/m² NIA
MECHANICAL & ELECTRICAL SERVICES		
Sanitaryware and above ground disposal installation		
Affordable	0 to 5	0 to 5
Private	25 to 35	50 to 70
Disposal installation		
Affordable	10 to 15	15 to 20
Private	5 to 15	9 to 16
Water installation		
Affordable	20 to 35	30 to 35
Private	30 to 60	40 to 60
Heat Source		
Affordable	0 to 15	10 to 15
Private	0 to 35	0 to 35
Space Heating & Air Treatment		
Affordable	5 to 10	40 to 60
Private	20 to 40	140 to 240
Ventilation		
Affordable to facade	10 to 20	10 to 20
Private (whole house vent)	25 to 35	30 to 50
Electrical Installations		
Affordable	35 to 40	40 to 60
Private	40 to 60	80 to 105
Gas Installations		
Affordable	1 to 5	5 to 10
Private	0 to 5	0 to 20
Protective Installations		
Affordable	10 to 35	10 to 40
Private	0 to 20	0 to 20
Communication Installations		
Affordable	20 to 30	30 to 45
Private	25 to 35	35 to 60
Special Installations		
Affordable	5 to 10	10 to 25
Private	5 to 10	5 to 10

Note: The range in cost differs due to the vast diversity in services strategies available. The lower end of the scale reflects all electric schemes (not always be possible due to Part L requirements) or local plant within apartment schemes, such as combi boilers, local ventilation etc. The high end of the scale is based on good quality apartments, which includes comfort cooling, sprinklers, home network installations, video entry, higher quality of sanitaryware and lighting.

ALL-IN-RATES

ALL-IN-RATES FOR PRICING MECHANICAL APPROXIMATE QUANTITIES	
ABOVE GROUND DRAINAGE	**Cost per point £**
Soil and Waste	335–380
WATER INSTALLATIONS	
Cold water	335–380
Hot water	380–430
HEAT SOURCE	**Cost per kW £**
Gas fired boilers including gas train and controls	30–35
Gas fired boilers including gas train, controls, flue, Plantroom pipework, valves and insulation, pumps and pressurization unit	65–115
SPACE HEATING AND AIR TREATMENT	
CHILLED WATER	**Cost per kW £**
Air cooled R134a refrigerant chiller including control panel, anti vibration mountings	115–135
Air cooled R134a refrigerant chiller including control panel, anti-vibration mountings, plantroom pipework, valves, insulation, pumps and pressurization units	190–240
Water cooled R134a refrigerant chiller including control panel, anti vibration mountings	65–85
Water cooled R134a refrigerant chiller including control panel, anti-vibration mountings, plantroom pipework, valves, insulation, pumps and pressurization units	140–180
Absorption steam medium chiller including control panel, anti-vibration mountings, plantroom pipework, valves, insulation, pumps and pressurization units	240–335
HEAT REJECTION	**Cost per kW (heat rejection) £**
Open circuit, forced draft cooling tower	50–65
Closed circuit, forced draft cooling tower	65–75
Dry Air	60–65
PUMPS	
Pumps including flexible connections, anti-vibration mountings	13–60
Pumps including flexible connections, anti-vibration mountings, plantroom pipework, valves, insulation and accessories	40–135
DUCTWORK	**Per m² of duct £**
The rates below allow for ductwork and for all other labour and material in fabrication, fittings, supports and jointing to equipment, stop and capped ends, elbows, bends, diminishing and transition pieces, regular and reducing couplings, volume control dampers, branch diffuser and 'snap on' grille connections, ties, 'Ys', crossover spigots, etc., turning vanes, regulating dampers, access doors and openings, hand-holes, test holes and covers, blanking plates, flanges, stiffeners, tie rods and all supports and brackets fixed to structure	
Rectangular galvanized mild steel ductwork as HVCA DW 144 up to 1000 mm longest side	50–60
Rectangular galvanized mild steel ductwork as HVCA DW 144 up to 2500 mm longest side	55–70
Rectangular galvanized mild steel ductwork as HVCA DW 144 3000 mm longest side and above	75–90
Circular galvanized mild steel ductwork as HVCA DW 144	55–70
Flat oval galvanized mild steel ductwork as HVCA DW 144 up to 545 mm wide	55–70

ALL-IN-RATES

ALL-IN-RATES FOR PRICING MECHANICAL APPROXIMATE QUANTITIES – cont'd	
SPACE HEATING AND AIR TREATMENT – cont'd	
DUCTWORK – cont'd	**Per m² of duct** **£**
Flat oval galvanized mild steel ductwork as HVCA DW 144 up to 880 mm wide	60–80
Flat oval galvanized mild steel ductwork as HVCA DW 144 up to 1785 mm wide	70–90
PACKAGED AIR HANDLING UNITS	**Cost per m³/s** **£**
Air handling unit including LPHW pre-heater coil, pre-filter panel, LPHW heater coils, chilled water coil, filter panels, inverter drive, motorized volume control dampers, sound attenuation, flexible connections to ductwork and all anti-vibration mountings	6,000–8,000
EXTRACT FANS	
Extract fan including inverter drive, sound attenuation, flexible connections to ductwork and all anti-vibration mountings	950–1,900
PROTECTIVE INSTALLATIONS	**£**
SPRINKLER INSTALLATION	
Recommended maximum area coverage per sprinkler head:	
Extra light hazard, 21 m² of floor area	
Ordinary hazard, 12 m² of floor area	
Extra high hazard, 9 m² of floor area	
Sprinkler equipment installation, pipework, valve sets, booster pumps and water storage	60,000–80,000
Price per sprinkler head; including pipework, valves and supports	180
HOSE REELS AND DRY RISERS	
Wall mounted concealed hose reel with 36 metre hose including approximately 15 metres of pipework and isolating valve:	
Price per hose reel	1,750
100 mm dry riser main including 2 way breeching valve and box,, 65 mm landing valve, complete with padlock and leather strap and automatic air vent and drain valve.	
Price per landing	1,500
COMMUNICATIONS INSTALLATIONS	
SECURITY	
ACCESS CONTROL SYSTEMS	
Door Mounted access control unit inclusive of door furniture, lock plus software. Including up to 50 meters of cable and termination. Including documentation testing and commissioning	
Internal single leaf door	1,100
Internal double door	1,200
External single leaf door	1,200
External double leaf door	1,300
Management control PC with printer software and commissioning up to 1000 users	12,000

ALL-IN-RATES

ALL-IN-RATES FOR PRICING MECHANICAL APPROXIMATE QUANTITIES – cont'd	
COMMUNICATIONS INSTALLATIONS – cont'd	
SECURITY – cont'd	
CCTV INSTALLATIONS	
CCTV Equipment inclusive of 50 m of cable including testing and commissioning	
Internal camera with Bracket	900
Internal camera with Housing	950
Internal PTZ camera with Bracket	1,700
External fixed camera with housing	1,200
External PTZ camera dome	2,400
External PTZ camera dome with power	2,900
IT INSTALLATIONS	**Cost per point £**
DATA CABLING	
Complete channel link including, patch leads, cable, panels, testing and documentation (excludes cabinets and/or frames, patch cords, backbone/harness connectivity as well as containment)	
Low Level	
Cat 5e (up to 5,000 outlets)	41.00
Cat 5e (5,000 to 15,000 outlets)	38.00
Cat 6 (up to 5,000 outlets)	53.00
Cat 6 (5,000 to 15,000 outlets)	50.00
Cat 6a (up to 5,000 outlets)	71.00
Cat 6a (5,000 to 15,000 outlets)	68.00
Cat 7 (up to 5,000 outlets)	86.00
Cat 7 (5,000 to 15,000 outlets)	83.00
Note: LSZH cable based on average of 50 metres false floor low level installation assuming 1 workstation in 2.5 m × 2.5 m (to 3.2 m × 3.2 m) density, with 4 data points per workstation. Not applicable for installations with less than 250 No. outlets.	
High Level	
Cat 5e (up to 500 outlets)	52.00
Cat 5e (over 500 outlets)	49.00
Cat 6 (up to 500 outlets)	64.00
Cat 6 (over 500 outlets)	61.00
Cat 6a (up to 500 outlets)	82.00
Cat 6a (over 500 outlets)	79.00
Cat 7 (up to 500 outlets)	97.00
Cat 7 (over 500 outlets)	94.00
Note: High level at 10 × 10 m grid.	

ALL-IN-RATES

ALL-IN-RATES FOR PRICING MECHANICAL APPROXIMATE QUANTITIES – cont'd	
PIPEWORK	
HOT AND COLD WATER *excludes insulation, valves and ancillaries etc.*	**Cost per metre £**
Light gauge copper tube to EN1057 R250 (TX) formerly BS 2871 part 1 table X with joints as described including allowance for waste, fittings and supports assuming average runs with capillary joints up to 54 mm and bronze welded thereafter	
Horizontal High Level Distribution	
15 mm	30.00
22 mm	31.00
28 mm	39.00
35 mm	43.00
42 mm	51.00
54 mm	64.00
67 mm	90.00
76 mm	106.00
108 mm	145.00
Risers	
15 mm	17.00
22 mm	20.00
28 mm	30.00
35 mm	34.00
42 mm	41.00
54 mm	50.00
67 mm	82.00
76 mm	90.00
108 mm	119.00
Toilet Areas etc at Low Level	
15 mm	51.00
22 mm	60.00
28 mm	85.00

LTHW AND CHILLED WATER *excludes insulation, valves and ancillaries etc.*	
Black heavy weight mild steel tube to BS1387 with joints in the running length, allowance for waste, fittings and supports assuming average runs	**Cost per metre LTHW/Chilled water**
Horizontal Distribution – Basements etc.	
15 mm	37.00
20 mm	40.00
25 mm	44.00
32 mm	51.00
40 mm	58.00
50 mm	70.00
65 mm	76.00

ALL-IN-RATES

ALL-IN-RATES FOR PRICING MECHANICAL APPROXIMATE QUANTITIES – cont'd	
PIPEWORK – cont'd	
LTHW AND CHILLED WATER – cont'd	
	Cost per metre LTHW/Chilled water
80 mm	102.00
100 mm	133.00
125 mm	190.00
150 mm	250.00
200 mm	335.00
250 mm	413.00
300 mm	484.00
Risers	
15 mm	23.00
20 mm	26.00
25 mm	29.00
32 mm	33.00
40 mm	37.00
50 mm	45.00
65 mm	56.00
80 mm	76.00
100 mm	97.00
125 mm	119.00
150 mm	156.00
200 mm	243.00
250 mm	314.00
300 mm	358.00
On Floor Distribution	
15 mm	35.00
20 mm	40.00
25 mm	43.00
32 mm	50.00
40 mm	56.00
50 mm	67.00
65 mm	-
Plantroom Areas etc.	
15 mm	37.00
20 mm	41.00
25 mm	46.00
32 mm	53.00
40 mm	60.00
50 mm	72.00
65 mm	78.00
80 mm	105.00

ALL-IN-RATES

ALL-IN-RATES FOR PRICING MECHANICAL APPROXIMATE QUANTITIES – cont'd	
PIPEWORK – cont'd	
LTHW AND CHILLED WATER – cont'd	
	Cost per metre **LTHW/Chilled water**
Plantroom Areas etc. – cont'd	
100 mm	138.00
125 mm	196.00
150 mm	265.00
200 mm	349.00
250 mm	430.00
300 mm	508.00

ALL-IN-RATES

ALL-IN-RATES FOR PRICING ELECTRICAL APPROXIMATE QUANTITIES	
HV/LV INSTALLATIONS	
The cost of HV/LV equipment will vary according to the electricity supplier's requirements, the duty required and the actual location of the site. For estimating purposes the items indicated below are typical of the equipment required in a HV substation incorporated into a building.	
RING MAIN UNIT	**Cost per unit £**
Ring Main Unit , 11kv including electrical terminations	12,000–18,000
TRANSFORMERS	**Cost per KVA £**
Oil filled transformers, 11 kv to 415 v including electrical terminations	14 to 18
Cast Resin transformers, 11 kv to 415 v including electrical terminations	15 to 20
Midal filled transformer, 11 kv to 415 v including electrical transformations	16 to 22
HV SWITCHGEAR	**Cost per section £**
Cubicle section HV switchpanel, Form 4 type 6 including air circuit breakers, meters and electrical terminations	20,000–25,000
LV SWITCHGEAR	**Cost per isolator £**
LV switchpanel, Form 3 including all isolators, fuses, meters and electrical terminations	1,800–2,800
LV switchpanel, Form 4 type 5 including all isolators, fuses, meters and electrical terminations	3,000–3,900
EXTERNAL PACKAGED SUB-STATION	
Extra over cost for prefabricated packaged sub station housing excludes base and protective security fencing	20,000–25,000
STANDBY GENERATING SETS	**Cost per KVA £**
Diesel powered including control panel, flue, oil day tank and attenuation	
Approximate installed cost, LV	220–340
Approximate installed cost, HV	240–390
UNINTERRUPTIBLE POWER SUPPLY	
Rotary UPS including control panel and choke transformer (excludes distribution)	
Approximate installed cost (range 1000 KVA to 2500 KVA)	390–490
Static UPS including control panel, automatic bypass, DC isolator and batteries for 30 minutes standby (excludes distribution)	
Approximate installed cost (range 500 KVA to 1000 KVA)	190–290

ALL-IN-RATES

ALL-IN-RATES FOR PRICING ELECTRICAL APPROXIMATE QUANTITIES – cont'd	
SMALL POWER	**Per point** **£**
Approximate prices for wiring of power points of length not exceeding 20 m, including accessories, wireways but excluding distribution boards.	
13 amp Accessories	
Wired in PVC insulated twin and earth cable in ring main circuit	
Domestic properties	55.00
Commercial properties	75.00
Industrial properties	75.00
Wired in PVC insulated twin and earth cable in radial circuit	
Domestic properties	75.00
Commercial properties	90.00
Industrial property	90.00
Wired in LSF insulated single cable in ring main circuit	
Commercial properties	90.00
Industrial property	90.00
Wired in LSF insulated single cable in radial circuit	
Commercial properties	110.00
Industrial property	110.00
45 amp wired in PVC insulated twin and earth cable	
Domestic properties	110.00
Low voltage power circuits Three phase four wire radial circuit feeding an individual load, wired in LSF insulated single cable including all wireways, isolator, *not exceeding 10 metres; in commercial properties.* Cable size mm²	
1.5	180.00
2.5	195.00
4	210.00
6	230.00
10	270.00
16	290.00
Three phase four core radial circuit feeding an individual load item, wired in LSF/SWA/XLPE insulated cable including terminations, isolator; clipped to surface, *not exceeding 10 metres in commercial properties.* Cable size mm²	
1.5	135.00
2.5	150.00
4	165.00
6	180.00
10	275.00
16	350.00

ALL-IN-RATES

ALL-IN-RATES FOR PRICING ELECTRICAL APPROXIMATE QUANTITIES – cont'd	
LIGHTING	**Per point £**
Approximate prices for wiring of lighting points including rose, wireways but excluding distribution boards, luminaires and switches	
Final Circuits	
Wired in PVC insulated twin and earth cable	
Domestic properties	40.00
Commercial properties	50.00
Industrial properties	50.00
Wired in LSF insulated single cable	
Commercial properties	65.00
Industrial property	65.00
ELECTRICAL WORKS IN CONNECTION WITH MECHANICAL SERVICES The cost of electrical connections to mechanical services equipment will vary depending on the type of building and complexity of the equipment. Includes for power wiring, isolators and associated wireways (£/m²)	5–10
FIRE ALARMS Cost per point for two core MICC insulated wired system including all terminations, supports and wireways	
Call point	240.00
Smoke detector	215.00
Smoke/heat detector	240.00
Heat detector	230.00
Heat detector and sounder	205.00
Input/output/relay units	290.00
Alarm sounder	230.00
Alarm sounder/beacon	270.00
Speakers/voice sounders	270.00
Speakers/voice sounders (weatherproof)	280.00
Beacon/strobe	215.00
Beacon/strobe (weatherproof)	310.00
Door release units	310.00
Beam detector	825.00
Call point	185.00
Smoke detector	185.00
Smoke/heat detector	225.00
Heat	195.00
Heat detector and sounder	390.00
Input/output/relay units	370.00
Alarm sounder	350.00
Alarm sounder/beacon	410.00
Speakers/voice sounders	370.00
Speakers/voice sounders (weatherproof)	445.00

ALL-IN-RATES

ALL-IN-RATES FOR PRICING ELECTRICAL APPROXIMATE QUANTITIES – cont'd	
FIRE ALARMS – cont'd Cost per point for two core MICC insulated wired system including all terminations, supports and wireways – cont'd	
Beacon/strobe	370.00
Beacon/strobe (weatherproof)	380.00
Door release units	290.00
Beam detector	970.00
For costs for zone control panel, battery chargers and batteries, see 'Prices for Measured Work' section.	
EXTERNAL LIGHTING	
Estate road lighting	
Post type road lighting lantern 70 watt CDM-T 3000k complete with 5m high column with hinged lockable door, control gear and cut-out including 2.5 mm two core butyl cable internal wiring, interconnections and earthing fed by 16 mm² four core XLPE/SWA /LSF cable and terminations. Approximate installed price *per metre road length* (based on 300 metres run) including time switch but excluding builder's work in connection	
Columns erected at 30 m intervals	£70.00 per m of road
Bollard lighting	
Bollard lighting fitting 26 watt TC-D 3500k including control gear, all internal wiring, interconnections, earthing and 25 metres of 2.5 mm² three core XLPE/SWA/LSF cable	
Approximate installed price excluding builder's work in connection	£970.00 *each*
Outdoor flood lighting	
Wall mounted outdoor flood light fitting complete with tungsten halogen lamp, mounting bracket, wire guard and all internal wiring and containment; fixed to brickwork or concrete and connected	
Installed price 500 watt	£165.00–£240.00
Installed price 1000 watt	£195.00–£290.00
Pedestal mounted outdoor floor light fitting complete with1000 watt MBF/U lamp, mounting bracket, control gear, contained in weatherproof steel box, all internal wiring and containment, interconnections and earthing; fixed to brickwork or concrete and connected	
Approximate installed price excluding builder's work in connection	£1,000.00 *each*

ALL-IN-RATES

ALL-IN-RATES FOR PRICING SPECIALIST APPROXIMATE QUANTITIES

LIFT INSTALLATIONS

The cost of lift installations will vary depending upon a variety of circumstances. The following prices assume a car height of 2.2 metres, manufacturers standard car finish, brushed stainless steel 2 panel centre opening doors to BSEN81 part 1 & 2 and Lift Regulations 1997.

Passenger Lifts Machine Room Above	8 Person £	10 Person £	13 Person £	17 Person £	21 Person £	26 Person £
Electrically operated AC drive serving 2 levels with directional collective controls and a speed of 1.0 m/s	58,000	63,000	66,000	75,000	85,000	97,000
As above serving 4 levels and a speed of 1.0 m/s	68,000	72,500	76,000	86,000	97,000	110,000
As above serving 6 levels and a speed of 1.0 m/s	77,000	82,000	86,000	96,000	108,000	124,000
As above serving 8 levels and a speed of 1.0 m/s	86,500	91,500	96,000	107,000	120,500	137,000
As above serving 10 levels and a speed of 1.0 m/s	96,000	101,000	105,000	118,000	132,000	150,000
As above serving 12 levels and a speed of 1.0 m/s	105,000	110,000	115,000	129,000	144,000	163,000
As above serving 14 levels and a speed of 1.0 m/s	116,500	120,000	126,000	141,000	157,000	177,000
Add to above for:						
Increase speed from 1.0 to 1.6 m/s	4,000	4,100	4,200	4,200	4,200	4,200
Increase speed from 1.6 m/s to 2.0 m/s	880	880	1,170	1,170	1,470	1,470
Increase speed from 2.0 m/s to 2.5 m/s	2,040	2,040	2,490	2,490	2,930	2,930
Enhanced finish to car – Centre mirror, flat ceiling, carpet	2,800	3,000	3,000	3,400	3,900	4,600
Bottom motor room	7,000	7,000	7,000	8,500	8,500	8,600
Fire fighting control	5,500	5,500	5,500	5,500	5,500	5,500
Glass back	2,500	2,800	3,300	4,000	4,000	4,000
Glass doors	19,000	19,000	20,900	22,500	21,500	21,500
Painting to entire pit	2,000	2,000	2,000	2,000	2,000	2,000
Dual seal shaft	4,000	4,000	4,000	4,800	4,800	4,800
Dust sealing machine room	800	800	1,300	1,300	1,300	1,300
Intercom to reception desk and security room	400	400	400	400	400	400
Heating, cooling and ventilation to machine room	1,000	1,000	1,000	1,000	1,000	1,000
Shaft lighting/small power	3,500	3,500	3,500	3,500	3,500	3,500
Motor room lighting/small power	1,300	1,300	1,500	1,600	1,600	1,600
Lifting beams	1,500	1,500	1,500	1,700	1,700	1,700

ALL-IN-RATES

ALL-IN-RATES FOR PRICING SPECIALIST APPROXIMATE QUANTITIES – cont'd

LIFT INSTALLATIONS – cont'd

Passenger Lifts Machine Room Above – cont'd	8 Person £	10 Person £	13 Person £	17 Person £	21 Person £	26 Person £
10 mm Equipotential bonding of all entrance metalwork	800	800	800	800	800	800
Shaft secondary steelwork	5,800	6,000	6,200	6,400	6,400	6,400
Independent insurance inspection	1,900	1,900	1,900	1,900	1,900	1,900
12 Month warranty service	2,000	2,000	2,000	2,000	2,000	2,000

Passenger Lifts Machine Room-less	8 Person £	10 Person £	13 Person £	17 Person £	21 Person £	26 Person £
Electrically operated AC drive serving 2 levels with directional collective controls and a speed of 1.0 m/s	52,500	57,800	61,500	74,500	81,500	90,000
As above serving 4 levels and a speed of 1.0 m/s	61,000	66,000	71,000	84,500	92,000	101,500
As above serving 6 levels and a speed of 1.0 m/s	70,500	75,000	80,000	94,500	102,000	112,500
As above serving 8 levels and a speed of 1.0 m/s	79,500	84,000	89,000	104,000	112,000	124,000
As above serving 10 levels and a speed of 1.0 m/s	88,500	93,000	98,500	114,000	122,000	135,000
As above serving 12 levels and a speed of 1.0 m/s	97,500	102,500	108,000	124,000	133,000	144,000
As above serving 14 levels and a speed of 1.0 m/s	109,000	111,500	119,000	136,000	146,000	160,000

Add to above for:						
Increase speed from 1.0 to 1.6 m/s	2,600	2,600	2,700	4,200	4,800	6,200
Enhanced finish to car – Centre mirror, flat ceiling, carpet	2,800	3,000	3,000	3,400	3,900	4,600
Fire fighting control	5,500	5,500	5,500	5,500	5,500	5,500
Painting to entire pit	2,000	2,000	2,000	2,000	2,000	2,000
Dual seal shaft	4,000	4,000	4,000	4,000	4,000	4,000
Shaft lighting/small power	3,500	3,500	3,500	3,500	3,500	3,500
Intercom to reception desk and security room	400	400	400	400	400	400
Lifting beams	1,500	1,500	1,500	1,700	1,700	1,700
10 mm Equipotential bonding of all entrance metalwork	800	800	800	800	800	800
Shaft secondary steelwork	5,800	6,000	6,200	6,400	6,400	6,400
Independent insurance inspection	1,900	1,900	1,900	1,900	1,900	1,900
12 Month warranty service	1,000	1,000	1,000	1,000	1,000	1,000

ALL-IN-RATES

ALL-IN-RATES FOR PRICING SPECIALIST APPROXIMATE QUANTITIES – cont'd				
LIFT INSTALLATIONS – cont'd				
Goods Lifts **Machine Room Above**	**2000 kg** **£**	**2250 kg** **£**	**2500 kg** **£**	**3000 kg** **£**
Electrically operated two speed serving 2 levels to take 1000 kg load, prime coated internal finish and a speed of 1.0 m/s	97,300	107,500	109,000	118,000
As above serving 4 levels and a speed of 1.0 m/s	111,000	121,500	123,000	138,500
A As above serving 6 levels and a speed of 1.0 m/s	124,000	135,000	137,000	152,500
As above serving 8 levels and a speed of 1.0 m/s	137,000	148,000	150,000	170,000
As above serving 10 levels and a speed of 1.0 m/s	150,000	162,000	164,000	187,000
As above serving 12 levels and a speed of 1.0 m/s	163,000	176,000	172,000	204,000
As above serving 14 levels and a speed of 1.0 m/s	176,500	189,000	191,000	221,000
Add to above for:				
Increased speed of travel from 1.0 to 1.6 metres per second	1,300	-	-	-
Enhanced finish to car – Centre mirror, flat ceiling, carpet	3,500	3,500	3,500	-
Bottom motor room	8,500	-	-	-
Painting to entire pit	2,000	2,000	2,000	2,000
Dual seal shaft	4,000	4,000	4,000	4,000
Intercom to reception desk and security room	400	400	400	400
Heating, cooling and ventilation to machine room	1,000	1,000	1,000	1,000
Lifting beams	1,500	1,500	1,500	1,500
10 mm Equipotential bonding of all entrance metalwork	800	800	800	800
Independent insurance inspection	1,900	1,900	1,900	1,900
12 Month warranty service	1,000	1,200	1,200	1,200
Goods Lifts **Machine Room-less**	**2000 kg** **£**	**2250 kg** **£**	**2500 kg** **£**	
Electrically operated two speed serving 2 levels to take 1000 kg load, prime coated internal finish and a speed of 1.0 m/s	88,000	94,000	100,000	
As above serving 4 levels and a speed of 1.0 m/s	99,000	106,000	112,000	
As above serving 6 levels and a speed of 1.0 m/s	110,000	110,000	125,000	
As above serving 8 levels and a speed of 1.0 m/s	121,000	129,000	135,000	
As above serving 10 levels and a speed of 1.0 m/s	133,000	140,000	147,500	
As above serving 12 levels and a speed of 1.0 m/s	144,000	152,000	159,000	
As above serving 14 levels and a speed of 1.0 m/s	158,000	165,000	171,000	
Add to above for:				
Increased speed of travel from 1.0 to 1.6 m/s	4,800	7,700	-	
Add to above for:				
Enhanced finish to car – Centre mirror, flat ceiling, carpet	4,100	5,400	8,000	

ALL-IN-RATES

ALL-IN-RATES FOR PRICING SPECIALIST APPROXIMATE QUANTITIES – cont'd			
LIFT INSTALLATIONS – cont'd			
Goods Lifts **Machine Room-less** – cont'd	**2000 kg** **£**	**2250 kg** **£**	**2500 kg** **£**
Add to above for:			
Painting to entire pit	2,000	2,000	2,000
Dual seal shaft	4,000	4,000	4,000
Intercom to reception desk and security room	400	400	400
Lifting beams	1,500	1,500	1,500
10 mm Equipotential bonding of all entrance metalwork	800	800	800
Independent insurance inspection	1,900	1,900	1,900
12 Month warranty service	1,000	1,000	1,000

ESCALATOR INSTALLATIONS

	Each £
30Ø Pitch escalator with a rise of 3 to 6 metres with standard balustrades	
1000 mm step width	80,000
Add to above for:	
Balustrade lighting	2,500
Skirting lighting	9,800
Emergency stop button pedestals	4,500
Truss cladding – Stainless steel	25,000
Truss cladding – Spray painted steel	22,000

ELEMENTAL COSTS

AIRPORT TERMINAL BUILDING

New build airport terminal building, premium quality, located in the South East, handling both domestic and international flights with a gross internal floor area (GIA) of 25,000 m². These costs exclude baggage handling, check-in systems, pre-check in and boarding security systems, vertical transportation and services to aircraft stands, with the heat source via district mains (excluded) and executed under landside access/logistics environment.

Cost Summary

El. Ref.	Element	Total cost £	Cost/m² £
5A	Sanitaryware	68,000	2.72
5C	Disposal Installations		
	Rainwater	136,000	5.44
	Soil and waste	180,000	7.20
	Condensate	34,000	1.36
5D	Water Installations		
	Hot and cold water services	407,000	16.28
5F	Space Heating and Air Treatment		
	LTHW Heating system	1,800,000	72.00
	Chilled water system	1,692,000	67.68
	Supply and extract air conditioning system	2,450,000	98.00
	Allowance for services to communications rooms	226,000	9.04
5G	Ventilating Services		
	Mechanical ventilation to baggage handling and plantrooms	564,000	22.56
	Toilet extract ventilation	203,000	8.16
	Smoke extract installation	226,000	9.04
	Kitchen extract system	67,000	2.68
5H	Electrical Installation		
	HV/LV Switchgear	1,498,000	59.92
	Standby generator	922,000	36.88
	Mains and sub mains installation	1,037,000	41.48
	Small power installation	345,000	13.80
	Lighting and luminaires	2,258,000	90.32
	Emergency lighting installation	276,000	11.04
	Power to mechanical services	161,000	6.44
5I	Gas Installation	46,000	1.84
5K	Protective Installations		
	Lightning protection	67,000	2.68
	Earthing and bonding	90,000	3.60
	Sprinkler installation	790,000	31.60
	Dry riser and hosereel installations	158,000	6.32
	Fire suppression installation to communications room	67,000	2.68
	Carried forward	15,768,000	630.72

ELEMENTAL COSTS

AIRPORT TERMINAL BUILDING – cont'd

El. Ref.	Element	Total cost £	Cost/m² £
	Brought forward	15,768,000	630.72
5L	Communications Installations		
	Fire and smoke detection and alarm system	692,000	27.68
	Voice/public address system	461,000	18.44
	Intruder detection	231,000	9.24
	Security, CCTV and access control	806,000	32.24
	Wireways for telephones, data and structured cable	345,000	13.80
	Structured cable installation	806,000	32.24
	Flight information display system	576,000	23.04
5M	Special Installations		
	BMS Installation	1,241,000	49.64
	Summary total	20,926,000	837.04

ELEMENTAL COSTS

SHOPPING MALL (TENANT'S FIT OUT EXCLUDED)

Natural ventilation shopping mall with approximately 33,000 m² two storey retail area and a 13,000 m² above ground covered car park, situated in a town centre in South East England.

Cost Summary

El. Ref.	Element	Total cost £	Cost/m² £
	RETAIL BUILDING 33,000m²		
5A	Sanitary appliances	24,500	0.74
5C	Disposal Installations		
	Rainwater	165,000	5.00
	Soil, waste and vent	193,800	5.87
5D	Water Installations		
	Cold water installation	184,600	5.59
	Hot water installation	174,400	5.28
5E	Heat Source	Included	
5F	Space Heating and Air Treatment		
	Condenser water system	967,000	29.30
	LTHW installation	128,500	3.89
	Air conditioning system	976,000	29.57
	Over-door heaters at entrances	29,500	0.89
5G	Ventilation Services		
	Public toilet ventilation	19,400	0.58
	Plant room ventilation	129,500	3.92
	Supply and extract systems to shop units	415,100	12.57
	Toilet extract systems to shop units	77,500	2.34
	Smoke ventilation system to Mall Area	313,000	9.48
	Service corridor ventilation	71,500	2.16
	Miscellaneous ventilation	531,400	16.10
5H	Electrical Installation		
	LV distribution	724,200	21.94
	Standby power	118,300	3.58
	General lighting	1,900,000	57.57
	External lighting	162,100	4.91
	Emergency lighting	365,000	11.06
	Small power	295,800	8.96
	Mechanical services power supplies	83,600	2.53
	General earthing	39,700	1.20
	UPS for security and CCTV	21,400	0.64
5I	Gas Installation		
	Gas supplies to boilers	16,300	0.49
	Gas supplies to Anchor (major) stores	10,200	0.30

ELEMENTAL COSTS

SHOPPING MALL (TENANT'S FIT OUT EXCLUDED) – cont'd

El. Ref.	Element	Total cost £	Cost/m² £
5K	Protective Installations		
	Lightning protection	38,800	1.17
	Sprinkler installation	415,000	12.57
	Dry Risers	Excluded	
	Hosereel installation	Excluded	
	Carried forward	8,591,100	260.33
	Brought forward	8,591,100	260.33
5L	Communications Installations		
	Fire alarm installation	246,800	7.47
	Public address/ voice alarm	162,100	4.91
	Security installation	355,000	10.75
	General containment	355,000	10.75
5M	Special Installations		
	BMS/Controls	531,000	16.09
	Summary total	10,241,000	310.33

	CAR PARK – 13,000m²		
5C	Disposal Installations		
	Car park drainage	70,000	5.38
5G	Ventilation Services		
	Car park ventilation (ducted system)	900,000	69.23
5H	Electrical Installation		
	LV distribution	140,000	10.76
	Standby power	Included	
	General lighting	260,000	20.00
	External lighting	Excluded	
	Emergency lighting	72,000	5.53
	Small power	65,000	5.00
	Mechanical services power supplies	48,000	3.69
	General earthing	14,500	1.11
	Ramp frost protection	19,000	1.46
5K	Protective Installations		
	Sprinkler installation	270,000	20.76
	Dry Riser and Hosereel Installation	Excluded	
5L	Communications Installations		
	Fire alarm installation	270,000	20.76
	Security installation	114,000	8.76
5M	Special Installations		
	BMS/Controls	71,000	5.46
	Entry/exit barriers, pay stations	66,000	5.07
	Summary total	2,379,500.00	183.03

ELEMENTAL COSTS

OFFICE BUILDING

Speculative 15 storey office in Central London for multiple tenant occupancy with a gross floor area of 19,300 m². A four pipe fan coil system, with roof mounted cooling towers, gas fired boilers, and basement mounted water cooled chillers.

Cost Summary

El. Ref.	Element	Total cost £	Cost/m² £
	SHELL AND CORE 19,300 m² GIA		
5A	Sanitaryware	125,000	6.48
5C	Disposal Installations		
	Rainwater/Soil and Waste	280,000	14.50
	Condensate	34,000	1.76
5D	Water Installations		
	Hot and cold water services	310,000	16.06
5E	Heat Source	165,000	8.55
5F	Space Heating and Air Treatment		
	LTHW Heating	287,000	14.87
	Chilled water	829,000	42.95
	Ductwork	781,800	40.51
5G	Ventilating Services		
	Toilet extract ventilation	110,000	5.69
	Basement extract	276,000	14.30
	Miscellaneous ventilation systems	350,000	18.13
5H	Electrical Installation		
	Generator	121,000	6.27
	HV/LV supply/distribution	1,100,000	56.99
	General lighting	395,000	20.47
	General power	76,000	3.94
	Electrical services for mechanical equipment	94,000	4.87
5I	Gas Installation	24,000	1.24
5K	Protection		
	Wet risers	181,000	9.37
	Sprinklers	316,000	16.37
	Earthing and bonding	37,000	1.91
	Lightning protection	37,000	1.91
5L	Communication Installation		
	Fire/Voice alarms	262,000	13.58
	Voice and data (wireways)	28,000	1.45
	Security (wireways)	24,000	1.24
	Disabled/refuge alarms	59,000	3.06
	CCTV/Access control/ Intruder detection	73,000	3.78
	Carried forward	6,374,800	330.30

ELEMENTAL COSTS

OFFICE BUILDING – cont'd

El. Ref.	Element	Total cost £	Cost/m² £
	Brought forward	6,374,800	330.30
5M	Special Installation		
	Building management systems	405,000	20.98
	Summary total (based on Gross Internal Area – GIA)	6,779,800	351.28

El. Ref.	Element	Total cost £	Cost/m² £
	CATEGORY 'A' FIT OUT – 14,500 m² NIA		
5F	Space Heating and Air Treatment		
	LTHW Heating	396,000	27.31
	Chilled water	470,000	32.41
	Ductwork and grilles	1,173,000	80.90
	Condensate	108,000	7.45
5H	Electrical Installation		
	Lighting installation	987,000	68.07
	Electrical services in connection	49,000	3.38
	Tenant distribution board	92,000	6.35
5K	Protection		
	Sprinkler installation	309,000	21.21
5L	Communication Installation		
	Fire/Voice alarms	182,000	12.55
5M	Special Installations		
	Building management system	275,100	18.97
	Summary total (based on Nett Internal Area – NIA)	4,041,000	278.70

ELEMENTAL COSTS

BUSINESS PARK

New build office in South East within the M25 part of a speculative business park with a gross floor area of 9,800 m². A full air displacement system with roof mounted air cooled chillers, gas fired boilers and air handling plant.

Cost Summary

El. Ref.	Element	Total cost £	Cost/m² £
	SHELL AND CORE – 9,800 m² GIA		
5A	Sanitaryware	63,700	6.50
5C	Disposal Installations		
	Rainwater	27,400	2.79
	Soil and waste	76,500	7.80
5D	Water Installations		
	Cold water services	89,000	9.08
	Hot water services	27,700	2.82
5E	Heat Source	62,800	6.40
5F	Space Heating and Air Treatment		
	LTHW Heating; plantroom and risers	107,300	10.94
	Chilled water; plantroom and risers	253,000	25.81
	Ductwork; Plantroom and risers	754,000	76.93
5G	Ventilating Services	69,900	7.13
	Toilet and miscellaneous ventilation		
5H	Electrical Installation		
	LV supply/distribution	211,700	21.60
	General lighting	180,600	18.42
	General power	51,100	5.21
	Electrical services in connection with mechanical	26,500	2.70
5I	Gas Installation	15,600	1.59
5K	Protective Installation		
	Earthing and bonding	16,000	1.63
	Lightning protection	21,000	2.14
	Dry risers	10,500	1.07
5L	Communication Installation		
	Fire alarms	81,000	8.27
	Security (wireways)	16,000	1.63
	Data and voice (wireways)	16,000	1.63
5M	Special Installation	182,300	18.60
	Building management systems		
	Summary total (based on Gross Internal Area – GIA)	2,359,600	240.77

ELEMENTAL COSTS

BUSINESS PARK – cont'd

El. Ref.	Element	Total cost £	Cost/m² £
	CATEGORY 'A' FIT OUT – 8,100 m² NIA		
5F	Space Heating and Air Treatment		
	LTHW Heating and perimeter heaters	324,300	40.03
	Floor swirl diffusers and supply ductwork	174,700	21.56
5H	Electrical Installation		
	Distribution boards	17,300	2.13
	General lighting, recessed including lighting controls	383,100	47.29
5K	Protective Installation		
	Earthing and bonding	10,000	1.23
5L	Communication Installation		
	Fire alarms	43,000	5.30
5M	Special Installations		
	Building management systems	46,000	5.67
	Summary total (Based on Nett Internal Area – NIA)	998,400	123.25

ELEMENTAL COSTS

PERFORMING ARTS CENTRE (MEDIUM SPECIFICATION)

Performing Arts centres with a Gross Internal Area (GIA) of approximately 8,000 m², upon which this cost analysis has been based, on a medium specification for the theatre systems and with cooling to the Auditorium.

The development comprises of dance studios and a theatre auditorium in the outer London area. The theatre would require all the necessary stage lighting, machinery and equipment installed in a modern professional theatre (these are excluded from the model, as assumed to be FF&E, but the containment and power wiring is included). Included are the staff call system/paging, audio and video recording, EPOS ticket system, production recording and relay to TV screens.

Cost Summary

El. Ref.	Element	Total cost £	Cost/m² £
5A	Sanitaryware	92,000	11.50
5C	Disposal Installations		
	Soil, Waste and Rainwater	89,000	11.13
5D	Water Installations		
	Cold water services	77,000	9.63
	Hot water services	72,000	9.00
5E	Heat Source	124,000	15.50
5F	Space Heating and Air Treatment		
	Heating	548,000	68.50
	Chilled water system	195,000	24.38
	Supply and extract air systems	911,000	113.88
5G	Ventilating Services		
	Ventilation and extract systems to toilets, kitchen and workshop	107,000	13.38
5H	Electrical Installation		
	LV supply/distribution	385,000	48.13
	General lighting	857,000	107.13
	Small power	225,000	28.13
5I	Gas Installation	29,000	3.63
5K	Protection		
	Lighting protection	10,000	1.25
5L	Communication Installation		
	Fire alarms and detection	184,000	23.00
	Voice and Data complete installation (excluding active equipment)	225,000	28.13
	Security, Access, Control, Disabled alarms, Staff paging	207,000	25.88
5M	Special Installation		
	Building management systems	337,000	42.13
	Theatre systems includes for containment and power wiring	160,000	20.00
	Summary total	4,834,000	604.25

ELEMENTAL COSTS

SPORTS HALL

Single storey sports hall, located in the South East, with a gross internal area of 1,200 m² (40 m × 30 m).

Cost Summary

El. Ref.	Element	Total cost £	Cost/m² £
5A	Sanitaryware	11,000	9.17
5C	Disposal Installations		
	Rainwater	3,500	2.92
	Soil and waste	7,500	6.25
5D	Water Installations		
	Hot and cold water services	17,000	14.17
5E	Heat Source		
	Boiler, flues, pumps and controls	14,000	11.66
5F	Space Heating and Air Treatment		
	Warm air heating to sports hall area	20,000	16.67
	Radiator heating to ancillary areas	23,000	19.17
5G	Ventilating Services		
	Ventilation to changing, fitness and sports hall areas	16,000	13.33
5H	Electrical Installations		
	Main switchgear and sub-mains	13,000	10.83
	Small power	12,000	10.00
	Lighting and luminaries to sports hall areas	19,500	16.25
	Lighting and luminaries to ancillary areas	22,000	18.33
5I	Gas Installation	Included in 5E	
5K	Protective Installations		
	Lightning protection	3,500	2.92
5L	Communications Installations		
	Fire, smoke detection and alarm system, intruder detection	13,500	11.25
	CCTV Installation	16,000	13.33
	Public address and music systems	7,500	6.25
	Wireways for telephone and data	2,500	20.08
	Summary total	221,500	184.58

ELEMENTAL COSTS

HOTELS

200 Bedroom, four star hotel, situated in Central London, with a gross internal floor area of 16,500 m².

The development comprises a ten storey building with large suites on each guest floor, together with banqueting, meeting rooms and leisure facilities.

Cost Summary

El. Ref.	Element	Total cost £	Cost/m² £
5A	Sanitaryware	522,000	31.63
5C	Disposal Installations		
	Rainwater, soil and waste	447,000	27.09
5D	Water Installations		
	Hot and cold water services	670,000	40.60
5E	Heat Source		
	Condensing boiler and pumps etc	179,000	10.84
5F	Space Heating and Air Treatment		
	Air conditioning system; chillers, pumps, air handling units, ductwork, fan coil units etc; to guest rooms, public areas, meeting and banquet rooms	2,085,000	126.36
5G	Ventilating Services		
	General toilet extract and ventilation to kitchens and bathrooms etc	670,000	40.60
5H	Electrical Installation		
	HV/LV Installation, standby power, lighting, emergency lighting and small power to guest floors and public areas including earthing and lightning protection	2,813,000	170.48
5I	Gas Installation	37,000	2.24
5K	Protective Installations		
	Dry risers and sprinkler installation	596,000	36.12
5L	Communications Installations		
	Fire, smoke detection and alarm system/security CCTV	699,000	42.36
	Background music, AV wireways	228,000	13.81
	Telecommunications, data and T.V. wiring (no hotel management and head end equipment)	396,000	24.00
5M	Special Installations		
	Building Management System	408,000	24.72
	Summary total	9,750,000	590.85

ELEMENTAL COSTS

STADIUM – NEW

A three storey stadium, located in Greater London with gross internal area of 85,000 m² and incorporating 60,000 spectator seats

Cost Summary

El. Ref.	Element	Total cost £	Cost/m² £
5A	Sanitaryware	825,000	9.71
5C	Disposal Installations		
	Rainwater	285,000	3.35
	Above ground drainage	890,000	10.47
5D	Water Installations		
	Hot and cold water	1,650,000	19.41
5E	Heat Source	735,000	8.65
5F	Space Heating and Air Treatment		
	Heating	500,000	5.88
	Cooling	1,100,000	12.94
5G	Ventilating Services		
	Ventilation	4,050,000	47.65
5H	Electrical Installation		
	HV/LV Supply	775,000	9.12
	LV Distribution	1,950,000	22.94
	General lighting	4,850,000	57.06
	Small power	1,150,000	13.53
	Earthing and bonding	85,000	1.00
	Power supply to mechanical equipment	100,000	1.18
	Pitch lighting	700,000	8.24
5I	Gas Installation	75,000	0.88
5K	Protective Installations		
	Lightning protection	100,000	1.18
	Hydrants	175,000	2.06
5L	Communications Installations		
	Wireways for data, TV, telecom and PA	630,000	7.41
	Public address	1,250,000	14.71
	Security	1,000,000	11.76
	Data voice installations	2,500,000	29.41
	Fire alarms	700,000	8.24
	Disabled/refuse alarm/call systems	225,000	2.65
5M	Special Installations		
	BMS/Controls	1,150,000	13.53
	Summary total	27,450,000	322.94

	Cost per seat	£457.50	

ELEMENTAL COSTS

PRIVATE HOSPITAL

New build project building. The works consist of a new 80 bed hospital of approximately 15,000 m², eight storey with a plant room.

All heat is provided from existing steam boiler plant, medical gases are also served from existing plant. The project includes the provision of additional standby electrical generation to serve the wider site requirements.

This hospital has six operating theatres, ITU/HDU department, pathology facilities, diagnostic imaging, out patient facilities and physiotherapy.

Cost Summary

El. Ref.	Element	Total cost £	Cost/m² £
5A	Sanitaryware	280,000	18.67
5C	Disposal Installations		
	Rainwater	36,000	2.40
	Soil and waste	324,000	21.60
	Specialist drainage (above ground)	20,000	1.33
5D	Water Installations	711,000	47.40
	Hot and cold water services		
5E	Heat Source	Included in 5F	
5F	Space Heating and Air Treatment		
	LPHW Heating	566,000	37.73
	Chilled Water	544,000	36.27
	Steam and condensate	329,000	21.93
5G	Ventilating Services	1,842,000	122.80
	Ventilation, comfort cooling and air conditioning		
5H	Electrical Installation		
	HV Distribution	33,000	2.20
	LV supply/distribution	391,000	26.07
	Standby Power	470,000	31.33
	UPS	324,000	21.60
	General lighting	563,000	37.53
	General power	560,000	37.33
	Emergency lighting	164,000	10.93
	Theatre lighting	200,000	13.33
	Specialist lighting	185,000	12.33
	External lighting	32,000	2.13
	Electrical supplies for mechanical equipment	169,000	11.27
5I	Gas Installation	46,000	3.07
	Oil Installations	93,000	6.20
5K	Protection		
	Dry risers	23,000	1.53
	Lightning Protection	6,200	0.41
	Carried forward	7,911,200	527.41

ELEMENTAL COSTS

PRIVATE HOSPITAL – cont'd

El. Ref.	Element	Total cost £	Cost/m² £
	Brought forward	7,911,200	527.39
5L	Communication Installation		
	Fire alarms and detection	267,000	17.80
	Voice and Data	165,000	11.00
	Security and CCTV	50,000	3.33
	Nurse call and cardiac alarm system	246,000	16.40
	Personnel paging	63,000	4.20
	Hospital radio (entertainment)	325,000	21.67
5M	Special Installation		
	Building management systems	558,000	37.20
	Pneumatic tube conveying system	39,000	2.60
	Group 1 Equipment	537,000	35.80
	Summary total (based on gross floor area)	10,161,200	677.39

ELEMENTAL COSTS

SCHOOL

New build secondary school (Academy) located in Southern England, with a gross internal floor area of 10,000 m².

The building comprises a three storey teaching block, including provision for music, drama, catering, sports hall, science laboratories, food technology, workshops and reception/admin (BB93 compliant). Excludes IT Cabling and sprinkler protection.

Cost Summary

El. Ref.	Element	Total cost £	Cost/m² £
5A	Sanitaryware		
	Toilet cores and changing facilities only	92,000	9.20
5C	Disposal Installations		
	Rainwater installations	46,500	4.65
	Soil and waste	110,600	11.06
5D	Water Installations		
	Potable hot and cold water services	276,500	27.65
	Non potable hot and cold water services to labs and art rooms	92,100	9.21
5E	Heat Source		
	Gas fired boiler installation	184,300	18.43
5F	Space Heating and Air Treatment		
	LTHW Heating system (primary)	368,600	36.86
	LTHW Heating system (secondary)	92,000	9.20
	DX Cooling system to ICT server rooms	46,500	4.65
	Mechanical supply and extract ventilation including DX type cooling to Music, Drama, Kitchen/Dining and Sports Hall	645,000	64.50
5G	Ventilating Services	46,500	4.65
	Toilet extract systems		
	Changing area extract systems	36,900	3.69
	Extract ventilation from design/food technology and science labs	73,800	7.38
5H	Electrical Installation		
	Mains and sub-mains distribution	282,300	28.23
	Lighting and luminaries; including emergency fittings	612,000	61.20
	Small power installation	282,300	28.23
	Earthing and bonding	18,500	1.85
5I	Gas Installation	92,000	9.20
5K	Protective Installations		
	Lightning protection	18,500	1.85
5L	Communications Installations		
	Containment for telephone, IT data, AV and security systems	28,000	2.80
	Fire, smoke detection and alarm system	141,600	14.16
	Security installations including CCTV, access control and intruder alarm	188,000	18.80
	Disabled toilet, refuge and induction loop systems	37,800	3.78
5M	Special Installations		
	Building Management system – To plant	202,700	20.27
	Building Management system – To opening vents/windows	73,700	7.37
	Summary total	4,088,700	408.87

ELEMENTAL COSTS

AFFORDABLE RESIDENTIAL DEVELOPMENT

A 12 storey, 50 apartment affordable residential development with a gross internal area of 3,400 m² and a net internal area of 2,400 m², situated within the London area. The development does not include a car park and is based on 71% efficiency.

Based on an individual radiator LTHW system within each apartment, with local gas combi boilers exhausting to building facade. Kitchens and bathrooms are also ventilated to the building facade, there are pendant light fittings, an audio entry system, telephone and satellite installation. Sanitaryware is of lower quality with plastic pipework runs. Full sprinkler installation installed throughout. Excludes remote metering and Local Controls only, no BMS.

Cost Summary

El. Ref.	Element	Total Cost £	Cost/m² £
	SHELL & CORE		
5A	Sanitaryware	5,300	1.56
5B	Services	-	-
5C	Disposal Installations	35,000	10.29
5D	Water Installations	58,000	17.06
5E	Heat Source	N/A	N/A
5F	Space Heating and Air Treatment	15,000	4.41
5G	Ventilating Services	61,000	17.94
5H	Electrical Installation	101,000	29.71
5I	Gas Installation	19,000	5.59
5K	Protective Installations	103,000	30.29
5L	Communications Installations	96,500	28.38
5M	Special Installations	28,500	8.38
	Summary total (based on Gross Internal Area – GIA)	522,300	153.62
	FITTING OUT		
5A	Sanitaryware	55,000	22.92
5B	Services	-	-
5C	Disposal Installations	20,000	8.33
5D	Water Installations	60,000	25.00
5E	Heat Source	75,000	31.25
5F	Space Heating and Air Treatment	84,000	35.00
5G	Ventilating Services	65,000	27.08
5H	Electrical Installation	125,000	52.08
5I	Gas Installation	15,000	6.25
5K	Protective Installations	75,000	31.25
5L	Communications Installations	40,000	16.67
5M	Special Installations	10,000	4.17
	Summary total (Net Internal Area – NIA)	624,000	260.00

	Cost per apartment – Shell & Core and Fit Out	£ 22,926	

ELEMENTAL COSTS

PRIVATE RESIDENTIAL DEVELOPMENT

A 20 storey, 250 apartment private residential developments with a gross internal area of 22,750 m² and a net internal area of 20,415 m², situated within the London area. The development does not include a car park and is based on 90% efficiency.

Included is a central boiler and hot water installation with apartment heat exchanger. Good quality sanitary appliances. Perimeter trench heating to each apartment and 4 pipe fan coil unit installation. Central air cooled chillers system. No gas to apartments. Whole house ventilation system discharging to local facade, wet riser with full sprinkler installation. Video entry and TV and satellite installation. Flood wiring for apartment home automation and sound system only. Public Health apartment runs in plastic. Local apartment controls only. Lutron or Creston type system excluded.

Cost Summary

El. Ref.	Element	Total cost £	Cost/m² £
	SHELL & CORE		
5A	Sanitaryware	9,000	0.40
5B	Services	-	-
5C	Disposal Installations	340,000	14.95
5D	Water Installations	580,000	25.49
5E	Heat Source	150,000	6.59
5F	Space Heating and Air Treatment	1,052,000	46.24
5G	Ventilating Services	120,000	5.27
5H	Electrical Installation	735,000	32.31
5I	Gas Installation	63,500	2.79
5K	Protective Installations	685,000	30.11
5L	Communications Installations	490,000	21.54
5M	Special Installations	360,630	15.85
	Summary total (based on Gross Internal Area – GIA)	4,585,130	201.54

El. Ref.	Element	Total cost £	Cost/m² £
	FITTING OUT		
5A	Sanitaryware	1,091,750	53.48
5B	Services	-	-
5C	Disposal Installations	167,500	8.20
5D	Water Installations	326,500	15.99
5E	Heat Source	488,500	23.93
5F	Space Heating and Air Treatment	4,500,000	220.43
5G	Ventilating Services	697,750	34.18
5H	Electrical Installation	2,005,500	98.24
5I	Gas Installation	N/A	N/A
5K	Protective Installations	430,250	21.08
5L	Communications Installations	1,000,000	48.98
5M	Special Installations	50,000	2.45
	Summary total (based on Gross Internal Area – GIA)	10,757,750	526.95

	Cost per apartment – Shell & Core and Fit Out	£ 61,372	

ELEMENTAL COSTS

SUPERMARKET

Supermarket located in the South East with a total gross floor area of 4,000 m², including a sales area of 2,350 m². The building is on one level and incorporates a main sales, coffee shop, bakery, offices and amenities areas and warehouse.

Cost Summary

El. Ref.	Element	Total Cost £	Cost/m² £
5A	Sanitaryware	4,587	1.15
5C	Disposal Installations		
	Soil and Waste	12,960	3.24
5D	Water Installations		
	Hot and Cold water services	101,403	25.35
5E	Heat Source – local heaters only used	N/A	N/A
5F	Space Heating and Air Treatment		
	Heating & ventilation with cooling via DX units	40,850	10.21
5G	Ventilating Services		
	Supply and extract system	9,052	2.26
5H	Electrical Installation	116,953	29.24
	Panels / Boards		
	Containment	6,050	1.51
	General lighting	41,120	10.28
	Small power	39,000	9.75
	Mechanical Services wiring	8,000	2.00
5I	Gas Installation		
	Gas mains services to plantroom	N/A	N/A
5K	Protection		
	Sprinklers	N/A	N/A
	Lightning protection	3,200	0.80
5L	Communication Installation		
	Fire alarms, detection and public address	9,650	2.41
	CCTV	9,300	2.33
	Intruder alarm, detection and store security	9,750	2.44
	Telecom and structured cabling	2,500	0.63
5M	Special Installations		
	BMS Installation	27,150	6.79
	Data Cabinet	9,300	2.33
	Controls Wiring	13,479	3.37
	Refrigeration		
	Installation	111,000	27.75
	Plant	111,000	27.75
	Cold Store	41,000	10.25
	Cabinets	303,000	75.75
	Summary total	1,030,304	257.59

ELEMENTAL COSTS

DISTRIBUTION CENTRE

Distribution centre located in London with a total gross floor area of 75,000 m², including a refrigerated cold box of 17,500 m².

The building is on one level and incorporates a office area, vehicle recovery unit, gate house and plantrooms.

Cost Summary

El. Ref.	Element	Total Cost £	Cost/m² £
5C	Disposal Installations		
	Soil and Waste	143,000	1.91
	Rainwater	464,000	6.19
5D	Water Installations		
	Hot and Cold water services	124,000	1.65
5F	Space Heating and Air Treatment		
	Heating with ventilation to offices, displacement system to main warehouse	1,600,000	21.33
5G	Ventilating Services	358,000	4.77
	Smoke extract system		
5H	Electrical Installation		
	Generator	1,109,000	14.79
	Main HV installation	1,415,000	18.87
	MV distribution	877,000	11.69
	Lighting installation	743,000	9.91
	Small power installation	902,000	12.03
5I	Gas Installation		
	Gas mains services to plantroom	41,000	0.55
5K	Protection		
	Sprinklers including racking protection	3,174,000	42.32
	Lightning protection	7,000	0.09
5L	Communication Installation		
	Fire alarms, detection and public address	825,000	11.00
	CCTV	530,000	7.07
5M	Special Installations		
	BMS Installation	404,000	5.39
	Refrigeration		
	Installation	2,580,000	34.40
	Summary total	15,296,000	203.95

ELEMENTAL COSTS

DATA CENTRE

New build data centre located in the London area/proximity to M25.

Net Technical Area (NTA) provided at 2,000 m² with typically other areas of 250 m² office space, 250 m² ancillary space and 1,000 m² internal plant. Total GIA 3,500 m².

Power and cooling to Technical Space @ 1,500 w/m². Cost/m² against Nett Technical Area – NTA.

Cost Summary

El. Ref.	Element	Total cost £	Cost/m² £
5C	Disposal Installations		
	Soil & Waste	28,000	14.00
	Rainwater	19,000	9.50
	Condensate	19,000	9.50
5D	Water Installations		
	Hot and Cold water services	48,000	24.00
5F	Space Heating and Air Treatment		
	Chilled water plant with redundancy of N+1 to provide 1,000w/m² net technical space cooling	729,000	364.50
	Chilled water distribution to data centre to free standing cooling units and distribution to ancillary office and workshop/build areas	1,347,000	673.50
	Freestanding cooling units with redundancy of N+20% to technical space and switchrooms. Based on single coil cooling units 30% of units with humidification	675,000	337.50
	Floor grilles	141,000	70.50
5G	Ventilating Services		
	Supply and extract ventilation systems to data centre, switchrooms and ancillary spaces including dedicated gas extract and hot aisle extract system	847,000	423.50
5H	Electrical Installation		
	Main HV installations including transformers with redundant capacity of N+1	1,280,000	640.00
	LV distribution including cabling and busbar installations to provide full system – System dual supplies/redundancy including supplies to mechanical services, office and ancillary areas	1,547,000	773.50
	Generator Installation		
	Standby rated containerised generators with redundant capacity of N+1 and including synchronization panel, 72 hours bulk fuel store and controls	1,803,000	901.50
	Uninterruptible Power Supplies		
	Static UPS to provide 2 × (N+1) system redundancy with 10 minute battery autonomy	1,976,000	988.00
	Carried forward	10,459,000	5,229.50

ELEMENTAL COSTS

DATA CENTRE – cont'd

El. Ref.	Element	Total cost £	Cost/m² £
	Brought Forward	10,459,000	5,229.50
5H	Electrical Installation *continued*		
	LV Switchgear		
	Incoming LV switchgear, UPS input and output boards for electrical and mechanical services systems	1,674,000	837.00
	Power Distribution Units (PDUs)		
	PDUs to provide a redundant capacity of 2 Nr PDU to include static transfer switch and isolating transformer	1,536,000	768.00
	Cabinet Supplies		
	A & B supply cables from PDUs to a BS4343 socket fixed under raised floor adjacent cabinet positions	534,000	267.00
	Lighting Installation		
	Lighting to technical, workshop and plant areas including office/ancillary spaces and external areas	197,000	98.50
5I	Gas Installation (Assumed not required)		
5K	Protective Installations		
	Gaseous suppression to technical areas and switchrooms	481,000	240.50
	Lightning protection	28,000	14.00
	Earthing and clean earth	58,000	29.00
	Leak detection	28,000	14.00
5L	Communications Installation		
	Fire, smoke detection and alarm system		
	Very early smoke detection alarm (VESDA) to technical areas	256,000	128.00
	CCTV installations	324,000	162.00
	Access control installations	275,000	137.50
5M	Special Installations		
	Building Management System	529,000	264.50
	PLC/Electrical monitoring system	481,000	240.50
	Summary total (Based on Nett Technical Area – NTA)	16,860,000	8,430.00

ELEMENTAL COSTS

	BUILDING MANAGEMENT INSTALLATIONS		
El. Ref.	Element	Total cost £	Cost/Point £
	Category A Fit Out		
	Option 1 – 189 nr four pipe fan coil – 756 points		
1.0	**Field Equipment**		
	Network devices		
	Valves/actuators		
	Sensing devices	51,000	67.46
2.0	**Cabling**		
	Power – from local isolator to DDC controller		
	Control – from DDC controller to field equipment	29,000	38.36
3.0	**Programming**		
	Software – central facility		
	Software – network devices		
	Graphics	16,000	21.16
4.0	**On site testing and commissioning**		
	Equipment		
	Programming/graphics		
	Power and control cabling	16,500	21.83
	Total Option 1 – Four pipe fan coil (On Point Basis)	112,500	148.81
	Category A Fit Out		
	Option 2 – 189 nr two pipe fan coil system with electric heating – 756 points		
1.0	**Field Equipment**		
	Network devices		
	Valves/actuators/thyristors		
	Sensing devices	66,000	87.30
2.0	**Cabling**		
	Power – from local isolator to DDC controller		
	Control – from DDC controller to field equipment	30,500	40.34
3.0	**Programming**		
	Software – central facility		
	Software – network devices		
	Graphics	16,000	21.16
4.0	**On site testing and commissioning**		
	Equipment		
	Programming/graphics		
	Power and control cabling	17,000	22.49
	Total Option 2 – Two pipe fan coil with electric heating (On Point Basis)	129,500	171.30

ELEMENTAL COSTS

El. Ref.	Element	Total cost £	Cost/Point £
BUILDING MANAGEMENT INSTALLATIONS – cont'd			
	Category A Fit Out		
	Option 3 – 180 Nr Chilled Beams with perimeter heating – 567 points		
1.0	**Field Equipment**		
	Network devices		
	Valves/actuators		
	Sensing devices	47,000	82.89
2.0	**Cabling**		
	Power – from local isolator to DDC controller		
	Control – from DDC controller to field equipment	32,000	56.44
3.0	**Programming**		
	Software – central facility		
	Software – network devices		
	Graphics	16,000	28.22
4.0	**On site testing and commissioning**		
	Equipment		
	Programming/graphics		
	Power and control cabling	17,000	29.98
	Total Option 3 – Chilled beams with perimeter heating (On Point Basis)	112,000	197.53

	Element	Total cost £	Cost/Point £
	Combined Shell & Core and Fit Out		
1.0	**Option 1 – 4 pipe fan coil – 1196 points**		
	Shell and core and Category A Fit out	388,000	324.41
2.0	**Option 2 – 2 pipe fan coil with electrical heating – 1196 points**		
	Shell and core and Category A Fit out	403,000	336.96
3.0	**Option 3 – Chilled beams with perimeter heating – 1007 points**		
	Shell and core and Category A Fit out	383,000	380.34

Spon's First Stage Estimating Handbook

3rd Edition

By **Bryan Spain**

Have you ever had to provide accurate costs for a new supermarket or a pub "just an idea…a ballpark figure…" ?

The earlier a pricing decision has to be made, the more difficult it is to estimate the cost and the more likely the design and the specs are to change. And yet a rough-and-ready estimate is more likely to get set in stone.

Spon's First Stage Estimating Handbook is the only comprehensive and reliable source of first stage estimating costs. Covering the whole spectrum of building costs and a wide range of related M&E work and landscaping work, vital cost data is presented as:

• Costs per square metre
• Principal rates

• Elemental cost analyses
• Composite rates

Compact and clear, *Spon's First Stage Estimating Handbook* is ideal for those key early meetings with clients. And with additional sections on whole life costing and general information, this is an essential reference for all construction professionals and clients making early judgements on the viability of new projects.

January 2010: 216x138: 244pp
Pb: 978-0-415-54715-4: **£45.00**

To Order: Tel: +44 (0) 1235 400524 **Fax:** +44 (0) 1235 400525
or Post: Taylor and Francis Customer Services,
Bookpoint Ltd, Unit T1, 200 Milton Park, Abingdon, Oxon, OX14 4TA UK
Email: book.orders@tandf.co.uk

For a complete listing of all our titles visit:
www.tandf.co.uk

Construction Contracts Questions and Answers

2nd Edition

By **David Chappell**

What they said about the first edition: "A fascinating concept, full of knowledgeable gems put in the most frank of styles... A book to sample when the time is right and to come back to when another time is right, maybe again and again." – *David A Simmonds,* Building Engineer *magazine*

• Is there a difference between inspecting and supervising?

• What does 'time-barred' mean?

• Is the contractor entitled to take possession of a section of the work even though it is the contractor's fault that possession is not practicable?

Construction law can be a minefield. Professionals need answers which are pithy and straightforward, as well as legally rigorous. The two hundred questions in the book are real questions, picked from the thousands of telephone enquiries David Chappell has received as a Specialist Adviser to the Royal Institute of British Architects. Although the enquiries were originally from architects, the answers to most of them are of interest to project managers, contractors, QSs, employers and others involved in construction.

The material is considerably updated from the first edition – weeded, extended and almost doubled in coverage. The questions range in content from extensions of time, liquidated damages and loss and/or expense to issues of warranties, bonds, novation, practical completion, defects, valuation, certificates and payment, architects' instructions, adjudication and fees. Brief footnotes and a table of cases will be retained for those who may wish to investigate further.

August 2010: 216x138: 352pp
Pb: 978-0-415-56650-6: **£34.99**

To Order: Tel: +44 (0) 1235 400524 **Fax:** +44 (0) 1235 400525
or Post: Taylor and Francis Customer Services,
Bookpoint Ltd, Unit T1, 200 Milton Park, Abingdon, Oxon, OX14 4TA UK
Email: book.orders@tandf.co.uk

For a complete listing of all our titles visit:
www.tandf.co.uk

The Impact of Building Information Modelling

R. Crotty

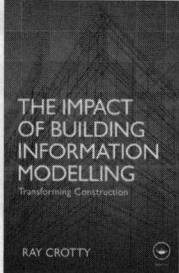

Construction projects involve a complex set of relationships, between parties with different professional backgrounds trying to achieve a very complex goal. Under these difficult circumstances, the quality of information on which projects are based should be of the highest possible standard. The line-based, two dimensional drawings on which conventional construction is based render this all but impossible. This is the source of some major shortcomings in the construction industry, and this book focuses on the two most fundamental of these: the failure to deliver projects predictably: to the required quality, on time and within budget; and the failure of most firms in the industry to make a survivable level of profit. By transforming the quality of information used in building, BIM aims to transform construction completely.

After describing and explaining these problems, the way in which BIM promises to provide solutions is examined in detail. A discussion of the theory and practice of BIM is also provided, followed by a review of various recent surveys of BIM usage in the US, UK and selected European economies. The way in which other industries, including retail and manufacturing, have been transformed by information are explored and compared with current developments in the deployment of BIM in construction. Five case studies from the UK show how BIM is being implemented, and the effects it is having on architects and contractors.

This book is perfect for any construction professional interested in improving the efficiency of their business, as well as undergraduate and postgraduate students wishing to understand the importance of BIM.

August 2011: 248pp
Hb: 978-0-415-60167-2: **£36.99**

To Order: Tel: +44 (0) 1235 400524 **Fax:** +44 (0) 1235 400525
or Post: Taylor and Francis Customer Services,
Bookpoint Ltd, Unit T1, 200 Milton Park, Abingdon, Oxon, OX14 4TA UK
Email: book.orders@tandf.co.uk

For a complete listing of all our titles visit:
www.tandf.co.uk

Managing the Brief for Better Design

2nd Edition

By **Alastair Blyth**, **John Worthington**

Briefing is not just presenting a set of documents to the design team; it is a process of developing a deep understanding about client needs. This book provides both inspiration to clients and a framework for practitioners. The coverage extends beyond new build, covering briefing for services and fit-outs. Written by an experienced and well-known team of authors, this new edition clearly explains how important the briefing process is to both the construction industry, in delivering well-designed buildings, and to their clients in achieving them. The text is illustrated by excellent examples of effective practice, drawn from DEGW experience, as well as five model briefs and invaluable process charts.

June 2010: 246x189: 272pp
Hb: 978-0-415-46030-9: **£95.00**
Pb: 978-0-415-46031-6: **£34.99**

PART 3

Material Costs/Measured Work Prices

Material Costs/Measured Work Prices

DIRECTIONS

The following explanations are given for each of the column headings and letter codes.

Unit	Prices for each unit are given as singular (i.e. 1 metre, 1 nr) unless stated otherwise.
Net price	Industry tender prices, plus nominal allowance for fixings (unless measured separately), waste and applicable trade discounts.
Material cost	Net price plus percentage allowance for overheads, profit and preliminaries.
Labour norms	In man-hours for each operation.
Labour cost	Labour constant multiplied by the appropriate all-in man-hour cost based on gang rate (See also relevant Rates of Wages Section) plus percentage allowance for overheads, profit and preliminaries.
Measured work Price (total rate)	Material cost plus Labour cost.

MATERIAL COSTS

The Material Costs given are based at Second Quarter 2011 but exclude any charges in respect of VAT. The average rate of copper during this quarter is US$9,000 / UK£5,625 per tonne. Users of the book are advised to register on the SPON's website www.pricebooks.co.uk/updates to receive the free quarterly updates - alerts will then be provided by e-mail as changes arise.

MEASURED WORK PRICES

These prices are intended to apply to new work in the London area. The prices are for reasonable quantities of work and the user should make suitable adjustments if the quantities are especially small or especially large. Adjustments may also be required for locality (e.g. outside London - refer to cost indices in approximate estimating section for details of adjustment factors) and for the market conditions e.g. volume of work secured or being tendered) at the time of use.

MECHANICAL INSTALLATIONS

The labour rate has been based on average gang rates per man hour effective from 4 October 2010. To this rate has been added 7.5% and 3% to cover preliminary items, site and head office overheads together with 2% for profit, resulting in an inclusive rate of £23.70 per man hour. The rate has been calculated on a working year of 2,007 hours; a detailed build-up of the rate is given at the end of these directions.

The rates and allowances under the National Agreement will be subject to an increase as referred to within JCC letter 100 - Revised. Future changes will be published in the free Spon's quarterly update by registering on their website.

DIRECTIONS

DUCTWORK INSTALLATIONS

The labour rate basis is as per Mechanical above and to this rate has been added 5.5% plus 12.5% to cover site and head office overheads only (factory overhead is included in the material rate) and preliminary items together with 3% for profit, resulting in an inclusive rate of £24.58 per man hour. The rate has been calculated on a working year of 2,007 hours; a detailed build-up of the rate is given at the end of these directions.

In calculating the 'Measured Work Prices' the following assumptions have been made:

(a)   That the work is carried out as a sub-contract under the Standard Form of Building Contract.
(b)   That, unless otherwise stated, the work is being carried out in open areas at a height which would not require more than simple scaffolding.
(c)   That the building in which the work is being carried out is no more than six storey's high.

Where these assumptions are not valid, as for example where work is carried out in ducts and similar confined spaces or in multi-storey structures when additional time is needed to get to and from upper floors, then an appropriate adjustment must be made to the prices. Such adjustment will normally be to the labour element only.

Note : The rates do not include for any uplift applied if the ductwork package is procured via the Mechanical Sub Contractor.

DIRECTIONS

LABOUR RATE – MECHANICAL

The annual cost of a notional twelve man gang of 12 men

	FOREMAN	SENIOR CRAFTSMAN (+2 Welding skill)	SENIOR CRAFTSMAN	CRAFTSMAN	INSTALLER	MATE (Over 18)	SUB TOTALS
	1 NR	1 NR	2 NR	4 NR	2 NR	2 NR	
Hourly Rate from 4th October 2010	**15.00**	**12.91**	**12.40**	**11.38**	**10.31**	**8.69**	
Working hours per annum per man	1,694.80	1,694.80	1,694.80	1,694.80	1,694.80	1,694.80	
x Hourly rate x nr of men = £ per annum	**25,422.00**	**21,879.87**	**42,031.04**	**77,147.30**	**34,946.78**	**29,455.62**	**230,882.60**
Overtime Rate	21.14	18.19	17.48	16.04	14.54	12.24	
Overtime hours per annum per man	312.20	312.20	312.20	312.20	312.20	312.20	
x Hourly rate x nr of men = £ per annum	**6,599.91**	**5,678.92**	**10,914.51**	**20,030.75**	**9,078.78**	**7,642.66**	**59,945.52**
Total	**32,021.91**	**27,558.79**	**52,945.55**	**97,178.05**	**44,025.55**	**37,098.28**	**290,828.13**
Incentive schemes 0.00%	**0.00**	**0.00**	**0.00**	**0.00**	**0.00**	**0.00**	**0.00**
Daily Travel Time Allowance (15-20 miles each way)	9.31	9.31	9.31	9.31	9.31	9.31	
Days per annum per man	223.00	223.00	223.00	223.00	223.00	223.00	
x nr of men = £ per annum	**2,076.13**	**2,076.13**	**4,152.26**	**8,304.52**	**4,152.26**	**4,152.26**	**24,913.56**
Daily Travel Fare (15-20 miles each way)	9.20	9.20	9.20	9.20	9.20	9.20	
Days per annum per man	223.00	223.00	223.00	223.00	223.00	223.00	
x nr of men = £ per annum	**2,051.60**	**2,051.60**	**4,103.20**	**8,206.40**	**4,103.20**	**4,103.20**	**24,619.20**
Employers Contributions to EasyBuild Stakeholder Pension (Death and accident cover is provided free):							
Number of weeks	52	52	52	52	52	52	
Total weekly £ contribution each	5.00	5.00	5.00	5.00	5.00	5.00	
£ Contributions/annum	**260.00**	**260.00**	**520.00**	**1,040.00**	**520.00**	**520.00**	**3,120.00**
National Insurance Contributions:							
Gross pay – subject to NI	36,149.64	31,686.52	61,201.01	113,688.97	52,281.01	45,353.74	
% of NI Contributions	13.80	13.80	13.80	13.80	13.80	13.80	
£ Contributions/annum	**3,835.44**	**3,219.53**	**6,139.32**	**11,076.23**	**4,908.36**	**3,952.39**	**33,131.27**
Holiday Credit and Welfare contributions:							
Number of weeks	52	52	52	52	52	52	
Total weekly £ contribution each	81.10	70.79	68.28	63.29	58.01	50.04	
X nr of men = £ Contributions/annum	**4,217.20**	**3,681.08**	**7,101.12**	**13,164.32**	**6,033.04**	**5,204.16**	**39,400.92**
Holiday Top-up Funding including overtime	14.60	12.59	12.10	11.08	10.07	8.48	
Cost	**759.03**	**654.70**	**1,258.77**	**2,305.15**	**1,046.85**	**881.68**	**6,906.18**

SUB-TOTAL		422,919.26
TRAINING (INCLUDING ANY TRADE REGISTRATIONS) – SAY	1.00%	4,229.19
SEVERANCE PAY AND SUNDRY COSTS – SAY	1.50%	6,407.23
EMPLOYER'S LIABILITY AND THIRD PARTY INSURANCE – SAY	2.00%	8,671.11
ANNUAL COST OF NOTIONAL GANG		442,226.80
MEN ACTUALLY WORKING =10.5 THEREFORE ANNUAL COST PER PRODUCTIVE MAN		42,116.84
AVERAGE NR OF HOURS WORKED PER MAN =2007 THEREFORE ALL IN MAN HOURS		20.98
PRELIMINARY ITEMS – SAY	7.50%	1.57
SITE AND HEAD OFFICE OVERHEADS – SAY	3.00%	0.68
PROFIT – SAY	2.00%	0.46
THEREFORE INCLUSIVE MAN HOUR RATE		23.70

Notes:

1) The following assumptions have been made in the above calculations:
   a) The working week of 38 hours i.e. the normal working week as defined by the National Agreement.
   b) The actual hours worked are five days of 9 hours each.
   c) A working year of 2007 hours.
   d) Five days in the year are lost through sickness or similar reason.
2) The incentive scheme addition of 5% is intended to reflect bonus schemes typically in use.
3) National insurance contributions are those effective from 6th April 2011. Calculation is based on employer making regular payment into the holiday pay scheme, allowing savings on NI.
4) Weekly Holiday Credit/Welfare Stamp values are those effective from 4th October 2010.
5) Rates are based from 4th October 2010.
6) Overtime rates are based on Premium Rate 1.
7) Easybuild Stakeholder Pension Contributions effective from 30th June 2009.
8) Fares with Oyster Card (New Malden to Waterloo + Zone 1 – Anytime fare) Current at April 2011 – 0845 748 4950 or 0845 330 9876.

DIRECTIONS

LABOUR RATE – DUCTWORK

The annual cost of notional eight man gang of 8 men

		FOREMAN	SENIOR CRAFTSMAN	CRAFTSMAN	INSTALLER	SUB TOTALS
		1 NR	1 NR	4 NR	2 NR	
Hourly Rate from 4th October 2010		15.00	12.40	11.38	10.31	
Working hours per annum per man		1,694.80	1,694.80	1,694.80	1,694.80	
x Hourly rate x nr of men = £ per annum		25,422.00	21,015.52	77,147.30	34,946.78	158,531.59
Overtime Rate		21.14	17.48	16.04	14.54	
Overtime hours per annum per man		312.20	312.20	312.20	312.20	
x hourly rate x nr of men = £ per annum		6,599.91	5,457.26	20,030.75	9,078.78	41,166.69
Total		32,021.91	26,472.78	97,178.05	44,025.55	199,698.28
Incentive schemes	0.00%	0.00	0.00	0.00	0.00	0.00
Daily Travel Time Allowance (15-20 miles each way)		9.31	9.31	9.31	9.31	
Days per annum per man		223.00	223.00	223.00	223.00	
x nr of men = £ per annum		2,076.13	2,076.13	8,304.52	4,152.26	16,609.04
Daily Travel Fare (15-20 miles each way)		9.20	9.20	9.20	9.20	
Days per annum per man		223.00	223.00	223.00	223.00	
x nr of men = £ per annum		2,051.60	2,051.60	8,206.40	4,103.20	16,412.80
Employers Contributions to EasyBuild Stakeholder Pension (Death and accident cover is provided free)						
Number of weeks		52	52	52	52	
Total weekly £ contribution each		5.00	5.00	5.00	5.00	
£ Contributions/annum		260.00	260.00	1,040.00	520.00	2,080.00
National Insurance Contributions:						
Gross pay - subject to NI		36,149.64	30,600.51	113,688.97	52,281.01	
% of NI Contributions		13.8	13.8	13.8	13.8	
£ Contributions/annum		3,835.44	3,069.66	11,076.23	4,908.36	22,889.69
Holiday Credit and Welfare contributions:						
Number of weeks		52	52	52	52	
Total weekly £ contribution each		81.10	68.28	63.29	58.01	
x nr of men = £ Contributions/annum		4,217.20	3,550.56	13,164.32	6,033.04	26,965.12
Holiday Top-up Funding including overtime		14.60	12.10	11.08	10.07	
Cost		759.03	629.38	2,305.15	1,046.85	4,740.42

SUB-TOTAL		289,395.35
TRAINING (INCLUDING ANY TRADE REGISTRATIONS) - SAY	1.00%	2,893.95
SEVERANCE PAY AND SUNDRY COSTS - SAY	1.50%	4,384.34
EMPLOYER'S LIABILITY AND THIRD PARTY INSURANCE - SAY	2.00%	5,933.47
ANNUAL COST OF NOTIONAL GANG		302,607.12
MEN ACTUALLY WORKING =7.5 THEREFORE ANNUAL COST PER PRODUCTIVE MAN		40,347.62
AVERAGE NR OF HOURS WORKED PER MAN =2007 THEREFORE ALL IN MAN HOURS		20.10
PRELIMINARY ITEMS - SAY	12.50%	2.51
SITE AND HEAD OFFICE OVERHEADS - SAY	5.50%	1.24
PROFIT - SAY	3.00%	0.72
THEREFORE INCLUSIVE MAN HOUR RATE		24.58

Notes:

1) The following assumptions have been made in the above calculations:-
   a) The working week of 38 hours i.e. the normal working week as defined by the National Agreement.
   b) The actual hours worked are five days of 9 hours each.
   c) A working year of 2007 hours.
   d) Five days in the year are lost through sickness or similar reason.
2) The incentive scheme addition of 5% is intended to reflect bonus schemes typically in use.
3) National insurance contributions are those effective from 6th April 2011. Calculation is based on employer making regular payment into the holiday pay scheme, allowing savings on NI.
4) Weekly Holiday Credit/Welfare Stamp values are those effective from 4th October 2010.
5) Rates are based from 4th October 2010.
6) Fares with Oyster Card (New Malden to Waterloo + Zone 1 – Anytime fare) current at April 2011 – 0845 748 4950 or 0845 330 9876.
7) Easybuild Stakeholder Pension Contributions effective from 30th June 2009.
8) Overtime rates are based on Premium Rate 1.

R: DISPOSAL SYSTEMS

Item	Net Price £	Material £	Labour hours	Labour £	Unit	Total rate £
R10: RAINWATER PIPEWORK/GUTTERS						
PVC-u gutters: push fit joints; fixed with brackets to backgrounds; BS 4576 BS EN 607						
Half round gutter, with brackets measured separately						
75 mm	3.02	3.41	0.69	16.35	m	**19.76**
100 mm	2.66	3.01	0.64	15.17	m	**18.18**
150 mm	7.01	7.92	0.82	19.42	m	**27.34**
Brackets: including fixing to backgrounds. For minimum fixing distances, refer to the Tables and Memoranda at the rear of the book						
75 mm; Fascia	1.18	1.34	0.15	3.56	nr	**4.90**
100 mm; Jointing	1.88	2.12	0.16	3.79	nr	**5.91**
100 mm; Support	0.77	0.87	0.16	3.79	nr	**4.66**
150 mm; Fascia	2.30	2.60	0.16	3.79	nr	**6.39**
Bracket supports: including fixing to backgrounds. For minimum fixing distances, refer to the Tables and Memoranda at the rear of the book						
Side rafter	3.00	3.39	0.16	3.79	nr	**7.18**
Top rafter	3.00	3.39	0.16	3.79	nr	**7.18**
Rise and fall	3.26	3.68	0.16	3.79	nr	**7.47**
Extra over fittings half round PVC-u gutter						
Union						
75 mm	1.76	1.99	0.19	4.51	nr	**6.50**
100 mm	1.88	2.12	0.24	5.69	nr	**7.81**
150 mm	7.00	7.91	0.28	6.63	nr	**14.54**
Rainwater pipe outlets						
Running: 75 × 53 mm dia.	4.05	4.57	0.12	2.85	nr	**7.42**
Running: 100 × 68 mm dia.	2.77	3.13	0.12	2.85	nr	**5.98**
Running: 150 × 110 mm dia.	12.96	14.64	0.12	2.85	nr	**17.49**
Stop end: 100 × 68 mm dia.	3.10	3.50	0.12	2.85	nr	**6.35**
Internal stop ends: short						
75 mm	1.76	2.97	0.09	2.13	nr	**5.10**
100 mm	0.89	1.99	0.09	2.13	nr	**4.12**
150 mm	4.67	5.27	0.09	2.13	nr	**7.40**
External stop ends: short						
75 mm	3.51	3.97	0.09	2.13	nr	**6.10**
100 mm	1.39	10.50	0.09	2.13	nr	**12.63**
150 mm	4.67	5.27	0.09	2.13	nr	**7.40**

R: DISPOSAL SYSTEMS

Item	Net Price £	Material £	Labour hours	Labour £	Unit	Total rate £
R10: RAINWATER PIPEWORK/GUTTERS – cont						
Extra over fittings half round PVC-u gutter – cont						
Angles						
75 mm; 45°	4.79	5.41	0.20	4.74	nr	**10.15**
75 mm; 90°	4.79	5.41	0.20	4.74	nr	**10.15**
100 mm; 90°	3.31	3.73	0.20	4.74	nr	**8.47**
100 mm; 120°	3.65	4.12	0.20	4.74	nr	**8.86**
100 mm; 135°	3.65	4.12	0.20	4.74	nr	**8.86**
100 mm; Prefabricated to special angle	20.63	23.30	0.23	5.46	nr	**28.76**
100 mm; Prefabricated to raked angle	22.32	25.20	0.23	5.46	nr	**30.66**
150 mm; 90°	11.84	13.37	0.20	4.74	nr	**18.11**
Gutter adaptors						
100 mm; Stainless steel clip	1.70	1.92	0.16	3.79	nr	**5.71**
100 mm; Cast iron spigot	4.66	5.26	0.23	5.46	nr	**10.72**
100 mm; Cast iron socket	4.66	5.26	0.23	5.46	nr	**10.72**
100 mm; Cast iron 'ogee' spigot	4.73	5.34	0.23	5.46	nr	**10.80**
100 mm; Cast iron 'ogee' socket	4.73	5.34	0.23	5.46	nr	**10.80**
100 mm; Half round to Square PVC-u	8.46	9.56	0.23	5.46	nr	**15.02**
100 mm; Gutter overshoot guard	10.47	11.82	0.58	13.74	nr	**25.56**
Square gutter, with brackets measured separately						
120 mm	2.86	3.23	0.82	19.42	m	**22.65**
Brackets: including fixing to backgrounds. For minimum fixing distances, refer to the Tables and Memoranda at the rear of the book						
Jointing	1.97	2.22	0.16	3.79	nr	**6.01**
Support	0.86	0.97	0.16	3.79	nr	**4.76**
Bracket support: including fixing to backgrounds. For minimum fixing distances, refer to the Tables and Memoranda at the rear of the book						
Side rafter	3.00	3.39	0.16	3.79	nr	**7.18**
Top rafter	3.00	3.39	0.16	3.79	nr	**7.18**
Rise and fall	3.26	3.68	0.16	3.79	nr	**7.47**
Extra over fittings square PVC-u gutter						
Rainwater pipe outlets						
Running: 62 mm square	2.78	3.14	0.12	2.85	nr	**5.99**
Stop end: 62 mm square	3.15	3.56	0.12	2.85	nr	**6.41**

R: DISPOSAL SYSTEMS

Item	Net Price £	Material £	Labour hours	Labour £	Unit	Total rate £
Stop ends: short						
External	1.44	1.62	0.09	2.13	nr	**3.75**
Angles						
90°	3.57	4.03	0.20	4.74	nr	**8.77**
120°	8.94	10.10	0.20	4.74	nr	**14.84**
135°	4.18	4.72	0.20	4.74	nr	**9.46**
Prefabricated to special angle	21.15	23.89	0.23	5.46	nr	**29.35**
Prefabricated to raked angle	26.08	29.46	0.23	5.46	nr	**34.92**
Gutter adaptors						
Cast iron	13.58	15.34	0.23	5.46	nr	**20.80**
High capacity square gutter, with brackets measured separately						
137 mm	6.52	7.36	0.82	19.42	m	**26.78**
Brackets: including fixing to backgrounds. For minimum fixing distances, refer to the Tables and Memoranda at the rear of the book						
Jointing	6.08	6.86	0.16	3.79	nr	**10.65**
Support	2.53	2.86	0.16	3.79	nr	**6.65**
Overslung	2.36	2.66	0.16	3.79	nr	**6.45**
Bracket supports: including fixing to backgrounds. For minimum fixing distances, refer to the Tables and Memoranda at the rear of the book						
Side rafter	3.62	4.09	0.16	3.79	nr	**7.88**
Top rafter	3.62	4.09	0.16	3.79	nr	**7.88**
Rise and fall	5.82	6.57	0.16	3.79	nr	**10.36**
Extra over fittings high capacity square PVC-u						
Rainwater pipe outlets						
Running: 75 mm square	10.16	11.48	0.12	2.85	nr	**14.33**
Running: 82 mm dia.	10.16	11.48	0.12	2.85	nr	**14.33**
Running: 110 mm dia.	8.89	10.04	0.12	2.85	nr	**12.89**
Screwed outlet adaptor						
75 mm square pipe	6.46	7.29	0.23	5.46	nr	**12.75**
Stop ends: short						
External	3.51	3.97	0.09	2.13	nr	**6.10**

R: DISPOSAL SYSTEMS

Item	Net Price £	Material £	Labour hours	Labour £	Unit	Total rate £
R10: RAINWATER PIPEWORK/GUTTERS – cont						
Extra over fittings high capacity square PVC-u – cont						
Angles						
90°	8.73	9.86	0.20	4.74	nr	**14.60**
135°	10.27	11.60	0.20	4.74	nr	**16.34**
Prefabricated to special angle	29.67	33.51	0.23	5.46	nr	**38.97**
Prefabricated to raked internal angle	51.66	58.34	0.23	5.46	nr	**63.80**
Prefabricated to raked external angle	51.66	58.34	0.23	5.46	nr	**63.80**
Deep eliptical gutter, with brackets measured separately						
137 mm	3.47	3.92	0.82	19.42	m	**23.34**
Brackets: including fixing to backgrounds. For minimum fixing distances, refer to the Tables and Memoranda at the rear of the book						
Jointing	2.75	3.10	0.16	3.79	nr	**6.89**
Support	1.17	1.33	0.16	3.79	nr	**5.12**
Bracket support: including fixing to backgrounds. For minimum fixing distances, refer to the Tables and Memoranda at the rear of the book						
Side rafter	3.81	4.30	0.16	3.79	nr	**8.09**
Top rafter	3.81	4.30	0.16	3.79	nr	**8.09**
Rise and fall	6.11	6.90	0.16	3.79	nr	**10.69**
Extra over fittings deep eliptical PVC-u gutter						
Rainwater pipe outlets						
Running: 68 mm dia.	4.36	4.93	0.12	2.85	nr	**7.78**
Running: 82 mm dia.	4.15	4.68	0.12	2.85	nr	**7.53**
Stop end: 68 mm dia.	4.44	5.02	0.12	2.85	nr	**7.87**
Stop ends: short						
External	2.13	2.41	0.09	2.13	nr	**4.54**
Angles						
90°	4.42	4.99	0.20	4.74	nr	**9.73**
135°	5.24	5.92	0.20	4.74	nr	**10.66**
Prefabricated to special angle	15.32	17.30	0.23	5.46	nr	**22.76**
Gutter adaptors						
Stainless steel clip	3.76	4.24	0.16	3.79	nr	**8.03**
Marley deepflow	3.03	3.42	0.23	5.46	nr	**8.88**

R: DISPOSAL SYSTEMS

Item	Net Price £	Material £	Labour hours	Labour £	Unit	Total rate £
Ogee profile PVC-u gutter, with brackets measured separately						
122 mm	3.65	4.12	0.82	19.42	m	**23.54**
Brackets: including fixing to backgrounds. For minimum fixing distances, refer to the Tables and Memoranda at the rear of the book						
Jointing	2.91	3.28	0.16	3.79	nr	**7.07**
Support	1.11	1.25	0.16	3.79	nr	**5.04**
Overslung	1.11	1.25	0.16	3.79	nr	**5.04**
Extra over fittings Ogee profile PVC-u gutter						
Rainwater pipe outlets						
Running: 68 mm dia.	4.15	4.68	0.12	2.85	nr	**7.53**
Stop ends: short						
Internal/External: left or right hand	2.10	2.38	0.09	2.13	nr	**4.51**
Angles						
90°: internal or external	4.21	4.75	0.20	4.74	nr	**9.49**
135°: internal or external	4.21	4.75	0.20	4.74	nr	**9.49**
PVC-u rainwater pipe: dry push fit joints; fixed with brackets to backgrounds; BS 4576/ BS EN 607						
Pipe: circular, with brackets measured separately						
53 mm	4.72	5.33	0.61	14.45	m	**19.78**
68 mm	4.91	5.55	0.61	14.45	m	**20.00**
Pipe clip: including fixing to backgrounds. For minimum fixing distances, refer to the Tables and Memoranda at the rear of the book						
68 mm	1.09	1.23	0.16	3.79	nr	**5.02**
Pipe clip adjustable: including fixing to backgrounds. For minimum fixing distances, refer to the Tables and Memoranda at the rear of the book						
53 mm	1.69	1.91	0.16	3.79	nr	**5.70**
68 mm	2.35	2.65	0.16	3.79	nr	**6.44**
Pipe clip drive in: including fixing to backgrounds. For minimum fixing distances, refer to the Tables and Memoranda at the rear of the book						
68 mm	2.65	2.99	0.16	3.79	nr	**6.78**

R: DISPOSAL SYSTEMS

Item	Net Price £	Material £	Labour hours	Labour £	Unit	Total rate £
R10: RAINWATER PIPEWORK/GUTTERS – cont						
Extra over fittings circular pipework PVC-u						
Pipe coupler: PVC-u to PVC-u						
68 mm	1.37	1.55	0.12	2.85	nr	4.40
Pipe coupler: PVC-u to Cast Iron						
68 mm: to 3' cast iron	5.38	6.08	0.17	4.03	nr	10.11
68 mm: to 3.¾' cast iron	18.48	20.87	0.17	4.03	nr	24.90
Access pipe: single socket						
68 mm	8.59	9.70	0.15	3.56	nr	13.26
Bend: short radius						
53 mm: 67.5°	2.00	2.25	0.20	4.74	nr	6.99
68 mm: 92.5°	2.23	2.52	0.20	4.74	nr	7.26
68 mm: 112.5°	1.97	2.22	0.20	4.74	nr	6.96
Bend: long radius						
68 mm: 112°	2.78	3.14	0.20	4.74	nr	7.88
Branch						
68 mm: 92°	18.05	20.38	0.23	5.46	nr	25.84
68 mm: 112°	18.11	20.45	0.23	5.46	nr	25.91
Double branch						
68 mm: 112°	35.73	40.35	0.24	5.69	nr	46.04
Shoe						
53 mm	2.97	3.36	0.12	2.85	nr	6.21
68 mm	1.59	1.80	0.12	2.85	nr	4.65
Rainwater head: including fixing to backgrounds						
68 mm	8.59	9.70	0.29	6.86	nr	16.56
Pipe: square, with brackets measured separately						
62 mm	3.00	3.39	0.45	10.66	m	14.05
75 mm	5.72	6.46	0.45	10.66	m	17.12
Pipe clip: including fixing to backgrounds. For minimum fixing distances, refer to the Tables and Memoranda at the rear of the book						
62 mm	1.01	1.14	0.16	3.79	nr	4.93
75 mm	2.23	2.52	0.16	3.79	nr	6.31

R: DISPOSAL SYSTEMS

Item	Net Price £	Material £	Labour hours	Labour £	Unit	Total rate £
Pipe clip adjustable: including fixing to backgrounds. For minimum fixing distances, refer to the Tables and Memoranda at the rear of the book						
62 mm	3.11	3.51	0.16	3.79	nr	7.30
Extra over fittings square pipework PVC-u						
Pipe coupler: PVC-u to PVC-u						
62 mm	1.47	1.66	0.20	4.74	nr	6.40
75 mm	2.30	2.60	0.20	4.74	nr	7.34
Square to circular adaptor: single socket						
62 mm to 68 mm	2.88	3.25	0.20	4.74	nr	7.99
Square to circular adaptor: single socket						
75 mm to 62 mm	3.71	4.19	0.20	4.74	nr	8.93
Access pipe						
62 mm	17.18	19.40	0.16	3.79	nr	23.19
75 mm	18.13	20.47	0.16	3.79	nr	24.26
Bends						
62 mm: 92.5°	2.66	3.01	0.20	4.74	nr	7.75
62 mm: 112.5°	1.94	2.19	0.20	4.74	nr	6.93
75 mm: 112.5°	4.84	5.47	0.20	4.74	nr	10.21
Bends: prefabricated special angle						
62 mm	17.26	19.49	0.23	5.46	nr	24.95
75 mm	23.98	27.08	0.23	5.46	nr	32.54
Offset						
62 mm	4.16	4.70	0.20	4.74	nr	9.44
75 mm	11.45	12.93	0.20	4.74	nr	17.67
Offset: prefabricated special angle						
62 mm	17.26	19.49	0.23	5.46	nr	24.95
Shoe						
62 mm	2.44	2.75	0.12	2.85	nr	5.60
75 mm	4.21	4.75	0.12	2.85	nr	7.60
Branch						
62 mm	6.04	6.82	0.23	5.46	nr	12.28
75 mm	29.43	33.23	0.23	5.46	nr	38.69
Double branch						
62 mm	32.38	36.57	0.24	5.69	nr	42.26

R: DISPOSAL SYSTEMS

Item	Net Price £	Material £	Labour hours	Labour £	Unit	Total rate £
R10: RAINWATER PIPEWORK/GUTTERS – cont						
Extra over fittings square pipework PVC-u – cont						
Rainwater head						
62 mm	7.79	8.80	0.29	6.86	nr	**15.66**
75 mm	28.57	32.26	3.45	81.71	nr	**113.97**
PVC-u rainwater pipe: solvent welded joints; fixed with brackets to backgrounds; BS 4576/ BS EN 607						
Pipe: circular, with brackets measured separately						
82 mm	9.26	10.45	0.35	8.29	m	**18.74**
Pipe clip: galvanized; including fixing to backgrounds. For minimum fixing distances, refer to the Tables and Memoranda at the rear of the book						
82 mm	3.77	4.25	0.58	13.74	nr	**17.99**
Pipe clip: galvanized plastic coated; including fixing to backgrounds. For minimum fixing distances, refer to the Tables and Memoranda at the rear of the book						
82 mm	5.21	5.89	0.58	13.74	nr	**19.63**
Pipe clip: PVC-u including fixing to backgrounds. For minimum fixing distances, refer to the Tables and Memoranda at the rear of the book						
82 mm	2.80	3.16	0.58	13.74	nr	**16.90**
Pipe clip: PVC-u adjustable: including fixing to backgrounds. For minimum fixing distances, refer to the Tables and Memoranda at the rear of the book						
82 mm	5.13	5.79	0.58	13.74	nr	**19.53**
Extra over fittings circular pipework PVC-u						
Pipe coupler: PVC-u to PVC-u						
82 mm	4.70	5.30	0.21	4.98	nr	**10.28**
Access pipe						
82 mm	28.92	32.66	0.23	5.46	nr	**38.12**

R: DISPOSAL SYSTEMS

Item	Net Price £	Material £	Labour hours	Labour £	Unit	Total rate £
Bend						
82 mm: 92, 112.5 and 135°	11.81	13.34	0.29	6.86	nr	**20.20**
Shoe						
82 mm	7.29	8.23	0.29	6.86	nr	**15.09**
110 mm	9.15	10.33	0.32	7.58	nr	**17.91**
Branch						
82 mm: 92, 112.5 and 135°	17.60	19.88	0.35	8.29	nr	**28.17**
Rainwater head						
82 mm	15.44	17.44	0.58	13.74	nr	**31.18**
110 mm	14.06	15.88	0.58	13.74	nr	**29.62**
Roof outlets: 178 dia.; Flat						
50 mm	15.43	17.42	1.15	27.25	nr	**44.67**
82 mm	15.43	17.42	1.15	27.25	nr	**44.67**
Roof outlets: 178 mm dia.; Domed						
50 mm	15.43	17.42	1.15	27.25	nr	**44.67**
82 mm	15.43	17.42	1.15	27.25	nr	**44.67**
Roof outlets: 406 mm dia.; Flat						
82 mm	30.16	34.06	1.15	27.25	nr	**61.31**
110 mm	30.16	34.06	1.15	27.25	nr	**61.31**
Roof outlets: 406 mm dia.; Domed						
82 mm	30.16	34.06	1.15	27.25	nr	**61.31**
110 mm	30.16	34.06	1.15	27.25	nr	**61.31**
Roof outlets: 406 mm dia.; Inverted						
82 mm	66.38	74.97	1.15	27.25	nr	**102.22**
110 mm	66.38	74.97	1.15	27.25	nr	**102.22**
Roof outlets: 406 mm dia.; Vent Pipe						
82 mm	43.60	49.24	1.15	27.25	nr	**76.49**
110 mm	43.60	49.24	1.15	27.25	nr	**76.49**
Balcony outlets: screed						
82 mm	25.80	29.14	1.15	27.25	nr	**56.39**
Balcony outlets: asphalt						
82 mm	25.67	28.99	1.15	27.25	nr	**56.24**
Adaptors						
82 mm × 62 mm square pipe	3.45	3.90	0.21	4.98	nr	**8.88**
82 mm × 68 mm circular pipe	3.45	3.90	0.21	4.98	nr	**8.88**

R: DISPOSAL SYSTEMS

Item	Net Price £	Material £	Labour hours	Labour £	Unit	Total rate £
R10: RAINWATER PIPEWORK/GUTTERS – cont						
For 110 mm diameter pipework and fittings refer to R11: Above Ground Drainage						
Cast iron gutters: mastic and bolted joints; BS 460; fixed with brackets to backgrounds						
Half round gutter, with brackets measured separately						
100 mm	12.13	13.70	0.85	20.13	m	**33.83**
115 mm	12.65	14.29	0.97	22.98	m	**37.27**
125 mm	14.80	16.72	0.97	22.98	m	**39.70**
150 mm	25.29	28.56	1.12	26.54	m	**55.10**
Brackets; fixed to backgrounds. For minimum fixing distances, refer to the Tables and Memoranda at the rear of the book						
Fascia						
100 mm	2.84	3.20	0.16	3.79	nr	**6.99**
115 mm	2.84	3.20	0.16	3.79	nr	**6.99**
125 mm	2.84	3.20	0.16	3.79	nr	**6.99**
150 mm	3.60	4.07	0.16	3.79	nr	**7.86**
Rise and fall						
100 mm	5.65	6.39	0.39	9.24	nr	**15.63**
115 mm	5.65	6.39	0.39	9.24	nr	**15.63**
125 mm	5.80	6.55	0.39	9.24	nr	**15.79**
150 mm	5.91	6.67	0.39	9.24	nr	**15.91**
Top rafter						
100 mm	3.48	3.93	0.16	3.79	nr	**7.72**
115 mm	3.48	3.93	0.16	3.79	nr	**7.72**
125 mm	4.71	5.31	0.16	3.79	nr	**9.10**
150 mm	6.40	7.23	0.16	3.79	nr	**11.02**
Side rafter						
100 mm	3.48	3.93	0.16	3.79	nr	**7.72**
115 mm	3.48	3.93	0.16	3.79	nr	**7.72**
125 mm	4.71	5.31	0.16	3.79	nr	**9.10**
150 mm	6.40	7.23	0.16	3.79	nr	**11.02**

R: DISPOSAL SYSTEMS

Item	Net Price £	Material £	Labour hours	Labour £	Unit	Total rate £
Extra over fittings half round gutter cast iron BS 460						
Union						
100 mm	6.67	7.54	0.39	9.24	nr	**16.78**
115 mm	8.32	9.39	0.48	11.37	nr	**20.76**
125 mm	9.41	10.63	0.48	11.37	nr	**22.00**
150 mm	10.55	11.91	0.55	13.04	nr	**24.95**
Stop end; internal						
100 mm	3.40	3.84	0.12	2.85	nr	**6.69**
115 mm	4.40	4.97	0.15	3.56	nr	**8.53**
125 mm	4.40	4.97	0.15	3.56	nr	**8.53**
150 mm	6.11	6.90	0.20	4.74	nr	**11.64**
Stop end; external						
100 mm	3.40	3.84	0.12	2.85	nr	**6.69**
115 mm	4.31	4.87	0.15	3.56	nr	**8.43**
125 mm	4.40	4.97	0.15	3.56	nr	**8.53**
150 mm	6.11	6.90	0.20	4.74	nr	**11.64**
90° angle; single socket						
100 mm	10.12	11.42	0.39	9.24	nr	**20.66**
115 mm	10.41	11.76	0.43	10.19	nr	**21.95**
125 mm	12.27	13.86	0.43	10.19	nr	**24.05**
150 mm	22.42	25.32	0.50	11.84	nr	**37.16**
90° angle; double socket						
100 mm	12.26	13.84	0.39	9.24	nr	**23.08**
115 mm	13.00	14.68	0.43	10.19	nr	**24.87**
125 mm	16.83	19.00	0.43	10.19	nr	**29.19**
135° angle; single socket						
100 mm	10.33	11.67	0.39	9.24	nr	**20.91**
115 mm	10.43	11.78	0.43	10.19	nr	**21.97**
125 mm	15.41	17.40	0.43	10.19	nr	**27.59**
150 mm	22.86	25.82	0.50	11.84	nr	**37.66**
Running outlet						
65 mm outlet						
100 mm	9.86	11.14	0.39	9.24	nr	**20.38**
115 mm	10.74	12.13	0.43	10.19	nr	**22.32**
125 mm	12.27	13.86	0.43	10.19	nr	**24.05**
75 mm outlet						
100 mm	9.86	11.14	0.39	9.24	nr	**20.38**
115 mm	10.74	12.13	0.43	10.19	nr	**22.32**
125 mm	12.27	13.86	0.43	10.19	nr	**24.05**
150 mm	21.25	24.00	0.50	11.84	nr	**35.84**

R: DISPOSAL SYSTEMS

Item	Net Price £	Material £	Labour hours	Labour £	Unit	Total rate £
R10: RAINWATER PIPEWORK/GUTTERS – cont						
Extra over fittings half round gutter cast iron BS 460 – cont						
100 mm outlet						
150 mm	21.25	24.00	0.50	11.84	nr	**35.84**
Stop end outlet; socket						
65 mm outlet						
100 mm	11.57	13.07	0.39	9.24	nr	**22.31**
115 mm	12.97	14.65	0.43	10.19	nr	**24.84**
75 mm outlet						
125 mm	11.57	13.07	0.43	10.19	nr	**23.26**
150 mm	24.31	51.45	0.50	11.84	nr	**63.29**
100 mm outlet						
150 mm	21.25	24.00	0.50	11.84	nr	**35.84**
Stop end outlet; spigot						
65 mm outlet						
100 mm	11.57	13.07	0.39	9.24	nr	**22.31**
115 mm	12.97	14.65	0.43	10.19	nr	**24.84**
75 mm outlet						
125 mm	11.57	13.07	0.43	10.19	nr	**23.26**
150 mm	24.31	27.46	0.50	11.84	nr	**39.30**
100 mm outlet						
150 mm	24.31	27.46	0.50	11.84	nr	**39.30**
Half round; 3 mm thick double beaded gutter, with brackets measured separately						
100 mm	12.35	13.94	0.85	20.13	m	**34.07**
115 mm	12.90	14.57	0.85	20.13	m	**34.70**
125 mm	14.79	16.71	0.97	22.98	m	**39.69**
Brackets; fixed to backgrounds. For minimum fixing distances, refer to the Tables and Memoranda at the rear of the book						
Fascia						
100 mm	2.84	3.20	0.16	3.79	nr	**6.99**
115 mm	2.84	3.20	0.16	3.79	nr	**6.99**
125 mm	2.84	3.20	0.16	3.79	nr	**6.99**

R: DISPOSAL SYSTEMS

Item	Net Price £	Material £	Labour hours	Labour £	Unit	Total rate £
Extra over fittings Half Round 3 mm thick Gutter BS 460						
Union						
100 mm	6.67	7.54	0.38	9.00	nr	**16.54**
115 mm	8.13	9.18	0.38	9.00	nr	**18.18**
125 mm	9.41	10.63	0.43	10.19	nr	**20.82**
Stop end; internal						
100 mm	3.40	3.84	0.12	2.85	nr	**6.69**
115 mm	4.40	4.97	0.12	2.85	nr	**7.82**
125 mm	4.42	4.99	0.15	3.56	nr	**8.55**
Stop end; external						
100 mm	3.40	3.84	0.12	2.85	nr	**6.69**
115 mm	4.40	4.97	0.12	2.85	nr	**7.82**
125 mm	4.42	4.99	0.15	3.56	nr	**8.55**
90° angle; single socket						
100 mm	10.50	11.86	0.38	9.00	nr	**20.86**
115 mm	10.65	12.03	0.38	9.00	nr	**21.03**
125 mm	12.27	13.86	0.43	10.19	nr	**24.05**
135° angle; single socket						
100 mm	10.33	11.67	0.38	9.00	nr	**20.67**
115 mm	10.41	11.76	0.38	9.00	nr	**20.76**
125 mm	12.98	14.66	0.43	10.19	nr	**24.85**
Running outlet						
65 mm outlet						
100 mm	10.59	11.96	0.38	9.00	nr	**20.96**
115 mm	10.95	12.36	0.38	9.00	nr	**21.36**
125 mm	12.71	14.35	0.43	10.19	nr	**24.54**
75 mm outlet						
115 mm	10.95	12.36	0.38	9.00	nr	**21.36**
125 mm	12.71	14.35	0.43	10.19	nr	**24.54**
Stop end outlet; socket						
65 mm outlet						
100 mm	11.57	13.07	0.38	9.00	nr	**22.07**
115 mm	12.97	14.65	0.38	9.00	nr	**23.65**
125 mm	14.50	16.37	0.43	10.19	nr	**26.56**
75 mm outlet						
125 mm	14.76	16.67	0.43	10.19	nr	**26.86**

R: DISPOSAL SYSTEMS

Item	Net Price £	Material £	Labour hours	Labour £	Unit	Total rate £
R10: RAINWATER PIPEWORK/GUTTERS – cont						
Extra over fittings Half Round 3 mm thick Gutter BS 460 – cont						
Stop end outlet; spigot						
65 mm outlet						
100 mm	11.57	13.07	0.38	9.00	nr	**22.07**
115 mm	12.97	14.65	0.38	9.00	nr	**23.65**
125 mm	14.50	16.37	0.43	10.19	nr	**26.56**
Deep half round gutter, with brackets measured separately						
100 × 75 mm	20.31	22.94	0.85	20.13	m	**43.07**
125 × 75 mm	26.25	29.64	0.97	22.98	m	**52.62**
Brackets; fixed to backgrounds. For minimum fixing distances, refer to the Tables and Memoranda at the rear of the book						
Fascia						
100 × 75 mm	9.59	10.83	0.16	3.79	nr	**14.62**
125 × 75 mm	11.82	13.35	0.16	3.79	nr	**17.14**
Extra over fittings Deep Half Round Gutter BS 460						
Union						
100 × 75 mm	11.16	12.61	0.38	9.00	nr	**21.61**
125 × 75 mm	11.82	13.35	0.43	10.19	nr	**23.54**
Stop end; internal						
100 × 75 mm	9.77	11.04	0.12	2.85	nr	**13.89**
125 × 75 mm	12.04	13.60	0.15	3.56	nr	**17.16**
Stop end; external						
100 × 75 mm	9.77	11.04	0.12	2.85	nr	**13.89**
125 × 75 mm	12.04	13.60	0.15	3.56	nr	**17.16**
90° angle; single socket						
100 × 75 mm	27.85	31.46	0.38	9.00	nr	**40.46**
125 × 75 mm	35.35	39.92	0.43	10.19	nr	**50.11**
135° angle; single socket						
100 × 75 mm	27.85	31.46	0.38	9.00	nr	**40.46**
125 × 75 mm	35.35	39.92	0.43	10.19	nr	**50.11**

R: DISPOSAL SYSTEMS

Item	Net Price £	Material £	Labour hours	Labour £	Unit	Total rate £
Running outlet						
65 mm outlet						
100 × 75 mm	27.85	31.46	0.38	9.00	nr	**40.46**
125 × 75 mm	35.35	39.92	0.43	10.19	nr	**50.11**
75 mm outlet						
100 × 75 mm	27.85	31.46	0.38	9.00	nr	**40.46**
125 × 75 mm	25.66	28.98	0.43	10.19	nr	**39.17**
Stop end outlet; socket						
65 mm outlet						
100 × 75 mm	37.62	42.48	0.38	9.00	nr	**51.48**
75 mm outlet						
100 × 75 mm	37.62	42.48	0.38	9.00	nr	**51.48**
125 × 75 mm	37.62	42.48	0.43	10.19	nr	**52.67**
Stop end outlet; spigot						
65 mm outlet						
100 × 75 mm	37.62	42.48	0.38	9.00	nr	**51.48**
75 mm outlet						
100 × 75 mm	37.62	42.48	0.38	9.00	nr	**51.48**
125 × 75 mm	37.62	42.48	0.43	10.19	nr	**52.67**
Ogee gutter, with brackets measured separately						
100 mm	13.53	15.28	0.85	20.13	m	**35.41**
115 mm	14.88	16.81	0.97	22.98	m	**39.79**
125 mm	15.61	17.63	0.97	22.98	m	**40.61**
Brackets; fixed to backgrounds.For minimum fixing distances, refer to the Tables and Memoranda at the rear of the book						
Fascia						
100 mm	3.09	3.49	0.16	3.79	nr	**7.28**
115 mm	3.09	3.49	0.16	3.79	nr	**7.28**
125 mm	3.49	3.94	0.16	3.79	nr	**7.73**

R: DISPOSAL SYSTEMS

Item	Net Price £	Material £	Labour hours	Labour £	Unit	Total rate £
R10: RAINWATER PIPEWORK/GUTTERS – cont						
Extra over fittings Ogee Cast Iron Gutter BS 460						
Union						
100 mm	6.68	7.55	0.38	9.00	nr	**16.55**
115 mm	8.13	9.18	0.43	10.19	nr	**19.37**
125 mm	9.41	10.63	0.43	10.19	nr	**20.82**
Stop end; internal						
100 mm	3.48	3.93	0.12	2.85	nr	**6.78**
115 mm	4.50	5.08	0.15	3.56	nr	**8.64**
125 mm	4.50	5.08	0.15	3.56	nr	**8.64**
Stop end; external						
100 mm	3.48	3.93	0.12	2.85	nr	**6.78**
115 mm	4.50	5.08	0.15	3.56	nr	**8.64**
125 mm	4.50	5.08	0.15	3.56	nr	**8.64**
90° angle; internal						
100 mm	10.55	11.91	0.38	9.00	nr	**20.91**
115 mm	11.43	12.91	0.43	10.19	nr	**23.10**
125 mm	12.48	14.10	0.43	10.19	nr	**24.29**
90° angle; external						
100 mm	10.55	11.91	0.38	9.00	nr	**20.91**
115 mm	11.43	12.91	0.43	10.19	nr	**23.10**
125 mm	12.48	14.10	0.43	10.19	nr	**24.29**
135° angle; internal						
100 mm	10.96	12.37	0.38	9.00	nr	**21.37**
115 mm	11.68	13.19	0.43	10.19	nr	**23.38**
125 mm	15.38	17.37	0.43	10.19	nr	**27.56**
135° angle; external						
100 mm	10.96	12.37	0.38	9.00	nr	**21.37**
115 mm	11.68	13.19	0.43	10.19	nr	**23.38**
125 mm	15.38	17.37	0.43	10.19	nr	**27.56**
Running outlet						
65 mm outlet						
100 mm	10.75	12.14	0.38	9.00	nr	**21.14**
115 mm	11.44	12.92	0.43	10.19	nr	**23.11**
125 mm	12.48	14.10	0.43	10.19	nr	**24.29**

R: DISPOSAL SYSTEMS

Item	Net Price £	Material £	Labour hours	Labour £	Unit	Total rate £
75 mm outlet						
125 mm	12.48	14.10	0.43	10.19	nr	**24.29**
Stop end outlet; socket						
65 mm outlet						
100 mm	16.98	19.18	0.38	9.00	nr	**28.18**
115 mm	16.98	19.18	0.43	10.19	nr	**29.37**
125 mm	16.98	19.18	0.43	10.19	nr	**29.37**
75 mm outlet						
125 mm	16.98	19.18	0.43	10.19	nr	**29.37**
Stop end outlet; spigot						
65 mm outlet						
100 mm	16.98	19.18	0.38	9.00	nr	**28.18**
115 mm	16.98	19.18	0.43	10.19	nr	**29.37**
125 mm	16.98	19.18	0.43	10.19	nr	**29.37**
75 mm outlet						
125 mm	16.98	19.18	0.43	10.19	nr	**29.37**
Notts Ogee Gutter, with brackets measured separately						
115 mm	24.03	27.14	0.85	20.13	m	**47.27**
Brackets; fixed to backgrounds. For minimum fixing distances, refer to the Tables and Memoranda at the rear of the book						
Fascia						
115 mm	9.37	10.58	0.16	3.79	nr	**14.37**
Extra over fittings Notts Ogee Cast Iron Gutter BS 460						
Union						
115 mm	11.39	12.86	0.38	9.00	nr	**21.86**
Stop end; internal						
115 mm	9.37	10.58	0.16	3.79	nr	**14.37**
Stop end; external						
115 mm	9.37	10.58	0.16	3.79	nr	**14.37**

R: DISPOSAL SYSTEMS

Item	Net Price £	Material £	Labour hours	Labour £	Unit	Total rate £
R10: RAINWATER PIPEWORK/GUTTERS – cont						
Extra over fittings Notts Ogee Cast Iron Gutter BS 460 – cont						
90° angle; internal						
115 mm	27.19	30.70	0.43	10.19	nr	**40.89**
90° angle; external						
115 mm	27.19	30.70	0.43	10.19	nr	**40.89**
135° angle; internal						
115 mm	27.70	31.28	0.43	10.19	nr	**41.47**
135° angle; external						
115 mm	27.70	31.28	0.43	10.19	nr	**41.47**
Running outlet						
65 mm outlet						
115 mm	32.60	36.81	0.43	10.19	nr	**47.00**
75 mm outlet						
115 mm	32.60	36.81	0.43	10.19	nr	**47.00**
Stop end outlet; socket						
65 mm outlet						
115 mm	41.97	47.40	0.43	10.19	nr	**57.59**
Stop end outlet; spigot						
65 mm outlet						
115 mm	41.97	47.40	0.43	10.19	nr	**57.59**
No 46 moulded Gutter, with brackets measured separately						
100 × 75 mm	23.25	26.25	0.85	20.13	m	**46.38**
125 × 100 mm	34.08	38.48	0.97	22.98	m	**61.46**
Brackets; fixed to backgrounds. For minimum fixing distances, refer to the Tables and Memoranda at the rear of the book						
Fascia						
100 × 75 mm	5.18	5.85	0.16	3.79	nr	**9.64**
125 × 100 mm	5.18	5.85	0.16	3.79	nr	**9.64**

R: DISPOSAL SYSTEMS

Item	Net Price £	Material £	Labour hours	Labour £	Unit	Total rate £
Extra over fittings						
Union						
100 × 75 mm	10.95	12.36	0.38	9.00	nr	**21.36**
125 × 100 mm	12.69	14.33	0.43	10.19	nr	**24.52**
Stop end; internal						
100 × 75 mm	9.81	11.08	0.12	2.85	nr	**13.93**
125 × 100 mm	12.69	14.33	0.15	3.56	nr	**17.89**
Stop end; external						
100 × 75 mm	9.81	11.08	0.12	2.85	nr	**13.93**
125 × 100 mm	12.69	14.33	0.15	3.56	nr	**17.89**
90° angle; internal						
100 × 75 mm	25.68	29.00	0.38	9.00	nr	**38.00**
125 × 100 mm	36.90	41.68	0.43	10.19	nr	**51.87**
90° angle; external						
100 × 75 mm	25.68	29.00	0.38	9.00	nr	**38.00**
125 × 100 mm	36.90	41.68	0.43	10.19	nr	**51.87**
135° angle; internal						
100 × 75 mm	26.17	29.56	0.38	9.00	nr	**38.56**
125 × 100 mm	36.90	41.68	0.43	10.19	nr	**51.87**
135° angle; external						
100 × 75 mm	26.17	29.56	0.38	9.00	nr	**38.56**
125 × 100 mm	36.90	41.68	0.43	10.19	nr	**51.87**
Running outlet						
65 mm outlet						
100 × 75 mm	26.17	29.56	0.38	9.00	nr	**38.56**
125 × 100 mm	36.90	41.68	0.43	10.19	nr	**51.87**
75 mm outlet						
100 × 75 mm	26.17	29.56	0.38	9.00	nr	**38.56**
125 × 100 mm	36.90	41.68	0.43	10.19	nr	**51.87**
100 mm outlet						
100 × 75 mm	26.17	29.56	0.38	9.00	nr	**38.56**
125 × 100 mm	36.90	41.68	0.43	10.19	nr	**51.87**
100 × 75 mm outlet						
125 × 100 mm	26.17	29.56	0.43	10.19	nr	**39.75**

R: DISPOSAL SYSTEMS

Item	Net Price £	Material £	Labour hours	Labour £	Unit	Total rate £
R10: RAINWATER PIPEWORK/GUTTERS – cont						
Extra over fittings – cont						
Stop end outlet; socket						
65 mm outlet						
100 × 75 mm	49.59	56.01	0.38	9.00	nr	**65.01**
75 mm outlet						
125 × 100 mm	49.59	56.01	0.43	10.19	nr	**66.20**
Stop end outlet; spigot						
65 mm outlet						
100 × 75 mm	49.59	56.01	0.38	9.00	nr	**65.01**
75 mm outlet						
125 × 100 mm	49.59	56.01	0.43	10.19	nr	**66.20**
Box gutter, with brackets measured separately						
100 × 75 mm	37.68	42.55	0.85	20.13	m	**62.68**
Brackets; fixed to backgrounds. For minimum fixing distances, refer to the Tables and Memoranda at the rear of the book						
Fascia						
100 × 75 mm	5.97	6.74	0.16	3.79	nr	**10.53**
Extra over fittings Box Cast Iron Gutter BS 460						
Union						
100 × 75 mm	7.93	8.96	0.38	9.00	nr	**17.96**
Stop end; external						
100 × 75 mm	5.97	6.74	0.12	2.85	nr	**9.59**
90° angle						
100 × 75 mm	30.12	34.02	0.38	9.00	nr	**43.02**
135° angle						
100 × 75 mm	30.12	34.02	0.38	9.00	nr	**43.02**

R: DISPOSAL SYSTEMS

Item	Net Price £	Material £	Labour hours	Labour £	Unit	Total rate £
Running outlet						
65 mm outlet						
100 × 75 mm	30.12	34.02	0.38	9.00	nr	**43.02**
75 mm outlet						
100 × 75 mm	30.12	34.02	0.38	9.00	nr	**43.02**
100 × 75 mm outlet						
100 × 75 mm	30.12	34.02	0.38	9.00	nr	**43.02**
Cast iron rainwater pipe; dry joints; BS 460; fixed to backgrounds						
Circular						
Plain socket pipe, with brackets measured separately						
65 mm	22.11	24.97	0.69	16.35	m	**41.32**
75 mm	22.11	24.97	0.69	16.35	m	**41.32**
100 mm	30.18	34.09	0.69	16.35	m	**50.44**
Bracket; fixed to backgrounds. For minimum fixing distances, refer to the Tables and Memoranda at the rear of the book						
65 mm	8.08	9.13	0.29	6.86	nr	**15.99**
75 mm	8.11	9.16	0.29	6.86	nr	**16.02**
100 mm	8.22	9.28	0.29	6.86	nr	**16.14**
Eared socket pipe, with wall spacers measured separately						
65 mm	23.63	26.68	0.62	14.69	m	**41.37**
75 mm	23.63	26.68	0.62	14.69	m	**41.37**
100 mm	31.73	35.83	0.62	14.69	m	**50.52**
Wall spacer plate; eared pipework						
65 mm	5.21	5.89	0.16	3.79	nr	**9.68**
75 mm	5.31	6.00	0.16	3.79	nr	**9.79**
100 mm	8.22	9.28	0.16	3.79	nr	**13.07**
Extra over fittings Circular Cast Iron Pipework BS 460						
Loose sockets						
Plain socket						
65 mm	6.39	7.22	0.23	5.46	nr	**12.68**
75 mm	6.39	7.22	0.23	5.46	nr	**12.68**
100 mm	9.88	11.16	0.23	5.46	nr	**16.62**

R: DISPOSAL SYSTEMS

Item	Net Price £	Material £	Labour hours	Labour £	Unit	Total rate £
R10: RAINWATER PIPEWORK/GUTTERS – cont						
Extra over fittings Circular Cast Iron Pipework BS 460 – cont						
Eared socket						
65 mm	9.24	10.43	0.29	6.86	nr	**17.29**
75 mm	9.24	10.43	0.29	6.86	nr	**17.29**
100 mm	12.48	14.10	0.29	6.86	nr	**20.96**
Shoe; front projection						
Plain socket						
65 mm	19.88	22.45	0.23	5.46	nr	**27.91**
75 mm	19.88	22.45	0.23	5.46	nr	**27.91**
100 mm	26.80	30.26	0.23	5.46	nr	**35.72**
Eared socket						
65 mm	23.04	26.02	0.29	6.86	nr	**32.88**
75 mm	23.04	26.02	0.29	6.86	nr	**32.88**
100 mm	30.58	34.54	0.29	6.86	nr	**41.40**
Access pipe						
65 mm	35.89	40.53	0.23	5.46	nr	**45.99**
75 mm	37.68	42.55	0.23	5.46	nr	**48.01**
100 mm	66.84	75.49	0.23	5.46	nr	**80.95**
100 mm; eared	75.47	85.23	0.29	6.86	nr	**92.09**
Bends; any degree						
65 mm	14.36	16.22	0.23	5.46	nr	**21.68**
75 mm	17.11	19.32	0.23	5.46	nr	**24.78**
100 mm	24.18	27.31	0.23	5.46	nr	**32.77**
Branch						
92.5°						
65 mm	27.70	31.28	0.29	6.86	nr	**38.14**
75 mm	30.55	34.50	0.29	6.86	nr	**41.36**
100 mm	35.61	40.22	0.29	6.86	nr	**47.08**
112.5°						
65 mm	27.70	31.28	0.29	6.86	nr	**38.14**
75 mm	30.55	34.50	0.29	6.86	nr	**41.36**
135°						
65 mm	27.70	31.28	0.29	6.86	nr	**38.14**
75 mm	30.55	34.50	0.29	6.86	nr	**41.36**

R: DISPOSAL SYSTEMS

Item	Net Price £	Material £	Labour hours	Labour £	Unit	Total rate £
Offsets						
75 to 150 mm projection						
65 mm	21.58	24.37	0.25	5.93	nr	**30.30**
75 mm	21.58	24.37	0.25	5.93	nr	**30.30**
100 mm	40.71	45.97	0.25	5.93	nr	**51.90**
225 mm projection						
65 mm	22.00	24.85	0.25	5.93	nr	**30.78**
75 mm	22.00	24.85	0.25	5.93	nr	**30.78**
100 mm	40.71	45.97	0.25	5.93	nr	**51.90**
305 mm projection						
65 mm	29.43	33.23	0.25	5.93	nr	**39.16**
75 mm	30.89	34.88	0.25	5.93	nr	**40.81**
100 mm	50.27	56.77	0.25	5.93	nr	**62.70**
380 mm projection						
65 mm	58.72	66.31	0.25	5.93	nr	**72.24**
75 mm	58.72	66.31	0.25	5.93	nr	**72.24**
100 mm	80.16	90.52	0.25	5.93	nr	**96.45**
455 mm projection						
65 mm	68.74	77.63	0.25	5.93	nr	**83.56**
75 mm	68.74	77.63	0.25	5.93	nr	**83.56**
100 mm	99.69	112.59	0.25	5.93	nr	**118.52**
Rectangular						
Plain socket						
100 × 75 mm	62.49	70.57	1.04	24.64	m	**95.21**
Bracket; fixed to backgrounds. For minimum fixing distances, refer to the Tables and Memoranda at the rear of the book						
100 × 75 mm; build in holdabat	35.56	40.16	0.35	8.29	nr	**48.45**
100 × 75 mm; trefoil earband	28.33	32.00	0.29	6.86	nr	**38.86**
100 × 75 mm; plain earband	27.40	30.95	0.29	6.86	nr	**37.81**
Eared socket, with wall spacers measured separately						
100 × 75 mm	63.52	71.74	1.16	27.49	m	**99.23**
Wall spacer plate; eared pipework						
100 × 75	8.22	9.28	0.16	3.79	nr	**13.07**

R: DISPOSAL SYSTEMS

Item	Net Price £	Material £	Labour hours	Labour £	Unit	Total rate £
R10: RAINWATER PIPEWORK/GUTTERS – cont						
Extra over fittings Rectangular Cast Iron Pipework BS 460						
Loose socket						
100 × 75 mm; plain	26.50	29.93	0.23	5.46	nr	**35.39**
100 × 75 mm; eared	43.92	49.60	0.29	6.86	nr	**56.46**
Shoe; front						
100 × 75 mm; plain	70.30	79.40	0.23	5.46	nr	**84.86**
100 × 75 mm; eared	85.90	97.01	0.29	6.86	nr	**103.87**
Shoe; side						
100 × 75 mm; plain	85.29	96.32	0.23	5.46	nr	**101.78**
100 × 75 mm; eared	106.64	120.43	0.29	6.86	nr	**127.29**
Bends; side; any degree						
100 × 75 mm; plain	64.69	73.05	0.25	5.93	nr	**78.98**
100 × 75 mm; 135°; plain	66.08	74.62	0.25	5.93	nr	**80.55**
Bends; side; any degree						
100 × 75 mm; eared	81.79	92.37	0.25	5.93	nr	**98.30**
Bends; front; any degree						
100 × 75 mm; plain	61.26	69.19	0.25	5.93	nr	**75.12**
100 × 75 mm; eared	69.42	78.40	0.25	5.93	nr	**84.33**
Offset; side						
Plain socket						
75 mm projection	85.22	96.25	0.25	5.93	nr	**102.18**
115 mm projection	88.64	100.10	0.25	5.93	nr	**106.03**
225 mm projection	110.44	124.73	0.25	5.93	nr	**130.66**
305 mm projection	127.28	143.74	0.25	5.93	nr	**149.67**
Offset; front						
Plain socket						
75 mm projection	64.85	73.24	0.25	5.93	nr	**79.17**
150 mm projection	71.67	80.94	0.25	5.93	nr	**86.87**
225 mm projection	88.71	100.18	0.25	5.93	nr	**106.11**
305 mm projection	105.54	119.19	0.25	5.93	nr	**125.12**

R: DISPOSAL SYSTEMS

Item	Net Price £	Material £	Labour hours	Labour £	Unit	Total rate £
Eared socket						
75 mm projection	82.95	93.68	0.25	5.93	nr	99.61
150 mm projection	89.45	101.02	0.25	5.93	nr	106.95
225 mm projection	105.81	119.49	0.25	5.93	nr	125.42
305 mm projection	123.10	139.03	0.25	5.93	nr	144.96
Offset; plinth						
115 mm projection; plain	66.93	75.58	0.25	5.93	nr	81.51
115 mm projection; eared	86.19	97.34	0.25	5.93	nr	103.27
Rainwater heads						
Flat hopper						
210 × 160 × 185 mm; 65 mm outlet	50.92	57.51	0.40	9.48	nr	66.99
210 × 160 × 185 mm; 75 mm outlet	50.92	57.51	0.40	9.48	nr	66.99
250 × 215 × 215 mm; 100 mm outlet	60.35	68.16	0.40	9.48	nr	77.64
Flat rectangular						
225 × 125 × 125 mm; 65 mm outlet	70.80	79.96	0.40	9.48	nr	89.44
225 × 125 × 125 mm; 75 mm outlet	70.80	79.96	0.40	9.48	nr	89.44
280 × 150 × 130 mm; 100 mm outlet	97.74	110.38	0.40	9.48	nr	119.86
Rectangular						
250 × 180 × 175 mm; 75 mm outlet	66.01	74.55	0.40	9.48	nr	84.03
250 × 180 × 175 mm; 100 mm outlet	66.01	74.55	0.40	9.48	nr	84.03
300 × 250 × 200 mm; 65 mm outlet	91.94	103.84	0.40	9.48	nr	113.32
300 × 250 × 200 mm; 75 mm outlet	91.94	103.84	0.40	9.48	nr	113.32
300 × 250 × 200 mm; 100 mm outlet	91.94	103.84	0.40	9.48	nr	113.32
300 × 250 × 200 mm; 100 × 75 mm outlet	91.94	103.84	0.40	9.48	nr	113.32
Castellated rectangular						
250 × 180 × 175 mm; 65 mm outlet	66.01	74.55	0.40	9.48	nr	84.03

R: DISPOSAL SYSTEMS

Item	Net Price £	Material £	Labour hours	Labour £	Unit	Total rate £
R11: ABOVE GROUND DRAINAGE						
Pricing note: degree angles are only indicated where material prices differ						
PVC-u overflow pipe; solvent welded joints; fixed with clips to backgrounds						
Pipe, with brackets measured separately						
19 mm	1.01	1.14	0.21	4.98	m	**6.12**
Fixings						
Pipe clip: including fixing to backgrounds. For minimum fixing distances, refer to the Tables and Memoranda at the rear of the book						
19 mm	0.40	0.45	0.18	4.27	nr	**4.72**
Extra over fittings overflow pipework PVC-u						
Straight coupler						
19 mm	1.08	1.22	0.17	4.03	nr	**5.25**
Bend						
19 mm: 91.25°	1.27	1.44	0.17	4.03	nr	**5.47**
19 mm: 135°	1.28	1.45	0.17	4.03	nr	**5.48**
Tee						
19 mm	1.39	1.57	0.18	4.27	nr	**5.84**
Reverse nut connector						
19 mm	0.50	0.56	0.15	3.56	nr	**4.12**
BSP adaptor: solvent welded socket to threaded socket						
19 mm × ¾'	1.84	2.08	0.14	3.33	nr	**5.41**
Straight tank connector						
19 mm	1.68	1.90	0.21	4.98	nr	**6.88**
32 mm	0.82	0.93	0.28	6.63	nr	**7.56**
40 mm	0.82	0.93	0.30	7.10	nr	**8.03**
Bent tank connector						
19 mm	1.99	2.24	0.21	4.98	nr	**7.22**
Tundish						
19 mm	26.23	29.62	0.38	9.00	nr	**38.62**

R: DISPOSAL SYSTEMS

Item	Net Price £	Material £	Labour hours	Labour £	Unit	Total rate £
MuPVC waste pipe; solvent welded joints; fixed with clips to backgrounds; BS 5255						
Pipe, with brackets measured separately						
32 mm	1.60	1.81	0.23	5.46	m	**7.27**
40 mm	1.99	2.24	0.23	5.46	m	**7.70**
50 mm	2.99	3.38	0.26	6.15	m	**9.53**
Fixings						
Pipe clip: including fixing to backgrounds. For minimum fixing distances, refer to the Tables and Memoranda at the rear of the book						
32 mm	0.33	0.38	0.13	3.08	nr	**3.46**
40 mm	0.38	0.43	0.13	3.08	nr	**3.51**
50 mm	0.50	0.56	0.13	3.08	nr	**3.64**
Pipe clip: expansion: including fixing to backgrounds. For minimum fixing distances, refer to the Tables and Memoranda at the rear of the book						
32 mm	0.40	0.45	0.13	3.08	nr	**3.53**
40 mm	0.42	0.48	0.13	3.08	nr	**3.56**
50 mm	0.96	1.08	0.13	3.08	nr	**4.16**
Pipe clip: metal; including fixing to backgrounds. For minimum fixing distances, refer to the Tables and Memoranda at the rear of the book						
32 mm	1.43	1.61	0.13	3.08	nr	**4.69**
40 mm	1.71	1.93	0.13	3.08	nr	**5.01**
50 mm	2.17	2.45	0.13	3.08	nr	**5.53**
Extra over fittings waste pipework MuPVC						
Screwed access plug						
32 mm	0.93	1.05	0.18	4.27	nr	**5.32**
40 mm	0.93	1.05	0.18	4.27	nr	**5.32**
50 mm	1.34	1.51	0.25	5.93	nr	**7.44**
Straight coupling						
32 mm	1.00	1.13	0.27	6.40	nr	**7.53**
40 mm	1.00	1.13	0.27	6.40	nr	**7.53**
50 mm	1.82	2.06	0.27	6.40	nr	**8.46**
Expansion coupling						
32 mm	1.76	1.99	0.27	6.40	nr	**8.39**
40 mm	2.12	2.40	0.27	6.40	nr	**8.80**
50 mm	2.88	3.25	0.27	6.40	nr	**9.65**

R: DISPOSAL SYSTEMS

Item	Net Price £	Material £	Labour hours	Labour £	Unit	Total rate £
R11: ABOVE GROUND DRAINAGE – cont						
Extra over fittings waste pipework MuPVC – cont						
MuPVC to copper coupling						
32 mm	1.76	1.99	0.27	6.40	nr	**8.39**
40 mm	2.12	2.40	0.27	6.40	nr	**8.80**
50 mm	2.88	3.25	0.27	6.40	nr	**9.65**
Spigot and socket coupling						
32 mm	1.76	1.99	0.27	6.40	nr	**8.39**
40 mm	2.12	2.40	0.27	6.40	nr	**8.80**
50 mm	2.88	3.25	0.27	6.40	nr	**9.65**
Union						
32 mm	4.19	4.73	0.28	6.63	nr	**11.36**
40 mm	5.48	6.19	0.28	6.63	nr	**12.82**
50 mm	6.25	7.06	0.28	6.63	nr	**13.69**
Reducer: socket						
32 × 19 mm	1.51	1.70	0.27	6.40	nr	**8.10**
40 × 32 mm	1.00	1.13	0.27	6.40	nr	**7.53**
50 × 32 mm	1.44	1.62	0.27	6.40	nr	**8.02**
50 × 40 mm	1.76	1.99	0.27	6.40	nr	**8.39**
Reducer: level invert						
40 × 32 mm	1.25	1.41	0.27	6.40	nr	**7.81**
50 × 32 mm	1.54	1.74	0.27	6.40	nr	**8.14**
50 × 40 mm	1.54	1.74	0.27	6.40	nr	**8.14**
Swept bend						
32 mm	1.02	1.15	0.27	6.40	nr	**7.55**
32 mm: 165°	1.06	1.19	0.27	6.40	nr	**7.59**
40 mm	1.13	1.28	0.27	6.40	nr	**7.68**
40 mm: 165°	2.00	2.25	0.27	6.40	nr	**8.65**
50 mm	1.99	2.24	0.30	7.10	nr	**9.34**
50 mm: 165°	2.64	2.98	0.30	7.10	nr	**10.08**
Knuckle bend						
32 mm	0.93	1.05	0.27	6.40	nr	**7.45**
40 mm	1.03	1.16	0.27	6.40	nr	**7.56**
Spigot and socket bend						
32 mm	1.67	1.89	0.27	6.40	nr	**8.29**
32 mm: 150°	1.72	1.94	0.27	6.40	nr	**8.34**
40 mm	1.91	2.15	0.27	6.40	nr	**8.55**
50 mm	2.71	3.06	0.30	7.10	nr	**10.16**

R: DISPOSAL SYSTEMS

Item	Net Price £	Material £	Labour hours	Labour £	Unit	Total rate £
Swept tee						
32 mm: 91.25°	1.37	1.55	0.31	7.34	nr	**8.89**
32 mm: 135°	1.64	1.86	0.31	7.34	nr	**9.20**
40 mm: 91.25°	1.73	1.96	0.31	7.34	nr	**9.30**
40 mm: 135°	2.16	2.44	0.31	7.34	nr	**9.78**
50 mm	3.38	3.81	0.31	7.34	nr	**11.15**
Swept cross						
40 mm: 91.25°	4.21	4.75	0.31	7.34	nr	**12.09**
50 mm: 91.25°	4.41	4.98	0.43	10.19	nr	**15.17**
50 mm: 135°	5.56	6.28	0.31	7.34	nr	**13.62**
Male iron adaptor						
32 mm	1.52	1.71	0.28	6.63	nr	**8.34**
40 mm	1.78	2.01	0.28	6.63	nr	**8.64**
Female iron adaptor						
32 mm	1.78	2.01	0.28	6.63	nr	**8.64**
40 mm	1.78	2.01	0.28	6.63	nr	**8.64**
50 mm	2.57	2.91	0.31	7.34	nr	**10.25**
Reverse nut adaptor						
32 mm	2.28	2.57	0.20	4.74	nr	**7.31**
40 mm	2.28	2.57	0.20	4.74	nr	**7.31**
Automatic air admittance valve						
32 mm	12.76	14.41	0.27	6.40	nr	**20.81**
40 mm	12.76	14.41	0.28	6.63	nr	**21.04**
50 mm	12.76	14.41	0.31	7.34	nr	**21.75**
MuPVC to metal adpator: including heat shrunk joint to metal						
50 mm	5.49	6.20	0.38	9.00	nr	**15.20**
Caulking bush: including joint to metal						
32 mm	2.39	2.70	0.31	7.34	nr	**10.04**
40 mm	2.39	2.70	0.31	7.34	nr	**10.04**
50 mm	2.39	2.70	0.32	7.58	nr	**10.28**
Weathering apron						
50 mm	2.11	2.39	0.65	15.40	nr	**17.79**
Vent Cowl						
50 mm	2.40	2.71	0.19	4.51	nr	**7.22**

R: DISPOSAL SYSTEMS

Item	Net Price £	Material £	Labour hours	Labour £	Unit	Total rate £
R11: ABOVE GROUND DRAINAGE – cont						
ABS waste pipe; solvent welded joints; fixed with clips to backgrounds; BS 5255						
Pipe, with brackets measured separately						
32 mm	1.57	1.77	0.23	5.46	m	**7.23**
40 mm	1.96	2.21	0.23	5.46	m	**7.67**
50 mm	2.47	2.78	0.26	6.15	m	**8.93**
Fixings						
Pipe clip: including fixing to backgrounds. For minimum fixing distances, refer to the Tables and Memoranda at the rear of the book						
32 mm	0.25	0.29	0.17	4.03	nr	**4.32**
40 mm	0.32	0.36	0.17	4.03	nr	**4.39**
50 mm	0.95	1.07	0.17	4.03	nr	**5.10**
Pipe clip: expansion: including fixing to backgrounds. For minimum fixing distances, refer to the Tables and Memoranda at the rear of the book						
32 mm	0.25	0.29	0.17	4.03	nr	**4.32**
40 mm	0.32	0.36	0.17	4.03	nr	**4.39**
50 mm	0.95	1.07	0.17	4.03	nr	**5.10**
Pipe clip: metal; including fixing to backgrounds. For minimum fixing distances, refer to the Tables and Memoranda at the rear of the book						
32 mm	1.43	1.61	0.17	4.03	nr	**5.64**
40 mm	1.71	1.93	0.17	4.03	nr	**5.96**
50 mm	2.17	2.45	0.17	4.03	nr	**6.48**
Extra over fittings waste pipework ABS						
Screwed access plug						
32 mm	0.98	1.11	0.18	4.27	nr	**5.38**
40 mm	0.98	1.11	0.18	4.27	nr	**5.38**
50 mm	2.03	2.30	0.25	5.93	nr	**8.23**
Straight coupling						
32 mm	0.98	1.11	0.27	6.40	nr	**7.51**
40 mm	0.98	1.11	0.27	6.40	nr	**7.51**
50 mm	2.03	2.30	0.27	6.40	nr	**8.70**

R: DISPOSAL SYSTEMS

Item	Net Price £	Material £	Labour hours	Labour £	Unit	Total rate £
Expansion coupling						
32 mm	2.03	2.30	0.27	6.40	nr	**8.70**
40 mm	2.03	2.30	0.27	6.40	nr	**8.70**
50 mm	3.10	3.50	0.27	6.40	nr	**9.90**
ABS to copper coupling						
32 mm	2.03	2.30	0.27	6.40	nr	**8.70**
40 mm	2.03	2.30	0.27	6.40	nr	**8.70**
50 mm	3.10	3.50	0.27	6.40	nr	**9.90**
Reducer: socket						
40 × 32 mm	0.98	1.11	0.27	6.40	nr	**7.51**
50 × 32 mm	2.23	2.52	0.27	6.40	nr	**8.92**
50 × 40 mm	2.23	2.52	0.27	6.40	nr	**8.92**
Swept bend						
32 mm	0.98	1.11	0.27	6.40	nr	**7.51**
40 mm	0.98	1.11	0.27	6.40	nr	**7.51**
50 mm	2.03	2.30	0.30	7.10	nr	**9.40**
Knuckle bend						
32 mm	0.98	1.11	0.27	6.40	nr	**7.51**
40 mm	0.98	1.11	0.27	6.40	nr	**7.51**
Swept tee						
32 mm	1.41	1.59	0.31	7.34	nr	**8.93**
40 mm	1.41	1.59	0.31	7.34	nr	**8.93**
50 mm	3.71	4.19	0.31	7.34	nr	**11.53**
Swept cross						
40 mm	4.73	5.34	0.23	5.46	nr	**10.80**
50 mm	5.40	6.10	0.43	10.19	nr	**16.29**
Male iron adaptor						
32 mm	2.03	2.30	0.28	6.63	nr	**8.93**
40 mm	2.03	2.30	0.28	6.63	nr	**8.93**
Female iron adapator						
32 mm	2.03	2.30	0.28	6.63	nr	**8.93**
40 mm	2.03	2.30	0.28	6.63	nr	**8.93**
50 mm	3.03	3.42	0.31	7.34	nr	**10.76**
Tank connectors						
32 mm	1.41	1.59	0.29	6.86	nr	**8.45**
40 mm	1.54	1.74	0.29	6.86	nr	**8.60**
Caulking bush: including joint to pipework						
50 mm	2.39	2.70	0.50	11.85	nr	**14.55**

R: DISPOSAL SYSTEMS

Item	Net Price £	Material £	Labour hours	Labour £	Unit	Total rate £
R11: ABOVE GROUND DRAINAGE – cont						
Fixings						
Polypropylene waste pipe; push fit joints; fixed with clips to backgrounds; BS 5254						
Pipe, with brackets measured separately						
32 mm	0.86	0.97	0.21	4.98	m	**5.95**
40 mm	1.07	1.20	0.21	4.98	m	**6.18**
50 mm	1.77	2.00	0.38	9.00	m	**11.00**
Pipe clip: saddle; including fixing to backgrounds. For minimum fixing distances, refer to the Tables and Memoranda at the rear of the book						
32 mm	0.29	0.33	0.17	4.03	nr	**4.36**
40 mm	0.29	0.33	0.17	4.03	nr	**4.36**
Pipe clip: including fixing to backgrounds. For minimum fixing distances, refer to the Tables and Memoranda at the rear of the book						
50 mm	0.70	0.80	0.17	4.03	nr	**4.83**
Extra over fittings waste pipework polypropylene						
Screwed access plug						
32 mm	0.82	0.93	0.16	3.79	nr	**4.72**
40 mm	0.82	0.93	0.16	3.79	nr	**4.72**
50 mm	1.41	1.59	0.20	4.74	nr	**6.33**
Straight coupling						
32 mm	0.82	0.93	0.19	4.51	nr	**5.44**
40 mm	0.82	0.93	0.19	4.51	nr	**5.44**
50 mm	1.41	1.59	0.20	4.74	nr	**6.33**
Universal waste pipe coupler						
32 mm dia.	1.40	1.58	0.20	4.74	nr	**6.32**
40 mm dia.	1.58	1.79	0.20	4.74	nr	**6.53**
Reducer						
40 × 32 mm	1.41	1.59	0.19	4.51	nr	**6.10**
50 × 32 mm	1.47	1.66	0.19	4.51	nr	**6.17**
50 × 40 mm	1.52	1.71	0.20	4.74	nr	**6.45**
Swept bend						
32 mm	0.82	0.93	0.19	4.51	nr	**5.44**
40 mm	0.82	0.93	0.19	4.51	nr	**5.44**
50 mm	1.41	1.59	0.20	4.74	nr	**6.33**

R: DISPOSAL SYSTEMS

Item	Net Price £	Material £	Labour hours	Labour £	Unit	Total rate £
Knuckle bend						
32 mm	0.82	0.93	0.19	4.51	nr	**5.44**
40 mm	0.82	0.93	0.19	4.51	nr	**5.44**
50 mm	1.41	1.59	0.20	4.74	nr	**6.33**
Spigot and socket bend						
32 mm	0.82	0.93	0.19	4.51	nr	**5.44**
40 mm	0.82	0.93	0.19	4.51	nr	**5.44**
Swept tee						
32 mm	0.90	1.02	0.22	5.22	nr	**6.24**
40 mm	0.90	1.02	0.22	5.22	nr	**6.24**
50 mm	1.52	1.71	0.23	5.46	nr	**7.17**
Male iron adaptor						
32 mm	0.82	0.93	0.13	3.08	nr	**4.01**
40 mm	0.82	0.93	0.19	4.51	nr	**5.44**
50 mm	1.41	1.59	0.15	3.56	nr	**5.15**
Tank connector						
32 mm	0.82	0.93	0.24	5.69	nr	**6.62**
40 mm	0.82	0.93	0.24	5.69	nr	**6.62**
50 mm	1.41	1.59	0.35	8.29	nr	**9.88**
Polypropylene traps; including fixing to appliance and connection to pipework; BS 3943						
Tubular P trap; 75 mm seal						
32 mm dia.	3.34	3.77	0.20	4.74	nr	**8.51**
40 mm dia.	3.86	4.36	0.20	4.74	nr	**9.10**
Tubular S trap; 75 mm seal						
32 mm dia.	4.22	4.76	0.20	4.74	nr	**9.50**
40 mm dia.	4.96	5.60	0.20	4.74	nr	**10.34**
Running tubular P trap; 75 mm seal						
32 mm dia.	5.13	5.79	0.20	4.74	nr	**10.53**
40 mm dia.	5.60	6.32	0.20	4.74	nr	**11.06**
Running tubular S trap; 75 mm seal						
32 mm dia.	6.16	6.96	0.20	4.74	nr	**11.70**
40 mm dia.	6.65	7.51	0.20	4.74	nr	**12.25**
Spigot and socket bend; converter from P to S Trap						
32 mm	1.31	1.48	0.20	4.74	nr	**6.22**
40 mm	1.41	1.59	0.21	4.98	nr	**6.57**

R: DISPOSAL SYSTEMS

Item	Net Price £	Material £	Labour hours	Labour £	Unit	Total rate £
R11: ABOVE GROUND DRAINAGE – cont						
Polypropylene traps – cont						
Bottle P trap; 75 mm seal						
32 mm dia.	3.72	4.20	0.20	4.74	nr	**8.94**
40 mm dia.	4.45	5.03	0.20	4.74	nr	**9.77**
Bottle S trap; 75 mm seal						
32 mm dia.	4.49	5.07	0.20	4.74	nr	**9.81**
40 mm dia.	5.46	6.17	0.25	5.93	nr	**12.10**
Bottle P trap; resealing; 75 mm seal						
32 mm dia.	4.63	5.23	0.20	4.74	nr	**9.97**
40 mm dia.	5.42	6.12	0.25	5.93	nr	**12.05**
Bottle S trap; resealing; 75 mm seal						
32 mm dia.	5.30	5.99	0.20	4.74	nr	**10.73**
40 mm dia.	6.14	6.94	0.25	5.93	nr	**12.87**
Bath trap, low level; 38 mm seal						
40 mm dia.	4.64	5.24	0.25	5.93	nr	**11.17**
Bath trap, low level; 38 mm seal complete with overflow hose						
40 mm dia.	7.18	8.11	0.25	5.93	nr	**14.04**
Bath trap; 75 mm seal complete with overlow hose						
40 mm dia.	7.15	8.08	0.25	5.93	nr	**14.01**
Bath trap; 75 mm seal complete with overflow hose and overflow outlet						
40 mm dia.	12.20	13.78	0.20	4.74	nr	**18.52**
Bath trap; 75 mm seal complete with overflow hose, overflow outlet and ABS chrome waste						
40 mm dia.	16.87	19.05	0.20	4.74	nr	**23.79**
Washing machine trap; 75 mm seal including stand pipe						
40 mm dia.	9.71	10.97	0.25	5.93	nr	**16.90**
Washing machine standpipe						
40 mm dia.	5.07	5.72	0.25	5.93	nr	**11.65**
Plastic unslotted chrome plated basin/sink waste including plug						
32 mm	6.10	6.88	0.34	8.05	nr	**14.93**
40 mm	8.16	9.22	0.34	8.05	nr	**17.27**

R: DISPOSAL SYSTEMS

Item	Net Price £	Material £	Labour hours	Labour £	Unit	Total rate £
Plastic slotted chrome plated basin/sink waste including plug						
32 mm	4.84	5.47	0.34	8.05	nr	**13.52**
40 mm	8.10	9.15	0.34	8.05	nr	**17.20**
Bath overflow outlet; plastic; white						
42 mm	4.24	4.78	0.37	8.76	nr	**13.54**
Bath overlow outlet; plastic; chrome plated						
42 mm	5.62	6.34	0.37	8.76	nr	**15.10**
Combined cistern and bath overflow outlet; plastic; white						
42 mm	8.73	9.86	0.39	9.24	nr	**19.10**
Combined cistern and bath overlow outlet; plastic; chrome plated						
42 mm	8.72	9.85	0.39	9.24	nr	**19.09**
Cistern overflow outlet; plastic; white						
42 mm	7.03	7.94	0.15	3.56	nr	**11.50**
Cistern overlow outlet; plastic; chrome plated						
42 mm	6.34	7.16	0.15	3.56	nr	**10.72**
PVC-u soil and waste pipe; solvent welded joints; fixed with clips to backgrounds; BS 4514/ BS EN 607						
Pipe, with brackets measured separately						
82 mm	7.27	8.21	0.35	8.29	m	**16.50**
110 mm	7.40	8.35	0.41	9.71	m	**18.06**
160 mm	19.20	21.69	0.51	12.09	m	**33.78**
Fixings						
Galvanized steel pipe clip: including fixing to backgrounds. For minimum fixing distances, refer to the Tables and Memoranda at the rear of the book						
82 mm	3.01	3.40	0.18	4.27	nr	**7.67**
110 mm	3.13	3.54	0.18	4.27	nr	**7.81**
160 mm	7.53	8.51	0.18	4.27	nr	**12.78**

R: DISPOSAL SYSTEMS

Item	Net Price £	Material £	Labour hours	Labour £	Unit	Total rate £
R11: ABOVE GROUND DRAINAGE – cont						
Fixings – cont						
Plastic coated steel pipe clip: including fixing to backgrounds. For minimum fixing distances, refer to the Tables and Memoranda at the rear of the book						
82 mm	4.22	4.76	0.18	4.27	nr	**9.03**
110 mm	5.75	6.50	0.18	4.27	nr	**10.77**
160 mm	7.26	8.20	0.18	4.27	nr	**12.47**
Plastic pipe clip: including fixing to backgrounds. For minimum fixing distances, refer to the Tables and Memoranda at the rear of the book						
82 mm	2.23	2.52	0.18	4.27	nr	**6.79**
110 mm	4.29	4.85	0.18	4.27	nr	**9.12**
Plastic coated steel pipe clip: adjustable; including fixing to backgrounds. For minimum fixing distances, refer to the Tables and Memoranda at the rear of the book						
82 mm	3.41	3.86	0.20	4.74	nr	**8.60**
110 mm	4.25	4.80	0.20	4.74	nr	**9.54**
Galvanized steel pipe clip: drive in; including fixing to backgrounds. For minimum fixing distances, refer to the Tables and Memoranda at the rear of the book						
110 mm	6.50	7.34	0.22	5.22	nr	**12.56**
Extra over fittings solvent welded pipework PVC-u						
Straight coupling						
82 mm	3.86	4.36	0.21	4.98	nr	**9.34**
110 mm	4.82	5.45	0.22	5.22	nr	**10.67**
160 mm	13.90	15.70	0.24	5.69	nr	**21.39**
Expansion coupling						
82 mm	5.77	6.52	0.21	4.98	nr	**11.50**
110 mm	5.90	6.66	0.22	5.22	nr	**11.88**
160 mm	17.74	20.03	0.24	5.69	nr	**25.72**
Slip coupling; double ring socket						
82 mm	11.98	13.53	0.21	4.98	nr	**18.51**
110 mm	15.00	16.94	0.22	5.22	nr	**22.16**
160 mm	23.82	26.90	0.24	5.69	nr	**32.59**

R: DISPOSAL SYSTEMS

Item	Net Price £	Material £	Labour hours	Labour £	Unit	Total rate £
Puddle flanges						
110 mm	117.06	132.20	0.45	10.66	nr	**142.86**
160 mm	189.22	213.70	0.55	13.04	nr	**226.74**
Socket reducer						
82 to 50 mm	5.86	6.62	0.18	4.27	nr	**10.89**
110 to 50 mm	7.31	8.25	0.18	4.27	nr	**12.52**
110 to 82 mm	7.53	8.51	0.22	5.22	nr	**13.73**
160 to 110 mm	15.26	17.24	0.26	6.15	nr	**23.39**
Socket plugs						
82 mm	4.54	5.13	0.15	3.56	nr	**8.69**
110 mm	6.65	7.51	0.20	4.74	nr	**12.25**
160 mm	12.22	13.80	0.27	6.40	nr	**20.20**
Access door; including cutting into pipe						
82 mm	12.88	14.55	0.28	6.63	nr	**21.18**
110 mm	12.88	14.55	0.34	8.05	nr	**22.60**
160 mm	23.01	25.99	0.46	10.89	nr	**36.88**
Screwed access cap						
82 mm	9.12	10.30	0.15	3.56	nr	**13.86**
110 mm	10.75	12.14	0.20	4.74	nr	**16.88**
160 mm	20.23	22.85	0.27	6.40	nr	**29.25**
Access pipe: spigot and socket						
110 mm	15.89	17.94	0.22	5.22	nr	**23.16**
Access pipe: double socket						
110 mm	15.89	17.94	0.22	5.22	nr	**23.16**
Swept bend						
82 mm	9.67	10.92	0.29	6.86	nr	**17.78**
110 mm	11.33	12.80	0.32	7.58	nr	**20.38**
160 mm	28.21	31.86	0.49	11.61	nr	**43.47**
Bend; special angle						
82 mm	18.63	21.04	0.29	6.86	nr	**27.90**
110 mm	22.22	25.09	0.32	7.58	nr	**32.67**
160 mm	37.51	42.36	0.49	11.61	nr	**53.97**
Spigot and socket bend						
82 mm	9.36	10.57	0.26	6.15	nr	**16.72**
110 mm	10.98	12.40	0.32	7.58	nr	**19.98**
110 mm: 135°	12.32	13.91	0.32	7.58	nr	**21.49**
160 mm: 135°	27.28	30.81	0.44	10.42	nr	**41.23**
Variable bend: single socket						
110 mm	21.23	23.98	0.33	7.81	nr	**31.79**

R: DISPOSAL SYSTEMS

Item	Net Price £	Material £	Labour hours	Labour £	Unit	Total rate £
R11: ABOVE GROUND DRAINAGE – cont						
Extra over fittings solvent welded pipework PVC-u – cont						
Variable bend: double socket						
110 mm	21.18	23.92	0.33	7.81	nr	**31.73**
Access bend						
110 mm	31.41	35.48	0.33	7.81	nr	**43.29**
Single branch: two bosses						
82 mm	13.52	15.27	0.35	8.29	nr	**23.56**
82 mm: 104°	14.42	16.29	0.35	8.29	nr	**24.58**
110 mm	14.98	16.92	0.42	9.95	nr	**26.87**
110 mm: 135°	15.62	17.65	0.42	9.95	nr	**27.60**
160 mm	31.80	35.91	0.50	11.85	nr	**47.76**
160 mm: 135°	32.53	36.74	0.50	11.85	nr	**48.59**
Single branch; four bosses						
110 mm	18.24	20.60	0.42	9.95	nr	**30.55**
Single access branch						
82 mm	43.83	49.50	0.35	8.29	nr	**57.79**
110 mm	25.65	28.97	0.42	9.95	nr	**38.92**
Unequal single branch						
160 × 160 × 110 mm	35.91	40.56	0.50	11.85	nr	**52.41**
160 × 160 × 110 mm: 135°	38.59	43.58	0.50	11.85	nr	**55.43**
Double branch						
110 mm	38.70	43.71	0.42	9.95	nr	**53.66**
110 mm: 135°	37.02	41.81	0.42	9.95	nr	**51.76**
Corner branch						
110 mm	65.39	73.85	0.42	9.95	nr	**83.80**
Unequal double branch						
160 × 160 × 110 mm	66.88	75.53	0.50	11.85	nr	**87.38**
Single boss pipe; single socket						
110 × 110 × 32 mm	4.05	4.57	0.24	5.69	nr	**10.26**
110 × 110 × 40 mm	4.05	4.57	0.24	5.69	nr	**10.26**
110 × 110 × 50 mm	4.26	4.81	0.24	5.69	nr	**10.50**
Single boss pipe; triple socket						
110 × 110 × 40 mm	6.57	7.42	0.24	5.69	nr	**13.11**

R: DISPOSAL SYSTEMS

Item	Net Price £	Material £	Labour hours	Labour £	Unit	Total rate £
Waste boss; including cutting into pipe						
82 to 32 mm	5.29	5.98	0.29	6.86	nr	**12.84**
82 to 40 mm	5.29	5.98	0.29	6.86	nr	**12.84**
110 to 32 mm	5.29	5.98	0.29	6.86	nr	**12.84**
110 to 40 mm	5.29	5.98	0.29	6.86	nr	**12.84**
110 to 50 mm	5.48	6.19	0.29	6.86	nr	**13.05**
160 to 32 mm	7.49	8.46	0.30	7.10	nr	**15.56**
160 to 40 mm	7.49	8.46	0.35	8.29	nr	**16.75**
160 to 50 mm	7.49	8.46	0.40	9.48	nr	**17.94**
Self locking waste boss; including cutting into pipe						
110 to 32 mm	7.07	7.99	0.30	7.10	nr	**15.09**
110 to 40 mm	7.38	8.33	0.30	7.10	nr	**15.43**
110 to 50 mm	9.54	10.77	0.30	7.10	nr	**17.87**
Adaptor saddle; including cutting to pipe						
82 to 32 mm	3.53	3.99	0.29	6.86	nr	**10.85**
110 to 40 mm	4.35	4.92	0.29	6.86	nr	**11.78**
160 to 50 mm	7.88	8.90	0.29	6.86	nr	**15.76**
Branch boss adaptor						
32 mm	2.10	2.38	0.26	6.15	nr	**8.53**
40 mm	2.10	2.38	0.26	6.15	nr	**8.53**
50 mm	3.00	3.39	0.26	6.15	nr	**9.54**
Branch boss adaptor bend						
32 mm	2.88	3.25	0.26	6.15	nr	**9.40**
40 mm	3.14	3.55	0.26	6.15	nr	**9.70**
50 mm	3.76	4.24	0.26	6.15	nr	**10.39**
Automatic air admittance valve						
82 to 110 mm	33.01	37.28	0.19	4.51	nr	**41.79**
PVC-u to metal adpator: including heat shrunk joint to metal						
110 mm	9.19	10.38	0.57	13.50	nr	**23.88**
Caulking bush: including joint to pipework						
82 mm	9.05	10.22	0.46	10.89	nr	**21.11**
110 mm	9.05	10.22	0.46	10.89	nr	**21.11**
Vent cowl						
82 mm	2.72	3.07	0.13	3.08	nr	**6.15**
110 mm	2.75	3.10	0.13	3.08	nr	**6.18**
160 mm	7.21	8.14	0.13	3.08	nr	**11.22**

R: DISPOSAL SYSTEMS

Item	Net Price £	Material £	Labour hours	Labour £	Unit	Total rate £
R11: ABOVE GROUND DRAINAGE – cont						
Extra over fittings solvent welded pipework PVC-u – cont						
Weathering apron; to lead slates						
82 mm	2.72	3.07	1.15	27.25	nr	**30.32**
110 mm	3.13	3.54	1.15	27.25	nr	**30.79**
160 mm	9.41	10.63	1.15	27.25	nr	**37.88**
Weathering apron; to asphalt						
82 mm	11.50	12.98	1.10	26.07	nr	**39.05**
110 mm	11.50	12.98	1.10	26.07	nr	**39.05**
Weathering slate; flat; 406 × 406 mm						
82 mm	32.41	36.61	1.04	24.64	nr	**61.25**
110 mm	32.41	36.61	1.04	24.64	nr	**61.25**
Weathering slate; flat; 457 × 457 mm						
82 mm	33.25	37.56	1.04	24.64	nr	**62.20**
110 mm	33.25	37.56	1.04	24.64	nr	**62.20**
Weathering slate; angled; 610 × 610 mm						
82 mm	44.93	50.75	1.04	24.64	nr	**75.39**
110 mm	44.93	50.75	1.04	24.64	nr	**75.39**
PVC-u soil and waste pipe; ring seal joints; fixed with clips to backgrounds; BS 4514/ BS EN 607						
Pipe, with brackets measured separately						
82 mm dia.	6.41	7.24	0.35	8.29	m	**15.53**
110 mm dia.	6.45	7.28	0.41	9.71	m	**16.99**
160 mm dia.	23.14	26.13	0.51	12.09	m	**38.22**
Fixings						
Galvanized steel pipe clip: including fixing to backgrounds. For minimum fixing distances, refer to the Tables and Memoranda at the rear of the book						
82 mm	3.01	3.40	0.18	4.27	nr	**7.67**
110 mm	3.13	3.54	0.18	4.27	nr	**7.81**
160 mm	7.53	8.51	0.18	4.27	nr	**12.78**

R: DISPOSAL SYSTEMS

Item	Net Price £	Material £	Labour hours	Labour £	Unit	Total rate £
Plastic coated steel pipe clip: including fixing to backgrounds. For minimum fixing distances, refer to the Tables and Memoranda at the rear of the book						
82 mm	4.22	4.76	0.18	4.27	nr	**9.03**
110 mm	5.75	6.50	0.18	4.27	nr	**10.77**
160 mm	7.26	8.20	0.18	4.27	nr	**12.47**
Plastic pipe clip: including fixing to backgrounds. For minimum fixing distances, refer to the Tables and Memoranda at the rear of the book						
82 mm	2.23	2.52	0.18	4.27	nr	**6.79**
110 mm	4.29	4.85	0.18	4.27	nr	**9.12**
Plastic coated steel pipe clip: adjustable; including fixing to backgrounds. For minimum fixing distances, refer to the Tables and Memoranda at the rear of the book						
82 mm	3.41	3.86	0.20	4.74	nr	**8.60**
110 mm	4.25	4.80	0.20	4.74	nr	**9.54**
Galvanized steel pipe clip: drive in; including fixing to backgrounds. For minimum fixing distances, refer to the Tables and Memoranda at the rear of the book						
110 mm	6.50	7.34	0.22	5.22	nr	**12.56**
Extra over fittings ring seal pipework PVC-u						
Straight coupling						
82 mm	4.20	4.74	0.21	4.98	nr	**9.72**
110 mm	4.71	5.32	0.22	5.22	nr	**10.54**
160 mm	15.89	17.94	0.24	5.69	nr	**23.63**
Straight coupling; double socket						
82 mm	7.40	8.35	0.21	4.98	nr	**13.33**
110 mm	6.72	7.59	0.22	5.22	nr	**12.81**
160 mm	15.89	17.94	0.24	5.69	nr	**23.63**
Reducer; socket						
82 to 50 mm	6.91	7.80	0.15	3.56	nr	**11.36**
110 to 50 mm	9.63	10.87	0.15	3.56	nr	**14.43**
110 to 82 mm	9.63	10.87	0.19	4.51	nr	**15.38**
160 to 110 mm	13.88	15.68	0.31	7.34	nr	**23.02**

R: DISPOSAL SYSTEMS

Item	Net Price £	Material £	Labour hours	Labour £	Unit	Total rate £
R11: ABOVE GROUND DRAINAGE – cont						
Extra over fittings ring seal pipework PVC-u – cont						
Access Cap						
82 mm	7.52	8.50	0.15	3.56	nr	**12.06**
110 mm	8.86	10.01	0.17	4.03	nr	**14.04**
Access Cap; pressure plug						
160 mm	20.82	23.51	0.33	7.81	nr	**31.32**
Access pipe						
82 mm	18.45	20.84	0.22	5.22	nr	**26.06**
110 mm	18.41	20.79	0.22	5.22	nr	**26.01**
160 mm	38.67	43.68	0.24	5.69	nr	**49.37**
Bend						
82 mm	9.38	10.60	0.29	6.86	nr	**17.46**
82 mm; adjustable radius	21.16	23.90	0.29	6.86	nr	**30.76**
110 mm	10.10	11.40	0.32	7.58	nr	**18.98**
110 mm; adjustable radius	18.84	21.28	0.32	7.58	nr	**28.86**
160 mm	30.31	34.23	0.49	11.61	nr	**45.84**
160 mm; adjustable radius	39.76	44.90	0.49	11.61	nr	**56.51**
Bend; spigot and socket						
110 mm	10.19	11.51	0.32	7.58	nr	**19.09**
Bend; offset						
82 mm	8.26	9.33	0.21	4.98	nr	**14.31**
110 mm	13.25	14.96	0.32	7.58	nr	**22.54**
160 mm	29.23	33.01	0.32	7.58	nr	**40.59**
Bend; access						
110 mm	27.80	31.40	0.33	7.81	nr	**39.21**
Single branch						
82 mm	14.73	16.64	0.35	8.29	nr	**24.93**
110 mm	13.61	15.37	0.42	9.95	nr	**25.32**
110 mm; 45°	14.01	15.82	0.31	7.34	nr	**23.16**
160 mm	35.12	39.67	0.50	11.85	nr	**51.52**
Single branch; access						
82 mm	22.07	24.93	0.35	8.29	nr	**33.22**
110 mm	31.96	36.10	0.42	9.95	nr	**46.05**
Unequal single branch						
160 × 160 × 110 mm	35.90	40.54	0.50	11.85	nr	**52.39**
160 × 160 × 110 mm; 45°	38.59	43.58	0.50	11.85	nr	**55.43**

R: DISPOSAL SYSTEMS

Item	Net Price £	Material £	Labour hours	Labour £	Unit	Total rate £
Double branch; 4 bosses						
110 mm	35.66	40.27	0.49	11.61	nr	**51.88**
Corner branch; 2 bosses						
110 mm	65.38	73.84	0.49	11.61	nr	**85.45**
Multibranch; 4 bosses						
110 mm	19.52	22.04	0.52	12.32	nr	**34.36**
Boss Branch						
110 × 32 mm	4.04	4.56	0.34	8.05	nr	**12.61**
110 × 40 mm	4.04	4.56	0.34	8.05	nr	**12.61**
Strap on boss						
110 × 32 mm	4.50	5.08	0.30	7.10	nr	**12.18**
110 × 40 mm	4.50	5.08	0.30	7.10	nr	**12.18**
110 × 50 mm	4.53	5.12	0.30	7.10	nr	**12.22**
Patch boss						
82 × 32 mm	5.29	5.98	0.31	7.34	nr	**13.32**
82 × 40 mm	5.29	5.98	0.31	7.34	nr	**13.32**
82 × 50 mm	5.29	5.98	0.31	7.34	nr	**13.32**
Boss Pipe; collar 4 boss						
110 mm	6.57	7.42	0.35	8.29	nr	**15.71**
Boss adaptor; rubber; push fit						
32 mm	2.46	2.77	0.26	6.15	nr	**8.92**
40 mm	2.46	2.77	0.26	6.15	nr	**8.92**
50 mm	2.78	3.14	0.26	6.15	nr	**9.29**
WC connector; cap and seal; solvent socket						
110 mm	6.84	7.72	0.23	5.46	nr	**13.18**
110 mm; 90°	11.00	12.42	0.27	6.40	nr	**18.82**
Vent terminal						
82 mm	2.72	3.07	0.13	3.08	nr	**6.15**
110 mm	2.75	3.10	0.13	3.08	nr	**6.18**
160 mm	7.22	8.15	0.13	3.08	nr	**11.23**
Weathering slate; inclined; 610 × 610 mm						
82 mm	44.93	50.75	1.04	24.64	nr	**75.39**
110 mm	44.93	50.75	1.04	24.64	nr	**75.39**
Weathering slate; inclined; 450 × 450 mm						
82 mm	33.25	37.56	1.04	24.64	nr	**62.20**
110 mm	33.25	37.56	1.04	24.64	nr	**62.20**

R: DISPOSAL SYSTEMS

Item	Net Price £	Material £	Labour hours	Labour £	Unit	Total rate £
R11: ABOVE GROUND DRAINAGE – cont						
Extra over fittings ring seal pipework PVC-u – cont						
Weathering slate; flat; 400 × 400 mm						
82 mm	32.43	36.63	1.04	24.64	nr	**61.27**
110 mm	32.43	36.63	1.04	24.64	nr	**61.27**
Air admittance valve						
82 mm	33.01	37.28	0.19	4.51	nr	**41.79**
110 mm	33.01	37.28	0.19	4.51	nr	**41.79**
Cast iron pipe; nitrile rubber gasket joint with continuity clip BSEN 877; fixed vertically to backgrounds						
Pipe, with brackets and jointing couplings measured separately						
50 mm	16.47	18.60	0.31	7.34	m	**25.94**
75 mm	19.25	21.74	0.34	8.05	m	**29.79**
100 mm	21.25	24.00	0.37	8.76	m	**32.76**
150 mm	44.37	50.11	0.60	14.22	m	**64.33**
Fixings						
Brackets; fixed to backgrounds. For minimum fixing distances, refer to the Tables and Memoranda at the rear of the book						
50 mm	5.97	6.74	0.15	3.56	nr	**10.30**
75 mm	6.53	7.37	0.18	4.27	nr	**11.64**
100 mm	6.70	7.57	0.18	4.27	nr	**11.84**
150 mm	12.41	14.01	0.20	4.74	nr	**18.75**
Extra over fittings nitrile gasket cast iron pipework BS 416/6087, with jointing couplings measured separately						
Standard coupling						
50 mm	7.91	8.94	0.17	4.03	nr	**12.97**
75 mm	8.72	9.85	0.17	4.03	nr	**13.88**
100 mm	11.40	12.87	0.17	4.03	nr	**16.90**
150 mm	22.75	25.69	0.17	4.03	nr	**29.72**
Conversion coupling						
65 × 75 mm	9.24	10.43	0.60	14.22	nr	**24.65**
70 × 75 mm	9.24	10.43	0.60	14.22	nr	**24.65**
90 × 100 mm	11.91	13.45	0.67	15.88	nr	**29.33**

R: DISPOSAL SYSTEMS

Item	Net Price £	Material £	Labour hours	Labour £	Unit	Total rate £
Access pipe; round door						
50 mm	23.31	26.33	0.41	9.71	nr	**36.04**
75 mm	33.51	37.84	0.46	10.89	nr	**48.73**
100 mm	35.22	39.78	0.67	15.90	nr	**55.68**
150 mm	58.59	66.17	0.83	19.69	nr	**85.86**
Access pipe; square door						
100 mm	69.40	78.38	0.67	15.90	nr	**94.28**
150 mm	106.23	119.97	0.83	19.69	nr	**139.66**
Taper reducer						
75 mm	19.16	21.63	0.60	14.22	nr	**35.85**
100 mm	24.31	27.46	0.67	15.88	nr	**43.34**
150 mm	47.34	53.47	0.83	19.69	nr	**73.16**
Blank cap						
50 mm	4.84	5.47	0.24	5.69	nr	**11.16**
75 mm	6.04	6.82	0.26	6.15	nr	**12.97**
100 mm	5.86	6.62	0.32	7.58	nr	**14.20**
150 mm	8.47	9.57	0.40	9.48	nr	**19.05**
Blank cap; 50 mm screwed tapping						
75 mm	11.90	13.44	0.26	6.15	nr	**19.59**
100 mm	13.81	15.60	0.32	7.58	nr	**23.18**
150 mm	15.34	17.33	0.40	9.48	nr	**26.81**
Universal connector						
50 × 56/48/40 mm	7.19	8.12	0.33	7.81	nr	**15.93**
Change piece; BS416						
100 mm	15.34	17.33	0.47	11.14	nr	**28.47**
WC connector						
100 mm	24.68	27.88	0.49	11.61	nr	**39.49**
Boss pipe; 2" BSPT socket						
50 mm	19.71	22.26	0.58	13.75	nr	**36.01**
75 mm	28.84	32.57	0.65	15.40	nr	**47.97**
100 mm	34.44	38.89	0.79	18.72	nr	**57.61**
150 mm	53.66	60.61	0.86	20.37	nr	**80.98**
Boss pipe; 2" BSPT socket; 135°						
100 mm	41.49	46.86	0.79	18.72	nr	**65.58**
Boss pipe; 2 × 2" BSPT socket; opposed						
75 mm	38.19	43.14	0.65	15.40	nr	**58.54**
100 mm	44.54	50.31	0.79	18.72	nr	**69.03**

R: DISPOSAL SYSTEMS

Item	Net Price £	Material £	Labour hours	Labour £	Unit	Total rate £
R11: ABOVE GROUND DRAINAGE – cont						
Fixings – cont						
Boss pipe; 2 × 2" BSPT socket; in line						
100 mm	44.98	50.80	0.79	18.72	nr	**69.52**
Boss pipe; 2 × 2" BSPT socket; 90°						
100 mm	44.54	50.31	0.79	18.72	nr	**69.03**
Bend; short radius						
50 mm	13.92	15.72	0.50	11.85	nr	**27.57**
75 mm	15.77	17.81	0.60	14.22	nr	**32.03**
100 mm	19.27	21.77	0.67	15.88	nr	**37.65**
100 mm; 11°	16.62	18.77	0.67	15.88	nr	**34.65**
100 mm; 67°	19.27	21.77	0.67	15.88	nr	**37.65**
150 mm	34.44	38.89	0.83	19.69	nr	**58.58**
Access bend; short radius						
50 mm	34.32	38.76	0.50	11.85	nr	**50.61**
75 mm	37.24	42.05	0.60	14.22	nr	**56.27**
100 mm	40.79	46.06	0.67	15.88	nr	**61.94**
100 mm; 45°	40.79	46.06	0.67	15.88	nr	**61.94**
150 mm	57.91	65.40	0.83	19.69	nr	**85.09**
150 mm; 45°	57.91	65.40	0.83	19.69	nr	**85.09**
Long radius bend						
75 mm	26.30	29.70	0.60	14.22	nr	**43.92**
100 mm	31.22	35.26	0.67	15.88	nr	**51.14**
100 mm; 5°	19.27	21.77	0.67	15.88	nr	**37.65**
150 mm	68.07	76.88	0.83	19.69	nr	**96.57**
150 mm; 22.5°	71.36	80.59	0.83	19.69	nr	**100.28**
Access bend; long radius						
75 mm	46.68	52.72	0.60	14.22	nr	**66.94**
100 mm	52.74	59.57	0.67	15.88	nr	**75.45**
150 mm	92.90	104.92	0.83	19.69	nr	**124.61**
Long tail bend						
100 × 250 mm long	24.90	28.12	0.70	16.60	nr	**44.72**
100 × 815 mm long	79.18	89.42	0.70	16.60	nr	**106.02**
Offset						
75 mm projection						
75 mm	20.56	23.23	0.53	12.56	nr	**35.79**
100 mm	20.28	22.91	0.66	15.65	nr	**38.56**

R: DISPOSAL SYSTEMS

Item	Net Price £	Material £	Labour hours	Labour £	Unit	Total rate £
115 mm projection						
75 mm	22.60	25.52	0.53	12.56	nr	**38.08**
100 mm	25.34	28.62	0.66	15.65	nr	**44.27**
150 mm projection						
75 mm	26.55	29.99	0.53	12.56	nr	**42.55**
100 mm	26.55	29.99	0.66	15.65	nr	**45.64**
225 mm projection						
100 mm	29.02	32.77	0.66	15.65	nr	**48.42**
300 mm projection						
100 mm	31.22	35.26	0.66	15.65	nr	**50.91**
Branch; equal and unequal						
50 mm	20.95	23.66	0.78	18.47	nr	**42.13**
75 mm	23.72	26.79	0.85	20.15	nr	**46.94**
100 mm	29.80	33.66	1.00	23.69	nr	**57.35**
150 mm	73.88	83.44	1.20	28.45	nr	**111.89**
150 × 100 mm; 87.5°	56.53	63.84	1.21	28.55	nr	**92.39**
150 × 100 mm; 45°	82.92	93.65	1.21	28.55	nr	**122.20**
Branch; 2' BSPT screwed socket						
100 mm	39.92	45.08	1.00	23.69	nr	**68.77**
Branch; long tail						
100 × 915 mm long	82.55	93.23	1.00	23.69	nr	**116.92**
Access branch; equal and unequal						
50 mm	45.64	51.54	0.78	18.47	nr	**70.01**
75 mm	45.64	51.54	0.85	20.15	nr	**71.69**
100 mm	51.31	57.95	1.02	24.18	nr	**82.13**
150 mm	109.24	123.38	1.20	28.45	nr	**151.83**
150 × 100 mm; 87.5°	91.81	103.69	1.20	28.45	nr	**132.14**
150 × 100 mm; 45°	99.52	112.39	1.20	28.45	nr	**140.84**
Parallel branch						
100 mm	31.22	35.26	1.00	23.69	nr	**58.95**
Double branch						
75 mm	35.22	39.78	0.95	22.51	nr	**62.29**
100 mm	36.86	41.63	1.30	30.81	nr	**72.44**
150 × 100 mm	103.80	117.23	1.56	36.96	nr	**154.19**
Double access branch						
100 mm	58.36	65.91	1.43	33.89	nr	**99.80**
Corner branch						
100 mm	52.40	59.18	1.30	30.81	nr	**89.99**

R: DISPOSAL SYSTEMS

Item	Net Price £	Material £	Labour hours	Labour £	Unit	Total rate £
R11: ABOVE GROUND DRAINAGE – cont						
Fixings – cont						
Puddle flange; grey epoxy coated						
100 mm	35.04	39.58	1.00	23.69	nr	**63.27**
Roof vent connector; asphalt						
75 mm	42.65	48.16	0.90	21.32	nr	**69.48**
100 mm	33.29	37.60	0.97	22.98	nr	**60.58**
P trap						
100 mm	30.90	34.89	1.00	23.69	nr	**58.58**
P trap with access						
50 mm	47.29	53.41	0.77	18.24	nr	**71.65**
75 mm	47.29	53.41	0.90	21.32	nr	**74.73**
100 mm	52.41	59.19	1.16	27.49	nr	**86.68**
150 mm	91.52	103.37	1.77	41.93	nr	**145.30**
Bellmouth gully inlet						
100 mm	44.73	50.52	1.08	25.59	nr	**76.11**
Balcony gully inlet						
100 mm	135.04	152.51	1.08	25.59	nr	**178.10**
Roof outlet						
Flat grate						
75 mm	83.36	94.15	0.83	19.67	nr	**113.82**
100 mm	117.28	132.46	1.08	25.59	nr	**158.05**
Dome grate						
75 mm	83.36	94.15	0.83	19.67	nr	**113.82**
100 mm	131.54	148.56	1.08	25.59	nr	**174.15**
Top Hat						
100 mm	178.09	201.13	1.08	25.59	nr	**226.72**
Cast iron pipe; EPDM rubber gasket joint with continuity clip; BS EN877; fixed to backgrounds						
Pipe, with brackets and jointing couplings measured separately						
50 mm	17.02	19.23	0.31	7.34	m	**26.57**
70 mm	19.90	22.47	0.34	8.05	m	**30.52**
100 mm	23.68	26.74	0.37	8.76	m	**35.50**

R: DISPOSAL SYSTEMS

Item	Net Price £	Material £	Labour hours	Labour £	Unit	Total rate £
125 mm	38.00	42.92	0.65	15.40	m	58.32
150 mm	46.92	52.99	0.70	16.60	m	69.59
200 mm	78.40	88.55	1.14	27.02	m	115.57
250 mm	109.55	123.73	1.25	29.62	m	153.35
300 mm	136.42	154.07	1.53	36.25	m	190.32
Fixings						
Brackets; fixed to backgrounds. For minimum fixing distances, refer to the Tables and Memoranda at the rear of the book						
Ductile iron						
50 mm	7.23	8.17	0.10	2.38	nr	10.55
70 mm	7.23	8.17	0.10	2.38	nr	10.55
100 mm	8.35	9.44	0.15	3.56	nr	13.00
150 mm	15.48	17.48	0.20	4.74	nr	22.22
200 mm	57.40	64.83	0.25	5.93	nr	70.76
Mild steel; vertical						
125 mm	14.08	15.90	0.15	3.56	nr	19.46
Mild steel; stand off						
250 mm	30.67	34.64	0.25	5.93	nr	40.57
300 mm	33.79	38.16	0.25	5.93	nr	44.09
Stack support; rubber seal						
70 mm	28.69	32.41	0.55	13.04	nr	45.45
100 mm	31.91	36.04	0.65	15.40	nr	51.44
125 mm	35.39	39.97	0.74	17.54	nr	57.51
150 mm	50.51	57.05	0.86	20.37	nr	77.42
Wall spacer plate; cast iron (eared sockets)						
50 mm	6.15	6.95	0.10	2.38	nr	9.33
70 mm	6.15	6.95	0.10	2.38	nr	9.33
100 mm	6.15	6.95	0.10	2.38	nr	9.33
Extra over fittings EPDM rubber jointed cast iron pipework BS EN 877, with jointing couplings measured separately						
Coupling						
50 mm	7.73	8.73	0.10	2.38	nr	11.11
70 mm	8.51	9.61	0.10	2.38	nr	11.99
100 mm	11.08	12.52	0.10	2.38	nr	14.90
125 mm	13.76	15.54	0.15	3.56	nr	19.10
150 mm	22.20	25.07	0.30	7.10	nr	32.17
200 mm	49.66	56.09	0.35	8.29	nr	64.38
250 mm	71.13	80.34	0.40	9.48	nr	89.82
300 mm	82.31	92.96	0.50	11.85	nr	104.81

R: DISPOSAL SYSTEMS

Item	Net Price £	Material £	Labour hours	Labour £	Unit	Total rate £
R11: ABOVE GROUND DRAINAGE – cont						
Extra over fittings EPDM rubber jointed cast iron pipework BS EN 877, with jointing couplings measured separately – cont						
Plain socket						
50 mm	19.42	21.93	0.10	2.38	nr	**24.31**
70 mm	19.42	21.93	0.10	2.38	nr	**24.31**
100 mm	22.30	25.18	0.10	2.38	nr	**27.56**
150 mm	37.62	42.48	0.10	2.38	nr	**44.86**
Eared socket						
50 mm	20.00	22.59	0.25	5.93	nr	**28.52**
70 mm	20.00	22.59	0.25	5.93	nr	**28.52**
100 mm	24.24	27.38	0.25	5.93	nr	**33.31**
150 mm	39.74	44.88	0.25	5.93	nr	**50.81**
Slip socket						
50 mm	25.16	28.42	0.25	5.93	nr	**34.35**
70 mm	25.16	28.42	0.25	5.93	nr	**34.35**
100 mm	29.38	33.18	0.25	5.93	nr	**39.11**
150 mm	44.23	49.95	0.25	5.93	nr	**55.88**
Stack support pipe						
70 mm	18.57	20.97	0.74	17.54	nr	**38.51**
100 mm	25.50	28.79	0.88	20.85	nr	**49.64**
125 mm	27.27	30.79	1.00	23.69	nr	**54.48**
150 mm	36.96	41.74	1.19	28.20	nr	**69.94**
Access pipe; round door						
50 mm	35.65	40.26	0.27	6.40	nr	**46.66**
70 mm	37.72	42.61	0.30	7.10	nr	**49.71**
100 mm	41.46	46.83	0.32	7.58	nr	**54.41**
150 mm	75.04	84.75	0.71	16.83	nr	**101.58**
Access pipe; square door						
100 mm	80.17	90.55	0.32	7.58	nr	**98.13**
125 mm	83.43	94.23	0.67	15.88	nr	**110.11**
150 mm	125.52	141.76	0.71	16.83	nr	**158.59**
200 mm	249.34	281.60	1.21	28.67	nr	**310.27**
250 mm	392.32	443.09	1.31	31.04	nr	**474.13**
300 mm	489.39	552.72	1.41	33.40	nr	**586.12**
Taper reducer						
70 mm	20.58	23.25	0.30	7.10	nr	**30.35**
100 mm	24.20	27.34	0.32	7.58	nr	**34.92**
125 mm	24.33	27.48	0.64	15.17	nr	**42.65**
150 mm	46.46	52.47	0.67	15.88	nr	**68.35**

R: DISPOSAL SYSTEMS

Item	Net Price £	Material £	Labour hours	Labour £	Unit	Total rate £
200 mm	75.45	85.21	1.15	27.25	nr	**112.46**
250 mm	155.84	176.00	1.25	29.62	nr	**205.62**
300 mm	214.24	241.96	1.37	32.46	nr	**274.42**
Blank cap						
50 mm	5.39	6.09	0.24	5.69	nr	**11.78**
70 mm	5.69	6.43	0.26	6.15	nr	**12.58**
100 mm	6.61	7.47	0.32	7.58	nr	**15.05**
125 mm	9.35	10.56	0.35	8.29	nr	**18.85**
150 mm	9.56	10.80	0.40	9.48	nr	**20.28**
200 mm	42.21	47.67	0.60	14.22	nr	**61.89**
250 mm	86.28	97.44	0.65	15.40	nr	**112.84**
300 mm	122.96	138.87	0.72	17.06	nr	**155.93**
Blank cap; 50 mm screwed tapping						
70 mm	12.68	14.32	0.26	6.15	nr	**20.47**
100 mm	13.69	15.46	0.32	7.58	nr	**23.04**
150 mm	16.46	18.59	0.40	9.48	nr	**28.07**
Universal connector; EPDM rubber						
50 × 56/48/40 mm	11.00	12.42	0.30	7.10	nr	**19.52**
Blank end; push fit						
100 × 38/32 mm	12.28	13.87	0.32	7.58	nr	**21.45**
Boss pipe; 2" BSPT socket						
50 mm	30.14	34.04	0.27	6.40	nr	**40.44**
75 mm	30.14	34.04	0.30	7.10	nr	**41.14**
100 mm	36.84	41.61	0.32	7.58	nr	**49.19**
150 mm	60.07	67.84	0.71	16.83	nr	**84.67**
Boss pipe; 2 × 2" BSPT socket; opposed						
100 mm	47.59	53.74	0.32	7.58	nr	**61.32**
Boss pipe; 2 × 2" BSPT socket; 90°						
100 mm	47.59	53.74	0.32	7.58	nr	**61.32**
Manifold connector						
100 mm	73.49	83.00	0.64	15.17	nr	**98.17**
150 mm	102.38	115.63	1.00	23.69	nr	**139.32**
Bend; short radius						
50 mm	13.38	15.12	0.27	6.40	nr	**21.52**
70 mm	15.06	17.01	0.30	7.10	nr	**24.11**
100 mm	17.83	20.13	0.32	7.58	nr	**27.71**
125 mm	31.62	35.71	0.62	14.70	nr	**50.41**
150 mm	32.02	36.16	0.67	15.88	nr	**52.04**
200 mm; 45°	95.33	107.66	1.21	28.67	nr	**136.33**
250 mm; 45°	186.01	210.08	1.31	31.04	nr	**241.12**
300 mm; 45°	261.66	295.51	1.43	33.88	nr	**329.39**

R: DISPOSAL SYSTEMS

Item	Net Price £	Material £	Labour hours	Labour £	Unit	Total rate £
R11: ABOVE GROUND DRAINAGE – cont						
Extra over fittings EPDM rubber jointed cast iron pipework BS EN 877, with jointing couplings measured separately – cont						
Access bend; short radius						
70 mm	29.19	32.97	0.30	7.10	nr	**40.07**
100 mm	42.64	48.15	0.32	7.58	nr	**55.73**
150 mm	75.03	84.74	0.67	15.88	nr	**100.62**
Bend; long radius bend						
100 mm; 88°	45.32	51.18	0.32	7.58	nr	**58.76**
100 mm; 22°	33.42	37.74	0.32	7.58	nr	**45.32**
150 mm; 88°	129.90	146.71	0.67	15.88	nr	**162.59**
Access bend; long radius						
100 mm	55.16	62.30	0.32	7.58	nr	**69.88**
150 mm	134.00	151.33	0.32	7.58	nr	**158.91**
Bend; long tail						
100 mm	31.24	35.28	0.32	7.58	nr	**42.86**
Bend; long tail double						
70 mm	48.42	54.68	0.30	7.10	nr	**61.78**
100 mm	53.41	60.32	0.32	7.58	nr	**67.90**
Offset; 75 mm projection						
100 mm	27.41	30.96	0.32	7.58	nr	**38.54**
Offset; 130 mm projection						
50 mm	22.67	25.60	0.27	6.40	nr	**32.00**
70 mm	34.40	38.85	0.30	7.10	nr	**45.95**
100 mm	45.18	51.03	0.32	7.58	nr	**58.61**
125 mm	57.26	64.67	0.67	15.88	nr	**80.55**
Branch; equal and unequal						
50 mm	21.46	24.24	0.37	8.76	nr	**33.00**
70 mm	22.66	25.59	0.40	9.48	nr	**35.07**
100 mm	31.09	35.11	0.42	9.95	nr	**45.06**
125 mm	62.31	70.37	0.76	18.00	nr	**88.37**
150 mm	67.80	76.57	0.97	22.98	nr	**99.55**
200 mm	173.91	196.41	1.51	35.78	nr	**232.19**
250 mm	222.26	251.02	1.63	38.63	nr	**289.65**
300 mm	366.74	414.19	1.77	41.93	nr	**456.12**
Branch; radius; equal and unequal						
70 mm	27.61	31.18	0.40	9.48	nr	**40.66**
100 mm	36.34	41.04	0.42	9.95	nr	**50.99**
150 mm	78.60	88.77	1.37	32.46	nr	**121.23**
200 mm	221.19	249.81	1.51	35.78	nr	**285.59**

R: DISPOSAL SYSTEMS

Item	Net Price £	Material £	Labour hours	Labour £	Unit	Total rate £
Branch; long tail						
100 mm	101.39	114.51	0.52	12.32	nr	**126.83**
Access branch; radius; equal and unequal						
70 mm	40.48	45.72	0.40	9.48	nr	**55.20**
100 mm	54.88	61.99	0.42	9.95	nr	**71.94**
150 mm	131.03	147.98	0.97	22.98	nr	**170.96**
Double branch; equal and unequal						
100 mm	30.61	34.57	0.52	12.32	nr	**46.89**
100 mm; 70°	45.66	51.57	0.52	12.32	nr	**63.89**
150 mm	131.03	147.98	1.37	32.46	nr	**180.44**
200 mm	224.08	253.07	1.51	35.78	nr	**288.85**
Double branch; radius; equal and unequal						
100 mm	39.48	44.58	0.52	12.32	nr	**56.90**
150 mm	161.62	182.53	1.37	32.46	nr	**214.99**
Corner branch						
100 mm	79.86	90.19	0.52	12.32	nr	**102.51**
150 mm	88.29	99.71	0.52	12.32	nr	**112.03**
Corner branch; long tail						
100 mm	119.14	134.56	0.52	12.32	nr	**146.88**
Roof vent connector; asphalt						
100 mm	50.24	56.74	0.32	7.58	nr	**64.32**
Movement connector						
100 mm	63.82	72.07	0.32	7.58	nr	**79.65**
150 mm	118.13	133.42	0.67	15.88	nr	**149.30**
Expansion plugs						
70 mm	14.20	16.03	0.32	7.58	nr	**23.61**
100 mm	17.69	19.98	0.39	9.24	nr	**29.22**
150 mm	31.76	35.87	0.55	13.04	nr	**48.91**
P trap						
100 mm dia.	33.10	37.38	0.32	7.58	nr	**44.96**
P trap with access						
50 mm	50.54	57.08	0.27	6.40	nr	**63.48**
70 mm	50.54	57.08	0.30	7.10	nr	**64.18**
100 mm	54.74	61.82	0.32	7.58	nr	**69.40**
150 mm	97.87	110.54	0.67	15.88	nr	**126.42**
Branch trap						
100 mm	119.95	135.47	0.42	9.95	nr	**145.42**

R: DISPOSAL SYSTEMS

Item	Net Price £	Material £	Labour hours	Labour £	Unit	Total rate £
R11: ABOVE GROUND DRAINAGE – cont						
Extra over fittings EPDM rubber jointed cast iron pipework BS EN 877, with jointing couplings measured separately – cont						
Stench trap						
100 mm	233.27	263.45	0.42	9.95	nr	**273.40**
Balcony gully inlet						
100 mm	135.04	152.51	1.00	23.69	nr	**176.20**
Roof outlet						
Flat grate						
70 mm	83.36	94.15	1.00	23.69	nr	**117.84**
100 mm	117.28	132.46	1.00	23.69	nr	**156.15**
Dome grate						
70 mm	83.36	94.15	1.00	23.69	nr	**117.84**
100 mm	131.54	148.56	1.00	23.69	nr	**172.25**
Top Hat						
100 mm	178.09	201.13	1.00	23.69	nr	**224.82**
Floor drains; for cast iron pipework BS 416 and BS EN877						
Adjustable clamp plate body						
100 mm; 165 mm nickel bronze grate and frame	79.39	89.66	0.50	11.85	nr	**101.51**
100 mm; 165 mm nickel bronze rodding eye	92.38	104.34	0.50	11.85	nr	**116.19**
100 mm; 150 × 150 mm nickel bronze grate and frame	89.08	100.60	0.50	11.85	nr	**112.45**
100 mm; 150 × 150 nickel bronze rodding eye	92.38	104.34	0.50	11.85	nr	**116.19**
Deck plate body						
100 mm; 165 mm nickel bronze grate and frame	79.39	89.66	0.50	11.85	nr	**101.51**
100 mm; 165 mm nickel bronze rodding eye	92.38	104.34	0.50	11.85	nr	**116.19**
100 mm; 150 × 150 mm nickel bronze grate and frame	89.08	100.60	0.50	11.85	nr	**112.45**
100 mm; 150 × 150 mm nickel bronze rodding eye	92.38	104.34	0.50	11.85	nr	**116.19**
Extra for						
100 mm; Srewed extension piece	30.96	34.97	0.30	7.10	nr	**42.07**
100 mm; Grating extension piece; screwed or spigot	23.49	26.53	0.30	7.10	nr	**33.63**
100 mm; Brewary trap	854.39	964.94	2.00	47.39	nr	**1012.33**

S: PIPED SUPPLY SYSTEMS

Item	Net Price £	Material £	Labour hours	Labour £	Unit	Total rate £
S10: COLD WATER						
Y10 – PIPELINES						
COPPER PIPEWORK						
Copper pipe; capillary or compression joints in the running length; EN1057 R250 (TX) formerly BS 2871 Table X						
Fixed vertically or at low level, with brackets measured separately						
12 mm dia.	1.27	1.44	0.39	9.24	m	**10.68**
15 mm dia.	1.43	1.61	0.40	9.48	m	**11.09**
22 mm dia.	2.88	3.25	0.47	11.14	m	**14.39**
28 mm dia.	3.64	4.11	0.51	12.09	m	**16.20**
35 mm dia.	8.63	9.75	0.58	13.75	m	**23.50**
42 mm dia.	10.51	11.87	0.66	15.65	m	**27.52**
54 mm dia.	13.52	15.27	0.72	17.06	m	**32.33**
67 mm dia.	17.68	19.97	0.75	17.77	m	**37.74**
76 mm dia.	25.01	28.24	0.76	18.00	m	**46.24**
108 mm dia.	35.98	40.64	0.78	18.47	m	**59.11**
133 mm dia.	46.69	52.73	1.05	24.88	m	**77.61**
159 mm dia.	73.98	83.55	1.15	27.25	m	**110.80**
Fixed horizontally at high level or suspended, with brackets measured separately						
12 mm dia.	1.27	1.44	0.45	10.66	m	**12.10**
15 mm dia.	1.43	1.61	0.46	10.89	m	**12.50**
22 mm dia.	2.88	3.25	0.54	12.80	m	**16.05**
28 mm dia.	3.64	8.23	0.59	13.98	m	**22.21**
35 mm dia.	8.63	9.75	0.67	15.88	m	**25.63**
42 mm dia.	10.51	11.87	0.76	18.00	m	**29.87**
54 mm dia.	13.52	15.27	0.83	19.67	m	**34.94**
67 mm dia.	17.68	19.97	0.86	20.37	m	**40.34**
76 mm dia.	25.01	28.24	0.87	20.61	m	**48.85**
108 mm dia.	35.98	40.64	0.90	21.32	m	**61.96**
133 mm dia.	46.69	52.73	1.21	28.67	m	**81.40**
159 mm dia.	73.98	83.55	1.32	31.27	m	**114.82**
Copper pipe; capillary or compression joints in the running length; EN1057 R250 (TY) formerly BS 2871 Table Y						
Fixed vertically or at low level with brackets measured separately (Refer to Copper Pipe Table X Section)						
12 mm dia.	1.65	1.87	0.41	9.71	m	**11.58**
15 mm dia.	2.35	2.65	0.43	10.19	m	**12.84**
22 mm dia.	4.14	4.67	0.50	11.85	m	**16.52**

S: PIPED SUPPLY SYSTEMS

Item	Net Price £	Material £	Labour hours	Labour £	Unit	Total rate £
S10: COLD WATER – cont						
COPPER PIPEWORK – cont						
Copper pipe; capillary or compression joints in the running length; EN1057 R250 (TY) formerly BS 2871 Table Y – cont						
Fixed vertically or at low level with brackets measured separately (Refer to Copper Pipe Table X Section) – cont						
28 mm dia.	5.35	6.04	0.54	12.80	m	**18.84**
35 mm dia.	7.80	8.81	0.62	14.70	m	**23.51**
42 mm dia.	9.46	10.68	0.71	16.83	m	**27.51**
54 mm dia.	16.12	18.21	0.78	18.47	m	**36.68**
67 mm dia.	21.41	24.18	0.82	19.42	m	**43.60**
76 mm dia.	31.29	35.34	0.60	14.22	m	**49.56**
108 mm dia.	43.46	49.08	0.88	20.85	m	**69.93**
Copper pipe; capillary or compression joints in the running length; EN1057 R250 (TX) formerly BS 2871 Table X						
Plastic coated gas and cold water service pipe for corrosive environments, fixed vertically or at low level with brackets measured separately						
15 mm dia. (white)	3.29	3.71	0.59	13.98	m	**17.69**
22 mm dia. (white)	5.82	6.57	0.68	16.12	m	**22.69**
28 mm dia. (white)	6.01	6.78	0.74	17.54	m	**24.32**
FIXINGS						
For copper pipework						
Saddle band						
6 mm dia.	0.05	0.07	0.05	1.20	nr	**1.27**
8 mm dia.	0.05	0.07	0.07	1.62	nr	**1.69**
10 mm dia.	0.07	0.09	0.09	2.02	nr	**2.11**
12 mm dia.	0.07	0.09	0.11	2.53	nr	**2.62**
15 mm dia.	0.07	0.09	0.13	3.08	nr	**3.17**
22 mm dia.	0.13	0.15	0.13	3.08	nr	**3.23**
28 mm dia.	0.08	0.10	0.16	3.79	nr	**3.89**
35 mm dia.	0.23	0.26	0.18	4.27	nr	**4.53**
42 mm dia.	0.49	0.55	0.21	4.98	nr	**5.53**
54 mm dia.	0.66	0.74	0.21	4.98	nr	**5.72**
Single spacing clip						
15 mm dia.	0.14	0.17	0.14	3.33	nr	**3.50**
22 mm dia.	0.14	0.17	0.15	3.56	nr	**3.73**
28 mm dia.	0.34	0.39	0.17	4.03	nr	**4.42**

S: PIPED SUPPLY SYSTEMS

Item	Net Price £	Material £	Labour hours	Labour £	Unit	Total rate £
Two piece spacing clip						
8 mm dia. Bottom	0.12	0.14	0.11	2.61	nr	**2.75**
8 mm dia. Top	0.12	0.14	0.11	2.61	nr	**2.75**
12 mm dia. Bottom	0.12	0.14	0.13	3.08	nr	**3.22**
12 mm dia. Top	0.12	0.14	0.13	3.08	nr	**3.22**
15 mm dia. Bottom	0.13	0.15	0.13	3.08	nr	**3.23**
15 mm dia. Top	0.13	0.15	0.13	3.08	nr	**3.23**
22 mm dia. Bottom	0.13	0.15	0.14	3.33	nr	**3.48**
22 mm dia. Top	0.14	0.17	0.14	3.33	nr	**3.50**
28 mm dia. Bottom	0.13	0.15	0.16	3.79	nr	**3.94**
28 mm dia. Top	0.22	0.25	0.16	3.79	nr	**4.04**
35 mm dia. Bottom	0.14	0.17	0.21	4.98	nr	**5.15**
35 mm dia. Top	0.33	0.38	0.21	4.98	nr	**5.36**
Single pipe bracket						
15 mm dia.	1.14	1.29	0.14	3.33	nr	**4.62**
22 mm dia.	1.16	1.31	0.14	3.33	nr	**4.64**
28 mm dia.	1.60	1.81	0.17	4.03	nr	**5.84**
Single pipe ring						
15 mm dia.	2.01	2.27	0.26	6.15	nr	**8.42**
22 mm dia.	2.18	2.46	0.26	6.15	nr	**8.61**
28 mm dia.	2.73	3.08	0.31	7.34	nr	**10.42**
35 mm dia.	3.85	4.35	0.32	7.58	nr	**11.93**
42 mm dia.	4.47	5.05	0.32	7.58	nr	**12.63**
54 mm dia.	5.72	6.46	0.34	8.05	nr	**14.51**
67 mm dia.	9.59	10.83	0.35	8.29	nr	**19.12**
76 mm dia.	19.71	22.26	0.42	9.95	nr	**32.21**
108 mm dia.	36.27	40.96	0.42	9.95	nr	**50.91**
Double pipe ring						
15 mm dia.	2.75	3.10	0.26	6.15	nr	**9.25**
22 mm dia.	3.44	3.89	0.26	6.15	nr	**10.04**
28 mm dia.	5.43	6.13	0.31	7.34	nr	**13.47**
35 mm dia.	7.24	8.18	0.32	7.58	nr	**15.76**
42 mm dia.	8.14	9.19	0.32	7.58	nr	**16.77**
54 mm dia.	8.39	9.48	0.34	8.05	nr	**17.53**
67 mm dia.	10.57	11.93	0.35	8.29	nr	**20.22**
76 mm dia.	16.04	18.12	0.42	9.95	nr	**28.07**
108 mm dia.	21.91	24.75	0.42	9.95	nr	**34.70**
Wall bracket						
15 mm dia.	2.28	2.57	0.05	1.18	nr	**3.75**
22 mm dia.	2.98	3.37	0.05	1.18	nr	**4.55**
28 mm dia.	3.57	4.03	0.05	1.18	nr	**5.21**
35 mm dia.	5.35	6.04	0.05	1.18	nr	**7.22**
42 mm dia.	7.06	7.98	0.05	1.18	nr	**9.16**
54 mm dia.	11.02	12.44	0.05	1.18	nr	**13.62**

S: PIPED SUPPLY SYSTEMS

Item	Net Price £	Material £	Labour hours	Labour £	Unit	Total rate £
S10: COLD WATER – cont						
FIXINGS – cont						
For copper pipework – cont						
Hospital bracket						
15 mm dia.	3.91	4.42	0.26	6.15	nr	**10.57**
22 mm dia.	4.63	5.23	0.26	6.15	nr	**11.38**
28 mm dia.	5.43	6.13	0.31	7.34	nr	**13.47**
35 mm dia.	5.75	6.50	0.32	7.58	nr	**14.08**
42 mm dia.	8.22	9.28	0.32	7.58	nr	**16.86**
54 mm dia.	11.00	12.42	0.34	8.05	nr	**20.47**
Screw on backplate, female						
All sizes 15 mm to 54 mm × 10 mm	1.26	1.43	0.10	2.38	nr	**3.81**
Screw on backplate, male						
All sizes 15 mm to 54 mm × 10 mm	1.67	1.89	0.10	2.38	nr	**4.27**
Pipe joist clips, single						
15 mm dia.	0.79	0.89	0.08	1.90	nr	**2.79**
22 mm dia.	0.79	0.89	0.08	1.90	nr	**2.79**
Pipe joist clips, double						
15 mm dia.	1.10	1.24	0.08	1.90	nr	**3.14**
22 mm dia.	0.60	0.67	0.08	1.90	nr	**2.57**
Extra Over channel sections for fabricated hangers and brackets						
Galvanized steel; including inserts, bolts, nuts, washers; fixed to backgrounds						
41 × 21 mm	4.92	5.56	0.29	6.86	m	**12.42**
41 × 41 mm	5.90	6.66	0.29	6.86	m	**13.52**
Threaded rods; metric thread; including nuts, washers etc						
10 mm dia. × 600 mm long for ring clips up to 54 mm	1.62	1.83	0.18	4.27	nr	**6.10**
12 mm dia. × 600 mm long for ring clips from 54 mm	2.49	2.82	0.18	4.27	nr	**7.09**
Extra over copper pipes; capillary fittings; BS 864						
Stop end						
15 mm dia.	1.15	1.30	0.13	3.08	nr	**4.38**
22 mm dia.	2.15	2.43	0.14	3.33	nr	**5.76**
28 mm dia.	3.85	4.35	0.17	4.03	nr	**8.38**

S: PIPED SUPPLY SYSTEMS

Item	Net Price £	Material £	Labour hours	Labour £	Unit	Total rate £
35 mm dia.	8.49	9.59	0.19	4.51	nr	14.10
42 mm dia.	14.62	16.51	0.22	5.22	nr	21.73
54 mm dia.	20.41	23.05	0.23	5.46	nr	28.51
Straight coupling; copper to copper						
6 mm dia.	1.21	1.37	0.23	5.46	nr	6.83
8 mm dia.	1.24	1.40	0.23	5.46	nr	6.86
10 mm dia.	0.63	0.71	0.23	5.46	nr	6.17
15 mm dia.	0.14	0.17	0.23	5.46	nr	5.63
22 mm dia.	0.37	0.42	0.26	6.15	nr	6.57
28 mm dia.	1.07	1.20	0.30	7.10	nr	8.30
35 mm dia.	3.47	3.92	0.34	8.05	nr	11.97
42 mm dia.	5.79	6.54	0.38	9.00	nr	15.54
54 mm dia.	10.69	12.08	0.42	9.95	nr	22.03
67 mm dia.	31.73	35.83	0.53	12.56	nr	48.39
Adaptor coupling; imperial to metric						
½' × 15 mm dia.	2.51	2.84	0.27	6.40	nr	9.24
¾' × 22 mm dia.	2.20	2.49	0.31	7.34	nr	9.83
1' × 28 mm dia.	4.32	4.88	0.36	8.53	nr	13.41
1¼' × 35 mm dia.	7.23	8.17	0.41	9.71	nr	17.88
1½' × 42 mm dia.	9.20	10.39	0.46	10.89	nr	21.28
Reducing coupling						
15 × 10 mm dia.	2.65	2.99	0.23	5.46	nr	8.45
22 × 10 mm dia.	3.88	4.39	0.26	6.15	nr	10.54
22 × 15 mm dia.	4.14	4.67	0.27	6.40	nr	11.07
28 × 15 mm dia.	4.77	5.39	0.28	6.63	nr	12.02
28 × 22 mm dia.	4.81	5.44	0.30	7.10	nr	12.54
35 × 28 mm dia.	6.88	7.77	0.34	8.05	nr	15.82
42 × 35 mm dia.	10.09	11.39	0.38	9.00	nr	20.39
54 × 35 mm dia.	17.70	19.99	0.42	9.95	nr	29.94
54 × 42 mm dia.	19.32	21.82	0.42	9.95	nr	31.77
Straight female connector						
15 mm × ½' dia.	2.63	2.97	0.27	6.40	nr	9.37
22 mm × ¾' dia.	3.80	4.29	0.31	7.34	nr	11.63
28 mm × 1' dia.	7.18	8.11	0.36	8.53	nr	16.64
35 mm × 1¼' dia.	12.41	14.01	0.41	9.71	nr	23.72
42 mm × 1½' dia.	16.11	18.20	0.46	10.89	nr	29.09
54 mm × 2' dia.	25.55	28.86	0.52	12.32	nr	41.18
Straight male connector						
15 mm × ½' dia.	2.23	2.52	0.27	6.40	nr	8.92
22 mm × ¾' dia.	3.99	4.51	0.31	7.34	nr	11.85
28 mm × 1' dia.	6.44	7.27	0.36	8.53	nr	15.80
35 mm × 1¼' dia.	11.32	12.78	0.41	9.71	nr	22.49
42 mm × 1½' dia.	14.57	16.45	0.46	10.89	nr	27.34
54 mm × 2' dia.	22.12	24.98	0.52	12.32	nr	37.30
67 mm × 2½' dia.	35.34	39.91	0.63	14.93	nr	54.84

S: PIPED SUPPLY SYSTEMS

Item	Net Price £	Material £	Labour hours	Labour £	Unit	Total rate £
S10: COLD WATER – cont						
FIXINGS – cont						
Extra over copper pipes; capillary fittings; BS 864 – cont						
Female reducing connector						
15 mm × ¾' dia.	6.52	7.36	0.27	6.40	nr	**13.76**
Male reducing connector						
15 mm × ¾' dia.	5.82	6.57	0.27	6.40	nr	**12.97**
22 mm × 1' dia.	8.88	10.03	0.31	7.34	nr	**17.37**
Flanged connector						
28 mm dia.	41.66	47.05	0.36	8.53	nr	**55.58**
35 mm dia.	52.74	59.57	0.41	9.71	nr	**69.28**
42 mm dia.	63.04	71.20	0.46	10.89	nr	**82.09**
54 mm dia.	95.30	107.63	0.52	12.32	nr	**119.95**
67 mm dia.	112.06	126.56	0.61	14.45	nr	**141.01**
Tank connector						
15 mm × ½' dia.	5.58	6.30	0.25	5.93	nr	**12.23**
22 mm × ¾' dia.	8.51	9.61	0.28	6.63	nr	**16.24**
28 mm × 1' dia.	11.18	12.63	0.32	7.58	nr	**20.21**
35 mm × 1¼' dia.	14.34	16.20	0.37	8.76	nr	**24.96**
42 mm × 1½' dia.	18.80	21.24	0.43	10.19	nr	**31.43**
54 mm × 2' dia.	28.73	32.45	0.46	10.89	nr	**43.34**
Tank connector with long thread						
15 mm × ½' dia.	7.23	8.17	0.30	7.10	nr	**15.27**
22 mm × ¾' dia.	10.29	11.62	0.33	7.81	nr	**19.43**
28 mm × 1' dia.	12.73	14.38	0.39	9.24	nr	**23.62**
Reducer						
15 × 10 mm dia.	0.89	1.01	0.23	5.46	nr	**6.47**
22 × 15 mm dia.	0.62	0.70	0.26	6.15	nr	**6.85**
28 × 15 mm dia.	2.35	2.65	0.28	6.63	nr	**9.28**
28 × 22 mm dia.	1.80	2.03	0.30	7.10	nr	**9.13**
35 × 22 mm dia.	6.68	7.55	0.34	8.05	nr	**15.60**
42 × 22 mm dia.	12.06	13.62	0.36	8.53	nr	**22.15**
42 × 35 mm dia.	9.33	10.54	0.38	9.00	nr	**19.54**
54 × 35 mm dia.	19.58	22.11	0.40	9.48	nr	**31.59**
54 × 42 mm dia.	16.89	19.07	0.42	9.95	nr	**29.02**
67 × 54 mm dia.	22.96	25.93	0.53	12.56	nr	**38.49**
Adaptor; copper to female iron						
15 mm × ½' dia.	4.52	5.10	0.27	6.40	nr	**11.50**
22 mm × ¾' dia.	6.89	7.78	0.31	7.34	nr	**15.12**
28 mm × 1' dia.	9.71	10.97	0.36	8.53	nr	**19.50**

S: PIPED SUPPLY SYSTEMS

Item	Net Price £	Material £	Labour hours	Labour £	Unit	Total rate £
35 mm × 1¼' dia.	17.57	19.84	0.41	9.71	nr	**29.55**
42 mm × 1½' dia.	22.12	24.98	0.46	10.89	nr	**35.87**
54 mm × 2' dia.	26.63	30.08	0.52	12.32	nr	**42.40**
Adaptor; copper to male iron						
15 mm × ½' dia.	4.61	5.20	0.27	6.40	nr	**11.60**
22 mm × ¾' dia.	5.90	6.66	0.31	7.34	nr	**14.00**
28 mm × 1' dia.	9.85	11.13	0.36	8.53	nr	**19.66**
35 mm × 1¼' dia.	14.35	16.21	0.41	9.71	nr	**25.92**
42 mm × 1½' dia.	19.83	22.40	0.46	10.89	nr	**33.29**
54 mm × 2' dia.	26.63	30.08	0.52	12.32	nr	**42.40**
Union coupling						
15 mm dia.	6.25	7.06	0.41	9.71	nr	**16.77**
22 mm dia.	10.00	11.29	0.45	10.66	nr	**21.95**
28 mm dia.	14.57	16.45	0.51	12.09	nr	**28.54**
35 mm dia.	19.13	21.60	0.64	15.17	nr	**36.77**
42 mm dia.	27.95	31.57	0.68	16.12	nr	**47.69**
54 mm dia.	53.18	60.06	0.78	18.47	nr	**78.53**
67 mm dia.	90.05	101.70	0.96	22.75	nr	**124.45**
Elbow						
15 mm dia.	0.24	0.28	0.23	5.46	nr	**5.74**
22 mm dia.	0.64	0.72	0.26	6.15	nr	**6.87**
28 mm dia.	1.71	1.93	0.31	7.34	nr	**9.27**
35 mm dia.	7.43	8.39	0.35	8.29	nr	**16.68**
42 mm dia.	12.28	13.87	0.41	9.71	nr	**23.58**
54 mm dia.	25.36	28.64	0.44	10.42	nr	**39.06**
67 mm dia.	65.82	74.34	0.54	12.80	nr	**87.14**
Backplate elbow						
15 mm dia.	4.69	5.29	0.51	12.09	nr	**17.38**
22 mm dia.	10.07	11.37	0.54	12.80	nr	**24.17**
Overflow bend						
22 mm dia.	14.19	16.02	0.26	6.15	nr	**22.17**
Return bend						
15 mm dia.	7.03	7.94	0.23	5.46	nr	**13.40**
22 mm dia.	13.80	15.59	0.26	6.15	nr	**21.74**
28 mm dia.	17.64	19.92	0.31	7.34	nr	**27.26**
Obtuse elbow						
15 mm dia.	0.90	1.02	0.23	5.46	nr	**6.48**
22 mm dia.	1.88	2.12	0.26	6.15	nr	**8.27**
28 mm dia.	3.60	4.07	0.31	7.34	nr	**11.41**
35 mm dia.	11.21	12.66	0.36	8.53	nr	**21.19**
42 mm dia.	19.96	22.54	0.41	9.71	nr	**32.25**
54 mm dia.	36.10	40.77	0.44	10.42	nr	**51.19**
67 mm dia.	65.50	73.97	0.54	12.80	nr	**86.77**

S: PIPED SUPPLY SYSTEMS

Item	Net Price £	Material £	Labour hours	Labour £	Unit	Total rate £
S10: COLD WATER – cont						
FIXINGS – cont						
Extra over copper pipes; capillary fittings; BS 864 – cont						
Straight tap connector						
15 mm × ½' dia.	1.20	1.36	0.13	3.08	nr	**4.44**
22 mm × ¾' dia.	1.79	2.02	0.14	3.33	nr	**5.35**
Bent tap connector						
15 mm × ½' dia.	1.78	2.01	0.13	3.08	nr	**5.09**
22 mm × ¾' dia.	5.47	6.18	0.14	3.33	nr	**9.51**
Bent male union connector						
15 mm × ½' dia.	9.14	10.32	0.41	9.71	nr	**20.03**
22 mm × ¾' dia.	11.87	13.40	0.45	10.66	nr	**24.06**
28 mm × 1' dia.	16.99	19.19	0.51	12.09	nr	**31.28**
35 mm × 1¼' dia.	27.71	31.29	0.64	15.17	nr	**46.46**
42 mm × 1½' dia.	45.05	50.88	0.68	16.12	nr	**67.00**
54 mm × 2' dia.	71.17	80.38	0.78	18.47	nr	**98.85**
Bent female union connector						
15 mm dia.	9.14	10.32	0.41	9.71	nr	**20.03**
22 mm × ¾' dia.	11.87	13.40	0.45	10.66	nr	**24.06**
28 mm × 1' dia.	16.99	19.19	0.51	12.09	nr	**31.28**
35 mm × 1¼' dia.	27.71	31.29	0.64	15.17	nr	**46.46**
42 mm × 1½' dia.	45.05	50.88	0.68	16.12	nr	**67.00**
54 mm × 2' dia.	71.17	80.38	0.78	18.47	nr	**98.85**
Straight union adaptor						
15 mm × ¾' dia.	3.91	4.42	0.41	9.71	nr	**14.13**
22 mm × 1' dia.	5.54	6.25	0.45	10.66	nr	**16.91**
28 mm × 1¼' dia.	8.96	10.12	0.51	12.09	nr	**22.21**
35 mm × 1½' dia.	13.80	15.59	0.64	15.17	nr	**30.76**
42 mm × 2' dia.	17.42	19.68	0.68	16.12	nr	**35.80**
54 mm × 2½' dia.	26.90	30.39	0.78	18.47	nr	**48.86**
Straight male union connector						
15 mm × ½' dia.	7.78	8.78	0.41	9.71	nr	**18.49**
22 mm × ¾' dia.	10.09	11.39	0.45	10.66	nr	**22.05**
28 mm × 1' dia.	15.04	16.98	0.51	12.09	nr	**29.07**
35 mm × 1¼' dia.	21.67	24.47	0.64	15.17	nr	**39.64**
42 mm × 1½' dia.	34.04	38.44	0.68	16.12	nr	**54.56**
54 mm × 2' dia.	48.92	55.25	0.78	18.47	nr	**73.72**

S: PIPED SUPPLY SYSTEMS

Item	Net Price £	Material £	Labour hours	Labour £	Unit	Total rate £
Straight female union connector						
15 mm × ½' dia.	7.78	8.78	0.41	9.71	nr	**18.49**
22 mm × ¾' dia.	10.09	11.39	0.45	10.66	nr	**22.05**
28 mm × 1' dia.	15.04	16.98	0.51	12.09	nr	**29.07**
35 mm × 1¼' dia.	21.67	24.47	0.64	15.17	nr	**39.64**
42 mm × 1½' dia.	34.04	38.44	0.68	16.12	nr	**54.56**
54 mm × 2' dia.	48.92	55.25	0.78	18.47	nr	**73.72**
Male nipple						
¾ × ½' dia.	1.84	2.08	0.24	5.62	nr	**7.70**
1 × ¾' dia.	2.20	2.49	0.32	7.58	nr	**10.07**
1¼ × 1' dia.	2.42	2.73	0.37	8.76	nr	**11.49**
1½ × 1¼' dia.	8.99	10.15	0.42	9.95	nr	**20.10**
2 × 1½' dia.	18.40	20.78	0.46	10.89	nr	**31.67**
2½ × 2' dia.	23.97	27.07	0.56	13.27	nr	**40.34**
Female nipple						
¾ × ½' dia.	2.85	3.22	0.19	4.51	nr	**7.73**
1 × ¾' dia.	4.46	5.04	0.32	7.58	nr	**12.62**
1¼ × 1' dia.	6.08	6.86	0.37	8.76	nr	**15.62**
1½ × 1¼' dia.	8.99	10.15	0.42	9.95	nr	**20.10**
2 × 1½' dia.	18.40	20.78	0.46	10.89	nr	**31.67**
2½ × 2' dia.	23.97	27.07	0.56	13.27	nr	**40.34**
Equal tee						
10 mm dia.	2.43	2.74	0.25	5.93	nr	**8.67**
15 mm dia.	0.24	0.28	0.36	8.53	nr	**8.81**
22 mm dia.	0.64	0.72	0.39	9.24	nr	**9.96**
28 mm dia.	4.75	5.37	0.43	10.19	nr	**15.56**
35 mm dia.	12.10	13.67	0.57	13.50	nr	**27.17**
42 mm dia.	19.41	21.92	0.60	14.22	nr	**36.14**
54 mm dia.	39.14	44.21	0.65	15.40	nr	**59.61**
67 mm dia.	53.25	60.14	0.78	18.47	nr	**78.61**
Female tee, reducing branch FI						
15 × 15 mm × ¼' dia.	5.85	6.61	0.36	8.53	nr	**15.14**
22 × 22 mm × ½' dia.	7.17	8.10	0.39	9.25	nr	**17.35**
28 × 28 mm × ¾' dia.	14.04	15.86	0.43	10.19	nr	**26.05**
35 × 35 mm × ¾' dia.	20.26	22.88	0.47	11.14	nr	**34.02**
42 × 42 mm × ½' dia.	24.34	27.49	0.60	14.22	nr	**41.71**
Backplate tee						
15 × 15 mm × ½' dia.	11.09	12.53	0.62	14.70	nr	**27.23**
Heater tee						
½ × ½' × 15 mm dia.	9.97	11.26	0.36	8.53	nr	**19.79**
Union heater tee						
½ × ½' × 15 mm dia.	9.97	11.26	0.36	8.53	nr	**19.79**

S: PIPED SUPPLY SYSTEMS

Item	Net Price £	Material £	Labour hours	Labour £	Unit	Total rate £
S10: COLD WATER – cont						
FIXINGS – cont						
Extra over copper pipes; capillary fittings; BS 864 – cont						
Sweep tee – equal						
15 mm dia.	7.93	8.96	0.36	8.53	nr	**17.49**
22 mm dia.	10.19	11.51	0.39	9.24	nr	**20.75**
28 mm dia.	17.16	19.38	0.43	10.19	nr	**29.57**
35 mm dia.	24.33	27.48	0.57	13.50	nr	**40.98**
42 mm dia.	36.08	40.75	0.60	14.22	nr	**54.97**
54 mm dia.	39.95	45.11	0.65	15.40	nr	**60.51**
67 mm dia.	54.47	61.52	0.78	18.47	nr	**79.99**
Sweep tee – reducing						
22 × 22 × 15 mm dia.	8.55	9.66	0.39	9.24	nr	**18.90**
28 × 28 × 22 mm dia.	14.52	16.40	0.43	10.19	nr	**26.59**
35 × 35 × 22 mm dia.	24.33	27.48	0.57	13.50	nr	**40.98**
Sweep tee – double						
15 mm dia.	8.96	10.12	0.36	8.53	nr	**18.65**
22 mm dia.	12.20	13.78	0.39	9.24	nr	**23.02**
28 mm dia.	18.54	20.94	0.43	10.19	nr	**31.13**
Cross						
15 mm dia.	11.86	13.39	0.48	11.37	nr	**24.76**
22 mm dia.	13.25	14.96	0.53	12.56	nr	**27.52**
28 mm dia.	19.01	21.47	0.61	14.45	nr	**35.92**
Extra over copper pipes; high duty capillary fittings; BS 864						
Stop end						
15 mm dia.	5.55	6.27	0.16	3.79	nr	**10.06**
Straight coupling; copper to copper						
15 mm dia.	2.54	2.87	0.27	6.40	nr	**9.27**
22 mm dia.	4.07	4.60	0.32	7.58	nr	**12.18**
28 mm dia.	5.44	6.14	0.37	8.76	nr	**14.90**
35 mm dia.	10.16	11.48	0.43	10.19	nr	**21.67**
42 mm dia.	11.11	12.55	0.50	11.85	nr	**24.40**
54 mm dia.	16.35	18.46	0.54	12.80	nr	**31.26**
Reducing coupling						
15 × 12 mm dia.	4.79	5.41	0.27	6.40	nr	**11.81**
22 × 15 mm dia.	5.55	6.27	0.32	7.58	nr	**13.85**
28 × 22 mm dia.	7.65	8.64	0.37	8.76	nr	**17.40**

S: PIPED SUPPLY SYSTEMS

Item	Net Price £	Material £	Labour hours	Labour £	Unit	Total rate £
Straight female connector						
15 mm × ½' dia.	6.25	7.06	0.32	7.58	nr	**14.64**
22 mm × ¾' dia.	7.04	7.96	0.36	8.53	nr	**16.49**
28 mm × 1' dia.	10.39	11.73	0.42	9.95	nr	**21.68**
Straight male connector						
15 mm × ½' dia.	6.08	6.86	0.32	7.58	nr	**14.44**
22 mm × ¾' dia.	7.04	7.96	0.36	8.53	nr	**16.49**
28 mm × 1' dia.	10.39	11.73	0.42	9.95	nr	**21.68**
42 mm × 1½' dia.	20.27	22.89	0.53	12.56	nr	**35.45**
54 mm × 2' dia.	32.95	37.21	0.62	14.70	nr	**51.91**
Reducer						
15 × 12 mm dia.	3.14	3.55	0.27	6.40	nr	**9.95**
22 × 15 mm dia.	3.07	3.47	0.32	7.58	nr	**11.05**
28 × 22 mm dia.	5.55	6.27	0.37	8.76	nr	**15.03**
35 × 28 mm dia.	7.04	7.96	0.43	10.19	nr	**18.15**
42 × 35 mm dia.	9.08	10.25	0.50	11.85	nr	**22.10**
54 × 42 mm dia.	14.64	16.53	0.39	9.24	nr	**25.77**
Straight union adaptor						
15 mm × ¾' dia.	5.08	5.73	0.27	6.40	nr	**12.13**
22 mm × 1' dia.	6.87	7.76	0.32	7.58	nr	**15.34**
28 mm × 1¼' dia.	9.08	10.25	0.37	8.76	nr	**19.01**
35 mm × 1½' dia.	16.45	18.57	0.43	10.19	nr	**28.76**
42 mm × 2' dia.	20.84	23.54	0.50	11.85	nr	**35.39**
Bent union adaptor						
15 mm × ¾' dia.	13.20	14.91	0.27	6.40	nr	**21.31**
22 mm × 1' dia.	17.81	20.11	0.32	7.58	nr	**27.69**
28 mm × 1¼' dia.	23.97	27.07	0.37	8.76	nr	**35.83**
Adaptor; male copper to FI						
15 mm × ½' dia.	5.94	6.71	0.27	6.40	nr	**13.11**
22 mm × ¾' dia.	10.19	11.51	0.32	7.58	nr	**19.09**
Union coupling						
15 mm dia.	11.43	12.91	0.54	12.80	nr	**25.71**
22 mm dia.	14.64	16.53	0.60	14.22	nr	**30.75**
28 mm dia.	20.32	22.95	0.68	16.12	nr	**39.07**
35 mm dia.	35.48	40.08	0.83	19.67	nr	**59.75**
42 mm dia.	41.78	47.19	0.89	21.08	nr	**68.27**
Elbow						
15 mm dia.	7.36	8.31	0.27	6.40	nr	**14.71**
22 mm dia.	7.87	8.88	0.32	7.58	nr	**16.46**
28 mm dia.	11.69	13.20	0.37	8.76	nr	**21.96**
35 mm dia.	18.28	20.64	0.43	10.19	nr	**30.83**
42 mm dia.	22.76	25.70	0.50	11.85	nr	**37.55**
54 mm dia.	39.59	44.72	0.52	12.32	nr	**57.04**

S: PIPED SUPPLY SYSTEMS

Item	Net Price £	Material £	Labour hours	Labour £	Unit	Total rate £
S10: COLD WATER – cont						
FIXINGS – cont						
Extra over copper pipes; high duty capillary fittings; BS 864 – cont						
Return bend						
28 mm dia.	17.64	19.92	0.37	8.76	nr	**28.68**
35 mm dia.	21.73	24.54	0.43	10.19	nr	**34.73**
Bent male union connector						
15 mm × ½' dia.	17.07	19.28	0.54	12.80	nr	**32.08**
22 mm × ¾' dia.	22.99	25.97	0.60	14.22	nr	**40.19**
28 mm × 1' dia.	41.78	47.19	0.68	16.12	nr	**63.31**
Composite flange						
35 mm dia.	3.45	3.90	0.38	9.00	nr	**12.90**
42 mm dia.	4.48	5.06	0.41	9.71	nr	**14.77**
54 mm dia.	5.71	6.45	0.43	10.19	nr	**16.64**
Equal tee						
15 mm dia.	8.46	9.56	0.44	10.42	nr	**19.98**
22 mm dia.	10.64	12.02	0.47	11.14	nr	**23.16**
28 mm dia.	14.02	15.83	0.53	12.56	nr	**28.39**
35 mm dia.	23.97	27.07	0.70	16.60	nr	**43.67**
42 mm dia.	30.52	34.47	0.84	19.90	nr	**54.37**
54 mm dia.	48.06	54.27	0.79	18.72	nr	**72.99**
Reducing tee						
15 × 12 mm dia.	11.60	13.10	0.44	10.42	nr	**23.52**
22 × 15 mm dia.	13.70	15.47	0.47	11.14	nr	**26.61**
28 × 22 mm dia.	19.56	22.09	0.53	12.56	nr	**34.65**
35 × 28 mm dia.	31.03	35.05	0.73	17.30	nr	**52.35**
42 × 28 mm dia.	39.72	44.86	0.84	19.90	nr	**64.76**
54 × 28 mm dia.	62.71	70.83	1.01	23.93	nr	**94.76**
Extra over copper pipes; compression fittings; BS 864						
Stop end						
15 mm dia.	1.30	1.47	0.10	2.38	nr	**3.85**
22 mm dia.	1.54	1.74	0.12	2.85	nr	**4.59**
28 mm dia.	1.77	2.00	0.15	3.56	nr	**5.56**
Straight connector; copper to copper						
15 mm dia.	2.69	3.04	0.18	4.27	nr	**7.31**
22 mm dia.	3.58	4.04	0.21	4.98	nr	**9.02**
28 mm dia.	4.74	5.36	0.24	5.69	nr	**11.05**

S: PIPED SUPPLY SYSTEMS

Item	Net Price £	Material £	Labour hours	Labour £	Unit	Total rate £
Straight connector; copper to imperial copper						
22 mm dia.	4.45	5.03	0.21	4.98	nr	**10.01**
Male coupling; copper to MI (BSP)						
15 mm dia.	0.77	0.87	0.19	4.51	nr	**5.38**
22 mm dia.	1.20	1.36	0.23	5.46	nr	**6.82**
28 mm dia.	2.80	3.16	0.26	6.15	nr	**9.31**
Male coupling with long thread and backnut						
15 mm dia.	4.93	5.57	0.19	4.51	nr	**10.08**
22 mm dia.	6.26	7.07	0.23	5.46	nr	**12.53**
Female coupling; copper to FI (BSP)						
15 mm dia.	0.94	1.06	0.19	4.51	nr	**5.57**
22 mm dia.	1.35	1.52	0.23	5.46	nr	**6.98**
28 mm dia.	3.91	4.42	0.27	6.40	nr	**10.82**
Elbow						
15 mm dia.	1.01	1.14	0.18	4.27	nr	**5.41**
22 mm dia.	1.72	1.94	0.21	4.98	nr	**6.92**
28 mm dia.	5.78	6.53	0.24	5.69	nr	**12.22**
Male elbow; copper to FI (BSP)						
15 mm × ½' dia.	1.93	2.18	0.19	4.51	nr	**6.69**
22 mm × ¾' dia.	2.50	2.83	0.23	5.46	nr	**8.29**
28 mm × 1' dia.	6.05	6.83	0.27	6.40	nr	**13.23**
Female elbow; copper to FI (BSP)						
15 mm × ½' dia.	2.96	3.35	0.19	4.51	nr	**7.86**
22 mm × ¾' dia.	4.27	4.82	0.23	5.46	nr	**10.28**
28 mm × 1' dia.	7.54	8.52	0.27	6.40	nr	**14.92**
Backplate elbow						
15 mm × ½' dia.	4.27	4.82	0.50	11.85	nr	**16.67**
Tank coupling; long thread						
22 mm dia.	6.26	7.07	0.46	10.89	nr	**17.96**
Tee equal						
15 mm dia.	1.42	1.60	0.28	6.63	nr	**8.23**
22 mm dia.	2.39	2.70	0.30	7.10	nr	**9.80**
28 mm dia.	10.90	12.31	0.34	8.05	nr	**20.36**
Tee reducing						
22 mm dia.	6.16	6.96	0.30	7.10	nr	**14.06**
Backplate tee						
15 mm dia.	15.05	16.99	0.62	14.70	nr	**31.69**

S: PIPED SUPPLY SYSTEMS

Item	Net Price £	Material £	Labour hours	Labour £	Unit	Total rate £
S10: COLD WATER – cont						
FIXINGS – cont						
Extra over fittings; silver brazed welded joints						
Reducer						
76 × 67 mm dia.	24.85	28.07	1.40	33.17	nr	**61.24**
108 × 76 mm dia.	52.06	58.79	1.80	42.65	nr	**101.44**
133 × 108 mm dia.	103.82	117.25	2.20	52.13	nr	**169.38**
159 × 133 mm dia.	132.78	149.96	2.60	61.61	nr	**211.57**
90° elbow						
76 mm dia.	60.02	67.79	1.60	37.91	nr	**105.70**
108 mm dia.	110.49	124.79	2.00	47.39	nr	**172.18**
133 mm dia.	233.56	263.78	2.40	56.86	nr	**320.64**
159 mm dia.	289.94	327.46	2.80	66.34	nr	**393.80**
45° elbow						
76 mm dia.	56.23	63.51	1.60	37.91	nr	**101.42**
108 mm dia.	88.19	99.60	2.00	47.39	nr	**146.99**
133 mm dia.	225.52	254.70	2.40	56.86	nr	**311.56**
159 mm dia.	324.60	366.60	2.80	66.34	nr	**432.94**
Equal tee						
76 mm dia.	73.09	82.55	2.40	56.86	nr	**139.41**
108 mm dia.	113.53	128.22	3.00	71.08	nr	**199.30**
133 mm dia.	267.38	301.98	3.60	85.30	nr	**387.28**
159 mm dia.	318.89	360.15	4.20	99.52	nr	**459.67**
Extra over copper pipes; dezincification resistant compression fittings; BS 864						
Stop end						
15 mm dia.	1.86	2.10	0.10	2.38	nr	**4.48**
22 mm dia.	2.69	3.04	0.13	3.08	nr	**6.12**
28 mm dia.	5.77	6.52	0.15	3.56	nr	**10.08**
35 mm dia.	9.04	10.21	0.18	4.27	nr	**14.48**
42 mm dia.	15.05	16.99	0.20	4.74	nr	**21.73**
Straight coupling; copper to copper						
15 mm dia.	1.49	1.68	0.18	4.27	nr	**5.95**
22 mm dia.	2.43	2.74	0.21	4.98	nr	**7.72**
28 mm dia.	5.51	6.22	0.24	5.69	nr	**11.91**
35 mm dia.	11.67	13.18	0.29	6.86	nr	**20.04**
42 mm dia.	15.34	17.33	0.33	7.81	nr	**25.14**
54 mm dia.	22.94	25.91	0.38	9.00	nr	**34.91**

S: PIPED SUPPLY SYSTEMS

Item	Net Price £	Material £	Labour hours	Labour £	Unit	Total rate £
Straight swivel connector; copper to imperial copper						
22 mm dia.	5.48	6.19	0.20	4.74	nr	10.93
Male coupling; copper to MI (BSP)						
15 mm × ½' dia.	1.33	1.50	0.19	4.51	nr	6.01
22 mm × ¾' dia.	2.02	2.28	0.23	5.46	nr	7.74
28 mm × 1' dia.	3.91	4.42	0.26	6.15	nr	10.57
35 mm × 1¼' dia.	8.87	10.02	0.32	7.58	nr	17.60
42 mm × 1½' dia.	13.29	15.01	0.37	8.76	nr	23.77
54 mm × 2' dia.	19.64	22.19	0.57	13.50	nr	35.69
Male coupling with long thread and backnuts						
22 mm dia.	7.54	8.52	0.23	5.46	nr	13.98
28 mm dia.	8.35	9.44	0.24	5.69	nr	15.13
Female coupling; copper to FI (BSP)						
15 mm × ½' dia.	1.60	1.81	0.19	4.51	nr	6.32
22 mm × ¾' dia.	2.34	2.64	0.23	5.46	nr	8.10
28 mm × 1' dia.	5.06	5.71	0.27	6.40	nr	12.11
35 mm × 1¼' dia.	10.65	12.03	0.32	7.58	nr	19.61
42 mm × 1½' dia.	14.31	16.16	0.37	8.76	nr	24.92
54 mm × 2' dia.	21.00	23.72	0.42	9.95	nr	33.67
Elbow						
15 mm dia.	1.79	2.02	0.18	4.27	nr	6.29
22 mm dia.	2.86	3.23	0.21	4.98	nr	8.21
28 mm dia.	7.11	8.03	0.24	5.69	nr	13.72
35 mm dia.	15.75	17.79	0.29	6.86	nr	24.65
42 mm dia.	21.33	24.09	0.33	7.81	nr	31.90
54 mm dia.	36.70	41.45	0.38	9.00	nr	50.45
Male elbow; copper to MI (BSP)						
15 mm × ½' dia.	3.11	3.51	0.19	4.51	nr	8.02
22 mm × ¾' dia.	3.48	3.93	0.23	5.46	nr	9.39
28 mm × 1' dia.	6.52	7.36	0.27	6.40	nr	13.76
Female elbow; copper to FI (BSP)						
15 mm × ½' dia.	3.33	3.76	0.19	4.51	nr	8.27
22 mm × ¾' dia.	4.80	5.42	0.23	5.46	nr	10.88
28 mm × 1' dia.	7.96	8.99	0.27	6.40	nr	15.39
Backplate elbow						
15 mm × ½' dia.	4.84	5.47	0.50	11.85	nr	17.32
Straight tap connector						
15 mm dia.	2.75	3.10	0.13	3.08	nr	6.18
22 mm dia.	5.97	6.74	0.15	3.56	nr	10.30

S: PIPED SUPPLY SYSTEMS

Item	Net Price £	Material £	Labour hours	Labour £	Unit	Total rate £
S10: COLD WATER – cont						
FIXINGS – cont						
Extra over copper pipes; dezincification resistant compression fittings; BS 864 – cont						
Tank coupling						
15 mm dia.	3.79	4.28	0.19	4.51	nr	8.79
22 mm dia.	4.20	4.74	0.23	5.46	nr	10.20
28 mm dia.	8.86	10.01	0.27	6.40	nr	16.41
35 mm dia.	15.64	17.67	0.32	7.58	nr	25.25
42 mm dia.	25.43	28.72	0.37	8.76	nr	37.48
54 mm dia.	32.75	36.99	0.31	7.34	nr	44.33
Tee equal						
15 mm dia.	2.52	2.85	0.28	6.63	nr	9.48
22 mm dia.	4.16	4.70	0.30	7.10	nr	11.80
28 mm dia.	11.35	12.82	0.34	8.05	nr	20.87
35 mm dia.	20.48	23.13	0.43	10.19	nr	33.32
42 mm dia.	32.20	36.36	0.46	10.89	nr	47.25
54 mm dia.	51.73	58.43	0.54	12.80	nr	71.23
Tee reducing						
22 mm dia.	6.65	7.51	0.30	7.10	nr	14.61
28 mm dia.	10.96	12.38	0.34	8.05	nr	20.43
35 mm dia.	20.01	22.60	0.43	10.19	nr	32.79
42 mm dia.	30.94	34.95	0.46	10.89	nr	45.84
54 mm dia.	51.73	58.43	0.54	12.80	nr	71.23
Extra over copper pipes; bronze one piece brazing flanges; metric, including jointing ring and bolts						
Bronze flange; PN6						
15 mm dia.	23.51	26.55	0.27	6.40	nr	32.95
22 mm dia.	28.02	31.65	0.32	7.58	nr	39.23
28 mm dia.	30.78	34.76	0.36	8.53	nr	43.29
35 mm dia.	36.77	41.52	0.47	11.14	nr	52.66
42 mm dia.	40.11	45.30	0.54	12.80	nr	58.10
54 mm dia.	46.53	52.55	0.63	14.93	nr	67.48
67 mm dia.	54.86	61.95	0.77	18.24	nr	80.19
76 mm dia.	66.11	74.66	0.93	22.03	nr	96.69
108 mm dia.	87.35	98.65	1.14	27.02	nr	125.67
133 mm dia.	119.68	135.17	1.41	33.40	nr	168.57
159 mm dia.	172.66	195.00	1.74	41.24	nr	236.24

S: PIPED SUPPLY SYSTEMS

Item	Net Price £	Material £	Labour hours	Labour £	Unit	Total rate £
Bronze flange; PN10						
15 mm dia.	26.92	30.41	0.27	6.40	nr	36.81
22 mm dia.	28.65	32.35	0.32	7.58	nr	39.93
28 mm dia.	31.51	35.59	0.38	9.00	nr	44.59
35 mm dia.	38.09	43.02	0.47	11.14	nr	54.16
42 mm dia.	42.70	48.23	0.54	12.80	nr	61.03
54 mm dia.	46.73	52.77	0.63	14.93	nr	67.70
67 mm dia.	54.89	62.00	0.77	18.24	nr	80.24
76 mm dia.	66.11	74.66	0.93	22.03	nr	96.69
108 mm dia.	83.56	94.37	1.14	27.02	nr	121.39
133 mm dia.	111.56	125.99	1.41	33.40	nr	159.39
159 mm dia.	141.40	159.70	1.74	41.24	nr	200.94
Bronze flange; PN16						
15 mm dia.	29.61	33.45	0.27	6.40	nr	39.85
22 mm dia.	31.52	35.60	0.32	7.58	nr	43.18
28 mm dia.	34.66	39.15	0.38	9.00	nr	48.15
35 mm dia.	41.90	47.32	0.47	11.14	nr	58.46
42 mm dia.	46.98	53.06	0.54	12.80	nr	65.86
54 mm dia.	51.40	58.05	0.63	14.93	nr	72.98
67 mm dia.	60.38	68.20	0.77	18.24	nr	86.44
76 mm dia.	72.71	82.12	0.93	22.03	nr	104.15
108 mm dia.	91.91	103.81	1.14	27.02	nr	130.83
133 mm dia.	122.71	138.59	1.41	33.40	nr	171.99
159 mm dia.	155.53	175.65	1.74	41.24	nr	216.89
Extra Over copper pipes; bronze blank flanges; metric, including jointing ring and bolts						
Gunmetal blank flange; PN6						
15 mm dia.	19.73	22.29	0.27	6.40	nr	28.69
22 mm dia.	25.18	28.44	0.27	6.40	nr	34.84
28 mm dia.	25.88	29.23	0.27	6.40	nr	35.63
35 mm dia.	42.35	47.83	0.32	7.58	nr	55.41
42 mm dia.	57.13	64.53	0.32	7.58	nr	72.11
54 mm dia.	63.00	71.16	0.34	8.05	nr	79.21
67 mm dia.	77.52	87.55	0.36	8.53	nr	96.08
76 mm dia.	99.86	112.78	0.37	8.76	nr	121.54
108 mm dia.	158.65	179.18	0.41	9.71	nr	188.89
133 mm dia.	187.17	211.38	0.58	13.75	nr	225.13
159 mm dia.	233.95	264.22	0.61	14.45	nr	278.67
Gunmetal blank flange; PN10						
15 mm dia.	23.87	26.96	0.27	6.40	nr	33.36
22 mm dia.	30.88	34.87	0.27	6.40	nr	41.27
28 mm dia.	34.20	38.63	0.27	6.40	nr	45.03
35 mm dia.	42.35	47.83	0.32	7.58	nr	55.41
42 mm dia.	79.38	89.65	0.32	7.58	nr	97.23
54 mm dia.	90.50	102.21	0.34	8.05	nr	110.26
67 mm dia.	97.08	109.64	0.46	10.89	nr	120.53

S: PIPED SUPPLY SYSTEMS

Item	Net Price £	Material £	Labour hours	Labour £	Unit	Total rate £
S10: COLD WATER – cont						
FIXINGS – cont						
Extra Over copper pipes; bronze blank flanges; metric, including jointing ring and bolts – cont						
Gunmetal blank flange – cont						
76 mm dia.	129.01	145.71	0.47	11.14	nr	**156.85**
108 mm dia.	151.42	171.01	0.51	12.09	nr	**183.10**
133 mm dia.	201.14	227.16	0.58	13.75	nr	**240.91**
159 mm dia.	273.94	309.39	0.71	16.83	nr	**326.22**
Gunmetal blank flange; PN16						
15 mm dia.	23.87	26.96	0.27	6.40	nr	**33.36**
22 mm dia.	31.35	35.40	0.27	6.40	nr	**41.80**
28 mm dia.	34.20	38.63	0.27	6.40	nr	**45.03**
35 mm dia.	42.35	47.83	0.32	7.58	nr	**55.41**
42 mm dia.	79.38	89.65	0.32	7.58	nr	**97.23**
54 mm dia.	90.50	102.21	0.34	8.05	nr	**110.26**
67 mm dia.	112.76	127.35	0.46	10.89	nr	**138.24**
76 mm dia.	129.01	145.71	0.47	11.14	nr	**156.85**
108 mm dia.	151.42	171.01	0.51	12.09	nr	**183.10**
133 mm dia.	347.75	392.75	0.58	13.75	nr	**406.50**
159 mm dia.	415.38	469.13	0.71	16.83	nr	**485.96**
Extra Over copper pipes; bronze screwed flanges; metric, including jointing ring and bolts						
Gunmetal screwed flange; 6 BSP						
15 mm dia.	19.73	22.29	0.35	8.29	nr	**30.58**
22 mm dia.	22.82	25.78	0.47	11.14	nr	**36.92**
28 mm dia.	23.78	26.86	0.52	12.32	nr	**39.18**
35 mm dia.	32.05	36.20	0.62	14.70	nr	**50.90**
42 mm dia.	38.48	43.46	0.70	16.60	nr	**60.06**
54 mm dia.	52.29	59.06	0.84	19.90	nr	**78.96**
67 mm dia.	65.63	74.12	1.03	24.41	nr	**98.53**
76 mm dia.	79.22	89.47	1.22	28.92	nr	**118.39**
108 mm dia.	125.45	141.68	1.41	33.40	nr	**175.08**
133 mm dia.	148.70	167.94	1.75	41.47	nr	**209.41**
159 mm dia.	189.30	213.79	2.21	52.37	nr	**266.16**
Gunmetal screwed flange; 10 BSP						
15 mm dia.	23.90	26.99	0.35	8.29	nr	**35.28**
22 mm dia.	27.83	31.43	0.47	11.14	nr	**42.57**
28 mm dia.	30.74	34.72	0.52	12.32	nr	**47.04**

S: PIPED SUPPLY SYSTEMS

Item	Net Price £	Material £	Labour hours	Labour £	Unit	Total rate £
35 mm dia.	43.80	49.47	0.62	14.70	nr	**64.17**
42 mm dia.	53.48	60.40	0.70	16.60	nr	**77.00**
54 mm dia.	76.05	85.89	0.84	19.90	nr	**105.79**
67 mm dia.	89.03	100.55	1.03	24.41	nr	**124.96**
76 mm dia.	100.50	113.51	1.22	28.92	nr	**142.43**
108 mm dia.	133.42	150.68	1.41	33.40	nr	**184.08**
133 mm dia.	161.69	182.61	1.75	41.47	nr	**224.08**
159 mm dia.	286.94	324.06	2.21	52.37	nr	**376.43**
Gunmetal screwed flange; 16 BSP						
15 mm dia.	18.78	21.21	0.35	8.29	nr	**29.50**
22 mm dia.	23.74	26.82	0.47	11.14	nr	**37.96**
28 mm dia.	27.48	31.04	0.52	12.32	nr	**43.36**
35 mm dia.	42.00	47.43	0.62	14.70	nr	**62.13**
42 mm dia.	50.54	57.08	0.70	16.60	nr	**73.68**
54 mm dia.	65.01	73.42	0.84	19.90	nr	**93.32**
67 mm dia.	93.97	106.13	1.03	24.41	nr	**130.54**
76 mm dia.	107.76	121.71	1.22	28.92	nr	**150.63**
108 mm dia.	126.09	142.40	1.41	33.40	nr	**175.80**
133 mm dia.	204.30	230.73	1.75	41.47	nr	**272.20**
159 mm dia.	252.73	285.44	2.21	52.37	nr	**337.81**
Extra over copper pipes; labour						
Made bend						
15 mm dia.	–	–	0.26	6.15	nr	**6.15**
22 mm dia.	–	–	0.28	6.63	nr	**6.63**
28 mm dia.	–	–	0.31	7.34	nr	**7.34**
35 mm dia.	–	–	0.42	9.95	nr	**9.95**
42 mm dia.	–	–	0.51	12.09	nr	**12.09**
54 mm dia.	–	–	0.58	13.75	nr	**13.75**
67 mm dia.	–	–	0.69	16.35	nr	**16.35**
76 mm dia.	–	–	0.80	18.95	nr	**18.95**
Bronze butt weld						
15 mm dia.	–	–	0.25	5.93	nr	**5.93**
22 mm dia.	–	–	0.31	7.34	nr	**7.34**
28 mm dia.	–	–	0.37	8.76	nr	**8.76**
35 mm dia.	–	–	0.49	11.61	nr	**11.61**
42 mm dia.	–	–	0.58	13.75	nr	**13.75**
54 mm dia.	–	–	0.72	17.06	nr	**17.06**
67 mm dia.	–	–	0.88	20.85	nr	**20.85**
76 mm dia.	–	–	1.08	25.59	nr	**25.59**
108 mm dia.	–	–	1.37	32.46	nr	**32.46**
133 mm dia.	–	–	1.73	40.99	nr	**40.99**
159 mm dia.	–	–	2.03	48.10	nr	**48.10**

S: PIPED SUPPLY SYSTEMS

Item	Net Price £	Material £	Labour hours	Labour £	Unit	Total rate £
S10: COLD WATER – cont						
PRESS FIT (COPPER FITTINGS)						
Mechanical press fit joints; butyl rubber O ring						
Coupler						
15 mm dia.	0.84	0.95	0.36	8.53	nr	9.48
22 mm dia.	1.33	1.50	0.36	8.53	nr	10.03
28 mm dia.	2.72	3.07	0.44	10.42	nr	13.49
35 mm dia.	3.43	3.88	0.44	10.42	nr	14.30
42 mm dia.	6.17	6.97	0.52	12.32	nr	19.29
54 mm dia.	7.88	8.90	0.60	14.22	nr	23.12
Stop end						
22 mm dia.	2.13	2.41	0.18	4.27	nr	6.68
28 mm dia.	3.36	3.79	0.22	5.22	nr	9.01
35 mm dia.	5.78	6.53	0.22	5.22	nr	11.75
42 mm dia.	8.68	9.80	0.26	6.15	nr	15.95
54 mm dia.	10.46	11.81	0.30	7.10	nr	18.91
Reducer						
22 × 15 mm dia.	0.98	1.11	0.36	8.53	nr	9.64
28 × 15 mm dia.	2.60	2.94	0.40	9.48	nr	12.42
28 × 22 mm dia.	2.68	3.03	0.40	9.48	nr	12.51
35 × 22 mm dia.	3.22	3.64	0.40	9.48	nr	13.12
35 × 28 mm dia.	3.57	4.03	0.44	10.42	nr	14.45
42 × 22 mm dia.	5.57	6.29	0.44	10.42	nr	16.71
42 × 28 mm dia.	5.30	5.99	0.48	11.37	nr	17.36
42 × 35 mm dia.	5.30	5.99	0.48	11.37	nr	17.36
54 × 35 mm dia.	7.18	8.11	0.52	12.32	nr	20.43
54 × 42 mm dia.	7.18	8.11	0.56	13.27	nr	21.38
90° elbow						
15 mm dia.	0.91	1.03	0.36	8.53	nr	9.56
22 mm dia.	1.53	1.72	0.36	8.53	nr	10.25
28 mm dia.	3.29	3.71	0.44	10.42	nr	14.13
35 mm dia.	6.79	7.67	0.44	10.42	nr	18.09
42 mm dia.	12.75	14.40	0.52	12.32	nr	26.72
54 mm dia.	17.69	19.98	0.60	14.22	nr	34.20
45° elbow						
15 mm dia.	1.12	1.26	0.36	8.53	nr	9.79
22 mm dia.	1.55	1.75	0.36	8.53	nr	10.28
28 mm dia.	4.66	5.26	0.44	10.42	nr	15.68
35 mm dia.	6.65	7.51	0.44	10.42	nr	17.93
42 mm dia.	11.07	12.51	0.52	12.32	nr	24.83
54 mm dia.	15.74	17.78	0.60	14.22	nr	32.00

S: PIPED SUPPLY SYSTEMS

Item	Net Price £	Material £	Labour hours	Labour £	Unit	Total rate £
Equal tee						
15 mm dia.	1.45	1.64	0.54	12.80	nr	**14.44**
22 mm dia.	2.65	2.99	0.54	12.80	nr	**15.79**
28 mm dia.	4.75	5.37	0.66	15.65	nr	**21.02**
35 mm dia.	8.21	9.27	0.66	15.65	nr	**24.92**
42 mm dia.	16.21	18.31	0.78	18.47	nr	**36.78**
54 mm dia.	20.23	22.85	0.90	21.32	nr	**44.17**
Reducing tee						
22 × 15 mm dia.	2.15	2.43	0.54	12.80	nr	**15.23**
28 × 15 mm dia.	4.18	4.72	0.62	14.70	nr	**19.42**
28 × 22 mm dia.	5.61	6.33	0.62	14.70	nr	**21.03**
35 × 22 mm dia.	7.31	8.25	0.62	14.70	nr	**22.95**
35 × 28 mm dia.	8.14	9.19	0.62	14.70	nr	**23.89**
42 × 28 mm dia.	14.72	16.63	0.70	16.60	nr	**33.23**
42 × 35 mm dia.	14.72	16.63	0.70	16.60	nr	**33.23**
54 × 35 mm dia.	24.87	28.09	0.82	19.42	nr	**47.51**
54 × 42 mm dia.	24.87	28.09	0.82	19.42	nr	**47.51**
Male iron connector; BSP thread						
15 mm dia.	3.07	3.47	0.18	4.27	nr	**7.74**
22 mm dia.	4.55	5.14	0.18	4.27	nr	**9.41**
28 mm dia.	6.09	6.87	0.22	5.22	nr	**12.09**
35 mm dia.	11.02	12.44	0.22	5.22	nr	**17.66**
42 mm dia.	14.77	16.68	0.26	6.15	nr	**22.83**
54 mm dia.	28.50	32.19	0.30	7.10	nr	**39.29**
90° elbow; male iron BSP thread						
15 mm dia.	4.97	5.61	0.36	8.53	nr	**14.14**
22 mm dia.	7.78	8.78	0.36	8.53	nr	**17.31**
28 mm dia.	11.92	13.46	0.44	10.42	nr	**23.88**
35 mm dia.	15.50	17.50	0.44	10.42	nr	**27.92**
42 mm dia.	20.21	22.83	0.52	12.32	nr	**35.15**
54 mm dia.	29.54	33.36	0.60	14.22	nr	**47.58**
Female iron connector; BSP thread						
15 mm dia.	3.51	3.97	0.18	4.27	nr	**8.24**
22 mm dia.	4.63	5.23	0.18	4.27	nr	**9.50**
28 mm dia.	6.24	7.05	0.22	5.22	nr	**12.27**
35 mm dia.	12.20	13.78	0.22	5.22	nr	**19.00**
42 mm dia.	17.43	19.69	0.26	6.15	nr	**25.84**
54 mm dia.	29.89	33.76	0.30	7.10	nr	**40.86**
90° elbow; female iron BSP thread						
15 mm dia.	4.20	4.74	0.36	8.53	nr	**13.27**
22 mm dia.	6.17	6.97	0.36	8.53	nr	**15.50**
28 mm dia.	10.19	11.51	0.44	10.42	nr	**21.93**
35 mm dia.	13.18	14.88	0.44	10.42	nr	**25.30**
42 mm dia.	17.93	20.25	0.52	12.32	nr	**32.57**
54 mm dia.	26.38	29.79	0.60	14.22	nr	**44.01**

S: PIPED SUPPLY SYSTEMS

Item	Net Price £	Material £	Labour hours	Labour £	Unit	Total rate £
S10: COLD WATER – cont						
STAINLESS STEEL PIPEWORK						
Stainless steel pipes; capillary or compression joints; BS 4127, vertical or at low level, with brackets measured separately						
Grade 304; satin finish						
15 mm dia.	3.51	3.97	0.41	9.71	m	**13.68**
22 mm dia.	4.92	5.56	0.51	12.09	m	**17.65**
28 mm dia.	6.71	7.58	0.58	13.75	m	**21.33**
35 mm dia.	10.13	11.44	0.65	15.40	m	**26.84**
42 mm dia.	12.86	14.52	0.71	16.83	m	**31.35**
54 mm dia.	17.92	20.24	0.80	18.95	m	**39.19**
Grade 316 satin finish						
15 mm dia.	4.51	5.09	0.61	14.45	m	**19.54**
22 mm dia.	8.44	9.54	0.76	18.02	m	**27.56**
28 mm dia.	10.01	11.30	0.87	20.62	m	**31.92**
35 mm dia.	18.16	20.51	0.98	23.24	m	**43.75**
42 mm dia.	23.50	26.54	1.06	25.13	m	**51.67**
54 mm dia.	27.38	30.93	1.16	27.49	m	**58.42**
FIXINGS						
For stainless steel pipework						
Single pipe ring						
15 mm dia.	9.93	11.21	0.26	6.15	nr	**17.36**
22 mm dia.	11.57	13.07	0.26	6.15	nr	**19.22**
28 mm dia.	12.12	13.69	0.31	7.34	nr	**21.03**
35 mm dia.	13.77	15.55	0.32	7.58	nr	**23.13**
42 mm dia.	15.64	17.67	0.32	7.58	nr	**25.25**
54 mm dia.	17.86	20.18	0.34	8.05	nr	**28.23**
Screw on backplate, female						
All sizes 15 mm to 54 mm dia.	8.95	10.11	0.10	2.38	nr	**12.49**
Screw on backplate, male						
All sizes 15 mm to 54 mm dia.	10.20	11.52	0.10	2.38	nr	**13.90**
Stainless steel threaded rods; metric thread; including nuts, washers etc						
10 mm dia. × 600 mm long	10.67	12.05	0.18	4.27	nr	**16.32**

S: PIPED SUPPLY SYSTEMS

Item	Net Price £	Material £	Labour hours	Labour £	Unit	Total rate £
Extra over stainless steel pipes; capillary fittings						
Straight coupling						
15 mm dia.	1.09	1.23	0.25	5.93	nr	7.16
22 mm dia.	1.77	2.00	0.28	6.63	nr	8.63
28 mm dia.	2.35	2.65	0.33	7.81	nr	10.46
35 mm dia.	5.40	6.10	0.37	8.76	nr	14.86
42 mm dia.	6.22	7.03	0.42	9.95	nr	16.98
54 mm dia.	9.36	10.57	0.45	10.66	nr	21.23
45° bend						
15 mm dia.	5.89	6.65	0.25	5.93	nr	12.58
22 mm dia.	7.74	8.74	0.30	7.03	nr	15.77
28 mm dia.	9.51	10.74	0.33	7.81	nr	18.55
35 mm dia.	11.26	12.72	0.37	8.76	nr	21.48
42 mm dia.	14.50	16.38	0.42	9.95	nr	26.33
54 mm dia.	17.92	20.24	0.45	10.66	nr	30.90
90° bend						
15 mm dia.	3.04	3.44	0.28	6.63	nr	10.07
22 mm dia.	4.11	4.64	0.28	6.63	nr	11.27
28 mm dia.	5.79	6.54	0.33	7.81	nr	14.35
35 mm dia.	14.13	15.96	0.37	8.76	nr	24.72
42 mm dia.	19.46	21.98	0.42	9.95	nr	31.93
54 mm dia.	26.38	29.79	0.45	10.66	nr	40.45
Reducer						
22 × 15 mm dia.	7.03	7.94	0.28	6.63	nr	14.57
28 × 22 mm dia.	7.83	8.84	0.33	7.81	nr	16.65
35 × 28 mm dia.	9.58	10.82	0.37	8.76	nr	19.58
42 × 35 mm dia.	10.33	11.67	0.42	9.95	nr	21.62
54 × 42 mm dia.	30.63	34.60	0.48	11.39	nr	45.99
Tap connector						
15 mm dia.	14.82	16.74	0.13	3.08	nr	19.82
22 mm dia.	19.58	22.11	0.14	3.33	nr	25.44
28 mm dia.	27.18	30.70	0.17	4.03	nr	34.73
Tank connector						
15 mm dia.	19.14	21.61	0.13	3.08	nr	24.69
22 mm dia.	28.48	32.16	0.13	3.08	nr	35.24
28 mm dia.	35.07	39.61	0.15	3.56	nr	43.17
35 mm dia.	50.71	57.27	0.18	4.27	nr	61.54
42 mm dia.	66.96	75.62	0.21	4.98	nr	80.60
54 mm dia.	101.38	114.50	0.24	5.69	nr	120.19

S: PIPED SUPPLY SYSTEMS

Item	Net Price £	Material £	Labour hours	Labour £	Unit	Total rate £
S10: COLD WATER – cont						
FIXINGS – cont						
Extra over stainless steel pipes; capillary fittings – cont						
Tee equal						
15 mm dia.	5.46	6.17	0.37	8.76	nr	14.93
22 mm dia.	6.79	7.67	0.40	9.48	nr	17.15
28 mm dia.	8.21	9.27	0.45	10.66	nr	19.93
35 mm dia.	19.71	22.26	0.59	13.98	nr	36.24
42 mm dia.	24.32	27.47	0.62	14.70	nr	42.17
54 mm dia.	49.12	55.48	0.67	15.88	nr	71.36
Unequal tee						
22 × 15 mm dia.	11.07	12.51	0.37	8.76	nr	21.27
28 × 15 mm dia.	12.46	14.08	0.45	10.66	nr	24.74
28 × 22 mm dia.	12.46	14.08	0.45	10.67	nr	24.75
35 × 22 mm dia.	21.77	24.58	0.59	13.98	nr	38.56
35 × 28 mm dia.	21.77	24.58	0.59	13.98	nr	38.56
42 × 28 mm dia.	26.76	30.22	0.62	14.70	nr	44.92
42 × 35 mm dia.	26.76	30.22	0.62	14.70	nr	44.92
54 × 35 mm dia.	55.42	62.59	0.67	15.88	nr	78.47
54 × 42 mm dia.	55.42	62.59	0.67	15.88	nr	78.47
Union, conical seat						
15 mm dia.	24.54	27.71	0.25	5.93	nr	33.64
22 mm dia.	38.64	43.64	0.28	6.63	nr	50.27
28 mm dia.	49.95	56.42	0.33	7.81	nr	64.23
35 mm dia.	65.58	74.06	0.37	8.76	nr	82.82
42 mm dia.	82.71	93.41	0.42	9.95	nr	103.36
54 mm dia.	109.43	123.59	0.45	10.66	nr	134.25
Union, flat seat						
15 mm dia.	25.62	28.94	0.25	5.93	nr	34.87
22 mm dia.	39.90	45.06	0.28	6.63	nr	51.69
28 mm dia.	51.58	58.25	0.33	7.81	nr	66.06
35 mm dia.	67.39	76.11	0.37	8.76	nr	84.87
42 mm dia.	84.90	95.89	0.42	9.95	nr	105.84
54 mm dia.	113.85	128.58	0.45	10.66	nr	139.24
Extra over stainless steel pipes; compression fittings						
Straight coupling						
15 mm dia.	21.33	24.09	0.18	4.27	nr	28.36
22 mm dia.	40.64	45.90	0.22	5.22	nr	51.12
28 mm dia.	54.70	61.78	0.25	5.93	nr	67.71
35 mm dia.	84.45	95.38	0.30	7.10	nr	102.48
42 mm dia.	98.58	111.33	0.40	9.48	nr	120.81

S: PIPED SUPPLY SYSTEMS

Item	Net Price £	Material £	Labour hours	Labour £	Unit	Total rate £
90° bend						
15 mm dia.	26.89	30.37	0.18	4.27	nr	**34.64**
22 mm dia.	53.41	60.32	0.22	5.22	nr	**65.54**
28 mm dia.	72.85	82.27	0.25	5.93	nr	**88.20**
35 mm dia.	147.50	166.59	0.33	7.81	nr	**174.40**
42 mm dia.	215.55	243.44	0.35	8.29	nr	**251.73**
Reducer						
22 × 15 mm dia.	38.70	43.71	0.28	6.63	nr	**50.34**
28 × 22 mm dia.	53.00	59.85	0.28	6.63	nr	**66.48**
35 × 28 mm dia.	77.45	87.48	0.30	7.10	nr	**94.58**
42 × 35 mm dia.	103.04	116.37	0.37	8.76	nr	**125.13**
Stud coupling						
15 mm dia.	22.17	25.04	0.42	9.95	nr	**34.99**
22 mm dia.	37.50	42.35	0.25	5.93	nr	**48.28**
28 mm dia.	52.06	58.79	0.25	5.93	nr	**64.72**
35 mm dia.	83.39	94.18	0.37	8.76	nr	**102.94**
42 mm dia.	98.58	111.33	0.42	9.95	nr	**121.28**
Equal tee						
15 mm dia.	37.86	42.76	0.37	8.76	nr	**51.52**
22 mm dia.	78.22	88.34	0.40	9.48	nr	**97.82**
28 mm dia.	106.98	120.82	0.45	10.66	nr	**131.48**
35 mm dia.	212.72	240.24	0.59	13.98	nr	**254.22**
42 mm dia.	294.96	333.12	0.62	14.70	nr	**347.82**
Running tee						
15 mm dia.	46.61	52.64	0.37	8.76	nr	**61.40**
22 mm dia.	83.96	94.82	0.40	9.48	nr	**104.30**
28 mm dia.	142.16	160.56	0.59	13.98	nr	**174.54**
PRESS FIT (STAINLESS STEEL)						
Press fit jointing system; butyl rubber O ring mechanical joint						
Pipework						
15 mm dia.	4.32	4.88	0.46	10.89	m	**15.77**
22 mm dia.	6.90	7.79	0.48	11.37	m	**19.16**
28 mm dia.	8.50	9.60	0.52	12.32	m	**21.92**
35 mm dia.	12.53	14.15	0.56	13.27	m	**27.42**
42 mm dia.	15.41	17.40	0.58	13.75	m	**31.15**
54 mm dia.	19.62	22.15	0.66	15.65	m	**37.80**

S: PIPED SUPPLY SYSTEMS

Item	Net Price £	Material £	Labour hours	Labour £	Unit	Total rate £
S10: COLD WATER – cont						
FIXINGS						
For stainless steel pipes						
Refer to fixings for stainless steel pipes; capillary or compression joints; BS 4127						
Extra over stainless steel pipes; Press fit jointing system						
Coupling						
15 mm dia.	4.31	4.87	0.36	8.53	nr	**13.40**
22 mm dia.	5.43	6.13	0.36	8.53	nr	**14.66**
28 mm dia.	6.10	6.88	0.44	10.42	nr	**17.30**
35 mm dia.	7.59	8.57	0.44	10.42	nr	**18.99**
42 mm dia.	10.36	11.70	0.52	12.32	nr	**24.02**
54 mm dia.	12.46	14.08	0.60	14.22	nr	**28.30**
Stop end						
22 mm dia.	4.10	4.63	0.18	4.27	nr	**8.90**
28 mm dia.	4.77	5.39	0.22	5.22	nr	**10.61**
35 mm dia.	7.80	8.81	0.22	5.22	nr	**14.03**
42 mm dia.	10.95	12.36	0.26	6.15	nr	**18.51**
54 mm dia.	12.67	14.31	0.30	7.10	nr	**21.41**
Reducer						
22 × 15 mm dia.	5.14	5.80	0.36	8.53	nr	**14.33**
28 × 15 mm dia.	5.81	6.56	0.40	9.48	nr	**16.04**
28 × 22 mm dia.	6.02	6.80	0.40	9.48	nr	**16.28**
35 × 22 mm dia.	7.34	8.29	0.40	9.48	nr	**17.77**
35 × 28 mm dia.	9.10	10.28	0.44	10.42	nr	**20.70**
42 × 35 mm dia.	9.59	10.83	0.48	11.37	nr	**22.20**
54 × 42 mm dia.	10.97	12.39	0.56	13.27	nr	**25.66**
90° bend						
15 mm dia.	6.16	6.96	0.36	8.53	nr	**15.49**
22 mm dia.	8.61	9.72	0.36	8.53	nr	**18.25**
28 mm dia.	10.87	12.28	0.44	10.42	nr	**22.70**
35 mm dia.	17.10	19.31	0.44	10.42	nr	**29.73**
42 mm dia.	28.56	32.25	0.52	12.32	nr	**44.57**
54 mm dia.	39.45	44.55	0.60	14.22	nr	**58.77**
45° bend						
15 mm dia.	8.36	9.45	0.36	8.53	nr	**17.98**
22 mm dia.	10.40	11.75	0.36	8.53	nr	**20.28**
28 mm dia.	12.09	13.66	0.44	10.42	nr	**24.08**
35 mm dia.	14.20	16.03	0.44	10.42	nr	**26.45**
42 mm dia.	22.85	25.81	0.52	12.32	nr	**38.13**
54 mm dia.	29.68	33.52	0.60	14.22	nr	**47.74**

S: PIPED SUPPLY SYSTEMS

Item	Net Price £	Material £	Labour hours	Labour £	Unit	Total rate £
Equal tee						
15 mm dia.	10.10	11.40	0.54	12.80	nr	**24.20**
22 mm dia.	12.40	14.00	0.54	12.80	nr	**26.80**
28 mm dia.	14.49	16.36	0.66	15.65	nr	**32.01**
35 mm dia.	18.35	20.73	0.66	15.65	nr	**36.38**
42 mm dia.	26.04	29.41	0.78	18.47	nr	**47.88**
54 mm dia.	31.17	35.20	0.90	21.32	nr	**56.52**
Reducing tee						
22 × 15 mm dia.	10.60	11.97	0.54	12.80	nr	**24.77**
28 × 15 mm dia.	12.85	14.51	0.62	14.70	nr	**29.21**
28 × 22 mm dia.	13.91	15.71	0.62	14.70	nr	**30.41**
35 × 22 mm dia.	16.49	18.63	0.62	14.70	nr	**33.33**
35 × 28 mm dia.	17.21	19.44	0.62	14.70	nr	**34.14**
42 × 28 mm dia.	24.47	27.63	0.70	16.60	nr	**44.23**
42 × 35 mm dia.	25.20	28.46	0.70	16.60	nr	**45.06**
54 × 35 mm dia.	28.47	32.15	0.82	19.42	nr	**51.57**
54 × 42 mm dia.	29.27	33.06	0.82	19.42	nr	**52.48**

FIXINGS

For stainless steel pipes

Refer to fixings for stainless steel pipes; capillary or compression joints; BS 4127

MEDIUM DENSITY POLYETHYLENE – BLUE

Note: MDPE is sized on Outside Diameter ie OD not ID.

Pipes for water distribution; laid underground; electrofusion joints in the running length; BS 6572

Item	Net Price £	Material £	Labour hours	Labour £	Unit	Total rate £
Coiled service pipe						
20 mm dia.	0.67	0.75	0.37	8.76	m	**9.51**
25 mm dia.	0.87	0.98	0.41	9.71	m	**10.69**
32 mm dia.	1.45	1.64	0.47	11.14	m	**12.78**
50 mm dia.	3.46	3.91	0.53	12.56	m	**16.47**
63 mm dia.	5.43	6.13	0.60	14.22	m	**20.35**
Mains service pipe						
90 mm dia.	8.19	9.25	0.90	21.32	m	**30.57**
110 mm dia.	12.27	13.86	1.10	26.07	m	**39.93**
125 mm dia.	15.52	17.52	1.20	28.45	m	**45.97**
160 mm dia.	24.73	27.93	1.48	35.07	m	**63.00**
180 mm dia.	32.22	36.39	1.50	35.58	m	**71.97**
225 mm dia.	48.95	55.28	1.77	41.93	m	**97.21**
250 mm dia.	61.79	69.79	1.75	41.49	m	**111.28**
315 mm dia.	95.42	107.76	1.90	45.01	m	**152.77**

S: PIPED SUPPLY SYSTEMS

Item	Net Price £	Material £	Labour hours	Labour £	Unit	Total rate £
S10: COLD WATER – cont						
MEDIUM DENSITY POLYETHYLENE – BLUE – cont						
Extra over fittings; MDPE blue; electrofusion joints						
Coupler						
20 mm dia.	5.57	6.29	0.36	8.53	nr	**14.82**
25 mm dia.	5.57	6.29	0.40	9.48	nr	**15.77**
32 mm dia.	5.57	6.29	0.44	10.42	nr	**16.71**
40 mm dia.	8.23	9.29	0.48	11.37	nr	**20.66**
50 mm dia.	7.91	8.94	0.52	12.32	nr	**21.26**
63 mm dia.	10.33	11.67	0.58	13.75	nr	**25.42**
90 mm dia.	15.18	17.15	0.67	15.88	nr	**33.03**
110 mm dia.	24.42	27.58	0.74	17.54	nr	**45.12**
125 mm dia.	27.55	31.11	0.83	19.67	nr	**50.78**
160 mm dia.	43.82	49.49	1.00	23.69	nr	**73.18**
180 mm dia.	51.52	58.19	1.25	29.62	nr	**87.81**
225 mm dia.	82.35	93.00	1.35	31.99	nr	**124.99**
250 mm dia.	120.57	136.17	1.50	35.55	nr	**171.72**
315 mm dia.	198.84	224.57	1.80	42.65	nr	**267.22**
Extra over fittings; MDPE blue; butt fused joints						
Cap						
25 mm dia.	10.57	11.93	0.20	4.74	nr	**16.67**
32 mm dia.	10.57	11.93	0.22	5.22	nr	**17.15**
40 mm dia.	11.21	12.66	0.24	5.69	nr	**18.35**
50 mm dia.	16.35	18.46	0.26	6.15	nr	**24.61**
63 mm dia.	18.75	21.18	0.32	7.58	nr	**28.76**
90 mm dia.	30.92	34.92	0.37	8.76	nr	**43.68**
110 mm dia.	62.27	70.33	0.40	9.48	nr	**79.81**
125 mm dia.	49.71	56.14	0.46	10.89	nr	**67.03**
160 mm dia.	56.83	64.19	0.50	11.85	nr	**76.04**
180 mm dia.	95.06	107.37	0.60	14.22	nr	**121.59**
225 mm dia.	112.28	126.81	0.68	16.12	nr	**142.93**
250 mm dia.	164.81	186.14	0.75	17.77	nr	**203.91**
315 mm dia.	212.50	240.00	0.90	21.32	nr	**261.32**
Reducer						
63 × 32 mm dia.	14.28	16.13	0.54	12.80	nr	**28.93**
63 × 50 mm dia.	16.72	18.88	0.60	14.22	nr	**33.10**
90 × 63 mm dia.	21.15	23.89	0.67	15.88	nr	**39.77**
110 × 90 mm dia.	29.14	32.92	0.74	17.54	nr	**50.46**
125 × 90 mm dia.	42.37	47.85	0.83	19.67	nr	**67.52**
125 × 110 mm dia.	46.69	52.73	1.00	23.69	nr	**76.42**
160 × 110 mm dia.	71.74	81.02	1.10	26.07	nr	**107.09**

S: PIPED SUPPLY SYSTEMS

Item	Net Price £	Material £	Labour hours	Labour £	Unit	Total rate £
180 × 125 mm dia.	77.91	88.00	1.25	29.62	nr	**117.62**
225 × 160 mm dia.	127.04	143.48	1.40	33.17	nr	**176.65**
250 × 180 mm dia.	97.61	110.24	1.80	42.65	nr	**152.89**
315 × 250 mm dia.	112.29	126.82	2.40	56.86	nr	**183.68**
Bend; 45°						
50 mm dia.	21.80	24.62	0.50	11.85	nr	**36.47**
63 mm dia.	26.38	29.79	0.58	13.75	nr	**43.54**
90 mm dia.	40.84	46.12	0.67	15.88	nr	**62.00**
110 mm dia.	59.88	67.63	0.74	17.54	nr	**85.17**
125 mm dia.	66.94	75.60	0.83	19.67	nr	**95.27**
160 mm dia.	124.05	140.10	1.00	23.69	nr	**163.79**
180 mm dia.	141.21	159.48	1.25	29.62	nr	**189.10**
225 mm dia.	179.99	203.28	1.40	33.16	nr	**236.44**
250 mm dia.	194.77	219.97	1.80	42.65	nr	**262.62**
315 mm dia.	242.67	274.07	2.40	56.86	nr	**330.93**
Bend; 90°						
50 mm dia.	21.80	24.62	0.50	11.85	nr	**36.47**
63 mm dia.	26.38	29.79	0.58	13.75	nr	**43.54**
90 mm dia.	40.84	46.12	0.67	15.88	nr	**62.00**
110 mm dia.	59.88	67.63	0.74	17.54	nr	**85.17**
125 mm dia.	66.94	75.60	0.83	19.67	nr	**95.27**
160 mm dia.	124.05	140.10	1.00	23.69	nr	**163.79**
180 mm dia.	240.93	272.11	1.25	29.62	nr	**301.73**
225 mm dia.	304.63	344.05	1.40	33.17	nr	**377.22**
250 mm dia.	333.22	376.34	1.80	42.65	nr	**418.99**
315 mm dia.	417.15	471.13	2.40	56.86	nr	**527.99**
Equal tee						
50 mm dia.	23.84	26.93	0.70	16.60	nr	**43.53**
63 mm dia.	26.02	29.39	0.75	17.77	nr	**47.16**
90 mm dia.	47.53	53.68	0.87	20.61	nr	**74.29**
110 mm dia.	70.86	80.03	1.00	23.69	nr	**103.72**
125 mm dia.	91.49	103.33	1.08	25.59	nr	**128.92**
160 mm dia.	150.44	169.90	1.35	31.99	nr	**201.89**
180 mm dia.	153.94	173.86	1.63	38.63	nr	**212.49**
225 mm dia.	186.56	210.70	1.90	45.01	nr	**255.71**
250 mm dia.	257.41	290.72	2.70	63.98	nr	**354.70**
315 mm dia.	641.72	724.75	3.60	85.30	nr	**810.05**
Extra over plastic fittings, compression joints						
Straight connector						
20 mm dia.	2.46	2.77	0.38	9.00	nr	**11.77**
25 mm dia.	2.59	2.93	0.45	10.66	nr	**13.59**
32 mm dia.	6.13	6.93	0.50	11.85	nr	**18.78**
50 mm dia.	14.13	15.96	0.68	16.12	nr	**32.08**
63 mm dia.	21.27	24.02	0.85	20.15	nr	**44.17**

S: PIPED SUPPLY SYSTEMS

Item	Net Price £	Material £	Labour hours	Labour £	Unit	Total rate £
S10: COLD WATER – cont						
MEDIUM DENSITY POLYETHYLENE – BLUE – cont						
Extra over plastic fittings, compression joints – cont						
Reducing connector						
25 mm dia.	5.07	5.72	0.38	9.00	nr	**14.72**
32 mm dia.	8.18	9.24	0.45	10.66	nr	**19.90**
50 mm dia.	22.66	25.59	0.50	11.85	nr	**37.44**
63 mm dia.	31.62	35.71	0.62	14.70	nr	**50.41**
Straight connector; polyethylene to MI						
20 mm dia.	2.23	2.52	0.31	7.35	nr	**9.87**
25 mm dia.	3.80	4.29	0.35	8.29	nr	**12.58**
32 mm dia.	4.12	4.65	0.40	9.48	nr	**14.13**
50 mm dia.	10.51	11.87	0.55	13.04	nr	**24.91**
63 mm dia.	14.82	16.74	0.65	15.40	nr	**32.14**
Straight connector; polyethylene to FI						
20 mm dia.	3.00	3.39	0.31	7.35	nr	**10.74**
25 mm dia.	3.25	3.67	0.35	8.29	nr	**11.96**
32 mm dia.	3.89	4.40	0.40	9.48	nr	**13.88**
50 mm dia.	12.35	13.94	0.55	13.04	nr	**26.98**
63 mm dia.	17.30	19.54	0.75	17.78	nr	**37.32**
Elbow						
20 mm dia.	3.28	3.70	0.38	9.00	nr	**12.70**
25 mm dia.	4.85	5.48	0.45	10.66	nr	**16.14**
32 mm dia.	7.07	7.99	0.50	11.85	nr	**19.84**
50 mm dia.	16.41	18.53	0.68	16.12	nr	**34.65**
63 mm dia.	22.32	25.20	0.80	18.95	nr	**44.15**
Elbow; polyethylene to MI						
25 mm dia.	4.18	4.72	0.35	8.29	nr	**13.01**
Elbow; polyethylene to FI						
20 mm dia.	2.98	3.37	0.31	7.35	nr	**10.72**
25 mm dia.	4.06	4.59	0.35	8.29	nr	**12.88**
32 mm dia.	6.09	6.87	0.42	9.95	nr	**16.82**
50 mm dia.	14.45	16.32	0.50	11.85	nr	**28.17**
63 mm dia.	18.94	21.39	0.55	13.04	nr	**34.43**
Tank coupling						
25 mm dia.	6.27	7.08	0.42	9.95	nr	**17.03**

S: PIPED SUPPLY SYSTEMS

Item	Net Price £	Material £	Labour hours	Labour £	Unit	Total rate £
Equal tee						
20 mm dia.	4.42	4.99	0.53	12.56	nr	**17.55**
25 mm dia.	6.91	7.80	0.55	13.04	nr	**20.84**
32 mm dia.	8.65	9.77	0.64	15.17	nr	**24.94**
50 mm dia.	20.19	22.81	0.75	17.78	nr	**40.59**
63 mm dia.	31.29	35.34	0.87	20.62	nr	**55.96**
Equal tee; FI branch						
20 mm dia.	4.28	4.83	0.45	10.66	nr	**15.49**
25 mm dia.	6.82	7.70	0.50	11.85	nr	**19.55**
32 mm dia.	8.30	9.37	0.60	14.22	nr	**23.59**
50 mm dia.	19.13	21.60	0.68	16.12	nr	**37.72**
63 mm dia.	26.85	30.32	0.81	19.20	nr	**49.52**
Equal tee; MI branch						
25 mm dia.	6.70	7.57	0.50	11.85	nr	**19.42**
ABS PIPEWORK						
Pipes; solvent welded joints in the running length, brackets measured separately						
Class C (9 bar pressure)						
1' dia.	4.47	5.05	0.30	7.10	m	**12.15**
1¼' dia.	7.50	8.47	0.33	7.81	m	**16.28**
1½' dia.	9.54	10.77	0.36	8.53	m	**19.30**
2' dia.	12.84	14.50	0.39	9.24	m	**23.74**
3' dia.	26.45	29.88	0.46	10.89	m	**40.77**
4' dia.	43.60	49.25	0.53	12.56	m	**61.81**
6' dia.	86.22	97.38	0.76	18.00	m	**115.38**
8' dia.	148.06	167.22	0.97	22.98	m	**190.20**
Class E (15 bar pressure)						
½' dia.	3.40	3.84	0.24	5.69	m	**9.53**
¾' dia.	5.24	5.92	0.27	6.40	m	**12.32**
1' dia.	6.91	7.80	0.30	7.10	m	**14.90**
1¼' dia.	10.31	11.65	0.33	7.81	m	**19.46**
1½' dia.	13.60	15.36	0.36	8.53	m	**23.89**
2' dia.	17.04	19.25	0.39	9.24	m	**28.49**
3' dia.	34.24	38.67	0.49	11.61	m	**50.28**
4' dia.	55.08	62.21	0.57	13.50	m	**75.71**

Fixings

Refer to steel pipes; galvanized iron. For minimum fixing dimensions, refer to the Tables and Memoranda at the rear of the book

S: PIPED SUPPLY SYSTEMS

Item	Net Price £	Material £	Labour hours	Labour £	Unit	Total rate £
S10: COLD WATER – cont						
ABS PIPEWORK – cont						
Extra over fittings; solvent welded joints						
Cap						
½' dia.	1.46	1.65	0.16	3.79	nr	**5.44**
¾' dia.	1.69	1.91	0.19	4.51	nr	**6.42**
1' dia.	1.94	2.19	0.22	5.22	nr	**7.41**
1¼' dia.	3.23	3.65	0.25	5.93	nr	**9.58**
1½' dia.	4.98	5.62	0.28	6.63	nr	**12.25**
2' dia.	6.31	7.13	0.31	7.34	nr	**14.47**
3' dia.	18.95	21.40	0.36	8.53	nr	**29.93**
4' dia.	28.98	32.73	0.44	10.42	nr	**43.15**
Elbow 90°						
½' dia.	2.03	2.30	0.29	6.86	nr	**9.16**
¾' dia.	2.43	2.74	0.34	8.05	nr	**10.79**
1' dia.	3.40	3.84	0.40	9.48	nr	**13.32**
1¼' dia.	5.76	6.51	0.45	10.66	nr	**17.17**
1½' dia.	7.48	8.45	0.51	12.09	nr	**20.54**
2' dia.	11.38	12.85	0.56	13.27	nr	**26.12**
3' dia.	32.68	36.90	0.65	15.40	nr	**52.30**
4' dia.	48.81	55.12	0.80	18.95	nr	**74.07**
6' dia.	196.49	221.91	1.21	28.67	nr	**250.58**
8' dia.	299.89	338.69	1.45	34.35	nr	**373.04**
Elbow 45°						
½' dia.	3.93	4.44	0.29	6.86	nr	**11.30**
¾' dia.	3.99	4.51	0.34	8.05	nr	**12.56**
1' dia.	4.98	5.62	0.40	9.48	nr	**15.10**
1¼' dia.	7.29	8.23	0.45	10.66	nr	**18.89**
1½' dia.	9.04	10.21	0.51	12.09	nr	**22.30**
2' dia.	12.56	14.19	0.56	13.27	nr	**27.46**
3' dia.	29.55	33.37	0.65	15.40	nr	**48.77**
4' dia.	61.27	69.20	0.80	18.95	nr	**88.15**
6' dia.	127.02	143.45	1.21	28.67	nr	**172.12**
8' dia.	273.34	308.71	1.45	34.35	nr	**343.06**
Reducing bush						
¾' × ½' dia.	1.50	1.69	0.42	9.95	nr	**11.64**
1' × ½' dia.	1.94	2.19	0.45	10.66	nr	**12.85**
1' × ¾' dia.	1.94	2.19	0.45	10.66	nr	**12.85**
1¼' × 1' dia.	2.61	2.95	0.48	11.37	nr	**14.32**
1½' × ¾' dia.	3.40	3.84	0.51	12.09	nr	**15.93**
1½' × 1' dia.	3.40	3.84	0.51	12.09	nr	**15.93**
1½' × 1¼' dia.	3.40	3.84	0.51	12.09	nr	**15.93**

S: PIPED SUPPLY SYSTEMS

Item	Net Price £	Material £	Labour hours	Labour £	Unit	Total rate £
2' × 1' dia.	4.47	5.05	0.56	13.27	nr	**18.32**
2' × 1¼' dia.	4.47	5.05	0.56	13.27	nr	**18.32**
2' × 1½' dia.	4.47	5.05	0.56	13.27	nr	**18.32**
3' × 1½' dia.	12.56	14.19	0.65	15.40	nr	**29.59**
3' × 2' dia.	12.56	14.19	0.65	15.40	nr	**29.59**
4' × 3' dia.	17.31	19.55	0.80	18.95	nr	**38.50**
6' × 4' dia.	53.31	60.21	1.21	28.67	nr	**88.88**
Union						
½' dia.	8.09	9.14	0.34	8.05	nr	**17.19**
¾' dia.	8.72	9.85	0.39	9.24	nr	**19.09**
1' dia.	11.75	13.27	0.43	10.19	nr	**23.46**
1¼' dia.	14.42	16.29	0.50	11.85	nr	**28.14**
1½' dia.	19.85	22.42	0.57	13.50	nr	**35.92**
2' dia.	25.88	29.23	0.62	14.70	nr	**43.93**
Sockets						
½' dia.	1.50	1.69	0.34	8.05	nr	**9.74**
¾' dia.	1.69	1.91	0.39	9.24	nr	**11.15**
1' dia.	1.94	2.19	0.43	10.19	nr	**12.38**
1¼' dia.	3.40	3.84	0.50	11.85	nr	**15.69**
1½' dia.	4.10	4.63	0.57	13.50	nr	**18.13**
2' dia.	5.76	6.51	0.62	14.70	nr	**21.21**
3' dia.	23.15	26.14	0.70	16.60	nr	**42.74**
4' dia.	32.86	37.11	0.70	16.60	nr	**53.71**
6' dia.	82.08	92.70	1.26	29.85	nr	**122.55**
8' dia.	163.96	185.17	1.55	36.73	nr	**221.90**
Barrel nipple						
½' dia.	2.83	3.19	0.34	8.05	nr	**11.24**
¾' dia.	3.68	4.15	0.39	9.24	nr	**13.39**
1' dia.	4.77	5.39	0.43	10.19	nr	**15.58**
1¼' dia.	6.61	7.47	0.50	11.85	nr	**19.32**
1½' dia.	7.79	8.80	0.57	13.50	nr	**22.30**
2' dia.	9.44	10.66	0.62	14.70	nr	**25.36**
3' dia.	25.09	28.34	0.70	16.60	nr	**44.94**
Tee, 90°						
½' dia.	2.32	2.62	0.41	9.71	nr	**12.33**
¾' dia.	3.23	3.65	0.47	11.14	nr	**14.79**
1' dia.	4.47	5.05	0.55	13.04	nr	**18.09**
1¼' dia.	6.41	7.24	0.64	15.17	nr	**22.41**
1½' dia.	9.44	10.66	0.71	16.83	nr	**27.49**
2' dia.	14.42	16.29	0.78	18.47	nr	**34.76**
3' dia.	42.03	47.47	0.91	21.56	nr	**69.03**
4' dia.	61.69	69.68	1.12	26.54	nr	**96.22**
6' dia.	215.56	243.45	1.69	40.05	nr	**283.50**
8' dia.	336.11	379.60	2.03	48.10	nr	**427.70**

S: PIPED SUPPLY SYSTEMS

Item	Net Price £	Material £	Labour hours	Labour £	Unit	Total rate £
S10: COLD WATER – cont						
ABS PIPEWORK – cont						
Extra over fittings; solvent welded joints – cont						
Full face flange						
½' dia.	25.11	28.36	0.10	2.38	nr	**30.74**
¾' dia.	25.70	29.03	0.13	3.08	nr	**32.11**
1' dia.	27.84	31.45	0.15	3.56	nr	**35.01**
1¼' dia.	30.94	34.95	0.18	4.27	nr	**39.22**
1½' dia.	37.21	42.02	0.21	4.98	nr	**47.00**
2' dia.	50.37	56.89	0.29	6.86	nr	**63.75**
3' dia.	86.36	97.53	0.37	8.76	nr	**106.29**
4' dia.	113.18	127.83	0.41	9.71	nr	**137.54**
PVC-U PIPEWORK						
Pipes; solvent welded joints in the running length, brackets Measured separately						
Class C (9 bar pressure)						
2' dia.	10.46	11.81	0.41	9.71	m	**21.52**
3' dia.	20.02	22.61	0.47	11.14	m	**33.75**
4' dia.	35.52	40.12	0.50	11.85	m	**51.97**
6' dia.	76.87	86.81	1.76	41.70	m	**128.51**
Class D (12 bar pressure)						
1¼' dia.	6.09	6.87	0.41	9.71	m	**16.58**
1½' dia.	8.37	9.46	0.42	9.95	m	**19.41**
2' dia.	12.97	14.65	0.45	10.66	m	**25.31**
3' dia.	27.79	31.39	0.48	11.37	m	**42.76**
4' dia.	46.53	52.55	0.53	12.56	m	**65.11**
6' dia.	86.35	97.52	0.58	13.75	m	**111.27**
Class E (15 bar pressure)						
½' dia.	2.97	3.36	0.38	9.00	m	**12.36**
¾' dia.	4.25	4.80	0.40	9.48	m	**14.28**
1' dia.	4.95	5.59	0.41	9.71	m	**15.30**
1¼' dia.	7.27	8.21	0.41	9.71	m	**17.92**
1½' dia.	9.46	10.68	0.42	9.95	m	**20.63**
2' dia.	14.78	16.70	0.45	10.66	m	**27.36**
3' dia.	32.01	36.15	0.47	11.14	m	**47.29**
4' dia.	52.55	59.35	0.50	11.85	m	**71.20**
6' dia.	113.84	128.57	0.53	12.56	m	**141.13**

S: PIPED SUPPLY SYSTEMS

Item	Net Price £	Material £	Labour hours	Labour £	Unit	Total rate £
Class 7						
½' dia.	5.27	5.96	0.32	7.58	m	**13.54**
¾' dia.	7.38	8.33	0.33	7.81	m	**16.14**
1' dia.	11.26	12.72	0.40	9.48	m	**22.20**
1¼' dia.	15.47	17.47	0.40	9.48	m	**26.95**
1½' dia.	19.16	21.63	0.41	9.71	m	**31.34**
2' dia.	31.84	35.95	0.43	10.19	m	**46.14**
Fixings						
Refer to steel pipes; galvanized iron. For minimum fixing dimensions, refer to the Tables and Memoranda at the rear of the book						
Extra over fittings; solvent welded joints						
End cap						
½' dia.	1.02	1.15	0.17	4.03	nr	**5.18**
¾' dia.	1.18	1.34	0.19	4.51	nr	**5.85**
1' dia.	1.33	1.50	0.22	5.22	nr	**6.72**
1¼' dia.	2.08	2.35	0.25	5.93	nr	**8.28**
1½' dia.	3.50	3.96	0.28	6.63	nr	**10.59**
2' dia.	4.28	4.83	0.31	7.34	nr	**12.17**
3' dia.	13.14	14.84	0.36	8.53	nr	**23.37**
4' dia.	20.29	22.92	0.44	10.42	nr	**33.34**
6' dia.	49.03	55.38	0.67	15.88	nr	**71.26**
Socket						
½' dia.	1.08	1.22	0.31	7.34	nr	**8.56**
¾' dia.	1.18	1.34	0.35	8.29	nr	**9.63**
1' dia.	1.39	1.57	0.42	9.95	nr	**11.52**
1¼' dia.	2.50	2.83	0.45	10.66	nr	**13.49**
1½' dia.	2.93	3.30	0.51	12.09	nr	**15.39**
2' dia.	4.16	4.70	0.56	13.27	nr	**17.97**
3' dia.	15.88	17.93	0.65	15.40	nr	**33.33**
4' dia.	23.02	26.00	0.80	18.95	nr	**44.95**
6' dia.	57.74	65.21	1.21	28.67	nr	**93.88**
Reducing socket						
¾ × ½' dia.	1.26	1.43	0.31	7.34	nr	**8.77**
1 × ¾' dia.	1.58	1.79	0.35	8.29	nr	**10.08**
1¼ × 1' dia.	3.01	3.40	0.42	9.95	nr	**13.35**
1½ × 1¼' dia.	3.36	3.79	0.45	10.66	nr	**14.45**
2 × 1½' dia.	5.08	5.73	0.51	12.09	nr	**17.82**
3 × 2' dia.	15.45	17.45	0.56	13.27	nr	**30.72**
4 × 3' dia.	22.87	25.83	0.65	15.40	nr	**41.23**
6 × 4' dia.	83.33	94.12	0.80	18.95	nr	**113.07**
8 × 6' dia.	129.07	145.77	1.21	28.67	nr	**174.44**

S: PIPED SUPPLY SYSTEMS

Item	Net Price £	Material £	Labour hours	Labour £	Unit	Total rate £
S10: COLD WATER – cont						
PVC-U PIPEWORK – cont						
Extra over fittings; solvent welded joints – cont						
Elbow, 90°						
½' dia.	1.42	1.60	0.31	7.34	nr	8.94
¾' dia.	1.71	1.93	0.35	8.29	nr	10.22
1' dia.	2.37	2.67	0.42	9.95	nr	12.62
1¼' dia.	4.16	4.70	0.45	10.66	nr	15.36
1½' dia.	5.35	6.04	0.45	10.66	nr	16.70
2' dia.	7.93	8.96	0.56	13.27	nr	22.23
3' dia.	22.87	25.83	0.65	15.40	nr	41.23
4' dia.	34.44	38.89	0.80	18.95	nr	57.84
6' dia.	136.36	154.00	1.21	28.67	nr	182.67
Elbow 45°						
½' dia.	2.70	3.05	0.31	7.34	nr	10.39
¾' dia.	2.89	3.26	0.35	8.29	nr	11.55
1' dia.	3.50	3.96	0.45	10.66	nr	14.62
1¼' dia.	5.01	5.66	0.45	10.66	nr	16.32
1½' dia.	6.29	7.10	0.51	12.09	nr	19.19
2' dia.	8.86	10.01	0.56	13.27	nr	23.28
3' dia.	20.86	23.56	0.65	15.40	nr	38.96
4' dia.	42.88	48.43	0.80	18.95	nr	67.38
6' dia.	88.47	99.92	1.21	28.67	nr	128.59
Bend 90° (long radius)						
3' dia.	63.75	72.00	0.65	15.40	nr	87.40
4' dia.	128.78	145.44	0.80	18.95	nr	164.39
6' dia.	283.00	319.62	1.21	28.67	nr	348.29
Bend 45° (long radius)						
1½' dia.	15.14	17.10	0.51	12.09	nr	29.19
2' dia.	24.74	27.94	0.56	13.27	nr	41.21
3' dia.	52.87	59.71	0.65	15.40	nr	75.11
4' dia.	102.90	116.22	0.80	18.95	nr	135.17
Socket union						
½' dia.	5.49	6.20	0.34	8.05	nr	14.25
¾' dia.	6.29	7.10	0.39	9.24	nr	16.34
1' dia.	8.16	9.22	0.45	10.66	nr	19.88
1¼' dia.	10.16	11.48	0.50	11.85	nr	23.33
1½' dia.	13.92	15.72	0.57	13.50	nr	29.22
2' dia.	18.00	20.33	0.62	14.70	nr	35.03
3' dia.	67.04	75.71	0.70	16.60	nr	92.31
4' dia.	90.76	102.50	0.89	21.08	nr	123.58

S: PIPED SUPPLY SYSTEMS

Item	Net Price £	Material £	Labour hours	Labour £	Unit	Total rate £
Saddle plain						
2' × 1¼' dia.	14.15	15.98	0.42	9.95	nr	**25.93**
3' × 1½' dia.	19.87	22.44	0.48	11.37	nr	**33.81**
4' × 2' dia.	22.39	25.29	0.68	16.12	nr	**41.41**
6' × 2' dia.	26.29	29.69	0.91	21.56	nr	**51.25**
Straight tank connector						
½' dia.	3.65	4.12	0.13	3.08	nr	**7.20**
¾' dia.	4.13	4.66	0.14	3.33	nr	**7.99**
1' dia.	8.78	9.91	0.14	3.33	nr	**13.24**
1¼' dia.	22.30	25.18	0.16	3.79	nr	**28.97**
1½' dia.	24.45	27.61	0.18	4.27	nr	**31.88**
2' dia.	29.29	33.08	0.24	5.69	nr	**38.77**
3' dia.	30.03	33.92	0.29	6.86	nr	**40.78**
Equal tee						
½' dia.	1.65	1.87	0.44	10.42	nr	**12.29**
¾' dia.	2.08	2.35	0.48	11.37	nr	**13.72**
1' dia.	3.14	3.55	0.54	12.80	nr	**16.35**
1¼' dia.	4.44	5.02	0.70	16.60	nr	**21.62**
1½' dia.	6.41	7.24	0.74	17.54	nr	**24.78**
2' dia.	10.16	11.48	0.80	18.95	nr	**30.43**
3' dia.	29.44	33.25	1.04	24.64	nr	**57.89**
4' dia.	43.17	48.76	1.28	30.32	nr	**79.08**
6' dia.	150.36	169.82	1.93	45.73	nr	**215.55**
PVC-C						
Pipes; solvent welded in the running length, brackets measured separately						
Pipe; 3 m long; PN25						
16 × 2.0 mm	3.42	3.87	0.20	4.74	m	**8.61**
20 × 2.3 mm	5.18	5.85	0.20	4.74	m	**10.59**
25 × 2.8 mm	6.70	7.57	0.20	4.74	m	**12.31**
32 × 3.6 mm	9.95	11.24	0.20	4.74	m	**15.98**
Pipe; 5 m long; PN25						
40 × 4.5 mm	12.13	13.70	0.20	4.74	m	**18.44**
50 × 5.6 mm	18.27	20.63	0.20	4.74	m	**25.37**
63 × 7.0 mm	28.21	31.86	0.20	4.74	m	**36.60**
Fixings						
Refer to steel pipes; galvanized iron. For minimum fixing dimensions, refer to the Tables and Memoranda at the rear of the book						

S: PIPED SUPPLY SYSTEMS

Item	Net Price £	Material £	Labour hours	Labour £	Unit	Total rate £
S10: COLD WATER – cont						
PVC-C – CONT						
Extra over fittings; solvent welded joints						
Straight coupling; PN25						
16 mm	0.45	0.51	0.20	4.74	nr	**5.25**
20 mm	0.63	0.71	0.20	4.74	nr	**5.45**
25 mm	0.78	0.88	0.20	4.74	nr	**5.62**
32 mm	2.41	2.72	0.20	4.74	nr	**7.46**
40 mm	3.10	3.50	0.20	4.74	nr	**8.24**
50 mm	4.16	4.70	0.20	4.74	nr	**9.44**
63 mm	7.34	8.29	0.20	4.74	nr	**13.03**
Elbow; 90°; PN25						
16 mm	0.73	0.83	0.20	4.74	nr	**5.57**
20 mm	1.11	1.25	0.20	4.74	nr	**5.99**
25 mm	1.39	1.57	0.20	4.74	nr	**6.31**
32 mm	2.89	3.26	0.20	4.74	nr	**8.00**
40 mm	4.45	5.03	0.20	4.74	nr	**9.77**
50 mm	6.16	6.96	0.20	4.74	nr	**11.70**
63 mm	10.53	11.89	0.20	4.74	nr	**16.63**
Elbow; 45°; PN25						
20 mm	1.11	1.25	0.20	4.74	nr	**5.99**
25 mm	1.39	1.57	0.20	4.74	nr	**6.31**
32 mm	2.89	3.26	0.20	4.74	nr	**8.00**
40 mm	4.45	5.03	0.20	4.74	nr	**9.77**
50 mm	6.16	6.96	0.20	4.74	nr	**11.70**
63 mm	10.53	11.89	0.20	4.74	nr	**16.63**
Reducer fitting; single stage reduction						
20/16 mm	0.78	0.88	0.20	4.74	nr	**5.62**
25/20 mm	0.95	1.07	0.20	4.74	nr	**5.81**
32/25 mm	1.90	2.14	0.20	4.74	nr	**6.88**
40/32 mm	2.50	2.83	0.20	4.74	nr	**7.57**
50/40 mm	2.89	3.26	0.20	4.74	nr	**8.00**
63/50 mm	4.38	4.95	0.20	4.74	nr	**9.69**
Equal tee; 90°; PN25						
16 mm	1.21	1.37	0.20	4.74	nr	**6.11**
20 mm	1.66	1.88	0.20	4.74	nr	**6.62**
25 mm	2.10	2.38	0.20	4.74	nr	**7.12**
32 mm	3.43	3.88	0.20	4.74	nr	**8.62**
40 mm	5.93	6.70	0.20	4.74	nr	**11.44**
50 mm	8.88	10.03	0.20	4.74	nr	**14.77**
63 mm	14.97	16.91	0.20	4.74	nr	**21.65**

S: PIPED SUPPLY SYSTEMS

Item	Net Price £	Material £	Labour hours	Labour £	Unit	Total rate £
Cap; PN25						
20 mm	0.83	0.94	0.20	4.74	nr	**5.68**
25 mm	1.11	1.25	0.20	4.74	nr	**5.99**
32 mm	1.62	1.83	0.20	4.74	nr	**6.57**
40 mm	2.22	2.51	0.20	4.74	nr	**7.25**
50 mm	3.10	3.50	0.20	4.74	nr	**8.24**
63 mm	4.94	5.58	0.20	4.74	nr	**10.32**
SCREWED STEEL PIPEWORK						
Galvanized steel pipes; screwed and socketed joints; BS 1387: 1985						
Galvanized; medium, fixed vertically, with brackets measured separately, screwed joints are within the running length, but any flanges are additional						
10 mm dia.	3.60	4.07	0.51	12.09	m	**16.16**
15 mm dia.	3.25	3.67	0.52	12.32	m	**15.99**
20 mm dia.	3.66	4.13	0.55	13.04	m	**17.17**
25 mm dia.	5.13	5.79	0.60	14.22	m	**20.01**
32 mm dia.	6.34	7.16	0.67	15.88	m	**23.04**
40 mm dia.	7.36	8.31	0.75	17.77	m	**26.08**
50 mm dia.	10.34	11.68	0.85	20.13	m	**31.81**
65 mm dia.	14.01	15.82	0.93	22.03	m	**37.85**
80 mm dia.	18.15	20.50	1.07	25.36	m	**45.86**
100 mm dia.	25.67	28.99	1.46	34.60	m	**63.59**
125 mm dia.	40.82	46.10	1.72	40.76	m	**86.86**
150 mm dia.	47.40	53.53	1.96	46.44	m	**99.97**
Galvanized; heavy, fixed vertically, with brackets measured separately, screwed joints are within the running length, but any flanges are additional						
15 mm dia.	3.85	4.35	0.52	12.32	m	**16.67**
20 mm dia.	4.36	4.93	0.55	13.04	m	**17.97**
25 mm dia.	6.22	7.03	0.60	14.22	m	**21.25**
32 mm dia.	7.71	8.71	0.67	15.88	m	**24.59**
40 mm dia.	9.00	10.17	0.75	17.77	m	**27.94**
50 mm dia.	12.47	14.09	0.85	20.13	m	**34.22**
65 mm dia.	16.94	19.14	0.93	22.03	m	**41.17**
80 mm dia.	21.52	24.31	1.07	25.36	m	**49.67**
100 mm dia.	29.99	33.87	1.46	34.60	m	**68.47**
125 mm dia.	43.46	49.08	1.72	40.76	m	**89.84**
150 mm dia.	50.82	57.40	1.96	46.44	m	**103.84**

S: PIPED SUPPLY SYSTEMS

Item	Net Price £	Material £	Labour hours	Labour £	Unit	Total rate £
S10: COLD WATER – cont						
SCREWED STEEL PIPEWORK – cont						
Galvanized steel pipes – cont						
Galvanized; medium, fixed horizontaly or suspended at high level, with brackets measured separately, screwed joints are within the running length, but any flanges are additional						
10 mm dia.	3.60	4.07	0.51	12.09	m	**16.16**
15 mm dia.	3.25	3.67	0.52	12.32	m	**15.99**
20 mm dia.	3.66	4.13	0.55	13.04	m	**17.17**
25 mm dia.	5.13	5.79	0.60	14.22	m	**20.01**
32 mm dia.	6.34	7.16	0.67	15.88	m	**23.04**
40 mm dia.	7.36	8.31	0.75	17.77	m	**26.08**
50 mm dia.	10.34	11.68	0.85	20.13	m	**31.81**
65 mm dia.	14.01	15.82	0.93	22.03	m	**37.85**
80 mm dia.	18.15	20.50	1.07	25.36	m	**45.86**
100 mm dia.	25.67	28.99	1.46	34.60	m	**63.59**
125 mm dia.	40.82	46.10	1.72	40.76	m	**86.86**
150 mm dia.	47.40	53.53	1.96	46.44	m	**99.97**
Galvanized; heavy, fixed horizontaly or suspended at high level, with brackets measured separately, screwed joints are within the running length, but any flanges are additional						
15 mm dia.	3.85	4.35	0.52	12.32	m	**16.67**
20 mm dia.	4.36	4.93	0.55	13.04	m	**17.97**
25 mm dia.	6.22	7.03	0.60	14.22	m	**21.25**
32 mm dia.	7.71	8.71	0.67	15.88	m	**24.59**
40 mm dia.	9.00	10.17	0.75	17.77	m	**27.94**
50 mm dia.	12.47	14.09	0.85	20.13	m	**34.22**
65 mm dia.	16.94	19.14	0.93	22.03	m	**41.17**
80 mm dia.	21.52	24.31	1.07	25.36	m	**49.67**
100 mm dia.	29.99	33.87	1.46	34.60	m	**68.47**
125 mm dia.	43.46	49.08	1.72	40.76	m	**89.84**
150 mm dia.	50.82	57.40	1.96	46.44	m	**103.84**

S: PIPED SUPPLY SYSTEMS

Item	Net Price £	Material £	Labour hours	Labour £	Unit	Total rate £
FIXINGS						
For steel pipes; galvanized iron. For minimum fixing dimensions, refer to the Tables and Memoranda at the rear of the book						
Single pipe bracket, screw on, galvanized iron; screwed to wood						
15 mm dia.	0.94	1.06	0.14	3.33	nr	**4.39**
20 mm dia.	1.05	1.18	0.14	3.33	nr	**4.51**
25 mm dia.	1.22	1.38	0.17	4.03	nr	**5.41**
32 mm dia.	1.67	1.89	0.19	4.51	nr	**6.40**
40 mm dia.	2.48	2.81	0.22	5.22	nr	**8.03**
50 mm dia.	3.28	3.70	0.22	5.22	nr	**8.92**
65 mm dia.	3.88	4.39	0.28	6.63	nr	**11.02**
80 mm dia.	6.08	6.86	0.32	7.58	nr	**14.44**
100 mm dia.	8.79	9.92	0.35	8.29	nr	**18.21**
Single pipe bracket, screw on, galvanized iron; plugged and screwed						
15 mm dia.	0.94	1.06	0.25	5.93	nr	**6.99**
20 mm dia.	1.05	1.18	0.25	5.93	nr	**7.11**
25 mm dia.	1.22	1.38	0.30	7.10	nr	**8.48**
32 mm dia.	1.67	1.89	0.32	7.58	nr	**9.47**
40 mm dia.	2.48	2.81	0.32	7.58	nr	**10.39**
50 mm dia.	3.28	3.70	0.32	7.58	nr	**11.28**
65 mm dia.	3.88	4.39	0.35	8.29	nr	**12.68**
80 mm dia.	6.08	6.86	0.42	9.95	nr	**16.81**
100 mm dia.	8.79	9.92	0.42	9.95	nr	**19.87**
Single pipe bracket for building in, galvanized iron						
15 mm dia.	1.00	1.13	0.10	2.38	nr	**3.51**
20 mm dia.	1.11	1.25	0.11	2.61	nr	**3.86**
25 mm dia.	1.22	1.38	0.12	2.85	nr	**4.23**
32 mm dia.	1.29	1.46	0.14	3.33	nr	**4.79**
40 mm dia.	1.67	1.89	0.15	3.56	nr	**5.45**
50 mm dia.	2.10	2.38	0.16	3.79	nr	**6.17**
Pipe ring, single socket, galvanized iron						
15 mm dia.	1.00	1.13	0.10	2.38	nr	**3.51**
20 mm dia.	1.11	1.25	0.11	2.61	nr	**3.86**
25 mm dia.	1.22	1.38	0.12	2.85	nr	**4.23**
32 mm dia.	1.29	1.46	0.15	3.56	nr	**5.02**
40 mm dia.	1.67	1.89	0.15	3.56	nr	**5.45**
50 mm dia.	2.10	2.38	0.16	3.79	nr	**6.17**
65 mm dia.	3.06	3.46	0.30	7.10	nr	**10.56**
80 mm dia.	3.66	4.13	0.35	8.29	nr	**12.42**
100 mm dia.	5.54	6.25	0.40	9.48	nr	**15.73**
125 mm dia.	11.21	12.66	0.60	14.22	nr	**26.88**
150 mm dia.	13.54	15.29	0.77	18.24	nr	**33.53**

S: PIPED SUPPLY SYSTEMS

Item	Net Price £	Material £	Labour hours	Labour £	Unit	Total rate £
S10: COLD WATER – cont						
FIXINGS – cont						
Pipe ring, double socket, galvanized iron						
15 mm dia.	8.83	9.98	0.10	2.38	nr	**12.36**
20 mm dia.	10.16	11.48	0.11	2.61	nr	**14.09**
25 mm dia.	11.19	12.64	0.12	2.85	nr	**15.49**
32 mm dia.	12.34	13.93	0.14	3.33	nr	**17.26**
40 mm dia.	15.58	17.59	0.15	3.56	nr	**21.15**
50 mm dia.	17.72	20.01	0.16	3.79	nr	**23.80**
Screw on backplate (Male), galvanized iron; plugged and screwed						
All sizes 15 mm to 50 mm × M12	0.78	0.88	0.10	2.38	nr	**3.26**
Screw on backplate (Female), galvanized iron; plugged and screwed						
All sizes 15 mm to 50 mm × M12	0.78	0.88	0.10	2.38	nr	**3.26**
Extra Over channel sections for fabricated hangers and brackets						
Galvanized steel; including inserts, bolts, nuts, washers; fixed to backgrounds						
41 × 21 mm	4.92	5.56	0.29	6.86	m	**12.42**
41 × 41 mm	5.90	6.66	0.29	6.86	m	**13.52**
Threaded rods; metric thread; including nuts, washers etc						
10 mm dia. × 600 mm long	1.62	1.83	0.18	4.27	nr	**6.10**
12 mm dia. × 600 mm long	2.49	2.82	0.18	4.27	nr	**7.09**
Extra over steel flanges, screwed and drilled; metric; BS 4504						
Screwed flanges; PN6						
15 mm dia.	12.35	13.94	0.35	8.29	nr	**22.23**
20 mm dia.	12.35	13.94	0.47	11.14	nr	**25.08**
25 mm dia.	12.35	13.94	0.53	12.56	nr	**26.50**
32 mm dia.	12.35	13.94	0.62	14.70	nr	**28.64**
40 mm dia.	12.35	13.94	0.70	16.60	nr	**30.54**
50 mm dia.	13.16	14.86	0.84	19.90	nr	**34.76**
65 mm dia.	18.30	20.67	1.03	24.41	nr	**45.08**
80 mm dia.	25.84	29.18	1.23	29.15	nr	**58.33**
100 mm dia.	30.56	34.52	1.41	33.40	nr	**67.92**
125 mm dia.	56.07	63.32	1.77	41.93	nr	**105.25**
150 mm dia.	56.07	63.32	2.21	52.37	nr	**115.69**

S: PIPED SUPPLY SYSTEMS

Item	Net Price £	Material £	Labour hours	Labour £	Unit	Total rate £
Screwed flanges; PN16						
15 mm dia.	15.70	17.73	0.35	8.29	nr	26.02
20 mm dia.	15.70	17.73	0.47	11.14	nr	28.87
25 mm dia.	15.70	17.73	0.53	12.56	nr	30.29
32 mm dia.	16.76	18.93	0.62	14.70	nr	33.63
40 mm dia.	16.76	18.93	0.70	16.60	nr	35.53
50 mm dia.	20.45	23.09	0.84	19.90	nr	42.99
65 mm dia.	25.54	28.85	1.03	24.41	nr	53.26
80 mm dia.	31.10	35.13	1.23	29.15	nr	64.28
100 mm dia.	34.88	39.39	1.41	33.40	nr	72.79
125 mm dia.	61.05	68.95	1.77	41.93	nr	110.88
150 mm dia.	60.08	67.85	2.21	52.37	nr	120.22
Extra over steel flanges, screwed and drilled; imperial; BS 10						
Screwed flanges; table E						
½' dia.	21.10	23.83	0.35	8.29	nr	32.12
¾' dia.	21.10	23.83	0.47	11.14	nr	34.97
1' dia.	21.10	23.83	0.53	12.56	nr	36.39
1¼' dia.	21.10	23.83	0.62	14.70	nr	38.53
1½' dia.	21.10	23.83	0.70	16.60	nr	40.43
2' dia.	21.10	23.83	0.84	19.90	nr	43.73
2½' dia.	25.10	28.35	1.03	24.41	nr	52.76
3' dia.	30.24	34.15	1.23	29.15	nr	63.30
4' dia.	38.48	43.46	1.41	33.40	nr	76.86
5' dia.	81.55	92.11	1.77	41.93	nr	134.04
Extra over steel flange connections						
Bolted connection between pair of flanges; including gasket, bolts, nuts and washers						
50 mm dia.	36.40	41.11	0.53	12.56	nr	53.67
65 mm dia.	45.84	51.78	0.53	12.56	nr	64.34
80 mm dia.	52.59	59.39	0.53	12.56	nr	71.95
100 mm dia.	62.76	70.88	0.53	12.56	nr	83.44
125 mm dia.	118.55	133.89	0.61	14.45	nr	148.34
150 mm dia.	121.93	137.71	0.90	21.32	nr	159.03
Extra over heavy steel tubular fittings; BS 1387						
Long screw connection with socket and backnut						
15 mm dia.	4.86	5.49	0.63	14.93	nr	20.42
20 mm dia.	6.10	6.88	0.84	19.90	nr	26.78
25 mm dia.	7.98	9.02	0.95	22.51	nr	31.53
32 mm dia.	10.52	11.88	1.11	26.31	nr	38.19
40 mm dia.	12.81	14.46	1.28	30.32	nr	44.78
50 mm dia.	18.90	21.35	1.53	36.25	nr	57.60
65 mm dia.	43.40	49.01	1.87	44.31	nr	93.32
80 mm dia.	56.13	63.39	2.21	52.37	nr	115.76
100 mm dia.	63.67	71.91	3.05	72.27	nr	144.18

S: PIPED SUPPLY SYSTEMS

Item	Net Price £	Material £	Labour hours	Labour £	Unit	Total rate £
S10: COLD WATER – cont						
FIXINGS – cont						
Extra over heavy steel tubular fittings; BS 1387 – cont						
Running nipple						
15 mm dia.	1.22	1.38	0.50	11.85	nr	13.23
20 mm dia.	1.52	1.71	0.68	16.12	nr	17.83
25 mm dia.	1.63	1.84	0.77	18.24	nr	20.08
32 mm dia.	2.63	2.97	0.90	21.32	nr	24.29
40 mm dia.	3.55	4.01	1.03	24.41	nr	28.42
50 mm dia.	5.40	6.10	1.23	29.15	nr	35.25
65 mm dia.	11.61	13.12	1.50	35.55	nr	48.67
80 mm dia.	18.10	20.44	1.78	42.17	nr	62.61
100 mm dia.	28.36	32.03	2.38	56.39	nr	88.42
Barrel nipple						
15 mm dia.	1.02	1.15	0.50	11.85	nr	13.00
20 mm dia.	1.54	1.74	0.68	16.12	nr	17.86
25 mm dia.	1.72	1.94	0.77	18.24	nr	20.18
32 mm dia.	2.86	3.23	0.90	21.32	nr	24.55
40 mm dia.	3.19	3.60	1.03	24.41	nr	28.01
50 mm dia.	4.56	5.15	1.23	29.15	nr	34.30
65 mm dia.	9.73	10.99	1.50	35.55	nr	46.54
80 mm dia.	13.59	15.35	1.78	42.17	nr	57.52
100 mm dia.	24.55	27.72	2.38	56.39	nr	84.11
125 mm dia.	45.63	51.53	2.87	68.00	nr	119.53
150 mm dia.	71.88	81.18	3.39	80.33	nr	161.51
Close taper nipple						
15 mm dia.	1.44	1.62	0.50	11.85	nr	13.47
20 mm dia.	1.87	2.11	0.68	16.12	nr	18.23
25 mm dia.	2.45	2.76	0.77	18.24	nr	21.00
32 mm dia.	3.66	4.13	0.90	21.32	nr	25.45
40 mm dia.	4.54	5.13	1.03	24.41	nr	29.54
50 mm dia.	6.98	7.88	1.23	29.15	nr	37.03
65 mm dia.	11.02	12.44	1.50	35.55	nr	47.99
80 mm dia.	18.05	20.39	1.78	42.17	nr	62.56
100 mm dia.	34.32	38.76	2.38	56.39	nr	95.15
90° bend with socket						
15 mm dia.	3.91	4.42	0.64	15.17	nr	19.59
20 mm dia.	5.25	5.93	0.85	20.13	nr	26.06
25 mm dia.	8.05	9.09	0.97	22.98	nr	32.07
32 mm dia.	11.53	13.03	1.12	26.54	nr	39.57
40 mm dia.	14.09	15.91	1.29	30.56	nr	46.47
50 mm dia.	21.91	24.75	1.55	36.73	nr	61.48
65 mm dia.	44.17	49.89	1.89	44.78	nr	94.67

S: PIPED SUPPLY SYSTEMS

Item	Net Price £	Material £	Labour hours	Labour £	Unit	Total rate £
80 mm dia.	65.57	74.05	2.24	53.08	nr	**127.13**
100 mm dia.	116.27	131.31	3.09	73.22	nr	**204.53**
125 mm dia.	284.78	321.63	3.92	92.88	nr	**414.51**
150 mm dia.	427.56	482.89	4.74	112.32	nr	**595.21**
Extra over heavy steel fittings; BS 1740						
Plug						
15 mm dia.	1.10	1.24	0.28	6.63	nr	**7.87**
20 mm dia.	1.72	1.94	0.38	9.00	nr	**10.94**
25 mm dia.	3.01	3.40	0.44	10.42	nr	**13.82**
32 mm dia.	4.68	5.28	0.51	12.09	nr	**17.37**
40 mm dia.	5.16	5.82	0.59	13.98	nr	**19.80**
50 mm dia.	7.38	8.33	0.70	16.60	nr	**24.93**
65 mm dia.	17.66	19.94	0.85	20.13	nr	**40.07**
80 mm dia.	33.06	37.34	1.00	23.69	nr	**61.03**
100 mm dia.	63.48	71.70	1.44	34.12	nr	**105.82**
Socket						
15 mm dia.	1.22	1.38	0.64	15.17	nr	**16.55**
20 mm dia.	1.39	1.57	0.85	20.13	nr	**21.70**
25 mm dia.	1.96	2.21	0.97	22.98	nr	**25.19**
32 mm dia.	2.83	3.19	1.12	26.54	nr	**29.73**
40 mm dia.	3.44	3.89	1.29	30.56	nr	**34.45**
50 mm dia.	5.30	5.99	1.55	36.73	nr	**42.72**
65 mm dia.	10.51	11.87	1.89	44.78	nr	**56.65**
80 mm dia.	13.60	15.36	2.24	53.08	nr	**68.44**
100 mm dia.	25.59	28.90	3.09	73.22	nr	**102.12**
150 mm dia.	61.09	68.99	4.74	112.32	nr	**181.31**
Elbow, female/female						
15 mm dia.	7.03	7.94	0.64	15.17	nr	**23.11**
20 mm dia.	9.16	10.34	0.85	20.13	nr	**30.47**
25 mm dia.	12.43	14.04	0.97	22.98	nr	**37.02**
32 mm dia.	23.15	26.14	1.12	26.54	nr	**52.68**
40 mm dia.	27.60	31.17	1.29	30.56	nr	**61.73**
50 mm dia.	45.24	51.09	1.55	36.73	nr	**87.82**
65 mm dia.	110.54	124.85	1.89	44.78	nr	**169.63**
80 mm dia.	131.81	148.87	2.24	53.08	nr	**201.95**
100 mm dia.	228.26	257.79	3.09	73.22	nr	**331.01**
Equal tee						
15 mm dia.	8.73	9.86	0.91	21.56	nr	**31.42**
20 mm dia.	10.15	11.46	1.22	28.92	nr	**40.38**
25 mm dia.	14.95	16.88	1.40	33.17	nr	**50.05**
32 mm dia.	30.88	34.87	1.62	38.39	nr	**73.26**
40 mm dia.	33.61	37.95	1.86	44.06	nr	**82.01**
50 mm dia.	54.66	61.73	2.21	52.37	nr	**114.10**
65 mm dia.	132.18	149.29	2.72	64.45	nr	**213.74**
80 mm dia.	141.85	160.20	3.21	76.06	nr	**236.26**
100 mm dia.	228.29	257.83	4.44	105.20	nr	**363.03**

S: PIPED SUPPLY SYSTEMS

Item	Net Price £	Material £	Labour hours	Labour £	Unit	Total rate £
S10: COLD WATER – cont						
FIXINGS – cont						
Extra over malleable iron fittings; BS 143						
Cap						
15 mm dia.	2.73	3.08	0.32	7.58	nr	**10.66**
20 mm dia.	2.87	3.24	0.43	10.19	nr	**13.43**
25 mm dia.	5.17	5.83	0.49	11.61	nr	**17.44**
32 mm dia.	9.82	11.09	0.58	13.75	nr	**24.84**
40 mm dia.	10.54	11.90	0.66	15.65	nr	**27.55**
50 mm dia.	16.76	18.93	0.78	18.47	nr	**37.40**
65 mm dia.	28.81	32.54	0.96	22.75	nr	**55.29**
80 mm dia.	47.53	53.68	1.13	26.78	nr	**80.46**
100 mm dia.	81.03	91.51	1.70	40.29	nr	**131.80**
Plain plug, hollow						
15 mm dia.	1.33	1.50	0.28	6.63	nr	**8.13**
20 mm dia.	2.06	2.33	0.38	9.00	nr	**11.33**
25 mm dia.	3.62	4.09	0.44	10.42	nr	**14.51**
32 mm dia.	5.61	6.33	0.51	12.09	nr	**18.42**
40 mm dia.	6.20	7.00	0.59	13.98	nr	**20.98**
50 mm dia.	8.87	10.02	0.70	16.60	nr	**26.62**
65 mm dia.	21.20	23.94	0.85	20.13	nr	**44.07**
80 mm dia.	39.67	44.80	1.00	23.69	nr	**68.49**
100 mm dia.	76.18	86.04	1.44	34.12	nr	**120.16**
Plain plug, solid						
15 mm dia.	1.65	1.87	0.29	6.86	nr	**8.73**
20 mm dia.	1.76	1.99	0.38	9.00	nr	**10.99**
25 mm dia.	2.34	2.64	0.44	10.42	nr	**13.06**
32 mm dia.	3.59	4.06	0.51	12.09	nr	**16.15**
40 mm dia.	4.84	5.47	0.59	13.98	nr	**19.45**
50 mm dia.	6.35	7.17	0.70	16.60	nr	**23.77**
Elbow, male/female						
15 mm dia.	0.86	0.97	0.64	15.17	nr	**16.14**
20 mm dia.	1.15	1.30	0.85	20.13	nr	**21.43**
25 mm dia.	1.92	2.17	0.97	22.98	nr	**25.15**
32 mm dia.	4.36	4.93	1.12	26.54	nr	**31.47**
40 mm dia.	6.02	6.80	1.29	30.56	nr	**37.36**
50 mm dia.	7.74	8.74	1.55	36.73	nr	**45.47**
65 mm dia.	17.27	19.50	1.89	44.78	nr	**64.28**
80 mm dia.	23.60	26.65	2.24	53.08	nr	**79.73**
100 mm dia.	41.28	46.62	3.09	73.22	nr	**119.84**
Elbow						
15 mm dia.	0.77	0.87	0.64	15.17	nr	**16.04**
20 mm dia.	1.06	1.19	0.85	20.13	nr	**21.32**
25 mm dia.	1.64	1.86	0.97	22.98	nr	**24.84**

S: PIPED SUPPLY SYSTEMS

Item	Net Price £	Material £	Labour hours	Labour £	Unit	Total rate £
32 mm dia.	3.41	3.86	1.12	26.54	nr	30.40
40 mm dia.	5.11	5.77	1.29	30.56	nr	36.33
50 mm dia.	5.98	6.75	1.55	36.73	nr	43.48
65 mm dia.	13.33	15.06	1.89	44.78	nr	59.84
80 mm dia.	19.58	22.11	2.24	53.08	nr	75.19
100 mm dia.	33.63	37.98	3.09	73.22	nr	111.20
125 mm dia.	80.69	91.13	4.44	105.20	nr	196.33
150 mm dia.	150.23	169.67	5.79	137.19	nr	306.86
45° elbow						
15 mm dia.	1.99	2.24	0.64	15.17	nr	17.41
20 mm dia.	2.45	2.76	0.85	20.13	nr	22.89
25 mm dia.	3.37	3.80	0.97	22.98	nr	26.78
32 mm dia.	7.86	8.87	1.12	26.54	nr	35.41
40 mm dia.	9.24	10.43	1.29	30.56	nr	40.99
50 mm dia.	12.67	14.31	1.55	36.73	nr	51.04
65 mm dia.	17.81	20.11	1.89	44.78	nr	64.89
80 mm dia.	26.78	30.24	2.24	53.08	nr	83.32
100 mm dia.	51.65	58.33	3.09	73.22	nr	131.55
150 mm dia.	157.18	177.52	5.79	137.19	nr	314.71
Bend, male/female						
15 mm dia.	1.52	1.71	0.64	15.17	nr	16.88
20 mm dia.	2.50	2.83	0.85	20.13	nr	22.96
25 mm dia.	3.51	3.97	0.97	22.98	nr	26.95
32 mm dia.	5.93	6.70	1.12	26.54	nr	33.24
40 mm dia.	8.69	9.81	1.29	30.56	nr	40.37
50 mm dia.	16.35	18.46	1.55	36.73	nr	55.19
65 mm dia.	25.02	28.25	1.89	44.78	nr	73.03
80 mm dia.	38.52	43.50	2.24	53.08	nr	96.58
100 mm dia.	95.40	107.74	3.09	73.22	nr	180.96
Bend, male						
15 mm dia.	3.48	3.93	0.64	15.17	nr	19.10
20 mm dia.	3.91	4.42	0.85	20.13	nr	24.55
25 mm dia.	5.74	6.49	0.97	22.98	nr	29.47
32 mm dia.	12.69	14.33	1.12	26.54	nr	40.87
40 mm dia.	17.80	20.10	1.29	30.56	nr	50.66
50 mm dia.	27.01	30.51	1.55	36.73	nr	67.24
Bend, female						
15 mm dia.	1.56	1.76	0.64	15.17	nr	16.93
20 mm dia.	2.23	2.52	0.85	20.13	nr	22.65
25 mm dia.	3.12	3.52	0.97	22.98	nr	26.50
32 mm dia.	6.09	6.87	1.12	26.54	nr	33.41
40 mm dia.	7.25	8.19	1.29	30.56	nr	38.75
50 mm dia.	11.42	12.89	1.55	36.73	nr	49.62
65 mm dia.	25.02	28.25	1.89	44.78	nr	73.03
80 mm dia.	37.09	41.89	2.24	53.08	nr	94.97
100 mm dia.	77.83	87.90	3.09	73.22	nr	161.12
125 mm dia.	157.03	177.35	4.44	105.20	nr	282.55
150 mm dia.	344.99	389.63	5.79	137.19	nr	526.82

S: PIPED SUPPLY SYSTEMS

Item	Net Price £	Material £	Labour hours	Labour £	Unit	Total rate £
S10: COLD WATER – cont						
FIXINGS – cont						
Extra over malleable iron fittings; BS 143 – cont						
Return bend						
15 mm dia.	7.97	9.00	0.64	15.17	nr	**24.17**
20 mm dia.	12.88	14.55	0.85	20.13	nr	**34.68**
25 mm dia.	16.07	18.15	0.97	22.98	nr	**41.13**
32 mm dia.	22.31	25.19	1.12	26.54	nr	**51.73**
40 mm dia.	30.22	34.13	1.29	30.56	nr	**64.69**
50 mm dia.	46.12	52.09	1.55	36.73	nr	**88.82**
Equal socket, parallel thread						
15 mm dia.	0.75	0.85	0.64	15.17	nr	**16.02**
20 mm dia.	0.98	1.11	0.85	20.13	nr	**21.24**
25 mm dia.	1.24	1.40	0.97	22.98	nr	**24.38**
32 mm dia.	2.75	3.10	1.12	26.54	nr	**29.64**
40 mm dia.	3.73	4.21	1.29	30.56	nr	**34.77**
50 mm dia.	5.40	6.10	1.55	36.73	nr	**42.83**
65 mm dia.	8.57	9.68	1.89	44.78	nr	**54.46**
80 mm dia.	12.14	13.71	2.24	53.08	nr	**66.79**
100 mm dia.	20.04	22.63	3.09	73.22	nr	**95.85**
Concentric reducing socket						
20 × 15 mm dia.	1.19	1.35	0.76	18.00	nr	**19.35**
25 × 15 mm dia.	1.56	1.76	0.86	20.37	nr	**22.13**
25 × 20 mm dia.	1.47	1.66	0.86	20.37	nr	**22.03**
32 × 25 mm dia.	2.83	3.19	1.01	23.93	nr	**27.12**
40 × 25 mm dia.	3.73	4.21	1.16	27.49	nr	**31.70**
40 × 32 mm dia.	4.14	4.67	1.16	27.49	nr	**32.16**
50 × 25 mm dia.	7.17	8.10	1.38	32.69	nr	**40.79**
50 × 40 mm dia.	5.79	6.54	1.38	32.69	nr	**39.23**
65 × 50 mm dia.	10.12	11.43	1.69	40.05	nr	**51.48**
80 × 50 mm dia.	12.61	14.24	2.00	47.39	nr	**61.63**
100 × 50 mm dia.	24.66	27.85	2.75	65.17	nr	**93.02**
100 × 80 mm dia.	73.75	83.29	2.75	65.17	nr	**148.46**
150 × 100 mm dia.	68.53	77.40	4.10	97.16	nr	**174.56**
Eccentric reducing socket						
20 × 15 mm dia.	2.70	3.05	0.76	18.00	nr	**21.05**
25 × 15 mm dia.	7.68	8.67	0.86	20.37	nr	**29.04**
25 × 20 mm dia.	8.72	9.85	0.86	20.37	nr	**30.22**
32 × 25 mm dia.	11.27	12.73	1.01	23.93	nr	**36.66**
40 × 25 mm dia.	12.92	14.60	1.16	27.49	nr	**42.09**
40 × 32 mm dia.	14.06	15.88	1.16	27.49	nr	**43.37**
50 × 25 mm dia.	14.10	15.92	1.18	27.97	nr	**43.89**

S: PIPED SUPPLY SYSTEMS

Item	Net Price £	Material £	Labour hours	Labour £	Unit	Total rate £
50 × 40 mm dia.	14.14	15.97	1.28	30.32	nr	**46.29**
65 × 50 mm dia.	14.32	16.18	1.69	40.05	nr	**56.23**
80 × 50 mm dia.	15.25	17.23	2.00	47.39	nr	**64.62**
Hexagon bush						
20 × 15 mm dia.	0.67	0.75	0.37	8.76	nr	**9.51**
25 × 15 mm dia.	0.92	1.04	0.43	10.19	nr	**11.23**
25 × 20 mm dia.	0.86	0.97	0.43	10.19	nr	**11.16**
32 × 25 mm dia.	1.16	1.31	0.51	12.09	nr	**13.40**
40 × 25 mm dia.	1.74	1.97	0.58	13.75	nr	**15.72**
40 × 32 mm dia.	1.74	1.97	0.58	13.75	nr	**15.72**
50 × 25 mm dia.	3.68	4.15	0.71	16.83	nr	**20.98**
50 × 40 mm dia.	3.44	3.89	0.71	16.83	nr	**20.72**
65 × 50 mm dia.	6.33	7.15	0.84	19.90	nr	**27.05**
80 × 50 mm dia.	9.55	10.78	1.00	23.69	nr	**34.47**
100 × 50 mm dia.	22.12	24.98	1.52	36.02	nr	**61.00**
100 × 80 mm dia.	18.41	20.79	1.52	36.02	nr	**56.81**
150 × 100 mm dia.	66.15	74.70	2.48	58.76	nr	**133.46**
Hexagon nipple						
15 mm dia.	0.73	0.83	0.28	6.63	nr	**7.46**
20 mm dia.	0.82	0.93	0.38	9.00	nr	**9.93**
25 mm dia.	1.15	1.30	0.44	10.42	nr	**11.72**
32 mm dia.	2.46	2.77	0.51	12.09	nr	**14.86**
40 mm dia.	2.83	3.19	0.59	13.98	nr	**17.17**
50 mm dia.	5.16	5.82	0.70	16.60	nr	**22.42**
65 mm dia.	8.63	9.75	0.85	20.13	nr	**29.88**
80 mm dia.	11.94	13.48	1.00	23.69	nr	**37.17**
100 mm dia.	21.18	23.92	1.44	34.12	nr	**58.04**
150 mm dia.	58.63	66.22	2.32	54.97	nr	**121.19**
Union, male/female						
15 mm dia.	3.62	4.09	0.64	15.17	nr	**19.26**
20 mm dia.	4.44	5.02	0.85	20.13	nr	**25.15**
25 mm dia.	5.16	5.82	0.97	22.98	nr	**28.80**
32 mm dia.	9.11	10.29	1.12	26.54	nr	**36.83**
40 mm dia.	11.66	13.17	1.29	30.56	nr	**43.73**
50 mm dia.	18.34	20.72	1.55	36.73	nr	**57.45**
65 mm dia.	41.04	46.35	1.89	44.78	nr	**91.13**
Union, female						
15 mm dia.	3.44	3.89	0.64	15.17	nr	**19.06**
20 mm dia.	3.77	4.25	0.85	20.13	nr	**24.38**
25 mm dia.	4.42	4.99	0.97	22.98	nr	**27.97**
32 mm dia.	7.62	8.61	1.12	26.54	nr	**35.15**
40 mm dia.	8.60	9.71	1.29	30.56	nr	**40.27**
50 mm dia.	12.82	14.47	1.55	36.73	nr	**51.20**
65 mm dia.	32.85	37.10	1.89	44.78	nr	**81.88**
80 mm dia.	43.43	49.05	2.24	53.08	nr	**102.13**
100 mm dia.	82.69	93.39	3.09	73.22	nr	**166.61**

S: PIPED SUPPLY SYSTEMS

Item	Net Price £	Material £	Labour hours	Labour £	Unit	Total rate £
S10: COLD WATER – cont						
FIXINGS – cont						
Extra over malleable iron fittings; BS 143 – cont						
Union elbow, male/female						
15 mm dia.	5.47	6.18	0.64	15.17	nr	**21.35**
20 mm dia.	6.85	7.73	0.85	20.13	nr	**27.86**
25 mm dia.	9.62	10.86	0.97	22.98	nr	**33.84**
Twin elbow						
15 mm dia.	5.26	5.94	0.91	21.56	nr	**27.50**
20 mm dia.	5.81	6.56	1.22	28.92	nr	**35.48**
25 mm dia.	9.42	10.64	1.39	32.94	nr	**43.58**
32 mm dia.	18.87	21.31	1.62	38.39	nr	**59.70**
40 mm dia.	23.89	26.98	1.86	44.06	nr	**71.04**
50 mm dia.	30.70	34.67	2.21	52.37	nr	**87.04**
65 mm dia.	49.61	56.03	2.72	64.45	nr	**120.48**
80 mm dia.	56.31	63.60	3.21	76.06	nr	**139.66**
Equal tee						
15 mm dia.	1.06	1.19	0.91	21.56	nr	**22.75**
20 mm dia.	1.54	1.74	1.22	28.92	nr	**30.66**
25 mm dia.	2.21	2.50	1.39	32.94	nr	**35.44**
32 mm dia.	4.67	5.27	1.62	38.39	nr	**43.66**
40 mm dia.	6.40	7.23	1.86	44.06	nr	**51.29**
50 mm dia.	9.21	10.40	2.21	52.37	nr	**62.77**
65 mm dia.	21.58	24.37	2.72	64.45	nr	**88.82**
80 mm dia.	25.16	28.42	3.21	76.06	nr	**104.48**
100 mm dia.	45.60	51.50	4.44	105.20	nr	**156.70**
125 mm dia.	84.58	95.52	5.38	127.48	nr	**223.00**
150 mm dia.	202.28	228.45	6.31	149.51	nr	**377.96**
Tee reducing on branch						
20 × 15 mm dia.	1.58	1.79	1.22	28.92	nr	**30.71**
25 × 15 mm dia.	2.15	2.43	1.39	32.94	nr	**35.37**
25 × 20 mm dia.	2.45	2.76	1.39	32.94	nr	**35.70**
32 × 25 mm dia.	4.75	5.37	1.62	38.39	nr	**43.76**
40 × 25 mm dia.	6.02	6.80	1.86	44.06	nr	**50.86**
40 × 32 mm dia.	8.83	9.98	1.86	44.06	nr	**54.04**
50 × 25 mm dia.	7.98	9.02	2.21	52.37	nr	**61.39**
50 × 40 mm dia.	12.41	14.01	2.21	52.37	nr	**66.38**
65 × 50 mm dia.	19.16	21.63	2.72	64.45	nr	**86.08**
80 × 50 mm dia.	25.90	29.25	3.21	76.06	nr	**105.31**
100 × 50 mm dia.	42.88	48.43	4.44	105.20	nr	**153.63**
100 × 80 mm dia.	66.15	74.70	4.44	105.20	nr	**179.90**
150 × 100 mm dia.	107.31	121.20	6.28	148.80	nr	**270.00**

S: PIPED SUPPLY SYSTEMS

Item	Net Price £	Material £	Labour hours	Labour £	Unit	Total rate £
Equal pitcher tee						
15 mm dia.	3.66	4.13	0.91	21.56	nr	**25.69**
20 mm dia.	4.52	5.10	1.22	28.92	nr	**34.02**
25 mm dia.	6.76	7.64	1.39	32.94	nr	**40.58**
32 mm dia.	10.48	11.83	1.62	38.39	nr	**50.22**
40 mm dia.	16.21	18.31	1.86	44.06	nr	**62.37**
50 mm dia.	22.76	25.70	2.21	52.37	nr	**78.07**
65 mm dia.	32.38	36.57	2.72	64.45	nr	**101.02**
80 mm dia.	50.53	57.07	3.21	76.06	nr	**133.13**
100 mm dia.	113.69	128.40	4.44	105.20	nr	**233.60**
Cross						
15 mm dia.	3.04	3.44	1.00	23.69	nr	**27.13**
20 mm dia.	4.76	5.38	1.33	31.51	nr	**36.89**
25 mm dia.	6.04	6.82	1.51	35.78	nr	**42.60**
32 mm dia.	8.97	10.13	1.77	41.93	nr	**52.06**
40 mm dia.	12.07	13.63	2.02	47.87	nr	**61.50**
50 mm dia.	18.76	21.19	2.42	57.34	nr	**78.53**
65 mm dia.	26.78	30.24	2.97	70.37	nr	**100.61**
80 mm dia.	35.62	40.23	3.50	82.94	nr	**123.17**
100 mm dia.	73.51	83.02	4.84	114.68	nr	**197.70**

Y11 – PIPELINE ANCILLARIES

VALVES

Regulators

Item	Net Price £	Material £	Labour hours	Labour £	Unit	Total rate £
Gunmetal; self-acting two port thermostat; single seat; screwed; normally closed; with adjustable or fixed bleed device						
25 mm dia.	536.34	605.74	1.46	34.60	nr	**640.34**
32 mm dia.	551.82	623.22	1.45	34.35	nr	**657.57**
40 mm dia.	589.83	666.15	1.55	36.74	nr	**702.89**
50 mm dia.	710.03	801.90	1.68	39.81	nr	**841.71**
Self acting temperature regulator for storage calorifier; integral sensing element and pocket; screwed ends						
15 mm dia.	527.70	595.99	1.32	31.27	nr	**627.26**
25 mm dia.	579.23	654.18	1.52	36.02	nr	**690.20**
32 mm dia.	748.00	844.78	1.79	42.41	nr	**887.19**
40 mm dia.	915.03	1033.43	1.99	47.15	nr	**1080.58**
50 mm dia.	1069.61	1208.02	2.26	53.54	nr	**1261.56**
Self acting temperature regulator for storage calorifier; integral sensing element and pocket; flanged ends; bolted connection						
15 mm dia.	774.66	874.89	0.61	14.45	nr	**889.34**
25 mm dia.	886.60	1001.32	0.72	17.06	nr	**1018.38**
32 mm dia.	1117.58	1262.19	0.94	22.27	nr	**1284.46**

S: PIPED SUPPLY SYSTEMS

Item	Net Price £	Material £	Labour hours	Labour £	Unit	Total rate £
S10: COLD WATER – cont						
VALVES – cont						
Regulators – cont						
Self acting temperature regulator for storage calorifier – cont						
40 mm dia.	1323.69	1494.97	1.03	24.41	nr	**1519.38**
50 mm dia.	1536.90	1735.76	1.18	27.97	nr	**1763.73**
Chrome plated thermostatic mixing valves including non-return valves and inlet swivel connections with strainers; copper compression fittings						
15 mm dia.	85.50	96.56	0.69	16.35	nr	**112.91**
Chrome plated thermostatic mixing valves including non-return valves and inlet swivel connections with angle pattern combined isolating valves and strainers; copper compression fittings						
15 mm dia.	140.31	158.47	0.69	16.35	nr	**174.82**
Gunmetal thermostatic mixing valves including non-return valves and inlet swivel connections with strainers; copper compression fittings						
15 mm dia.	177.44	200.40	0.69	16.35	nr	**216.75**
Gunmetal thermostatic mixing valves including non-return valves and inlet swivel connections with angle pattern combined isolating valves and strainers; copper compression fittings						
15 mm dia.	186.56	210.70	0.69	16.35	nr	**227.05**
Ball float valves						
Bronze, equilibrium; copper float; working pressure cold services up to 16 bar; flanged ends; BS 4504 Table 16/21; bolted connections						
25 mm dia.	115.79	130.77	1.04	24.64	nr	**155.41**
32 mm dia.	160.64	181.43	1.22	28.92	nr	**210.35**
40 mm dia.	217.87	246.06	1.38	32.69	nr	**278.75**
50 mm dia.	342.38	386.68	1.66	39.34	nr	**426.02**
65 mm dia.	367.16	414.67	1.93	45.73	nr	**460.40**
80 mm dia.	409.39	462.37	2.16	51.18	nr	**513.55**

S: PIPED SUPPLY SYSTEMS

Item	Net Price £	Material £	Labour hours	Labour £	Unit	Total rate £
Heavy, equilibrium; with long tail and backnut; copper float; screwed for iron						
25 mm dia.	106.67	120.47	1.58	37.44	nr	**157.91**
32 mm dia.	146.44	165.39	1.78	42.17	nr	**207.56**
40 mm dia.	206.45	233.16	1.90	45.01	nr	**278.17**
50 mm dia.	271.10	306.18	2.65	62.79	nr	**368.97**
Brass, ball valve; BS 1212; copper float; screwed						
15 mm dia.	5.77	6.52	0.25	5.93	nr	**12.45**
22 mm dia.	10.48	11.83	0.29	6.86	nr	**18.69**
28 mm dia.	45.65	51.56	0.35	8.29	nr	**59.85**
Gate valves						
DZR copper alloy wedge non-rising stem; capillary joint to copper						
15 mm dia.	12.00	13.56	0.84	19.90	nr	**33.46**
22 mm dia.	14.73	16.64	1.01	23.93	nr	**40.57**
28 mm dia.	19.92	22.50	1.19	28.20	nr	**50.70**
35 mm dia.	35.74	40.36	1.38	32.69	nr	**73.05**
42 mm dia.	60.73	68.58	1.62	38.39	nr	**106.97**
54 mm dia.	85.08	96.08	1.94	45.97	nr	**142.05**
Cocks; capillary joints to copper						
Stopcock; brass head with gun metal body						
15 mm dia.	5.30	5.99	0.45	10.66	nr	**16.65**
22 mm dia.	9.89	11.17	0.46	10.89	nr	**22.06**
28 mm dia.	28.12	31.76	0.54	12.80	nr	**44.56**
Lockshield stop cocks; brass head with gun metal body						
15 mm dia.	6.12	6.92	0.45	10.66	nr	**17.58**
22 mm dia.	6.12	6.92	0.46	10.89	nr	**17.81**
28 mm dia.	6.12	6.92	0.54	12.80	nr	**19.72**
DZR stopcock; brass head with gun metal body						
15 mm dia.	14.52	16.40	0.45	10.66	nr	**27.06**
22 mm dia.	25.14	28.40	0.46	10.89	nr	**39.29**
28 mm dia.	41.89	47.31	0.54	12.80	nr	**60.11**
Gunmetal stopcock						
35 mm dia.	54.76	61.84	0.69	16.35	nr	**78.19**
42 mm dia.	72.72	82.13	0.71	16.83	nr	**98.96**
54 mm dia.	108.63	122.69	0.81	19.19	nr	**141.88**

S: PIPED SUPPLY SYSTEMS

Item	Net Price £	Material £	Labour hours	Labour £	Unit	Total rate £
S10: COLD WATER – cont						
VALVES – cont						
Cocks; capillary joints to copper – cont						
Double union stopcock						
15 mm dia.	15.60	17.62	0.60	14.22	nr	**31.84**
22 mm dia.	21.93	24.77	0.60	14.22	nr	**38.99**
28 mm dia.	39.02	44.06	0.69	16.35	nr	**60.41**
Double union DZR stopcock						
15 mm dia.	25.47	28.76	0.60	14.22	nr	**42.98**
22 mm dia.	31.30	35.35	0.61	14.45	nr	**49.80**
28 mm dia.	57.87	65.36	0.69	16.35	nr	**81.71**
Double union gun metal stopcock						
35 mm dia.	96.41	108.88	0.63	14.93	nr	**123.81**
42 mm dia.	132.31	149.43	0.67	15.88	nr	**165.31**
54 mm dia.	208.15	235.08	0.85	20.13	nr	**255.21**
Double union stopcock with easy clean cover						
15 mm dia.	27.03	30.53	0.60	14.22	nr	**44.75**
22 mm dia.	33.72	38.09	0.61	14.45	nr	**52.54**
28 mm dia.	62.69	70.80	0.69	16.35	nr	**87.15**
Combined stopcock and drain						
15 mm dia.	26.81	30.28	0.67	15.88	nr	**46.16**
22 mm dia.	32.72	36.95	0.68	16.12	nr	**53.07**
Combined DZR stopcock and drain						
15 mm dia.	35.43	40.01	0.67	15.88	nr	**55.89**
Gate valve						
DZR copper alloy wedge non-rising stem; compression joint to copper						
15 mm dia.	12.00	13.56	0.84	19.90	nr	**33.46**
22 mm dia.	14.73	16.64	1.01	23.93	nr	**40.57**
28 mm dia.	19.92	22.50	1.19	28.20	nr	**50.70**
35 mm dia.	35.74	40.36	1.38	32.69	nr	**73.05**
42 mm dia.	60.73	68.58	1.62	38.39	nr	**106.97**
54 mm dia.	85.08	96.08	1.94	45.97	nr	**142.05**
Cocks; compression joints to copper						
Stopcock; brass head gun metal body						
15 mm dia.	4.82	5.45	0.42	9.95	nr	**15.40**
22 mm dia.	8.99	10.15	0.42	9.95	nr	**20.10**
28 mm dia.	25.57	28.88	0.45	10.66	nr	**39.54**

S: PIPED SUPPLY SYSTEMS

Item	Net Price £	Material £	Labour hours	Labour £	Unit	Total rate £
Lockshield stopcock; brass head gun metal body						
15 mm dia.	5.10	5.76	0.42	9.95	nr	**15.71**
22 mm dia.	7.30	8.24	0.42	9.95	nr	**18.19**
28 mm dia.	13.00	14.68	0.45	10.66	nr	**25.34**
DZR Stopcock						
15 mm dia.	12.10	13.67	0.38	9.00	nr	**22.67**
22 mm dia.	20.96	23.67	0.39	9.24	nr	**32.91**
28 mm dia.	34.91	39.42	0.40	9.48	nr	**48.90**
35 mm dia.	54.76	61.84	0.52	12.32	nr	**74.16**
42 mm dia.	72.72	82.13	0.54	12.80	nr	**94.93**
54 mm dia.	108.63	122.69	0.63	14.93	nr	**137.62**
DZR Lockshield stopcock						
15 mm dia.	5.10	5.76	0.38	9.00	nr	**14.76**
22 mm dia.	8.05	9.09	0.39	9.24	nr	**18.33**
Combined stop/draincock						
15 mm dia.	26.81	30.28	0.22	5.22	nr	**35.50**
22 mm dia.	34.49	38.95	0.45	10.66	nr	**49.61**
DZR Combined stop/draincock						
15 mm dia.	35.43	40.01	0.41	9.71	nr	**49.72**
22 mm dia.	45.91	51.85	0.42	9.95	nr	**61.80**
Stopcock to polyethylene						
15 mm dia.	18.52	20.92	0.38	9.00	nr	**29.92**
20 mm dia.	23.34	26.36	0.39	9.24	nr	**35.60**
25 mm dia.	31.03	35.05	0.40	9.48	nr	**44.53**
Draw off coupling						
15 mm dia.	11.25	12.71	0.38	9.00	nr	**21.71**
DZR Draw off coupling						
15 mm dia.	11.25	12.71	0.38	9.00	nr	**21.71**
22 mm dia.	13.02	14.71	0.39	9.24	nr	**23.95**
Draw off elbow						
15 mm dia.	12.35	13.94	0.38	9.00	nr	**22.94**
22 mm dia.	15.22	17.19	0.39	9.24	nr	**26.43**
Lockshield drain cock						
15 mm dia.	12.91	14.58	0.41	9.71	nr	**24.29**

S: PIPED SUPPLY SYSTEMS

Item	Net Price £	Material £	Labour hours	Labour £	Unit	Total rate £
S10: COLD WATER – cont						
VALVES – cont						
Check valves						
DZR copper alloy and bronze, WRC approved cartridge double check valve; BS 6282; working pressure cold services up to 10 bar at 65°C; screwed ends						
32 mm dia.	78.45	88.60	1.38	32.69	nr	**121.29**
40 mm dia.	89.39	100.96	1.62	38.39	nr	**139.35**
50 mm dia.	146.58	165.55	1.94	45.97	nr	**211.52**
Y20 – PUMPS						
Packaged cold water pressure booster set; fully automatic; 3 phase supply; includes fixing in position; electrical work elsewhere						
Pressure booster set						
0.75 l/s @ 30 m head	3962.17	4474.85	9.38	222.26	nr	**4697.11**
1.5 l/s @ 30 m head	4705.08	5313.89	9.38	222.26	nr	**5536.15**
3 l/s @ 30 m head	5748.68	6492.53	10.38	245.95	nr	**6738.48**
6 l/s @ 30 m head	12983.19	14663.15	10.38	245.95	nr	**14909.10**
12 l/s @ 30 m head	16220.13	18318.93	12.38	293.34	nr	**18612.27**
0.75 l/s @ 50 m head	4510.52	5094.16	9.38	222.26	nr	**5316.42**
1.5 l/s @ 50 m head	5748.68	6492.53	9.38	222.26	nr	**6714.79**
3 l/s @ 50 m head	6350.10	7171.77	10.38	245.95	nr	**7417.72**
6 l/s @ 50 m head	14500.84	16377.18	10.38	245.95	nr	**16623.13**
12 l/s @ 50 m head	17812.07	20116.86	12.38	293.34	nr	**20410.20**
0.75 l/s @ 70 m head	4935.02	5573.59	9.38	222.26	nr	**5795.85**
1.5 l/s @ 70 m head	6346.55	7167.76	9.38	222.26	nr	**7390.02**
3 l/s @ 70 m head	6756.92	7631.23	10.38	245.95	nr	**7877.18**
6 l/s @ 70 m head	15830.99	17879.44	10.38	245.95	nr	**18125.39**
12 l/s @ 70 m head	19244.81	21735.00	12.38	293.34	nr	**22028.34**
Automatic sump pump for clear and drainage water; single stage centrifugal pump, presure tight electric motor; single phase supply; includes fixing in position; electrical work elsewhere						
Single pump						
1 l/s @ 2.68 m total head	209.41	236.51	3.50	82.94	nr	**319.45**
1 l/s @ 4.68 m total head	230.50	260.32	3.50	82.94	nr	**343.26**
1 l/s @ 6.68 m total head	304.32	343.70	3.50	82.94	nr	**426.64**
2 l/s @ 4.38 m total head	304.32	343.70	4.00	94.78	nr	**438.48**
2 l/s @ 6.38 m total head	304.32	343.70	4.00	94.78	nr	**438.48**
2 l/s @ 8.38 m total head	391.69	442.37	4.00	94.78	nr	**537.15**
3 l/s @ 3.7 m total head	304.32	343.70	4.50	106.63	nr	**450.33**

S: PIPED SUPPLY SYSTEMS

Item	Net Price £	Material £	Labour hours	Labour £	Unit	Total rate £
3 l/s @ 5.7 m total head	391.69	442.37	4.50	106.63	nr	549.00
4 l/s @ 2.9 m total head	304.32	343.70	5.00	118.47	nr	462.17
4 l/s @ 4.9 m total head	391.69	442.37	5.00	118.47	nr	560.84
4 l/s @ 6.9 m total head	1086.20	1226.74	5.00	118.47	nr	1345.21
Extra for high level alarm box with single float switch, local alarm and volt free contacts for remote alarm.	318.74	359.98	–	–	nr	359.98
Duty/standby pump unit						
1 l/s @ 2.68 m total head	397.71	449.17	5.00	118.47	nr	567.64
1 l/s @ 4.68 m total head	436.89	493.42	5.00	118.47	nr	611.89
1 l/s @ 6.68 m total head	590.56	666.98	5.00	118.47	nr	785.45
2 l/s @ 4.38 m total head	590.56	666.98	5.50	130.33	nr	797.31
2 l/s @ 6.38 m total head	590.56	666.98	5.50	130.33	nr	797.31
2 l/s @ 8.38 m total head	756.27	854.13	5.50	130.33	nr	984.46
3 l/s @ 3.7 m total head	590.56	666.98	6.00	142.17	nr	809.15
3 l/s @ 5.7 m total head	756.27	854.13	6.00	142.17	nr	996.30
4 l/s @ 2.9 m total head	590.56	666.98	6.50	154.02	nr	821.00
4 l/s @ 4.9 m total head	756.27	854.13	6.50	154.02	nr	1008.15
4 /s @ 6.9 m total head	2009.69	2269.73	7.00	165.86	nr	2435.59
Extra for 4 nr float switches to give pump on, off and high level alarm	333.02	376.11	–	–	nr	376.11
Extra for dual pump control panel, internal wall mounted IP54, including volt free contacts	1585.81	1791.01	4.00	94.78	nr	1885.79

Y21 – TANKS

Cisterns; fibreglass; complete with ball valve, fixing plate and fitted covers

Item	Net Price £	Material £	Labour hours	Labour £	Unit	Total rate £
Rectangular						
70 litres capacity	217.75	245.92	1.33	31.51	nr	277.43
110 litres capacity	226.35	255.64	1.40	33.17	nr	288.81
170 litres capacity	239.24	270.20	1.61	38.15	nr	308.35
280 litres capacity	389.65	440.07	1.61	38.15	nr	478.22
420 litres capacity	408.28	461.11	1.99	47.15	nr	508.26
710 litres capacity	673.31	760.43	3.31	78.43	nr	838.86
840 litres capacity	695.74	785.77	3.60	85.30	nr	871.07
1590 litres capacity	859.71	970.95	13.32	315.61	nr	1286.56
2275 litres capacity	901.78	1018.47	20.18	478.17	nr	1496.64
3365 litres capacity	987.97	1115.81	24.50	580.52	nr	1696.33
4545 litres capacity	1035.68	1169.70	29.91	708.71	nr	1878.41

S: PIPED SUPPLY SYSTEMS

Item	Net Price £	Material £	Labour hours	Labour £	Unit	Total rate £
S10: COLD WATER – cont						
Y21 – TANKS – cont						
Cisterns; polypropylene; complete with ball valve, fixing plate and cover; includes placing in position						
Rectangular						
18 litres capacity	54.19	61.20	1.00	23.69	nr	**84.89**
68 litres capacity	73.01	82.46	1.00	23.69	nr	**106.15**
91 litres capacity	73.49	83.00	1.00	23.69	nr	**106.69**
114 litres capacity	81.76	92.34	1.00	23.69	nr	**116.03**
182 litres capacity	107.99	121.96	1.00	23.69	nr	**145.65**
227 litres capacity	137.54	155.34	1.00	23.69	nr	**179.03**
Circular						
114 litres capacity	76.49	86.38	1.00	23.69	nr	**110.07**
227 litres capacity	77.99	88.08	1.00	23.69	nr	**111.77**
318 litres capacity	129.73	146.51	1.00	23.69	nr	**170.20**
455 litres capacity	149.11	168.40	1.00	23.69	nr	**192.09**
Steel sectional water storage tank; hot pressed steel tank to BS 1564 TYPE 1; 5 mm plate; pre-insulated and complete with all connections and fittings to comply with BSEN 13280; 2001 and WRAS water supply (water fittings) regulations 1999; externally flanged base and sides; cost of erection (on prepared base) is included within the net price, labour cost allows for offloading and positioning materials						
Note: Prices are based on the most economical tank size for each volume, and the cost will vary with differing tank dimensions, for the same volume						
Volume, size						
4,900 litres, 3.66 m × 1.22 m × 1,22 m (h)	5602.50	6327.44	6.00	142.17	nr	**6469.61**
20,300 litres, 3.66 m × 2.4 m × 2.4 m (h)	11934.21	13478.43	12.00	284.34	nr	**13762.77**
52,000 litres, 6.1 m × 3.6 m × 2.4 m (h)	21279.28	24032.71	19.00	450.20	nr	**24482.91**
94,000 litres, 7.3 m × 3.6 m × 3.6 m (h)	32013.26	36155.61	28.00	663.45	nr	**36819.06**
140,000 litres, 9.7 m × 6.1 m × 2.44 m (h)	41809.47	47219.41	28.00	663.45	nr	**47882.86**

S: PIPED SUPPLY SYSTEMS

Item	Net Price £	Material £	Labour hours	Labour £	Unit	Total rate £
GRP sectional water storage tank; pre-insulated and complete with all connections and fittings to comply with BSEN 13280; 2001 and WRAS water supply (water fittings) regulations 1999; externally flanged base and sides; cost of erection (on prepared base) is included within the net price, labour cost allows for offloading and positioning materials						
Note: Prices are based on the most economical tank size for each volume, and the cost will vary with differing tank dimensions, for the same volume.						
Volume, size						
4,500 litres, 3 m × 1 m × 1.5 m (h)	4010.61	4529.56	5.00	118.47	nr	**4648.03**
10,000 litres, 2.5 m × 2 m × 2 m (h)	5331.07	6020.89	7.00	165.86	nr	**6186.75**
20,000 litres, 4 m × 2.5 m × 2 m (h)	7668.19	8660.41	10.00	236.95	nr	**8897.36**
30,000 litres 5 m × 3 m × 2 m (h)	9568.83	10806.99	12.00	284.34	nr	**11091.33**
40,000 litres, 5 m × 4 m × 2 m (h)	11212.20	12663.00	12.00	284.34	nr	**12947.34**
50,000 litres, 5 m × 4 m × 2.5 m (h)	13823.00	15611.63	14.00	331.72	nr	**15943.35**
60,000 litres, 6 m × 4 m × 2.5 m (h)	15926.14	17986.90	16.00	379.11	nr	**18366.01**
70,000 litres, 7 m × 4 m × 2.5 m (h)	17647.51	19931.01	16.00	379.11	nr	**20310.12**
80,000 litres, 8 m × 4 m × 2.5 m (h)	20058.54	22654.02	16.00	379.11	nr	**23033.13**
90,000 litres, 6 m × 5 m × 3 m (h)	20441.68	23086.73	16.00	379.11	nr	**23465.84**
105,000 litres, 7 m × 5 m × 3 m (h)	22595.44	25519.18	24.00	568.67	nr	**26087.85**
120,000 litres, 8 m × 5 m × 3 m (h)	23842.00	26927.03	24.00	568.67	nr	**27495.70**
135,000 litres, 9 m × 6 m × 2.5 m (h)	27480.41	31036.23	24.00	568.67	nr	**31604.90**
144,000 litres, 8 m × 6 m × 3 m (h)	29345.45	33142.61	24.00	568.67	nr	**33711.28**
Y25 – CLEANING AND CHEMICAL TREATMENT						
Electromagnetic water conditioner, complete with control box; maximum inlet pressure 16 bar; electrical work elsewhere						
Connection size, nominal flow rate at 50 mbar						
20 mm dia., 0.3l/s	1545.56	1745.55	1.25	29.62	nr	**1775.17**
25 mm dia., 0.6l/s	2030.06	2292.74	1.45	34.35	nr	**2327.09**
32 mm dia., 1.2l/s	2892.47	3266.74	1.55	36.73	nr	**3303.47**
40 mm dia., 1.7l/s	3464.18	3912.42	1.65	39.10	nr	**3951.52**
50 mm dia., 3.4l/s	4530.07	5116.24	1.75	41.47	nr	**5157.71**
65 mm dia., 5.2l/s	4966.13	5608.73	1.90	45.01	nr	**5653.74**
100 mm dia., 30.5l/s	9444.38	10666.44	3.00	71.08	nr	**10737.52**

S: PIPED SUPPLY SYSTEMS

Item	Net Price £	Material £	Labour hours	Labour £	Unit	Total rate £
S10: COLD WATER – cont						
Y25 – CLEANING AND CHEMICAL TREATMENT – cont						
Ultraviolet water sterillising unit, complete with control unit; UV lamp housed in quartz tube; unit complete with UV intensity sensor, flushing and discharge valve and facilities for remote alarm; electrical work elsewhere						
Maximum flow rate (@ 250J/m2 exposure), connection size						
0.82l/s, 40 mm dia.	2336.99	2639.38	1.98	46.92	nr	**2686.30**
1.28l/s, 40 mm dia.	2683.40	3030.61	1.98	46.92	nr	**3077.53**
2.00l/s, 40 mm dia.	3509.48	3963.59	1.98	46.92	nr	**4010.51**
4.14l/s, 50 mm dia.	3855.89	4354.82	2.10	49.77	nr	**4404.59**
1.28l/s, 40 mm dia.	3829.25	4324.74	1.98	46.92	nr	**4371.66**
2.00l/s, 40 mm dia.	5321.51	6010.08	1.98	46.92	nr	**6057.00**
4.14l/s, 50 mm dia.	6094.28	6882.85	2.10	49.77	nr	**6932.62**
7.4l/s, 80 mm dia.	10216.65	11538.64	3.60	85.30	nr	**11623.94**
16.8l/s 100 mm dia.	12742.83	14391.69	3.60	85.30	nr	**14476.99**
32.3l/s 100 mm dia.	14315.04	16167.34	3.60	85.30	nr	**16252.64**
Base exchange water softener complete with resin tank, brine tank and consumption data monitoring facilities						
Capacities of softeners are based on 300ppm hardness and quoted in m³ of softened water produced. Design flow rates are recommended for continuous use						
Simplex configuration						
Design flow rate, min-max softenend water produced						
1l/s, 5.8 m³-11.2 m³	1684.12	1902.03	8.00	189.56	nr	**2091.59**
1.3l/s, 11.7 m³-21.4 m³	2302.34	2600.26	8.00	189.56	nr	**2789.82**
1.3l/s, 15.5 m³-28.5 m³	2622.11	2961.40	10.00	236.95	nr	**3198.35**
1.6l/s, 23.3 m³-42.7 m³	3330.94	3761.94	10.00	236.95	nr	**3998.89**
1.6l/s, 38.8 m³-71.2 m³	3751.97	4237.46	12.00	284.34	nr	**4521.80**
1.9l/s, 11.7 m³-21.4 m³	3783.95	4273.58	12.00	284.34	nr	**4557.92**
3.2l/s, 19.4 m³-35.6 m³	4061.08	4586.56	12.00	284.34	nr	**4870.90**
4.4l/s, 31 m³-57 m³	4785.89	5405.16	15.00	355.42	nr	**5760.58**
5.1l/s, 46.6 m³-85.4 m³	7130.87	8053.57	15.00	355.42	nr	**8408.99**
5.1l/s, 77.7 m³-142.4 m³	8719.06	9847.26	18.00	426.50	nr	**10273.76**

S: PIPED SUPPLY SYSTEMS

Item	Net Price £	Material £	Labour hours	Labour £	Unit	Total rate £
Duplex configuration						
Design flow rate, min-max softenend water produced						
2l/s, 5.8 m³-22.4 m³	2803.32	3166.06	12.00	284.34	nr	**3450.40**
2.6l/s, 11.7 m³-42.8 m³	3805.26	4297.64	12.00	284.34	nr	**4581.98**
2.6l/s, 15.5 m³-57 m³	3986.47	4502.30	15.00	355.42	nr	**4857.72**
3.2l/s, 23.3 m³-85.4 m³	5148.30	5814.47	15.00	355.42	nr	**6169.89**
3.2l/s, 38.8 m³-142.4 m³	6150.24	6946.05	18.00	426.50	nr	**7372.55**
3.8l/s, 11.7 m³-42.8 m³	6629.90	7487.78	18.00	426.50	nr	**7914.28**
6.4l/s, 19.4 m³-71.2 m³	7130.87	8053.57	18.00	426.50	nr	**8480.07**
8.8l/s, 31.1 m³-114 m³	8175.45	9233.32	23.00	544.98	nr	**9778.30**
10.2l/s, 46.6 m³-170.8 m³	13099.91	14794.98	23.00	544.98	nr	**15339.96**
10.2l/s, 77.7 m³-284.8 m³	16244.32	18346.25	27.00	639.75	nr	**18986.00**
Triplex configuration						
Design flow rate, min-max softenend water produced						
3l/s, 5.8 m³-33.6 m³	3895.86	4399.96	15.00	355.42	nr	**4755.38**
3.9l/s, 11.7 m³-64.2 m³	5521.36	6235.80	15.00	355.42	nr	**6591.22**
3.9l/s, 15.5 m³-85.5 m³	5830.47	6584.91	18.00	426.50	nr	**7011.41**
4.8l/s, 23.3 m³-128.1 m³	7589.21	8571.21	18.00	426.50	nr	**8997.71**
4.8l/s, 38.8 m³-213.6 m³	8719.06	9847.26	22.00	521.28	nr	**10368.54**
5.7l/s, 11.7 m³-64.2 m³	9763.64	11027.01	22.00	521.28	nr	**11548.29**
9.6l/s, 19.4 m³-106.8 m³	10584.39	11953.96	22.00	521.28	nr	**12475.24**
13.2l/s, 31.1 m³-171.0 m³	12140.60	13711.53	27.00	639.75	nr	**14351.28**
15.3l/s, 46.6 m³-256.2 m³	19527.29	22054.02	27.00	639.75	nr	**22693.77**
15.3l/s, 77.7 m³-427.2 m³	24398.45	27555.48	32.00	758.23	nr	**28313.71**
Y50 –THERMAL INSULATION						
Flexible closed cell walled insulation; Class 1/Class O; adhesive joints; including around fittings						
6 mm wall thickness						
15 mm diameter	1.00	1.13	0.15	3.56	m	**4.69**
22 mm diameter	1.18	1.34	0.15	3.56	m	**4.90**
28 mm diameter	1.49	1.68	0.15	3.56	m	**5.24**
9 mm wall thickness						
15 mm diameter	1.14	1.29	0.15	3.56	m	**4.85**
22 mm diameter	1.44	1.62	0.15	3.56	m	**5.18**
28 mm diameter	1.53	1.72	0.15	3.56	m	**5.28**
35 mm diameter	1.73	1.96	0.15	3.56	m	**5.52**
42 mm diameter	1.99	2.24	0.15	3.56	m	**5.80**
54 mm diameter	2.05	2.32	0.15	3.56	m	**5.88**

S: PIPED SUPPLY SYSTEMS

Item	Net Price £	Material £	Labour hours	Labour £	Unit	Total rate £
S10: COLD WATER – cont						
Y50 –THERMAL INSULATION – cont						
Flexible closed cell walled insulation – cont						
13 mm wall thickness						
15 mm diameter	1.46	1.65	0.15	3.56	m	**5.21**
22 mm diameter	1.80	2.03	0.15	3.56	m	**5.59**
28 mm diameter	2.14	2.42	0.15	3.56	m	**5.98**
35 mm diameter	2.31	2.61	0.15	3.56	m	**6.17**
42 mm diameter	2.71	3.06	0.15	3.56	m	**6.62**
54 mm diameter	3.78	4.27	0.15	3.56	m	**7.83**
67 mm diameter	5.58	6.30	0.15	3.56	m	**9.86**
76 mm diameter	6.57	7.42	0.15	3.56	m	**10.98**
108 mm diameter	6.88	7.77	0.15	3.56	m	**11.33**
19 mm wall thickness						
15 mm diameter	2.38	2.69	0.15	3.56	m	**6.25**
22 mm diameter	2.90	3.27	0.15	3.56	m	**6.83**
28 mm diameter	3.93	4.44	0.15	3.56	m	**8.00**
35 mm diameter	4.55	5.14	0.15	3.56	m	**8.70**
42 mm diameter	5.38	6.08	0.15	3.56	m	**9.64**
54 mm diameter	6.80	7.68	0.15	3.56	m	**11.24**
67 mm diameter	8.16	9.22	0.15	3.56	m	**12.78**
76 mm diameter	9.42	10.64	0.22	5.22	m	**15.86**
108 mm diameter	12.86	14.52	0.22	5.22	m	**19.74**
25 mm wall thickness						
15 mm diameter	4.67	5.27	0.15	3.56	m	**8.83**
22 mm diameter	5.12	5.78	0.15	3.56	m	**9.34**
28 mm diameter	5.78	6.53	0.15	3.56	m	**10.09**
35 mm diameter	6.41	7.24	0.15	3.56	m	**10.80**
42 mm diameter	6.84	7.72	0.15	3.56	m	**11.28**
54 mm diameter	8.04	9.08	0.15	3.56	m	**12.64**
67 mm diameter	10.83	12.23	0.15	3.56	m	**15.79**
76 mm diameter	12.21	13.79	0.22	5.22	m	**19.01**
32 mm wall thickness						
15 mm diameter	5.85	6.61	0.15	3.56	m	**10.17**
22 mm diameter	6.41	7.24	0.15	3.56	m	**10.80**
28 mm diameter	7.44	8.40	0.15	3.56	m	**11.96**
35 mm diameter	7.68	8.67	0.15	3.56	m	**12.23**
42 mm diameter	8.93	10.09	0.15	3.56	m	**13.65**
54 mm diameter	11.25	12.71	0.15	3.56	m	**16.27**
76 mm diameter	16.82	18.99	0.22	5.22	m	**24.21**

S: PIPED SUPPLY SYSTEMS

Item	Net Price £	Material £	Labour hours	Labour £	Unit	Total rate £
Note: For mineral fibre sectional insulation; bright class O foil faced; bright class O foil taped joints; 19 mm aluminium bands rates, refer to Section T31 – Low Temperature Hot Water Heating, Y50 Thermal Insulation						
Note: For mineral fibre sectional insulation; bright class O foil faced; bright class O foil taped joints; 22 swg plain/embossed aluminium cladding; pop riveted rates, refer to Section T31 – Low Temperature Hot Water Heating, Y50 Thermal Insulation						
Note: For mineral fibre sectional insulation; bright class O foil faced; bright class O foil taped joints; 0.8 mm polyisobutylene sheeting; welded joints rates, refer to Section T31 – Low Temperature Hot Water Heating, Y50 Thermal Insulation						

S: PIPED SUPPLY SYSTEMS

Item	Net Price £	Material £	Labour hours	Labour £	Unit	Total rate £
S11 – HOT WATER						
Y10 – PIPELINES						
Note: For pipework prices refer to Section S10 – Cold Water.						
Y11 – PIPELINE ANCILLARIES						
Note: For prices for ancillaries refer to Section S10 – Cold Water.						
Y23 – STORAGE CYLINDERS/ CALORIFIERS						
CYLINDERS						
Insulated copper storage cylinders; BS 699; includes placing in position						
Grade 3 (maximum 10 m working head)						
BS size 6; 115 litres capacity; 400 mm dia.; 1050 mm high	169.55	191.48	1.50	35.58	nr	**227.06**
BS size 7; 120 litres capacity; 450 mm dia.; 900 mm high	200.46	226.40	2.00	47.39	nr	**273.79**
BS size 8; 144 litres capacity; 450 mm dia.; 1050 mm high	213.09	240.66	2.80	66.37	nr	**307.03**
Grade 4 (maximum 6 m working head)						
BS size 2; 96 litres capacity; 400 mm dia.; 900 mm high	131.27	148.26	1.50	35.58	nr	**183.84**
BS size 7; 120 litres capacity; 450 mm dia.; 900 mm high	137.96	155.82	1.50	35.58	nr	**191.40**
BS size 8; 144 litres capacity; 450 mm dia.; 1050 mm high	164.68	185.99	1.50	35.58	nr	**221.57**
BS size 9; 166 litres capacity; 450 mm dia.; 1200 mm high	240.80	271.96	1.50	35.58	nr	**307.54**
Storage cylinders; brazed copper construction; to BS 699; screwed bosses; includes placing in position						
Tested to 2.2 bar, 15 m maximum head						
144 litres	746.29	842.86	3.00	71.16	nr	**914.02**
160 litres	843.66	952.82	3.00	71.16	nr	**1023.98**
200 litres	869.60	982.12	3.76	89.08	nr	**1071.20**
255 litres	989.65	1117.71	3.76	89.08	nr	**1206.79**
290 litres	1336.83	1509.81	3.76	89.08	nr	**1598.89**
370 litres	1531.54	1729.72	4.50	106.73	nr	**1836.45**
450 litres	2086.28	2356.23	5.00	118.47	nr	**2474.70**

S: PIPED SUPPLY SYSTEMS

Item	Net Price £	Material £	Labour hours	Labour £	Unit	Total rate £
Tested to 2.55 bar, 17 m maximum head						
550 litres	2259.02	2551.33	5.00	118.47	nr	**2669.80**
700 litres	2641.01	2982.75	6.02	142.75	nr	**3125.50**
800 litres	3054.66	3449.92	6.54	154.87	nr	**3604.79**
900 litres	3305.45	3733.16	8.00	189.56	nr	**3922.72**
1000 litres	3488.18	3939.54	8.00	189.56	nr	**4129.10**
1250 litres	3820.35	4314.68	13.16	311.77	nr	**4626.45**
1500 litres	5851.57	6608.73	15.15	359.01	nr	**6967.74**
2000 litres	7022.40	7931.06	17.24	408.52	nr	**8339.58**
3000 litres	9864.06	11140.42	24.39	577.92	nr	**11718.34**
Indirect cylinders; copper; bolted top; up to 5 tappings for connections; BS 1586; includes placing in position						
Grade 3, tested to 1.45 bar, 10 m maximum head						
74 litres capacity	342.17	386.45	1.50	35.58	nr	**422.03**
96 litres capacity	348.37	393.44	1.50	35.58	nr	**429.02**
114 litres capacity	357.70	403.98	1.50	35.58	nr	**439.56**
117 litres capacity	371.26	419.30	2.00	47.39	nr	**466.69**
140 litres capacity	382.57	432.07	2.50	59.24	nr	**491.31**
162 litres capacity	534.99	604.22	3.00	71.16	nr	**675.38**
190 litres capacity	584.79	660.46	3.51	83.13	nr	**743.59**
245 litres capacity	684.32	772.86	3.80	90.10	nr	**862.96**
280 litres capacity	1213.07	1370.03	4.00	94.78	nr	**1464.81**
360 litres capacity	1312.62	1482.47	4.50	106.73	nr	**1589.20**
440 litres capacity	1524.13	1721.34	4.50	106.73	nr	**1828.07**
Grade 2, tested to 2.2 bar, 15 m maximum head						
117 litres capacity	494.55	558.54	2.00	47.39	nr	**605.93**
140 litres capacity	538.12	607.75	2.50	59.24	nr	**666.99**
162 litres capacity	615.87	695.56	2.80	66.37	nr	**761.93**
190 litres capacity	715.45	808.02	3.00	71.16	nr	**879.18**
245 litres capacity	864.74	976.63	4.00	94.78	nr	**1071.41**
280 litres capacity	1381.04	1559.74	4.00	94.78	nr	**1654.52**
360 litres capacity	1524.13	1721.34	4.50	106.73	nr	**1828.07**
440 litres capacity	1804.06	2037.50	4.50	106.73	nr	**2144.23**
Grade 1, tested 3.65 bar, 25 m maximum head						
190 litres capacity	1063.81	1201.46	3.00	71.16	nr	**1272.62**
245 litres capacity	1209.94	1366.50	3.00	71.16	nr	**1437.66**
280 litres capacity	1710.77	1932.13	4.00	94.78	nr	**2026.91**
360 litres capacity	2164.93	2445.06	4.50	106.73	nr	**2551.79**
440 litres capacity	2628.34	2968.43	4.50	106.73	nr	**3075.16**

S: PIPED SUPPLY SYSTEMS

Item	Net Price £	Material £	Labour hours	Labour £	Unit	Total rate £
S11 – HOT WATER – cont						
CYLINDERS – cont						
Indirect cylinders, including manhole; BS 853						
Grade 3, tested to 1.5 bar, 10 m maximum head						
550 litres capacity	2212.42	2498.69	5.21	123.41	nr	**2622.10**
700 litres capacity	2449.49	2766.44	6.02	142.75	nr	**2909.19**
800 litres capacity	2844.56	3212.63	6.54	154.87	nr	**3367.50**
1000 litres capacity	3555.73	4015.82	7.04	166.87	nr	**4182.69**
1500 litres capacity	4108.82	4640.48	10.00	236.95	nr	**4877.43**
2000 litres capacity	5689.16	6425.31	16.13	382.17	nr	**6807.48**
Grade 2, tested to 2.55 bar, 15 m maximum head						
550 litres capacity	2452.89	2770.28	5.21	123.41	nr	**2893.69**
700 litres capacity	3069.96	3467.19	6.02	142.75	nr	**3609.94**
800 litres capacity	3239.65	3658.84	6.54	154.87	nr	**3813.71**
1000 litres capacity	4011.00	4530.00	7.04	166.87	nr	**4696.87**
1500 litres capacity	4936.62	5575.39	10.00	236.95	nr	**5812.34**
2000 litres capacity	6170.80	6969.27	16.13	382.17	nr	**7351.44**
Grade 1, tested to 4 bar, 25 m maximum head						
550 litres capacity	2853.98	3223.27	5.21	123.41	nr	**3346.68**
700 litres capacity	3239.65	3658.84	6.02	142.75	nr	**3801.59**
800 litres capacity	3471.07	3920.21	6.54	154.87	nr	**4075.08**
1000 litres capacity	4628.11	5226.96	7.04	166.87	nr	**5393.83**
1500 litres capacity	5553.67	6272.29	10.00	236.95	nr	**6509.24**
2000 litres capacity	6787.85	7666.17	16.13	382.17	nr	**8048.34**
Storage calorifiers; copper; heater battery capable of raising temperature of contents from 10°C to 65°C in one hour; static head not exceeding 1.35 bar; BS 853; includes fixing in position on cradles or legs						
Horizontal; primary LPHW at 82°C/71°C						
400 litres capacity	3167.19	3577.01	7.04	166.87	nr	**3743.88**
1000 litres capacity	5067.47	5723.18	8.00	189.56	nr	**5912.74**
2000 litres capacity	10135.01	11446.43	14.08	333.72	nr	**11780.15**
3000 litres capacity	12510.38	14129.16	25.00	592.37	nr	**14721.53**
4000 litres capacity	15202.48	17169.61	40.00	947.78	nr	**18117.39**
4500 litres capacity	17132.52	19349.38	50.00	1184.74	nr	**20534.12**

S: PIPED SUPPLY SYSTEMS

Item	Net Price £	Material £	Labour hours	Labour £	Unit	Total rate £
Vertical; primary LPHW at 82°C/71°C						
400 litres capacity	3569.43	4031.30	7.04	166.87	nr	**4198.17**
1000 litres capacity	5736.61	6478.90	8.00	189.56	nr	**6668.46**
2000 litres capacity	10926.80	12340.67	14.08	333.72	nr	**12674.39**
3000 litres capacity	13658.51	15425.86	25.00	592.37	nr	**16018.23**
4000 litres capacity	16754.38	18922.32	40.00	947.78	nr	**19870.10**
4500 litres capacity	18939.76	21390.47	50.00	1184.74	nr	**22575.21**
Storage calorifiers; galvanized mild steel; heater battery capable of raising temperature of contents from 10°C to 65°C in one hour; static head not exceeding 1.35 bar; BS 853; includes fixing in position on cradles or legs						
Horizontal; primary LPHW at 82°C/71°C						
400 litres capacity	3167.19	3577.01	7.04	166.87	nr	**3743.88**
1000 litres capacity	5067.47	5723.18	8.00	189.56	nr	**5912.74**
2000 litres capacity	10135.01	11446.43	14.08	333.72	nr	**11780.15**
3000 litres capacity	12510.38	14129.16	25.00	592.37	nr	**14721.53**
4000 litres capacity	15202.48	17169.61	40.00	947.78	nr	**18117.39**
4500 litres capacity	17132.52	19349.38	50.00	1184.74	nr	**20534.12**
Vertical; primary LPHW at 82°C/71°C						
400 litres capacity	3569.43	4031.30	7.04	166.87	nr	**4198.17**
1000 litres capacity	5736.61	6478.90	8.00	189.56	nr	**6668.46**
2000 litres capacity	10926.80	12340.67	14.08	333.72	nr	**12674.39**
3000 litres capacity	13658.51	15425.86	25.00	592.37	nr	**16018.23**
4000 litres capacity	16754.38	18922.32	40.00	947.78	nr	**19870.10**
4500 litres capacity	18939.76	21390.47	50.00	1184.74	nr	**22575.21**
Indirect cylinders; mild steel, welded throughout, galvanized; with bolted connections; includes placing in position						
3.2 mm plate						
136 litres capacity	1222.86	1381.09	2.50	59.24	nr	**1440.33**
159 litres capacity	1316.95	1487.35	2.80	66.37	nr	**1553.72**
182 litres capacity	1618.00	1827.36	3.00	71.16	nr	**1898.52**
227 litres capacity	1994.23	2252.27	3.00	71.16	nr	**2323.43**
273 litres capacity	2326.27	2627.28	4.00	94.78	nr	**2722.06**
364 litres capacity	2709.14	3059.69	4.50	106.73	nr	**3166.42**
455 litres capacity	2798.21	3160.29	5.00	118.47	nr	**3278.76**
683 litres capacity	4528.35	5114.30	6.02	142.75	nr	**5257.05**
910 litres capacity	5259.32	5939.85	7.04	166.87	nr	**6106.72**

S: PIPED SUPPLY SYSTEMS

Item	Net Price £	Material £	Labour hours	Labour £	Unit	Total rate £
S11 – HOT WATER – cont						
Y50 – INSULATION						
Refer to Sections S10 – Cold Water and T31 – Low Temperature Hot Water Heating for details						
Local Electric Hot Water Heaters						
Unvented multi-point water heater; providing hot water for one or more outlets: Used with conventional taps or mixers: Factory fitted temperature and pressure relief valve: Externally adjustable thermostat: Elemental 'on' indicator: Fitted with 1 metre of 3 core cable: Electrical supply and connection excluded						
5 litre capacity, 2.2 kW rating	168.87	190.72	1.50	35.55	nr	**226.27**
10 litre capacity, 2.2 kW rating	188.33	212.70	1.50	35.55	nr	**248.25**
15 litre capacity, 2.2 kW rating	209.28	236.36	1.50	35.55	nr	**271.91**
30 litre capacity, 3 kW rating	372.99	421.25	2.00	47.39	nr	**468.64**
50 litre capacity, 3 kW rating	392.69	443.51	2.00	47.39	nr	**490.90**
80 litre capacity, 3 kW rating	772.68	872.66	2.00	47.39	nr	**920.05**
100 litre capacity, 3 kW rating	829.52	936.86	2.00	47.39	nr	**984.25**
Accessories						
Pressure reducing valve and expansion kit	137.03	154.76	2.00	47.39	nr	**202.15**
Thermostatic blending valve	72.92	82.35	1.00	23.69	nr	**106.04**

S: PIPED SUPPLY SYSTEMS

Item	Net Price £	Material £	Labour hours	Labour £	Unit	Total rate £
S32: NATURAL GAS						
Y10 – PIPELINES						
MEDIUM DENSITY POLYETHYLENE – YELLOW						
Pipe; laid underground; electrofusion joints in the running length; BS 6572; BGT PL2 standards						
Coiled service pipe						
20 mm dia.	1.21	1.37	0.37	8.76	m	**10.13**
25 mm dia.	1.58	1.79	0.41	9.71	m	**11.50**
32 mm dia.	2.60	2.94	0.47	11.14	m	**14.08**
63 mm dia.	9.92	11.20	0.60	14.22	m	**25.42**
90 mm dia.	13.02	14.71	0.90	21.32	m	**36.03**
Mains service pipe						
63 mm dia.	9.63	10.87	0.60	14.22	m	**25.09**
90 mm dia.	12.65	14.29	0.90	21.32	m	**35.61**
125 mm dia.	24.42	27.58	1.20	28.44	m	**56.02**
180 mm dia.	50.42	56.95	1.50	35.55	m	**92.50**
250 mm dia.	92.80	104.80	1.75	41.47	m	**146.27**
Extra over fittings, electrofusion joints						
Straight connector						
32 mm dia.	7.55	8.53	0.47	11.14	nr	**19.67**
63 mm dia.	14.19	16.02	0.58	13.75	nr	**29.77**
90 mm dia.	20.93	23.63	0.67	15.88	nr	**39.51**
125 mm dia.	39.19	44.26	0.83	19.67	nr	**63.93**
180 mm dia.	70.56	79.69	1.25	29.62	nr	**109.31**
Reducing connector						
90 × 63 mm dia.	29.22	33.00	0.67	15.88	nr	**48.88**
125 × 90 mm dia.	58.60	66.18	0.83	19.67	nr	**85.85**
180 × 125 mm dia.	107.51	121.42	1.25	29.62	nr	**151.04**
Bend; 45°						
90 mm dia.	56.47	63.78	0.67	15.88	nr	**79.66**
125 mm dia.	92.37	104.33	0.83	19.67	nr	**124.00**
180 mm dia.	209.56	236.68	1.25	29.62	nr	**266.30**
Bend; 90°						
63 mm dia.	36.59	41.32	0.58	13.75	nr	**55.07**
90 mm dia.	56.47	63.78	0.67	15.88	nr	**79.66**
125 mm dia.	92.37	104.33	0.83	19.67	nr	**124.00**
180 mm dia.	209.56	236.68	1.25	29.62	nr	**266.30**

S: PIPED SUPPLY SYSTEMS

Item	Net Price £	Material £	Labour hours	Labour £	Unit	Total rate £
S32: NATURAL GAS – cont						
MEDIUM DENSITY POLYETHYLENE – YELLOW – cont						
Extra over malleable iron fittings, compression joints						
Straight connector						
20 mm dia.	10.78	12.18	0.38	9.00	nr	**21.18**
25 mm dia.	11.74	13.26	0.45	10.66	nr	**23.92**
32 mm dia.	13.17	14.87	0.50	11.85	nr	**26.72**
63 mm dia.	26.45	29.88	0.85	20.15	nr	**50.03**
Straight connector; polyethylene to MI						
20 mm dia.	9.16	10.34	0.31	7.35	nr	**17.69**
25 mm dia.	9.98	11.27	0.35	8.29	nr	**19.56**
32 mm dia.	11.15	12.60	0.40	9.48	nr	**22.08**
63 mm dia.	18.68	21.09	0.65	15.40	nr	**36.49**
Straight connector; polyethylene to FI						
20 mm dia.	9.16	10.34	0.31	7.35	nr	**17.69**
25 mm dia.	11.15	12.60	0.35	8.29	nr	**20.89**
32 mm dia.	11.15	12.60	0.40	9.48	nr	**22.08**
63 mm dia.	18.68	21.09	0.75	17.78	nr	**38.87**
Elbow						
20 mm dia.	14.01	15.82	0.38	9.00	nr	**24.82**
25 mm dia.	15.28	17.26	0.45	10.66	nr	**27.92**
32 mm dia.	17.12	19.34	0.50	11.85	nr	**31.19**
63 mm dia.	34.40	38.85	0.80	18.95	nr	**57.80**
Equal tee						
20 mm dia.	18.68	21.09	0.53	12.56	nr	**33.65**
25 mm dia.	21.84	24.66	0.55	13.04	nr	**37.70**
32 mm dia.	27.50	31.06	0.64	15.17	nr	**46.23**

S: PIPED SUPPLY SYSTEMS

Item	Net Price £	Material £	Labour hours	Labour £	Unit	Total rate £
SCREWED STEEL						
For prices for steel pipework refer to Section T31 – Low Temperature Hot Water Heating						
PIPE IN PIPE						
Note: for pipe in pipe, a sleeve size two pipe sizes bigger than actual pipe size has been allowed. All rates refer to actual pipe size.						
Black steel pipes – Screwed and socketed joints; BS 1387: 1985 upto 50 mm pipe size. Butt welded joints; BS 1387: 1985 65 mm pipe size and above						
Pipe						
25 mm	11.65	13.16	1.73	40.99	m	54.15
32 mm	15.47	17.47	1.95	46.21	m	63.68
40 mm	19.90	22.47	2.16	51.18	m	73.65
50 mm	26.12	29.50	2.44	57.81	m	87.31
65 mm	35.83	40.46	2.95	69.90	m	110.36
80 mm	43.30	48.90	3.42	81.04	m	129.94
100 mm	54.09	61.09	4.00	94.78	m	155.87
Extra over black steel pipes – Screwed pipework; black malleable iron fittings; BS 143. Welded pipework; butt welded steel fittings; BS 1965						
Bend, 90°						
25 mm	6.72	7.59	2.91	68.95	m	76.54
32 mm	10.24	11.57	3.45	81.74	m	93.31
40 mm	14.34	16.20	5.34	126.53	m	142.73
50 mm	16.16	18.25	6.53	154.72	m	172.97
65 mm	21.84	24.66	8.84	209.46	m	234.12
80 mm	37.22	42.03	10.73	254.25	m	296.28
100 mm	49.43	55.82	12.76	302.34	m	358.16
Bend, 45°						
25 mm	8.50	9.60	2.91	68.95	m	78.55
32 mm	11.02	12.44	3.45	81.74	m	94.18
40 mm	12.31	13.90	5.34	126.53	m	140.43
50 mm	14.68	16.58	6.53	154.72	m	171.30
65 mm	19.17	21.65	8.84	209.46	m	231.11
80 mm	32.80	37.05	10.73	254.25	m	291.30
100 mm	42.02	47.46	12.76	302.34	m	349.80

S: PIPED SUPPLY SYSTEMS

Item	Net Price £	Material £	Labour hours	Labour £	Unit	Total rate £
S32: NATURAL GAS – cont						
PIPE IN PIPE – cont						
Extra over black steel pipes – Screwed pipework – cont						
Equal tee						
25 mm	56.59	63.91	4.18	99.05	m	**162.96**
32 mm	58.09	65.61	4.94	117.06	m	**182.67**
40 mm	71.47	80.72	7.28	172.49	m	**253.21**
50 mm	73.33	82.81	8.48	200.93	m	**283.74**
65 mm	97.64	110.27	11.47	271.78	m	**382.05**
80 mm	167.52	189.20	14.23	337.18	m	**526.38**
100 mm	189.59	214.12	17.92	424.61	m	**638.73**
Copper pipe; capillary or compression joints in the running length; EN1057 R250 (TX) formerly BS 2871 Table X						
Plastic coated gas service pipe for corrosive environments, fixed vertically or at low level with brackets measured separtely						
15 mm dia. (yellow)	6.81	7.69	0.85	20.15	m	**27.84**
22 mm dia. (yellow)	13.44	15.18	0.96	22.76	m	**37.94**
28 mm dia. (yellow)	17.06	19.27	1.06	25.12	m	**44.39**
Copper pipe; capillary or compression joints in the running length; EN1057 R250 (TY) formerly BS 2871 Table Y						
Plastic coated gas and cold water service pipe for corrosive environments, fixed vertically or at low level with brackets measured separately (Refer to Copper Pipe Table X Section)						
15 mm dia. (yellow)	7.88	8.90	0.61	14.45	m	**23.35**
22 mm dia. (yellow)	14.32	16.18	0.69	16.35	m	**32.53**
FIXINGS						
Refer to Section S10 Cold Water						
Extra over copper pipes; capillary fittings; BS 864						
Refer to Section S10 – Cold Water						

S: PIPED SUPPLY SYSTEMS

Item	Net Price £	Material £	Labour hours	Labour £	Unit	Total rate £
GAS BOOSTERS						
Complete skid mounted gas booster set, including AV mounts, flexible connections, low pressure switch,control panel and NRV (for run/standby unit); 3 phase supply; in accordance with IGE/UP/2; includes delivery, offloading and positioning						
Single unit						
Flow, pressure range						
0-200 m³/hour, 0.1-2.6 kPa	3017.82	3408.31	10.00	236.95	nr	**3645.26**
0-200 m³/hour, 0.1-4.0 kPa	3443.41	3888.97	10.00	236.95	nr	**4125.92**
0-200 m³/hour, 0.1-7 kPa	4007.15	4525.66	10.00	236.95	nr	**4762.61**
0-200 m³/hour, 0.1-9.5 kPa	4172.97	4712.93	10.00	236.95	nr	**4949.88**
0-200 m³/hour 0.1-11.0 kPa	4662.14	5265.39	10.00	236.95	nr	**5502.34**
0-400 m³/hour, 0.1-4.0 kPa	3802.65	4294.69	10.00	236.95	nr	**4531.64**
0-1000 m³/hour, 0.1-7.4 kPa	5267.33	5948.89	10.00	236.95	nr	**6185.84**
50-1000 m³/hour, 0.1-16.0 kPa	9841.04	11114.42	20.00	473.89	nr	**11588.31**
50-1000 m³/hour, 0.1-24.5 kPa	11126.07	12565.73	20.00	473.89	nr	**13039.62**
50-1000 m³/hour, 0.1-31.0 kPa	12654.33	14291.74	20.00	473.89	nr	**14765.63**
50-1000 m³/hour, 0.1-41.0 kPa	13967.03	15774.29	20.00	473.89	nr	**16248.18**
50-1000 m³/hour, 0.1-51.0 kPa	14492.07	16367.27	20.00	473.89	nr	**16841.16**
100-1800 m³/hour, 3.5-23.5 kPa	14870.72	16794.91	20.00	473.89	nr	**17268.80**
100-1800 m³/hour, 4.5-27.0 kPa	16039.69	18115.15	20.00	473.89	nr	**18589.04**
100-1800 m³/hour, 6.0-32.5 kPa	17764.16	20062.76	20.00	473.89	nr	**20536.65**
100-1800 m³/hour, 7.2-39.0 kPa	20436.51	23080.90	20.00	473.89	nr	**23554.79**
100-1800 m³/hour, 9.0-42.0 kPa	21481.14	24260.69	20.00	473.89	nr	**24734.58**
Run/Standby unit						
Flow, pressure range						
0-200 m³/hour, 0.1-2.6 kPa	15521.81	17530.25	16.00	379.11	nr	**17909.36**
0-200 m³/hour, 0.1-4.0 kPa	15994.54	18064.15	16.00	379.11	nr	**18443.26**
0-200 m³/hour, 0.1-7 kPa	16364.99	18482.54	16.00	379.11	nr	**18861.65**
0-200 m³/hour, 0.1-9.5 kPa	16770.50	18940.52	16.00	379.11	nr	**19319.63**
0-200 m³/hour 0.1-11.0 kPa	17048.33	19254.30	16.00	379.11	nr	**19633.41**
0-400 m³/hour, 0.1-4.0 kPa	18800.47	21233.16	16.00	379.11	nr	**21612.27**
0-1000 m³/hour, 0.1-7.4 kPa	23964.28	27065.14	25.00	592.37	nr	**27657.51**
50-1000 m³/hour, 0.1-16.0 kPa	32845.11	37095.11	25.00	592.37	nr	**37687.48**
50-1000 m³/hour, 0.1-24.5 kPa	37132.87	41937.68	25.00	592.37	nr	**42530.05**
50-1000 m³/hour, 0.1-31.0 kPa	41410.57	46768.89	25.00	592.37	nr	**47361.26**
50-1000 m³/hour, 0.1-41.0 kPa	45700.82	51614.27	25.00	592.37	nr	**52206.64**
50-1000 m³/hour, 0.1-51.0 kPa	45906.07	51846.09	25.00	592.37	nr	**52438.46**
100-1800 m³/hour, 3.5-23.5 kPa	47761.87	53942.02	25.00	592.37	nr	**54534.39**
100-1800 m³/hour, 4.5-27.0 kPa	49071.33	55420.91	25.00	592.37	nr	**56013.28**
100-1800 m³/hour, 6.0-32.5 kPa	52133.65	58879.48	25.00	592.37	nr	**59471.85**
100-1800 m³/hour, 7.2-39.0 kPa	55412.67	62582.79	25.00	592.37	nr	**63175.16**
100-1800 m³/hour, 9.0-42.0 kPa	58246.15	65782.91	25.00	592.37	nr	**66375.28**

S: PIPED SUPPLY SYSTEMS

Item	Net Price £	Material £	Labour hours	Labour £	Unit	Total rate £
S41: FUEL OIL STORAGE/DISTRIBUTION						
Y10 – PIPELINES						
For pipework prices refer to Section T31 – Low Temperature Hot Water Heating						
Y21 – TANKS						
Fuel storage tanks; mild steel; with all necessary screwed bosses; oil resistant joint rings; includes placing in position						
Rectangular						
1360 litres (300 gallon) capacity; 2 mm plate	334.83	378.15	12.03	285.05	nr	**663.20**
2730 litres (600 gallon) capacity; 2.5 mm plate	447.42	505.32	18.60	440.72	nr	**946.04**
4550 litres (1000 gallon) capacity; 3 mm plate	938.03	1059.40	25.00	592.37	nr	**1651.77**
Fuel storage tanks; 5 mm plate mild steel to BS 799 type J; complete with raised neck manhole with bolted cover, screwed connections, vent and fill connections, drain valve, gauge and overfill alarm; includes placing in position; excludes pumps and control panel						
Nominal capacity, size						
5,600 litres, 2.5 m × 1.5 m × 1.5 m high	2189.43	2472.73	20.00	473.89	nr	**2946.62**
Extra for bund unit (internal use)	1259.09	1422.01	30.00	710.84	nr	**2132.85**
Extra for external use with bund (watertight)	797.42	900.60	2.00	47.39	nr	**947.99**
10,200 litres, 3.05 m × 1.83 m × 1.83 m high	2721.04	3073.13	30.00	710.84	nr	**3783.97**
Extra for bund unit (internal use)	1832.68	2069.81	40.00	947.78	nr	**3017.59**
Extra for external use with bund (watertight)	979.29	1106.01	2.00	47.39	nr	**1153.40**
15,000 litres, 3.75 m × 2 m × 2 m high	3441.52	3886.83	40.00	947.78	nr	**4834.61**
Extra for bund unit (internal use)	2462.23	2780.83	55.00	1303.21	nr	**4084.04**
Extra for external use with bund (watertight)	1147.17	1295.60	2.00	47.39	nr	**1342.99**
20,000 litres, 4 m × 2.5 m × 2 m high	4539.74	5127.16	50.00	1184.74	nr	**6311.90**
Extra for bund unit (internal use)	2986.87	3373.35	65.00	1540.16	nr	**4913.51**
Extra for external use with bund (watertight)	1357.02	1532.61	2.00	47.39	nr	**1580.00**
Extra for BMS output (all tank sizes)	475.65	537.19	–	–	nr	**537.19**

S: PIPED SUPPLY SYSTEMS

Item	Net Price £	Material £	Labour hours	Labour £	Unit	Total rate £
Fuel storage tanks; plastic; with all necessary screwed bosses; oil resistant joint rings; includes placing in position						
Cylindrical; horizontal						
1250 litres (285 gallon) capacity	270.90	305.95	3.73	88.38	nr	**394.33**
1350 litres (300 gallon) capacity	276.77	312.58	4.30	101.89	nr	**414.47**
2500 litres (550 gallon) capacity	426.20	481.35	4.88	115.63	nr	**596.98**
Cylindrical; vertical						
1365 litres (300 gallon) capacity	171.83	194.07	3.73	88.38	nr	**282.45**
2600 litres (570 gallon) capacity	261.52	295.36	4.88	115.63	nr	**410.99**
3635 litres (800 gallon) capacity	407.72	460.48	4.88	115.63	nr	**576.11**
5455 litres (1200 gallon) capacity	593.95	670.80	5.95	140.98	nr	**811.78**
Bunded tanks						
1135 litres (250 gallon) capacity	550.86	622.14	4.30	101.89	nr	**724.03**
1590 litres (350 gallon) capacity	655.33	740.12	4.88	115.63	nr	**855.75**
2500 litres (550 gallon) capacity	778.94	879.73	5.95	140.98	nr	**1020.71**
5000 litres (1100 gallon) capacity	1571.83	1775.22	6.53	154.72	nr	**1929.94**

S: PIPED SUPPLY SYSTEMS

Item	Net Price £	Material £	Labour hours	Labour £	Unit	Total rate £
S60: FIRE HOSE REELS						
Y10 – PIPELINES						
For pipework prices refer to Section S10 – Cold Water						
Y11 – PIPELINE ANCILLARIES						
For prices for ancillaries refer to Section S10 – Cold Water						
Hose reels; automatic; connection to 25 mm screwed joint; reel with 30.5 m, 19 mm rubber hose; suitable for working pressure up to 7 bar						
Reels						
Non-swing pattern	197.42	222.95	3.75	88.84	nr	**311.79**
Recessed non-swing pattern	253.19	285.94	3.75	88.84	nr	**374.78**
Swinging pattern	263.64	297.74	3.75	88.84	nr	**386.58**
Recessed swinging pattern	271.98	307.16	3.75	88.84	nr	**396.00**
Hose reels; manual; connection to 25 mm screwed joint; reel with 30.5 m, 19 mm rubber hose; suitable for working pressure up to 7 bar						
Reels						
Non-swing pattern	198.84	224.56	3.25	77.01	nr	**301.57**
Recessed non-swing pattern	233.39	263.58	3.25	77.01	nr	**340.59**
Swinging pattern	251.85	284.43	3.25	77.01	nr	**361.44**
Recessed swinging pattern	260.91	294.66	3.25	77.01	nr	**371.67**

S: PIPED SUPPLY SYSTEMS

Item	Net Price £	Material £	Labour hours	Labour £	Unit	Total rate £
S61: DRY RISERS						
Y10 – PIPELINES						
For pipework prices refer to Section S10 – Cold Water						
Y11 – PIPELINE ANCILLARIES						
VALVES (BS 5041, PARTS 2 AND 3)						
Bronze/gunmetal inlet breeching for pumping in with 65 mm dia. instantaneous male coupling; with cap, chain and 25 mm drain valve						
Double inlet with back pressure valve, flanged to steel	250.40	282.81	1.75	41.47	nr	**324.28**
Quadruple inlet with back pressure valve, flanged to steel	560.65	633.20	1.75	41.47	nr	**674.67**
Bronze/gunmetal gate type outlet valve with 65 mm dia. instantaneous female coupling; cap and chain; wheel head secured by padlock and leather strap						
Flanged to BS 4504 PN6 (bolted connection to counter flanges measured separately)	200.81	226.80	1.75	41.47	nr	**268.27**
Bronze/gunmetal landing type outlet valve, with 65 mm dia. instantaneous female coupling; cap and chain; wheelhead secured by padlock and leather strap; bolted connections to counter flanges measured separately						
Horizontal, flanged to BS 4504 PN6	218.70	247.00	1.50	35.55	nr	**282.55**
Oblique, flanged to BS 4504 PN6	218.70	247.00	1.50	35.55	nr	**282.55**
Air valve, screwed joints to steel						
25 mm dia.	43.47	49.09	0.55	13.04	nr	**62.13**
INLET BOXES (BS 5041, PART 5)						
Steel dry riser inlet box with hinged wire glazed door suitably lettered (fixing by others)						
610 × 460 × 325 mm; double inlet	266.49	300.97	3.00	71.08	nr	**372.05**
610 × 610 × 356 mm; quadruple inlet	497.37	561.72	3.00	71.08	nr	**632.80**
OUTLET BOXES (BS 5041, PART 5)						
Steel dry riser outlet box with hinged wire glazed door suitably lettered (fixing by others)						
610 × 460 × 325 mm; single outlet	260.24	293.91	3.00	71.08	nr	**364.99**

S: PIPED SUPPLY SYSTEMS

Item	Net Price £	Material £	Labour hours	Labour £	Unit	Total rate £
S63: SPRINKLERS						
Y10 – PIPELINES						
Prefabricated black steel pipework; screwed joints, including all coupliings, unions and the like to BS 1387:1985; includes fixing to backgrounds, with brackets measured separately						
Heavy weight						
25 mm dia.	4.76	5.38	0.47	11.14	m	**16.52**
32 mm dia.	5.91	6.67	0.53	12.56	m	**19.23**
40 mm dia.	6.89	7.78	0.58	13.75	m	**21.53**
50 mm dia.	9.56	10.80	0.63	14.93	m	**25.73**
FIXINGS						
For steel pipes; black malleable iron. For minimum fixing distances, refer to the Tables and Memoranda at the rear of the book						
Pipe ring, single socket, black malleable iron						
25 mm dia.	0.80	0.91	0.12	2.85	nr	**3.76**
32 mm dia.	0.85	0.96	0.14	3.33	nr	**4.29**
40 mm dia.	1.10	1.24	0.15	3.56	nr	**4.80**
50 mm dia.	1.40	1.58	0.16	3.79	nr	**5.37**
Extra Over channel sections for fabricated hangers and brackets						
Galvanized steel; including inserts, bolts, nuts, washers; fixed to backgrounds						
41 × 21 mm	4.92	5.56	0.29	6.86	m	**12.42**
41 × 41 mm	5.90	6.66	0.29	6.86	m	**13.52**
Threaded rods; metric thread; including nuts, washers etc						
12 mm dia. × 600 mm long	2.49	2.82	0.18	4.27	nr	**7.09**
Extra over for black malleable iron fittings; BS 143						
Plain plug, solid						
25 mm dia.	1.70	1.92	0.40	9.48	nr	**11.40**
32 mm dia.	2.44	2.75	0.44	10.42	nr	**13.17**
40 mm dia.	3.44	3.89	0.48	11.37	nr	**15.26**
50 mm dia.	4.52	5.10	0.56	13.27	nr	**18.37**

S: PIPED SUPPLY SYSTEMS

Item	Net Price £	Material £	Labour hours	Labour £	Unit	Total rate £
Concentric reducing socket						
32 mm dia.	2.02	2.28	0.48	11.37	nr	**13.65**
40 mm dia.	2.63	2.97	0.55	13.04	nr	**16.01**
50 mm dia.	3.67	4.14	0.60	14.22	nr	**18.36**
Elbow; 90° female/female						
25 mm dia.	1.18	1.34	0.44	10.42	nr	**11.76**
32 mm dia.	2.16	2.44	0.53	12.56	nr	**15.00**
40 mm dia.	3.63	4.10	0.60	14.22	nr	**18.32**
50 mm dia.	4.25	4.80	0.65	15.40	nr	**20.20**
Tee						
25 mm dia. equal	1.61	1.82	0.51	12.09	nr	**13.91**
32 mm dia. reducing to 25 mm dia.	2.94	3.33	0.54	12.80	nr	**16.13**
40 mm dia.	4.52	5.10	0.65	15.40	nr	**20.50**
50 mm dia.	6.49	7.33	0.78	18.47	nr	**25.80**
Cross tee						
25 mm dia. equal	4.38	4.95	1.16	27.49	nr	**32.44**
32 mm dia.	6.39	7.22	1.40	33.17	nr	**40.39**
40 mm dia.	8.22	9.28	1.60	37.91	nr	**47.19**
50 mm dia.	13.34	15.07	1.68	39.81	nr	**54.88**
Prefabricated black steel pipework; welded joints, including all couplings, unions and the like to BS 1387:1985; fixing to backgrounds						
Heavy weight						
65 mm dia.	13.01	14.70	0.65	15.40	m	**30.10**
80 mm dia.	16.56	18.71	0.70	16.60	m	**35.31**
100 mm dia.	23.11	26.10	0.85	20.13	m	**46.23**
150 mm dia.	31.27	35.31	1.15	27.25	m	**62.56**
FIXINGS						
For steel pipes; black malleable iron. For minimum fixing distances, refer to the Tables and Memoranda at the rear of the book						
Pipe ring, single socket, black malleable iron						
65 mm dia.	2.02	2.28	0.30	7.10	nr	**9.38**
80 mm dia.	2.41	2.72	0.35	8.29	nr	**11.01**
100 mm dia.	3.65	4.12	0.40	9.48	nr	**13.60**
150 mm dia.	8.29	9.36	0.77	18.24	nr	**27.60**

S: PIPED SUPPLY SYSTEMS

Item	Net Price £	Material £	Labour hours	Labour £	Unit	Total rate £
S63: SPRINKLERS – cont						
FIXINGS – cont						
Extra over fittings						
Reducer (one size down)						
65 mm dia.	6.68	7.55	2.70	63.98	nr	**71.53**
80 mm dia.	8.69	9.81	2.86	67.76	nr	**77.57**
100 mm dia.	17.34	19.58	3.22	76.31	nr	**95.89**
150 mm dia.	48.16	54.40	4.20	99.52	nr	**153.92**
Elbow; 90°						
65 mm dia.	9.19	10.38	3.06	72.51	nr	**82.89**
80 mm dia.	13.50	15.25	3.40	80.56	nr	**95.81**
100 mm dia.	16.87	19.05	3.70	87.68	nr	**106.73**
150 mm dia.	25.30	28.57	5.20	123.22	nr	**151.79**
Branch bend						
65 mm dia.	9.35	10.56	3.60	85.30	nr	**95.86**
80 mm dia.	10.95	12.36	3.80	90.04	nr	**102.40**
100 mm dia.	16.85	19.03	5.10	120.85	nr	**139.88**
150 mm dia.	43.59	49.24	7.50	177.71	nr	**226.95**
Prefabricated black steel pipe; victaulic joints; including all couplings and the like to BS 1387: 1985; fixing to backgrounds						
Heavy weight						
65 mm dia.	11.23	12.68	0.70	16.60	m	**29.28**
80 mm dia.	14.58	16.46	0.78	18.47	m	**34.93**
100 mm dia.	21.26	24.01	0.93	22.03	m	**46.04**
150 mm dia.	42.34	47.82	1.25	29.62	m	**77.44**
FIXINGS						
For fixings refer to For steel pipes; black malleable iron.						
Extra over fittings						
Coupling						
65 mm dia.	8.22	9.28	0.26	6.15	nr	**15.43**
80 mm dia.	8.81	9.95	0.26	6.15	nr	**16.10**
100 mm dia.	11.54	13.04	0.32	7.58	nr	**20.62**
150 mm dia.	20.32	22.95	0.35	8.29	nr	**31.24**

S: PIPED SUPPLY SYSTEMS

Item	Net Price £	Material £	Labour hours	Labour £	Unit	Total rate £
Reducer						
65 mm dia.	6.68	7.55	0.48	11.37	nr	**18.92**
80 mm dia.	8.69	9.81	0.43	10.19	nr	**20.00**
100 mm dia.	17.34	19.58	0.46	10.89	nr	**30.47**
150 mm dia.	26.00	29.37	0.45	10.66	nr	**40.03**
Elbow; any degree						
65 mm dia.	9.19	10.38	0.56	13.27	nr	**23.65**
80 mm dia.	13.50	15.25	0.63	14.93	nr	**30.18**
100 mm dia.	26.03	29.40	0.71	16.83	nr	**46.23**
150 mm dia.	39.04	44.09	0.80	18.95	nr	**63.04**
Equal tee						
65 mm dia.	14.23	16.08	0.74	17.54	nr	**33.62**
80 mm dia.	17.14	19.36	0.83	19.67	nr	**39.03**
100 mm dia.	31.41	35.48	0.94	22.27	nr	**57.75**
150 mm dia.	47.12	53.21	1.05	24.88	nr	**78.09**

Y11 – PIPELINE ANCILLARIES

SPRINKLER HEADS

Item	Net Price £	Material £	Labour hours	Labour £	Unit	Total rate £
Sprinkler heads; brass body; frangible glass bulb; manufactured to standard operating temperature of 57-141°C; quick response; RTI<50						
Conventional pattern; 15 mm dia.	3.76	4.24	0.15	3.56	nr	**7.80**
Sidewall pattern; 15 mm dia.	5.52	6.23	0.15	3.56	nr	**9.79**
Conventional pattern; 15 mm dia.; satin chrome plated	4.41	4.98	0.15	3.56	nr	**8.54**
Sidewall pattern; 15 mm dia.; satin chrome plated	5.74	6.49	0.15	3.56	nr	**10.05**
Fully concealed; fusible link; 15 mm dia.	13.79	15.58	0.15	3.56	nr	**19.14**

VALVES

Item	Net Price £	Material £	Labour hours	Labour £	Unit	Total rate £
Wet system alarm valves; including internal non-return valve; working pressure up to 12.5 bar; BS4504 PN16 flanged ends; bolted connections						
100 mm dia.	1193.24	1347.63	25.00	592.37	nr	**1940.00**
150 mm dia.	1457.56	1646.16	25.00	592.37	nr	**2238.53**
Wet system by-pass alarm valves; including internal non-return valve; working pressure up to 12.5 bar; BS4504 PN16 flanged ends; bolted connections						
100 mm dia.	2103.34	2375.50	25.00	592.37	nr	**2967.87**
150 mm dia.	2659.33	3003.43	25.00	592.37	nr	**3595.80**

S: PIPED SUPPLY SYSTEMS

Item	Net Price £	Material £	Labour hours	Labour £	Unit	Total rate £
S63: SPRINKLERS – cont						
VALVES – cont						
Alternate system wet/dry alarm station; including butterfly valve, wet alarm valve, dry pipe differential pressure valve and pressure gauges; working pressure up to 10.5 bar; BS4505 PN16 flanged ends; bolted connections						
100 mm dia.	2606.77	2944.08	40.00	947.78	nr	**3891.86**
150 mm dia.	3037.90	3430.98	40.00	947.78	nr	**4378.76**
Alternate system wet/dry alarm station; including electrically operated butterfly valve, water supply accelerator set, wet alarm valve, dry pipe differential pressure valve and pressure gauges; working pressure up to 10.5 bar; BS4505 PN16 flanged ends; bolted connections						
100 mm dia.	2971.40	3355.88	45.00	1066.26	nr	**4422.14**
150 mm dia.	3406.63	3847.43	45.00	1066.26	nr	**4913.69**
ALARM/GONGS						
Water operated motor alarm and gong; stainless steel and aluminum body and gong; screwed connections						
Connection to sprinkler system and drain pipework	363.71	410.77	6.00	142.17	nr	**552.94**
Y21 – WATER TANKS						
Note: Prices are based on the most economical tank size for each volume, and the cost will vary with differing tank dimensions, for the same volume						
Steel sectional sprinkler tank; ordinary hazard, life safety classification; two compartment tank, complete with all fittings and accessories to comply with LPCB type A requirements; cost of erection (on prepared supports) is included within net price, labour cost allows for offloading and positioning of materials						
Volume, size						
70 m³, 6.1 m × 4.88 m × 2.44 m (h)	22574.38	25495.39	24.00	568.67	nr	**26064.06**
105 m³, 7.3 m × 6.1 m × 2.44 m (h)	26489.75	29917.40	28.00	663.45	nr	**30580.85**
168 m³, 9.76 m × 4.88 m x 3.66 m (h)	32728.78	36963.72	28.00	663.45	nr	**37627.17**
211 m³, 9.76 m × 6.1 m × 3.66 m (h)	36132.94	40808.36	32.00	758.23	nr	**41566.59**

S: PIPED SUPPLY SYSTEMS

Item	Net Price £	Material £	Labour hours	Labour £	Unit	Total rate £
Steel sectional sprinkler tank; ordinary hazard, property protection classification; single compartment tank, complete with all fittings and accessories to comply with LPCB type A requirements; cost of erection (on prepared supports) is included within net price, labour cost allows for offloading and positioning of materials						
Volume, size						
70 m³, 6.1 m × 4.88 m × 2.44 m (h)	15917.08	17976.67	24.00	568.67	nr	**18545.34**
105 m³, 7.3 m × 6.1 m × 2.44 m (h)	19663.99	22208.41	28.00	663.45	nr	**22871.86**
168 m³, 9.76 m × 4.88 m × 3.66 m (h)	25409.24	28697.07	28.00	663.45	nr	**29360.52**
211 m³, 9.76 m × 6.1 m × 3.66 m (h)	28511.33	32200.55	32.00	758.23	nr	**32958.78**
GRP sectional sprinkler tank; ordinary hazard, life safety classification; two compartment tank, complete with all fittings and accessories to comply with LPCB type A requirements; cost of erection (on prepared supports) is included within net price, labour cost allows for offloading and positioning of materials						
Volume, size						
55 m³, 6 m × 4 m × 3 m (h)	25191.05	28450.65	14.00	331.72	nr	**28782.37**
70 m³, 6 m × 5 m × 3 m (h)	28165.43	31809.89	16.00	379.11	nr	**32189.00**
80 m³, 8 m × 4 m × 3 m (h)	29189.46	32966.43	16.00	379.11	nr	**33345.54**
105 m³, 10 m × 4 m × 3 m (h)	33030.15	37304.08	24.00	568.67	nr	**37872.75**
125 m³, 10 m × 5 m × 3 m (h)	37250.69	42070.75	24.00	568.67	nr	**42639.42**
140 m³, 8 m × 7 m × 3 m (h)	38793.41	43813.08	24.00	568.67	nr	**44381.75**
135 m³, 9 m × 6 m × 3 m (h)	39966.27	45137.70	24.00	568.67	nr	**45706.37**
160 m³, 13 m × 5 m × 3 m (h)	43879.15	49556.89	24.00	568.67	nr	**50125.56**
185 m³, 12 m × 6 m × 3 m (h)	46415.91	52421.90	24.00	568.67	nr	**52990.57**
GRP sectional sprinkler tank; ordinary hazard, property protection classification; single compartment tank, complete with all fittings and accessories to comply with LPCB type A requirements; cost of erection (on prepared supports) is within net price, labour cost allows for offloading and positioning of materials						
Volume, size						
55 m³, 6 m × 4 m × 3 m (h)	19380.05	21887.73	14.00	331.72	nr	**22219.45**
70 m³, 6 m × 5 m × 3 m (h)	21894.60	24727.66	16.00	379.11	nr	**25106.77**
80 m³, 8 m × 4 m × 3 m (h)	23369.56	26393.47	16.00	379.11	nr	**26772.58**
105 m³, 10 m × 4 m × 3 m (h)	27377.96	30920.53	24.00	568.67	nr	**31489.20**
125 m³, 10 m × 5 m × 3 m (h)	30958.76	34964.67	24.00	568.67	nr	**35533.34**
140 m³, 8 m × 7 m × 3 m (h)	32049.43	36196.46	24.00	568.67	nr	**36765.13**
135 m³, 9 m × 6 m × 3 m (h)	33673.39	38030.56	24.00	568.67	nr	**38599.23**
160 m³, 13 m × 5 m × 3 m (h)	37750.49	42635.21	24.00	568.67	nr	**43203.88**
185 m³, 12 m × 6 m × 3 m (h)	39657.51	44789.00	24.00	568.67	nr	**45357.67**

S: PIPED SUPPLY SYSTEMS

Item	Net Price £	Material £	Labour hours	Labour £	Unit	Total rate £
S65: FIRE HYDRANTS						
EXTINGUISHERS						
Fire extinguishers; hand held; BS 5423; placed in position						
Water type; cartridge operated; for Class A fires						
Water type, 9 litres capacity; 55 gm CO_2 cartridge; Class A fires (fire-rating 13A)	79.60	89.90	1.00	23.69	nr	**113.59**
Foam type, 9 litres capacity; 75 gm CO_2 cartridge; Class A & B fires (fire-rating 13A:183B)	94.40	106.61	1.00	23.69	nr	**130.30**
Dry powder type; cartridge operated; for Class A, B & C fires and electrical equipment fires						
Dry powder type, 1 kg capacity; 12 gm CO_2 cartridge; Class A, B & C fires (fire-rating 5A:34B)	33.66	38.02	1.00	23.69	nr	**61.71**
Dry powder type, 2 kg capacity; 28 gm CO_2 cartridge; Class A, B & C fires (fire-rating 13A:55B)	45.33	51.19	1.00	23.69	nr	**74.88**
Dry powder type, 4 kg capacity; 90 gm CO_2 cartridge; Class A, B & C fires (fire-rating 21A:183B)	80.75	91.20	1.00	23.69	nr	**114.89**
Dry powder type, 9 kg capacity; 190 gm CO_2 cartridge; Class A, B & C fires (fire-rating 43A:233B)	108.19	122.19	1.00	23.69	nr	**145.88**
Dry powder type; stored pressure type; for Class A, B & C fires and electrical equipment fires						
Dry powder type, 1 kg capacity; Class A, B & C fires (fire-rating 5A:34B)	31.01	35.03	1.00	23.69	nr	**58.72**
Dry powder type, 2 kg capacity; Class A, B & C fires (fire-rating 13A:55B)	36.81	41.58	1.00	23.69	nr	**65.27**
Dry powder type, 4 kg capacity; Class A, B & C fires (fire-rating 21A:183B)	68.76	77.65	1.00	23.69	nr	**101.34**
Dry powder type, 9 kg capacity; Class A, B & C fires (fire-rating 43A:233B)	89.09	100.61	1.00	23.69	nr	**124.30**
Carbon dioxide type; for Class B fires and electrical equipment fires						
CO_2 type with hose and horn, 2 kg capacity, Class B fires (fire-rating 34B)	85.17	96.19	1.00	23.69	nr	**119.88**
CO_2 type with hose and horn, 5 kg capacity, Class B fires (fire-rating 55B)	133.45	150.72	1.00	23.69	nr	**174.41**

S: PIPED SUPPLY SYSTEMS

Item	Net Price £	Material £	Labour hours	Labour £	Unit	Total rate £
Glass fibre blanket, in GRP container						
1100 × 1100 mm	27.03	30.53	0.50	11.85	nr	**42.38**
1200 × 1200 mm	29.62	33.46	0.50	11.85	nr	**45.31**
1800 × 1200 mm	39.51	44.63	0.50	11.85	nr	**56.48**
HYDRANTS						
Fire hydrants; bolted connections						
Underground hydrants, complete with frost plug to BS 750						
sluice valve pattern type 1	222.26	251.02	4.50	106.63	nr	**357.65**
screw down pattern type 2	161.44	182.32	4.50	106.63	nr	**288.95**
Stand pipe for underground hydrant; screwed base; light alloy						
Single outlet	127.83	144.37	1.00	23.69	nr	**168.06**
Double outlet	187.04	211.24	1.00	23.69	nr	**234.93**
64 mm diameter bronze/gunmetal outlet valves						
Oblique flanged landing valve	139.91	158.02	1.00	23.69	nr	**181.71**
Oblique screwed landing valve	139.91	158.02	1.00	23.69	nr	**181.71**
Cast iron surface box; fixing by others						
400 × 200 × 100 mm	122.51	138.36	1.00	23.69	nr	**162.05**
500 × 200 × 150 mm	165.07	186.43	1.00	23.69	nr	**210.12**
Frost plug	29.73	33.58	0.25	5.93	nr	**39.51**

T: MECHANICAL/COOLING/HEATING SYSTEMS

Item	Net Price £	Material £	Labour hours	Labour £	Unit	Total rate £
T10: GAS/OIL FIRED BOILERS						
DOMESTIC						
Domestic water boilers; stove enamelled casing; electric controls; placing in position; assembling and connecting; electrical work elsewhere						
Gas fired; floor standing; connected to conventional flue						
9 to 12 kW	544.29	614.72	8.59	203.54	nr	**818.26**
12 to 15 kW	572.59	646.68	8.59	203.54	nr	**850.22**
15 to 18 kW	609.87	688.79	8.88	210.41	nr	**899.20**
18 to 21 kW	708.53	800.21	9.92	235.05	nr	**1035.26**
21 to 23 kW	785.48	887.11	10.66	252.59	nr	**1139.70**
23 to 29 kW	1018.76	1150.58	11.81	279.83	nr	**1430.41**
29 to 37 kW	1204.94	1360.85	11.81	279.83	nr	**1640.68**
37 to 41 kW	1253.09	1415.23	12.68	300.45	nr	**1715.68**
Gas fired; wall hung; connected to conventional flue						
9 to 12 kW	478.38	540.28	8.59	203.54	nr	**743.82**
12 to 15 kW	617.61	697.53	8.59	203.54	nr	**901.07**
13 to 18 kW	783.80	885.22	8.59	203.54	nr	**1088.76**
Gas fired; floor standing; connected to balanced flue						
9 to 12 kW	679.91	767.89	9.16	217.05	nr	**984.94**
12 to 15 kW	707.41	798.95	10.95	259.46	nr	**1058.41**
15 to 18 kW	761.47	860.00	11.98	283.87	nr	**1143.87**
18 to 21 kW	897.32	1013.43	12.78	302.82	nr	**1316.25**
21 to 23 kW	1034.94	1168.86	12.78	302.82	nr	**1471.68**
23 to 29 kW	1319.15	1489.84	15.45	366.08	nr	**1855.92**
29 to 37 kW	2535.93	2864.07	17.65	418.21	nr	**3282.28**
Gas fired; wall hung; connected to balanced flue						
6 to 9 kW	513.36	579.79	9.16	217.05	nr	**796.84**
9 to 12 kW	587.45	663.46	9.16	217.05	nr	**880.51**
12 to 15 kW	665.20	751.27	9.45	223.91	nr	**975.18**
15 to 18 kW	797.61	900.81	9.74	230.80	nr	**1131.61**
18 to 22 kW	846.63	956.18	9.74	230.80	nr	**1186.98**
Gas fired; wall hung; connected to fan flue (including flue kit)						
6 to 9 kW	587.13	663.10	9.16	217.05	nr	**880.15**
9 to 12 kW	654.30	738.96	9.16	217.05	nr	**956.01**
12 to 15 kW	710.27	802.18	10.95	259.46	nr	**1061.64**
15 to 18 kW	765.07	864.06	11.98	283.87	nr	**1147.93**
18 to 23 kW	996.86	1125.85	12.78	302.82	nr	**1428.67**
23 to 29 kW	1290.91	1457.95	15.45	366.08	nr	**1824.03**
29 to 35 kW	1584.87	1789.95	17.65	418.21	nr	**2208.16**

T: MECHANICAL/COOLING/HEATING SYSTEMS

Item	Net Price £	Material £	Labour hours	Labour £	Unit	Total rate £
Oil fired; floor standing; connected to conventional flue						
12 to 15	1019.43	1151.34	10.38	245.95	nr	**1397.29**
15 to 19	1063.42	1201.02	12.20	289.08	nr	**1490.10**
21 to 25	1213.66	1370.71	14.30	338.83	nr	**1709.54**
26 to 32	1330.62	1502.80	15.80	374.37	nr	**1877.17**
35 to 50	1502.32	1696.71	20.46	484.80	nr	**2181.51**
Fire place mounted natural gas fire and back boiler; cast iron water boiler; electric control box; fire output 3 kW with wood surround						
10.50 kW	297.41	335.90	8.88	210.41	nr	**546.31**
FORCED DRAFT						
Commercial cast iron sectional floor standing boilers; pressure jet burner; including controls, enamelled jacket, insulation, assembly and commissioning; electrical work elsewhere						
Gas fired (on/off type), connected to conventional flue						
16-26 kW; 3 sections; 125 mm dia. flue	2163.52	2443.47	8.00	189.56	nr	**2633.03**
26-33 kW; 4 sections; 125 mm dia. flue	2306.69	2605.16	8.00	189.56	nr	**2794.72**
33-40 kW; 5 sections; 125 mm dia. flue	2608.95	2946.54	8.00	189.56	nr	**3136.10**
35-50 kW; 3 sections; 153 mm dia. flue	2863.48	3234.00	8.00	189.56	nr	**3423.56**
50-65 kW; 4 sections; 153 mm dia. flue	3086.20	3485.53	8.00	189.56	nr	**3675.09**
65-80 kW; 5 sections; 153 mm dia. flue	3356.64	3790.97	8.00	189.56	nr	**3980.53**
80-100 kW; 6 sections; 180 mm dia. flue	3900.70	4405.43	8.00	189.56	nr	**4594.99**
100-120 kW; 7 sections; 180 mm dia. flue	4931.55	5569.67	8.00	189.56	nr	**5759.23**
105-140 kW; 5 sections; 180 mm dia. flue	5726.96	6468.00	8.00	189.56	nr	**6657.56**
140-180 kW; 6 sections; 180 mm dia. flue	6522.38	7366.35	8.00	189.56	nr	**7555.91**
180-230 kW; 7 sections; 200 mm dia. flue	7317.78	8264.66	8.00	189.56	nr	**8454.22**
230-280 kW; 8 sections; 200 mm dia. flue	7795.04	8803.68	8.00	189.56	nr	**8993.24**
280-330 kW; 9 sections; 200 mm dia. flue	8749.53	9881.68	8.00	189.56	nr	**10071.24**
Gas fired (high/low type), connected to conventional flue						
105-140 kW; 5 sections; 180 mm dia. flue	6840.54	7725.67	8.00	189.56	nr	**7915.23**
140-180 kW; 6 sections; 180 mm dia. flue	7349.61	8300.62	8.00	189.56	nr	**8490.18**
180-230 kW; 7 sections; 200 mm dia. flue	8335.92	9414.55	8.00	189.56	nr	**9604.11**
230-280 kW; 8 sections; 200 mm dia. flue	8908.61	10061.34	8.00	189.56	nr	**10250.90**
280-330 kW; 9 sections; 200 mm dia. flue	9544.94	10780.00	8.00	189.56	nr	**10969.56**
300-390 kW; 8 sections; 250 mm dia. flue	11453.93	12936.01	12.00	284.34	nr	**13220.35**
390-450 kW; 9 sections; 250 mm dia. flue	12615.23	14247.57	12.00	284.34	nr	**14531.91**
450-540 kW; 10 sections; 250 mm dia. flue	13840.17	15631.02	12.00	284.34	nr	**15915.36**
540-600 kW; 11 sections; 300 mm dia. flue	14158.34	15990.36	12.00	284.34	nr	**16274.70**
600-670 kW; 12 sections; 300 mm dia. flue	17380.32	20679.81	12.00	284.34	nr	**20964.15**
670-720 kW; 13 sections; 300 mm dia. flue	17698.49	21039.16	12.00	284.34	nr	**21323.50**

T: MECHANICAL/COOLING/HEATING SYSTEMS

Item	Net Price £	Material £	Labour hours	Labour £	Unit	Total rate £
T10: GAS/OIL FIRED BOILERS – cont						
FORCED DRAFT – cont						
Commercial cast iron sectional floor standing boilers – cont						
Gas fired (high/low type), connected to conventional flue – cont						
720-780 kW; 14 sections; 300 mm dia. flue	17937.11	21308.65	12.00	284.34	nr	**21592.99**
754-812 kW; 14 sections; 400 mm dia. flue	18096.19	21488.31	12.00	284.34	nr	**21772.65**
812-870 kW; 15 sections; 400 mm dia. flue	22868.66	26878.32	12.00	284.34	nr	**27162.66**
870-928 kW; 16 sections; 400 mm dia. flue	24777.66	29034.33	12.00	284.34	nr	**29318.67**
928-986 kW; 17 sections; 400 mm dia. flue	25413.99	29752.99	12.00	284.34	nr	**30037.33**
986-1044 kW; 18 sections; 400 mm dia. flue	26527.57	31010.67	12.00	284.34	nr	**31295.01**
1044-1102 kW; 19 sections; 400 mm dia. flue	27163.89	31729.32	12.00	284.34	nr	**32013.66**
1102-1160 kW; 20 sections; 400 mm dia. flue	27863.86	32519.86	12.00	284.34	nr	**32804.20**
1160-1218 kW; 21 sections; exceeding 400 mm dia. flue	29391.05	34244.66	12.00	284.34	nr	**34529.00**
1218-1276 kW; 22 sections; exceeding 400 mm dia. flue	30897.72	36097.34	12.00	284.34	nr	**36381.68**
1276-1334 kW; 23 sections; exceeding 400 mm dia. flue	31995.39	37337.04	12.00	284.34	nr	**37621.38**
1334-1392 kW; 24 sections; exceeding 400 mm dia. flue	33427.13	38954.04	12.00	284.34	nr	**39238.38**
1392-1450 kW; 25 sections; exceeding 400 mm dia. flue	34381.63	40032.05	12.00	284.34	nr	**40316.39**
Oil fired (on/off type), connected to conventional flue						
16-26 kW; 3 sections; 125 mm dia. flue	1590.83	1796.68	8.00	189.56	nr	**1986.24**
26-33 kW; 4 sections; 125 mm dia. flue	1749.90	1976.33	8.00	189.56	nr	**2165.89**
33-40 kW; 5 sections; 125 mm dia. flue	1908.99	2156.00	8.00	189.56	nr	**2345.56**
35-50 kW; 3 sections; 153 mm dia. flue	2243.06	2533.30	8.00	189.56	nr	**2722.86**
50-65 kW; 4 sections; 153 mm dia. flue	2354.42	2659.07	8.00	189.56	nr	**2848.63**
65-80 kW; 5 sections; 153 mm dia. flue	2449.87	2766.87	8.00	189.56	nr	**2956.43**
80-100 kW; 6 sections; 180 mm dia. flue	3658.90	4132.35	8.00	189.56	nr	**4321.91**
100-120 kW; 7 sections; 180 mm dia. flue	4136.14	4671.33	8.00	189.56	nr	**4860.89**
105-140 kW; 5 sections; 180 mm dia. flue	4772.47	5390.01	8.00	189.56	nr	**5579.57**
140-180 kW; 6 sections; 180 mm dia. flue	5408.80	6108.67	8.00	189.56	nr	**6298.23**
180-230 kW; 7 sections; 200 mm dia. flue	6681.46	7546.01	8.00	189.56	nr	**7735.57**
230-280 kW; 8 sections; 200 mm dia. flue	7317.78	8264.66	8.00	189.56	nr	**8454.22**
280-330 kW; 9 sections; 200 mm dia. flue	8113.19	9163.00	8.00	189.56	nr	**9352.56**
Oil fired (high/low type), connected to conventional flue						
105-140 kW; 5 sections; 180 mm dia. flue	3833.89	4329.97	8.00	189.56	nr	**4519.53**
140-180 kW; 6 sections; 180 mm dia. flue	4136.14	4671.33	8.00	189.56	nr	**4860.89**
180-230 kW; 7 sections; 200 mm dia. flue	4613.39	5210.34	8.00	189.56	nr	**5399.90**

T: MECHANICAL/COOLING/HEATING SYSTEMS

Item	Net Price £	Material £	Labour hours	Labour £	Unit	Total rate £
230-280 kW; 8 sections; 200 mm dia. flue	5440.62	6144.61	8.00	189.56	nr	**6334.17**
280-330 kW; 9 sections; 200 mm dia. flue	8590.46	9702.03	8.00	189.56	nr	**9891.59**
300-390 kW; 8 sections; 250 mm dia. flue	9544.94	10780.00	12.00	284.34	nr	**11064.34**
390-450 kW; 9 sections; 250 mm dia. flue	10260.81	11588.51	12.00	284.34	nr	**11872.85**
450-540 kW; 10 sections; 250 mm dia. flue	11613.01	13115.68	12.00	284.34	nr	**13400.02**
540-600 kW; 11 sections; 300 mm dia. flue	12201.62	13780.44	12.00	284.34	nr	**14064.78**
600-670 kW; 12 sections; 300 mm dia. flue	13044.76	14732.69	12.00	284.34	nr	**15017.03**
670-720 kW; 13 sections; 300 mm dia. flue	14794.66	16709.02	12.00	284.34	nr	**16993.36**
720-780 kW; 14 sections; 300 mm dia. flue	15033.28	16978.51	12.00	284.34	nr	**17262.85**
754-812 kW; 14 sections; 400 mm dia. flue	15240.09	17212.08	12.00	284.34	nr	**17496.42**
812-870 kW; 15 sections; 400 mm dia. flue	20362.55	22997.36	12.00	284.34	nr	**23281.70**
870-928 kW; 16 sections; 400 mm dia. flue	21142.05	23877.72	12.00	284.34	nr	**24162.06**
928-986 kW; 17 sections; 400 mm dia. flue	21873.82	24704.19	12.00	284.34	nr	**24988.53**
986-1044 kW; 18 sections; 400 mm dia. flue	22748.78	25692.36	12.00	284.34	nr	**25976.70**
1044-1102 kW; 19 sections; 400 mm dia. flue	23862.36	26950.03	12.00	284.34	nr	**27234.37**
1102-1160 kW; 20 sections; 400 mm dia. flue	25198.65	28459.23	12.00	284.34	nr	**28743.57**
1160-1218 kW; 21 sections; exceeding 400 mm dia. flue	26598.58	30040.31	12.00	284.34	nr	**30324.65**
1218-1276 kW; 22 sections; exceeding 400 mm dia. flue	27918.96	31531.54	12.00	284.34	nr	**31815.88**
1276-1334 kW; 23 sections; exceeding 400 mm dia. flue	28905.27	32645.47	12.00	284.34	nr	**32929.81**
1334-1392 kW; 24 sections; exceeding 400 mm dia. flue	30599.49	34558.92	12.00	284.34	nr	**34843.26**
1392-1450 kW; 25 sections; exceeding 400 mm dia. flue	31721.03	35825.57	12.00	284.34	nr	**36109.91**
Commercial steel shell floor standing boilers; pressure jet burner; including controls, enamelled jacket, insulation, placing in position and commissioning; electrical work elsewhere						
Gas fired (on/off type), connected to conventional flue						
130-190 kW	5703.80	6441.84	8.00	189.56	nr	**6631.40**
200-250 kW	6460.59	7296.56	8.00	189.56	nr	**7486.12**
280-360 kW	7737.89	8739.14	8.00	189.56	nr	**8928.70**
375-500 kW	9430.66	10650.94	8.00	189.56	nr	**10840.50**
Gas fired (high/low type), connected to conventional flue						
130-190 kW	6947.11	7846.03	8.00	189.56	nr	**8035.59**
200-250 kW	7419.72	8379.79	8.00	189.56	nr	**8569.35**
280-360 kW	8868.46	10015.99	8.00	189.56	nr	**10205.55**
375-500 kW	9894.00	11174.23	8.00	189.56	nr	**11363.79**
580-730 kW	12488.74	14104.72	10.00	236.95	nr	**14341.67**
655-820 kW	12743.58	14392.54	10.00	236.95	nr	**14629.49**
830-1040 kW	12788.37	14443.12	12.00	284.34	nr	**14727.46**

T: MECHANICAL/COOLING/HEATING SYSTEMS

Item	Net Price £	Material £	Labour hours	Labour £	Unit	Total rate £
T10: GAS/OIL FIRED BOILERS – cont						
FORCED DRAFT – cont						
Commercial steel shell floor standing boilers – cont						
Gas fired (high/low type), connected to conventional flue – cont						
1070-1400 kW	16913.70	19102.24	12.00	284.34	nr	**19386.58**
1420-1850 kW	21062.20	23787.54	12.00	284.34	nr	**24071.88**
1850-2350 kW	23922.59	27018.06	14.00	331.72	nr	**27349.78**
2300-3000 kW	28227.09	31879.54	14.00	331.72	nr	**32211.26**
2800-3500 kW	36607.48	41344.30	14.00	331.72	nr	**41676.02**
Oil fired (on/off type), connected to conventional flue						
130-190 kW	5221.92	5897.61	8.00	189.56	nr	**6087.17**
200-250 kW	5702.25	6440.10	8.00	189.56	nr	**6629.66**
Oil fired (high/low type), connected to conventional flue						
130-190 kW	5683.72	6419.17	8.00	189.56	nr	**6608.73**
200-250 kW	6164.06	6961.66	8.00	189.56	nr	**7151.22**
280-360 kW	7106.20	8025.71	8.00	189.56	nr	**8215.27**
375-500 kW	8496.24	9595.61	8.00	189.56	nr	**9785.17**
580-730 kW	9450.72	10673.60	10.00	236.95	nr	**10910.55**
655-820 kW	9704.02	10959.68	10.00	236.95	nr	**11196.63**
830-1040 kW	11226.89	12679.59	12.00	284.34	nr	**12963.93**
1070-1400 kW	13718.15	15493.21	12.00	284.34	nr	**15777.55**
1420-1850 kW	16610.98	18760.36	12.00	284.34	nr	**19044.70**
1850-2350 kW	21272.24	24024.76	14.00	331.72	nr	**24356.48**
2300-3000 kW	24847.74	28062.91	14.00	331.72	nr	**28394.63**
2800-3500 kW	33228.13	37527.69	14.00	331.72	nr	**37859.41**
ATMOSPHERIC						
Commercial cast iron sectional floor standing boilers; atmospheric; including controls, enamelled jacket, insulation, assembly and commissioning; electrical work elsewhere						
Gas (on/off type), connected to conventional flue						
30-40 kW	2038.10	2301.82	8.00	189.56	nr	**2491.38**
40-50 kW	2234.46	2523.59	8.00	189.56	nr	**2713.15**
50-60 kW	2374.96	2682.26	8.00	189.56	nr	**2871.82**

T: MECHANICAL/COOLING/HEATING SYSTEMS

Item	Net Price £	Material £	Labour hours	Labour £	Unit	Total rate £
60-70 kW	2592.14	2927.55	8.00	189.56	nr	**3117.11**
70-80 kW	2747.40	3102.90	8.00	189.56	nr	**3292.46**
80-90 kW	3210.11	3625.48	8.00	189.56	nr	**3815.04**
90-100 kW	3339.49	3771.60	8.00	189.56	nr	**3961.16**
100-110 kW	3538.89	3996.81	8.00	189.56	nr	**4186.37**
110-120 kW	3742.85	4227.16	8.00	189.56	nr	**4416.72**
Gas (high/low type), connected to conventional flue						
30-40 kW	2313.60	2612.96	8.00	189.56	nr	**2802.52**
40-50 kW	2453.63	2771.12	8.00	189.56	nr	**2960.68**
50-60 kW	2637.81	2979.13	8.00	189.56	nr	**3168.69**
60-70 kW	2876.78	3249.02	8.00	189.56	nr	**3438.58**
70-80 kW	3045.74	3439.85	8.00	189.56	nr	**3629.41**
80-90 kW	3385.16	3823.18	8.00	189.56	nr	**4012.74**
90-100 kW	3569.34	4031.19	8.00	189.56	nr	**4220.75**
100-110 kW	3779.39	4268.42	8.00	189.56	nr	**4457.98**
110-120 kW	4015.31	4534.87	8.00	189.56	nr	**4724.43**
120-140 kW	6022.96	6802.30	8.00	189.56	nr	**6991.86**
140-160 kW	6175.18	6974.22	8.00	189.56	nr	**7163.78**
160-180 kW	6280.22	7092.85	8.00	189.56	nr	**7282.41**
180-200 kW	6867.74	7756.40	8.00	189.56	nr	**7945.96**
200-220 kW	7164.54	8091.60	10.00	236.95	nr	**8328.55**
220-260 kW	7785.57	8792.98	10.00	236.95	nr	**9029.93**
260-300 kW	8490.30	9588.90	10.00	236.95	nr	**9825.85**
300-340 kW	9055.00	10226.67	12.00	284.34	nr	**10511.01**

CONDENSING

Low Nox wall mounted condensing boiler with high efficiency modulating pre-mix burner: Aluminium heat exchanger: Placing in position

Maximum output						
35 kW	1701.18	1921.30	11.00	260.64	nr	**2181.94**
45 kW	1758.26	1985.77	11.00	260.64	nr	**2246.41**
60 kW	1986.58	2243.63	11.00	260.64	nr	**2504.27**
80 kW	2390.60	2699.93	11.00	260.64	nr	**2960.57**

T: MECHANICAL/COOLING/HEATING SYSTEMS

Item	Net Price £	Material £	Labour hours	Labour £	Unit	Total rate £
T10: GAS/OIL FIRED BOILERS – cont						
CONDENSING – cont						
Low Nox floor standing condensing boiler with high efficiency modulating pre-mix burner: Stainless steel heat exchanger: Including controls: Placing in position						
Maximum output						
50 kW	2795.75	3157.50	8.00	189.56	nr	**3347.06**
60 kW	2988.25	3374.91	8.00	189.56	nr	**3564.47**
80 kW	3262.46	3684.61	8.00	189.56	nr	**3874.17**
100 kW	3751.54	4236.97	8.00	189.56	nr	**4426.53**
125 kW	4662.57	5265.88	10.00	236.95	nr	**5502.83**
150 kW	5159.49	5827.11	10.00	236.95	nr	**6064.06**
200 kW	7132.63	8055.55	10.00	236.95	nr	**8292.50**
250 kW	7808.63	8819.03	10.00	236.95	nr	**9055.98**
300 kW	9885.85	11165.03	10.00	236.95	nr	**11401.98**
350 kW	10771.13	12164.86	10.00	236.95	nr	**12401.81**
400 kW	11587.03	13086.33	10.00	236.95	nr	**13323.28**
450 kW	11780.65	13305.00	10.00	236.95	nr	**13541.95**
500 kW	11974.27	13523.68	10.00	236.95	nr	**13760.63**
650 kW	12590.94	14220.15	10.00	236.95	nr	**14457.10**
700 kW	20114.17	22716.84	10.00	236.95	nr	**22953.79**
800 kW	21509.81	24293.07	12.00	284.34	nr	**24577.41**
900 kW	22250.72	25129.85	12.00	284.34	nr	**25414.19**
1000 kW	22993.87	25969.16	12.00	284.34	nr	**26253.50**
1300 kW	24392.86	27549.17	12.00	284.34	nr	**27833.51**
FLUE SYSTEMS						
Flues; suitable for domestic, medium sized industrial and commercial oil and gas appliances; stainless steel, twin wall, insulated; for use internally or externally						
Straight length; 120 mm long; including one locking band						
127 mm dia.	40.62	45.88	0.49	11.61	nr	**57.49**
152 mm dia.	45.52	51.41	0.51	12.09	nr	**63.50**
175 mm dia.	52.84	59.68	0.54	12.80	nr	**72.48**
203 mm dia.	60.12	67.90	0.58	13.75	nr	**81.65**
254 mm dia.	71.10	80.30	0.70	16.60	nr	**96.90**
304 mm dia.	89.07	100.59	0.74	17.54	nr	**118.13**
355 mm dia.	127.55	144.05	0.80	18.95	nr	**163.00**

T: MECHANICAL/COOLING/HEATING SYSTEMS

Item	Net Price £	Material £	Labour hours	Labour £	Unit	Total rate £
Straight length; 300 mm long; including one locking band						
127 mm dia.	62.67	70.78	0.52	12.32	nr	**83.10**
152 mm dia.	70.84	80.01	0.52	12.32	nr	**92.33**
178 mm dia.	81.40	91.93	0.55	13.04	nr	**104.97**
203 mm dia.	92.12	104.04	0.64	15.17	nr	**119.21**
254 mm dia.	102.16	115.38	0.79	18.72	nr	**134.10**
304 mm dia.	122.56	138.41	0.86	20.37	nr	**158.78**
355 mm dia.	134.49	151.89	0.94	22.27	nr	**174.16**
400 mm dia.	143.92	162.55	1.03	24.41	nr	**186.96**
450 mm dia.	164.61	185.91	1.03	24.41	nr	**210.32**
500 mm dia.	176.56	199.41	1.10	26.07	nr	**225.48**
550 mm dia.	194.85	220.06	1.10	26.07	nr	**246.13**
600 mm dia.	215.00	242.82	1.10	26.07	nr	**268.89**
Straight length; 500 mm long; including one locking band						
127 mm dia.	73.65	83.18	0.55	13.04	nr	**96.22**
152 mm dia.	82.11	92.74	0.55	13.04	nr	**105.78**
178 mm dia.	92.47	104.44	0.63	14.93	nr	**119.37**
203 mm dia.	108.09	122.07	0.63	14.93	nr	**137.00**
254 mm dia.	125.54	141.78	0.86	20.37	nr	**162.15**
304 mm dia.	150.41	169.87	0.95	22.51	nr	**192.38**
355 mm dia.	168.91	190.77	1.03	24.41	nr	**215.18**
400 mm dia.	184.44	208.30	1.12	26.54	nr	**234.84**
450 mm dia.	213.01	240.58	1.12	26.54	nr	**267.12**
500 mm dia.	229.52	259.22	1.19	28.20	nr	**287.42**
550 mm dia.	253.00	285.73	1.19	28.20	nr	**313.93**
600 mm dia.	262.08	295.99	1.19	28.20	nr	**324.19**
Straight length; 1000 mm long; including one locking band						
127 mm dia.	131.66	148.70	0.62	14.70	nr	**163.40**
152 mm dia.	146.74	165.73	0.68	16.12	nr	**181.85**
178 mm dia.	165.10	186.47	0.74	17.54	nr	**204.01**
203 mm dia.	194.47	219.64	0.80	18.95	nr	**238.59**
254 mm dia.	220.39	248.91	0.87	20.61	nr	**269.52**
304 mm dia.	254.39	287.30	1.06	25.12	nr	**312.42**
355 mm dia.	291.92	329.69	1.16	27.49	nr	**357.18**
400 mm dia.	312.62	353.07	1.26	29.85	nr	**382.92**
450 mm dia.	329.87	372.55	1.26	29.85	nr	**402.40**
500 mm dia.	357.98	404.30	1.33	31.51	nr	**435.81**
550 mm dia.	393.70	444.64	1.33	31.51	nr	**476.15**
600 mm dia.	413.61	467.13	1.33	31.51	nr	**498.64**

T: MECHANICAL/COOLING/HEATING SYSTEMS

Item	Net Price £	Material £	Labour hours	Labour £	Unit	Total rate £
T10: GAS/OIL FIRED BOILERS – cont						
FLUE SYSTEMS – cont						
Flues – cont						
Adjustable length; boiler removal; internal use only; including one locking band						
127 mm dia.	60.35	68.16	0.52	12.32	nr	**80.48**
152 mm dia.	68.20	77.02	0.55	13.04	nr	**90.06**
178 mm dia.	77.29	87.29	0.59	13.98	nr	**101.27**
203 mm dia.	88.63	100.10	0.64	15.17	nr	**115.27**
254 mm dia.	131.42	148.42	0.79	18.72	nr	**167.14**
304 mm dia.	157.40	177.77	0.86	20.37	nr	**198.14**
355 mm dia.	176.51	199.35	0.99	23.46	nr	**222.81**
400 mm dia.	308.88	348.85	0.91	21.56	nr	**370.41**
450 mm dia.	330.63	373.41	0.91	21.56	nr	**394.97**
500 mm dia.	360.60	407.26	0.99	23.46	nr	**430.72**
550 mm dia.	393.39	444.29	0.99	23.46	nr	**467.75**
600 mm dia.	411.82	465.11	0.99	23.46	nr	**488.57**
Inspection length; 500 mm long; including one locking band						
127 mm dia.	155.12	175.20	0.55	13.04	nr	**188.24**
152 mm dia.	160.62	181.41	0.55	13.04	nr	**194.45**
178 mm dia.	169.18	191.07	0.63	14.93	nr	**206.00**
203 mm dia.	179.15	202.33	0.63	14.93	nr	**217.26**
254 mm dia.	231.63	261.60	0.86	20.37	nr	**281.97**
304 mm dia.	251.04	283.52	0.95	22.51	nr	**306.03**
355 mm dia.	283.45	320.13	1.03	24.41	nr	**344.54**
400 mm dia.	454.76	513.60	1.12	26.54	nr	**540.14**
450 mm dia.	467.14	527.58	1.12	26.54	nr	**554.12**
500 mm dia.	512.58	578.90	1.19	28.20	nr	**607.10**
550 mm dia.	536.00	605.36	1.19	28.20	nr	**633.56**
600 mm dia.	545.11	615.64	1.19	28.20	nr	**643.84**
Adapters						
127 mm dia.	12.97	14.65	0.49	11.61	nr	**26.26**
152 mm dia.	14.20	16.03	0.51	12.09	nr	**28.12**
178 mm dia.	15.18	17.15	0.54	12.80	nr	**29.95**
203 mm dia.	17.28	19.51	0.58	13.75	nr	**33.26**
254 mm dia.	19.01	21.47	0.70	16.60	nr	**38.07**
304 mm dia.	23.72	26.79	0.74	17.54	nr	**44.33**
355 mm dia.	28.81	32.54	0.80	18.95	nr	**51.49**
400 mm dia.	32.32	36.51	0.89	21.08	nr	**57.59**
450 mm dia.	34.52	38.98	0.89	21.08	nr	**60.06**
500 mm dia.	36.63	41.37	0.96	22.75	nr	**64.12**
550 mm dia.	42.60	48.11	0.96	22.75	nr	**70.86**
600 mm dia.	51.18	57.80	0.96	22.75	nr	**80.55**

T: MECHANICAL/COOLING/HEATING SYSTEMS

Item	Net Price £	Material £	Labour hours	Labour £	Unit	Total rate £
Fittings for flue system						
90° insulated tee; including two locking bands						
127 mm dia.	150.82	170.34	1.89	44.78	nr	**215.12**
152 mm dia.	174.06	196.58	2.04	48.34	nr	**244.92**
178 mm dia.	190.24	214.85	2.39	56.63	nr	**271.48**
203 mm dia.	222.75	251.57	2.56	60.66	nr	**312.23**
254 mm dia.	225.66	254.86	2.95	69.90	nr	**324.76**
304 mm dia.	280.62	316.93	3.41	80.79	nr	**397.72**
355 mm dia.	357.76	404.05	3.77	89.32	nr	**493.37**
400 mm dia.	471.43	532.43	4.25	100.70	nr	**633.13**
450 mm dia.	491.36	554.94	4.76	112.78	nr	**667.72**
500 mm dia.	556.01	627.95	5.12	121.32	nr	**749.27**
550 mm dia.	597.67	675.01	5.61	132.93	nr	**807.94**
600 mm dia.	621.88	702.35	5.98	141.70	nr	**844.05**
135° insulated tee; including two locking bands						
127 mm dia.	195.60	220.91	1.89	44.78	nr	**265.69**
152 mm dia.	211.49	238.85	2.04	48.34	nr	**287.19**
178 mm dia.	231.23	261.15	2.39	56.63	nr	**317.78**
203 mm dia.	293.46	331.43	2.56	60.66	nr	**392.09**
254 mm dia.	333.45	376.59	2.95	69.90	nr	**446.49**
304 mm dia.	251.24	283.75	3.41	80.79	nr	**364.54**
355 mm dia.	494.35	558.32	3.77	89.32	nr	**647.64**
400 mm dia.	650.76	734.96	4.25	100.70	nr	**835.66**
450 mm dia.	704.48	795.64	4.76	112.78	nr	**908.42**
500 mm dia.	820.71	926.90	5.12	121.32	nr	**1048.22**
550 mm dia.	842.05	951.01	5.61	132.93	nr	**1083.94**
600 mm dia.	888.70	1003.69	5.98	141.70	nr	**1145.39**
Wall sleeve; for 135° tee through wall						
127 mm dia.	20.02	22.61	1.89	44.78	nr	**67.39**
152 mm dia.	26.86	30.33	2.04	48.34	nr	**78.67**
178 mm dia.	28.13	31.77	2.39	56.63	nr	**88.40**
203 mm dia.	31.65	35.74	2.56	60.66	nr	**96.40**
254 mm dia.	35.29	39.85	2.95	69.90	nr	**109.75**
304 mm dia.	41.13	46.45	3.41	80.79	nr	**127.24**
355 mm dia.	45.70	51.61	3.77	89.32	nr	**140.93**
15° insulated elbow; including two locking bands						
127 mm dia.	104.66	118.20	1.57	37.20	nr	**155.40**
152 mm dia.	116.32	131.38	1.79	42.41	nr	**173.79**
178 mm dia.	124.40	140.49	2.05	48.57	nr	**189.06**
203 mm dia.	131.88	148.94	2.33	55.20	nr	**204.14**
254 mm dia.	135.83	153.41	2.45	58.05	nr	**211.46**
304 mm dia.	171.73	193.95	3.43	81.27	nr	**275.22**
355 mm dia.	229.74	259.47	4.71	111.61	nr	**371.08**

T: MECHANICAL/COOLING/HEATING SYSTEMS

Item	Net Price £	Material £	Labour hours	Labour £	Unit	Total rate £
T10: GAS/OIL FIRED BOILERS – cont						
FLUE SYSTEMS – cont						
Fittings for flue system – cont						
30° insulated elbow; including two locking bands						
127 mm dia.	104.66	118.20	1.44	34.12	nr	**152.32**
152 mm dia.	116.32	131.38	1.62	38.39	nr	**169.77**
178 mm dia.	124.40	140.49	1.89	44.78	nr	**185.27**
203 mm dia.	131.88	148.94	2.17	51.42	nr	**200.36**
254 mm dia.	135.83	153.41	2.16	51.18	nr	**204.59**
304 mm dia.	171.73	193.95	2.74	64.93	nr	**258.88**
355 mm dia.	229.30	258.97	3.17	75.11	nr	**334.08**
400 mm dia.	229.74	259.47	3.53	83.64	nr	**343.11**
450 mm dia.	266.25	300.71	3.88	91.93	nr	**392.64**
500 mm dia.	279.14	315.26	4.24	100.47	nr	**415.73**
550 mm dia.	299.58	338.34	4.61	109.23	nr	**447.57**
600 mm dia.	326.66	368.92	4.96	117.52	nr	**486.44**
45° insulated elbow; including two locking bands						
127 mm dia.	104.66	118.20	1.44	34.12	nr	**152.32**
152 mm dia.	116.32	131.38	1.51	35.78	nr	**167.16**
178 mm dia.	124.40	140.49	1.58	37.44	nr	**177.93**
203 mm dia.	131.88	148.94	1.66	39.34	nr	**188.28**
254 mm dia.	135.83	153.41	1.72	40.76	nr	**194.17**
304 mm dia.	171.73	193.95	1.80	42.65	nr	**236.60**
355 mm dia.	229.74	259.47	1.94	45.97	nr	**305.44**
400 mm dia.	292.97	330.88	2.01	47.62	nr	**378.50**
450 mm dia.	306.87	346.58	2.09	49.52	nr	**396.10**
500 mm dia.	329.17	371.76	2.16	51.18	nr	**422.94**
550 mm dia.	359.00	405.45	2.23	52.85	nr	**458.30**
600 mm dia.	368.10	415.73	2.30	54.49	nr	**470.22**
Flue supports						
Wall support, galvanized; including plate and brackets						
127 mm dia.	63.54	71.76	2.24	53.08	nr	**124.84**
152 mm dia.	69.86	78.90	2.44	57.81	nr	**136.71**
178 mm dia.	77.12	87.10	2.52	59.71	nr	**146.81**
203 mm dia.	80.31	90.70	2.77	65.63	nr	**156.33**
254 mm dia.	95.78	108.17	2.98	70.61	nr	**178.78**
304 mm dia.	108.29	122.30	3.46	81.99	nr	**204.29**
355 mm dia.	144.94	163.69	4.08	96.68	nr	**260.37**

T: MECHANICAL/COOLING/HEATING SYSTEMS

Item	Net Price £	Material £	Labour hours	Labour £	Unit	Total rate £
400 mm dia.; with 300 mm support length and collar	372.32	420.49	4.80	113.73	nr	**534.22**
450 mm dia.; with 300 mm support length and collar	396.95	448.31	5.62	133.17	nr	**581.48**
500 mm dia.; with 300 mm support length and collar	433.66	489.77	6.24	147.86	nr	**637.63**
550 mm dia.; with 300 mm support length and collar	474.45	535.84	6.97	165.15	nr	**700.99**
600 mm dia.; with 300 mm support length and collar	500.86	565.67	7.49	177.47	nr	**743.14**
Ceiling/floor support						
127 mm dia.	19.67	22.22	1.86	44.06	nr	**66.28**
152 mm dia.	21.86	24.68	2.14	50.71	nr	**75.39**
178 mm dia.	26.18	29.57	1.93	45.73	nr	**75.30**
203 mm dia.	39.42	44.52	2.74	64.93	nr	**109.45**
254 mm dia.	44.97	50.79	3.21	76.06	nr	**126.85**
304 mm dia.	51.40	58.05	3.68	87.20	nr	**145.25**
355 mm dia.	61.25	69.18	4.28	101.41	nr	**170.59**
400 mm dia.	78.24	88.36	4.86	115.16	nr	**203.52**
450 mm dia.	84.96	95.95	5.46	129.38	nr	**225.33**
500 mm dia.	103.15	116.49	6.04	143.12	nr	**259.61**
550 mm dia.	126.19	142.51	6.65	157.57	nr	**300.08**
600 mm dia.	181.10	204.53	7.24	171.55	nr	**376.08**
Ceiling/floor firestop spacer						
127 mm dia.	3.83	4.32	0.66	15.65	nr	**19.97**
152 mm dia.	4.26	4.81	0.69	16.35	nr	**21.16**
178 mm dia.	4.82	5.45	0.70	16.60	nr	**22.05**
203 mm dia.	5.71	6.45	0.87	20.61	nr	**27.06**
254 mm dia.	5.89	6.65	0.91	21.56	nr	**28.21**
304 mm dia.	7.08	8.00	0.95	22.51	nr	**30.51**
355 mm dia.	12.90	14.57	0.99	23.46	nr	**38.03**
Wall band; internal or external use						
127 mm dia.	23.17	26.16	1.03	24.41	nr	**50.57**
152 mm dia.	24.19	27.32	1.07	25.36	nr	**52.68**
178 mm dia.	25.15	28.41	1.11	26.31	nr	**54.72**
203 mm dia.	26.54	29.98	1.18	27.97	nr	**57.95**
254 mm dia.	27.63	31.20	1.30	30.79	nr	**61.99**
304 mm dia.	29.85	33.71	1.45	34.35	nr	**68.06**
355 mm dia.	31.76	35.87	1.65	39.10	nr	**74.97**
400 mm dia.	39.39	44.48	1.85	43.83	nr	**88.31**
450 mm dia.	41.97	47.40	2.39	56.63	nr	**104.03**
500 mm dia.	50.45	56.98	2.25	53.31	nr	**110.29**
550 mm dia.	52.84	59.68	2.45	58.05	nr	**117.73**
600 mm dia.	55.86	63.09	2.66	63.04	nr	**126.13**

T: MECHANICAL/COOLING/HEATING SYSTEMS

Item	Net Price £	Material £	Labour hours	Labour £	Unit	Total rate £
T10: GAS/OIL FIRED BOILERS – cont						
FLUE SYSTEMS – cont						
Flashings and terminals						
Insulated top stub; including one locking band						
127 mm dia.	57.06	64.44	1.49	35.30	nr	**99.74**
152 mm dia.	64.53	72.88	1.90	45.01	nr	**117.89**
178 mm dia.	69.47	78.46	1.92	45.49	nr	**123.95**
203 mm dia.	73.92	83.49	2.20	52.13	nr	**135.62**
254 mm dia.	78.49	88.65	2.49	59.00	nr	**147.65**
304 mm dia.	107.30	121.19	2.79	66.11	nr	**187.30**
355 mm dia.	141.65	159.98	3.19	75.59	nr	**235.57**
400 mm dia.	136.57	154.24	3.59	85.07	nr	**239.31**
450 mm dia.	148.37	167.57	3.97	94.06	nr	**261.63**
500 mm dia.	171.65	193.86	4.38	103.79	nr	**297.65**
550 mm dia.	179.60	202.84	4.78	113.26	nr	**316.10**
600 mm dia.	185.87	209.92	5.17	122.50	nr	**332.42**
Rain cap; including one locking band						
127 mm dia.	30.50	34.45	1.49	35.30	nr	**69.75**
152 mm dia.	31.88	36.01	1.54	36.50	nr	**72.51**
178 mm dia.	35.11	39.66	1.72	40.76	nr	**80.42**
203 mm dia.	41.98	47.41	2.00	47.39	nr	**94.80**
254 mm dia.	55.19	62.33	2.49	59.00	nr	**121.33**
304 mm dia.	37.79	42.68	2.80	66.34	nr	**109.02**
355 mm dia.	99.70	112.60	3.19	75.59	nr	**188.19**
400 mm dia.	99.47	112.34	3.45	81.74	nr	**194.08**
450 mm dia.	108.18	122.18	3.97	94.06	nr	**216.24**
500 mm dia.	116.97	132.11	4.38	103.79	nr	**235.90**
550 mm dia.	125.50	141.74	4.78	113.26	nr	**255.00**
600 mm dia.	134.18	151.54	5.17	122.50	nr	**274.04**
Round top; including one locking band						
127 mm dia.	58.32	65.86	1.49	35.30	nr	**101.16**
152 mm dia.	63.60	71.83	1.65	39.10	nr	**110.93**
178 mm dia.	72.50	81.89	1.92	45.49	nr	**127.38**
203 mm dia.	85.32	96.36	2.20	52.13	nr	**148.49**
254 mm dia.	100.87	113.92	2.49	59.00	nr	**172.92**
304 mm dia.	132.49	149.63	2.80	66.34	nr	**215.97**
355 mm dia.	176.76	199.63	3.19	75.59	nr	**275.22**
Coping cap; including one locking band						
127 mm dia.	32.66	36.88	1.49	35.30	nr	**72.18**
152 mm dia.	34.22	38.65	1.65	39.10	nr	**77.75**
178 mm dia.	37.65	42.52	1.92	45.49	nr	**88.01**
203 mm dia.	45.17	51.01	2.20	52.13	nr	**103.14**
254 mm dia.	55.19	62.33	2.49	59.00	nr	**121.33**
304 mm dia.	74.42	84.05	2.79	66.11	nr	**150.16**
355 mm dia.	99.70	112.60	3.19	75.59	nr	**188.19**

T: MECHANICAL/COOLING/HEATING SYSTEMS

Item	Net Price £	Material £	Labour hours	Labour £	Unit	Total rate £
Storm collar						
127 mm dia.	6.19	6.99	0.52	12.32	nr	**19.31**
152 mm dia.	6.63	7.49	0.55	13.04	nr	**20.53**
178 mm dia.	7.34	8.29	0.57	13.50	nr	**21.79**
203 mm dia.	7.69	8.68	0.66	15.65	nr	**24.33**
254 mm dia.	9.63	10.87	0.66	15.65	nr	**26.52**
304 mm dia.	10.04	11.34	0.72	17.06	nr	**28.40**
355 mm dia.	10.73	12.12	0.77	18.24	nr	**30.36**
400 mm dia.	28.45	32.13	0.82	19.42	nr	**51.55**
450 mm dia.	31.32	35.37	0.87	20.61	nr	**55.98**
500 mm dia.	34.15	38.57	0.92	21.80	nr	**60.37**
550 mm dia.	37.01	41.80	0.98	23.23	nr	**65.03**
600 mm dia.	39.85	45.00	1.03	24.41	nr	**69.41**
Flat flashing; including storm collar and sealant						
127 mm dia.	34.78	39.28	1.49	35.30	nr	**74.58**
152 mm dia.	35.94	40.59	1.65	39.10	nr	**79.69**
178 mm dia.	37.70	42.57	1.92	45.49	nr	**88.06**
203 mm dia.	41.29	46.63	2.20	52.13	nr	**98.76**
254 mm dia.	56.42	63.72	2.49	59.00	nr	**122.72**
304 mm dia.	67.41	76.13	2.80	66.34	nr	**142.47**
355 mm dia.	106.24	119.98	3.20	75.83	nr	**195.81**
400 mm dia.	148.21	167.39	3.59	85.07	nr	**252.46**
450 mm dia.	170.26	192.29	3.97	94.06	nr	**286.35**
500 mm dia.	184.45	208.31	4.38	103.79	nr	**312.10**
550 mm dia.	195.84	221.18	4.78	113.26	nr	**334.44**
600 mm dia.	202.92	229.17	5.17	122.50	nr	**351.67**
5°–30° rigid adjustable flashing; including storm collar and sealant						
127 mm dia.	59.23	66.89	1.49	35.30	nr	**102.19**
152 mm dia.	62.41	70.48	1.65	39.10	nr	**109.58**
178 mm dia.	66.37	74.96	1.92	45.49	nr	**120.45**
203 mm dia.	69.79	78.82	2.20	52.13	nr	**130.95**
254 mm dia.	73.46	82.97	2.49	59.00	nr	**141.97**
304 mm dia.	90.91	102.67	2.80	66.34	nr	**169.01**
355 mm dia.	103.39	116.77	3.19	75.59	nr	**192.36**
400 mm dia.	333.34	376.47	3.59	85.07	nr	**461.54**
450 mm dia.	389.01	439.34	3.97	94.06	nr	**533.40**
500 mm dia.	415.30	469.04	4.38	103.79	nr	**572.83**
550 mm dia.	434.59	490.82	4.77	113.02	nr	**603.84**
600 mm dia.	472.83	534.01	5.17	122.50	nr	**656.51**

T: MECHANICAL/COOLING/HEATING SYSTEMS

Item	Net Price £	Material £	Labour hours	Labour £	Unit	Total rate £
T10: GAS/OIL FIRED BOILERS – cont						
FLUE SYSTEMS – cont						
Domestic and small commercial; twin walled gas vent system suitable for gas fired appliances; domestic gas boilers; small commercial boilers with internal or external flues						
152 mm long						
100 mm dia.	6.34	7.16	0.52	12.32	nr	**19.48**
125 mm dia.	7.79	8.80	0.52	12.32	nr	**21.12**
150 mm dia.	8.43	9.52	0.52	12.32	nr	**21.84**
305 mm long						
100 mm dia.	9.61	10.85	0.52	12.32	nr	**23.17**
125 mm dia.	11.28	12.74	0.52	12.32	nr	**25.06**
150 mm dia.	13.38	15.12	0.52	12.32	nr	**27.44**
457 mm long						
100 mm dia.	10.64	12.02	0.55	13.04	nr	**25.06**
125 mm dia.	11.96	13.50	0.55	13.04	nr	**26.54**
150 mm dia.	14.80	16.72	0.55	13.04	nr	**29.76**
914 mm long						
100 mm dia.	19.00	21.46	0.62	14.70	nr	**36.16**
125 mm dia.	22.15	25.02	0.62	14.70	nr	**39.72**
150 mm dia.	25.40	28.68	0.62	14.70	nr	**43.38**
1524 mm long						
100 mm dia.	27.45	31.00	0.82	19.42	nr	**50.42**
125 mm dia.	33.78	38.15	0.84	19.90	nr	**58.05**
150 mm dia.	36.27	40.96	0.84	19.90	nr	**60.86**
Adjustable length 305 mm long						
100 mm dia.	12.17	13.75	0.56	13.27	nr	**27.02**
125 mm dia.	13.66	15.43	0.56	13.27	nr	**28.70**
150 mm dia.	17.23	19.46	0.56	13.27	nr	**32.73**
Adjustable length 457 mm long						
100 mm dia.	16.41	18.53	0.56	13.27	nr	**31.80**
125 mm dia.	19.89	22.46	0.56	13.27	nr	**35.73**
150 mm dia.	22.13	24.99	0.56	13.27	nr	**38.26**
Adjustable elbow 0°–90°						
100 mm dia.	13.88	15.68	0.48	11.37	nr	**27.05**
125 mm dia.	16.41	18.53	0.48	11.37	nr	**29.90**
150 mm dia.	20.55	23.21	0.48	11.37	nr	**34.58**

T: MECHANICAL/COOLING/HEATING SYSTEMS

Item	Net Price £	Material £	Labour hours	Labour £	Unit	Total rate £
Draughthood connector						
100 mm dia.	4.28	4.83	0.48	11.37	nr	**16.20**
125 mm dia.	4.83	5.46	0.48	11.37	nr	**16.83**
150 mm dia.	5.24	5.92	0.48	11.37	nr	**17.29**
Adaptor						
100 mm dia.	10.39	11.73	0.48	11.37	nr	**23.10**
125 mm dia.	10.61	11.98	0.48	11.37	nr	**23.35**
150 mm dia.	10.83	12.23	0.48	11.37	nr	**23.60**
Support plate						
100 mm dia.	7.50	8.47	0.48	11.37	nr	**19.84**
125 mm dia.	7.97	9.00	0.48	11.37	nr	**20.37**
150 mm dia.	8.54	9.65	0.48	11.37	nr	**21.02**
Wall band						
100 mm dia.	6.77	7.65	0.48	11.37	nr	**19.02**
125 mm dia.	7.23	8.17	0.48	11.37	nr	**19.54**
150 mm dia.	9.16	10.34	0.48	11.37	nr	**21.71**
Firestop						
100 mm dia.	2.94	3.33	0.48	11.37	nr	**14.70**
125 mm dia.	2.94	3.33	0.48	11.37	nr	**14.70**
150 mm dia.	3.37	3.80	0.48	11.37	nr	**15.17**
Flat flashing						
125 mm dia.	19.98	22.56	0.55	13.04	nr	**35.60**
150 mm dia.	27.79	31.39	0.55	13.04	nr	**44.43**
Adjustable flashing 5°–30°						
100 mm dia.	52.06	58.79	0.55	13.04	nr	**71.83**
125 mm dia.	81.28	91.80	0.55	13.04	nr	**104.84**
Storm collar						
100 mm dia.	4.15	4.69	0.55	13.04	nr	**17.73**
125 mm dia.	4.24	4.78	0.55	13.04	nr	**17.82**
150 mm dia.	4.35	4.92	0.55	13.04	nr	**17.96**
Gas vent terminal						
100 mm dia.	15.84	17.89	0.55	13.04	nr	**30.93**
125 mm dia.	17.41	19.67	0.55	13.04	nr	**32.71**
150 mm dia.	22.34	25.23	0.55	13.04	nr	**38.27**
Twin wall galvanized steel flue box, 125 mm dia.; fitted for gas fire, where no chimney exists						
Free standing	100.68	113.71	2.15	50.95	nr	**164.66**
Recess	100.68	113.71	2.15	50.95	nr	**164.66**
Back boiler	74.18	83.78	2.40	56.86	nr	**140.64**

T: MECHANICAL/COOLING/HEATING SYSTEMS

Item	Net Price £	Material £	Labour hours	Labour £	Unit	Total rate £
T13: PACKAGED STEAM GENERATORS						
Packaged steam boilers; boiler mountings centrifugal water feed pump; insulation; and sheet steel wrap around casing; plastic coated						
Gas fired						
293 kW rating	17860.71	20171.80	86.45	2048.40	nr	**22220.20**
1465 kW rating	38096.45	43025.94	148.22	3512.03	nr	**46537.97**
2930 kW rating	54707.89	61786.82	207.50	4916.66	nr	**66703.48**
Oil fired						
293 kW rating	16199.58	18295.72	86.45	2048.40	nr	**20344.12**
1465 kW rating	35240.94	39800.94	148.22	3512.03	nr	**43312.97**
2930 kW rating	53335.05	60236.33	207.50	4916.66	nr	**65152.99**

T: MECHANICAL/COOLING/HEATING SYSTEMS

Item	Net Price £	Material £	Labour hours	Labour £	Unit	Total rate £
T31: LOW TEMPERATURE HOT WATER HEATING						
Y10 – PIPELINES						
SCREWED STEEL						
Black steel pipes; screwed and socketed joints; BS 1387: 1985. Fixed vertically, brackets measured separately. Screwed joints are within the running length, but any flanges are additional						
Medium weight						
10 mm dia.	3.65	4.12	0.37	8.76	m	**12.88**
15 mm dia.	2.31	2.61	0.37	8.76	m	**11.37**
20 mm dia.	2.71	3.06	0.37	8.76	m	**11.82**
25 mm dia.	3.90	4.41	0.41	9.71	m	**14.12**
32 mm dia.	4.83	5.46	0.48	11.37	m	**16.83**
40 mm dia.	5.61	6.33	0.52	12.32	m	**18.65**
50 mm dia.	7.89	8.91	0.62	14.70	m	**23.61**
65 mm dia.	10.71	12.10	0.65	15.40	m	**27.50**
80 mm dia.	13.91	15.71	1.10	26.07	m	**41.78**
100 mm dia.	19.71	22.26	1.31	31.04	m	**53.30**
125 mm dia.	25.08	28.33	1.66	39.34	m	**67.67**
150 mm dia.	29.13	32.90	1.88	44.54	m	**77.44**
Heavy weight						
15 mm dia.	2.74	3.09	0.37	8.76	m	**11.85**
20 mm dia.	3.26	3.68	0.37	8.76	m	**12.44**
25 mm dia.	4.76	5.38	0.41	9.71	m	**15.09**
32 mm dia.	5.91	6.67	0.48	11.37	m	**18.04**
40 mm dia.	6.89	7.78	0.52	12.32	m	**20.10**
50 mm dia.	9.56	10.80	0.62	14.70	m	**25.50**
65 mm dia.	13.01	14.70	0.64	15.17	m	**29.87**
80 mm dia.	16.56	18.71	1.10	26.07	m	**44.78**
100 mm dia.	23.11	26.10	1.31	31.04	m	**57.14**
125 mm dia.	26.74	30.20	1.66	39.34	m	**69.54**
150 mm dia.	31.27	35.31	1.88	44.54	m	**79.85**
200 mm dia.	49.28	55.66	2.99	70.85	m	**126.51**
250 mm dia.	61.78	69.78	3.49	82.69	m	**152.47**
300 mm dia.	73.57	83.09	3.91	92.65	m	**175.74**

T: MECHANICAL/COOLING/HEATING SYSTEMS

Item	Net Price £	Material £	Labour hours	Labour £	Unit	Total rate £
T31: LOW TEMPERATURE HOT WATER HEATING – cont						
Black steel pipes; screwed and socketed joints; BS 1387: 1985. Fixed at high level or suspended, brackets measured separately. Screwed joints are within the running length, but any flanges are additional						
Medium weight						
10 mm dia.	3.65	4.12	0.58	13.75	m	**17.87**
15 mm dia.	2.31	2.61	0.58	13.75	m	**16.36**
20 mm dia.	2.71	3.06	0.58	13.75	m	**16.81**
25 mm dia.	3.90	4.41	0.60	14.22	m	**18.63**
32 mm dia.	4.83	5.46	0.68	16.12	m	**21.58**
40 mm dia.	5.61	6.33	0.73	17.30	m	**23.63**
50 mm dia.	7.89	8.91	0.85	20.13	m	**29.04**
65 mm dia.	10.71	12.10	0.88	20.85	m	**32.95**
80 mm dia.	13.91	15.71	1.45	34.35	m	**50.06**
100 mm dia.	19.71	22.26	1.74	41.24	m	**63.50**
125 mm dia.	25.08	28.33	2.21	52.37	m	**80.70**
150 mm dia.	29.13	32.90	2.50	59.24	m	**92.14**
Heavy weight						
15 mm dia.	2.74	3.09	0.58	13.75	m	**16.84**
20 mm dia.	3.26	3.68	0.58	13.75	m	**17.43**
25 mm dia.	4.76	5.38	0.60	14.22	m	**19.60**
32 mm dia.	5.91	6.67	0.68	16.12	m	**22.79**
40 mm dia.	6.89	7.78	0.73	17.30	m	**25.08**
50 mm dia.	9.56	10.80	0.85	20.13	m	**30.93**
65 mm dia.	13.01	14.70	0.88	20.85	m	**35.55**
80 mm dia.	16.56	18.71	1.45	34.35	m	**53.06**
100 mm dia.	23.11	26.10	1.74	41.24	m	**67.34**
125 mm dia.	26.74	30.20	2.21	52.37	m	**82.57**
150 mm dia.	31.27	35.31	2.50	59.24	m	**94.55**
200 mm dia.	49.28	55.66	2.99	70.85	m	**126.51**
250 mm dia.	61.78	69.78	3.49	82.69	m	**152.47**
300 mm dia.	73.57	83.09	3.91	92.65	m	**175.74**
FIXINGS						
For steel pipes; black malleable iron. For minimum fixing distances, refer to the Tables and Memoranda to the rear of the book						
Single pipe bracket, screw on, black malleable iron; screwed to wood						
15 mm dia.	0.57	0.64	0.14	3.33	nr	**3.97**
20 mm dia.	0.64	0.72	0.14	3.33	nr	**4.05**
25 mm dia.	0.75	0.85	0.17	4.03	nr	**4.88**

T: MECHANICAL/COOLING/HEATING SYSTEMS

Item	Net Price £	Material £	Labour hours	Labour £	Unit	Total rate £
32 mm dia.	1.36	1.54	0.19	4.51	nr	**6.05**
40 mm dia.	1.29	1.46	0.22	5.22	nr	**6.68**
50 mm dia.	1.80	2.03	0.22	5.22	nr	**7.25**
65 mm dia.	2.37	2.67	0.28	6.63	nr	**9.30**
80 mm dia.	3.26	3.68	0.32	7.58	nr	**11.26**
100 mm dia.	4.75	5.37	0.35	8.29	nr	**13.66**
Single pipe bracket, screw on, black malleable iron; plugged and screwed						
15 mm dia.	0.57	0.64	0.25	5.93	nr	**6.57**
20 mm dia.	0.64	0.72	0.25	5.93	nr	**6.65**
25 mm dia.	0.75	0.85	0.30	7.10	nr	**7.95**
32 mm dia.	1.36	1.54	0.32	7.58	nr	**9.12**
40 mm dia.	1.29	1.46	0.32	7.58	nr	**9.04**
50 mm dia.	1.80	2.03	0.32	7.58	nr	**9.61**
65 mm dia.	2.37	2.67	0.35	8.29	nr	**10.96**
80 mm dia.	3.26	3.68	0.42	9.95	nr	**13.63**
100 mm dia.	4.75	5.37	0.42	9.95	nr	**15.32**
Single pipe bracket for building in, black malleable iron						
15 mm dia.	1.68	1.90	0.10	2.38	nr	**4.28**
20 mm dia.	1.68	1.90	0.11	2.61	nr	**4.51**
25 mm dia.	1.68	1.90	0.12	2.85	nr	**4.75**
32 mm dia.	1.90	2.14	0.14	3.33	nr	**5.47**
40 mm dia.	1.92	2.17	0.15	3.56	nr	**5.73**
50 mm dia.	2.00	2.25	0.16	3.79	nr	**6.04**
Pipe ring, single socket, black malleable iron						
15 mm dia.	0.66	0.74	0.10	2.38	nr	**3.12**
20 mm dia.	0.74	0.84	0.11	2.61	nr	**3.45**
25 mm dia.	0.80	0.91	0.12	2.85	nr	**3.76**
32 mm dia.	0.85	0.96	0.14	3.33	nr	**4.29**
40 mm dia.	1.10	1.24	0.15	3.56	nr	**4.80**
50 mm dia.	1.40	1.58	0.16	3.79	nr	**5.37**
65 mm dia.	2.02	2.28	0.30	7.10	nr	**9.38**
80 mm dia.	2.41	2.72	0.35	8.29	nr	**11.01**
100 mm dia.	3.65	4.12	0.40	9.48	nr	**13.60**
125 mm dia.	7.39	8.34	0.60	14.22	nr	**22.56**
150 mm dia.	8.29	9.36	0.77	18.24	nr	**27.60**
200 mm dia.	22.91	25.88	0.90	21.32	nr	**47.20**
250 mm dia.	28.63	32.33	1.10	26.07	nr	**58.40**
300 mm dia.	34.37	38.82	1.25	29.62	nr	**68.44**
350 mm dia.	40.11	45.30	1.50	35.55	nr	**80.85**
400 mm dia.	47.46	53.60	1.75	41.47	nr	**95.07**

T: MECHANICAL/COOLING/HEATING SYSTEMS

Item	Net Price £	Material £	Labour hours	Labour £	Unit	Total rate £
T31: LOW TEMPERATURE HOT WATER HEATING – cont						
FIXINGS – cont						
For steel pipes – cont						
Pipe ring, double socket, black malleable iron						
15 mm dia.	0.77	0.87	0.10	2.38	nr	**3.25**
20 mm dia.	0.88	0.99	0.11	2.61	nr	**3.60**
25 mm dia.	0.99	1.12	0.12	2.85	nr	**3.97**
32 mm dia.	1.17	1.33	0.14	3.33	nr	**4.66**
40 mm dia.	1.26	1.43	0.15	3.56	nr	**4.99**
50 mm dia.	1.98	2.23	0.16	3.79	nr	**6.02**
Screw on backplate (Male), black malleable iron; plugged and screwed						
M12	0.42	0.48	0.10	2.38	nr	**2.86**
Screw on backplate (Female), black malleable iron; plugged and screwed						
M12	0.42	0.48	0.10	2.38	nr	**2.86**
Extra Over channel sections for fabricated hangers and brackets						
Galvanized steel; including inserts, bolts, nuts, washers; fixed to backgrounds						
41 × 21 mm	4.92	5.56	0.29	6.86	m	**12.42**
41 × 41 mm	5.90	6.66	0.29	6.86	m	**13.52**
Threaded rods; metric thread; including nuts, washers etc						
12 mm dia. × 600 mm long	2.49	2.82	0.18	4.27	nr	**7.09**
Pipe roller and chair						
Roller and chair; black malleable						
Up to 50 mm dia.	12.70	14.34	0.20	4.74	nr	**19.08**
65 mm dia.	13.08	14.77	0.20	4.74	nr	**19.51**
80 mm dia.	14.21	16.04	0.20	4.74	nr	**20.78**
100 mm dia.	14.45	16.32	0.20	4.74	nr	**21.06**
125 mm dia.	20.24	22.86	0.20	4.74	nr	**27.60**
150 mm dia.	20.24	22.86	0.30	7.10	nr	**29.96**
175 mm dia.	34.17	38.59	0.30	7.10	nr	**45.69**
200 mm dia.	52.33	59.10	0.30	7.10	nr	**66.20**
250 mm dia.	68.27	77.10	0.30	7.10	nr	**84.20**
300 mm dia.	85.96	97.08	0.30	7.10	nr	**104.18**

T: MECHANICAL/COOLING/HEATING SYSTEMS

Item	Net Price £	Material £	Labour hours	Labour £	Unit	Total rate £
Roller bracket; black malleable						
25 mm dia.	2.66	3.01	0.20	4.74	nr	**7.75**
32 mm dia.	2.80	3.16	0.20	4.74	nr	**7.90**
40 mm dia.	2.99	3.38	0.20	4.74	nr	**8.12**
50 mm dia.	3.16	3.57	0.20	4.74	nr	**8.31**
65 mm dia.	4.16	4.70	0.20	4.74	nr	**9.44**
80 mm dia.	5.99	6.76	0.20	4.74	nr	**11.50**
100 mm dia.	6.66	7.52	0.20	4.74	nr	**12.26**
125 mm dia.	10.99	12.41	0.20	4.74	nr	**17.15**
150 mm dia.	10.99	12.41	0.30	7.10	nr	**19.51**
175 mm dia.	24.49	27.66	0.30	7.10	nr	**34.76**
200 mm dia.	24.49	27.66	0.30	7.10	nr	**34.76**
250 mm dia.	32.55	36.76	0.30	7.10	nr	**43.86**
300 mm dia.	39.99	45.17	0.30	7.10	nr	**52.27**
350 mm dia.	64.29	72.61	0.30	7.10	nr	**79.71**
400 mm dia.	73.49	83.00	0.30	7.10	nr	**90.10**
Extra over black steel screwed pipes; black steel flanges, screwed and drilled; metric; BS 4504						
Screwed flanges; PN6						
15 mm dia.	6.79	7.67	0.35	8.29	nr	**15.96**
20 mm dia.	6.79	7.67	0.47	11.14	nr	**18.81**
25 mm dia.	6.79	7.67	0.53	12.56	nr	**20.23**
32 mm dia.	6.79	7.67	0.62	14.70	nr	**22.37**
40 mm dia.	6.79	7.67	0.70	16.60	nr	**24.27**
50 mm dia.	7.24	8.18	0.84	19.90	nr	**28.08**
65 mm dia.	10.07	11.37	1.03	24.41	nr	**35.78**
80 mm dia.	14.22	16.07	1.23	29.15	nr	**45.22**
100 mm dia.	16.81	18.98	1.41	33.40	nr	**52.38**
125 mm dia.	30.83	34.82	1.77	41.93	nr	**76.75**
150 mm dia.	30.83	34.82	2.21	52.37	nr	**87.19**
Screwed flanges; PN16						
15 mm dia.	8.63	9.75	0.35	8.29	nr	**18.04**
20 mm dia.	8.63	9.75	0.47	11.14	nr	**20.89**
25 mm dia.	8.63	9.75	0.53	12.56	nr	**22.31**
32 mm dia.	9.22	10.41	0.62	14.70	nr	**25.11**
40 mm dia.	9.22	10.41	0.70	16.60	nr	**27.01**
50 mm dia.	11.25	12.71	0.84	19.90	nr	**32.61**
65 mm dia.	14.05	15.87	1.03	24.41	nr	**40.28**
80 mm dia.	17.11	19.33	1.23	29.15	nr	**48.48**
100 mm dia.	19.19	21.68	1.41	33.40	nr	**55.08**
125 mm dia.	33.58	37.92	1.77	41.93	nr	**79.85**
150 mm dia.	33.04	37.31	2.21	52.37	nr	**89.68**

T: MECHANICAL/COOLING/HEATING SYSTEMS

Item	Net Price £	Material £	Labour hours	Labour £	Unit	Total rate £
T31: LOW TEMPERATURE HOT WATER HEATING – cont						
FIXINGS – cont						
Extra over black steel screwed pipes; black steel flanges, screwed and drilled; imperial; BS 10						
Screwed flanges; Table E						
½' dia.	12.90	14.57	0.35	8.29	nr	22.86
¾' dia.	12.90	14.57	0.47	11.14	nr	25.71
1' dia.	12.88	14.55	0.53	12.56	nr	27.11
1¼' dia.	12.90	14.57	0.62	14.70	nr	29.27
1½' dia.	12.90	14.57	0.70	16.60	nr	31.17
2' dia.	12.90	14.57	0.84	19.90	nr	34.47
2½' dia.	15.34	17.33	1.03	24.41	nr	41.74
3' dia.	17.38	19.62	1.23	29.15	nr	48.77
4' dia.	23.52	26.56	1.41	33.40	nr	59.96
5' dia.	49.84	56.29	1.77	41.93	nr	98.22
6' dia.	49.84	56.29	2.21	52.37	nr	108.66
Extra over black steel screwed pipes; black steel flange connections						
Bolted connection between pair of flanges; including gasket, bolts, nuts and washers						
50 mm dia.	6.25	7.06	0.53	12.56	nr	19.62
65 mm dia.	7.39	8.34	0.53	12.56	nr	20.90
80 mm dia.	11.10	12.54	0.53	12.56	nr	25.10
100 mm dia.	12.05	13.61	0.61	14.45	nr	28.06
125 mm dia.	14.18	16.01	0.61	14.45	nr	30.46
150 mm dia.	20.26	22.88	0.90	21.32	nr	44.20
Extra over black steel screwed pipes; black heavy steel tubular fittings; BS 1387						
Long screw connection with socket and backnut						
15 mm dia.	4.58	5.17	0.63	14.93	nr	20.10
20 mm dia.	5.77	6.52	0.84	19.90	nr	26.42
25 mm dia.	7.56	8.54	0.95	22.51	nr	31.05
32 mm dia.	10.14	11.45	1.11	26.31	nr	37.76
40 mm dia.	12.19	13.77	1.28	30.32	nr	44.09
50 mm dia.	17.96	20.29	1.53	36.25	nr	56.54
65 mm dia.	25.12	28.37	1.87	44.31	nr	72.68
80 mm dia.	57.42	64.85	2.21	52.37	nr	117.22
100 mm dia.	93.03	105.07	3.05	72.27	nr	177.34

T: MECHANICAL/COOLING/HEATING SYSTEMS

Item	Net Price £	Material £	Labour hours	Labour £	Unit	Total rate £
Running nipple						
15 mm dia.	1.12	1.26	0.50	11.85	nr	**13.11**
20 mm dia.	1.42	1.60	0.68	16.12	nr	**17.72**
25 mm dia.	1.75	1.98	0.77	18.24	nr	**20.22**
32 mm dia.	2.45	2.76	0.90	21.32	nr	**24.08**
40 mm dia.	3.30	3.72	1.03	24.41	nr	**28.13**
50 mm dia.	5.03	5.68	1.23	29.15	nr	**34.83**
65 mm dia.	10.83	12.23	1.50	35.55	nr	**47.78**
80 mm dia.	16.89	19.07	1.78	42.17	nr	**61.24**
100 mm dia.	26.46	29.89	2.38	56.39	nr	**86.28**
Barrel nipple						
15 mm dia.	0.64	0.72	0.50	11.85	nr	**12.57**
20 mm dia.	1.02	1.15	0.68	16.12	nr	**17.27**
25 mm dia.	1.35	1.52	0.77	18.24	nr	**19.76**
32 mm dia.	2.02	2.28	0.90	21.32	nr	**23.60**
40 mm dia.	2.48	2.81	1.03	24.41	nr	**27.22**
50 mm dia.	3.54	4.00	1.23	29.15	nr	**33.15**
65 mm dia.	7.57	8.55	1.50	35.55	nr	**44.10**
80 mm dia.	10.55	11.91	1.78	42.17	nr	**54.08**
100 mm dia.	19.12	21.59	2.38	56.39	nr	**77.98**
125 mm dia.	35.48	40.08	2.87	68.00	nr	**108.08**
150 mm dia.	55.92	63.16	3.39	80.33	nr	**143.49**
Close taper nipple						
15 mm dia.	1.35	1.52	0.50	11.85	nr	**13.37**
20 mm dia.	1.75	1.98	0.68	16.12	nr	**18.10**
25 mm dia.	2.28	2.57	0.77	18.24	nr	**20.81**
32 mm dia.	3.42	3.87	0.90	21.32	nr	**25.19**
40 mm dia.	4.23	4.77	1.03	24.41	nr	**29.18**
50 mm dia.	6.52	7.36	1.23	29.15	nr	**36.51**
65 mm dia.	12.66	14.30	1.50	35.55	nr	**49.85**
80 mm dia.	15.60	17.62	1.78	42.17	nr	**59.79**
100 mm dia.	32.04	36.19	2.38	56.39	nr	**92.58**
Extra over black steel screwed pipes; black malleable iron fittings; BS 143						
Cap						
15 mm dia.	0.54	0.61	0.32	7.58	nr	**8.19**
20 mm dia.	0.62	0.70	0.43	10.19	nr	**10.89**
25 mm dia.	0.78	0.88	0.49	11.61	nr	**12.49**
32 mm dia.	1.37	1.55	0.58	13.75	nr	**15.30**
40 mm dia.	1.62	1.83	0.66	15.65	nr	**17.48**
50 mm dia.	3.37	3.80	0.78	18.47	nr	**22.27**
65 mm dia.	5.63	6.35	0.96	22.75	nr	**29.10**
80 mm dia.	6.37	7.19	1.13	26.78	nr	**33.97**
100 mm dia.	13.94	15.75	1.70	40.29	nr	**56.04**

T: MECHANICAL/COOLING/HEATING SYSTEMS

Item	Net Price £	Material £	Labour hours	Labour £	Unit	Total rate £
T31: LOW TEMPERATURE HOT WATER HEATING – cont						
FIXINGS – cont						
Extra over black steel screwed pipes; black malleable iron fittings; BS 143 – cont						
Plain plug, hollow						
15 mm dia.	0.38	0.43	0.28	6.63	nr	**7.06**
20 mm dia.	0.48	0.54	0.38	9.00	nr	**9.54**
25 mm dia.	0.66	0.74	0.44	10.42	nr	**11.16**
32 mm dia.	0.92	1.04	0.51	12.09	nr	**13.13**
40 mm dia.	1.73	1.96	0.59	13.98	nr	**15.94**
50 mm dia.	2.42	2.73	0.70	16.60	nr	**19.33**
65 mm dia.	3.96	4.47	0.85	20.13	nr	**24.60**
80 mm dia.	6.18	6.98	1.00	23.69	nr	**30.67**
100 mm dia.	11.39	12.86	1.44	34.12	nr	**46.98**
Plain plug, solid						
15 mm dia.	1.18	1.34	0.28	6.63	nr	**7.97**
20 mm dia.	1.14	1.29	0.38	9.00	nr	**10.29**
25 mm dia.	1.70	1.92	0.44	10.42	nr	**12.34**
32 mm dia.	2.44	2.75	0.51	12.09	nr	**14.84**
40 mm dia.	3.44	3.89	0.59	13.98	nr	**17.87**
50 mm dia.	4.52	5.10	0.70	16.60	nr	**21.70**
90° Elbow, male/female						
15 mm dia.	0.62	0.70	0.64	15.17	nr	**15.87**
20 mm dia.	0.84	0.95	0.85	20.13	nr	**21.08**
25 mm dia.	1.40	1.58	0.97	22.98	nr	**24.56**
32 mm dia.	2.55	2.88	1.12	26.54	nr	**29.42**
40 mm dia.	4.28	4.83	1.29	30.56	nr	**35.39**
50 mm dia.	5.50	6.21	1.55	36.73	nr	**42.94**
65 mm dia.	11.90	13.44	1.89	44.78	nr	**58.22**
80 mm dia.	16.26	18.36	2.24	53.08	nr	**71.44**
100 mm dia.	28.43	32.11	3.09	73.22	nr	**105.33**
90° Elbow						
15 mm dia.	0.56	0.63	0.64	15.17	nr	**15.80**
20 mm dia.	0.77	0.87	0.85	20.13	nr	**21.00**
25 mm dia.	1.18	1.34	0.97	22.98	nr	**24.32**
32 mm dia.	2.16	2.44	1.12	26.54	nr	**28.98**
40 mm dia.	3.63	4.10	1.29	30.56	nr	**34.66**
50 mm dia.	4.25	4.80	1.55	36.73	nr	**41.53**
65 mm dia.	9.19	10.38	1.89	44.78	nr	**55.16**
80 mm dia.	13.50	15.25	2.24	53.08	nr	**68.33**
100 mm dia.	26.03	29.40	3.09	73.22	nr	**102.62**
125 mm dia.	63.30	71.49	4.44	105.20	nr	**176.69**
150 mm dia.	117.85	133.10	5.79	137.19	nr	**270.29**

T: MECHANICAL/COOLING/HEATING SYSTEMS

Item	Net Price £	Material £	Labour hours	Labour £	Unit	Total rate £
45° Elbow						
15 mm dia.	1.34	1.51	0.64	15.17	nr	**16.68**
20 mm dia.	1.64	1.86	0.85	20.13	nr	**21.99**
25 mm dia.	2.44	2.75	0.97	22.98	nr	**25.73**
32 mm dia.	4.99	5.64	1.12	26.54	nr	**32.18**
40 mm dia.	6.12	6.92	1.29	30.56	nr	**37.48**
50 mm dia.	8.39	9.48	1.55	36.73	nr	**46.21**
65 mm dia.	12.28	13.87	1.89	44.78	nr	**58.65**
80 mm dia.	18.45	20.84	2.24	53.08	nr	**73.92**
100 mm dia.	40.47	45.71	3.09	73.22	nr	**118.93**
150 mm dia.	113.00	127.62	5.79	137.19	nr	**264.81**
90° Bend, male/female						
15 mm dia.	1.09	1.23	0.64	15.17	nr	**16.40**
20 mm dia.	1.60	1.81	0.85	20.13	nr	**21.94**
25 mm dia.	2.36	2.66	0.97	22.98	nr	**25.64**
32 mm dia.	3.76	4.24	1.12	26.54	nr	**30.78**
40 mm dia.	5.93	6.70	1.29	30.56	nr	**37.26**
50 mm dia.	10.36	11.70	1.55	36.73	nr	**48.43**
65 mm dia.	17.24	19.47	1.89	44.78	nr	**64.25**
80 mm dia.	26.53	29.97	2.24	53.08	nr	**83.05**
100 mm dia.	65.74	74.25	3.09	73.22	nr	**147.47**
90° Bend, male						
15 mm dia.	2.53	2.86	0.64	15.17	nr	**18.03**
20 mm dia.	2.83	3.19	0.85	20.13	nr	**23.32**
25 mm dia.	4.16	4.70	0.97	22.98	nr	**27.68**
32 mm dia.	9.02	10.19	1.12	26.54	nr	**36.73**
40 mm dia.	12.66	14.30	1.29	30.56	nr	**44.86**
50 mm dia.	19.23	21.72	1.55	36.73	nr	**58.45**
90° Bend, female						
15 mm dia.	1.01	1.14	0.64	15.17	nr	**16.31**
20 mm dia.	1.43	1.61	0.85	20.13	nr	**21.74**
25 mm dia.	2.03	2.30	0.97	22.98	nr	**25.28**
32 mm dia.	3.83	4.32	1.12	26.54	nr	**30.86**
40 mm dia.	5.11	5.77	1.29	30.56	nr	**36.33**
50 mm dia.	7.19	8.12	1.55	36.73	nr	**44.85**
65 mm dia.	15.36	17.35	1.89	44.78	nr	**62.13**
80 mm dia.	24.47	27.63	2.24	53.08	nr	**80.71**
100 mm dia.	53.63	60.57	3.09	73.22	nr	**133.79**
125 mm dia.	126.51	142.88	4.44	105.20	nr	**248.08**
150 mm dia.	230.09	259.87	5.79	137.19	nr	**397.06**
Return bend						
15 mm dia.	5.78	6.53	0.64	15.17	nr	**21.70**
20 mm dia.	9.34	10.55	0.85	20.13	nr	**30.68**
25 mm dia.	11.65	13.16	0.97	22.98	nr	**36.14**
32 mm dia.	18.05	20.39	1.12	26.54	nr	**46.93**
40 mm dia.	21.50	24.29	1.29	30.56	nr	**54.85**
50 mm dia.	32.81	37.06	1.55	36.73	nr	**73.79**

T: MECHANICAL/COOLING/HEATING SYSTEMS

Item	Net Price £	Material £	Labour hours	Labour £	Unit	Total rate £
T31: LOW TEMPERATURE HOT WATER HEATING – cont						
FIXINGS – cont						
Extra over black steel screwed pipes; black malleable iron fittings; BS 143 – cont						
Equal socket, parallel thread						
15 mm dia.	0.54	0.61	0.64	15.17	nr	**15.78**
20 mm dia.	0.66	0.74	0.85	20.13	nr	**20.87**
25 mm dia.	0.91	1.03	0.97	22.98	nr	**24.01**
32 mm dia.	1.82	2.06	1.12	26.54	nr	**28.60**
40 mm dia.	2.66	3.01	1.29	30.56	nr	**33.57**
50 mm dia.	3.84	4.33	1.55	36.73	nr	**41.06**
65 mm dia.	5.90	6.66	1.89	44.78	nr	**51.44**
80 mm dia.	8.36	9.45	2.24	53.08	nr	**62.53**
100 mm dia.	13.80	15.59	3.09	73.22	nr	**88.81**
Concentric reducing socket						
20 × 15 mm dia.	0.85	0.96	0.76	18.00	nr	**18.96**
25 × 15 mm dia.	1.01	1.14	0.85	20.13	nr	**21.27**
25 × 20 mm dia.	1.07	1.20	0.86	20.37	nr	**21.57**
32 × 25 mm dia.	2.02	2.28	1.01	23.93	nr	**26.21**
40 × 25 mm dia.	2.53	2.86	1.16	27.49	nr	**30.35**
40 × 32 mm dia.	2.63	2.97	1.16	27.49	nr	**30.46**
50 × 25 mm dia.	4.89	5.52	1.38	32.69	nr	**38.21**
50 × 40 mm dia.	3.67	4.14	1.38	32.69	nr	**36.83**
65 × 50 mm dia.	6.68	7.55	1.69	40.05	nr	**47.60**
80 × 50 mm dia.	8.69	9.81	2.00	47.39	nr	**57.20**
100 × 50 mm dia.	17.34	19.58	2.75	65.17	nr	**84.75**
100 × 80 mm dia.	16.08	18.16	2.75	65.17	nr	**83.33**
150 × 100 mm dia.	48.16	54.40	4.10	97.16	nr	**151.56**
Eccentric reducing socket						
20 × 15 mm dia.	1.53	1.72	0.73	17.30	nr	**19.02**
25 × 15 mm dia.	4.38	4.95	0.85	20.13	nr	**25.08**
25 × 20 mm dia.	4.96	5.60	0.85	20.13	nr	**25.73**
32 × 25 mm dia.	6.78	7.66	1.01	23.93	nr	**31.59**
40 × 25 mm dia.	7.75	8.75	1.16	27.49	nr	**36.24**
40 × 32 mm dia.	3.90	4.41	1.16	27.49	nr	**31.90**
50 × 25 mm dia.	5.03	5.68	1.38	32.69	nr	**38.37**
50 × 40 mm dia.	4.67	5.27	1.38	32.69	nr	**37.96**
65 × 50 mm dia.	8.07	9.12	1.69	40.05	nr	**49.17**
80 × 50 mm dia.	14.13	15.96	2.00	47.39	nr	**63.35**
Hexagon bush						
20 × 15 mm dia.	0.48	0.54	0.37	8.76	nr	**9.30**
25 × 15 mm dia.	0.59	0.66	0.43	10.19	nr	**10.85**
25 × 20 mm dia.	0.62	0.70	0.43	10.19	nr	**10.89**

T: MECHANICAL/COOLING/HEATING SYSTEMS

Item	Net Price £	Material £	Labour hours	Labour £	Unit	Total rate £
32 × 25 mm dia.	0.80	0.91	0.51	12.09	nr	**13.00**
40 × 25 mm dia.	1.27	1.44	0.58	13.75	nr	**15.19**
40 × 32 mm dia.	1.16	1.31	0.58	13.75	nr	**15.06**
50 × 25 mm dia.	2.63	2.97	0.71	16.83	nr	**19.80**
50 × 40 mm dia.	2.42	2.73	0.71	16.83	nr	**19.56**
65 × 50 mm dia.	4.06	4.59	0.85	20.13	nr	**24.72**
80 × 50 mm dia.	6.59	7.45	1.00	23.69	nr	**31.14**
100 × 50 mm dia.	15.23	17.20	1.52	36.02	nr	**53.22**
100 × 80 mm dia.	12.68	14.32	1.52	36.02	nr	**50.34**
150 × 100 mm dia.	45.57	51.47	2.57	60.89	nr	**112.36**
Hexagon nipple						
15 mm dia.	0.52	0.59	0.28	6.63	nr	**7.22**
20 mm dia.	0.59	0.66	0.38	9.00	nr	**9.66**
25 mm dia.	0.84	0.95	0.44	10.42	nr	**11.37**
32 mm dia.	1.73	1.96	0.51	12.09	nr	**14.05**
40 mm dia.	2.02	2.28	0.59	13.98	nr	**16.26**
50 mm dia.	3.64	4.11	0.70	16.60	nr	**20.71**
65 mm dia.	5.95	6.72	0.85	20.13	nr	**26.85**
80 mm dia.	8.60	9.71	1.00	23.69	nr	**33.40**
100 mm dia.	14.59	16.47	1.44	34.12	nr	**50.59**
150 mm dia.	32.56	36.77	2.32	54.97	nr	**91.74**
Union, male/female						
15 mm dia.	2.74	3.09	0.64	15.17	nr	**18.26**
20 mm dia.	3.35	3.78	0.85	20.13	nr	**23.91**
25 mm dia.	4.20	4.74	0.97	22.98	nr	**27.72**
32 mm dia.	6.46	7.29	1.12	26.54	nr	**33.83**
40 mm dia.	8.28	9.35	1.29	30.56	nr	**39.91**
50 mm dia.	13.03	14.72	1.55	36.73	nr	**51.45**
65 mm dia.	24.87	28.09	1.89	44.78	nr	**72.87**
80 mm dia.	34.20	38.63	2.24	53.08	nr	**91.71**
Union, female						
15 mm dia.	2.24	2.53	0.64	15.17	nr	**17.70**
20 mm dia.	2.44	2.75	0.85	20.13	nr	**22.88**
25 mm dia.	2.85	3.22	0.97	22.98	nr	**26.20**
32 mm dia.	5.37	6.07	1.12	26.54	nr	**32.61**
40 mm dia.	6.07	6.85	1.29	30.56	nr	**37.41**
50 mm dia.	10.05	11.35	1.55	36.73	nr	**48.08**
65 mm dia.	21.67	24.47	1.89	44.78	nr	**69.25**
80 mm dia.	29.92	33.79	2.24	53.08	nr	**86.87**
100 mm dia.	56.97	64.34	3.09	73.22	nr	**137.56**
Union elbow, male/female						
15 mm dia.	3.97	4.49	0.55	13.04	nr	**17.53**
20 mm dia.	4.97	5.61	0.85	20.13	nr	**25.74**
25 mm dia.	6.98	7.88	0.97	22.98	nr	**30.86**

T: MECHANICAL/COOLING/HEATING SYSTEMS

Item	Net Price £	Material £	Labour hours	Labour £	Unit	Total rate £
T31: LOW TEMPERATURE HOT WATER HEATING – cont						
FIXINGS – cont						
Extra over black steel screwed pipes; black malleable iron fittings; BS 143 – cont						
Twin elbow						
15 mm dia.	3.22	3.64	0.91	21.56	nr	**25.20**
20 mm dia.	3.56	4.02	1.22	28.92	nr	**32.94**
25 mm dia.	5.76	6.51	1.39	32.94	nr	**39.45**
32 mm dia.	11.82	13.35	1.62	38.39	nr	**51.74**
40 mm dia.	14.97	16.91	1.86	44.06	nr	**60.97**
50 mm dia.	19.24	21.73	2.21	52.37	nr	**74.10**
65 mm dia.	34.19	38.62	2.72	64.45	nr	**103.07**
80 mm dia.	58.24	65.78	3.21	76.06	nr	**141.84**
Equal tee						
15 mm dia.	0.77	0.87	0.91	21.56	nr	**22.43**
20 mm dia.	1.11	1.25	1.22	28.92	nr	**30.17**
25 mm dia.	1.61	1.82	1.39	32.94	nr	**34.76**
32 mm dia.	2.94	3.33	1.62	38.39	nr	**41.72**
40 mm dia.	4.52	5.10	1.86	44.06	nr	**49.16**
50 mm dia.	6.49	7.33	2.21	52.37	nr	**59.70**
65 mm dia.	14.23	16.08	2.72	64.45	nr	**80.53**
80 mm dia.	17.34	19.58	3.21	76.06	nr	**95.64**
100 mm dia.	31.41	35.48	4.44	105.20	nr	**140.68**
125 mm dia.	91.27	103.08	5.38	127.48	nr	**230.56**
150 mm dia.	145.43	164.25	6.31	149.51	nr	**313.76**
Tee reducing on branch						
20 × 15 mm dia.	1.01	1.14	1.22	28.92	nr	**30.06**
25 × 15 mm dia.	1.40	1.58	1.39	32.94	nr	**34.52**
25 × 20 mm dia.	1.66	1.88	1.39	32.94	nr	**34.82**
32 × 25 mm dia.	2.89	3.26	1.62	38.39	nr	**41.65**
40 × 25 mm dia.	3.82	4.31	1.86	44.06	nr	**48.37**
40 × 32 mm dia.	5.00	5.65	1.86	44.06	nr	**49.71**
50 × 25 mm dia.	5.68	6.42	2.21	52.37	nr	**58.79**
50 × 40 mm dia.	8.46	9.56	2.21	52.37	nr	**61.93**
65 × 50 mm dia.	13.20	14.91	2.72	64.45	nr	**79.36**
80 × 50 mm dia.	17.85	20.16	3.21	76.06	nr	**96.22**
100 × 50 mm dia.	29.54	33.36	4.44	105.20	nr	**138.56**
100 × 80 mm dia.	45.57	51.47	4.44	105.20	nr	**156.67**
150 × 100 mm dia.	107.12	120.98	6.31	149.51	nr	**270.49**
Equal pitcher tee						
15 mm dia.	2.35	2.65	0.91	21.56	nr	**24.21**
20 mm dia.	2.92	3.29	1.22	28.92	nr	**32.21**
25 mm dia.	4.38	4.95	1.39	32.94	nr	**37.89**

T: MECHANICAL/COOLING/HEATING SYSTEMS

Item	Net Price £	Material £	Labour hours	Labour £	Unit	Total rate £
32 mm dia.	6.65	7.51	1.62	38.39	nr	**45.90**
40 mm dia.	10.27	11.60	1.86	44.06	nr	**55.66**
50 mm dia.	14.43	16.30	2.21	52.37	nr	**68.67**
65 mm dia.	22.31	25.19	2.72	64.45	nr	**89.64**
80 mm dia.	34.82	39.32	3.21	76.06	nr	**115.38**
100 mm dia.	78.33	88.46	4.44	105.20	nr	**193.66**
Cross						
15 mm dia.	2.30	2.60	1.00	23.69	nr	**26.29**
20 mm dia.	3.45	3.90	1.33	31.51	nr	**35.41**
25 mm dia.	4.38	4.95	1.51	35.78	nr	**40.73**
32 mm dia.	6.39	7.22	1.77	41.93	nr	**49.15**
40 mm dia.	8.22	9.28	2.02	47.87	nr	**57.15**
50 mm dia.	13.34	15.07	2.42	57.34	nr	**72.41**
65 mm dia.	18.45	20.84	2.97	70.37	nr	**91.21**
80 mm dia.	24.54	27.71	3.50	82.94	nr	**110.65**
100 mm dia.	50.64	57.19	4.84	114.68	nr	**171.87**

BLACK WELDED STEEL

Black steel pipes; butt welded joints; BS 1387: 1985; including protective painting. Fixed Vertical with brackets measured separately (Refer to Screwed Steel Section). Welded butt joints are within the running length, but any flanges are additional

Item	Net Price £	Material £	Labour hours	Labour £	Unit	Total rate £
Medium weight						
10 mm dia.	2.16	2.44	0.37	8.76	m	**11.20**
15 mm dia.	2.31	2.61	0.37	8.76	m	**11.37**
20 mm dia.	2.71	3.06	0.37	8.76	m	**11.82**
25 mm dia.	3.90	4.41	0.41	9.71	m	**14.12**
32 mm dia.	4.83	5.46	0.48	11.37	m	**16.83**
40 mm dia.	5.61	6.33	0.52	12.32	m	**18.65**
50 mm dia.	7.89	8.91	0.62	14.70	m	**23.61**
65 mm dia.	10.71	12.10	0.64	15.17	m	**27.27**
80 mm dia.	13.91	15.71	1.10	26.07	m	**41.78**
100 mm dia.	19.71	22.26	1.31	31.04	m	**53.30**
125 mm dia.	25.08	28.33	1.66	39.34	m	**67.67**
150 mm dia.	29.13	32.90	1.88	44.54	m	**77.44**
Heavy weight						
15 mm dia.	2.74	3.09	0.37	8.76	m	**11.85**
20 mm dia.	3.26	3.68	0.37	8.76	m	**12.44**
25 mm dia.	4.76	5.38	0.41	9.71	m	**15.09**
32 mm dia.	5.91	6.67	0.48	11.37	m	**18.04**
40 mm dia.	6.89	7.78	0.52	12.32	m	**20.10**
50 mm dia.	9.56	10.80	0.62	14.70	m	**25.50**
65 mm dia.	13.01	14.70	0.64	15.17	m	**29.87**

T: MECHANICAL/COOLING/HEATING SYSTEMS

Item	Net Price £	Material £	Labour hours	Labour £	Unit	Total rate £
T31: LOW TEMPERATURE HOT WATER HEATING – cont						
BLACK WELDED STEEL – cont						
Black steel pipes – cont						
Heavy weight – cont						
80 mm dia.	16.56	18.71	1.10	26.07	m	**44.78**
100 mm dia.	23.11	26.10	1.31	31.04	m	**57.14**
125 mm dia.	26.74	30.20	1.66	39.34	m	**69.54**
150 mm dia.	31.27	35.31	1.88	44.54	m	**79.85**
Black steel pipes; butt welded joints; BS 1387: 1985; including protective painting. Fixed at High Level or Suspended with brackets measured separately (Refer to Screwed Steel Section). Welded butt joints are within the running length, but any flanges are additional						
Medium weight						
10 mm dia.	2.16	2.44	0.58	13.75	m	**16.19**
15 mm dia.	2.31	2.61	0.58	13.75	m	**16.36**
20 mm dia.	2.71	3.06	0.58	13.75	m	**16.81**
25 mm dia.	3.90	4.41	0.60	14.22	m	**18.63**
32 mm dia.	4.83	5.46	0.68	16.12	m	**21.58**
40 mm dia.	5.61	6.33	0.73	17.30	m	**23.63**
50 mm dia.	7.89	8.91	0.85	20.13	m	**29.04**
65 mm dia.	10.71	12.10	0.88	20.85	m	**32.95**
80 mm dia.	13.91	15.71	1.45	34.35	m	**50.06**
100 mm dia.	19.71	22.26	1.74	41.24	m	**63.50**
125 mm dia.	25.08	28.33	2.21	52.37	m	**80.70**
150 mm dia.	29.13	32.90	2.50	59.24	m	**92.14**
Heavy weight						
15 mm dia.	2.74	3.09	0.58	13.75	m	**16.84**
20 mm dia.	3.26	3.68	0.58	13.75	m	**17.43**
25 mm dia.	4.76	5.38	0.60	14.22	m	**19.60**
32 mm dia.	5.91	6.67	0.68	16.12	m	**22.79**
40 mm dia.	6.89	7.78	0.73	17.30	m	**25.08**
50 mm dia.	9.56	10.80	0.85	20.13	m	**30.93**
65 mm dia.	13.01	14.70	0.88	20.85	m	**35.55**
80 mm dia.	16.56	18.71	1.45	34.35	m	**53.06**
100 mm dia.	23.11	26.10	1.74	41.24	m	**67.34**
125 mm dia.	26.74	30.20	2.21	52.37	m	**82.57**
150 mm dia.	31.27	35.31	2.50	59.24	m	**94.55**

T: MECHANICAL/COOLING/HEATING SYSTEMS

Item	Net Price £	Material £	Labour hours	Labour £	Unit	Total rate £
FIXINGS						
Refer to steel pipes; black malleable iron. For minimum fixing distances, refer to the Tables and Memoranda to the rear of the book						
Extra over black steel butt welded pipes; black steel flanges, welded and drilled; metric; BS 4504						
Welded flanges; PN6						
15 mm dia.	2.86	3.23	0.59	13.98	nr	**17.21**
20 mm dia.	2.86	3.23	0.69	16.35	nr	**19.58**
25 mm dia.	2.86	3.23	0.84	19.90	nr	**23.13**
32 mm dia.	4.75	5.37	1.00	23.69	nr	**29.06**
40 mm dia.	5.06	5.71	1.11	26.31	nr	**32.02**
50 mm dia.	5.20	5.88	1.37	32.46	nr	**38.34**
65 mm dia.	7.66	8.65	1.54	36.50	nr	**45.15**
80 mm dia.	11.49	12.97	1.67	39.58	nr	**52.55**
100 mm dia.	12.56	14.19	2.22	52.61	nr	**66.80**
125 mm dia.	16.46	18.59	2.61	61.84	nr	**80.43**
150 mm dia.	16.63	18.78	2.99	70.85	nr	**89.63**
Welded flanges; PN16						
15 mm dia.	5.73	6.47	0.59	13.98	nr	**20.45**
20 mm dia.	5.73	6.47	0.69	16.35	nr	**22.82**
25 mm dia.	5.73	6.47	0.84	19.90	nr	**26.37**
32 mm dia.	7.73	8.73	1.00	23.69	nr	**32.42**
40 mm dia.	7.73	8.73	1.11	26.31	nr	**35.04**
50 mm dia.	10.16	11.48	1.37	32.46	nr	**43.94**
65 mm dia.	11.72	13.24	1.54	36.50	nr	**49.74**
80 mm dia.	14.89	16.82	1.67	39.58	nr	**56.40**
100 mm dia.	15.19	17.16	2.22	52.61	nr	**69.77**
125 mm dia.	24.13	27.25	2.61	61.84	nr	**89.09**
150 mm dia.	28.80	32.53	2.99	70.85	nr	**103.38**
Blank flanges, slip on for welding; PN6						
15 mm dia.	1.79	2.02	0.48	11.37	nr	**13.39**
20 mm dia.	1.79	2.02	0.55	13.04	nr	**15.06**
25 mm dia.	1.79	2.02	0.64	15.17	nr	**17.19**
32 mm dia.	3.68	4.15	0.76	18.00	nr	**22.15**
40 mm dia.	3.68	4.15	0.84	19.90	nr	**24.05**
50 mm dia.	3.34	3.77	1.01	23.93	nr	**27.70**
65 mm dia.	6.63	7.49	1.30	30.79	nr	**38.28**
80 mm dia.	6.92	7.81	1.41	33.40	nr	**41.21**
100 mm dia.	7.12	8.04	1.78	42.17	nr	**50.21**
125 mm dia.	14.21	16.04	2.06	48.82	nr	**64.86**
150 mm dia.	13.44	15.18	2.35	55.68	nr	**70.86**

T: MECHANICAL/COOLING/HEATING SYSTEMS

Item	Net Price £	Material £	Labour hours	Labour £	Unit	Total rate £
T31: LOW TEMPERATURE HOT WATER HEATING – cont						
FIXINGS – cont						
Extra over black steel butt welded pipes; black steel flanges, welded and drilled; metric; BS 4504 – cont						
Blank flanges, slip on for welding; PN16						
15 mm dia.	1.63	1.84	0.48	11.37	nr	**13.21**
20 mm dia.	2.50	2.83	0.55	13.04	nr	**15.87**
25 mm dia.	3.04	3.44	0.64	15.17	nr	**18.61**
32 mm dia.	3.11	3.51	0.76	18.00	nr	**21.51**
40 mm dia.	4.02	4.54	0.84	19.90	nr	**24.44**
50 mm dia.	6.24	7.05	1.01	23.93	nr	**30.98**
65 mm dia.	7.66	8.65	1.30	30.79	nr	**39.44**
80 mm dia.	9.93	11.21	1.41	33.40	nr	**44.61**
100 mm dia.	12.31	13.90	1.78	42.17	nr	**56.07**
125 mm dia.	18.06	20.40	2.06	48.82	nr	**69.22**
150 mm dia.	23.30	26.32	2.35	55.68	nr	**82.00**
Extra over black steel butt welded pipes; black steel flanges, welding and drilled; imperial; BS 10						
Welded flanges; Table E						
15 mm dia.	5.87	6.63	0.59	13.98	nr	**20.61**
20 mm dia.	5.87	6.63	0.69	16.35	nr	**22.98**
25 mm dia.	5.87	6.63	0.84	19.90	nr	**26.53**
32 mm dia.	5.87	6.63	1.00	23.69	nr	**30.32**
40 mm dia.	6.05	6.83	1.11	26.31	nr	**33.14**
50 mm dia.	7.72	8.72	1.37	32.46	nr	**41.18**
65 mm dia.	9.40	10.62	1.54	36.50	nr	**47.12**
80 mm dia.	12.35	13.94	1.67	39.58	nr	**53.52**
100 mm dia.	18.23	20.59	2.22	52.61	nr	**73.20**
125 mm dia.	35.46	40.05	2.61	61.84	nr	**101.89**
150 mm dia.	35.46	40.05	2.99	70.85	nr	**110.90**
Blank flanges, slip on for welding; Table E						
15 mm dia.	4.04	4.56	0.48	11.37	nr	**15.93**
20 mm dia.	4.04	4.56	0.55	13.04	nr	**17.60**
25 mm dia.	4.04	4.56	0.64	15.17	nr	**19.73**
32 mm dia.	5.09	5.75	0.76	18.00	nr	**23.75**
40 mm dia.	5.52	6.23	0.84	19.90	nr	**26.13**
50 mm dia.	6.10	6.88	1.01	23.93	nr	**30.81**
65 mm dia.	7.03	7.94	1.30	30.79	nr	**38.73**
80 mm dia.	15.36	17.35	1.41	33.40	nr	**50.75**
100 mm dia.	17.44	19.70	1.78	42.17	nr	**61.87**
125 mm dia.	28.69	32.41	2.06	48.82	nr	**81.23**
150 mm dia.	36.89	41.67	2.35	55.68	nr	**97.35**

T: MECHANICAL/COOLING/HEATING SYSTEMS

Item	Net Price £	Material £	Labour hours	Labour £	Unit	Total rate £
Extra over black steel butt welded pipes; black steel flange connections						
Bolted connection between pair of flanges; including gasket, bolts, nuts and washers						
50 mm dia.	6.25	7.06	0.50	11.85	nr	**18.91**
65 mm dia.	7.39	8.34	0.50	11.85	nr	**20.19**
80 mm dia.	11.10	12.54	0.50	11.85	nr	**24.39**
100 mm dia.	12.05	13.61	0.50	11.85	nr	**25.46**
125 mm dia.	14.18	16.01	0.50	11.85	nr	**27.86**
150 mm dia.	20.26	22.88	0.88	20.85	nr	**43.73**
Extra over fittings; BS 1965; butt welded						
Cap						
25 mm dia.	13.87	15.67	0.47	11.14	nr	**26.81**
32 mm dia.	13.87	15.67	0.59	13.98	nr	**29.65**
40 mm dia.	13.94	15.75	0.70	16.60	nr	**32.35**
50 mm dia.	16.30	18.41	0.99	23.46	nr	**41.87**
65 mm dia.	19.15	21.62	1.35	31.99	nr	**53.61**
80 mm dia.	19.49	22.01	1.66	39.34	nr	**61.35**
100 mm dia.	25.48	28.77	2.23	52.85	nr	**81.62**
125 mm dia.	35.84	40.47	3.03	71.80	nr	**112.27**
150 mm dia.	41.45	46.82	3.79	89.80	nr	**136.62**
Concentric reducer						
20 × 15 mm dia.	7.72	8.72	0.69	16.35	nr	**25.07**
25 × 15 mm dia.	9.63	10.87	0.87	20.61	nr	**31.48**
25 × 20 mm dia.	7.45	8.42	0.87	20.61	nr	**29.03**
32 × 25 mm dia.	10.27	11.60	1.08	25.59	nr	**37.19**
40 × 25 mm dia.	13.38	15.12	1.38	32.69	nr	**47.81**
40 × 32 mm dia.	9.17	10.35	1.38	32.69	nr	**43.04**
50 × 25 mm dia.	12.73	14.38	1.82	43.12	nr	**57.50**
50 × 40 mm dia.	9.02	10.19	1.82	43.12	nr	**53.31**
65 × 50 mm dia.	12.45	14.07	2.52	59.71	nr	**73.78**
80 × 50 mm dia.	12.66	14.30	3.24	76.78	nr	**91.08**
100 × 50 mm dia.	20.97	23.68	4.08	96.68	nr	**120.36**
100 × 80 mm dia.	14.43	16.30	4.08	96.68	nr	**112.98**
125 × 80 mm dia.	31.51	35.59	4.71	111.61	nr	**147.20**
150 × 100 mm dia.	34.08	38.49	5.33	126.29	nr	**164.78**
Eccentric reducer						
20 × 15 mm dia.	11.47	12.95	0.69	16.35	nr	**29.30**
25 × 15 mm dia.	15.03	16.97	0.87	20.61	nr	**37.58**
25 × 20 mm dia.	12.61	14.24	0.87	20.61	nr	**34.85**
32 × 25 mm dia.	13.94	15.75	1.08	25.59	nr	**41.34**
40 × 25 mm dia.	17.22	19.45	1.38	32.69	nr	**52.14**
40 × 32 mm dia.	16.44	18.56	1.38	32.69	nr	**51.25**
50 × 25 mm dia.	19.61	22.14	1.82	43.12	nr	**65.26**

T: MECHANICAL/COOLING/HEATING SYSTEMS

Item	Net Price £	Material £	Labour hours	Labour £	Unit	Total rate £
T31: LOW TEMPERATURE HOT WATER HEATING – cont						
FIXINGS – cont						
Extra over fittings – cont						
Eccentric reducer – cont						
50 × 40 mm dia.	14.14	15.97	1.82	43.12	nr	**59.09**
65 × 50 mm dia.	16.68	18.84	2.52	59.71	nr	**78.55**
80 × 50 mm dia.	20.42	23.06	3.24	76.78	nr	**99.84**
100 × 50 mm dia.	34.58	39.06	4.08	96.68	nr	**135.74**
100 × 80 mm dia.	25.51	28.82	4.08	96.68	nr	**125.50**
125 × 80 mm dia.	68.16	76.98	4.71	111.61	nr	**188.59**
150 × 100 mm dia.	51.62	58.30	5.33	126.29	nr	**184.59**
45° elbow, long radius						
15 mm dia.	2.90	3.27	0.56	13.27	nr	**16.54**
20 mm dia.	2.94	3.33	0.75	17.77	nr	**21.10**
25 mm dia.	3.82	4.31	0.93	22.03	nr	**26.34**
32 mm dia.	4.63	5.23	1.17	27.72	nr	**32.95**
40 mm dia.	4.68	5.28	1.46	34.60	nr	**39.88**
50 mm dia.	6.41	7.24	1.97	46.68	nr	**53.92**
65 mm dia.	8.27	9.34	2.70	63.98	nr	**73.32**
80 mm dia.	7.63	8.62	3.32	78.66	nr	**87.28**
100 mm dia.	11.54	13.04	4.09	96.91	nr	**109.95**
125 mm dia.	24.54	27.71	4.94	117.06	nr	**144.77**
150 mm dia.	30.48	34.43	5.78	136.96	nr	**171.39**
90° elbow, long radius						
15 mm dia.	2.99	3.38	0.56	13.27	nr	**16.65**
20 mm dia.	2.93	3.30	0.75	17.77	nr	**21.07**
25 mm dia.	3.89	4.40	0.93	22.03	nr	**26.43**
32 mm dia.	4.64	5.24	1.17	27.72	nr	**32.96**
40 mm dia.	4.69	5.29	1.46	34.60	nr	**39.89**
50 mm dia.	6.41	7.24	1.97	46.68	nr	**53.92**
65 mm dia.	8.27	9.34	2.70	63.98	nr	**73.32**
80 mm dia.	8.97	10.13	3.32	78.66	nr	**88.79**
100 mm dia.	13.58	15.34	4.09	96.91	nr	**112.25**
125 mm dia.	28.25	31.91	4.94	117.06	nr	**148.97**
150 mm dia.	35.85	40.48	5.78	136.96	nr	**177.44**
Branch bend (based on branch and pipe sizes being the same)						
15 mm dia.	11.94	13.48	0.85	20.13	nr	**33.61**
20 mm dia.	11.83	13.36	0.85	20.13	nr	**33.49**
25 mm dia.	11.96	13.50	1.02	24.17	nr	**37.67**
32 mm dia.	11.29	12.75	1.11	26.31	nr	**39.06**
40 mm dia.	11.14	12.58	1.36	32.22	nr	**44.80**
50 mm dia.	10.83	12.23	1.70	40.29	nr	**52.52**
65 mm dia.	16.01	18.08	1.78	42.17	nr	**60.25**

T: MECHANICAL/COOLING/HEATING SYSTEMS

Item	Net Price £	Material £	Labour hours	Labour £	Unit	Total rate £
80 mm dia.	25.01	28.24	1.82	43.12	nr	**71.36**
100 mm dia.	32.62	36.84	1.87	44.31	nr	**81.15**
125 mm dia.	57.48	64.91	2.21	52.37	nr	**117.28**
150 mm dia.	88.38	99.82	2.65	62.79	nr	**162.61**
Equal tee						
15 mm dia.	28.29	31.95	0.82	19.42	nr	**51.37**
20 mm dia.	28.29	31.95	1.10	26.07	nr	**58.02**
25 mm dia.	28.29	31.95	1.35	31.99	nr	**63.94**
32 mm dia.	28.29	31.95	1.63	38.63	nr	**70.58**
40 mm dia.	28.29	31.95	2.14	50.71	nr	**82.66**
50 mm dia.	29.80	33.66	3.02	71.56	nr	**105.22**
65 mm dia.	43.47	49.09	3.61	85.54	nr	**134.63**
80 mm dia.	43.54	49.17	4.18	99.05	nr	**148.22**
100 mm dia.	54.17	61.18	5.24	124.16	nr	**185.34**
125 mm dia.	123.98	140.03	6.70	158.76	nr	**298.79**
150 mm dia.	135.43	152.95	8.45	200.22	nr	**353.17**
Extra over black steel butt welded pipes; labour						
Made bend						
15 mm dia.	–	–	0.42	9.95	nr	**9.95**
20 mm dia.	–	–	0.42	9.95	nr	**9.95**
25 mm dia.	–	–	0.50	11.85	nr	**11.85**
32 mm dia.	–	–	0.62	14.70	nr	**14.70**
40 mm dia.	–	–	0.74	17.54	nr	**17.54**
50 mm dia.	–	–	0.89	21.08	nr	**21.08**
65 mm dia.	–	–	1.05	24.88	nr	**24.88**
80 mm dia.	–	–	1.13	26.78	nr	**26.78**
100 mm dia.	–	–	2.90	68.72	nr	**68.72**
125 mm dia.	–	–	3.56	84.35	nr	**84.35**
150 mm dia.	–	–	4.18	99.05	nr	**99.05**
Splay cut end						
15 mm dia.	–	–	0.14	3.33	nr	**3.33**
20 mm dia.	–	–	0.16	3.79	nr	**3.79**
25 mm dia.	–	–	0.18	4.27	nr	**4.27**
32 mm dia.	–	–	0.25	5.93	nr	**5.93**
40 mm dia.	–	–	0.27	6.40	nr	**6.40**
50 mm dia.	–	–	0.31	7.34	nr	**7.34**
65 mm dia.	–	–	0.35	8.29	nr	**8.29**
80 mm dia.	–	–	0.40	9.48	nr	**9.48**
100 mm dia.	–	–	0.48	11.37	nr	**11.37**
125 mm dia.	–	–	0.56	13.27	nr	**13.27**
150 mm dia.	–	–	0.64	15.17	nr	**15.17**

T: MECHANICAL/COOLING/HEATING SYSTEMS

Item	Net Price £	Material £	Labour hours	Labour £	Unit	Total rate £
T31: LOW TEMPERATURE HOT WATER HEATING – cont						
FIXINGS – cont						
Extra over black steel butt welded pipes; labour – cont						
Screwed joint to fitting						
15 mm dia.	–	–	0.30	7.10	nr	**7.10**
20 mm dia.	–	–	0.40	9.48	nr	**9.48**
25 mm dia.	–	–	0.46	10.89	nr	**10.89**
32 mm dia.	–	–	0.53	12.56	nr	**12.56**
40 mm dia.	–	–	0.61	14.45	nr	**14.45**
50 mm dia.	–	–	0.73	17.30	nr	**17.30**
65 mm dia.	–	–	0.89	21.08	nr	**21.08**
80 mm dia.	–	–	1.05	24.88	nr	**24.88**
100 mm dia.	–	–	1.46	34.60	nr	**34.60**
125 mm dia.	–	–	2.10	49.77	nr	**49.77**
150 mm dia.	–	–	2.73	64.69	nr	**64.69**
Straight butt weld						
15 mm dia.	–	–	0.31	7.34	nr	**7.34**
20 mm dia.	–	–	0.42	9.95	nr	**9.95**
25 mm dia.	–	–	0.52	12.32	nr	**12.32**
32 mm dia.	–	–	0.69	16.35	nr	**16.35**
40 mm dia.	–	–	0.83	19.67	nr	**19.67**
50 mm dia.	–	–	1.22	28.92	nr	**28.92**
65 mm dia.	–	–	1.57	37.20	nr	**37.20**
80 mm dia.	–	–	1.95	46.21	nr	**46.21**
100 mm dia.	–	–	2.38	56.39	nr	**56.39**
125 mm dia.	–	–	2.83	67.05	nr	**67.05**
150 mm dia.	–	–	3.27	77.48	nr	**77.48**
CARBON WELDED STEEL						
Hot finished seamless carbon steel pipe; BS 806 and BS 3601; wall thickness to BS 3600; butt welded joints; including protective painting, fixed vertically or at low level, brackets measured separately (Refer to Screwed Pipework Section). Welded butt joints are within the running length, but any flanges are additional						
Pipework						
200 mm dia.	71.50	80.75	2.04	48.34	m	**129.09**
250 mm dia.	64.38	72.71	2.59	61.37	m	**134.08**
300 mm dia.	151.37	170.95	2.99	70.85	m	**241.80**
350 mm dia.	159.75	180.42	3.52	83.41	m	**263.83**
400 mm dia.	278.52	314.56	4.08	96.68	m	**411.24**

T: MECHANICAL/COOLING/HEATING SYSTEMS

Item	Net Price £	Material £	Labour hours	Labour £	Unit	Total rate £
Hot finished seamless carbon steel pipe; BS 806 and BS 3601; wall thickness to BS 3600; butt welded joints; including protective painting, fixed at high level or suspended, brackets measured separately (Refer to Screwed Pipework Section). Welded butt joints are within the running length, but any flanges are additional						
Pipework						
200 mm dia.	71.50	80.75	3.70	87.68	m	**168.43**
250 mm dia.	64.38	72.71	4.73	112.08	m	**184.79**
300 mm dia.	151.37	170.95	5.65	133.88	m	**304.83**
350 mm dia.	159.75	180.42	6.68	158.28	m	**338.70**
400 mm dia.	278.52	314.56	7.70	182.46	m	**497.02**
FIXINGS						
Refer to steel pipes; black malleable iron. For minimum fixing distances, refer to the Tables and Memoranda to the rear of the book						
Extra over fittings; BS 1965 part 1; butt welded						
Cap						
200 mm dia.	58.55	66.13	3.70	87.68	nr	**153.81**
250 mm dia.	112.35	126.89	4.73	112.08	nr	**238.97**
300 mm dia.	122.89	138.79	5.65	133.88	nr	**272.67**
350 mm dia.	189.33	213.83	6.68	158.28	nr	**372.11**
400 mm dia.	218.99	247.33	7.70	182.46	nr	**429.79**
Concentric reducer						
200 mm × 150 mm dia.	63.96	72.24	7.27	172.26	nr	**244.50**
250 mm × 150 mm dia.	92.75	104.75	9.05	214.43	nr	**319.18**
250 mm × 200 mm dia.	56.76	64.11	9.10	215.63	nr	**279.74**
300 mm × 150 mm dia.	107.13	120.99	10.75	254.71	nr	**375.70**
300 mm × 200 mm dia.	110.95	125.31	10.75	254.71	nr	**380.02**
300 mm × 250 mm dia.	98.63	111.39	11.15	264.20	nr	**375.59**
350 mm × 200 mm dia.	168.25	190.02	12.50	296.19	nr	**486.21**
350 mm × 250 mm dia.	141.07	159.32	12.70	300.93	nr	**460.25**
350 mm × 300 mm dia.	141.07	159.32	13.00	308.03	nr	**467.35**
400 mm × 250 mm dia.	206.26	232.95	14.46	342.63	nr	**575.58**
400 mm × 300 mm dia.	169.99	191.98	14.51	343.81	nr	**535.79**
400 mm × 350 mm dia.	222.65	251.46	15.16	359.21	nr	**610.67**

T: MECHANICAL/COOLING/HEATING SYSTEMS

Item	Net Price £	Material £	Labour hours	Labour £	Unit	Total rate £
T31: LOW TEMPERATURE HOT WATER HEATING – cont						
FIXINGS – cont						
Extra over fittings – cont						
Eccentric reducer						
200 mm × 150 mm dia.	115.51	130.46	7.27	172.26	nr	**302.72**
250 mm × 150 mm dia.	153.93	173.85	9.05	214.43	nr	**388.28**
250 mm × 200 mm dia.	100.72	113.75	9.10	215.63	nr	**329.38**
300 mm × 150 mm dia.	184.65	208.54	10.75	254.71	nr	**463.25**
300 mm × 200 mm dia.	212.98	240.54	10.75	254.71	nr	**495.25**
300 mm × 250 mm dia.	171.44	193.63	11.15	264.20	nr	**457.83**
350 mm × 200 mm dia.	195.61	220.92	12.50	296.19	nr	**517.11**
350 mm × 250 mm dia.	162.63	183.67	12.70	300.93	nr	**484.60**
350 mm × 300 mm dia.	162.63	183.67	13.00	308.03	nr	**491.70**
400 mm × 250 mm dia.	297.15	335.60	14.46	342.63	nr	**678.23**
400 mm × 300 mm dia.	244.59	276.24	14.51	343.81	nr	**620.05**
400 mm × 350 mm dia.	216.97	245.04	15.16	359.21	nr	**604.25**
45° elbow						
200 mm dia.	61.17	69.08	7.75	183.63	nr	**252.71**
250 mm dia.	114.89	129.75	10.05	238.13	nr	**367.88**
300 mm dia.	167.86	189.58	12.20	289.08	nr	**478.66**
350 mm dia.	135.84	153.42	14.65	347.13	nr	**500.55**
400 mm dia.	225.78	254.99	17.12	405.65	nr	**660.64**
90° elbow						
200 mm dia.	71.97	81.28	7.75	183.63	nr	**264.91**
250 mm dia.	135.19	152.68	10.05	238.13	nr	**390.81**
300 mm dia.	197.49	223.04	12.20	289.08	nr	**512.12**
350 mm dia.	271.50	306.63	14.65	347.13	nr	**653.76**
400 mm dia.	346.79	391.66	17.12	405.65	nr	**797.31**
Equal tee						
200 mm dia.	190.14	214.74	11.25	266.57	nr	**481.31**
250 mm dia.	326.02	368.21	14.53	344.28	nr	**712.49**
300 mm dia.	404.61	456.96	17.55	415.84	nr	**872.80**
350 mm dia.	192.90	217.86	20.98	497.12	nr	**714.98**
400 mm dia.	217.75	245.92	24.38	577.68	nr	**823.60**
Extra over black steel butt welded pipes; labour						
Straight butt weld						
200 mm dia.	–	–	4.08	96.68	nr	**96.68**
250 mm dia.	–	–	5.20	123.22	nr	**123.22**
300 mm dia.	–	–	6.22	147.39	nr	**147.39**
350 mm dia.	–	–	7.33	173.68	nr	**173.68**
400 mm dia.	–	–	8.41	199.27	nr	**199.27**

T: MECHANICAL/COOLING/HEATING SYSTEMS

Item	Net Price £	Material £	Labour hours	Labour £	Unit	Total rate £
Branch weld						
100 mm dia.	–	–	3.46	81.99	nr	**81.99**
125 mm dia.	–	–	4.23	100.24	nr	**100.24**
150 mm dia.	–	–	5.00	118.47	nr	**118.47**
Extra over black steel butt welded pipes; black steel flanges, welding and drilled; metric; BS 4504						
Welded flanges; PN16						
200 mm dia.	39.15	44.22	4.10	97.16	nr	**141.38**
250 mm dia.	56.23	63.51	5.33	126.29	nr	**189.80**
300 mm dia.	78.77	88.96	6.40	151.64	nr	**240.60**
350 mm dia.	152.35	172.06	7.43	176.05	nr	**348.11**
400 mm dia.	201.14	227.16	8.45	200.22	nr	**427.38**
Welded flanges; PN25						
200 mm dia.	129.98	146.80	4.10	97.16	nr	**243.96**
250 mm dia.	155.81	175.97	5.33	126.29	nr	**302.26**
300 mm dia.	210.70	237.97	6.40	151.64	nr	**389.61**
Blank flanges, slip on for welding; PN16						
200 mm dia.	90.51	102.22	2.70	63.98	nr	**166.20**
250 mm dia.	137.56	155.36	3.48	82.46	nr	**237.82**
300 mm dia.	146.98	165.99	4.20	99.52	nr	**265.51**
350 mm dia.	230.02	259.78	4.78	113.26	nr	**373.04**
400 mm dia.	304.15	343.51	5.35	126.77	nr	**470.28**
Blank flanges, slip on for welding; PN25						
200 mm dia.	90.51	102.22	2.70	63.98	nr	**166.20**
250 mm dia.	137.56	155.36	3.48	82.46	nr	**237.82**
300 mm dia.	146.98	165.99	4.20	99.52	nr	**265.51**
Extra over black steel butt welded pipes; black steel flange connections						
Bolted connection between pair of flanges; including gasket, bolts, nuts and washers						
200 mm dia.	75.96	85.79	3.83	90.75	nr	**176.54**
250 mm dia.	117.56	132.77	4.93	116.81	nr	**249.58**
300 mm dia.	160.28	181.02	5.90	139.80	nr	**320.82**

T: MECHANICAL/COOLING/HEATING SYSTEMS

Item	Net Price £	Material £	Labour hours	Labour £	Unit	Total rate £
T31: LOW TEMPERATURE HOT WATER HEATING – cont						
PRESS FIT						
Press fit jointing system; operating temperature -20°C to +120°C; operating pressure 16 bar; butyl rubber 'O' ring mechanical joint. With brackets measured separately (Refer to Screwed Steel Section)						
Carbon steel						
Pipework						
15 mm dia.	1.24	1.40	0.46	10.89	m	**12.29**
20 mm dia.	1.99	2.24	0.48	11.37	m	**13.61**
25 mm dia.	2.79	3.15	0.52	12.32	m	**15.47**
32 mm dia.	3.60	4.07	0.56	13.27	m	**17.34**
40 mm dia.	4.93	5.57	0.58	13.75	m	**19.32**
50 mm dia.	6.40	7.23	0.66	15.65	m	**22.88**
Extra over for Carbon Steel press fit fittings						
Coupling						
15 mm dia.	0.84	0.95	0.36	8.53	nr	**9.48**
20 mm dia.	1.02	1.15	0.36	8.53	nr	**9.68**
25 mm dia.	1.29	1.46	0.44	10.42	nr	**11.88**
32 mm dia.	2.16	2.44	0.44	10.42	nr	**12.86**
40 mm dia.	2.89	3.26	0.52	12.32	nr	**15.58**
50 mm dia.	3.41	3.86	0.60	14.22	nr	**18.08**
Reducer						
20 × 15 mm dia.	0.78	0.88	0.36	8.53	nr	**9.41**
25 × 15 mm dia.	1.02	1.15	0.40	9.48	nr	**10.63**
25 × 20 mm dia.	1.08	1.22	0.40	9.48	nr	**10.70**
32 × 20 mm dia.	1.19	1.35	0.40	9.48	nr	**10.83**
32 × 25 mm dia.	1.28	1.45	0.44	10.42	nr	**11.87**
40 × 32 mm dia.	2.77	3.13	0.48	11.37	nr	**14.50**
50 × 20 mm dia.	7.99	9.03	0.48	11.37	nr	**20.40**
50 × 25 mm dia.	8.05	9.09	0.52	12.32	nr	**21.41**
50 × 40 mm dia.	8.47	9.57	0.56	13.27	nr	**22.84**
90° Elbow						
15 mm dia.	1.22	1.38	0.36	8.53	nr	**9.91**
20 mm dia.	1.60	1.81	0.36	8.53	nr	**10.34**
25 mm dia.	2.19	2.47	0.44	10.42	nr	**12.89**
32 mm dia.	5.47	6.18	0.44	10.42	nr	**16.60**
40 mm dia.	8.74	9.87	0.52	12.32	nr	**22.19**
50 mm dia.	10.45	11.80	0.60	14.22	nr	**26.02**

T: MECHANICAL/COOLING/HEATING SYSTEMS

Item	Net Price £	Material £	Labour hours	Labour £	Unit	Total rate £
45° Elbow						
15 mm dia.	1.45	1.64	0.36	8.53	nr	**10.17**
20 mm dia.	1.62	1.83	0.36	8.53	nr	**10.36**
25 mm dia.	2.20	2.49	0.44	10.42	nr	**12.91**
32 mm dia.	4.33	4.89	0.44	10.42	nr	**15.31**
40 mm dia.	5.44	6.14	0.52	12.32	nr	**18.46**
50 mm dia.	6.14	6.94	0.60	14.22	nr	**21.16**
Equal tee						
15 mm dia.	2.33	2.63	0.54	12.80	nr	**15.43**
20 mm dia.	2.69	3.04	0.54	12.80	nr	**15.84**
25 mm dia.	3.60	4.07	0.66	15.65	nr	**19.72**
32 mm dia.	5.62	6.34	0.66	15.65	nr	**21.99**
40 mm dia.	8.29	9.36	0.78	18.47	nr	**27.83**
50 mm dia.	9.95	11.24	0.90	21.32	nr	**32.56**
Reducing tee						
20 × 15 mm dia.	2.65	2.99	0.54	12.80	nr	**15.79**
25 × 15 mm dia.	3.57	4.03	0.62	14.70	nr	**18.73**
25 × 20 mm dia.	3.88	4.39	0.62	14.70	nr	**19.09**
32 × 15 mm dia.	5.23	5.91	0.62	14.70	nr	**20.61**
32 × 20 mm dia.	5.66	6.40	0.62	14.70	nr	**21.10**
32 × 25 mm dia.	5.75	6.50	0.62	14.70	nr	**21.20**
40 × 20 mm dia.	7.58	8.56	0.70	16.60	nr	**25.16**
40 × 25 mm dia.	7.86	8.87	0.70	16.60	nr	**25.47**
40 × 32 mm dia.	7.67	8.66	0.70	16.60	nr	**25.26**
50 × 20 mm dia.	9.03	10.20	0.82	19.42	nr	**29.62**
50 × 25 mm dia.	9.21	10.40	0.82	19.42	nr	**29.82**
50 × 32 mm dia.	9.48	10.71	0.82	19.42	nr	**30.13**
50 × 40 mm dia.	9.94	11.23	0.82	19.42	nr	**30.65**

MECHANICAL GROOVED

Mechanical grooved jointing system; working temperature not exceeding 82°C BS 5750; pipework complete with grooved joints; painted finish. With brackets measured separately (Refer to Screwed Steel Section)

Item	Net Price £	Material £	Labour hours	Labour £	Unit	Total rate £
Grooved Joints						
65 mm	11.31	12.77	0.58	13.75	m	**26.52**
80 mm	11.85	13.38	0.68	16.12	m	**29.50**
100 mm	15.38	17.37	0.79	18.72	m	**36.09**
125 mm	31.21	35.25	1.02	24.17	m	**59.42**
150 mm	25.19	28.45	1.15	27.25	m	**55.70**

T: MECHANICAL/COOLING/HEATING SYSTEMS

Item	Net Price £	Material £	Labour hours	Labour £	Unit	Total rate £
T31: LOW TEMPERATURE HOT WATER HEATING – cont						
MECHANICAL GROOVED – cont						
Extra over mechanical grooved system fittings						
Couplings						
65 mm	11.31	12.77	0.41	9.71	nr	**22.48**
80 mm	11.85	13.38	0.41	9.71	nr	**23.09**
100 mm	15.38	17.37	0.66	15.65	nr	**33.02**
125 mm	31.21	35.25	0.68	16.12	nr	**51.37**
150 mm	25.19	28.45	0.80	18.95	nr	**47.40**
Concentric reducers (one size down)						
80 mm	18.15	20.50	0.59	13.98	nr	**34.48**
100 mm	18.19	20.54	0.71	16.83	nr	**37.37**
125 mm	31.36	35.41	0.85	20.13	nr	**55.54**
150 mm	40.65	45.91	0.98	23.23	nr	**69.14**
Short radius elbow; 90°						
65 mm	20.78	23.47	0.53	12.56	nr	**36.03**
80 mm	21.21	23.95	0.61	14.45	nr	**38.40**
100 mm	28.39	32.06	0.80	18.95	nr	**51.01**
125 mm	46.82	52.88	0.90	21.32	nr	**74.20**
150 mm	60.78	68.65	0.94	22.27	nr	**90.92**
Short radius elbow; 45°						
65 mm	17.79	20.09	0.53	12.56	nr	**32.65**
80 mm	19.99	22.57	0.61	14.45	nr	**37.02**
100 mm	24.85	28.07	0.80	18.95	nr	**47.02**
125 mm	42.03	47.47	0.90	21.32	nr	**68.79**
150 mm	46.27	52.25	0.94	22.27	nr	**74.52**
Equal tee						
65 mm	37.41	42.25	0.83	19.67	nr	**61.92**
80 mm	39.64	44.77	0.93	22.03	nr	**66.80**
100 mm	44.32	50.05	1.18	27.97	nr	**78.02**
125 mm	117.55	132.76	1.37	32.46	nr	**165.22**
150 mm	108.98	123.08	1.43	33.88	nr	**156.96**

T: MECHANICAL/COOLING/HEATING SYSTEMS

Item	Net Price £	Material £	Labour hours	Labour £	Unit	Total rate £
PLASTIC PIPEWORK						
Polypropylene PP-R 80 pipe, mechanically stabilized by fibre compound mixture in middle layer; suitable for continuous working temperatures of 0–90°C; thermally fused joints in the running length						
Pipe; 4 m long; PN 20						
20 mm dia.	2.20	2.49	0.35	8.29	m	**10.78**
25 mm dia.	3.31	3.73	0.39	9.24	m	**12.97**
32 mm dia.	3.78	4.27	0.43	10.19	m	**14.46**
40 mm dia.	5.05	5.70	0.47	11.14	m	**16.84**
50 mm dia.	7.34	8.29	0.51	12.09	m	**20.38**
63 mm dia.	12.12	13.69	0.52	12.32	m	**26.01**
75 mm dia.	15.68	17.71	0.60	14.22	m	**31.93**
90 mm dia.	24.18	27.31	0.69	16.35	m	**43.66**
110 mm dia.	36.36	41.07	0.69	16.35	m	**57.42**
125 mm dia.	38.95	43.99	0.85	20.13	m	**64.12**
FIXINGS						
Refer to steel pipes; black malleable iron. For minimum fixing distances, refer to the Tables and Memoranda to the rear of the book						
Extra over fittings; thermally fused joints						
Overbridge bow						
20 mm dia.	1.46	1.65	0.51	12.09	nr	**13.74**
25 mm dia.	2.68	3.03	0.56	13.27	nr	**16.30**
32 mm dia.	5.37	6.07	0.65	15.40	nr	**21.47**
Elbow 90°						
20 mm dia.	0.49	0.55	0.44	10.42	nr	**10.97**
25 mm dia.	0.65	0.73	0.52	12.32	nr	**13.05**
32 mm dia.	0.94	1.06	0.59	13.98	nr	**15.04**
40 mm dia.	1.47	1.66	0.66	15.65	nr	**17.31**
50 mm dia.	3.18	3.59	0.73	17.30	nr	**20.89**
63 mm dia.	4.86	5.49	0.85	20.13	nr	**25.62**
75 mm dia.	10.77	12.17	0.85	20.13	nr	**32.30**
90 mm dia.	19.88	22.45	1.04	24.64	nr	**47.09**
110 mm dia.	28.28	31.94	1.04	24.64	nr	**56.58**
125 mm dia.	43.56	49.19	1.30	30.79	nr	**79.98**
Long bend 90°						
20 mm dia.	2.62	2.96	0.48	11.37	nr	**14.33**
25 mm dia.	2.73	3.08	0.57	13.50	nr	**16.58**
32 mm dia.	3.21	3.62	0.65	15.40	nr	**19.02**
40 mm dia.	5.87	6.63	0.73	17.30	nr	**23.93**

T: MECHANICAL/COOLING/HEATING SYSTEMS

Item	Net Price £	Material £	Labour hours	Labour £	Unit	Total rate £
T31: LOW TEMPERATURE HOT WATER HEATING – cont						
FIXINGS – cont						
Extra over fittings – cont						
Elbow 90°, female/male						
20 mm dia.	0.50	0.56	0.44	10.42	nr	**10.98**
25 mm dia.	0.66	0.74	0.52	12.32	nr	**13.06**
32 mm dia.	0.95	1.07	0.59	13.98	nr	**15.05**
Elbow 45°						
20 mm dia.	0.50	0.56	0.44	10.42	nr	**10.98**
25 mm dia.	0.66	0.74	0.52	12.32	nr	**13.06**
32 mm dia.	0.95	1.07	0.59	13.98	nr	**15.05**
40 mm dia.	1.48	1.67	0.66	15.65	nr	**17.32**
50 mm dia.	3.18	3.59	0.73	17.30	nr	**20.89**
63 mm dia.	4.87	5.50	0.85	20.13	nr	**25.63**
75 mm dia.	10.78	12.18	0.85	20.13	nr	**32.31**
90 mm dia.	19.88	22.45	1.04	24.64	nr	**47.09**
110 mm dia.	28.29	31.95	1.04	24.64	nr	**56.59**
125 mm dia.	43.57	49.20	1.30	30.79	nr	**79.99**
Elbow 45°, female/male						
20 mm dia.	0.50	0.56	0.44	10.42	nr	**10.98**
25 mm dia.	0.66	0.74	0.52	12.32	nr	**13.06**
32 mm dia.	0.95	1.07	0.59	13.98	nr	**15.05**
T-Piece 90°						
20 mm dia.	0.69	0.78	0.61	14.45	nr	**15.23**
25 mm dia.	0.93	1.05	0.72	17.06	nr	**18.11**
32 mm dia.	1.21	1.37	0.83	19.67	nr	**21.04**
40 mm dia.	1.86	2.10	0.92	21.80	nr	**23.90**
50 mm dia.	5.29	5.98	1.01	23.93	nr	**29.91**
63 mm dia.	7.59	8.57	1.11	26.31	nr	**34.88**
75 mm dia.	12.65	14.29	1.18	27.97	nr	**42.26**
90 mm dia.	23.28	26.30	1.46	34.60	nr	**60.90**
110 mm dia.	36.32	41.02	1.46	34.60	nr	**75.62**
125 mm dia.	48.26	54.51	1.82	43.12	nr	**97.63**
T-Piece 90° reducing						
25 × 20 × 20 mm	0.95	1.07	0.72	17.06	nr	**18.13**
25 × 20 × 25 mm	0.95	1.07	0.72	17.06	nr	**18.13**
32 × 20 × 32 mm	1.21	1.37	0.83	19.67	nr	**21.04**
32 × 25 × 32 mm	1.21	1.37	0.83	19.67	nr	**21.04**
40 × 20 × 40 mm	1.86	2.10	0.92	21.80	nr	**23.90**
40 × 25 × 40 mm	1.86	2.10	0.92	21.80	nr	**23.90**
40 × 32 × 40 mm	1.86	2.10	0.92	21.80	nr	**23.90**
50 × 25 × 50 mm	5.29	5.98	1.01	23.93	nr	**29.91**
50 × 32 × 50 mm	5.29	5.98	1.01	23.93	nr	**29.91**

T: MECHANICAL/COOLING/HEATING SYSTEMS

Item	Net Price £	Material £	Labour hours	Labour £	Unit	Total rate £
50 × 40 × 50 mm	5.29	5.98	1.01	23.93	nr	**29.91**
63 × 20 × 63 mm	7.14	8.07	1.11	26.31	nr	**34.38**
63 × 25 × 63 mm	7.14	8.07	1.11	26.31	nr	**34.38**
63 × 32 × 63 mm	7.14	8.07	1.11	26.31	nr	**34.38**
63 × 40 × 63 mm	7.14	8.07	1.01	23.93	nr	**32.00**
63 × 50 × 63 mm	7.14	8.07	1.01	23.93	nr	**32.00**
75 × 20 × 75 mm	11.61	13.12	1.18	27.97	nr	**41.09**
75 × 25 × 75 mm	11.61	13.12	1.18	27.97	nr	**41.09**
75 × 32 × 75 mm	11.61	13.12	1.18	27.97	nr	**41.09**
75 × 40 × 75 mm	11.61	13.12	1.18	27.97	nr	**41.09**
75 × 50 × 75 mm	11.61	13.12	1.18	27.97	nr	**41.09**
75 × 63 × 75 mm	11.61	13.12	1.18	27.97	nr	**41.09**
90 × 63 × 90 mm	23.28	26.30	1.46	34.60	nr	**60.90**
110 × 75 × 110 mm	36.32	41.02	1.46	34.60	nr	**75.62**
110 × 90 × 110 mm	36.32	41.02	1.46	34.60	nr	**75.62**
125 × 90 × 125 mm	43.20	48.79	1.82	43.12	nr	**91.91**
125 × 110 × 125 mm	44.08	49.79	1.82	43.12	nr	**92.91**
Reducer						
25 × 20 mm	0.54	0.61	0.59	13.98	nr	**14.59**
32 × 20 mm	0.70	0.80	0.62	14.70	nr	**15.50**
32 × 25 mm	0.70	0.80	0.62	14.70	nr	**15.50**
40 × 20 mm	1.11	1.25	0.66	15.65	nr	**16.90**
40 × 25 mm	1.11	1.25	0.66	15.65	nr	**16.90**
40 × 32 mm	1.11	1.25	0.66	15.65	nr	**16.90**
50 × 20 mm	1.78	2.01	0.73	17.30	nr	**19.31**
50 × 25 mm	1.78	2.01	0.73	17.30	nr	**19.31**
50 × 32 mm	1.78	2.01	0.73	17.30	nr	**19.31**
50 × 40 mm	1.78	2.01	0.73	17.30	nr	**19.31**
63 × 40 mm	3.60	4.07	0.78	18.47	nr	**22.54**
63 × 25 mm	3.60	4.07	0.78	18.47	nr	**22.54**
63 × 32 mm	3.60	4.07	0.78	18.47	nr	**22.54**
63 × 50 mm	3.60	4.07	0.78	18.47	nr	**22.54**
75 × 50 mm	4.02	4.54	0.85	20.13	nr	**24.67**
75 × 63 mm	4.02	4.54	0.85	20.13	nr	**24.67**
90 × 63 mm	8.98	10.14	1.04	24.64	nr	**34.78**
90 × 75 mm	8.98	10.14	1.04	24.64	nr	**34.78**
110 × 90 mm	14.52	16.40	1.17	27.72	nr	**44.12**
125 × 110 mm	22.68	25.61	1.43	33.88	nr	**59.49**
Socket						
20 mm dia.	0.48	0.54	0.51	12.09	nr	**12.63**
25 mm dia.	0.54	0.61	0.56	13.27	nr	**13.88**
32 mm dia.	0.70	0.80	0.65	15.40	nr	**16.20**
40 mm dia.	0.86	0.97	0.74	17.54	nr	**18.51**
50 mm dia.	1.78	2.01	0.81	19.19	nr	**21.20**
63 mm dia.	3.60	4.07	0.86	20.37	nr	**24.44**
75 mm dia.	4.02	4.54	0.91	21.56	nr	**26.10**
90 mm dia.	10.38	11.72	0.91	21.56	nr	**33.28**
110 mm dia.	17.64	19.92	0.91	21.56	nr	**41.48**
125 mm dia.	24.57	27.75	1.30	30.79	nr	**58.54**

T: MECHANICAL/COOLING/HEATING SYSTEMS

Item	Net Price £	Material £	Labour hours	Labour £	Unit	Total rate £
T31: LOW TEMPERATURE HOT WATER HEATING – cont						
FIXINGS – cont						
Extra over fittings – cont						
End Cap						
20 mm dia.	0.76	0.86	0.25	5.93	nr	6.79
25 mm dia.	0.93	1.05	0.29	6.86	nr	7.91
32 mm dia.	1.14	1.29	0.33	7.81	nr	9.10
40 mm dia.	1.83	2.07	0.36	8.53	nr	10.60
50 mm dia.	2.54	2.87	0.40	9.48	nr	12.35
63 mm dia.	4.27	4.82	0.44	10.42	nr	15.24
75 mm dia.	6.15	6.95	0.47	11.14	nr	18.09
90 mm dia.	13.94	15.75	0.57	13.50	nr	29.25
110 mm dia.	16.75	18.92	0.57	13.50	nr	32.42
125 mm dia.	25.53	28.84	0.85	20.13	nr	48.97
Stub flange with gasket						
32 mm dia.	16.94	19.14	0.23	5.46	nr	24.60
40 mm dia.	21.31	24.07	0.27	6.40	nr	30.47
50 mm dia.	25.78	29.11	0.38	9.00	nr	38.11
63 mm dia.	30.95	34.96	0.43	10.19	nr	45.15
75 mm dia.	36.32	41.02	0.48	11.37	nr	52.39
90 mm dia.	49.15	55.51	0.53	12.56	nr	68.07
110 mm dia.	68.82	77.72	0.53	12.56	nr	90.28
125 mm dia.	98.63	111.39	0.75	17.77	nr	129.16
Weld in saddle with female thread						
40 – ½'	1.23	1.39	0.36	8.53	nr	9.92
50 – ½'	1.23	1.39	0.36	8.53	nr	9.92
63 – ½'	1.23	1.39	0.40	9.48	nr	10.87
75 – ½'	1.23	1.39	0.40	9.48	nr	10.87
90 – ½'	1.23	1.39	0.46	10.89	nr	12.28
110 – ½'	1.23	1.39	0.46	10.89	nr	12.28
Weld in saddle with male thread						
50 – ½'	1.23	1.39	0.36	8.53	nr	9.92
63 – ½'	1.23	1.39	0.40	9.48	nr	10.87
75 – ½'	1.23	1.39	0.40	9.48	nr	10.87
90 – ½'	1.23	1.39	0.46	10.89	nr	12.28
110 – ½'	1.23	1.39	0.46	10.89	nr	12.28
Transition piece, round with female thread						
20 × ½'	2.85	3.22	0.29	6.86	nr	10.08
20 × ¾'	3.75	4.23	0.29	6.86	nr	11.09
25 × ½'	2.85	3.22	0.33	7.81	nr	11.03
25 × ¾'	3.75	4.23	0.33	7.81	nr	12.04

T: MECHANICAL/COOLING/HEATING SYSTEMS

Item	Net Price £	Material £	Labour hours	Labour £	Unit	Total rate £
Transition piece, hexagon with female thread						
32 × 1'	10.60	11.97	0.36	8.53	nr	**20.50**
40 × 1¼'	16.77	18.94	0.36	8.53	nr	**27.47**
50 × 1½'	19.47	21.99	0.36	8.53	nr	**30.52**
63 × 2'	30.16	34.06	0.40	9.48	nr	**43.54**
75 × 2'	31.46	35.53	0.40	9.48	nr	**45.01**
125 × 5'	178.66	201.78	0.51	12.09	nr	**213.87**
Stop valve for surface assembly						
20 mm dia.	11.71	13.23	0.25	5.93	nr	**19.16**
25 mm dia.	11.71	13.23	0.29	6.86	nr	**20.09**
32 mm dia.	22.04	24.89	0.33	7.81	nr	**32.70**
Ball valve						
20 mm dia.	41.22	46.55	0.25	5.93	nr	**52.48**
25 mm dia.	44.17	49.89	0.29	6.86	nr	**56.75**
32 mm dia.	53.06	59.93	0.33	7.81	nr	**67.74**
40 mm dia.	67.73	76.49	0.36	8.53	nr	**85.02**
50 mm dia.	92.99	105.02	0.40	9.48	nr	**114.50**
63 mm dia.	104.84	118.40	0.44	10.42	nr	**128.82**
Floor or ceiling cover plates						
Plastic						
15 mm dia.	0.40	0.45	0.16	3.79	nr	**4.24**
20 mm dia.	0.45	0.51	0.22	5.22	nr	**5.73**
25 mm dia.	0.47	0.53	0.22	5.22	nr	**5.75**
32 mm dia.	0.53	0.60	0.24	5.69	nr	**6.29**
40 mm dia.	1.17	1.33	0.26	6.15	nr	**7.48**
50 mm dia.	1.30	1.47	0.26	6.15	nr	**7.62**
Chromium plated						
15 mm dia.	2.72	3.07	0.16	3.79	nr	**6.86**
20 mm dia.	2.91	3.28	0.17	4.03	nr	**7.31**
25 mm dia.	3.01	3.40	0.21	4.98	nr	**8.38**
32 mm dia.	3.08	3.48	0.22	5.22	nr	**8.70**
40 mm dia.	3.48	3.93	0.26	6.15	nr	**10.08**
50 mm dia.	4.16	4.70	0.26	6.15	nr	**10.85**

T: MECHANICAL/COOLING/HEATING SYSTEMS

Item	Net Price £	Material £	Labour hours	Labour £	Unit	Total rate £
T31: LOW TEMPERATURE HOT WATER HEATING – cont						
Y11 – PIPELINE ANCILLARIES						
EXPANSION JOINTS						
Axial movement bellows expansion joints; stainless steel						
Screwed ends for steel pipework; up to 6 bar G at 100°C						
15 mm dia.	65.87	74.39	0.68	16.12	nr	**90.51**
20 mm dia.	70.36	79.47	0.81	19.19	nr	**98.66**
25 mm dia.	71.66	80.94	0.93	22.03	nr	**102.97**
32 mm dia.	75.87	85.69	1.06	25.12	nr	**110.81**
40 mm dia.	80.36	90.76	1.16	27.49	nr	**118.25**
50 mm dia.	81.66	92.23	1.19	28.20	nr	**120.43**
Screwed ends for steel pipework; aluminium and steel outer sleeves; up to 16 bar G at 120°C						
20 mm dia.	70.36	79.47	1.32	31.27	nr	**110.74**
25 mm dia.	71.66	80.94	1.52	36.02	nr	**116.96**
32 mm dia.	75.87	85.69	1.80	42.65	nr	**128.34**
40 mm dia.	80.36	90.76	2.03	48.10	nr	**138.86**
50 mm dia.	81.66	92.23	2.26	53.54	nr	**145.77**
Flanged ends for steel pipework; aluminium and steel outer sleeves; up to 16 bar G at 120°C						
20 mm dia.	63.49	71.71	0.53	12.56	nr	**84.27**
25 mm dia.	68.02	76.83	0.64	15.17	nr	**92.00**
32 mm dia.	93.72	105.85	0.74	17.54	nr	**123.39**
40 mm dia.	101.28	114.38	0.82	19.42	nr	**133.80**
50 mm dia.	120.18	135.73	0.89	21.08	nr	**156.81**
Flanged ends for steel pipework; up to 16 bar G at 120°C						
65 mm dia.	111.11	125.49	1.10	26.07	nr	**151.56**
80 mm dia.	119.42	134.87	1.31	31.04	nr	**165.91**
100 mm dia.	136.80	154.50	1.78	42.17	nr	**196.67**
150 mm dia.	209.36	236.45	3.08	72.98	nr	**309.43**
Screwed ends for non-ferrous pipework; up to 6 bar G at 100°C						
20 mm dia.	70.36	79.47	0.72	17.06	nr	**96.53**
25 mm dia.	71.66	80.94	0.84	19.90	nr	**100.84**
32 mm dia.	75.87	85.69	1.02	24.17	nr	**109.86**
40 mm dia.	80.36	90.76	1.11	26.31	nr	**117.07**
50 mm dia.	81.66	92.23	1.18	27.97	nr	**120.20**

T: MECHANICAL/COOLING/HEATING SYSTEMS

Item	Net Price £	Material £	Labour hours	Labour £	Unit	Total rate £
Flanged ends for steel, copper or non-ferrous pipework; up to 16 bar G at 120°C						
65 mm dia.	142.85	161.33	0.87	20.61	nr	**181.94**
80 mm dia.	147.38	166.45	0.95	22.51	nr	**188.96**
100 mm dia.	170.06	192.07	1.15	27.25	nr	**219.32**
150 mm dia.	278.90	314.99	1.36	32.22	nr	**347.21**
Angular movement bellows expansion joints; stainless steel						
Flanged ends for steel pipework; up to 16 bar G at 120°C						
50 mm dia.	164.67	185.98	0.71	16.83	nr	**202.81**
65 mm dia.	199.85	225.71	0.83	19.67	nr	**245.38**
80 mm dia.	233.53	263.75	0.91	21.56	nr	**285.31**
100 mm dia.	301.49	340.50	0.97	22.98	nr	**363.48**
125 mm dia.	398.23	449.76	1.16	27.49	nr	**477.25**
150 mm dia.	473.53	534.81	1.18	27.97	nr	**562.78**
VALVES						
Isolating valves						
Bronze gate valve; non-rising stem; BS 5154, series B, PN 32; working pressure up to 14 bar for saturated steam, 32 bar from -10°C to 100°C; screwed ends to steel						
15 mm dia.	34.32	38.76	1.11	26.31	nr	**65.07**
20 mm dia.	44.53	50.30	1.28	30.32	nr	**80.62**
25 mm dia.	60.31	68.12	1.49	35.30	nr	**103.42**
32 mm dia.	83.30	94.07	1.88	44.54	nr	**138.61**
40 mm dia.	111.85	126.33	2.31	54.73	nr	**181.06**
50 mm dia.	157.16	177.50	2.80	66.34	nr	**243.84**
Bronze gate valve; non-rising stem; BS 5154, series B, PN 20; working pressure up to 9 bar for saturated steam, 20 bar from -10°C to 100°C; screwed ends to steel						
15 mm dia.	23.41	26.44	0.84	19.90	nr	**46.34**
20 mm dia.	33.27	37.58	1.01	23.93	nr	**61.51**
25 mm dia.	43.04	48.61	1.19	28.20	nr	**76.81**
32 mm dia.	61.42	69.37	1.38	32.69	nr	**102.06**
40 mm dia.	84.06	94.94	1.62	38.39	nr	**133.33**
50 mm dia.	120.12	135.66	1.94	45.97	nr	**181.63**

T: MECHANICAL/COOLING/HEATING SYSTEMS

Item	Net Price £	Material £	Labour hours	Labour £	Unit	Total rate £
T31: LOW TEMPERATURE HOT WATER HEATING – cont						
VALVES – cont						
Isolating valves – cont						
Bronze gate valve; non-rising stem; BS 5154, series B, PN 16; working pressure up to 7 bar for saturated steam, 16 bar from -10°C to 100°C; BS4504 flanged ends; bolted connections						
15 mm dia.	111.63	126.07	1.18	27.97	nr	**154.04**
20 mm dia.	116.24	131.28	1.24	29.39	nr	**160.67**
25 mm dia.	153.08	172.89	1.31	31.04	nr	**203.93**
32 mm dia.	192.07	216.92	1.43	33.88	nr	**250.80**
40 mm dia.	232.21	262.25	1.53	36.25	nr	**298.50**
50 mm dia.	321.02	362.56	1.63	38.63	nr	**401.19**
65 mm dia.	498.56	563.07	1.71	40.52	nr	**603.59**
80 mm dia.	705.64	796.95	1.88	44.54	nr	**841.49**
100 mm dia.	1282.35	1448.28	2.03	48.10	nr	**1496.38**
Cast iron gate valve; bronze trim; non rising stem; BS 5150, PN6; working pressure 6 bar from -10°C to 120°C; BS4504 flanged ends; bolted connections						
50 mm dia.	208.79	235.80	1.85	43.83	nr	**279.63**
65 mm dia.	218.23	246.47	2.00	47.39	nr	**293.86**
80 mm dia.	252.20	284.83	2.27	53.78	nr	**338.61**
100 mm dia.	319.01	360.28	2.76	65.39	nr	**425.67**
125 mm dia.	456.65	515.74	6.05	143.35	nr	**659.09**
150 mm dia.	525.46	593.46	8.03	190.27	nr	**783.73**
200 mm dia.	1000.87	1130.37	9.17	217.28	nr	**1347.65**
250 mm dia.	1540.04	1739.31	10.72	254.01	nr	**1993.32**
300 mm dia.	1826.60	2062.95	11.75	278.41	nr	**2341.36**
Cast iron gate valve; bronze trim; non rising stem; BS 5150, PN10; working pressure up to 8.4 bar for saturated steam, 10 bar from -10°C to 120°C; BS4504 flanged ends; bolted connections						
50 mm dia.	219.00	247.34	1.85	43.83	nr	**291.17**
65 mm dia.	272.16	307.38	2.00	47.39	nr	**354.77**
80 mm dia.	303.01	342.22	2.27	53.78	nr	**396.00**
100 mm dia.	403.36	455.55	2.76	65.39	nr	**520.94**
125 mm dia.	563.95	636.92	6.05	143.35	nr	**780.27**
150 mm dia.	665.35	751.44	8.03	190.27	nr	**941.71**
200 mm dia.	1222.45	1380.63	9.17	217.28	nr	**1597.91**
250 mm dia.	1838.47	2076.36	10.72	254.01	nr	**2330.37**
300 mm dia.	2068.82	2336.51	11.75	278.41	nr	**2614.92**
350 mm dia.	2406.27	2717.63	12.67	300.22	nr	**3017.85**

T: MECHANICAL/COOLING/HEATING SYSTEMS

Item	Net Price £	Material £	Labour hours	Labour £	Unit	Total rate £
Cast iron gate valve; bronze trim; non rising stem; BS 5163 series A, PN16; working pressure for cold water services up to 16 bar; BS4504 flanged ends; bolted connections						
50 mm dia.	219.00	247.34	1.85	43.83	nr	**291.17**
65 mm dia.	272.16	307.38	2.00	47.39	nr	**354.77**
80 mm dia.	303.01	342.22	2.27	53.78	nr	**396.00**
100 mm dia.	403.36	455.55	2.76	65.39	nr	**520.94**
125 mm dia.	563.95	636.92	6.05	143.35	nr	**780.27**
150 mm dia.	665.35	751.44	8.03	190.27	nr	**941.71**
Ball valves						
Malleable iron body; lever operated stainless steel ball and stem; working pressure up to 12 bar; flanged ends to BS 4504 16/11; bolted connections						
40 mm dia.	291.07	328.74	1.54	36.50	nr	**365.24**
50 mm dia.	366.78	414.24	1.64	38.86	nr	**453.10**
80 mm dia.	593.30	670.07	1.92	45.49	nr	**715.56**
100 mm dia.	613.43	692.80	2.80	66.34	nr	**759.14**
150 mm dia.	1133.87	1280.59	12.05	285.52	nr	**1566.11**
Malleable iron body; lever operated stainless steel ball and stem; working pressure up to 16 bar; screwed ends to steel						
20 mm dia.	34.57	39.05	1.34	31.76	nr	**70.81**
25 mm dia.	35.55	40.15	1.40	33.18	nr	**73.33**
32 mm dia.	48.97	55.30	1.46	34.64	nr	**89.94**
40 mm dia.	48.97	55.30	1.54	36.52	nr	**91.82**
50 mm dia.	58.57	66.15	1.64	38.90	nr	**105.05**
Carbon steel body; lever operated stainless steel ball and stem; Class 150; working pressure up to 19 bar; screwed ends to steel						
15 mm dia.	63.96	72.24	0.84	19.91	nr	**92.15**
20 mm dia.	71.90	81.20	1.14	27.02	nr	**108.22**
25 mm dia.	77.85	87.92	1.30	30.81	nr	**118.73**
Globe valves						
Bronze; rising stem; renewable disc; BS 5154 series B, PN32; working pressure up to 14 bar for saturated steam, 32 bar from -10°C to 100°C; screwed ends to steel						
15 mm dia.	35.22	39.78	0.77	18.24	nr	**58.02**
20 mm dia.	48.16	54.40	1.03	24.41	nr	**78.81**
25 mm dia.	69.09	78.03	1.19	28.20	nr	**106.23**
32 mm dia.	97.28	109.86	1.38	32.69	nr	**142.55**
40 mm dia.	135.47	153.00	1.62	38.39	nr	**191.39**
50 mm dia.	207.93	234.83	1.61	38.15	nr	**272.98**

T: MECHANICAL/COOLING/HEATING SYSTEMS

Item	Net Price £	Material £	Labour hours	Labour £	Unit	Total rate £
T31: LOW TEMPERATURE HOT WATER HEATING – cont						
VALVES – cont						
Globe valves – cont						
Bronze; needle valve; rising stem; BS 5154, series B, PN32; working pressure up to 14 bar for saturated steam, 32 bar from -10°C to 100°C; screwed ends to steel						
15 mm dia.	26.77	30.23	1.07	25.36	nr	**55.59**
20 mm dia.	45.32	51.18	1.18	27.97	nr	**79.15**
25 mm dia.	63.91	72.18	1.27	30.09	nr	**102.27**
32 mm dia.	133.02	150.24	1.35	31.99	nr	**182.23**
40 mm dia.	209.58	236.70	1.47	34.83	nr	**271.53**
50 mm dia.	265.44	299.79	1.61	38.15	nr	**337.94**
Bronze; rising stem; renewable disc; BS 5154, series B, PN16; working pressure up to 7 bar for saturated steam, 16 bar from -10°C to 100°C; BS4504 flanged ends; bolted connections						
15 mm dia.	71.16	80.37	1.16	27.49	nr	**107.86**
20 mm dia.	82.31	92.96	1.26	29.85	nr	**122.81**
25 mm dia.	144.33	163.01	1.38	32.69	nr	**195.70**
32 mm dia.	182.20	205.77	1.47	34.83	nr	**240.60**
40 mm dia.	207.10	233.90	1.56	36.96	nr	**270.86**
50 mm dia.	297.14	335.59	1.71	40.52	nr	**376.11**
Bronze; rising stem; renewable disc; BS 2060, class 250; working pressure up to 24 bar for saturated steam, 38 bar from -10°C to 100°C; flanged ends (BS 10 table H); bolted connections						
15 mm dia.	181.22	204.67	1.16	27.49	nr	**232.16**
20 mm dia.	210.61	237.86	1.26	29.85	nr	**267.71**
25 mm dia.	289.42	326.87	1.38	32.69	nr	**359.56**
32 mm dia.	378.08	427.00	1.47	34.83	nr	**461.83**
40 mm dia.	452.76	511.35	1.56	36.96	nr	**548.31**
50 mm dia.	704.02	795.12	1.71	40.52	nr	**835.64**
65 mm dia.	841.94	950.88	1.88	44.54	nr	**995.42**
80 mm dia.	1069.89	1208.33	2.03	48.10	nr	**1256.43**

T: MECHANICAL/COOLING/HEATING SYSTEMS

Item	Net Price £	Material £	Labour hours	Labour £	Unit	Total rate £
Check valves						
Bronze; swing pattern; BS 5154 series B, PN 25; working pressure up to 10.5 bar for saturated steam, 25 bar from -10°C to 100°C; screwed ends to steel						
15 mm dia.	16.21	18.31	0.77	18.24	nr	**36.55**
20 mm dia.	19.25	21.74	1.03	24.41	nr	**46.15**
25 mm dia.	26.70	30.15	1.19	28.20	nr	**58.35**
32 mm dia.	45.22	51.07	1.38	32.69	nr	**83.76**
40 mm dia.	56.27	63.55	1.62	38.39	nr	**101.94**
50 mm dia.	86.29	97.45	1.94	45.97	nr	**143.42**
65 mm dia.	179.74	203.00	2.45	58.05	nr	**261.05**
80 mm dia.	254.15	287.04	2.83	67.05	nr	**354.09**
Bronze; vertical lift pattern; BS 5154 series B, PN32; working pressure up to 14 bar for saturated steam, 32 bar from -10°C to 100°C; screwed ends to steel						
15 mm dia.	20.57	23.24	0.96	22.75	nr	**45.99**
20 mm dia.	29.25	33.04	1.07	25.36	nr	**58.40**
25 mm dia.	42.96	48.52	1.17	27.72	nr	**76.24**
32 mm dia.	65.43	73.90	1.33	31.51	nr	**105.41**
40 mm dia.	85.25	96.28	1.41	33.40	nr	**129.68**
50 mm dia.	127.75	144.28	1.55	36.73	nr	**181.01**
65 mm dia.	392.44	443.22	1.80	42.65	nr	**485.87**
80 mm dia.	586.69	662.60	1.99	47.15	nr	**709.75**
Bronze; oblique swing pattern; BS 5154 series A, PN32; working pressure up to 14 bar for saturated steam, 32 bar from -10°C to 120°C; screwed connections to steel						
15 mm dia.	26.12	29.50	0.96	22.75	nr	**52.25**
20 mm dia.	31.05	35.07	1.07	25.36	nr	**60.43**
25 mm dia.	43.03	48.59	1.17	27.72	nr	**76.31**
32 mm dia.	64.35	72.67	1.33	31.51	nr	**104.18**
40 mm dia.	82.95	93.69	1.41	33.40	nr	**127.09**
50 mm dia.	137.40	155.18	1.55	36.73	nr	**191.91**
Cast iron; swing pattern; BS 5153 PN6; working pressure up to 6 bar from -10°C to 120°C; BS 4504 flanged ends; bolted connections						
50 mm dia.	456.11	515.13	1.86	44.06	nr	**559.19**
65 mm dia.	499.39	564.01	2.00	47.39	nr	**611.40**
80 mm dia.	562.11	634.85	2.56	60.66	nr	**695.51**
100 mm dia.	720.07	813.25	2.76	65.39	nr	**878.64**
125 mm dia.	1074.25	1213.25	6.05	143.35	nr	**1356.60**
150 mm dia.	1207.86	1364.15	8.11	192.17	nr	**1556.32**
200 mm dia.	2671.16	3016.79	9.26	219.41	nr	**3236.20**
250 mm dia.	3338.86	3770.89	10.72	254.01	nr	**4024.90**
300 mm dia.	4006.65	4525.09	11.75	278.41	nr	**4803.50**

T: MECHANICAL/COOLING/HEATING SYSTEMS

Item	Net Price £	Material £	Labour hours	Labour £	Unit	Total rate £
T31: LOW TEMPERATURE HOT WATER HEATING – cont						
VALVES – cont						
Check valves – cont						
Cast iron; horizontal lift pattern; BS 5153 PN16; working pressure up to 13 bar for saturated steam, 16 bar from -10°C to 120°C; BS 4504 flanged ends; bolted connections						
50 mm dia.	411.88	465.17	1.86	44.06	nr	**509.23**
65 mm dia.	411.88	465.17	2.00	47.39	nr	**512.56**
80 mm dia.	509.95	575.93	2.56	60.66	nr	**636.59**
100 mm dia.	593.85	670.69	2.96	70.14	nr	**740.83**
125 mm dia.	886.01	1000.65	7.76	183.87	nr	**1184.52**
150 mm dia.	996.14	1125.04	10.50	248.80	nr	**1373.84**
Cast iron; semi lugged butterfly valve; BS5155 PN16; working pressure 16 bar from -10°C to 120°C; BS 4504 flanged ends; bolted connections						
50 mm dia.	132.23	149.34	2.20	52.13	nr	**201.47**
65 mm dia.	136.46	154.12	2.31	54.73	nr	**208.85**
80 mm dia.	160.75	181.55	2.88	68.24	nr	**249.79**
100 mm dia.	223.47	252.39	3.11	73.70	nr	**326.09**
125 mm dia.	326.16	368.36	5.02	118.95	nr	**487.31**
150 mm dia.	374.37	422.81	6.98	165.39	nr	**588.20**
200 mm dia.	597.61	674.93	8.25	195.48	nr	**870.41**
250 mm dia.	882.06	996.19	10.47	248.08	nr	**1244.27**
300 mm dia.	1323.11	1494.31	11.48	272.01	nr	**1766.32**
Commissioning valves						
Bronze commissioning set; metering station; double regulating valve; BS5154 PN20 Series B; working pressure 20 bar from -10°C to 100°C; screwed ends to steel						
15 mm dia.	69.00	77.93	1.08	25.59	nr	**103.52**
20 mm dia.	110.87	125.22	1.46	34.60	nr	**159.82**
25 mm dia.	131.84	148.90	1.68	39.81	nr	**188.71**
32 mm dia.	176.77	199.64	1.95	46.21	nr	**245.85**
40 mm dia.	256.36	289.53	2.27	53.78	nr	**343.31**
50 mm dia.	386.88	436.94	2.73	64.69	nr	**501.63**

T: MECHANICAL/COOLING/HEATING SYSTEMS

Item	Net Price £	Material £	Labour hours	Labour £	Unit	Total rate £
Cast iron commissioning set; metering station; double regulating valve; BS5152 PN16; working pressure 16 bar from -10°C to 90°C; flanged ends (BS 4504, Part 1, Table 16); bolted connections						
65 mm dia.	586.13	661.97	1.80	42.65	nr	704.62
80 mm dia.	701.83	792.64	2.56	60.66	nr	853.30
100 mm dia.	977.05	1103.48	2.30	54.49	nr	1157.97
125 mm dia.	1431.15	1616.33	2.44	57.81	nr	1674.14
150 mm dia.	1885.26	2129.20	2.90	68.72	nr	2197.92
200 mm dia.	4825.51	5449.91	8.26	195.72	nr	5645.63
250 mm dia.	6036.65	6817.76	10.49	248.55	nr	7066.31
300 mm dia.	6555.93	7404.23	11.49	272.25	nr	7676.48
Cast iron variable orifice double regulating valve; orifice valve; BS5152 PN16; working pressure 16 bar from -10° to 90°C; flanged ends (BS 4504, Part 1, Table 16); bolted connections						
65 mm dia.	429.13	484.65	2.00	47.39	nr	532.04
80 mm dia.	525.26	593.22	2.56	60.66	nr	653.88
100 mm dia.	720.83	814.10	2.96	70.14	nr	884.24
125 mm dia.	1079.44	1219.11	7.76	183.87	nr	1402.98
150 mm dia.	1386.51	1565.91	10.50	248.80	nr	1814.71
200 mm dia.	3633.89	4104.09	8.26	195.72	nr	4299.81
250 mm dia.	5525.31	6240.26	10.49	248.55	nr	6488.81
300 mm dia.	9911.29	11193.77	11.49	272.25	nr	11466.02
Cast iron globe valve with double regulating feature; BS5152 PN16; working pressure 16 bar from -10°C to 120°C; flanged ends (BS 4504, Part 1, Table 16); bolted connections						
65 mm dia.	511.21	577.36	2.00	47.39	nr	624.75
80 mm dia.	618.21	698.20	2.56	60.66	nr	758.86
100 mm dia.	814.37	919.74	2.96	70.14	nr	989.88
125 mm dia.	1135.45	1282.37	7.76	183.87	nr	1466.24
150 mm dia.	1437.06	1623.00	10.50	248.80	nr	1871.80
200 mm dia.	3720.43	4201.84	8.26	195.72	nr	4397.56
250 mm dia.	5628.92	6357.27	10.49	248.55	nr	6605.82
300 mm dia.	8843.84	9988.19	11.49	272.25	nr	10260.44
Bronze autoflow commissioning valve; PN25; working pressure 25 bar up to 100°C; screwed ends to steel						
15 mm dia.	78.24	88.36	0.82	19.42	nr	107.78
20 mm dia.	82.85	93.57	1.08	25.59	nr	119.16
25 mm dia.	103.16	116.50	1.27	30.09	nr	146.59
32 mm dia.	161.11	181.96	1.50	35.55	nr	217.51
40 mm dia.	176.42	199.25	1.76	41.70	nr	240.95
50 mm dia.	260.84	294.60	2.13	50.47	nr	345.07

T: MECHANICAL/COOLING/HEATING SYSTEMS

Item	Net Price £	Material £	Labour hours	Labour £	Unit	Total rate £
T31: LOW TEMPERATURE HOT WATER HEATING – cont						
VALVES – cont						
Commissioning valves – cont						
Ductile iron autoflow commissioning valves; PN16; working pressure 16 bar from -10°C to 120°C; for ANSI 150 flanged ends						
65 mm dia.	686.89	775.77	2.31	54.73	nr	**830.50**
80 mm dia.	759.45	857.72	2.88	68.24	nr	**925.96**
100 mm dia.	970.17	1095.70	3.11	73.70	nr	**1169.40**
150 mm dia.	1998.56	2257.17	6.98	165.39	nr	**2422.56**
200 mm dia.	2911.10	3287.79	8.26	195.72	nr	**3483.51**
250 mm dia.	4043.08	4566.23	10.49	248.55	nr	**4814.78**
300 mm dia.	5163.65	5831.80	11.49	272.25	nr	**6104.05**
Strainers						
Bronze strainer; Y type; PN32 ; working pressure 32 bar from -10°C to 100°C; screwed ends to steel						
15 mm dia.	31.36	35.41	0.82	19.42	nr	**54.83**
20 mm dia.	39.84	44.99	1.08	25.59	nr	**70.58**
25 mm dia.	52.84	59.68	1.27	30.09	nr	**89.77**
32 mm dia.	85.34	96.38	1.50	35.55	nr	**131.93**
40 mm dia.	110.38	124.66	1.76	41.70	nr	**166.36**
50 mm dia.	184.88	208.80	2.13	50.47	nr	**259.27**
Cast iron strainer; Y type; PN16; working pressure 16 bar from -10°C to 120°C; BS 4504 flanged ends						
65 mm dia.	195.05	220.29	2.31	54.73	nr	**275.02**
80 mm dia.	230.87	260.74	2.88	68.24	nr	**328.98**
100 mm dia.	337.79	381.50	3.11	73.70	nr	**455.20**
125 mm dia.	622.72	703.30	5.02	118.95	nr	**822.25**
150 mm dia.	798.95	902.33	6.98	165.39	nr	**1067.72**
200 mm dia.	1341.00	1514.52	8.26	195.72	nr	**1710.24**
250 mm dia.	1939.16	2190.07	10.49	248.55	nr	**2438.62**
300 mm dia.	3157.89	3566.50	11.49	272.25	nr	**3838.75**
Regulators						
Gunmetal; self-acting two port thermostatic regulator; single seat; screwed ends; complete with sensing element, 2 m long capillary tube						
15 mm dia.	546.98	617.75	1.37	32.46	nr	**650.21**
20 mm dia.	564.57	637.62	1.24	29.39	nr	**667.01**
25 mm dia.	579.39	654.36	1.34	31.74	nr	**686.10**

T: MECHANICAL/COOLING/HEATING SYSTEMS

Item	Net Price £	Material £	Labour hours	Labour £	Unit	Total rate £
Gunmetal; self-acting two port thermostatic regulator; double seat; flanged ends (BS 4504 PN25); with sensing element, 2 m long capillary tube; steel body						
65 mm dia.	2028.50	2290.98	1.23	33.60	nr	**2324.58**
80 mm dia.	2394.21	2704.01	1.62	44.26	nr	**2748.27**
Control valves; electrically operated (electrical work elsewhere)						
Cast iron; butterfly type; two position electrically controlled 240 V motor and linkage mechanism; for low pressure hot water; maximum pressure 6 bar at 120°C; flanged ends						
25 mm dia.	679.94	767.92	1.47	34.83	nr	**802.75**
32 mm dia.	702.56	793.47	1.52	36.02	nr	**829.49**
40 mm dia.	981.71	1108.74	1.61	38.15	nr	**1146.89**
50 mm dia.	1009.62	1140.26	1.71	40.52	nr	**1180.78**
65 mm dia.	1034.90	1168.81	2.51	59.48	nr	**1228.29**
80 mm dia.	1075.37	1214.51	2.69	63.74	nr	**1278.25**
100 mm dia.	1128.51	1274.53	2.81	66.58	nr	**1341.11**
125 mm dia.	1249.96	1411.70	2.94	69.67	nr	**1481.37**
150 mm dia.	1354.54	1529.81	3.33	78.90	nr	**1608.71**
200 mm dia.	1680.08	1897.48	3.67	86.97	nr	**1984.45**
Cast iron; three way 240 V motorized; for low pressure hot water; maximum pressure 6 bar 120°C; flanged ends, drilled (BS 10, Table F)						
25 mm dia.	759.93	858.26	1.99	47.15	nr	**905.41**
40 mm dia.	797.05	900.18	2.13	50.47	nr	**950.65**
50 mm dia.	786.92	888.75	3.21	76.06	nr	**964.81**
65 mm dia.	879.25	993.02	3.23	76.54	nr	**1069.56**
80 mm dia.	985.21	1112.69	3.50	82.94	nr	**1195.63**
Two port normally closed motorized valve; electric actuator; spring return; domestic usage						
22 mm dia.	107.74	121.69	1.18	27.97	nr	**149.66**
28 mm dia.	146.04	164.93	1.35	31.99	nr	**196.92**
Two port on/off motorized valve; electric actuator; spring return; domestic usage						
22 mm dia.	107.74	121.69	1.18	27.97	nr	**149.66**
Three port motorized valve; electric actuator; spring return; domestic usage						
22 mm dia.	157.21	177.55	1.18	27.97	nr	**205.52**

T: MECHANICAL/COOLING/HEATING SYSTEMS

Item	Net Price £	Material £	Labour hours	Labour £	Unit	Total rate £
T31: LOW TEMPERATURE HOT WATER HEATING – cont						
VALVES – cont						
Safety and relief valves						
Bronze relief valve; spring type; side outlet; working pressure up to 20.7 bar at 120°C; screwed ends to steel						
15 mm dia.	120.16	135.71	0.26	6.15	nr	**141.86**
20 mm dia.	131.78	148.83	0.36	8.53	nr	**157.36**
Bronze relief valve; spring type; side outlet; working pressure up to 17.2 bar at 120°C; screwed ends to steel						
25 mm dia.	180.23	203.55	0.38	9.00	nr	**212.55**
32 mm dia.	244.19	275.79	0.48	11.37	nr	**287.16**
Bronze relief valve; spring type; side outlet; working pressure up to 13.8 bar at 120°C; screwed ends to steel						
40 mm dia.	318.80	360.05	0.64	15.17	nr	**375.22**
50 mm dia.	415.70	469.49	0.76	18.00	nr	**487.49**
65 mm dia.	769.68	869.27	0.94	22.27	nr	**891.54**
80 mm dia.	1009.80	1140.46	1.10	26.07	nr	**1166.53**
Cocks; screwed joints to steel						
Bronze gland cock; complete with malleable iron lever; working pressure up to 10 bar at 100°C; screwed ends to steel						
15 mm dia.	42.22	47.68	0.77	18.24	nr	**65.92**
20 mm dia.	60.07	67.84	1.03	24.41	nr	**92.25**
25 mm dia.	80.95	91.42	1.19	28.20	nr	**119.62**
32 mm dia.	122.32	138.15	1.38	32.69	nr	**170.84**
40 mm dia.	170.79	192.89	1.62	38.39	nr	**231.28**
50 mm dia.	260.66	294.39	1.94	45.97	nr	**340.36**
Bronze three-way plug cock; complete with malleable iron lever; working pressure up to 10 bar at 100°C; screwed ends to steel						
15 mm dia.	85.60	96.68	0.77	18.24	nr	**114.92**
20 mm dia.	99.08	111.90	1.03	24.41	nr	**136.31**
25 mm dia.	138.51	156.44	1.19	28.20	nr	**184.64**
32 mm dia.	196.45	221.87	1.38	32.69	nr	**254.56**
40 mm dia.	236.85	267.50	1.62	38.39	nr	**305.89**

T: MECHANICAL/COOLING/HEATING SYSTEMS

Item	Net Price £	Material £	Labour hours	Labour £	Unit	Total rate £
Air vents; including regulating, adjusting and testing						
Automatic air vent; maximum pressure up to 7 bar at 93°C; screwed ends to steel						
15 mm dia.	10.78	12.18	0.80	18.95	nr	**31.13**
Automatic air vent; maximum pressure up to 7 bar at 93°C; lockhead isolating valve; screwed ends to steel						
15 mm dia.	10.78	12.18	0.83	19.67	nr	**31.85**
Automatic air vent; maximum pressure up to 17 bar at 200°C; flanged ends (BS10, Table H); bolted connections to counter flange (measured separately)						
15 mm dia.	387.50	437.64	0.83	19.67	nr	**457.31**
Radiator valves						
Bronze; wheelhead or lockshield; chromium plated finish; screwed joints to steel						
Straight						
15 mm dia.	13.75	15.52	0.59	13.98	nr	**29.50**
20 mm dia.	21.40	24.17	0.73	17.30	nr	**41.47**
25 mm dia.	26.86	30.33	0.85	20.13	nr	**50.46**
Angled						
15 mm dia.	46.52	52.54	0.59	13.98	nr	**66.52**
20 mm dia.	61.34	69.28	0.73	17.30	nr	**86.58**
25 mm dia.	78.86	89.07	0.85	20.13	nr	**109.20**
Bronze; wheelhead or lockshield; chromium plated finish; compression joints to copper						
Straight						
15 mm dia.	24.62	27.81	0.59	13.98	nr	**41.79**
20 mm dia.	31.03	35.05	0.73	17.30	nr	**52.35**
25 mm dia.	38.99	44.03	0.85	20.13	nr	**64.16**
Angled						
15 mm dia.	18.59	20.99	0.59	13.98	nr	**34.97**
20 mm dia.	19.50	22.02	0.73	17.30	nr	**39.32**
25 mm dia.	24.17	27.30	0.85	20.13	nr	**47.43**
Twin entry						
8 mm dia.	25.83	29.17	0.23	5.46	nr	**34.63**
10 mm dia.	29.31	33.10	0.23	5.46	nr	**38.56**

T: MECHANICAL/COOLING/HEATING SYSTEMS

Item	Net Price £	Material £	Labour hours	Labour £	Unit	Total rate £
T31: LOW TEMPERATURE HOT WATER HEATING – cont						
VALVES – cont						
Bronze; thermostatic head; chromium plated finish; compression joints to copper						
Straight						
15 mm dia.	10.17	11.49	0.59	13.98	nr	**25.47**
20 mm dia.	14.93	16.86	0.73	17.30	nr	**34.16**
Angled						
15 mm dia.	9.48	10.71	0.59	13.98	nr	**24.69**
20 mm dia.	13.89	15.69	0.73	17.30	nr	**32.99**
GAUGES						
Thermometers and pressure gauges						
Dial thermometer; coated steel case and dial; glass window; brass pocket; BS 5235; pocket length 100 mm; screwed end						
Back/bottom entry						
100 mm dia. face	51.82	58.53	0.81	19.20	nr	**77.73**
150 mm dia. face	84.89	95.87	0.81	19.20	nr	**115.07**
Dial pressure/altitude gauge; bronze bourdon tube type; coated steel case and dial; glass window BS 1780; screwed end						
100 mm dia. face	76.81	86.75	0.81	19.20	nr	**105.95**
150 mm dia. face	86.03	97.17	0.81	19.20	nr	**116.37**
EQUIPMENT						
PRESSURISATION UNITS						
LTHW pressurization unit complete with expansion vessel(s), interconnecting pipework and all necessary isolating and drain valves; includes placing in position; electrical work elsewhere. Selection based on a final working pressure of 4 bar, a 3 m static head and system operating temperatures of 82/71°C						
System volume						
2,400 litres	1906.95	2153.70	15.00	355.42	nr	**2509.12**
6,000–20,000 litres	2090.84	2361.38	22.00	521.28	nr	**2882.66**
25,000 litres	2595.91	2931.81	22.00	521.28	nr	**3453.09**

T: MECHANICAL/COOLING/HEATING SYSTEMS

Item	Net Price £	Material £	Labour hours	Labour £	Unit	Total rate £
DIRT SEPARATORS						
Dirt seperator; maximum operating temperature and pressure of 110°C and 10 bar; fitted with drain valve						
Bore size, flow rate (at 1.0 m/s velocity); threaded connections						
32 mm dia., 3.7 m³/h	72.65	82.05	2.29	54.25	nr	**136.30**
40 mm dia., 5.0 m³/h	86.48	97.67	2.45	58.05	nr	**155.72**
Bore size, flow rate (at 1.5 m/s velocity); flanged connections to PN16						
50 mm dia., 13.0 m³/h	488.92	552.19	3.00	71.08	nr	**623.27**
65 mm dia., 21.0 m³/h	509.67	575.62	3.00	71.08	nr	**646.70**
80 mm dia., 29.0 m³/h	717.23	810.03	3.84	90.98	nr	**901.01**
100 mm dia., 49.0 m³/h	758.75	856.93	4.44	105.20	nr	**962.13**
125 mm dia., 74.0 m³/h	1454.07	1642.22	11.64	275.81	nr	**1918.03**
150 mm dia., 109.0 m³/h	1516.34	1712.55	15.75	373.20	nr	**2085.75**
200 mm dia., 181.0 m³/h	2285.46	2581.19	15.75	373.20	nr	**2954.39**
250 mm dia., 288.0 m³/h	4050.88	4575.05	15.75	373.20	nr	**4948.25**
300 mm dia., 407.0 m³/h	5921.22	6687.40	17.24	408.50	nr	**7095.90**
MICROBUBBLE DEAERATORS						
Microbubble deaerator; maximum operating temperature and pressure of 110°C and 10 bar; fitted with drain valve						
Bore size, flow rate (at 1.0 m/s velocity); threaded connections						
32 mm dia., 3.7 m³/h	66.88	75.53	2.29	54.25	nr	**129.78**
40 mm dia., 5.0 m³/h	79.56	89.85	2.45	58.05	nr	**147.90**
Bore size, flow rate (at 1.5 m/s velocity); flanged connections to PN16						
50 mm dia., 13.0 m³/h	488.92	552.19	3.00	71.08	nr	**623.27**
65 mm dia., 21.0 m³/h	508.71	574.54	3.00	71.08	nr	**645.62**
80 mm dia., 29.0 m³/h	717.23	810.03	3.84	90.98	nr	**901.01**
100 mm dia., 49.0 m³/h	758.75	856.93	4.44	105.20	nr	**962.13**
125 mm dia., 74.0 m³/h	1454.07	1642.22	11.64	275.81	nr	**1918.03**
150 mm dia., 109.0 m³/h	1516.34	1712.55	15.75	373.20	nr	**2085.75**
200 mm dia., 181.0 m³/h	2285.46	2581.19	15.75	373.20	nr	**2954.39**
250 mm dia., 288.0 m³/h	4050.88	4575.05	15.75	373.20	nr	**4948.25**
300 mm dia., 407.0 m³/h	5941.98	6710.85	17.24	408.50	nr	**7119.35**

T: MECHANICAL/COOLING/HEATING SYSTEMS

Item	Net Price £	Material £	Labour hours	Labour £	Unit	Total rate £
T31: LOW TEMPERATURE HOT WATER HEATING – cont						
COMBINED MICROBUBBLE DEAERATORS AND DIRT SEPARATORS						
Combined deaerator and dirt separators; maximum operating temperature and pressure of 110°C and 10 bar; fitted with drain valve						
Bore size, flow rate (at 1.5 m/s velocity); threaded connections						
25 mm dia., 2.0 m³/h	92.25	104.18	2.75	65.17	nr	**169.35**
Bore size, flow rate (at 1.5 m/s velocity); flanged connections to PN16						
50 mm dia., 13.0 m³/h	623.83	704.55	3.60	85.30	nr	**789.85**
65 mm dia., 21.0 m³/h	644.59	727.99	3.60	85.30	nr	**813.29**
80 mm dia., 29.0 m³/h	872.90	985.85	4.61	109.23	nr	**1095.08**
100 mm dia., 49.0 m³/h	914.42	1032.74	5.33	126.29	nr	**1159.03**
125 mm dia., 74.0 m³/h	1744.66	1970.41	13.97	331.01	nr	**2301.42**
150 mm dia., 109.0 m³/h	1808.08	2042.04	18.90	447.83	nr	**2489.87**
200 mm dia., 181.0 m³/h	2804.36	3167.23	18.90	447.83	nr	**3615.06**
250 mm dia., 288.0 m³/h	4944.54	5584.34	18.90	447.83	nr	**6032.17**
300 mm dia., 407.0 m³/h	7582.85	8564.03	20.69	490.25	nr	**9054.28**
Y20 – PUMPS						
Centrifugal heating and chilled water pump; belt drive; 3 phase, 1450 rpm motor; max. pressure 1000 kN/m²; max. temperature 125°C; bed plate; coupling guard; bolted connections; supply only mating flanges;includes fixing on prepared concrete base; electrical work elsewhere						
40 mm pump size; 4.0 l/s at 70 kPa max. head; 0.25 kW max. motor rating	1167.56	1318.64	7.59	179.85	nr	**1498.49**
40 mm pump size; 4.0 l/s at 130 kPa max. head; 1.5 kW max. motor rating	1461.18	1650.25	8.09	191.69	nr	**1841.94**
50 mm pump size; 8.5 l/s at 90 kPa max. head; 2.2 kW max. motor rating	1506.88	1701.86	8.67	205.44	nr	**1907.30**
50 mm pump size; 8.5 l/s at 190 kPa max. head; 3 kW max. motor rating	2619.04	2957.93	11.20	265.38	nr	**3223.31**
50 mm pump size; 8.5 l/s at 215 kPa max. head; 4 kW max. motor rating	1711.87	1933.38	11.70	277.24	nr	**2210.62**
65 mm pump size; 14.0 l/s at 90 kPa max. head; 3 kW max. motor rating	1501.35	1695.62	11.70	277.24	nr	**1972.86**
65 mm pump size; 14.0 l/s at 160 kPa max. head; 4 kW max. motor rating	1664.78	1880.20	11.70	277.24	nr	**2157.44**

T: MECHANICAL/COOLING/HEATING SYSTEMS

Item	Net Price £	Material £	Labour hours	Labour £	Unit	Total rate £
80 mm pump size; 14.5 l/s at 210 kPa max. head; 5.5 kW max. motor rating	2553.95	2884.42	11.70	277.24	nr	**3161.66**
80 mm pump size; 22.0 l/s at 130 kPa max. head; 5.5 kW max. motor rating	2553.95	2884.42	13.64	323.20	nr	**3207.62**
80 mm pump size; 22.0 l/s at 200 kPa max. head; 7.5 kW max. motor rating	2674.44	3020.50	13.64	323.20	nr	**3343.70**
100 mm pump size; 22.0 l/s at 250 kPa max. head; 11 kW max. motor rating	3753.36	4239.03	13.64	323.20	nr	**4562.23**
100 mm pump size; 30.0 l/s at 100 kPa max. head; 4.0 kW max. motor rating	2101.05	2372.92	19.15	453.76	nr	**2826.68**
100 mm pump size; 36.0 l/s at 250 kPa max. head; 15.0 kW max. motor rating	3974.96	4489.30	19.15	453.76	nr	**4943.06**
100 mm pump size; 36.0 l/s at 550 kPa max. head; 30.0 kW max. motor rating	5394.59	6092.62	19.15	453.76	nr	**6546.38**
Centrifugal heating and chilled water pump; TWIN HEAD BELT DRIVE; 3 phase, 1450 rpm motor; max. pressure 1000 kN/m²; max. temperature 125°C; bed plate; coupling guard; bolted connections; supply only mating flanges; includes fixing on prepared concrete base; electrical work elsewhere						
40 mm pump size; 4.0 l/s at 70 kPa max. head; 0.75 kW max. motor rating	2501.32	2824.98	7.59	179.85	nr	**3004.83**
40 mm pump size; 4.0 l/s at 130 kPa max. head; 1.5 kW max. motor rating	3010.99	3400.60	8.09	191.69	nr	**3592.29**
50 mm pump size; 8.5 l/s at 90 kPa max. head; 2.2 kW max. motor rating	3076.10	3474.13	8.67	205.44	nr	**3679.57**
50 mm pump size; 8.5 l/s at 190 kPa max. head; 4 kW max. motor rating	3545.61	4004.40	11.20	265.38	nr	**4269.78**
65 mm pump size; 8.5 l/s at 215 kPa max. head; 4 kW max. motor rating	3697.97	4176.47	11.70	277.24	nr	**4453.71**
65 mm pump size; 14.0 l/s at 90 kPa max. head; 3 kW max. motor rating	3311.55	3740.04	11.70	277.24	nr	**4017.28**
65 mm pump size; 14.0 l/s at 160 kPa max. head; 4 kW max. motor rating	3697.97	4176.47	11.70	277.24	nr	**4453.71**
80 mm pump size; 14.5 l/s at 210 kPa max. head; 7.5 kW max. motor rating	5555.25	6274.07	13.64	323.20	nr	**6597.27**

T: MECHANICAL/COOLING/HEATING SYSTEMS

Item	Net Price £	Material £	Labour hours	Labour £	Unit	Total rate £
T31: LOW TEMPERATURE HOT WATER HEATING – cont						
Y20 – PUMPS – cont						
Centrifugal heating and chilled water pump; CLOSE COUPLED; 3 phase, 1450 rpm motor; max. pressure 1000kN/m²; max. temperature 110°C; bed plate; coupling guard; bolted connections; supply only mating flanges; includes fixing on prepared concrete base; electrical work elsewhere						
40 mm pump size; 4.0 l/s at 23 kPa max. head; 0.55 kW max. motor rating	702.20	793.06	7.31	173.21	nr	**966.27**
50 mm pump size; 4.0 l/s at 75 kPa max. head; 0.75 kW max. motor rating	778.38	879.10	7.31	173.21	nr	**1052.31**
50 mm pump size; 7.0 l/s at 65 kPa max. head; 0.75 kW max. motor rating	778.38	879.10	8.01	189.79	nr	**1068.89**
65 mm pump size; 10.0 l/s at 33 kPa max. head; 0.75 kW max. motor rating	882.26	996.42	8.01	189.79	nr	**1186.21**
50 mm pump size; 4.0 l/s at 120 kPa max. head; 1.5 kW max. motor rating	1062.30	1199.75	8.01	189.79	nr	**1389.54**
80 mm pump size; 16.0 l/s at 80 kPa max. head; 2.2 kW max. motor rating	1375.31	1553.27	12.35	292.63	nr	**1845.90**
80 mm pump size; 16.0 l/s at 120 kPa max. head; 4.0 kW max. motor rating	1340.68	1514.16	12.35	292.63	nr	**1806.79**
100 mm pump size; 28.0 l/s at 40 kPa max. head; 2.2 kW max. motor rating	1358.69	1534.50	17.86	423.19	nr	**1957.69**
100 mm pump size; 28.0 l/s at 90 kPa max. head; 4.0 kW max. motor rating	1499.95	1694.04	17.86	423.19	nr	**2117.23**
125 mm pump size; 40.0 l/s at 50 kPa max. head; 3.0 kW max. motor rating	1430.71	1615.83	25.85	612.50	nr	**2228.33**
125 mm pump size; 40.0 l/s at 120 kPa max. head; 7.5 kW max. motor rating	1765.88	1994.38	25.85	612.50	nr	**2606.88**
150 mm pump size; 70.0 l/s at 75 kPa max. head; 11 kW max. motor rating	2595.50	2931.35	30.43	721.03	nr	**3652.38**
150 mm pump size; 70.0 l/s at 120 kPa max. head; 15.0 kW max. motor rating	2774.16	3133.12	30.43	721.03	nr	**3854.15**
150 mm pump size; 70.0 l/s at 150 kPa max. head; 15.0 kW max. motor rating	2774.16	3133.12	30.43	721.03	nr	**3854.15**

T: MECHANICAL/COOLING/HEATING SYSTEMS

Item	Net Price £	Material £	Labour hours	Labour £	Unit	Total rate £
Centrifugal heating and chilled water pump; close coupled; 3 phase, VARIABLE SPEED motor; max. system pressure 1000 kN/m²; max. temperature 110°C; bed plate; coupling guard; bolted connections; supply only mating flanges; includes fixing on prepared concrete base; electrical work elsewhere						
40 mm pump size; 4.0 l/s at 23 kPa max. head; 0.55 kW max. motor rating	1267.28	1431.26	7.31	173.21	nr	**1604.47**
40 mm pump size; 4.0 l/s at 75 kPa max. head; 0.75 kW max. motor rating	1398.86	1579.87	7.31	173.21	nr	**1753.08**
50 mm pump size; 7.0 l/s at 65 kPa max. head; 1.5 kW max. motor rating	1716.02	1938.06	8.01	189.79	nr	**2127.85**
50 mm pump size; 10.0 l/s at 33 kPa max. head; 1.5 kW max. motor rating	1716.02	1938.06	8.01	189.79	nr	**2127.85**
50 mm pump size; 4.0 l/s at 120 kPa max. head; 1.5 kW max. motor rating	1716.02	1938.06	8.01	189.79	nr	**2127.85**
80 mm pump size; 16.0 l/s at 80 kPa max. head; 2.2 kW max. motor rating	2236.78	2526.20	12.35	292.63	nr	**2818.83**
80 mm pump size; 16.0 l/s at 120 kPa max. head; 3.0 kW max. motor rating	2412.68	2724.87	12.35	292.63	nr	**3017.50**
100 mm pump size; 28.0 l/s at 40 kPa max. head; 2.2 kW max. motor rating	2292.18	2588.78	17.86	423.19	nr	**3011.97**
100 mm pump size; 28.0 l/s at 90 kPa max. head; 4.0.kW max. motor rating	2734.00	3087.76	17.86	423.19	nr	**3510.95**
125 mm pump size; 40.0 l/s at 50 kPa max. head; 3.0 kW max. motor rating	2624.59	2964.20	25.85	612.50	nr	**3576.70**
125 mm pump size; 40.0 l/s at 120 kPa max. head; 7.5 kW max. motor rating	3693.80	4171.76	25.85	612.50	nr	**4784.26**
150 mm pump size; 70.0 l/s at 75 kPa max. head; 7.5 kW max. motor rating	4269.97	4822.48	30.43	721.03	nr	**5543.51**
Glandless domestic heating pump; for low pressure domestic hot water heating systems; 240 volt; 50 Hz electric motor; max. working pressure 1000 N/m² and max. temperature of 130°C; includes fixing in position; electrical work elsewhere						
1' BSP unions – 2 speed	126.32	142.67	1.58	37.44	nr	**180.11**
1.25' BSP unions – 3 speed	186.14	210.22	1.58	37.44	nr	**247.66**
Glandless pumps; for hot water secondary supply; silent running; 3 phase; max. pressure 1000kN/m²; max. temperature 130°C ; bolted connections; supply only mating flanges; including fixing in position; electrical elsewhere						
1' BSP unions – 3 speed	245.15	276.87	1.58	37.44	nr	**314.31**

T: MECHANICAL/COOLING/HEATING SYSTEMS

Item	Net Price £	Material £	Labour hours	Labour £	Unit	Total rate £
T31: LOW TEMPERATURE HOT WATER HEATING – cont						
Y20 – PUMPS – cont						
Pipeline mounted circulator; for heating and chilled water; silent running; 3 phase; 1450 rpm motor; max. pressure 1000 kN/m²; max. temperature 120°C; bolted connections; supply only mating flanges; includes fixing in position; electrical elsewhere						
32 mm pump size; 2.0 l/s at 17 kPa max. head; 0.2 kW max. motor rating	490.29	553.73	6.44	152.59	nr	**706.32**
50 mm pump size; 3.0 l/s at 20 kPa max. head; 0.2 kW max. motor rating	487.52	550.61	6.86	162.55	nr	**713.16**
65 mm pump size; 5.0 l/s at 30 kPa max. head; 0.37 kW max. motor rating	775.61	875.97	7.48	177.24	nr	**1053.21**
65 mm pump size; 8.0 l/s at 37 kPa max. head; 0.75 kW max. motor rating	868.40	980.77	7.48	177.24	nr	**1158.01**
80 mm pump size; 12.0 l/s at 42 kPa max. head; 1.1 kW max. motor rating	1009.67	1140.32	8.01	189.79	nr	**1330.11**
100 mm pump size; 25.0 l/s at 37 kPa max. head; 2.2 kW max. motor rating	1439.02	1625.22	9.11	215.86	nr	**1841.08**
Dual pipeline mounted circulator; for heating and chilled water; silent running; 3 phase; 1450 rpm motor; max. pressure 1000 kN/m²; max. temperature 120°C; bolted connections; supply only mating flanges; includes fixing in position; electrical work elsewhere						
40 mm pump size; 2.0 l/s at 17 kPa max. head; 0.8 kW max. motor rating	904.42	1021.45	7.88	186.71	nr	**1208.16**
50 mm pump size; 3.0 l/s at 20 kPa max. head; 0.2 kW max. motor rating	907.18	1024.57	8.01	189.79	nr	**1214.36**
65 mm pump size; 5.0 l/s at 30 kPa max. head; 0.37 kW max. motor rating	1472.26	1662.76	9.20	217.99	nr	**1880.75**
65 mm pump size; 8.0 l/s at 37 kPa max. head; 0.75 kW max. motor rating	1702.17	1922.42	9.20	217.99	nr	**2140.41**
100 mm pump size; 12.0 l/s at 42 kPa max. head; 1.1 kW max. motor rating	1925.16	2174.26	9.45	223.91	nr	**2398.17**

T: MECHANICAL/COOLING/HEATING SYSTEMS

Item	Net Price £	Material £	Labour hours	Labour £	Unit	Total rate £
Glandless accelerator pumps; for low and medium pressure heating services; silent running; 3 phase; 1450 rpm motor; max. pressure 1000 kN/m²; max. temperature 130°C; bolted connections; supply only mating flanges; includes fixing in position; electrical work elsewhere						
40 mm pump size; 4.0 l/s at 15 kPa max. head; 0.35 kW max. motor rating	400.62	452.46	6.94	164.44	nr	**616.90**
50 mm pump size; 6.0 l/s at 20 kPa max. head; 0.45 kW max. motor rating	426.95	482.19	7.35	174.15	nr	**656.34**
80 mm pump size; 13.0 l/s at 28 kPa max. head; 0.58 kW max. motor rating	841.31	950.17	7.76	183.87	nr	**1134.04**
Y22 – HEAT EXCHANGERS						
Plate heat exchanger; for use in LTHW systems; painted carbon steel frame; stainless steel plates, nitrile rubber gaskets; design pressure of 10 bar and operating temperature of 110/135°C						
Primary side; 80°C in, 69°C out; secondary side; 82°C in, 71°C out						
107 KW, 2.38 l/s	1850.79	2090.28	10.00	236.95	nr	**2327.23**
245 kW, 5.46 l/s	2844.01	3212.01	10.00	236.95	nr	**3448.96**
287 kW, 6.38 l/s	3157.00	3565.50	10.00	236.95	nr	**3802.45**
328 kW, 7.31 l/s	3444.80	3890.53	10.00	236.95	nr	**4127.48**
364 kW, 8.11 l/s	4088.21	4617.20	10.00	236.95	nr	**4854.15**
403 kW, 8.96 l/s	4291.70	4847.02	10.00	236.95	nr	**5083.97**
453 kW, 10.09 l/s	4465.15	5042.92	10.00	236.95	nr	**5279.87**
490 kW, 10.89 l/s	4628.91	5227.87	10.00	236.95	nr	**5464.82**
1000 kW, 21.7 l/s	3443.83	3889.44	12.00	284.34	nr	**4173.78**
1500 kW, 32.6 l/s	4387.63	4955.36	12.00	284.34	nr	**5239.70**
2000 kW, 43.4 l/s	5263.61	5944.69	15.00	355.42	nr	**6300.11**
2500 kW, 54.3 l/s	6470.01	7307.20	15.00	355.42	nr	**7662.62**
Note: For temperature conditions different to those above, the cost of the units can vary significantly, and so manufacturers advice should be sought.						

T: MECHANICAL/COOLING/HEATING SYSTEMS

Item	Net Price £	Material £	Labour hours	Labour £	Unit	Total rate £
T31: LOW TEMPERATURE HOT WATER HEATING – cont						
Y23 – CALORIFIERS						
Non-storage calorifiers; mild steel; heater battery duty 116°C/90°C to BS 853, maximum test on shell 11.55 bar, tubes 26.25 bar						
Horizontal or vertical; primary water at 116°C on, 90°C off						
40 kW capacity	769.37	868.92	3.00	71.08	nr	**940.00**
88 kW capacity	896.33	1012.31	5.00	118.47	nr	**1130.78**
176 kW capacity	1027.14	1160.05	7.04	166.87	nr	**1326.92**
293 kW capacity	1424.43	1608.74	9.01	213.47	nr	**1822.21**
586 kW capacity	2122.11	2396.70	22.22	526.54	nr	**2923.24**
879 kW capacity	2955.45	3337.87	28.57	676.99	nr	**4014.86**
1465 kW capacity	4748.10	5362.48	50.00	1184.74	nr	**6547.22**
2000 kW capacity	8642.17	9760.42	60.00	1421.69	nr	**11182.11**
HEAT EMITTERS						
Perimeter convector heating; metal casing with standard finish; aluminium extruded grille; including backplates						
Top/sloping/flat front outlet						
60 × 200 mm	30.86	34.85	2.00	47.39	m	**82.24**
60 × 300 mm	33.57	37.91	2.00	47.39	m	**85.30**
60 × 450 mm	41.61	46.99	2.00	47.39	m	**94.38**
60 × 525 mm	44.29	50.02	2.00	47.39	m	**97.41**
60 × 600 mm	48.31	54.56	2.00	47.39	m	**101.95**
90 × 260 mm	33.57	37.91	2.00	47.39	m	**85.30**
90 × 300 mm	34.89	39.40	2.00	47.39	m	**86.79**
90 × 450 mm	42.97	48.53	2.00	47.39	m	**95.92**
90 × 525 mm	46.99	53.07	2.00	47.39	m	**100.46**
90 × 600 mm	49.67	56.10	2.00	47.39	m	**103.49**
Extra over for dampers						
Damper	14.76	16.67	0.25	5.93	nr	**22.60**
Extra over for fittings						
60 mm End caps	11.81	13.34	0.25	5.93	nr	**19.27**
90 mm End caps	18.59	20.99	0.25	5.93	nr	**26.92**
60 mm Corners	25.11	28.36	0.25	5.93	nr	**34.29**
90 mm Corners	36.92	41.70	0.25	5.93	nr	**47.63**

T: MECHANICAL/COOLING/HEATING SYSTEMS

Item	Net Price £	Material £	Labour hours	Labour £	Unit	Total rate £
Radiant Strip Heaters						
Suitable for connection to hot water system; aluminium sheet panels with steel pipe clamped to upper surface; including insulation, sliding brackets, cover plates, end closures; weld or screwed BSP ends						
One pipe						
1500 mm long	62.41	70.48	3.11	73.70	nr	**144.18**
3000 mm long	98.72	111.50	3.11	73.70	nr	**185.20**
4500 mm long	133.91	151.24	3.11	73.70	nr	**224.94**
6000 mm long	184.59	208.48	3.11	73.70	nr	**282.18**
Two pipe						
1500 mm long	116.52	131.60	4.15	98.34	nr	**229.94**
3000 mm long	184.05	207.87	4.15	98.34	nr	**306.21**
4500 mm long	251.34	283.87	4.15	98.34	nr	**382.21**
6000 mm long	338.62	382.44	4.15	98.34	nr	**480.78**
Pressed steel panel type radiators; fixed with and including brackets; taking down once for decoration; refixing						
300 mm high; single panel						
500 mm length	17.04	19.25	2.03	48.10	nr	**67.35**
1000 mm length	34.05	38.45	2.03	48.10	nr	**86.55**
1500 mm length	44.61	50.38	2.03	48.10	nr	**98.48**
2000 mm length	50.05	56.53	2.47	58.53	nr	**115.06**
2500 mm length	55.48	62.66	2.97	70.37	nr	**133.03**
3000 mm length	66.39	74.98	3.22	76.31	nr	**151.29**
300 mm high; double panel; convector						
500 mm length	32.75	36.99	2.13	50.47	nr	**87.46**
1000 mm length	65.55	74.03	2.13	50.47	nr	**124.50**
1500 mm length	98.31	111.03	2.13	50.47	nr	**161.50**
2000 mm length	131.10	148.06	2.57	60.89	nr	**208.95**
2500 mm length	163.85	185.05	3.07	72.75	nr	**257.80**
3000 mm length	196.64	222.08	3.31	78.43	nr	**300.51**
450 mm high; single panel						
500 mm length	15.89	17.94	2.08	49.29	nr	**67.23**
1000 mm length	31.81	35.92	2.08	49.29	nr	**85.21**
1600 mm length	50.89	57.48	2.53	59.95	nr	**117.43**
2000 mm length	63.62	71.85	2.97	70.37	nr	**142.22**
2400 mm length	76.33	86.21	3.47	82.22	nr	**168.43**
3000 mm length	95.42	107.76	3.82	90.51	nr	**198.27**

T: MECHANICAL/COOLING/HEATING SYSTEMS

Item	Net Price £	Material £	Labour hours	Labour £	Unit	Total rate £
T31: LOW TEMPERATURE HOT WATER HEATING – cont						
HEAT EMITTERS – cont						
Pressed steel panel type radiators – cont						
450 mm high; double panel; convector						
500 mm length	29.13	32.90	2.18	51.66	nr	**84.56**
1000 mm length	58.25	65.79	2.18	51.66	nr	**117.45**
1600 mm length	106.68	120.48	2.63	62.32	nr	**182.80**
2000 mm length	179.34	202.54	3.06	72.51	nr	**275.05**
2400 mm length	215.22	243.07	3.37	79.85	nr	**322.92**
3000 mm length	269.02	303.83	3.92	92.88	nr	**396.71**
600 mm high; single panel						
500 mm length	21.30	24.05	2.18	51.66	nr	**75.71**
1000 mm length	42.61	48.12	2.43	57.58	nr	**105.70**
1600 mm length	68.16	76.98	3.13	74.16	nr	**151.14**
2000 mm length	85.19	96.22	3.77	89.32	nr	**185.54**
2400 mm length	102.25	115.48	4.07	96.44	nr	**211.92**
3000 mm length	127.81	144.35	5.11	121.08	nr	**265.43**
600 mm high; double panel; convector						
500 mm length	36.68	41.42	2.28	54.02	nr	**95.44**
1000 mm length	73.34	82.83	2.28	54.02	nr	**136.85**
1600 mm length	134.34	151.73	3.23	76.54	nr	**228.27**
2000 mm length	225.84	255.06	3.87	91.70	nr	**346.76**
2400 mm length	271.01	306.08	4.17	98.81	nr	**404.89**
3000 mm length	338.75	382.58	5.24	124.16	nr	**506.74**
700 mm high; single panel						
500 mm length	24.92	28.14	2.23	52.85	nr	**80.99**
1000 mm length	49.83	56.27	2.83	67.05	nr	**123.32**
1600 mm length	79.73	90.05	3.73	88.38	nr	**178.43**
2000 mm length	99.67	112.57	4.46	105.68	nr	**218.25**
2400 mm length	119.62	135.10	4.48	106.15	nr	**241.25**
3000 mm length	149.50	168.84	5.24	124.16	nr	**293.00**
700 mm high; double panel; convector						
500 mm length	47.67	53.84	2.33	55.20	nr	**109.04**
1000 mm length	128.24	144.83	3.08	72.98	nr	**217.81**
1600 mm length	205.17	231.71	3.83	90.75	nr	**322.46**
2000 mm length	256.46	289.65	4.17	98.81	nr	**388.46**
2400 mm length	307.76	347.59	4.37	103.54	nr	**451.13**
3000 mm length	384.69	434.47	4.82	114.21	nr	**548.68**

T: MECHANICAL/COOLING/HEATING SYSTEMS

Item	Net Price £	Material £	Labour hours	Labour £	Unit	Total rate £
Flat panel type steel radiators; fixed with and including brackets; taking down once for decoration; refixing						
300 mm high; single panel (44 mm deep)						
500 mm length	143.24	161.77	2.03	48.10	nr	**209.87**
1000 mm length	215.04	242.86	2.03	48.10	nr	**290.96**
1500 mm length	286.84	323.95	2.03	48.10	nr	**372.05**
2000 mm length	358.65	405.06	2.47	58.53	nr	**463.59**
2400 mm length	416.09	469.93	2.97	70.37	nr	**540.30**
3000 mm length	620.28	700.55	3.22	76.31	nr	**776.86**
300 mm high; double panel (100 mm deep)						
500 mm length	330.66	373.44	2.03	48.10	nr	**421.54**
1000 mm length	518.45	585.53	2.03	48.10	nr	**633.63**
1500 mm length	706.25	797.64	2.03	48.10	nr	**845.74**
2000 mm length	894.04	1009.73	2.47	58.53	nr	**1068.26**
2400 mm length	1044.27	1179.40	2.97	70.37	nr	**1249.77**
3000 mm length	1387.65	1567.21	3.22	76.31	nr	**1643.52**
500 mm high; single panel (44 mm deep)						
500 mm length	157.05	177.37	2.13	50.47	nr	**227.84**
1000 mm length	242.66	274.06	2.13	50.47	nr	**324.53**
1500 mm length	328.27	370.75	2.13	50.47	nr	**421.22**
2000 mm length	413.88	467.44	2.57	60.89	nr	**528.33**
2400 mm length	482.37	544.78	3.07	72.75	nr	**617.53**
3000 mm length	703.13	794.11	3.31	78.43	nr	**872.54**
500 mm high; double panel (100 mm deep)						
500 mm length	383.13	432.70	2.08	49.29	nr	**481.99**
1000 mm length	623.40	704.07	2.08	49.29	nr	**753.36**
1500 mm length	863.66	975.42	2.53	59.95	nr	**1035.37**
2000 mm length	1103.92	1246.77	2.97	70.37	nr	**1317.14**
2400 mm length	1296.13	1463.84	3.47	82.22	nr	**1546.06**
3000 mm length	1702.47	1922.76	3.82	90.51	nr	**2013.27**
600 mm high; single panel (44 mm deep)						
500 mm length	185.40	209.39	2.18	51.66	nr	**261.05**
1000 mm length	290.34	327.91	2.18	51.66	nr	**379.57**
1500 mm length	395.28	774.34	2.63	62.32	nr	**836.66**
2000 mm length	500.23	564.96	3.06	72.51	nr	**637.47**
2400 mm length	584.18	659.77	3.37	79.85	nr	**739.62**
3000 mm length	828.14	935.30	3.92	92.88	nr	**1028.18**
600 mm high; double panel (100 mm deep)						
500 mm length	437.08	493.64	2.18	51.66	nr	**545.30**
1000 mm length	713.24	805.52	2.43	57.58	nr	**863.10**
1500 mm length	989.41	1117.43	3.13	74.16	nr	**1191.59**
2000 mm length	1265.57	1429.33	3.77	89.32	nr	**1518.65**
2400 mm length	1486.50	1678.85	4.07	96.44	nr	**1775.29**
3000 mm length	1935.93	2186.43	5.11	121.08	nr	**2307.51**

T: MECHANICAL/COOLING/HEATING SYSTEMS

Item	Net Price £	Material £	Labour hours	Labour £	Unit	Total rate £
T31: LOW TEMPERATURE HOT WATER HEATING – cont						
HEAT EMITTERS – cont						
Flat panel type steel radiators – cont						
700 mm high; single panel (44 mm deep)						
500 mm length	196.45	221.87	2.28	54.02	nr	**275.89**
1000 mm length	312.43	352.86	2.28	54.02	nr	**406.88**
1500 mm length	428.42	483.86	3.23	76.54	nr	**560.40**
2000 mm length	544.41	614.86	3.87	91.70	nr	**706.56**
2400 mm length	637.20	719.65	4.17	98.81	nr	**818.46**
3000 mm length	894.42	1010.16	5.24	124.16	nr	**1134.32**
700 mm high; double panel (100 mm deep)						
500 mm length	472.98	534.18	2.23	52.85	nr	**587.03**
1000 mm length	785.04	886.62	2.83	67.05	nr	**953.67**
1500 mm length	1097.11	1239.08	3.73	88.38	nr	**1327.46**
2000 mm length	1409.18	1591.52	4.46	105.68	nr	**1697.20**
2400 mm length	1658.83	1873.47	4.48	106.15	nr	**1979.62**
3000 mm length	1867.27	2108.88	5.24	124.16	nr	**2233.04**
Fan convector; sheet metal casing with lockable access panel; centrifugal fan; air filter; LPHW heating coil; extruded aluminium grilles; 3 speed; includes fixing in position; electrical work elsewhere						
Free standing flat top, 695 mm high, medium speed rating						
Entering air temperature, 18°C						
695 mm long, 1 row 1.94 kW, 75 l/sec	613.45	692.82	2.73	64.69	nr	**757.51**
695 mm long, 2 row 2.64 kW, 75 l/sec	613.45	692.82	2.73	64.69	nr	**757.51**
895 mm long, 1 row 4.02 kW, 150 l/sec	691.27	780.72	2.73	64.69	nr	**845.41**
895 mm long, 2 row 5.62 kW, 150 l/sec	691.27	780.72	2.73	64.69	nr	**845.41**
1195 mm long, 1 row 6.58 kW, 250 l/sec	787.01	888.85	3.00	71.08	nr	**959.93**
1195 mm long, 2 row 9.27 kW, 250 l/sec	787.01	888.85	3.00	71.08	nr	**959.93**
1495 mm long, 1 row 9.04 kW, 340 l/sec	878.29	991.94	3.26	77.23	nr	**1069.17**
1495 mm long, 2 row 12.73 kW, 340 l/sec	878.29	991.94	3.26	77.23	nr	**1069.17**
Free standing flat top, 695 mm high, medium speed rating, c/w floor plinth						
695 mm long, 1 row 1.94 kW, 75 l/sec	644.12	727.46	2.73	64.69	nr	**792.15**
695 mm long, 2 row 2.64 kW, 75 l/sec	644.12	727.46	2.73	64.69	nr	**792.15**
895 mm long, 1 row 4.02 kW, 150 l/sec	725.82	819.73	2.73	64.69	nr	**884.42**
895 mm long, 2 row 5.62 kW, 150 l/sec	725.82	819.73	2.73	64.69	nr	**884.42**
1195 mm long, 1 row 6.58 kW, 250 l/sec	826.37	933.30	3.00	71.08	nr	**1004.38**
1195 mm long, 2 row 9.27 kW, 250 l/sec	826.37	933.30	3.00	71.08	nr	**1004.38**
1495 mm long, 1 row 9.04 kW, 340 l/sec	922.20	1041.53	3.26	77.23	nr	**1118.76**
1495 mm long, 2 row 12.73 kW, 340 l/sec	922.20	1041.53	3.26	77.23	nr	**1118.76**

T: MECHANICAL/COOLING/HEATING SYSTEMS

Item	Net Price £	Material £	Labour hours	Labour £	Unit	Total rate £
Free standing sloping top, 695 mm high, medium speed rating, c/w floor plinth						
695 mm long, 1 row 1.94 kW, 75 l/sec	666.57	752.82	2.73	64.69	nr	**817.51**
695 mm long, 2 row 2.64 kW, 75 l/sec	666.57	752.82	2.73	64.69	nr	**817.51**
895 mm long, 1 row 4.02 kW, 150 l/sec	748.26	845.08	2.73	64.69	nr	**909.77**
895 mm long, 2 row 5.62 kW, 150 l/sec	748.26	845.08	2.73	64.69	nr	**909.77**
1195 mm long, 1 row 6.58 kW, 250 l/sec	848.81	958.64	3.00	71.08	nr	**1029.72**
1195 mm long, 2 row 9.27 kW, 250 l/sec	848.81	958.64	3.00	71.08	nr	**1029.72**
1495 mm long, 1 row 9.04 kW, 340 l/sec	944.64	1066.87	3.26	77.23	nr	**1144.10**
1495 mm long, 2 row 12.73 kW, 340 l/sec	944.64	1066.87	3.26	77.23	nr	**1144.10**
Wall mounted high level sloping discharge						
695 mm long, 1 row 1.94 kW, 75 l/sec	673.30	760.42	2.73	64.69	nr	**825.11**
695 mm long, 2 row 2.64 kW, 75 l/sec	673.30	760.42	2.73	64.69	nr	**825.11**
895 mm long, 1 row 4.02 kW, 150 l/sec	692.77	782.41	2.73	64.69	nr	**847.10**
895 mm long, 2 row 5.62 kW, 150 l/sec	692.77	782.41	2.73	64.69	nr	**847.10**
1195 mm long, 1 row 6.58 kW, 250 l/sec	843.11	952.20	3.00	71.08	nr	**1023.28**
1195 mm long, 2 row 9.27 kW, 250 l/sec	843.11	952.20	3.00	71.08	nr	**1023.28**
1495 mm long, 1 row 9.04 kW, 340 l/sec	914.18	1032.47	3.26	77.23	nr	**1109.70**
1495 mm long, 2 row 12.73 kW, 340 l/sec	914.18	1032.47	3.26	77.23	nr	**1109.70**
Ceiling mounted sloping inlet/outlet 665 mm wide						
895 mm long, 1 row 4.02 kW, 150 l/sec	761.58	860.13	4.15	98.34	nr	**958.47**
895 mm long, 2 row 5.62 kW, 150 l/sec	761.58	860.13	4.15	98.34	nr	**958.47**
1195 mm long, 1 row 6.58 kW, 250 l/sec	855.85	966.59	4.15	98.34	nr	**1064.93**
1195 mm long, 2 row 9.27 kW, 250 l/sec	855.85	966.59	4.15	98.34	nr	**1064.93**
1495 mm long, 1 row 9.04 kW, 340 l/sec	939.63	1061.22	4.15	98.34	nr	**1159.56**
1495 mm long, 2 row 12.73 kW, 340 l/sec	939.63	1061.22	4.15	98.34	nr	**1159.56**
Free standing unit, extended height 1700/1900/2100 mm						
895 mm long, 1 row 4.02 kW, 150 l/sec	881.58	995.65	3.11	73.70	nr	**1069.35**
895 mm long, 2 row 5.62 kW, 150 l/sec	881.29	995.33	3.11	73.70	nr	**1069.03**
1195 mm long, 1 row 6.58 kW, 250 l/sec	1029.41	1162.61	3.11	73.70	nr	**1236.31**
1195 mm long, 2 row 9.27 kW, 250 l/sec	1029.41	1162.61	3.11	73.70	nr	**1236.31**
1495 mm long, 1 row 9.04 kW, 340 l/sec	1132.65	1279.21	3.11	73.70	nr	**1352.91**
1495 mm long, 2 row 12.73 kW, 340 l/sec	1132.65	1279.21	3.11	73.70	nr	**1352.91**

T: MECHANICAL/COOLING/HEATING SYSTEMS

Item	Net Price £	Material £	Labour hours	Labour £	Unit	Total rate £
T31: LOW TEMPERATURE HOT WATER HEATING – cont						
HEAT EMITTERS – cont						
LTHW trench heating; water temperatures 90°C/70°C; room air temperature 20°C; convector with copper tubes and aluminium fins within steel duct; Includes fixing within floor screed; electrical work elsewhere						
Natural convection type						
Normal capacity, 182 mm width, complete with linear, natural anodized aluminium grille (grille also costed separately below)						
92 mm deep						
1250 mm long, 234 W output	215.45	243.33	2.00	47.39	nr	290.72
2250 mm long, 471 W output	344.71	389.31	4.00	94.78	nr	484.09
3250 mm long, 709 W output	471.53	532.54	5.00	118.47	nr	651.01
4250 mm long, 946 W output	602.02	679.92	7.00	165.86	nr	845.78
5000 mm long, 1124 W output	822.38	928.79	8.00	189.56	nr	1118.35
120 mm deep						
1250 mm long, 294 W output	251.15	283.65	2.00	47.39	nr	331.04
2250 mm long, 471 W output	395.20	446.34	4.00	94.78	nr	541.12
3250 mm long, 891 W output	538.00	607.61	5.00	118.47	nr	726.08
4250 mm long, 1190 W output	680.81	768.91	7.00	165.86	nr	934.77
5000 mm long, 1414 W output	917.19	1035.87	8.00	189.56	nr	1225.43
150 mm deep						
1250 mm long, 329 W output	257.31	290.61	2.00	47.39	nr	338.00
2250 mm long, 664 W output	406.27	458.84	4.00	94.78	nr	553.62
3250 mm long, 998 W output	552.78	624.31	5.00	118.47	nr	742.78
4250 mm long, 1333 W output	702.97	793.93	7.00	165.86	nr	959.79
5000 mm long, 1584 W output	943.03	1065.05	8.00	189.56	nr	1254.61
200 mm deep						
1250 mm long, 396 W output	268.38	303.10	2.00	47.39	nr	350.49
2250 mm long, 799 W output	491.22	554.78	4.00	94.78	nr	649.56
3250 mm long, 1201 W output	584.78	660.45	5.00	118.47	nr	778.92
4250 mm long, 1603 W output	748.51	845.37	7.00	165.86	nr	1011.23
5000 mm long, 1905 W output	994.74	1123.46	8.00	189.56	nr	1313.02

T: MECHANICAL/COOLING/HEATING SYSTEMS

Item	Net Price £	Material £	Labour hours	Labour £	Unit	Total rate £
Fan assisted type (outputs assume fan at 50%)						
Normal capacity, 182 mm width, complete with Natural anodized aluminium grille						
112 mm deep						
1250 mm long, 437 W output	568.77	642.37	2.00	47.39	nr	**689.76**
2250 mm long, 1019 W output	688.20	777.25	4.00	94.78	nr	**872.03**
3250 mm long, 1488 W output	808.84	913.50	5.00	118.47	nr	**1031.97**
4250 mm long, 1845 W output	928.26	1048.38	7.00	165.86	nr	**1214.24**
5000 mm long, 2038 W output	1047.68	1183.24	8.00	189.56	nr	**1372.80**
Linear grille anodized aluminium, 170 mm width (if supplied as a separate item)	134.20	151.56	–	–	m	**151.56**
Roll up grille, natural anodized aluminium	134.20	151.56	–	–	m	**151.56**
Thermostatic valve with remote regulator (c/w valve body)	55.41	62.58	4.00	94.78	nr	**157.36**
Fan speed controller	27.08	30.58	2.00	47.39	nr	**77.97**
Note: as an alternative to thermostatic control, the system can be controlled via two port valves. Refer to valve section in T31 for valve prices						
LTHW underfloor heating; water flow and return temperatures of 60°C and 70°C; pipework at 300 mm centres; pipe fixings; flow and return manifolds and zone actuators; wiring block; insulation; includes fixing in position; excludes secondary pump, mixing valve, zone thermostats and floor finishes; electrical work elsewhere						
Note: All rates are expressed on a m² basis, for the following example areas						
Screeded floor with 15–25 mm stone/marble finish (producing 80–100 W/m²)						
250 m² area (single zone)	23.53	26.57	0.14	3.33	m²	**29.90**
1000 m² area (single zone)	22.92	25.89	0.12	2.85	m²	**28.74**
5000 m² area (multi-zone)	22.87	25.83	0.10	2.38	m²	**28.21**
Screeded floor with 10 mm carpet tile (producing 80–100 W/m²)						
250 m² area (single zone)	23.53	26.57	0.14	3.33	m²	**29.90**
1000 m² area (single zone)	22.92	25.89	0.12	2.85	m²	**28.74**
5000 m² area (multi-zone)	22.87	25.83	0.10	2.38	m²	**28.21**

T: MECHANICAL/COOLING/HEATING SYSTEMS

Item	Net Price £	Material £	Labour hours	Labour £	Unit	Total rate £
T31: LOW TEMPERATURE HOT WATER HEATING – cont						
HEAT EMITTERS – cont						
Floating timber floor with 20 mm timber finish (producing 70–80 Wm²)						
250 m² are (single zone)	29.65	33.49	0.14	3.33	m²	**36.82**
1000 m² area(single zone)	31.78	35.89	0.12	2.85	m²	**38.74**
5000 m² area (multi-zone)	30.36	34.29	0.10	2.38	m²	**36.67**
Floating timber floor with 10 mm carpet tile (producing 70–80 W/m²)						
250 m² are (single zone)	38.59	43.58	0.16	3.79	m²	**47.37**
1000 m² area (single zone)	35.79	40.42	0.12	2.85	m²	**43.27**
5000 m² area (multi-zone)	34.23	38.66	0.10	2.38	m²	**41.04**
PIPE FREEZING						
Freeze isolation of carbon steel or copper pipelines containing static water, either side of work location, freeze duration not exceeding 4 hours assuming that flow and return circuits are treated concurrently and activities undertaken during normal working hours						
Up to 4 freezes						
50 mm dia.	368.47	416.15	1.00	23.69	nr	**439.84**
65 mm dia.	368.47	416.15	1.00	23.69	nr	**439.84**
80 mm dia.	421.24	475.75	1.00	23.69	nr	**499.44**
100 mm dia.	474.03	535.37	1.00	23.69	nr	**559.06**
150 mm dia.	788.75	890.81	1.00	23.69	nr	**914.50**
200 mm dia.	1209.01	1365.45	1.00	23.69	nr	**1389.14**

T: MECHANICAL/COOLING/HEATING SYSTEMS

Item	Net Price £	Material £	Labour hours	Labour £	Unit	Total rate £
ENERGY METERS						
Ultrasonic						
Energy meter for measuring energy use in LTHW systems; includes ultrasonic flow meter (with sensor and signal converter), energy calculator, pair of temperature sensors with brass pockets, and 3 m of interconnecting cable; includes fixing in position; electrical work elsewhere						
Pipe size (flanged connections to PN16); maximum flow rate						
50 mm, 36 m³/hr	1088.03	1228.81	1.80	42.65	nr	**1271.46**
65 mm, 60 m³/hr	1199.34	1354.53	2.32	54.97	nr	**1409.50**
80 mm, 100 m³/hr	1344.37	1518.32	2.56	60.66	nr	**1578.98**
125 mm, 250 m³/hr	1557.58	1759.12	3.60	85.30	nr	**1844.42**
150 mm, 360 m³/hr	1690.84	1909.62	4.80	113.73	nr	**2023.35**
200 mm, 600 m³/hr	1889.16	2133.61	6.24	147.86	nr	**2281.47**
250 mm, 1000 m³/hr	2179.98	2462.06	9.60	227.47	nr	**2689.53**
300 mm, 1500 m³/hr	2562.52	2894.10	10.80	255.90	nr	**3150.00**
350 mm, 2000 m³/hr	3087.72	3487.26	13.20	312.77	nr	**3800.03**
400 mm, 2500 m³/hr	3534.53	3991.88	15.60	369.64	nr	**4361.52**
500 mm, 3000 m³/hr	4011.13	4530.15	24.00	568.67	nr	**5098.82**
600 mm, 3500 m³/hr	4506.55	5089.68	28.00	663.45	nr	**5753.13**
Y53 – CONTROL COMPONENTS – MECHANICAL						
Room thermostats; light and medium duty; installed and connected						
Range 3°C to 27°C; 240 Volt						
1 amp; on/off type	25.06	28.30	0.30	7.10	nr	**35.40**
Range 0°C to +15°C; 240 Volt						
6 amp; frost thermostat	16.90	19.08	0.30	7.10	nr	**26.18**
Range 3°C to 27°C; 250 Volt						
2 amp; changeover type; dead zone	41.92	47.35	0.30	7.10	nr	**54.45**
2 amp; changeover type	19.45	21.97	0.30	7.10	nr	**29.07**
2 amp; changeover type; concealed setting	24.53	27.70	0.30	7.10	nr	**34.80**
6 amp; on/off type	15.14	17.10	0.30	7.10	nr	**24.20**
6 amp; temperature set-back	31.33	35.38	0.30	7.10	nr	**42.48**
16 amp; on/off type	23.31	26.33	0.30	7.10	nr	**33.43**
16 amp; on/off type; concealed setting	25.42	28.71	0.30	7.10	nr	**35.81**
20 amp; on/off type; concealed setting	21.65	24.45	0.30	7.10	nr	**31.55**
20 amp; indicated 'off' position	27.30	30.83	0.30	7.10	nr	**37.93**
20 amp; manual; double pole on/off and neon indicator	49.69	56.12	0.30	7.10	nr	**63.22**
20 amp; indicated 'off' position	32.52	36.73	0.30	7.10	nr	**43.83**

T: MECHANICAL/COOLING/HEATING SYSTEMS

Item	Net Price £	Material £	Labour hours	Labour £	Unit	Total rate £
T31: LOW TEMPERATURE HOT WATER HEATING – cont						
Y53 – CONTROL COMPONENTS – MECHANICAL – cont						
Room thermostats – cont						
Range 10°C to 40°C; 240 Volt						
20 amp; changeover contacts	29.63	33.47	0.30	7.10	nr	**40.57**
2 amp; 'heating-cooling' switch	64.09	72.38	0.30	7.10	nr	**79.48**
Surface thermostats						
Cylinder thermostat						
6 amp; changeover type; with cable	17.49	19.76	0.25	5.93	nr	**25.69**
Electrical thermostats; installed and connected						
Range 5°C to 30°C; 230 Volt standard port single time						
10 amp with sensor	24.29	27.44	0.30	7.10	nr	**34.54**
Range 5°C to 30°C; 230 Volt standard port double time						
10 amp with sensor	27.53	31.09	0.30	7.10	nr	**38.19**
10 amp with sensor and on/off switch	42.00	47.43	0.30	7.10	nr	**54.53**
Radiator thermostats						
Angled valve body; thermostatic head; built in sensor						
15 mm; liquid filled	15.15	17.11	0.84	19.91	nr	**37.02**
15 mm; wax filled	15.15	17.11	0.84	19.91	nr	**37.02**
Immersion thermostats; stem type; domestic water boilers; fitted; electrical work elsewhere						
Temperature range 0°C to 40°C						
Non standard; 280 mm stem	9.22	10.41	0.25	5.93	nr	**16.34**
Temperature range 18°C to 88°C						
13 amp; 178 mm stem	6.12	6.92	0.25	5.93	nr	**12.85**
20 amp; 178 mm stem	9.52	10.75	0.25	5.93	nr	**16.68**
Non standard; pocket clip; 280 mm stem	8.83	9.98	0.25	5.93	nr	**15.91**

T: MECHANICAL/COOLING/HEATING SYSTEMS

Item	Net Price £	Material £	Labour hours	Labour £	Unit	Total rate £
Temperature range 40°C to 80°C						
13 amp; 178 mm stem	3.36	3.79	0.25	5.93	nr	9.72
20 amp; 178 mm stem	6.36	7.18	0.25	5.93	nr	13.11
Non standard; pocket clip; 280 mm stem	9.89	11.17	0.25	5.93	nr	17.10
13 amp; 457 mm stem	3.98	4.50	0.25	5.93	nr	10.43
20 amp; 457 mm stem	6.96	7.86	0.25	5.93	nr	13.79
Temperature range 50°C to 100°C						
Non standard; 1780 mm stem	8.32	9.39	0.25	5.93	nr	15.32
Non standard; 280 mm stem	8.68	9.80	0.25	5.93	nr	15.73
Pockets for thermostats						
For 178 mm stem	10.89	12.30	0.25	5.93	nr	18.23
For 280 mm stem	11.08	12.52	0.25	5.93	nr	18.45
Immersion thermostats; stem type; industrial installations; fitted; electrical work elsewhere						
Temperature range 5°C to 105°C						
For 305 mm stem	136.80	154.50	0.50	11.85	nr	166.35

T: MECHANICAL/COOLING/HEATING SYSTEMS

Item	Net Price £	Material £	Labour hours	Labour £	Unit	Total rate £
T31: THERMAL INSULATION						
Y50 -THERMAL INSULATION						
For flexible closed cell insulation see Section S10 – Cold Water						
Mineral fibre sectional insulation; bright class O foil faced; bright class O foil taped joints; 19 mm aluminium bands						
Concealed pipework						
20 mm thick						
15 mm diameter	2.68	3.03	0.15	3.56	m	**6.59**
20 mm diameter	2.86	3.23	0.15	3.56	m	**6.79**
25 mm diameter	3.07	3.47	0.15	3.56	m	**7.03**
32 mm diameter	3.43	3.88	0.15	3.56	m	**7.44**
40 mm diameter	3.67	4.14	0.15	3.56	m	**7.70**
50 mm diameter	4.20	4.74	0.15	3.56	m	**8.30**
Extra over for fittings concealed insulation						
Flange/union						
15 mm diameter	1.34	1.51	0.13	3.08	nr	**4.59**
20 mm diameter	1.43	1.61	0.13	3.08	nr	**4.69**
25 mm diameter	1.54	1.74	0.13	3.08	nr	**4.82**
32 mm diameter	1.72	1.94	0.13	3.08	nr	**5.02**
40 mm diameter	1.83	2.07	0.13	3.08	nr	**5.15**
50 mm diameter	2.09	2.36	0.13	3.08	nr	**5.44**
Valves						
15 mm diameter	2.68	3.03	0.15	3.56	nr	**6.59**
20 mm diameter	3.07	3.47	0.15	3.56	nr	**7.03**
25 mm diameter	3.07	3.47	0.15	3.56	nr	**7.03**
32 mm diameter	3.43	3.88	0.15	3.56	nr	**7.44**
40 mm diameter	3.67	4.14	0.15	3.56	nr	**7.70**
50 mm diameter	4.20	4.74	0.15	3.56	nr	**8.30**
Expansion bellows						
15 mm diameter	5.36	6.05	0.22	5.22	nr	**11.27**
20 mm diameter	5.73	6.47	0.22	5.22	nr	**11.69**
25 mm diameter	6.15	6.95	0.22	5.22	nr	**12.17**
32 mm diameter	6.85	7.73	0.22	5.22	nr	**12.95**
40 mm diameter	7.34	8.29	0.22	5.22	nr	**13.51**
50 mm diameter	8.38	9.47	0.22	5.22	nr	**14.69**

T: MECHANICAL/COOLING/HEATING SYSTEMS

Item	Net Price £	Material £	Labour hours	Labour £	Unit	Total rate £
25 mm thick						
15 mm diameter	2.96	3.35	0.15	3.56	m	6.91
20 mm diameter	3.18	3.59	0.15	3.56	m	7.15
25 mm diameter	3.57	4.03	0.15	3.56	m	7.59
32 mm diameter	3.88	4.39	0.15	3.56	m	7.95
40 mm diameter	4.15	4.69	0.15	3.56	m	8.25
50 mm diameter	4.75	5.37	0.15	3.56	m	8.93
65 mm diameter	5.43	6.13	0.15	3.56	m	9.69
80 mm diameter	5.95	6.72	0.22	5.22	m	11.94
100 mm diameter	7.85	8.86	0.22	5.22	m	14.08
125 mm diameter	9.07	10.24	0.22	5.22	m	15.46
150 mm diameter	10.80	12.20	0.22	5.22	m	17.42
200 mm diameter	15.27	17.25	0.25	5.93	m	23.18
250 mm diameter	18.29	20.65	0.25	5.93	m	26.58
300 mm diameter	19.54	22.07	0.25	5.93	m	28.00
Extra over for fittings concealed insulation						
Flange/union						
15 mm diameter	1.47	1.66	0.13	3.08	nr	4.74
20 mm diameter	1.59	1.80	0.13	3.08	nr	4.88
25 mm diameter	1.78	2.01	0.13	3.08	nr	5.09
32 mm diameter	1.95	2.20	0.13	3.08	nr	5.28
40 mm diameter	2.06	2.33	0.13	3.08	nr	5.41
50 mm diameter	2.37	2.67	0.13	3.08	nr	5.75
65 mm diameter	2.71	3.06	0.13	3.08	nr	6.14
80 mm diameter	2.98	3.37	0.18	4.27	nr	7.64
100 mm diameter	3.92	4.43	0.18	4.27	nr	8.70
125 mm diameter	4.53	5.12	0.18	4.27	nr	9.39
150 mm diameter	5.41	6.11	0.18	4.27	nr	10.38
200 mm diameter	7.64	8.63	0.22	5.22	nr	13.85
250 mm diameter	9.15	10.33	0.22	5.22	nr	15.55
300 mm diameter	9.78	11.05	0.22	5.22	nr	16.27
Valves						
15 mm diameter	2.96	3.35	0.15	3.56	nr	6.91
20 mm diameter	3.18	3.59	0.15	3.56	nr	7.15
25 mm diameter	3.57	4.03	0.15	3.56	nr	7.59
32 mm diameter	3.88	4.39	0.15	3.56	nr	7.95
40 mm diameter	4.15	4.69	0.15	3.56	nr	8.25
50 mm diameter	4.75	5.37	0.15	3.56	nr	8.93
65 mm diameter	5.43	6.13	0.15	3.56	nr	9.69
80 mm diameter	5.95	6.72	0.20	4.74	nr	11.46
100 mm diameter	7.85	8.86	0.20	4.74	nr	13.60
125 mm diameter	9.07	10.24	0.20	4.74	nr	14.98
150 mm diameter	10.80	12.20	0.20	4.74	nr	16.94
200 mm diameter	15.27	17.25	0.25	5.93	nr	23.18
250 mm diameter	18.29	20.65	0.25	5.93	nr	26.58
300 mm diameter	19.54	22.07	0.25	5.93	nr	28.00

T: MECHANICAL/COOLING/HEATING SYSTEMS

Item	Net Price £	Material £	Labour hours	Labour £	Unit	Total rate £
T31: THERMAL INSULATION – cont						
Extra over for fittings concealed insulation – cont						
Expansion bellows						
15 mm diameter	5.90	6.66	0.22	5.22	nr	**11.88**
20 mm diameter	6.37	7.19	0.22	5.22	nr	**12.41**
25 mm diameter	7.10	8.02	0.22	5.22	nr	**13.24**
32 mm diameter	7.74	8.74	0.22	5.22	nr	**13.96**
40 mm diameter	8.30	9.37	0.22	5.22	nr	**14.59**
50 mm diameter	9.52	10.75	0.22	5.22	nr	**15.97**
65 mm diameter	10.85	12.25	0.22	5.22	nr	**17.47**
80 mm diameter	11.89	13.43	0.29	6.86	nr	**20.29**
100 mm diameter	15.68	17.71	0.29	6.86	nr	**24.57**
125 mm diameter	18.14	20.49	0.29	6.86	nr	**27.35**
150 mm diameter	21.62	24.42	0.29	6.86	nr	**31.28**
200 mm diameter	30.53	34.48	0.36	8.53	nr	**43.01**
250 mm diameter	36.57	41.30	0.36	8.53	nr	**49.83**
300 mm diameter	39.10	44.16	0.36	8.53	nr	**52.69**
30 mm thick						
15 mm diameter	3.84	4.33	0.15	3.56	m	**7.89**
20 mm diameter	4.11	4.64	0.15	3.56	m	**8.20**
25 mm diameter	4.35	4.92	0.15	3.56	m	**8.48**
32 mm diameter	4.74	5.36	0.15	3.56	m	**8.92**
40 mm diameter	5.02	5.67	0.15	3.56	m	**9.23**
50 mm diameter	5.74	6.49	0.15	3.56	m	**10.05**
65 mm diameter	6.49	7.33	0.15	3.56	m	**10.89**
80 mm diameter	7.09	8.01	0.22	5.22	m	**13.23**
100 mm diameter	9.17	10.35	0.22	5.22	m	**15.57**
125 mm diameter	10.56	11.92	0.22	5.22	m	**17.14**
150 mm diameter	12.40	14.00	0.22	5.22	m	**19.22**
200 mm diameter	17.34	19.58	0.25	5.93	m	**25.51**
250 mm diameter	20.62	23.29	0.25	5.93	m	**29.22**
300 mm diameter	21.85	24.67	0.25	5.93	m	**30.60**
350 mm diameter	24.01	27.12	0.25	5.93	m	**33.05**
Extra over for fittings concealed insulation						
Flange/union						
15 mm diameter	1.91	2.15	0.13	3.08	nr	**5.23**
20 mm diameter	2.05	2.32	0.13	3.08	nr	**5.40**
25 mm diameter	2.17	2.45	0.13	3.08	nr	**5.53**
32 mm diameter	2.37	2.67	0.13	3.08	nr	**5.75**
40 mm diameter	2.51	2.84	0.13	3.08	nr	**5.92**
50 mm diameter	2.88	3.25	0.13	3.08	nr	**6.33**
65 mm diameter	3.26	3.68	0.13	3.08	nr	**6.76**

T: MECHANICAL/COOLING/HEATING SYSTEMS

Item	Net Price £	Material £	Labour hours	Labour £	Unit	Total rate £
80 mm diameter	3.54	4.00	0.18	4.27	nr	**8.27**
100 mm diameter	4.59	5.18	0.18	4.27	nr	**9.45**
125 mm diameter	5.27	5.96	0.18	4.27	nr	**10.23**
150 mm diameter	6.21	7.02	0.18	4.27	nr	**11.29**
200 mm diameter	8.67	9.79	0.22	5.22	nr	**15.01**
250 mm diameter	10.32	11.66	0.22	5.22	nr	**16.88**
300 mm diameter	10.94	12.35	0.22	5.22	nr	**17.57**
350 mm diameter	12.01	13.57	0.22	5.22	nr	**18.79**
Valves						
15 mm diameter	4.11	4.64	0.15	3.56	nr	**8.20**
20 mm diameter	4.11	4.64	0.15	3.56	nr	**8.20**
25 mm diameter	4.35	4.92	0.15	3.56	nr	**8.48**
32 mm diameter	4.74	5.36	0.15	3.56	nr	**8.92**
40 mm diameter	5.02	5.67	0.15	3.56	nr	**9.23**
50 mm diameter	5.74	6.49	0.15	3.56	nr	**10.05**
65 mm diameter	6.49	7.33	0.15	3.56	nr	**10.89**
80 mm diameter	7.09	8.01	0.20	4.74	nr	**12.75**
100 mm diameter	9.17	10.35	0.20	4.74	nr	**15.09**
125 mm diameter	10.56	11.92	0.20	4.74	nr	**16.66**
150 mm diameter	12.40	14.00	0.20	4.74	nr	**18.74**
200 mm diameter	17.34	19.58	0.25	5.93	nr	**25.51**
250 mm diameter	20.62	23.29	0.25	5.93	nr	**29.22**
300 mm diameter	21.85	24.67	0.25	5.93	nr	**30.60**
350 mm diameter	24.01	27.12	0.25	5.93	nr	**33.05**
Expansion bellows						
15 mm diameter	7.68	8.67	0.22	5.22	nr	**13.89**
20 mm diameter	8.21	9.27	0.22	5.22	nr	**14.49**
25 mm diameter	8.69	9.81	0.22	5.22	nr	**15.03**
32 mm diameter	9.50	10.73	0.22	5.22	nr	**15.95**
40 mm diameter	10.04	11.34	0.22	5.22	nr	**16.56**
50 mm diameter	11.48	12.96	0.22	5.22	nr	**18.18**
65 mm diameter	12.99	14.67	0.22	5.22	nr	**19.89**
80 mm diameter	14.18	16.01	0.29	6.86	nr	**22.87**
100 mm diameter	18.35	20.73	0.29	6.86	nr	**27.59**
125 mm diameter	21.10	23.83	0.29	6.86	nr	**30.69**
150 mm diameter	24.80	28.01	0.29	6.86	nr	**34.87**
200 mm diameter	34.67	39.16	0.36	8.53	nr	**47.69**
250 mm diameter	41.24	46.57	0.36	8.53	nr	**55.10**
300 mm diameter	43.73	49.39	0.36	8.53	nr	**57.92**
350 mm diameter	48.01	54.22	0.36	8.53	nr	**62.75**
40 mm thick						
15 mm diameter	4.91	5.55	0.15	3.56	m	**9.11**
20 mm diameter	5.07	5.72	0.15	3.56	m	**9.28**
25 mm diameter	5.46	6.17	0.15	3.56	m	**9.73**
32 mm diameter	5.81	6.56	0.15	3.56	m	**10.12**
40 mm diameter	6.10	6.88	0.15	3.56	m	**10.44**
50 mm diameter	6.91	7.80	0.15	3.56	m	**11.36**
65 mm diameter	7.75	8.75	0.15	3.56	m	**12.31**

T: MECHANICAL/COOLING/HEATING SYSTEMS

Item	Net Price £	Material £	Labour hours	Labour £	Unit	Total rate £
T31: THERMAL INSULATION – cont						
40 mm thick – cont						
80 mm diameter	8.43	9.52	0.22	5.22	m	**14.74**
100 mm diameter	10.95	12.36	0.22	5.22	m	**17.58**
125 mm diameter	12.37	13.97	0.22	5.22	m	**19.19**
150 mm diameter	14.44	16.31	0.22	5.22	m	**21.53**
200 mm diameter	19.95	22.53	0.25	5.93	m	**28.46**
250 mm diameter	23.38	26.41	0.25	5.93	m	**32.34**
300 mm diameter	25.00	28.23	0.25	5.93	m	**34.16**
350 mm diameter	27.65	31.23	0.25	5.93	m	**37.16**
400 mm diameter	30.80	34.78	0.25	5.93	m	**40.71**
Extra over for fittings concealed insulation						
Flange/union						
15 mm diameter	2.46	2.77	0.13	3.08	nr	**5.85**
20 mm diameter	2.53	2.86	0.13	3.08	nr	**5.94**
25 mm diameter	2.72	3.07	0.13	3.08	nr	**6.15**
32 mm diameter	2.90	3.27	0.13	3.08	nr	**6.35**
40 mm diameter	3.05	3.45	0.13	3.08	nr	**6.53**
50 mm diameter	3.46	3.91	0.13	3.08	nr	**6.99**
65 mm diameter	3.88	4.39	0.13	3.08	nr	**7.47**
80 mm diameter	4.22	4.76	0.18	4.27	nr	**9.03**
100 mm diameter	5.47	6.18	0.18	4.27	nr	**10.45**
125 mm diameter	6.17	6.97	0.18	4.27	nr	**11.24**
150 mm diameter	7.23	8.17	0.18	4.27	nr	**12.44**
200 mm diameter	9.98	11.27	0.22	5.22	nr	**16.49**
250 mm diameter	11.69	13.20	0.22	5.22	nr	**18.42**
300 mm diameter	12.49	14.11	0.22	5.22	nr	**19.33**
350 mm diameter	13.83	15.62	0.22	5.22	nr	**20.84**
400 mm diameter	15.42	17.41	0.22	5.22	nr	**22.63**
Valves						
15 mm diameter	4.91	5.55	0.15	3.56	nr	**9.11**
20 mm diameter	5.07	5.72	0.15	3.56	nr	**9.28**
25 mm diameter	5.46	6.17	0.15	3.56	nr	**9.73**
32 mm diameter	5.81	6.56	0.15	3.56	nr	**10.12**
40 mm diameter	6.10	6.88	0.15	3.56	nr	**10.44**
50 mm diameter	6.91	7.80	0.15	3.56	nr	**11.36**
65 mm diameter	7.75	8.75	0.15	3.56	nr	**12.31**
80 mm diameter	8.43	9.52	0.20	4.74	nr	**14.26**
100 mm diameter	10.95	12.36	0.20	4.74	nr	**17.10**
125 mm diameter	12.37	13.97	0.20	4.74	nr	**18.71**
150 mm diameter	14.44	16.31	0.20	4.74	nr	**21.05**
200 mm diameter	19.95	22.53	0.25	5.93	nr	**28.46**
250 mm diameter	23.38	26.41	0.25	5.93	nr	**32.34**
300 mm diameter	25.00	28.23	0.25	5.93	nr	**34.16**
350 mm diameter	27.65	31.23	0.25	5.93	nr	**37.16**
400 mm diameter	30.80	34.78	0.25	5.93	nr	**40.71**

T: MECHANICAL/COOLING/HEATING SYSTEMS

Item	Net Price £	Material £	Labour hours	Labour £	Unit	Total rate £
Expansion bellows						
15 mm diameter	9.85	11.13	0.22	5.22	nr	**16.35**
20 mm diameter	10.13	11.44	0.22	5.22	nr	**16.66**
25 mm diameter	10.91	12.32	0.22	5.22	nr	**17.54**
32 mm diameter	11.62	13.13	0.22	5.22	nr	**18.35**
40 mm diameter	12.22	13.80	0.22	5.22	nr	**19.02**
50 mm diameter	13.83	15.62	0.22	5.22	nr	**20.84**
65 mm diameter	15.49	17.49	0.22	5.22	nr	**22.71**
80 mm diameter	16.88	19.06	0.29	6.86	nr	**25.92**
100 mm diameter	21.89	24.72	0.29	6.86	nr	**31.58**
125 mm diameter	24.74	27.94	0.29	6.86	nr	**34.80**
150 mm diameter	28.89	32.63	0.29	6.86	nr	**39.49**
200 mm diameter	39.90	45.06	0.36	8.53	nr	**53.59**
250 mm diameter	46.76	52.82	0.36	8.53	nr	**61.35**
300 mm diameter	50.00	56.47	0.36	8.53	nr	**65.00**
350 mm diameter	27.65	31.23	0.36	8.53	nr	**39.76**
400 mm diameter	61.63	69.60	0.36	8.53	nr	**78.13**
50 mm thick						
15 mm diameter	6.84	7.72	0.15	3.56	m	**11.28**
20 mm diameter	7.19	8.12	0.15	3.56	m	**11.68**
25 mm diameter	7.65	8.64	0.15	3.56	m	**12.20**
32 mm diameter	7.98	9.02	0.15	3.56	m	**12.58**
40 mm diameter	8.40	9.49	0.15	3.56	m	**13.05**
50 mm diameter	9.42	10.64	0.15	3.56	m	**14.20**
65 mm diameter	10.30	11.63	0.15	3.56	m	**15.19**
80 mm diameter	11.02	12.44	0.22	5.22	m	**17.66**
100 mm diameter	14.09	15.91	0.22	5.22	m	**21.13**
125 mm diameter	15.81	17.86	0.22	5.22	m	**23.08**
150 mm diameter	18.26	20.62	0.22	5.22	m	**25.84**
200 mm diameter	24.91	28.13	0.25	5.93	m	**34.06**
250 mm diameter	28.78	32.51	0.25	5.93	m	**38.44**
300 mm diameter	30.49	34.44	0.25	5.93	m	**40.37**
350 mm diameter	33.65	38.01	0.25	5.93	m	**43.94**
400 mm diameter	37.33	42.16	0.25	5.93	m	**48.09**
Extra over for fittings concealed insulation						
Flange/union						
15 mm diameter	3.42	3.87	0.13	3.08	nr	**6.95**
20 mm diameter	3.60	4.07	0.13	3.08	nr	**7.15**
25 mm diameter	3.83	4.32	0.13	3.08	nr	**7.40**
32 mm diameter	4.00	4.52	0.13	3.08	nr	**7.60**
40 mm diameter	4.21	4.75	0.13	3.08	nr	**7.83**
50 mm diameter	4.70	5.30	0.13	3.08	nr	**8.38**
65 mm diameter	5.15	5.81	0.13	3.08	nr	**8.89**
80 mm diameter	5.51	6.22	0.18	4.27	nr	**10.49**
100 mm diameter	7.05	7.97	0.18	4.27	nr	**12.24**
125 mm diameter	7.91	8.94	0.18	4.27	nr	**13.21**
150 mm diameter	9.13	10.31	0.18	4.27	nr	**14.58**

T: MECHANICAL/COOLING/HEATING SYSTEMS

Item	Net Price £	Material £	Labour hours	Labour £	Unit	Total rate £
T31: THERMAL INSULATION – cont						
Extra over for fittings concealed insulation – cont						
Flange/union – cont						
200 mm diameter	12.46	14.08	0.22	5.22	nr	**19.30**
250 mm diameter	14.39	16.25	0.22	5.22	nr	**21.47**
300 mm diameter	15.25	17.23	0.22	5.22	nr	**22.45**
350 mm diameter	16.84	19.02	0.22	5.22	nr	**24.24**
400 mm diameter	18.67	21.08	0.22	5.22	nr	**26.30**
Valves						
15 mm diameter	6.84	7.72	0.15	3.56	nr	**11.28**
20 mm diameter	7.19	8.12	0.15	3.56	nr	**11.68**
25 mm diameter	7.65	8.64	0.15	3.56	nr	**12.20**
32 mm diameter	7.98	9.02	0.15	3.56	nr	**12.58**
40 mm diameter	8.40	9.49	0.15	3.56	nr	**13.05**
50 mm diameter	9.42	10.64	0.15	3.56	nr	**14.20**
65 mm diameter	10.30	11.63	0.15	3.56	nr	**15.19**
80 mm diameter	11.02	12.44	0.20	4.74	nr	**17.18**
100 mm diameter	14.09	15.91	0.20	4.74	nr	**20.65**
125 mm diameter	15.81	17.86	0.20	4.74	nr	**22.60**
150 mm diameter	18.26	20.62	0.20	4.74	nr	**25.36**
200 mm diameter	24.91	28.13	0.25	5.93	nr	**34.06**
250 mm diameter	28.78	32.51	0.25	5.93	nr	**38.44**
300 mm diameter	30.49	34.44	0.25	5.93	nr	**40.37**
350 mm diameter	33.65	38.01	0.25	5.93	nr	**43.94**
400 mm diameter	37.33	42.16	0.25	5.93	nr	**48.09**
Expansion bellows						
15 mm diameter	13.65	15.41	0.22	5.22	nr	**20.63**
20 mm diameter	14.38	16.24	0.22	5.22	nr	**21.46**
25 mm diameter	15.28	17.26	0.22	5.22	nr	**22.48**
32 mm diameter	15.99	18.05	0.22	5.22	nr	**23.27**
40 mm diameter	16.80	18.97	0.22	5.22	nr	**24.19**
50 mm diameter	18.84	21.28	0.22	5.22	nr	**26.50**
65 mm diameter	20.58	23.25	0.22	5.22	nr	**28.47**
80 mm diameter	22.04	24.89	0.29	6.86	nr	**31.75**
100 mm diameter	28.17	31.81	0.29	6.86	nr	**38.67**
125 mm diameter	31.62	35.71	0.29	6.86	nr	**42.57**
150 mm diameter	36.51	41.24	0.29	6.86	nr	**48.10**
200 mm diameter	49.86	56.31	0.36	8.53	nr	**64.84**
250 mm diameter	57.55	64.99	0.36	8.53	nr	**73.52**
300 mm diameter	61.01	68.90	0.36	8.53	nr	**77.43**
350 mm diameter	67.33	76.04	0.36	8.53	nr	**84.57**
400 mm diameter	74.65	84.31	0.36	8.53	nr	**92.84**

T: MECHANICAL/COOLING/HEATING SYSTEMS

Item	Net Price £	Material £	Labour hours	Labour £	Unit	Total rate £
Mineral fibre sectional insulation; bright class O foil faced; bright class O foil taped joints; 22 swg plain/embossed aluminium cladding; pop riveted						
Plantroom pipework						
20 mm thick						
15 mm diameter	4.38	4.95	0.44	10.42	m	**15.37**
20 mm diameter	4.63	5.23	0.44	10.42	m	**15.65**
25 mm diameter	4.95	5.59	0.44	10.42	m	**16.01**
32 mm diameter	5.37	6.07	0.44	10.42	m	**16.49**
40 mm diameter	5.66	6.40	0.44	10.42	m	**16.82**
50 mm diameter	6.30	7.12	0.44	10.42	m	**17.54**
Extra over for fittings plantroom insulation						
Flange/union						
15 mm diameter	4.89	5.52	0.58	13.75	nr	**19.27**
20 mm diameter	5.19	5.87	0.58	13.75	nr	**19.62**
25 mm diameter	5.56	6.28	0.58	13.75	nr	**20.03**
32 mm diameter	6.09	6.87	0.58	13.75	nr	**20.62**
40 mm diameter	6.42	7.25	0.58	13.75	nr	**21.00**
50 mm diameter	7.21	8.14	0.58	13.75	nr	**21.89**
Bends						
15 mm diameter	2.40	2.71	0.44	10.42	nr	**13.13**
20 mm diameter	2.56	2.89	0.44	10.42	nr	**13.31**
25 mm diameter	2.73	3.08	0.44	10.42	nr	**13.50**
32 mm diameter	2.96	3.35	0.44	10.42	nr	**13.77**
40 mm diameter	3.10	3.50	0.44	10.42	nr	**13.92**
50 mm diameter	3.47	3.92	0.44	10.42	nr	**14.34**
Tees						
15 mm diameter	1.44	1.62	0.44	10.42	nr	**12.04**
20 mm diameter	1.53	1.72	0.44	10.42	nr	**12.14**
25 mm diameter	1.64	1.86	0.44	10.42	nr	**12.28**
32 mm diameter	1.77	2.00	0.44	10.42	nr	**12.42**
40 mm diameter	1.87	2.11	0.44	10.42	nr	**12.53**
50 mm diameter	2.07	2.34	0.44	10.42	nr	**12.76**
Valves						
15 mm diameter	2.68	3.03	0.78	18.47	nr	**21.50**
20 mm diameter	3.07	3.47	0.78	18.47	nr	**21.94**
25 mm diameter	3.07	3.47	0.78	18.47	nr	**21.94**
32 mm diameter	3.43	3.88	0.78	18.47	nr	**22.35**
40 mm diameter	3.67	4.14	0.78	18.47	nr	**22.61**
50 mm diameter	4.20	4.74	0.78	18.47	nr	**23.21**

T: MECHANICAL/COOLING/HEATING SYSTEMS

Item	Net Price £	Material £	Labour hours	Labour £	Unit	Total rate £
T31: THERMAL INSULATION – cont						
Extra over for fittings plantroom insulation – cont						
Pumps						
15 mm diameter	14.39	16.25	2.34	55.44	nr	**71.69**
20 mm diameter	15.26	17.24	2.34	55.44	nr	**72.68**
25 mm diameter	16.39	18.51	2.34	55.44	nr	**73.95**
32 mm diameter	17.88	20.20	2.34	55.44	nr	**75.64**
40 mm diameter	18.89	21.34	2.34	55.44	nr	**76.78**
50 mm diameter	21.23	23.98	2.34	55.44	nr	**79.42**
Expansion bellows						
15 mm diameter	11.50	12.98	1.05	24.88	nr	**37.86**
20 mm diameter	12.21	13.79	1.05	24.88	nr	**38.67**
25 mm diameter	13.11	14.81	1.05	24.88	nr	**39.69**
32 mm diameter	14.29	16.14	1.05	24.88	nr	**41.02**
40 mm diameter	15.10	17.05	1.05	24.88	nr	**41.93**
50 mm diameter	16.98	19.18	1.05	24.88	nr	**44.06**
25 mm thick						
15 mm diameter	4.81	5.44	0.44	10.42	m	**15.86**
20 mm diameter	5.19	5.87	0.44	10.42	m	**16.29**
25 mm diameter	5.70	6.44	0.44	10.42	m	**16.86**
32 mm diameter	6.07	6.85	0.44	10.42	m	**17.27**
40 mm diameter	6.52	7.36	0.44	10.42	m	**17.78**
50 mm diameter	7.31	8.25	0.44	10.42	m	**18.67**
65 mm diameter	8.26	9.33	0.44	10.42	m	**19.75**
80 mm diameter	8.95	10.11	0.52	12.32	m	**22.43**
100 mm diameter	11.11	12.55	0.52	12.32	m	**24.87**
125 mm diameter	12.90	14.57	0.52	12.32	m	**26.89**
150 mm diameter	15.02	16.96	0.52	12.32	m	**29.28**
200 mm diameter	20.51	23.16	0.60	14.22	m	**37.38**
250 mm diameter	24.37	27.52	0.60	14.22	m	**41.74**
300 mm diameter	27.29	30.82	0.60	14.22	m	**45.04**
Extra over for fittings plantroom insulation						
Flange/union						
15 mm diameter	5.37	6.07	0.58	13.75	nr	**19.82**
20 mm diameter	5.81	6.56	0.58	13.75	nr	**20.31**
25 mm diameter	6.41	7.24	0.58	13.75	nr	**20.99**
32 mm diameter	6.88	7.77	0.58	13.75	nr	**21.52**
40 mm diameter	7.38	8.33	0.58	13.75	nr	**22.08**
50 mm diameter	8.32	9.39	0.58	13.75	nr	**23.14**
65 mm diameter	9.44	10.66	0.58	13.75	nr	**24.41**
80 mm diameter	10.26	11.59	0.67	15.88	nr	**27.47**
100 mm diameter	12.99	14.67	0.67	15.88	nr	**30.55**
125 mm diameter	15.09	17.04	0.67	15.88	nr	**32.92**
150 mm diameter	17.68	19.97	0.67	15.88	nr	**35.85**

T: MECHANICAL/COOLING/HEATING SYSTEMS

Item	Net Price £	Material £	Labour hours	Labour £	Unit	Total rate £
200 mm diameter	24.43	27.59	0.87	20.61	nr	**48.20**
250 mm diameter	29.11	32.87	0.87	20.61	nr	**53.48**
300 mm diameter	32.16	36.32	0.87	20.61	nr	**56.93**
Bends						
15 mm diameter	2.64	2.98	0.44	10.42	nr	**13.40**
20 mm diameter	2.86	3.23	0.44	10.42	nr	**13.65**
25 mm diameter	3.13	3.54	0.44	10.42	nr	**13.96**
32 mm diameter	3.34	3.77	0.44	10.42	nr	**14.19**
40 mm diameter	3.60	4.07	0.44	10.42	nr	**14.49**
50 mm diameter	4.02	4.54	0.44	10.42	nr	**14.96**
65 mm diameter	4.53	5.12	0.44	10.42	nr	**15.54**
80 mm diameter	4.92	5.56	0.52	12.32	nr	**17.88**
100 mm diameter	6.10	6.88	0.52	12.32	nr	**19.20**
125 mm diameter	7.10	8.02	0.52	12.32	nr	**20.34**
150 mm diameter	8.26	9.33	0.52	12.32	nr	**21.65**
200 mm diameter	11.28	12.74	0.60	14.22	nr	**26.96**
250 mm diameter	13.41	15.15	0.60	14.22	nr	**29.37**
300 mm diameter	15.01	16.95	0.60	14.22	nr	**31.17**
Tees						
15 mm diameter	1.59	1.80	0.44	10.42	nr	**12.22**
20 mm diameter	1.71	1.93	0.44	10.42	nr	**12.35**
25 mm diameter	1.88	2.12	0.44	10.42	nr	**12.54**
32 mm diameter	2.01	2.27	0.44	10.42	nr	**12.69**
40 mm diameter	2.15	2.43	0.44	10.42	nr	**12.85**
50 mm diameter	2.40	2.71	0.44	10.42	nr	**13.13**
65 mm diameter	2.72	3.07	0.44	10.42	nr	**13.49**
80 mm diameter	2.96	3.35	0.52	12.32	nr	**15.67**
100 mm diameter	3.66	4.13	0.52	12.32	nr	**16.45**
125 mm diameter	4.26	4.81	0.52	12.32	nr	**17.13**
150 mm diameter	4.95	5.59	0.52	12.32	nr	**17.91**
200 mm diameter	6.76	7.64	0.60	14.22	nr	**21.86**
250 mm diameter	8.05	9.09	0.60	14.22	nr	**23.31**
300 mm diameter	9.01	10.18	0.60	14.22	nr	**24.40**
Valves						
15 mm diameter	8.54	9.65	0.78	18.47	nr	**28.12**
20 mm diameter	9.22	10.41	0.78	18.47	nr	**28.88**
25 mm diameter	10.18	11.50	0.78	18.47	nr	**29.97**
32 mm diameter	10.93	12.34	0.78	18.47	nr	**30.81**
40 mm diameter	11.72	13.24	0.78	18.47	nr	**31.71**
50 mm diameter	13.21	14.92	0.78	18.47	nr	**33.39**
65 mm diameter	15.00	16.94	0.78	18.47	nr	**35.41**
80 mm diameter	16.31	18.42	0.92	21.80	nr	**40.22**
100 mm diameter	20.63	23.30	0.92	21.80	nr	**45.10**
125 mm diameter	23.97	27.07	0.92	21.80	nr	**48.87**
150 mm diameter	28.08	31.71	0.92	21.80	nr	**53.51**
200 mm diameter	38.81	43.83	1.12	26.54	nr	**70.37**
250 mm diameter	46.23	52.21	1.12	26.54	nr	**78.75**
300 mm diameter	51.08	57.69	1.12	26.54	nr	**84.23**

T: MECHANICAL/COOLING/HEATING SYSTEMS

Item	Net Price £	Material £	Labour hours	Labour £	Unit	Total rate £
T31: THERMAL INSULATION – cont						
Extra over for fittings plantroom insulation – cont						
Pumps						
15 mm diameter	15.80	17.84	2.34	55.44	nr	**73.28**
20 mm diameter	17.08	19.29	2.34	55.44	nr	**74.73**
25 mm diameter	18.87	21.31	2.34	55.44	nr	**76.75**
32 mm diameter	20.22	22.84	2.34	55.44	nr	**78.28**
40 mm diameter	21.72	24.53	2.34	55.44	nr	**79.97**
50 mm diameter	24.48	27.65	2.34	55.44	nr	**83.09**
65 mm diameter	27.76	31.35	2.34	55.44	nr	**86.79**
80 mm diameter	30.20	34.11	2.76	65.39	nr	**99.50**
100 mm diameter	38.21	43.16	2.76	65.39	nr	**108.55**
125 mm diameter	44.37	50.11	2.76	65.39	nr	**115.50**
150 mm diameter	52.01	58.74	2.76	65.39	nr	**124.13**
200 mm diameter	71.86	81.16	3.36	79.61	nr	**160.77**
250 mm diameter	85.61	96.69	3.36	79.61	nr	**176.30**
300 mm diameter	94.57	106.80	3.36	79.61	nr	**186.41**
Expansion bellows						
15 mm diameter	12.64	14.28	1.05	24.88	nr	**39.16**
20 mm diameter	13.66	15.43	1.05	24.88	nr	**40.31**
25 mm diameter	15.09	17.04	1.05	24.88	nr	**41.92**
32 mm diameter	16.19	18.29	1.05	24.88	nr	**43.17**
40 mm diameter	17.37	19.61	1.05	24.88	nr	**44.49**
50 mm diameter	19.58	22.11	1.05	24.88	nr	**46.99**
65 mm diameter	22.21	25.08	1.05	24.88	nr	**49.96**
80 mm diameter	24.17	27.30	1.26	29.85	nr	**57.15**
100 mm diameter	30.56	34.52	1.26	29.85	nr	**64.37**
125 mm diameter	35.48	40.08	1.26	29.85	nr	**69.93**
150 mm diameter	41.59	46.97	1.26	29.85	nr	**76.82**
200 mm diameter	57.50	64.94	1.53	36.25	nr	**101.19**
250 mm diameter	68.50	77.37	1.53	36.25	nr	**113.62**
300 mm diameter	75.66	85.45	1.53	36.25	nr	**121.70**
30 mm thick						
15 mm diameter	5.92	6.68	0.44	10.42	m	**17.10**
20 mm diameter	6.27	7.08	0.44	10.42	m	**17.50**
25 mm diameter	6.64	7.50	0.44	10.42	m	**17.92**
32 mm diameter	7.27	8.21	0.44	10.42	m	**18.63**
40 mm diameter	7.57	8.55	0.44	10.42	m	**18.97**
50 mm diameter	8.40	9.49	0.44	10.42	m	**19.91**
65 mm diameter	9.48	10.71	0.44	10.42	m	**21.13**
80 mm diameter	10.36	11.70	0.52	12.32	m	**24.02**
100 mm diameter	12.72	14.36	0.52	12.32	m	**26.68**
125 mm diameter	14.54	16.42	0.52	12.32	m	**28.74**
150 mm diameter	16.95	19.15	0.52	12.32	m	**31.47**

T: MECHANICAL/COOLING/HEATING SYSTEMS

Item	Net Price £	Material £	Labour hours	Labour £	Unit	Total rate £
200 mm diameter	22.77	25.71	0.60	14.22	m	**39.93**
250 mm diameter	26.94	30.43	0.60	14.22	m	**44.65**
300 mm diameter	29.66	33.50	0.60	14.22	m	**47.72**
350 mm diameter	33.02	37.29	0.60	14.22	m	**51.51**
Extra over for fittings plantroom insulation						
Flange/union						
15 mm diameter	6.73	7.60	0.58	13.75	nr	**21.35**
20 mm diameter	7.15	8.08	0.58	13.75	nr	**21.83**
25 mm diameter	7.58	8.56	0.58	13.75	nr	**22.31**
32 mm diameter	8.30	9.37	0.58	13.75	nr	**23.12**
40 mm diameter	8.68	9.80	0.58	13.75	nr	**23.55**
50 mm diameter	9.73	10.99	0.58	13.75	nr	**24.74**
65 mm diameter	10.99	12.41	0.58	13.75	nr	**26.16**
80 mm diameter	12.01	13.57	0.67	15.88	nr	**29.45**
100 mm diameter	15.00	16.94	0.67	15.88	nr	**32.82**
125 mm diameter	17.16	19.38	0.67	15.88	nr	**35.26**
150 mm diameter	20.08	22.67	0.67	15.88	nr	**38.55**
200 mm diameter	27.34	30.88	0.87	20.61	nr	**51.49**
250 mm diameter	32.41	36.61	0.87	20.61	nr	**57.22**
300 mm diameter	35.26	39.82	0.87	20.61	nr	**60.43**
350 mm diameter	39.12	44.19	0.87	20.61	nr	**64.80**
Bends						
15 mm diameter	3.26	3.68	0.44	10.42	nr	**14.10**
20 mm diameter	3.44	3.89	0.44	10.42	nr	**14.31**
25 mm diameter	3.65	4.12	0.44	10.42	nr	**14.54**
32 mm diameter	3.99	4.51	0.44	10.42	nr	**14.93**
40 mm diameter	4.16	4.70	0.44	10.42	nr	**15.12**
50 mm diameter	4.63	5.23	0.44	10.42	nr	**15.65**
65 mm diameter	5.21	5.89	0.44	10.42	nr	**16.31**
80 mm diameter	5.69	6.43	0.52	12.32	nr	**18.75**
100 mm diameter	7.00	7.91	0.52	12.32	nr	**20.23**
125 mm diameter	7.99	9.03	0.52	12.32	nr	**21.35**
150 mm diameter	9.33	10.54	0.52	12.32	nr	**22.86**
200 mm diameter	12.53	14.15	0.60	14.22	nr	**28.37**
250 mm diameter	14.83	16.75	0.60	14.22	nr	**30.97**
300 mm diameter	16.31	18.42	0.60	14.22	nr	**32.64**
350 mm diameter	18.17	20.52	0.60	14.22	nr	**34.74**
Tees						
15 mm diameter	1.96	2.21	0.44	10.42	nr	**12.63**
20 mm diameter	2.06	2.33	0.44	10.42	nr	**12.75**
25 mm diameter	2.19	2.47	0.44	10.42	nr	**12.89**
32 mm diameter	2.39	2.70	0.44	10.42	nr	**13.12**
40 mm diameter	2.50	2.83	0.44	10.42	nr	**13.25**
50 mm diameter	2.77	3.13	0.44	10.42	nr	**13.55**
65 mm diameter	3.12	3.52	0.44	10.42	nr	**13.94**

T: MECHANICAL/COOLING/HEATING SYSTEMS

Item	Net Price £	Material £	Labour hours	Labour £	Unit	Total rate £
T31: THERMAL INSULATION – cont						
Extra over for fittings plantroom insulation – cont						
Tees – cont						
80 mm diameter	3.42	3.87	0.52	12.32	nr	**16.19**
100 mm diameter	4.21	4.75	0.52	12.32	nr	**17.07**
125 mm diameter	4.81	5.44	0.52	12.32	nr	**17.76**
150 mm diameter	5.60	6.32	0.52	12.32	nr	**18.64**
200 mm diameter	7.52	8.50	0.60	14.22	nr	**22.72**
250 mm diameter	8.90	10.05	0.60	14.22	nr	**24.27**
300 mm diameter	9.79	11.06	0.60	14.22	nr	**25.28**
350 mm diameter	10.91	12.32	0.60	14.22	nr	**26.54**
Valves						
15 mm diameter	10.71	12.10	0.78	18.47	nr	**30.57**
20 mm diameter	11.36	12.83	0.78	18.47	nr	**31.30**
25 mm diameter	12.04	13.60	0.78	18.47	nr	**32.07**
32 mm diameter	13.19	14.89	0.78	18.47	nr	**33.36**
40 mm diameter	13.80	15.59	0.78	18.47	nr	**34.06**
50 mm diameter	15.45	17.45	0.78	18.47	nr	**35.92**
65 mm diameter	17.46	19.72	0.78	18.47	nr	**38.19**
80 mm diameter	19.06	21.52	0.92	21.80	nr	**43.32**
100 mm diameter	23.81	26.89	0.92	21.80	nr	**48.69**
125 mm diameter	27.27	30.79	0.92	21.80	nr	**52.59**
150 mm diameter	31.88	36.01	0.92	21.80	nr	**57.81**
200 mm diameter	43.41	49.03	1.12	26.54	nr	**75.57**
250 mm diameter	51.47	58.13	1.12	26.54	nr	**84.67**
300 mm diameter	56.00	63.25	1.12	26.54	nr	**89.79**
350 mm diameter	62.13	70.17	1.12	26.54	nr	**96.71**
Pumps						
15 mm diameter	19.83	22.40	2.34	55.44	nr	**77.84**
20 mm diameter	21.04	23.77	2.34	55.44	nr	**79.21**
25 mm diameter	22.29	25.17	2.34	55.44	nr	**80.61**
32 mm diameter	24.42	27.58	2.34	55.44	nr	**83.02**
40 mm diameter	25.54	28.85	2.34	55.44	nr	**84.29**
50 mm diameter	28.61	32.31	2.34	55.44	nr	**87.75**
65 mm diameter	32.32	36.51	2.34	55.44	nr	**91.95**
80 mm diameter	35.29	39.85	2.76	65.39	nr	**105.24**
100 mm diameter	44.09	49.80	2.76	65.39	nr	**115.19**
125 mm diameter	50.49	57.03	2.76	65.39	nr	**122.42**
150 mm diameter	59.04	66.68	2.76	65.39	nr	**132.07**
200 mm diameter	80.40	90.80	3.36	79.61	nr	**170.41**
250 mm diameter	95.34	107.68	3.36	79.61	nr	**187.29**
300 mm diameter	103.72	117.14	3.36	79.61	nr	**196.75**
350 mm diameter	115.05	129.94	3.36	79.61	nr	**209.55**

T: MECHANICAL/COOLING/HEATING SYSTEMS

Item	Net Price £	Material £	Labour hours	Labour £	Unit	Total rate £
Expansion bellows						
15 mm diameter	15.86	17.91	1.05	24.88	nr	**42.79**
20 mm diameter	16.83	19.01	1.05	24.88	nr	**43.89**
25 mm diameter	17.83	20.13	1.05	24.88	nr	**45.01**
32 mm diameter	19.54	22.07	1.05	24.88	nr	**46.95**
40 mm diameter	20.44	23.08	1.05	24.88	nr	**47.96**
50 mm diameter	22.88	25.84	1.05	24.88	nr	**50.72**
65 mm diameter	25.86	29.20	1.05	24.88	nr	**54.08**
80 mm diameter	28.25	31.91	1.26	29.85	nr	**61.76**
100 mm diameter	35.27	39.83	1.26	29.85	nr	**69.68**
125 mm diameter	40.41	45.63	1.26	29.85	nr	**75.48**
150 mm diameter	47.24	53.36	1.26	29.85	nr	**83.21**
200 mm diameter	64.32	72.64	1.53	36.25	nr	**108.89**
250 mm diameter	76.26	86.13	1.53	36.25	nr	**122.38**
300 mm diameter	82.98	93.72	1.53	36.25	nr	**129.97**
350 mm diameter	92.05	103.96	1.53	36.25	nr	**140.21**
40 mm thick						
15 mm diameter	7.25	8.19	0.44	10.42	m	**18.61**
20 mm diameter	7.61	8.60	0.44	10.42	m	**19.02**
25 mm diameter	8.10	9.15	0.44	10.42	m	**19.57**
32 mm diameter	8.49	9.59	0.44	10.42	m	**20.01**
40 mm diameter	9.02	10.19	0.44	10.42	m	**20.61**
50 mm diameter	9.99	11.28	0.44	10.42	m	**21.70**
65 mm diameter	11.13	12.57	0.44	10.42	m	**22.99**
80 mm diameter	11.98	13.53	0.52	12.32	m	**25.85**
100 mm diameter	17.54	19.81	0.52	12.32	m	**32.13**
125 mm diameter	16.43	18.55	0.52	12.32	m	**30.87**
150 mm diameter	19.17	21.65	0.52	12.32	m	**33.97**
200 mm diameter	25.34	28.62	0.60	14.22	m	**42.84**
250 mm diameter	29.69	33.53	0.60	14.22	m	**47.75**
300 mm diameter	32.83	37.08	0.60	14.22	m	**51.30**
350 mm diameter	36.69	41.44	0.60	14.22	m	**55.66**
400 mm diameter	41.25	46.58	0.60	14.22	m	**60.80**
Extra over for fittings plantroom insulation						
Flange/union						
15 mm diameter	8.36	9.45	0.58	13.75	nr	**23.20**
20 mm diameter	8.73	9.86	0.58	13.75	nr	**23.61**
25 mm diameter	9.34	10.55	0.58	13.75	nr	**24.30**
32 mm diameter	9.83	11.10	0.58	13.75	nr	**24.85**
40 mm diameter	10.43	11.78	0.58	13.75	nr	**25.53**
50 mm diameter	11.62	13.13	0.58	13.75	nr	**26.88**
65 mm diameter	12.98	14.66	0.58	13.75	nr	**28.41**
80 mm diameter	14.00	15.81	0.67	15.88	nr	**31.69**
100 mm diameter	17.54	19.81	0.67	15.88	nr	**35.69**
125 mm diameter	19.64	22.19	0.67	15.88	nr	**38.07**
150 mm diameter	22.92	25.89	0.67	15.88	nr	**41.77**

T: MECHANICAL/COOLING/HEATING SYSTEMS

Item	Net Price £	Material £	Labour hours	Labour £	Unit	Total rate £
T31: THERMAL INSULATION – cont						
Extra over for fittings plantroom insulation – cont						
Flange/union – cont						
200 mm diameter	30.75	34.73	0.87	20.61	nr	**55.34**
250 mm diameter	36.04	40.71	0.87	20.61	nr	**61.32**
300 mm diameter	39.44	44.54	0.87	20.61	nr	**65.15**
350 mm diameter	43.95	49.63	0.87	20.61	nr	**70.24**
400 mm diameter	49.28	55.66	0.87	20.61	nr	**76.27**
Bends						
15 mm diameter	3.98	4.50	0.44	10.42	nr	**14.92**
20 mm diameter	4.18	4.72	0.44	10.42	nr	**15.14**
25 mm diameter	4.46	5.04	0.44	10.42	nr	**15.46**
32 mm diameter	4.66	5.26	0.44	10.42	nr	**15.68**
40 mm diameter	4.96	5.60	0.44	10.42	nr	**16.02**
50 mm diameter	5.49	6.20	0.44	10.42	nr	**16.62**
65 mm diameter	6.12	6.92	0.44	10.42	nr	**17.34**
80 mm diameter	6.58	7.44	0.52	12.32	nr	**19.76**
100 mm diameter	8.12	9.17	0.52	12.32	nr	**21.49**
125 mm diameter	9.03	10.20	0.52	12.32	nr	**22.52**
150 mm diameter	10.54	11.90	0.52	12.32	nr	**24.22**
200 mm diameter	13.92	15.72	0.60	14.22	nr	**29.94**
250 mm diameter	16.22	18.32	0.60	14.22	nr	**32.54**
300 mm diameter	18.06	20.40	0.60	14.22	nr	**34.62**
350 mm diameter	20.18	22.79	0.60	14.22	nr	**37.01**
400 mm diameter	22.69	25.62	0.60	14.22	nr	**39.84**
Tees						
15 mm diameter	2.39	2.70	0.44	10.42	nr	**13.12**
20 mm diameter	2.50	2.83	0.44	10.42	nr	**13.25**
25 mm diameter	2.67	3.02	0.44	10.42	nr	**13.44**
32 mm diameter	2.80	3.16	0.44	10.42	nr	**13.58**
40 mm diameter	2.98	3.37	0.44	10.42	nr	**13.79**
50 mm diameter	3.29	3.71	0.44	10.42	nr	**14.13**
65 mm diameter	3.67	4.14	0.44	10.42	nr	**14.56**
80 mm diameter	3.95	4.46	0.52	12.32	nr	**16.78**
100 mm diameter	4.87	5.50	0.52	12.32	nr	**17.82**
125 mm diameter	5.43	6.13	0.52	12.32	nr	**18.45**
150 mm diameter	6.33	7.15	0.52	12.32	nr	**19.47**
200 mm diameter	8.36	9.45	0.60	14.22	nr	**23.67**
250 mm diameter	9.81	11.08	0.60	14.22	nr	**25.30**
300 mm diameter	10.82	12.22	0.60	14.22	nr	**26.44**
350 mm diameter	12.11	13.68	0.60	14.22	nr	**27.90**
400 mm diameter	13.61	15.37	0.60	14.22	nr	**29.59**

T: MECHANICAL/COOLING/HEATING SYSTEMS

Item	Net Price £	Material £	Labour hours	Labour £	Unit	Total rate £
Valves						
15 mm diameter	13.29	15.01	0.78	18.47	nr	**33.48**
20 mm diameter	13.87	15.67	0.78	18.47	nr	**34.14**
25 mm diameter	14.83	16.75	0.78	18.47	nr	**35.22**
32 mm diameter	15.61	17.63	0.78	18.47	nr	**36.10**
40 mm diameter	16.55	18.69	0.78	18.47	nr	**37.16**
50 mm diameter	18.45	20.84	0.78	18.47	nr	**39.31**
65 mm diameter	20.61	23.28	0.78	18.47	nr	**41.75**
80 mm diameter	22.24	25.12	0.92	21.80	nr	**46.92**
100 mm diameter	27.88	31.49	0.92	21.80	nr	**53.29**
125 mm diameter	31.17	35.20	0.92	21.80	nr	**57.00**
150 mm diameter	36.40	41.11	0.92	21.80	nr	**62.91**
200 mm diameter	48.83	55.15	1.12	26.54	nr	**81.69**
250 mm diameter	57.24	64.65	1.12	26.54	nr	**91.19**
300 mm diameter	62.64	70.75	1.12	26.54	nr	**97.29**
350 mm diameter	69.82	78.86	1.12	26.54	nr	**105.40**
400 mm diameter	78.29	88.42	1.12	26.54	nr	**114.96**
Pumps						
15 mm diameter	24.59	27.77	2.34	55.44	nr	**83.21**
20 mm diameter	25.68	29.00	2.34	55.44	nr	**84.44**
25 mm diameter	27.46	31.02	2.34	55.44	nr	**86.46**
32 mm diameter	28.90	32.64	2.34	55.44	nr	**88.08**
40 mm diameter	30.66	34.63	2.34	55.44	nr	**90.07**
50 mm diameter	34.18	38.61	2.34	55.44	nr	**94.05**
65 mm diameter	38.17	43.11	2.34	55.44	nr	**98.55**
80 mm diameter	41.19	46.52	2.76	65.39	nr	**111.91**
100 mm diameter	51.61	58.29	2.76	65.39	nr	**123.68**
125 mm diameter	57.73	65.20	2.76	65.39	nr	**130.59**
150 mm diameter	67.40	76.12	2.76	65.39	nr	**141.51**
200 mm diameter	90.42	102.12	3.36	79.61	nr	**181.73**
250 mm diameter	106.00	119.72	3.36	79.61	nr	**199.33**
300 mm diameter	151.66	171.29	3.36	79.61	nr	**250.90**
350 mm diameter	129.27	145.99	3.36	79.61	nr	**225.60**
400 mm diameter	144.97	163.73	3.36	79.61	nr	**243.34**
Expansion bellows						
15 mm diameter	19.67	22.22	1.05	24.88	nr	**47.10**
20 mm diameter	20.55	23.21	1.05	24.88	nr	**48.09**
25 mm diameter	21.97	24.82	1.05	24.88	nr	**49.70**
32 mm diameter	23.12	26.11	1.05	24.88	nr	**50.99**
40 mm diameter	24.53	27.70	1.05	24.88	nr	**52.58**
50 mm diameter	27.34	30.88	1.05	24.88	nr	**55.76**
65 mm diameter	30.53	34.48	1.05	24.88	nr	**59.36**
80 mm diameter	32.96	37.22	1.26	29.85	nr	**67.07**
100 mm diameter	41.29	46.63	1.26	29.85	nr	**76.48**
125 mm diameter	46.17	52.14	1.26	29.85	nr	**81.99**
150 mm diameter	53.92	60.89	1.26	29.85	nr	**90.74**
200 mm diameter	72.33	81.69	1.53	36.25	nr	**117.94**

T: MECHANICAL/COOLING/HEATING SYSTEMS

Item	Net Price £	Material £	Labour hours	Labour £	Unit	Total rate £
T31: THERMAL INSULATION – cont						
Extra over for fittings plantroom insulation – cont						
Expansion bellows – cont						
250 mm diameter	84.81	95.79	1.53	36.25	nr	**132.04**
300 mm diameter	92.78	104.78	1.53	36.25	nr	**141.03**
350 mm diameter	103.41	116.79	1.53	36.25	nr	**153.04**
400 mm diameter	115.98	130.99	1.53	36.25	nr	**167.24**
50 mm thick						
15 mm diameter	9.45	10.67	0.44	10.42	m	**21.09**
20 mm diameter	9.96	11.25	0.44	10.42	m	**21.67**
25 mm diameter	10.55	11.91	0.44	10.42	m	**22.33**
32 mm diameter	11.10	12.54	0.44	10.42	m	**22.96**
40 mm diameter	11.60	13.10	0.44	10.42	m	**23.52**
50 mm diameter	12.80	14.45	0.44	10.42	m	**24.87**
65 mm diameter	13.76	15.54	0.44	10.42	m	**25.96**
80 mm diameter	14.78	16.70	0.52	12.32	m	**29.02**
100 mm diameter	18.26	20.62	0.52	12.32	m	**32.94**
125 mm diameter	20.20	22.82	0.52	12.32	m	**35.14**
150 mm diameter	23.21	26.21	0.52	12.32	m	**38.53**
200 mm diameter	30.43	34.36	0.60	14.22	m	**48.58**
250 mm diameter	35.16	39.71	0.60	14.22	m	**53.93**
300 mm diameter	38.28	43.24	0.60	14.22	m	**57.46**
350 mm diameter	42.67	48.20	0.60	14.22	m	**62.42**
400 mm diameter	47.72	53.90	0.60	14.22	m	**68.12**
Extra over for fittings plantroom insulation						
Flange/union						
15 mm diameter	11.13	12.57	0.58	13.75	nr	**26.32**
20 mm diameter	11.74	13.26	0.58	13.75	nr	**27.01**
25 mm diameter	12.44	14.05	0.58	13.75	nr	**27.80**
32 mm diameter	13.06	14.75	0.58	13.75	nr	**28.50**
40 mm diameter	13.68	15.45	0.58	13.75	nr	**29.20**
50 mm diameter	15.19	17.16	0.58	13.75	nr	**30.91**
65 mm diameter	16.41	18.53	0.58	13.75	nr	**32.28**
80 mm diameter	17.59	19.87	0.67	15.88	nr	**35.75**
100 mm diameter	22.00	24.85	0.67	15.88	nr	**40.73**
125 mm diameter	24.47	27.63	0.67	15.88	nr	**43.51**
150 mm diameter	28.15	31.79	0.67	15.88	nr	**47.67**
200 mm diameter	37.42	42.26	0.87	20.61	nr	**62.87**
250 mm diameter	43.25	48.85	0.87	20.61	nr	**69.46**
300 mm diameter	46.66	52.69	0.87	20.61	nr	**73.30**
350 mm diameter	51.88	58.59	0.87	20.61	nr	**79.20**
400 mm diameter	57.88	65.37	0.87	20.61	nr	**85.98**

T: MECHANICAL/COOLING/HEATING SYSTEMS

Item	Net Price £	Material £	Labour hours	Labour £	Unit	Total rate £
Bend						
15 mm diameter	5.20	5.88	0.44	10.42	nr	**16.30**
20 mm diameter	5.48	6.19	0.44	10.42	nr	**16.61**
25 mm diameter	5.80	6.55	0.44	10.42	nr	**16.97**
32 mm diameter	6.10	6.88	0.44	10.42	nr	**17.30**
40 mm diameter	6.37	7.19	0.44	10.42	nr	**17.61**
50 mm diameter	7.04	7.96	0.44	10.42	nr	**18.38**
65 mm diameter	7.57	8.55	0.44	10.42	nr	**18.97**
80 mm diameter	8.12	9.17	0.52	12.32	nr	**21.49**
100 mm diameter	10.03	11.33	0.52	12.32	nr	**23.65**
125 mm diameter	11.11	12.55	0.52	12.32	nr	**24.87**
150 mm diameter	12.77	14.42	0.52	12.32	nr	**26.74**
200 mm diameter	16.73	18.89	0.60	14.22	nr	**33.11**
250 mm diameter	19.34	21.84	0.60	14.22	nr	**36.06**
300 mm diameter	21.05	23.78	0.60	14.22	nr	**38.00**
350 mm diameter	23.46	26.50	0.60	14.22	nr	**40.72**
400 mm diameter	26.24	29.63	0.60	14.22	nr	**43.85**
Tee						
15 mm diameter	3.12	3.52	0.44	10.42	nr	**13.94**
20 mm diameter	3.28	3.70	0.44	10.42	nr	**14.12**
25 mm diameter	3.48	3.93	0.44	10.42	nr	**14.35**
32 mm diameter	3.66	4.13	0.44	10.42	nr	**14.55**
40 mm diameter	3.83	4.32	0.44	10.42	nr	**14.74**
50 mm diameter	4.22	4.76	0.44	10.42	nr	**15.18**
65 mm diameter	4.53	5.12	0.44	10.42	nr	**15.54**
80 mm diameter	4.87	5.50	0.52	12.32	nr	**17.82**
100 mm diameter	6.03	6.81	0.52	12.32	nr	**19.13**
125 mm diameter	6.67	7.54	0.52	12.32	nr	**19.86**
150 mm diameter	7.66	8.65	0.52	12.32	nr	**20.97**
200 mm diameter	10.03	11.33	0.60	14.22	nr	**25.55**
250 mm diameter	11.61	13.12	0.60	14.22	nr	**27.34**
300 mm diameter	12.64	14.28	0.60	14.22	nr	**28.50**
350 mm diameter	14.08	15.90	0.60	14.22	nr	**30.12**
400 mm diameter	15.75	17.79	0.60	14.22	nr	**32.01**
Valves						
15 mm diameter	17.69	19.98	0.78	18.47	nr	**38.45**
20 mm diameter	18.64	21.05	0.78	18.47	nr	**39.52**
25 mm diameter	19.76	22.32	0.78	18.47	nr	**40.79**
32 mm diameter	20.75	23.44	0.78	18.47	nr	**41.91**
40 mm diameter	21.73	24.54	0.78	18.47	nr	**43.01**
50 mm diameter	24.12	27.24	0.78	18.47	nr	**45.71**
65 mm diameter	26.06	29.43	0.78	18.47	nr	**47.90**
80 mm diameter	27.95	31.57	0.92	21.80	nr	**53.37**
100 mm diameter	34.94	39.46	0.92	21.80	nr	**61.26**
125 mm diameter	38.85	43.88	0.92	21.80	nr	**65.68**
150 mm diameter	44.71	50.50	0.92	21.80	nr	**72.30**
200 mm diameter	59.45	67.15	1.12	26.54	nr	**93.69**

T: MECHANICAL/COOLING/HEATING SYSTEMS

Item	Net Price £	Material £	Labour hours	Labour £	Unit	Total rate £
T31: THERMAL INSULATION – cont						
Extra over for fittings plantroom insulation – cont						
Valves – cont						
250 mm diameter	68.69	77.58	1.12	26.54	nr	**104.12**
300 mm diameter	74.41	84.04	1.12	26.54	nr	**110.58**
350 mm diameter	82.39	93.05	1.12	26.54	nr	**119.59**
400 mm diameter	91.93	103.83	1.12	26.54	nr	**130.37**
Pumps						
15 mm diameter	32.76	37.00	2.34	55.44	nr	**92.44**
20 mm diameter	34.54	39.00	2.34	55.44	nr	**94.44**
25 mm diameter	36.58	41.31	2.34	55.44	nr	**96.75**
32 mm diameter	38.42	43.39	2.34	55.44	nr	**98.83**
40 mm diameter	40.25	45.46	2.34	55.44	nr	**100.90**
50 mm diameter	44.64	50.42	2.34	55.44	nr	**105.86**
65 mm diameter	36.04	40.71	2.34	55.44	nr	**96.15**
80 mm diameter	51.76	58.46	2.76	65.39	nr	**123.85**
100 mm diameter	64.70	73.07	2.76	65.39	nr	**138.46**
125 mm diameter	71.96	81.27	2.76	65.39	nr	**146.66**
150 mm diameter	82.80	93.51	2.76	65.39	nr	**158.90**
200 mm diameter	110.08	124.33	3.36	79.61	nr	**203.94**
250 mm diameter	127.20	143.66	3.36	79.61	nr	**223.27**
300 mm diameter	137.26	155.02	3.36	79.61	nr	**234.63**
350 mm diameter	152.58	172.32	3.36	79.61	nr	**251.93**
400 mm diameter	170.24	192.27	3.36	79.61	nr	**271.88**
Expansion bellows						
15 mm diameter	26.20	29.59	1.05	24.88	nr	**54.47**
20 mm diameter	27.62	31.19	1.05	24.88	nr	**56.07**
25 mm diameter	29.26	33.05	1.05	24.88	nr	**57.93**
32 mm diameter	30.73	34.71	1.05	24.88	nr	**59.59**
40 mm diameter	32.19	36.35	1.05	24.88	nr	**61.23**
50 mm diameter	35.73	40.35	1.05	24.88	nr	**65.23**
65 mm diameter	38.60	43.59	1.05	24.88	nr	**68.47**
80 mm diameter	41.42	46.78	1.26	29.85	nr	**76.63**
100 mm diameter	51.75	58.45	1.26	29.85	nr	**88.30**
125 mm diameter	57.56	65.00	1.26	29.85	nr	**94.85**
150 mm diameter	66.25	74.83	1.26	29.85	nr	**104.68**
200 mm diameter	88.07	99.47	1.53	36.25	nr	**135.72**
250 mm diameter	101.75	114.91	1.53	36.25	nr	**151.16**
300 mm diameter	109.82	124.03	1.53	36.25	nr	**160.28**
350 mm diameter	122.06	137.85	1.53	36.25	nr	**174.10**
400 mm diameter	136.18	153.81	1.53	36.25	nr	**190.06**

T: MECHANICAL/COOLING/HEATING SYSTEMS

Item	Net Price £	Material £	Labour hours	Labour £	Unit	Total rate £
Mineral fibre sectional insulation; bright class O foil faced; bright class O foil taped joints; 0.8 mm polyisobutylene sheeting; welded joints						
External pipework						
20 mm thick						
15 mm diameter	4.03	4.55	0.30	7.10	m	11.65
20 mm diameter	4.29	4.85	0.30	7.10	m	11.95
25 mm diameter	4.63	5.23	0.30	7.10	m	12.33
32 mm diameter	5.09	5.75	0.30	7.10	m	12.85
40 mm diameter	5.43	6.13	0.30	7.10	m	13.23
50 mm diameter	6.12	6.92	0.30	7.10	m	14.02
Extra over for fittings external insulation						
Flange/union						
15 mm diameter	6.17	6.97	0.75	17.77	nr	24.74
20 mm diameter	6.55	7.39	0.75	17.77	nr	25.16
25 mm diameter	7.04	7.96	0.75	17.77	nr	25.73
32 mm diameter	7.67	8.66	0.75	17.77	nr	26.43
40 mm diameter	8.10	9.15	0.75	17.77	nr	26.92
50 mm diameter	9.08	10.25	0.75	17.77	nr	28.02
Bends						
15 mm diameter	1.02	1.15	0.30	7.10	nr	8.25
20 mm diameter	1.08	1.22	0.30	7.10	nr	8.32
25 mm diameter	1.16	1.31	0.30	7.10	nr	8.41
32 mm diameter	1.27	1.44	0.30	7.10	nr	8.54
40 mm diameter	1.36	1.54	0.30	7.10	nr	8.64
50 mm diameter	1.53	1.72	0.30	7.10	nr	8.82
Tees						
15 mm diameter	1.02	1.15	0.30	7.10	nr	8.25
20 mm diameter	1.08	1.22	0.30	7.10	nr	8.32
25 mm diameter	1.16	1.31	0.30	7.10	nr	8.41
32 mm diameter	1.27	1.44	0.30	7.10	nr	8.54
40 mm diameter	1.36	1.54	0.30	7.10	nr	8.64
50 mm diameter	1.53	1.72	0.30	7.10	nr	8.82
Valves						
15 mm diameter	9.81	11.08	1.03	24.41	nr	35.49
20 mm diameter	10.42	11.77	1.03	24.41	nr	36.18
25 mm diameter	11.18	12.63	1.03	24.41	nr	37.04
32 mm diameter	12.17	13.75	1.03	24.41	nr	38.16
40 mm diameter	12.86	14.52	1.03	24.41	nr	38.93
50 mm diameter	14.41	16.28	1.03	24.41	nr	40.69

T: MECHANICAL/COOLING/HEATING SYSTEMS

Item	Net Price £	Material £	Labour hours	Labour £	Unit	Total rate £
T31: THERMAL INSULATION – cont						
Extra over for fittings external insulation – cont						
Expansion bellows						
15 mm diameter	14.53	16.41	1.42	33.64	nr	**50.05**
20 mm diameter	15.44	17.44	1.42	33.64	nr	**51.08**
25 mm diameter	16.59	18.74	1.42	33.64	nr	**52.38**
32 mm diameter	18.05	20.39	1.42	33.64	nr	**54.03**
40 mm diameter	19.05	21.51	1.42	33.64	nr	**55.15**
50 mm diameter	21.35	24.11	1.42	33.64	nr	**57.75**
25 mm thick						
15 mm diameter	4.45	5.03	0.30	7.10	m	**12.13**
20 mm diameter	4.77	5.39	0.30	7.10	m	**12.49**
25 mm diameter	5.25	5.93	0.30	7.10	m	**13.03**
32 mm diameter	5.69	6.43	0.30	7.10	m	**13.53**
40 mm diameter	6.05	6.83	0.30	7.10	m	**13.93**
50 mm diameter	6.83	7.71	0.30	7.10	m	**14.81**
65 mm diameter	7.74	8.74	0.30	7.10	m	**15.84**
80 mm diameter	8.47	9.57	0.40	9.48	m	**19.05**
100 mm diameter	10.70	12.09	0.40	9.48	m	**21.57**
125 mm diameter	12.32	13.91	0.40	9.48	m	**23.39**
150 mm diameter	14.45	16.32	0.40	9.48	m	**25.80**
200 mm diameter	19.59	22.12	0.50	11.85	m	**33.97**
250 mm diameter	23.39	26.42	0.50	11.85	m	**38.27**
300 mm diameter	25.45	28.74	0.50	11.85	m	**40.59**
Extra over for fittings external insulation						
Flange/union						
15 mm diameter	6.79	7.67	0.75	17.77	nr	**25.44**
20 mm diameter	7.33	8.28	0.75	17.77	nr	**26.05**
25 mm diameter	8.04	9.08	0.75	17.77	nr	**26.85**
32 mm diameter	8.60	9.71	0.75	17.77	nr	**27.48**
40 mm diameter	9.21	10.40	0.75	17.77	nr	**28.17**
50 mm diameter	10.32	11.66	0.75	17.77	nr	**29.43**
65 mm diameter	11.67	13.18	0.75	17.77	nr	**30.95**
80 mm diameter	12.69	14.33	0.89	21.08	nr	**35.41**
100 mm diameter	15.78	17.82	0.89	21.08	nr	**38.90**
125 mm diameter	18.27	20.63	0.89	21.08	nr	**41.71**
150 mm diameter	21.25	24.00	0.89	21.08	nr	**45.08**
200 mm diameter	28.76	32.48	1.15	27.25	nr	**59.73**
250 mm diameter	34.23	38.66	1.15	27.25	nr	**65.91**
300 mm diameter	38.03	42.95	1.15	27.25	nr	**70.20**
Bends						
15 mm diameter	1.10	1.24	0.30	7.10	nr	**8.34**
20 mm diameter	1.20	1.36	0.30	7.10	nr	**8.46**
25 mm diameter	1.30	1.47	0.30	7.10	nr	**8.57**

T: MECHANICAL/COOLING/HEATING SYSTEMS

Item	Net Price £	Material £	Labour hours	Labour £	Unit	Total rate £
32 mm diameter	1.41	1.59	0.30	7.10	nr	**8.69**
40 mm diameter	1.51	1.70	0.30	7.10	nr	**8.80**
50 mm diameter	1.71	1.93	0.30	7.10	nr	**9.03**
65 mm diameter	1.95	2.20	0.30	7.10	nr	**9.30**
80 mm diameter	2.11	2.39	0.40	9.48	nr	**11.87**
100 mm diameter	2.67	3.02	0.40	9.48	nr	**12.50**
125 mm diameter	3.07	3.47	0.40	9.48	nr	**12.95**
150 mm diameter	3.61	4.08	0.40	9.48	nr	**13.56**
200 mm diameter	4.89	5.52	0.50	11.85	nr	**17.37**
250 mm diameter	5.85	6.61	0.50	11.85	nr	**18.46**
300 mm diameter	6.36	7.18	0.50	11.85	nr	**19.03**
Tees						
15 mm diameter	1.10	1.24	0.30	7.10	nr	**8.34**
20 mm diameter	1.20	1.36	0.30	7.10	nr	**8.46**
25 mm diameter	1.30	1.47	0.30	7.10	nr	**8.57**
32 mm diameter	1.41	1.59	0.30	7.10	nr	**8.69**
40 mm diameter	1.51	1.70	0.30	7.10	nr	**8.80**
50 mm diameter	1.71	1.93	0.30	7.10	nr	**9.03**
65 mm diameter	1.95	2.20	0.30	7.10	nr	**9.30**
80 mm diameter	2.11	2.39	0.40	9.48	nr	**11.87**
100 mm diameter	2.67	3.02	0.40	9.48	nr	**12.50**
125 mm diameter	3.07	3.47	0.40	9.48	nr	**12.95**
150 mm diameter	3.61	4.08	0.40	9.48	nr	**13.56**
200 mm diameter	4.89	5.52	0.50	11.85	nr	**17.37**
250 mm diameter	5.85	6.61	0.50	11.85	nr	**18.46**
300 mm diameter	6.36	7.18	0.50	11.85	nr	**19.03**
Valves						
15 mm diameter	10.79	12.19	1.03	24.41	nr	**36.60**
20 mm diameter	11.64	13.15	1.03	24.41	nr	**37.56**
25 mm diameter	12.76	14.41	1.03	24.41	nr	**38.82**
32 mm diameter	13.67	15.44	1.03	24.41	nr	**39.85**
40 mm diameter	14.61	16.50	1.03	24.41	nr	**40.91**
50 mm diameter	16.41	18.53	1.03	24.41	nr	**42.94**
65 mm diameter	18.55	20.95	1.03	24.41	nr	**45.36**
80 mm diameter	20.16	22.77	1.25	29.62	nr	**52.39**
100 mm diameter	25.07	28.32	1.25	29.62	nr	**57.94**
125 mm diameter	29.00	32.75	1.25	29.62	nr	**62.37**
150 mm diameter	33.77	38.14	1.25	29.62	nr	**67.76**
200 mm diameter	45.69	51.60	1.55	36.73	nr	**88.33**
250 mm diameter	54.37	61.40	1.55	36.73	nr	**98.13**
300 mm diameter	60.40	68.22	1.55	36.73	nr	**104.95**
Expansion bellows						
15 mm diameter	16.00	18.07	1.42	33.64	nr	**51.71**
20 mm diameter	17.24	19.47	1.42	33.64	nr	**53.11**
25 mm diameter	18.91	21.36	1.42	33.64	nr	**55.00**
32 mm diameter	20.27	22.89	1.42	33.64	nr	**56.53**
40 mm diameter	21.66	24.46	1.42	33.64	nr	**58.10**
50 mm diameter	24.28	27.42	1.42	33.64	nr	**61.06**
65 mm diameter	27.47	31.03	1.42	33.64	nr	**64.67**

T: MECHANICAL/COOLING/HEATING SYSTEMS

Item	Net Price £	Material £	Labour hours	Labour £	Unit	Total rate £
T31: THERMAL INSULATION – cont						
Extra over for fittings external insulation – cont						
Expansion bellows – cont						
80 mm diameter	29.86	33.72	1.75	41.47	nr	**75.19**
100 mm diameter	37.12	41.92	1.75	41.47	nr	**83.39**
125 mm diameter	42.97	48.53	1.75	41.47	nr	**90.00**
150 mm diameter	50.02	56.49	1.75	41.47	nr	**97.96**
200 mm diameter	67.68	76.44	2.17	51.42	nr	**127.86**
250 mm diameter	80.54	90.96	2.17	51.42	nr	**142.38**
300 mm diameter	89.48	101.06	3.17	75.11	nr	**176.17**
30 mm thick						
15 mm diameter	6.44	7.27	0.30	7.10	m	**14.37**
20 mm diameter	5.82	6.57	0.30	7.10	m	**13.67**
25 mm diameter	6.17	6.97	0.30	7.10	m	**14.07**
32 mm diameter	6.70	7.57	0.30	7.10	m	**14.67**
40 mm diameter	7.06	7.98	0.30	7.10	m	**15.08**
50 mm diameter	7.95	8.98	0.30	7.10	m	**16.08**
65 mm diameter	8.93	10.09	0.30	7.10	m	**17.19**
80 mm diameter	9.72	10.98	0.40	9.48	m	**20.46**
100 mm diameter	12.15	13.72	0.40	9.48	m	**23.20**
125 mm diameter	13.91	15.71	0.40	9.48	m	**25.19**
150 mm diameter	16.15	18.24	0.40	9.48	m	**27.72**
200 mm diameter	21.75	24.56	0.50	11.85	m	**36.41**
250 mm diameter	25.81	29.15	0.50	11.85	m	**41.00**
300 mm diameter	27.84	31.45	0.50	11.85	m	**43.30**
350 mm diameter	30.45	34.39	0.50	11.85	m	**46.24**
Extra over for fittings external insulation						
Flange/union						
15 mm diameter	8.31	9.38	0.75	17.77	nr	**27.15**
20 mm diameter	8.81	9.95	0.75	17.77	nr	**27.72**
25 mm diameter	9.34	10.55	0.75	17.77	nr	**28.32**
32 mm diameter	10.18	11.50	0.75	17.77	nr	**29.27**
40 mm diameter	10.64	12.02	0.75	17.77	nr	**29.79**
50 mm diameter	11.88	13.41	0.75	17.77	nr	**31.18**
65 mm diameter	13.37	15.10	0.75	17.77	nr	**32.87**
80 mm diameter	14.57	16.45	0.89	21.08	nr	**37.53**
100 mm diameter	17.94	20.26	0.89	21.08	nr	**41.34**
125 mm diameter	20.49	23.14	0.89	21.08	nr	**44.22**
150 mm diameter	23.81	26.89	0.89	21.08	nr	**47.97**
200 mm diameter	31.81	35.92	1.15	27.25	nr	**63.17**
250 mm diameter	37.67	42.54	1.15	27.25	nr	**69.79**
300 mm diameter	41.28	46.62	1.15	27.25	nr	**73.87**
350 mm diameter	45.60	51.50	1.15	27.25	nr	**78.75**

T: MECHANICAL/COOLING/HEATING SYSTEMS

Item	Net Price £	Material £	Labour hours	Labour £	Unit	Total rate £
Bends						
15 mm diameter	1.37	1.55	0.30	7.10	nr	8.65
20 mm diameter	1.45	1.64	0.30	7.10	nr	8.74
25 mm diameter	1.55	1.75	0.30	7.10	nr	8.85
32 mm diameter	1.67	1.89	0.30	7.10	nr	8.99
40 mm diameter	1.76	1.99	0.30	7.10	nr	9.09
50 mm diameter	2.00	2.25	0.30	7.10	nr	9.35
65 mm diameter	2.24	2.53	0.30	7.10	nr	9.63
80 mm diameter	2.43	2.74	0.40	9.48	nr	12.22
100 mm diameter	3.04	3.44	0.40	9.48	nr	12.92
125 mm diameter	3.48	3.93	0.40	9.48	nr	13.41
150 mm diameter	4.04	4.56	0.40	9.48	nr	14.04
200 mm diameter	5.45	6.15	0.50	11.85	nr	18.00
250 mm diameter	6.46	7.29	0.50	11.85	nr	19.14
300 mm diameter	6.96	7.86	0.50	11.85	nr	19.71
350 mm diameter	7.61	8.60	0.50	11.85	nr	20.45
Tees						
15 mm diameter	1.37	1.55	0.30	7.10	nr	8.65
20 mm diameter	1.45	1.64	0.30	7.10	nr	8.74
25 mm diameter	1.55	1.75	0.30	7.10	nr	8.85
32 mm diameter	1.67	1.89	0.30	7.10	nr	8.99
40 mm diameter	1.76	1.99	0.30	7.10	nr	9.09
50 mm diameter	2.00	2.25	0.30	7.10	nr	9.35
65 mm diameter	2.24	2.53	0.30	7.10	nr	9.63
80 mm diameter	2.43	2.74	0.40	9.48	nr	12.22
100 mm diameter	3.04	3.44	0.40	9.48	nr	12.92
125 mm diameter	3.48	3.93	0.40	9.48	nr	13.41
150 mm diameter	4.04	4.56	0.40	9.48	nr	14.04
200 mm diameter	5.45	6.15	0.50	11.85	nr	18.00
250 mm diameter	6.46	7.29	0.50	11.85	nr	19.14
300 mm diameter	6.96	7.86	0.50	11.85	nr	19.71
350 mm diameter	7.61	8.60	0.50	11.85	nr	20.45
Valves						
15 mm diameter	13.21	14.92	1.03	24.41	nr	39.33
20 mm diameter	13.99	15.80	1.03	24.41	nr	40.21
25 mm diameter	14.84	16.76	1.03	24.41	nr	41.17
32 mm diameter	16.16	18.25	1.03	24.41	nr	42.66
40 mm diameter	16.92	19.10	1.03	24.41	nr	43.51
50 mm diameter	18.87	21.31	1.03	24.41	nr	45.72
65 mm diameter	21.23	23.98	1.03	24.41	nr	48.39
80 mm diameter	23.14	26.13	1.25	29.62	nr	55.75
100 mm diameter	28.47	32.15	1.25	29.62	nr	61.77
125 mm diameter	32.54	36.75	1.25	29.62	nr	66.37
150 mm diameter	37.80	42.69	1.25	29.62	nr	72.31
200 mm diameter	50.53	57.07	1.55	36.73	nr	93.80
250 mm diameter	59.84	67.59	1.55	36.73	nr	104.32
300 mm diameter	65.56	74.04	1.55	36.73	nr	110.77
350 mm diameter	72.42	81.79	1.55	36.73	nr	118.52

T: MECHANICAL/COOLING/HEATING SYSTEMS

Item	Net Price £	Material £	Labour hours	Labour £	Unit	Total rate £
T31: THERMAL INSULATION – cont						
Extra over for fittings external insulation – cont						
Expansion bellows						
15 mm diameter	19.55	22.08	1.42	33.64	nr	**55.72**
20 mm diameter	20.74	23.42	1.42	33.64	nr	**57.06**
25 mm diameter	21.98	24.83	1.42	33.64	nr	**58.47**
32 mm diameter	23.95	27.05	1.42	33.64	nr	**60.69**
40 mm diameter	25.07	28.32	1.42	33.64	nr	**61.96**
50 mm diameter	27.94	31.56	1.42	33.64	nr	**65.20**
65 mm diameter	31.46	35.53	1.42	33.64	nr	**69.17**
80 mm diameter	34.29	38.73	1.75	41.47	nr	**80.20**
100 mm diameter	42.18	47.63	1.75	41.47	nr	**89.10**
125 mm diameter	48.21	54.45	1.75	41.47	nr	**95.92**
150 mm diameter	56.00	63.25	1.75	41.47	nr	**104.72**
200 mm diameter	74.86	84.55	2.17	51.42	nr	**135.97**
250 mm diameter	88.65	100.12	2.17	51.42	nr	**151.54**
300 mm diameter	97.12	109.69	2.17	51.42	nr	**161.11**
350 mm diameter	107.30	121.19	2.17	51.42	nr	**172.61**
40 mm thick						
15 mm diameter	6.85	7.73	0.30	7.10	m	**14.83**
20 mm diameter	7.08	8.00	0.30	7.10	m	**15.10**
25 mm diameter	7.57	8.55	0.30	7.10	m	**15.65**
32 mm diameter	8.05	9.09	0.30	7.10	m	**16.19**
40 mm diameter	8.43	9.52	0.30	7.10	m	**16.62**
50 mm diameter	9.40	10.62	0.30	7.10	m	**17.72**
65 mm diameter	10.48	11.83	0.30	7.10	m	**18.93**
80 mm diameter	11.36	12.83	0.40	9.48	m	**22.31**
100 mm diameter	14.19	16.02	0.40	9.48	m	**25.50**
125 mm diameter	15.99	18.05	0.40	9.48	m	**27.53**
150 mm diameter	18.45	20.84	0.40	9.48	m	**30.32**
200 mm diameter	24.61	27.80	0.50	11.85	m	**39.65**
250 mm diameter	28.82	32.55	0.50	11.85	m	**44.40**
300 mm diameter	31.20	35.24	0.50	11.85	m	**47.09**
350 mm diameter	34.29	38.73	0.50	11.85	m	**50.58**
400 mm diameter	38.17	43.11	0.50	11.85	m	**54.96**
Extra over for fittings external insulation						
Flange/union						
15 mm diameter	10.23	11.56	0.75	17.77	nr	**29.33**
20 mm diameter	10.69	12.08	0.75	17.77	nr	**29.85**
25 mm diameter	11.40	12.87	0.75	17.77	nr	**30.64**
32 mm diameter	12.01	13.57	0.75	17.77	nr	**31.34**
40 mm diameter	12.68	14.32	0.75	17.77	nr	**32.09**
50 mm diameter	14.05	15.87	0.75	17.77	nr	**33.64**
65 mm diameter	15.65	17.68	0.75	17.77	nr	**35.45**

T: MECHANICAL/COOLING/HEATING SYSTEMS

Item	Net Price £	Material £	Labour hours	Labour £	Unit	Total rate £
80 mm diameter	16.87	19.05	0.89	21.08	nr	**40.13**
100 mm diameter	20.78	23.47	0.89	21.08	nr	**44.55**
125 mm diameter	23.24	26.24	0.89	21.08	nr	**47.32**
150 mm diameter	26.93	30.42	0.89	21.08	nr	**51.50**
200 mm diameter	35.51	40.11	1.15	27.25	nr	**67.36**
250 mm diameter	41.62	47.00	1.15	27.25	nr	**74.25**
300 mm diameter	45.74	51.66	1.15	27.25	nr	**78.91**
350 mm diameter	50.73	57.29	1.15	27.25	nr	**84.54**
400 mm diameter	56.81	64.16	1.15	27.25	nr	**91.41**
Bends						
15 mm diameter	1.71	1.93	0.30	7.10	nr	**9.03**
20 mm diameter	1.77	2.00	0.30	7.10	nr	**9.10**
25 mm diameter	1.89	2.13	0.30	7.10	nr	**9.23**
32 mm diameter	2.02	2.28	0.30	7.10	nr	**9.38**
40 mm diameter	2.11	2.39	0.30	7.10	nr	**9.49**
50 mm diameter	2.36	2.66	0.30	7.10	nr	**9.76**
65 mm diameter	2.63	2.97	0.30	7.10	nr	**10.07**
80 mm diameter	2.85	3.22	0.40	9.48	nr	**12.70**
100 mm diameter	3.55	4.01	0.40	9.48	nr	**13.49**
125 mm diameter	3.99	4.51	0.40	9.48	nr	**13.99**
150 mm diameter	4.62	5.22	0.40	9.48	nr	**14.70**
200 mm diameter	6.15	6.95	0.50	11.85	nr	**18.80**
250 mm diameter	7.20	8.13	0.50	11.85	nr	**19.98**
300 mm diameter	7.80	8.81	0.50	11.85	nr	**20.66**
350 mm diameter	8.57	9.68	0.50	11.85	nr	**21.53**
400 mm diameter	9.54	10.77	0.50	11.85	nr	**22.62**
Tees						
15 mm diameter	1.71	1.93	0.30	7.10	nr	**9.03**
20 mm diameter	1.77	2.00	0.30	7.10	nr	**9.10**
25 mm diameter	1.89	2.13	0.30	7.10	nr	**9.23**
32 mm diameter	2.02	2.28	0.30	7.10	nr	**9.38**
40 mm diameter	2.11	2.39	0.30	7.10	nr	**9.49**
50 mm diameter	2.36	2.66	0.30	7.10	nr	**9.76**
65 mm diameter	2.63	2.97	0.30	7.10	nr	**10.07**
80 mm diameter	2.85	3.22	0.40	9.48	nr	**12.70**
100 mm diameter	3.55	4.01	0.40	9.48	nr	**13.49**
125 mm diameter	3.99	4.51	0.40	9.48	nr	**13.99**
150 mm diameter	4.62	5.22	0.40	9.48	nr	**14.70**
200 mm diameter	6.15	6.95	0.50	11.85	nr	**18.80**
250 mm diameter	7.20	8.13	0.50	11.85	nr	**19.98**
300 mm diameter	7.80	8.81	0.50	11.85	nr	**20.66**
350 mm diameter	8.57	9.68	0.50	11.85	nr	**21.53**
400 mm diameter	9.54	10.77	0.50	11.85	nr	**22.62**

T: MECHANICAL/COOLING/HEATING SYSTEMS

Item	Net Price £	Material £	Labour hours	Labour £	Unit	Total rate £
T31: THERMAL INSULATION – cont						
Extra over for fittings external insulation – cont						
Valves						
15 mm diameter	16.24	18.34	1.03	24.41	nr	42.75
20 mm diameter	16.97	19.17	1.03	24.41	nr	43.58
25 mm diameter	18.09	20.43	1.03	24.41	nr	44.84
32 mm diameter	19.06	21.52	1.03	24.41	nr	45.93
40 mm diameter	20.15	22.76	1.03	24.41	nr	47.17
50 mm diameter	22.33	25.21	1.03	24.41	nr	49.62
65 mm diameter	24.85	28.07	1.03	24.41	nr	52.48
80 mm diameter	26.79	30.25	1.25	29.62	nr	59.87
100 mm diameter	33.00	37.27	1.25	29.62	nr	66.89
125 mm diameter	36.91	41.69	1.25	29.62	nr	71.31
150 mm diameter	42.78	48.32	1.25	29.62	nr	77.94
200 mm diameter	56.40	63.70	1.55	36.73	nr	100.43
250 mm diameter	66.08	74.63	1.55	36.73	nr	111.36
300 mm diameter	72.65	82.05	1.55	36.73	nr	118.78
350 mm diameter	80.57	90.99	1.55	36.73	nr	127.72
400 mm diameter	90.21	101.89	1.55	36.73	nr	138.62
Expansion bellows						
15 mm diameter	24.06	27.17	1.42	33.64	nr	60.81
20 mm diameter	25.15	28.41	1.42	33.64	nr	62.05
25 mm diameter	26.80	30.26	1.42	33.64	nr	63.90
32 mm diameter	28.23	31.89	1.42	33.64	nr	65.53
40 mm diameter	29.85	33.71	1.42	33.64	nr	67.35
50 mm diameter	33.08	37.36	1.42	33.64	nr	71.00
65 mm diameter	36.82	41.59	1.42	33.64	nr	75.23
80 mm diameter	39.69	44.83	1.75	41.47	nr	86.30
100 mm diameter	48.90	55.22	1.75	41.47	nr	96.69
125 mm diameter	54.68	61.75	1.75	41.47	nr	103.22
150 mm diameter	63.39	71.59	1.75	41.47	nr	113.06
200 mm diameter	83.56	94.37	2.17	51.42	nr	145.79
250 mm diameter	97.90	110.57	2.17	51.42	nr	161.99
300 mm diameter	107.63	121.55	2.17	51.42	nr	172.97
350 mm diameter	119.37	134.81	2.17	51.42	nr	186.23
400 mm diameter	133.65	150.94	2.17	51.42	nr	202.36
50 mm thick						
15 mm diameter	9.02	10.19	0.30	7.10	m	17.29
20 mm diameter	9.47	10.70	0.30	7.10	m	17.80
25 mm diameter	10.01	11.30	0.30	7.10	m	18.40
32 mm diameter	10.80	12.20	0.30	7.10	m	19.30
40 mm diameter	10.98	12.40	0.30	7.10	m	19.50
50 mm diameter	12.16	13.73	0.30	7.10	m	20.83
65 mm diameter	13.26	14.97	0.30	7.10	m	22.07

T: MECHANICAL/COOLING/HEATING SYSTEMS

Item	Net Price £	Material £	Labour hours	Labour £	Unit	Total rate £
80 mm diameter	14.19	16.02	0.40	9.48	m	**25.50**
100 mm diameter	17.56	19.83	0.40	9.48	m	**29.31**
125 mm diameter	19.65	22.20	0.40	9.48	m	**31.68**
150 mm diameter	22.46	25.37	0.40	9.48	m	**34.85**
200 mm diameter	29.75	33.60	0.50	11.85	m	**45.45**
250 mm diameter	34.36	38.81	0.50	11.85	m	**50.66**
300 mm diameter	36.85	41.62	0.50	11.85	m	**53.47**
350 mm diameter	40.44	45.68	0.50	11.85	m	**57.53**
400 mm diameter	44.80	50.59	0.50	11.85	m	**62.44**
Extra over for fittings external insulation						
Flange/union						
15 mm diameter	13.29	15.01	0.75	17.77	nr	**32.78**
20 mm diameter	13.99	15.80	0.75	17.77	nr	**33.57**
25 mm diameter	14.79	16.71	0.75	17.77	nr	**34.48**
32 mm diameter	15.52	17.52	0.75	17.77	nr	**35.29**
40 mm diameter	16.24	18.34	0.75	17.77	nr	**36.11**
50 mm diameter	17.91	20.23	0.75	17.77	nr	**38.00**
65 mm diameter	19.36	21.87	0.75	17.77	nr	**39.64**
80 mm diameter	20.76	23.45	0.89	21.08	nr	**44.53**
100 mm diameter	25.51	28.82	0.89	21.08	nr	**49.90**
125 mm diameter	28.36	32.03	0.89	21.08	nr	**53.11**
150 mm diameter	32.48	36.68	0.89	21.08	nr	**57.76**
200 mm diameter	42.49	47.99	1.15	27.25	nr	**75.24**
250 mm diameter	49.10	55.46	1.15	27.25	nr	**82.71**
300 mm diameter	53.28	60.17	1.15	27.25	nr	**87.42**
350 mm diameter	58.95	66.58	1.15	27.25	nr	**93.83**
400 mm diameter	65.69	74.19	1.15	27.25	nr	**101.44**
Bend						
15 mm diameter	2.26	2.55	0.30	7.10	nr	**9.65**
20 mm diameter	2.37	2.67	0.30	7.10	nr	**9.77**
25 mm diameter	2.50	2.83	0.30	7.10	nr	**9.93**
32 mm diameter	2.63	2.97	0.30	7.10	nr	**10.07**
40 mm diameter	2.75	3.10	0.30	7.10	nr	**10.20**
50 mm diameter	3.04	3.44	0.30	7.10	nr	**10.54**
65 mm diameter	3.31	3.73	0.30	7.10	nr	**10.83**
80 mm diameter	3.54	4.00	0.40	9.48	nr	**13.48**
100 mm diameter	4.39	4.96	0.40	9.48	nr	**14.44**
125 mm diameter	4.90	5.54	0.40	9.48	nr	**15.02**
150 mm diameter	5.62	6.34	0.40	9.48	nr	**15.82**
200 mm diameter	7.44	8.40	0.50	11.85	nr	**20.25**
250 mm diameter	8.60	9.71	0.50	11.85	nr	**21.56**
300 mm diameter	9.22	10.41	0.50	11.85	nr	**22.26**
350 mm diameter	10.12	11.43	0.50	11.85	nr	**23.28**
400 mm diameter	11.19	12.64	0.50	11.85	nr	**24.49**

T: MECHANICAL/COOLING/HEATING SYSTEMS

Item	Net Price £	Material £	Labour hours	Labour £	Unit	Total rate £
T31: THERMAL INSULATION – cont						
Extra over for fittings external insulation – cont						
Tee						
15 mm diameter	2.26	2.55	0.30	7.10	nr	**9.65**
20 mm diameter	2.37	2.67	0.30	7.10	nr	**9.77**
25 mm diameter	2.50	2.83	0.30	7.10	nr	**9.93**
32 mm diameter	2.63	2.97	0.30	7.10	nr	**10.07**
40 mm diameter	2.75	3.10	0.30	7.10	nr	**10.20**
50 mm diameter	3.04	3.44	0.30	7.10	nr	**10.54**
65 mm diameter	3.31	3.73	0.30	7.10	nr	**10.83**
80 mm diameter	3.54	4.00	0.40	9.48	nr	**13.48**
100 mm diameter	4.39	4.96	0.40	9.48	nr	**14.44**
125 mm diameter	4.90	5.54	0.40	9.48	nr	**15.02**
150 mm diameter	5.62	6.34	0.40	9.48	nr	**15.82**
200 mm diameter	7.44	8.40	0.50	11.85	nr	**20.25**
250 mm diameter	8.60	9.71	0.50	11.85	nr	**21.56**
300 mm diameter	9.22	10.41	0.50	11.85	nr	**22.26**
350 mm diameter	10.12	11.43	0.50	11.85	nr	**23.28**
400 mm diameter	11.19	12.64	0.50	11.85	nr	**24.49**
Valves						
15 mm diameter	21.12	23.86	1.03	24.41	nr	**48.27**
20 mm diameter	22.22	25.09	1.03	24.41	nr	**49.50**
25 mm diameter	23.49	26.53	1.03	24.41	nr	**50.94**
32 mm diameter	24.66	27.85	1.03	24.41	nr	**52.26**
40 mm diameter	25.80	29.14	1.03	24.41	nr	**53.55**
50 mm diameter	28.45	32.13	1.03	24.41	nr	**56.54**
65 mm diameter	30.77	34.75	1.03	24.41	nr	**59.16**
80 mm diameter	32.97	37.24	1.25	29.62	nr	**66.86**
100 mm diameter	40.53	45.78	1.25	29.62	nr	**75.40**
125 mm diameter	45.05	50.88	1.25	29.62	nr	**80.50**
150 mm diameter	51.57	58.24	1.25	29.62	nr	**87.86**
200 mm diameter	67.48	76.21	1.55	36.73	nr	**112.94**
250 mm diameter	77.98	88.07	1.55	36.73	nr	**124.80**
300 mm diameter	84.60	95.54	1.55	36.73	nr	**132.27**
350 mm diameter	93.62	105.73	1.55	36.73	nr	**142.46**
400 mm diameter	104.32	117.82	1.55	36.73	nr	**154.55**
Expansion bellows						
15 mm diameter	31.30	35.35	1.42	33.64	nr	**68.99**
20 mm diameter	32.92	37.18	1.42	33.64	nr	**70.82**
25 mm diameter	34.80	39.30	1.42	33.64	nr	**72.94**
32 mm diameter	36.55	41.28	1.42	33.64	nr	**74.92**
40 mm diameter	38.22	43.17	1.42	33.64	nr	**76.81**
50 mm diameter	42.14	47.59	1.42	33.64	nr	**81.23**
65 mm diameter	45.58	51.48	1.42	33.64	nr	**85.12**

T: MECHANICAL/COOLING/HEATING SYSTEMS

Item	Net Price £	Material £	Labour hours	Labour £	Unit	Total rate £
80 mm diameter	48.85	55.17	1.75	41.47	nr	**96.64**
100 mm diameter	60.05	67.82	1.75	41.47	nr	**109.29**
125 mm diameter	66.74	75.38	1.75	41.47	nr	**116.85**
150 mm diameter	76.42	86.31	1.75	41.47	nr	**127.78**
200 mm diameter	99.97	112.90	2.17	51.42	nr	**164.32**
250 mm diameter	115.52	130.47	2.17	51.42	nr	**181.89**
300 mm diameter	125.34	141.56	2.17	51.42	nr	**192.98**
350 mm diameter	138.70	156.65	2.17	51.42	nr	**208.07**
400 mm diameter	154.56	174.56	2.17	51.42	nr	**225.98**

T: MECHANICAL/COOLING/HEATING SYSTEMS

Item	Net Price £	Material £	Labour hours	Labour £	Unit	Total rate £
T33: STEAM HEATING						
Y10 – PIPELINES						
For pipework prices refer to Section T31 – Low Temperature Hot Water Heating						
Y11 – PIPELINE ANCILLARIES						
Steam traps and accessories						
Cast iron; inverted bucket type; steam trap pressure range up to 17 bar at 210°C; screwed ends						
½' dia.	94.43	106.65	0.85	20.13	nr	**126.78**
¾' dia.	94.43	106.65	1.13	26.78	nr	**133.43**
1' dia.	159.11	179.69	1.35	31.99	nr	**211.68**
1½' dia.	491.23	554.79	1.80	42.65	nr	**597.44**
2' dia.	549.86	621.01	2.18	51.66	nr	**672.67**
Cast iron; inverted bucket type; steam trap pressure range up to 17 bar at 210°C; flanged ends to BS 4504 PN16; bolted connections						
15 mm dia.	158.16	178.62	1.15	27.25	nr	**205.87**
20 mm dia.	191.14	215.87	1.25	29.62	nr	**245.49**
25 mm dia.	237.26	267.96	1.33	31.51	nr	**299.47**
40 mm dia.	503.35	568.48	1.46	34.60	nr	**603.08**
50 mm dia.	561.97	634.68	1.60	37.91	nr	**672.59**
Steam traps and strainers						
Stainless steel; thermodynamic trap with pressure range up to 42 bar; temperature range to 400°C; screwed ends to steel						
15 mm dia.	63.23	71.41	0.84	19.91	nr	**91.32**
20 mm dia.	71.56	80.81	1.14	27.02	nr	**107.83**
Stainless steel; thermodynamic trap with pressure range up to 24 bar; temperature range to 288°C; flanged ends to DIN 2456 PN64; bolted connections						
15 mm dia.	132.38	149.51	1.24	29.40	nr	**178.91**
20 mm dia.	145.11	163.88	1.34	31.76	nr	**195.64**
25 mm dia.	167.73	189.43	1.40	33.18	nr	**222.61**

T: MECHANICAL/COOLING/HEATING SYSTEMS

Item	Net Price £	Material £	Labour hours	Labour £	Unit	Total rate £
Malleable iron pipeline strainer; max. steam working pressure 14 bar and temperature range to 230°C; screwed ends to steel						
½' dia.	18.41	20.79	0.84	19.91	nr	40.70
¾' dia.	23.01	25.99	1.14	27.02	nr	53.01
1' dia.	35.90	40.54	1.30	30.81	nr	71.35
1½' dia.	73.74	83.28	1.50	35.58	nr	118.86
2' dia.	103.54	116.93	1.74	41.28	nr	158.21
Bronze pipeline strainer; max. steam working pressure 25 bar; flanged ends to BS 4504 PN25; bolted connections						
15 mm dia.	159.59	180.24	1.24	29.40	nr	209.64
20 mm dia.	194.65	219.84	1.34	31.76	nr	251.60
25 mm dia.	223.19	252.07	1.40	33.18	nr	285.25
32 mm dia.	346.47	391.30	1.46	34.64	nr	425.94
40 mm dia.	393.19	444.07	1.54	36.52	nr	480.59
50 mm dia.	604.70	682.94	1.64	38.90	nr	721.84
65 mm dia.	669.58	756.22	2.50	59.24	nr	815.46
80 mm dia.	833.08	940.88	2.91	68.88	nr	1009.76
100 mm dia.	1442.98	1629.69	3.51	83.13	nr	1712.82
Balanced pressure thermostatic steam trap and strainer; max. working pressure up to 13 bar; screwed ends to steel						
½' dia.	131.04	147.99	1.26	29.89	nr	177.88
¾' dia.	131.39	148.39	1.71	40.57	nr	188.96
Bimetallic thermostatic steam trap and strainer; max. working pressure up to 21 bar; flanged ends						
15 mm	170.47	192.52	1.24	29.40	nr	221.92
20 mm	187.27	211.50	1.34	31.76	nr	243.26
Sight glasses						
Pressed brass; straight; single window; screwed ends to steel						
15 mm dia.	40.22	45.42	0.84	19.91	nr	65.33
20 mm dia.	44.64	50.42	1.14	27.02	nr	77.44
25 mm dia.	55.79	63.01	1.30	30.81	nr	93.82
Gunmetal; straight; double window; screwed ends to steel						
15 mm dia.	64.87	73.27	0.84	19.91	nr	93.18
20 mm dia.	71.37	80.60	1.14	27.02	nr	107.62
25 mm dia.	88.23	99.64	1.30	30.81	nr	130.45
32 mm dia.	145.34	164.15	1.35	32.02	nr	196.17
40 mm dia.	145.34	164.15	1.74	41.28	nr	205.43
50 mm dia.	176.48	199.32	2.08	49.37	nr	248.69

T: MECHANICAL/COOLING/HEATING SYSTEMS

Item	Net Price £	Material £	Labour hours	Labour £	Unit	Total rate £
T33: STEAM HEATING – cont						
Sight glasses – cont						
SG Iron flanged; BS 4504, PN 25						
15 mm dia.	132.36	149.49	1.00	23.69	nr	**173.18**
20 mm dia.	155.73	175.88	1.25	29.62	nr	**205.50**
25 mm dia.	198.56	224.26	1.50	35.58	nr	**259.84**
32 mm dia.	219.29	247.67	1.70	40.30	nr	**287.97**
40 mm dia.	286.79	323.90	2.00	47.39	nr	**371.29**
50 mm dia.	346.47	391.30	2.30	54.59	nr	**445.89**
Check valve and sight glass; gun metal; screwed						
15 mm dia.	65.51	73.99	0.84	19.91	nr	**93.90**
20 mm dia.	69.41	78.39	1.14	27.02	nr	**105.41**
25 mm dia.	116.78	131.89	1.30	30.81	nr	**162.70**
Pressure reducing valves						
Pressure reducing valve for steam; maximum range of 17 bar and 232°C; screwed ends to steel						
15 mm dia.	490.51	553.98	0.87	20.61	nr	**574.59**
20 mm dia.	530.73	599.40	0.91	21.56	nr	**620.96**
25 mm dia.	572.25	646.29	1.35	31.99	nr	**678.28**
Pressure reducing valve for steam; maximum range of 17 bar and 232°C; flanged ends to BS 4504 PN 25						
25 mm dia.	689.05	778.21	1.70	40.29	nr	**818.50**
32 mm dia.	782.49	883.74	1.87	44.31	nr	**928.05**
40 mm dia.	934.31	1055.20	2.12	50.23	nr	**1105.43**
50 mm dia.	1078.35	1217.88	2.57	60.89	nr	**1278.77**
Safety and relief valves						
Bronze safety valve; 'pop' type; side outlet; including easing lever; working pressure saturated steam up to 20.7 bar; screwed ends to steel						
15 mm dia.	149.23	168.53	0.32	7.58	nr	**176.11**
20 mm dia.	168.68	190.51	0.40	9.48	nr	**199.99**
Bronze safety valve; 'pop' type; side outlet; including easing lever; working pressure saturated steam up to 17.2 bar; screwed ends to steel						
25 mm dia.	240.05	271.12	0.47	11.14	nr	**282.26**
32 mm dia.	286.79	323.90	0.56	13.27	nr	**337.17**

T: MECHANICAL/COOLING/HEATING SYSTEMS

Item	Net Price £	Material £	Labour hours	Labour £	Unit	Total rate £
Bronze safety valve; 'pop' type; side outlet; including easing lever; working pressure saturated steam up to 13.8 bar; screwed ends to steel						
40 mm dia.	371.13	419.15	0.64	15.17	nr	**434.32**
50 mm dia.	516.47	583.30	0.76	18.00	nr	**601.30**
65 mm dia.	734.48	829.52	0.94	22.27	nr	**851.79**
80 mm dia.	952.50	1075.75	1.10	26.07	nr	**1101.82**
EQUIPMENT						
Y23 – CALORIFIERS						
Non-storage calorifiers; mild steel shell construction with indirect steam heating for secondary LPHW at 82°C flow and 71°C return to BS 853; maximum test on shell 11 bar, tubes 26 bar						
Horizontal/vertical, for steam at 3 bar–5.5 bar						
88 kW capacity	580.43	655.53	8.00	189.56	nr	**845.09**
176 kW capacity	844.97	954.30	12.05	285.48	nr	**1239.78**
293 kW capacity	920.55	1039.67	14.08	333.72	nr	**1373.39**
586 kW capacity	1344.00	1517.90	37.04	877.59	nr	**2395.49**
879 kW capacity	1673.46	1890.00	40.00	947.78	nr	**2837.78**
1465 kW capacity	2040.71	2304.77	45.45	1077.04	nr	**3381.81**

T: MECHANICAL/COOLING/HEATING SYSTEMS

Item	Net Price £	Material £	Labour hours	Labour £	Unit	Total rate £
T42: LOCAL HEATING UNITS						
Warm air unit heater for connection to LTHW or steam supplies; suitable for heights up to 3 m; recirculating type; mild steel casing; heating coil; adjustable discharge louvre; axial fan; horizontal or vertical discharge; normal speed; entering air temperature 15°C; complete with enclosures; includes fixing in position; includes connections to primary heating supply; electrical work elsewhere						
Low pressure hot water						
7.5 kW, 265 l/sec	336.74	380.32	6.53	154.72	nr	**535.04**
15.4 kW, 575 l/sec	407.62	460.37	7.54	178.66	nr	**639.03**
26.9 kW, 1040 l/sec	552.34	623.81	8.65	204.96	nr	**828.77**
48.0 kW, 1620 l/sec	728.09	822.30	9.35	221.54	nr	**1043.84**
Steam, 2 Bar						
9.2 kW, 265 l/sec	481.45	543.75	6.53	154.72	nr	**698.47**
18.8 kW, 575 l/sec	521.33	588.78	6.82	161.60	nr	**750.38**
34.8 kW, 1040 l/sec	599.61	677.20	6.82	161.60	nr	**838.80**
51.6 kW, 1625 l/sec	812.29	917.40	7.10	168.24	nr	**1085.64**

T: MECHANICAL/COOLING/HEATING SYSTEMS

Item	Net Price £	Material £	Labour hours	Labour £	Unit	Total rate £
T60: CENTRAL REFRIGERATION PLANT						
CHILLERS						
Air cooled						
Selection of air cooled chillers based on chilled water flow and return temperatures 6°C and 12°C, and an outdoor temperature of 35°C						
Air cooled liquid chiller; refrigerant 407C; scroll compressors; twin circuit; integral controls; includes placing in position; electrical work elsewhere						
Cooling load						
100 kW	18426.04	20810.27	8.00	189.56	nr	**20999.83**
150 kW	20785.72	23475.29	8.00	189.56	nr	**23664.85**
200 kW	26385.12	29799.22	8.00	189.56	nr	**29988.78**
Air cooled liquid chiller; refrigerant 407C; reciprocating compressors; twin circuit; integral controls; includes placing in position; electrical work elsewhere						
Cooling load						
250 kW	34678.19	39165.38	8.00	189.56	nr	**39354.94**
400 kW	48806.65	55121.98	8.00	189.56	nr	**55311.54**
550 kW	62178.18	70223.73	8.00	189.56	nr	**70413.29**
700 kW	78051.88	88151.40	9.00	213.25	nr	**88364.65**
Air cooled liquid chiller; refrigerant R134a; screw compressors; twin circuit; integral controls; includes placing in position; electrical work elsewhere						
Cooling load						
250 kW	39365.65	44459.37	8.00	189.56	nr	**44648.93**
400 kW	49662.75	56088.86	8.00	189.56	nr	**56278.42**
600 kW	64512.79	72860.43	8.00	189.56	nr	**73049.99**
800 kW	95926.19	108338.56	9.00	213.25	nr	**108551.81**
1000 kW	111574.15	126011.29	9.00	213.25	nr	**126224.54**
1200 kW	129152.10	145863.73	10.00	236.95	nr	**146100.68**

T: MECHANICAL/COOLING/HEATING SYSTEMS

Item	Net Price £	Material £	Labour hours	Labour £	Unit	Total rate £
T60: CENTRAL REFRIGERATION PLANT – cont						
CHILLERS – cont						
Air cooled – cont						
Air cooled liquid chiller; ductable for indoor installation; refrigerant 407C; scroll compressors; integral controls; includes placing in position; electrical work elsewhere						
Cooling load						
40 kW	12940.63	14615.08	6.00	142.17	nr	**14757.25**
80 kW	18793.10	21224.83	6.00	142.17	nr	**21367.00**
Higher efficiency air cooled						
Selection of air cooled chillers based on chilled water flow and return temperatures of 6°C and 12°C and an outdoor temperature of 25°C						
These machines have significantly higher part load operating efficiencies than conventional air cooled machines						
Air cooled liquid chiller, refrigerant R410A; scroll compressors; complete with free cooling facility; integral controls; including placing in position; electrical work elsewhere						
Cooling load						
250 kW	35998.19	40656.18	8.00	189.56	nr	**40845.74**
300 kW	38398.08	43366.60	8.00	189.56	nr	**43556.16**
350 kW	40797.95	46077.00	8.00	189.56	nr	**46266.56**
400 kW	44397.78	50142.63	8.00	189.56	nr	**50332.19**
450 kW	47997.59	54208.24	8.00	189.56	nr	**54397.80**
500 kW	51597.41	58273.85	8.00	189.56	nr	**58463.41**
600 kW	55197.23	62339.47	8.00	189.56	nr	**62529.03**
650 kW	59996.99	67760.30	9.00	213.25	nr	**67973.55**
700 kW	71996.39	81312.36	9.00	213.25	nr	**81525.61**
750 kW	76796.15	86733.19	9.00	213.25	nr	**86946.44**

T: MECHANICAL/COOLING/HEATING SYSTEMS

Item	Net Price £	Material £	Labour hours	Labour £	Unit	Total rate £
Water cooled						
Selection of water cooled chillers based on chilled water flow and return temperatures of 6°C and 12°C, and condenser entering and leaving temperatures of 27°C and 33°C						
Water cooled liquid chiller; refrigerant 407C; reciprocating compressors; twin circuit; integral controls; includes placing in position; electrical work elsewhere						
Cooling load						
200 kw	19342.56	21845.39	8.00	189.56	nr	**22034.95**
350 kW	32438.21	36635.55	8.00	189.56	nr	**36825.11**
500 kW	41520.14	46892.64	8.00	189.56	nr	**47082.20**
650 kW	53559.06	60489.34	9.00	213.25	nr	**60702.59**
750 kW	58004.84	65510.38	9.00	213.25	nr	**65723.63**
Water cooled condenserless liquid chiller; refrigerant 407C; reciprocating compressors; twin circuit; integral controls; includes placing in position; electrical work elsewhere						
Cooling load						
200 kW	17027.34	19230.59	8.00	189.56	nr	**19420.15**
350 kW	29890.44	33758.11	8.00	189.56	nr	**33947.67**
500 kW	37777.71	42665.96	8.00	189.56	nr	**42855.52**
650 kW	46203.02	52181.46	9.00	213.25	nr	**52394.71**
750 kW	52574.16	59376.99	9.00	213.25	nr	**59590.24**
Water cooled liquid chiller; refrigerant R134a; screw compressors; twin circuit; integral controls; includes placing in position; electrical work elsewhere						
Cooling load						
300 kW	31888.76	36015.01	8.00	189.56	nr	**36204.57**
500 kW	42042.24	47482.30	8.00	189.56	nr	**47671.86**
700 kW	61526.13	69487.31	9.00	213.25	nr	**69700.56**
900 kW	70696.97	79844.80	9.00	213.25	nr	**80058.05**
1100 kW	85270.00	96303.51	10.00	236.95	nr	**96540.46**
1300 kW	95003.98	107297.02	10.00	236.95	nr	**107533.97**

T: MECHANICAL/COOLING/HEATING SYSTEMS

Item	Net Price £	Material £	Labour hours	Labour £	Unit	Total rate £
T60: CENTRAL REFRIGERATION PLANT – cont						
CHILLERS – cont						
Water cooled – cont						
Water cooled liquid chiller; refrigerant R134a; centrifugal compressors; twin circuit; integral controls; includes placing in position; electrical work elsewhere						
Cooling load						
700 kW	60644.95	68492.10	9.00	213.25	nr	**68705.35**
1000 kW	85495.71	96558.42	10.00	236.95	nr	**96795.37**
1300 kW	112455.36	127006.52	10.00	236.95	nr	**127243.47**
1600 kW	134969.20	152433.54	11.00	260.64	nr	**152694.18**
1900 kW	164379.80	185649.72	11.00	260.64	nr	**185910.36**
2200 kW	195842.17	221183.16	13.00	308.03	nr	**221491.19**
2500 kW	202339.89	228521.66	13.00	308.03	nr	**228829.69**
3000 kW	246227.65	278088.28	15.00	355.42	nr	**278443.70**
3500 kW	335143.18	378509.04	15.00	355.42	nr	**378864.46**
4000 kW	355951.68	402010.05	20.00	473.89	nr	**402483.94**
4500 kW	379762.82	428902.23	20.00	473.89	nr	**429376.12**
5000 kW	463155.35	523085.34	25.00	592.37	nr	**523677.71**
Absorption						
Absorption chiller, for operation using low pressure steam; selection based on chilled water flow and return temperatures of 6°C and 12°C, steam at 1 bar gauge and condenser entering and leaving temperatures of 27°C and 33°C; integral controls; includes placing in position; electrical work elsewhere						
Cooling load						
400 kW	58150.77	65675.19	8.00	189.56	nr	**65864.75**
700 kW	73837.52	83391.72	9.00	213.25	nr	**83604.97**
1000 kW	84238.36	95138.38	10.00	236.95	nr	**95375.33**
1300 kW	100529.29	113537.28	12.00	284.34	nr	**113821.62**
1600 kW	118975.85	134370.73	14.00	331.72	nr	**134702.45**
2000 kW	141008.69	159254.51	15.00	355.42	nr	**159609.93**

T: MECHANICAL/COOLING/HEATING SYSTEMS

Item	Net Price £	Material £	Labour hours	Labour £	Unit	Total rate £
Absorption chiller, for operation using low pressure hot water; selection based on chilled water flow and return temperatures of 6°C and 12°C, cooling water temperatures of 27°C and 33°C and hot water at 90°C; integral controls; includes placing in position; electrical work elsewhere						
Cooling load						
700 kW	84238.36	95138.38	9.00	213.25	nr	**95351.63**
1000 kW	105301.04	118926.47	10.00	236.95	nr	**119163.42**
1300 kW	122472.01	138319.27	12.00	284.34	nr	**138603.61**
1600 kW	141008.69	159254.51	14.00	331.72	nr	**159586.23**
HEAT REJECTION						
Dry air liquid coolers						
Dry air liquid cooler; selection based on fluid temperatures 45°C on, 40°C off at 32°C dry bulb ambient temperature; includes 20% ethylene glycol; includes placing in postion; electrical work elsewhere						
Flat coil configuration						
500 kW	13121.71	14819.59	15.00	355.42	nr	**15175.01**
Extra for inverter panels (factory wired and mounted on units)	6974.99	7877.52	15.00	355.42	nr	**8232.94**
800 kW	22199.00	25071.44	15.00	355.42	nr	**25426.86**
Extra for inverter panels (factory wired and mounted on units)	12818.15	14476.76	15.00	355.42	nr	**14832.18**
1100 kW	28503.48	32191.69	15.00	355.42	nr	**32547.11**
Extra for inverter panels (factory wired and mounted on units)	13948.78	15753.69	15.00	355.42	nr	**16109.11**
1400 kW	36206.34	40891.26	15.00	355.42	nr	**41246.68**
Extra over for inverter panels (factory wired and mounted on units)	19227.81	21715.79	15.00	355.42	nr	**22071.21**
1700 kW	42754.65	48286.89	15.00	355.42	nr	**48642.31**
Extra for inverter panels (factory wired and mounted on units)	19549.33	22078.92	15.00	355.42	nr	**22434.34**
2000 kW	50014.08	56485.65	15.00	355.42	nr	**56841.07**
Extra for inverter panels (factory wired and mounted on units)	25729.52	29058.79	15.00	355.42	nr	**29414.21**
Note: heat rejection capacities above 500 kW require multiple units. Prices are therefore for total number of units						

T: MECHANICAL/COOLING/HEATING SYSTEMS

Item	Net Price £	Material £	Labour hours	Labour £	Unit	Total rate £
T60: CENTRAL REFRIGERATION PLANT – cont						
HEAT REJECTION – cont						
Dry air liquid coolers – cont						
'Vee' type coil configuration						
500 kW	12678.31	14318.82	15.00	355.42	nr	**14674.24**
Extra for inverter panels (factory wired and mounted on units)	4906.15	5540.98	15.00	355.42	nr	**5896.40**
800 kW	19447.79	21964.24	15.00	355.42	nr	**22319.66**
Extra for inverter panels (factory wired and mounted on units)	11361.92	12832.10	15.00	355.42	nr	**13187.52**
1100 kW	27500.76	31059.22	15.00	355.42	nr	**31414.64**
Extra for inverter panels (factory wired and mounted on units)	13738.72	15516.44	15.00	355.42	nr	**15871.86**
1400 kW	35482.97	40074.29	15.00	355.42	nr	**40429.71**
Extra for inverter panels (factory wired and mounted on units)	17083.10	19293.57	15.00	355.42	nr	**19648.99**
1700 kW	40761.71	46036.07	15.00	355.42	nr	**46391.49**
Extra for inverter panels (factory wired and mounted on units)	22723.09	25663.34	15.00	355.42	nr	**26018.76**
2000 kW	53224.45	60111.43	15.00	355.42	nr	**60466.85**
Extra for inverter panels (factory wired and mounted on units)	25624.63	28940.33	15.00	355.42	nr	**29295.75**
Note: Heat rejection capacities above 1100 kW require multiple units. Prices are for total number of units.						
Air cooled condensers						
Air cooled condenser; refrigerant 407C; selection based on condensing temperature of 45°C at 32°C dry bulb ambient; includes placing in position; electrical work elsewhere						
Flat coil configuration						
500 kW	14006.73	15819.13	15.00	355.42	nr	**16174.55**
Extra for inverter panels (factory wired and mounted on units)	6409.08	7238.38	15.00	355.42	nr	**7593.80**
800 kW	22979.44	25952.86	15.00	355.42	nr	**26308.28**
Extra for inverter panels (factory wired and mounted on units)	9811.71	11081.30	15.00	355.42	nr	**11436.72**
1100 kW	30763.56	34744.21	15.00	355.42	nr	**35099.63**
Extra for inverter panels (factory wired and mounted on units)	13948.47	15753.33	15.00	355.42	nr	**16108.75**
1400 kW	38407.84	43377.62	15.00	355.42	nr	**43733.04**

T: MECHANICAL/COOLING/HEATING SYSTEMS

Item	Net Price £	Material £	Labour hours	Labour £	Unit	Total rate £
Extra for inverter panels (factory wired and mounted on units)	17153.02	19372.53	15.00	355.42	nr	**19727.95**
1700 kW	50014.08	56485.65	15.00	355.42	nr	**56841.07**
Extra for inverter panels (factory wired and mounted on units)	25729.52	29058.79	15.00	355.42	nr	**29414.21**
2000 kW	57611.75	65066.42	15.00	355.42	nr	**65421.84**
Extra for inverter panels (factory wired and mounted on units)	25729.52	29058.79	15.00	355.42	nr	**29414.21**
Note: Heat rejection capacities above 500 kW require multiple units. Prices are for total number of units.						
'Vee' type coil configuration						
500 kW	13750.37	15529.60	15.00	355.42	nr	**15885.02**
Extra for inverter panels (factory wired and mounted on units)	4905.86	5540.65	15.00	355.42	nr	**5896.07**
800 kW	20683.83	23360.21	15.00	355.42	nr	**23715.63**
Extra for inverter panels (factory wired and mounted on units)	10044.77	11344.51	15.00	355.42	nr	**11699.93**
1100 kW	30355.72	34283.60	15.00	355.42	nr	**34639.02**
Extra for inverter panels (factory wired and mounted on units)	17083.10	19293.57	15.00	355.42	nr	**19648.99**
1400 kW	35343.13	39916.35	15.00	355.42	nr	**40271.77**
Extra for inverter panels (factory wired and mounted on units)	20089.53	22689.01	15.00	355.42	nr	**23044.43**
1700 kW	45533.56	51425.37	15.00	355.42	nr	**51780.79**
Extra for inverter panels (factory wired and mounted on units)	25636.30	28953.51	15.00	355.42	nr	**29308.93**
2000 kW	53014.71	59874.55	15.00	355.42	nr	**60229.97**
Extra for inverter panels (factory wired and mounted on units)	30134.30	34033.52	15.00	355.42	nr	**34388.94**
Note: Heat rejection capacities above 1100 kW require multiple units. Prices are for total number of units.						

T: MECHANICAL/COOLING/HEATING SYSTEMS

Item	Net Price £	Material £	Labour hours	Labour £	Unit	Total rate £
T60: CENTRAL REFRIGERATION PLANT – cont						
HEAT REJECTION – cont						
Cooling towers						
Cooling towers; forced draught, centrifugal fan, conterflow design; based on water temperatures of 35°C on and 29°C off at 21°C wet bulb ambient temperature; includes placing in position; electrical work elsewhere						
Open circuit type						
900 kW	9330.00	10537.25	20.00	473.89	nr	**11011.14**
Extra for stainless steel construction	4182.94	4724.19	–	–	nr	**4724.19**
Extra for intake and discharge sound attenuation	4501.94	5084.47	–	–	nr	**5084.47**
Extra for fan dampers for capacity control	1044.54	1179.70	–	–	nr	**1179.70**
1500 kW	14475.74	16348.83	20.00	473.89	nr	**16822.72**
Extra for stainless steel construction	6700.41	7567.41	–	–	nr	**7567.41**
Extra for intake and discharge sound attenuation	6746.28	7619.22	–	–	nr	**7619.22**
Extra for fan dampers for capacity control	1060.20	1197.39	–	–	nr	**1197.39**
2100 kW	20536.08	23193.34	20.00	473.89	nr	**23667.23**
Extra for stainless steel construction	9503.19	10732.86	–	–	nr	**10732.86**
Extra for intake and discharge sound attenuation	10017.14	11313.31	–	–	nr	**11313.31**
Extra for fan dampers for capacity control	1210.63	1367.28	–	–	nr	**1367.28**
2700 kW	24800.94	28010.06	20.00	473.89	nr	**28483.95**
Extra for stainless steel construction	11592.74	13092.78	–	–	nr	**13092.78**
Extra for intake and discharge sound attenuation	13494.97	15241.16	–	–	nr	**15241.16**
Extra for fan dampers for capacity control	1884.53	2128.38	–	–	nr	**2128.38**
3300 kW	30562.83	34517.50	20.00	473.89	nr	**34991.39**
Extra for stainless steel construction	14148.77	15979.55	–	–	nr	**15979.55**
Extra for intake and discharge sound attenuation	12727.19	14374.02	–	–	nr	**14374.02**
Extra for fan dampers for capacity control	1541.56	1741.03	–	–	nr	**1741.03**
3900 kW	34447.42	38904.75	20.00	473.89	nr	**39378.64**
Extra for stainless steel construction	15947.87	18011.45	–	–	nr	**18011.45**
Extra for intake and discharge sound attenuation	12971.48	14649.92	–	–	nr	**14649.92**
Extra for fan dampers for capacity control	1541.56	1741.03	–	–	nr	**1741.03**
4500 kW	38767.64	43783.98	23.00	544.98	nr	**44328.96**
Extra for stainless steel construction	18078.53	20417.80	–	–	nr	**20417.80**
Extra for intake and discharge sound attenuation	17203.86	19429.95	–	–	nr	**19429.95**

T: MECHANICAL/COOLING/HEATING SYSTEMS

Item	Net Price £	Material £	Labour hours	Labour £	Unit	Total rate £
Extra for fan dampers for capacity control	2426.07	2740.00	–	–	nr	2740.00
5100 kW	44782.25	50576.85	23.00	544.98	nr	51121.83
Extra for stainless steel construction	20633.29	23303.13	–	–	nr	23303.13
Extra for intake and discharge sound attenuation	17158.12	19378.30	–	–	nr	19378.30
Extra for fan dampers for capacity control	2426.07	2740.00	–	–	nr	2740.00
5700 kW	47344.28	53470.39	30.00	710.84	nr	54181.23
Extra for stainless steel construction	22107.28	24967.86	–	–	nr	24967.86
Extra for intake and discharge sound attenuation	25408.65	28696.40	–	–	nr	28696.40
Extra for fan dampers for capacity control	2667.95	3013.17	–	–	nr	3013.17
6300 kW	57084.63	64471.10	30.00	710.84	nr	65181.94
Extra for stainless steel construction	26427.72	29847.33	–	–	nr	29847.33
Extra for intake and discharge sound attenuation	25775.69	29110.93	–	–	nr	29110.93
Extra for fan dampers for capacity control	2667.95	3013.17	–	–	nr	3013.17
Closed circuit type (includes 20% ethylene glycol)						
900 kW	26009.16	29374.61	20.00	473.89	nr	29848.50
Extra for stainless steel construction	21178.14	23918.49	–	–	nr	23918.49
Extra for intake and discharge sound attenuation	7826.49	8839.20	–	–	nr	8839.20
Extra for fan dampers for capacity control	1060.20	1197.39	–	–	nr	1197.39
1500 kW	48355.14	54612.05	20.00	473.89	nr	55085.94
Extra for stainless steel construction	38552.46	43540.95	–	–	nr	43540.95
Extra for intake and discharge sound attenuation	12492.53	14109.00	–	–	nr	14109.00
Extra for fan dampers for capacity control	1541.56	1741.03	–	–	nr	1741.03
2100 kW	58928.23	66553.25	20.00	473.89	nr	67027.14
Extra for stainless steel construction	52376.09	59153.30	–	–	nr	59153.30
Extra for intake and discharge sound attenuation	15037.73	16983.54	–	–	nr	16983.54
Extra for fan dampers for capacity control	1541.56	1741.03	–	–	nr	1741.03
2700 kW	63534.87	71755.96	25.00	592.37	nr	72348.33
Extra for stainless steel construction	64896.64	73293.94	–	–	nr	73293.94
Extra for intake and discharge sound attenuation	22444.69	25348.92	–	–	nr	25348.92
Extra for fan dampers for capacity control	2332.20	2633.98	–	–	nr	2633.98
3300 kW	97890.82	110557.40	25.00	592.37	nr	111149.77
Extra for stainless steel construction	80960.16	91436.00	–	–	nr	91436.00
Extra for intake and discharge sound attenuation	29856.43	33719.70	–	–	nr	33719.70
Extra for fan dampers for capacity control	2667.95	3013.17	–	–	nr	3013.17
3900 kW	107782.78	121729.33	25.00	592.37	nr	122321.70
Extra for stainless steel construction	92712.27	104708.77	–	–	nr	104708.77
Extra for intake and discharge sound attenuation	30345.01	34271.50	–	–	nr	34271.50
Extra for fan dampers for capacity control	2667.95	3013.17	–	–	nr	3013.17
4500 kW	129037.26	145734.04	40.00	947.78	nr	146681.82
Extra for stainless steel construction	96181.09	108626.44	–	–	nr	108626.44

T: MECHANICAL/COOLING/HEATING SYSTEMS

Item	Net Price £	Material £	Labour hours	Labour £	Unit	Total rate £
T60: CENTRAL REFRIGERATION PLANT – cont						
HEAT REJECTION – cont						
Closed circuit type (includes 20% ethylene glycol) – cont						
Extra for intake and discharge sound attenuation	43371.85	48983.95	–	–	nr	**48983.95**
Extra for fan dampers for capacity control	4664.39	5267.94	–	–	nr	**5267.94**
5100 kW	143917.41	162539.60	40.00	947.78	nr	**163487.38**
Extra for stainless steel construction	124293.58	140376.55	–	–	nr	**140376.55**
Extra for intake and discharge sound attenuation	43371.85	48983.95	–	–	nr	**48983.95**
Extra for fan dampers for capacity control	4664.39	5267.94	–	–	nr	**5267.94**
5700 kW	177458.66	200420.92	40.00	947.78	nr	**201368.70**
Extra for stainless steel construction	128386.62	144999.20	–	–	nr	**144999.20**
Extra for intake and discharge sound attenuation	48486.31	54760.20	–	–	nr	**54760.20**
Extra for fan dampers for capacity control	5335.90	6026.34	–	–	nr	**6026.34**
6300 kW	189555.24	214082.74	40.00	947.78	nr	**215030.52**
Extra for stainless steel construction	146853.50	165855.61	–	–	nr	**165855.61**
Extra for intake and discharge sound attenuation	48486.31	54760.20	–	–	nr	**54760.20**
Extra for fan dampers for capacity control	5335.90	6026.34	–	–	nr	**6026.34**

T: MECHANICAL/COOLING/HEATING SYSTEMS

Item	Net Price £	Material £	Labour hours	Labour £	Unit	Total rate £
T61: CHILLED WATER						
SCREWED STEEL						
Y10 – PIPELINES						
For pipework prices refer to Section T31 – Low Temperature Hot Water Heating, with the exception of chilled water blocks within brackets as detailed hereafter. For minimum fixing distances, refer to the Tables and Memoranda to the rear of the book						
FIXINGS						
For steel pipes; black malleable iron						
Oversized pipe clip, to contain 30 mm insulation block for vapour barrier						
15 mm dia.	3.11	3.51	0.10	2.38	nr	**5.89**
20 mm dia.	3.14	3.55	0.11	2.61	nr	**6.16**
25 mm dia.	3.30	3.72	0.12	2.85	nr	**6.57**
32 mm dia.	3.40	3.84	0.14	3.33	nr	**7.17**
40 mm dia.	3.60	4.07	0.15	3.56	nr	**7.63**
50 mm dia.	5.13	5.79	0.16	3.79	nr	**9.58**
65 mm dia.	5.53	6.24	0.30	7.10	nr	**13.34**
80 mm dia.	5.87	6.63	0.35	8.29	nr	**14.92**
100 mm dia.	9.31	10.52	0.40	9.48	nr	**20.00**
125 mm dia.	10.80	12.20	0.60	14.22	nr	**26.42**
150 mm dia.	16.90	19.08	0.77	18.24	nr	**37.32**
200 mm dia.	21.57	24.36	0.90	21.32	nr	**45.68**
250 mm dia.	26.63	30.08	1.10	26.07	nr	**56.15**
300 mm dia.	31.70	35.80	1.25	29.62	nr	**65.42**
350 mm dia.	36.76	41.51	1.50	35.55	nr	**77.06**
400 mm dia.	41.17	46.50	1.75	41.47	nr	**87.97**
Screw on backplate (Male), black malleable iron; plugged and screwed						
M12	0.42	0.48	0.10	2.38	nr	**2.86**
Screw on backplate (Female), black malleable iron; plugged and screwed						
M12	0.44	0.96	0.10	2.38	nr	**3.34**

T: MECHANICAL/COOLING/HEATING SYSTEMS

Item	Net Price £	Material £	Labour hours	Labour £	Unit	Total rate £
T61: CHILLED WATER – cont						
FIXINGS – cont						
Extra Over channel sections for fabricated hangers and brackets						
Galvanized steel; including inserts, bolts, nuts, washers; fixed to backgrounds						
41 × 21 mm	4.92	5.56	0.29	6.86	m	**12.42**
41 × 41 mm	5.90	6.66	0.29	6.86	m	**13.52**
Threaded rods; metric thread; including nuts, washers etc.						
12 mm dia. × 600 mm long	2.49	2.82	0.18	4.27	nr	**7.09**
For plastic pipework suitable for chilled water systems, refer to ABS pipework details in Section S10 – Cold Water with the exception of chilled water blocks within brackets as detailed for the aforementioned steel pipe. For minimum fixing distances, refer to the Tables and Memoranda to the rear of the book						
For copper pipework, refer to Section S10 – Cold Water with the exception of chilled water blocks within brackets as detailed hereafter. For minimum fixing distances, refer to the Tables and Memoranda to the rear of the book						
FIXINGS						
For copper pipework						
Oversized pipe clip, to contain 30 mm insulation block for vapour barrier						
15 mm dia.	3.11	3.51	0.10	2.38	nr	**5.89**
22 mm dia.	3.14	3.55	0.11	2.61	nr	**6.16**
28 mm dia.	3.30	3.72	0.12	2.85	nr	**6.57**
35 mm dia.	3.40	3.84	0.14	3.33	nr	**7.17**
42 mm dia.	3.60	4.07	0.15	3.56	nr	**7.63**
54 mm dia.	5.13	5.79	0.16	3.79	nr	**9.58**
67 mm dia.	5.53	6.24	0.30	7.10	nr	**13.34**
76 mm dia.	5.87	6.63	0.35	8.29	nr	**14.92**
108 mm dia.	2.41	13.24	0.40	9.48	nr	**22.72**
133 mm dia.	10.80	12.20	0.60	14.22	nr	**26.42**
159 mm dia.	16.90	19.08	0.77	18.24	nr	**37.32**

T: MECHANICAL/COOLING/HEATING SYSTEMS

Item	Net Price £	Material £	Labour hours	Labour £	Unit	Total rate £
Screw on backplate, female						
All sizes 15 mm to 54 mm × 10 mm	1.26	1.43	0.10	2.38	nr	3.81
Screw on backplate, male						
All sizes 15 mm to 54 mm × 10 mm	1.67	1.89	0.10	2.38	nr	4.27
Extra Over channel sections for fabricated hangers and brackets						
Galvanized steel; including inserts, bolts, nuts, washers; fixed to backgrounds						
41 × 21 mm	4.92	5.56	0.29	6.86	m	12.42
41 × 41 mm	5.90	6.66	0.29	6.86	m	13.52
Threaded rods; metric thread; including nuts, washers etc.						
10 mm dia. × 600 mm long for ring clips up to 54 mm	1.62	1.83	0.18	4.27	nr	6.10
12 mm dia. × 600 mm long for ring clips from 54 mm	2.49	2.82	0.18	4.27	nr	7.09
Y11 – PIPELINE ANCILLARIES						
For prices for ancillaries refer to Section T31 – Low Temperature Hot Water Heating						
Y22 – HEAT EXCHANGERS						
Plate heat exchanger; for use in CHW systems; painted carbon steel frame, stainless steel plates, nitrile rubber gaskets, design pressure of 10 bar and operating temperature of 110/135°C						
Primary side; 13°C in, 8°C out; secondary side; 6°C in, 11°C out						
264 kW, 12.60 l/s	4621.56	5219.56	10.00	236.95	nr	5456.51
290 kW, 13.85 l/s	4871.55	5501.90	12.00	284.34	nr	5786.24
316 kW, 15.11 l/s	5071.32	5727.52	12.00	284.34	nr	6011.86
350 kW, 16.69 l/s	5453.78	6159.47	16.00	379.11	nr	6538.58
395 kW, 18.88 l/s	5836.24	6591.42	16.00	379.11	nr	6970.53
454 kW, 21.69 l/s	6339.42	7159.71	10.00	236.95	nr	7396.66
475 kW, 22.66 l/s	6462.28	7298.47	10.00	236.95	nr	7535.42
527 kW, 25.17 l/s	7002.85	7908.99	10.00	236.95	nr	8145.94
554 kW, 26.43 l/s	7188.74	8118.92	10.00	236.95	nr	8355.87
580 kW, 27.68 l/s	7435.05	8397.11	10.00	236.95	nr	8634.06
633 kW, 30.19 l/s	7882.08	8901.98	10.00	236.95	nr	9138.93
661 kW, 31.52 l/s	8072.24	9116.75	10.00	236.95	nr	9353.70
713 kW, 34.04 l/s	8617.08	9732.09	12.00	284.34	nr	10016.43

T: MECHANICAL/COOLING/HEATING SYSTEMS

Item	Net Price £	Material £	Labour hours	Labour £	Unit	Total rate £
T61: CHILLED WATER – cont						
Y22 – HEAT EXCHANGERS – cont						
Plate heat exchanger – cont						
Primary side – cont						
740 kW, 35.28 l/s	8805.11	9944.45	12.00	284.34	nr	**10228.79**
804 kW, 38.33 l/s	9375.60	10588.75	12.00	284.34	nr	**10873.09**
1925 kW, 91.82 l/s	15738.53	17775.02	15.00	355.42	nr	**18130.44**
2710 kW, 129.26 l/s	20904.93	23609.92	15.00	355.42	nr	**23965.34**
3100 kW, 147.87 l/s	23289.43	26302.96	15.00	355.42	nr	**26658.38**
Note: For temperature conditions different to those above, the cost of the units can vary significantly, and therefore the manufacturers advice should be sought.						
Y24 – TRACE HEATING						
Trace heating; for freeze protection or temperature maintainance of pipework; to BS 6351; including fixing to parent structures by plastic pull ties						
Straight laid						
15 mm	19.68	22.23	0.27	6.40	m	**28.63**
25 mm	19.68	22.23	0.27	6.40	m	**28.63**
28 mm	19.68	22.23	0.27	6.40	m	**28.63**
32 mm	19.68	22.23	0.30	7.10	m	**29.33**
35 mm	19.68	22.23	0.31	7.34	m	**29.57**
50 mm	19.68	22.23	0.34	8.05	m	**30.28**
100 mm	19.68	22.23	0.40	9.48	m	**31.71**
150 mm	19.68	22.23	0.40	9.48	m	**31.71**
Helically wound						
15 mm	25.07	28.32	1.00	23.69	m	**52.01**
25 mm	25.07	28.32	1.00	23.69	m	**52.01**
28 mm	25.07	28.32	1.00	23.69	m	**52.01**
32 mm	25.07	28.32	1.00	23.69	m	**52.01**
35 mm	25.07	28.32	1.00	23.69	m	**52.01**
50 mm	25.07	28.32	1.00	23.69	m	**52.01**
100 mm	25.07	28.32	1.00	23.69	m	**52.01**
150 mm	25.07	28.32	1.00	23.69	m	**52.01**

T: MECHANICAL/COOLING/HEATING SYSTEMS

Item	Net Price £	Material £	Labour hours	Labour £	Unit	Total rate £
Accessories for trace heating; weatherproof; polycarbonate enclosure to IP standards; fully installed						
Connection junction box						
100 × 100 × 75 mm	43.08	48.65	1.40	33.17	nr	**81.82**
Single air thermostat						
150 × 150 × 75 mm	94.39	106.60	1.42	33.64	nr	**140.24**
Single capillary thermostat						
150 × 150 × 75 mm	135.81	153.39	1.46	34.60	nr	**187.99**
Twin capillary thermostat						
150 × 150 × 75 mm	243.64	275.17	1.46	34.60	nr	**309.77**
EQUIPMENT						
PRESSURIZATION UNITS						
Chilled water packaged pressurization unit complete with expansion vessel(s), interconnecting pipework and necessary isolating and drain valves; includes placing in position; electrical work elasewhere						
Selection based on a final working pressure of 4 bar, a 3 m static head and system operating temperatures of 6°/12°C						
System volume						
1800 litres	1544.25	1744.07	8.00	189.56	nr	**1933.63**
4500 litres	1544.25	1744.07	8.00	189.56	nr	**1933.63**
7200 litres	1618.97	1828.45	10.00	236.95	nr	**2065.40**
9900 litres	1618.97	1828.45	10.00	236.95	nr	**2065.40**
15300 litres	1681.23	1898.77	13.00	308.03	nr	**2206.80**
22500 litres	1782.11	2012.70	20.00	473.89	nr	**2486.59**
27000 litres	1782.11	2012.70	20.00	473.89	nr	**2486.59**

T: MECHANICAL/COOLING/HEATING SYSTEMS

Item	Net Price £	Material £	Labour hours	Labour £	Unit	Total rate £
T61: CHILLED WATER – cont						
CHILLED BEAMS						
Static (passive) beams; based on water at 14°C flow and 16°C return, 24°C room temperature; 600 mm wide coil providing 350-400 W/m output						
Static cooled beam for exposed installation with standard casing	114.31	129.10	4.00	94.78	m	**223.88**
Static cooled beam for installation above open grid or perforated ceiling	78.38	88.53	4.00	94.78	m	**183.31**
Ventilated (active) beams; based on water at 14°C flow and 16°C return, 24°C room temperature; air supply at 10l/s/linear metre; 300 mm wide beam providing 250-350 W/m output unless stated otherwise; all exposed beams c/w standard casing; electrical work elsewhere						
Ventilated cooled beam flush mounted within a false ceiling; closed type with integrated secondary air circulation; 600 mm wide beam providing 400 W/m output	177.45	200.41	4.50	106.63	m	**307.04**
Ventilated cooled beam flush mounted within a false ceiling; open type	156.76	177.04	4.00	94.78	m	**271.82**
Ventilated cooled beam for exposed mounting with standard casing	156.76	177.04	4.00	94.78	m	**271.82**
Ventilated cooled beam flush mounted within a false ceiling; open type; with recessed integrated flush mounted 28 W or 35 W T5 light fittings	284.14	320.90	4.00	94.78	m	**415.68**
Ventilated cooled beam for exposed mounting with recessed integrated flush mounted 28 W or 35 W T5 light fittings	284.14	320.90	4.00	94.78	m	**415.68**
Ventilated cooled beam for exposed mounting with recessed integrated flush mounted direct and indirect 28 W or 35 W T5 light fittings	310.27	350.42	4.00	94.78	m	**445.20**

T: MECHANICAL/COOLING/HEATING SYSTEMS

Item	Net Price £	Material £	Labour hours	Labour £	Unit	Total rate £
LEAK DETECTION						
Leak detection system consisting of a central control module connected by a leader cable to water sensing cables						
Control modules						
Alarm only	311.80	352.14	4.00	109.28	nr	**461.42**
Alarm and location	2044.88	2309.47	8.00	218.57	nr	**2528.04**
Cables						
Sensing – 3 m length	85.40	96.45	4.00	109.28	nr	**205.73**
Sensing – 7.5 m length	125.82	142.10	4.00	109.28	nr	**251.38**
Sensing – 15 m length	217.54	245.69	8.00	218.57	nr	**464.26**
Leader – 3.5 m length	38.32	43.28	2.00	54.64	nr	**97.92**
End terminal						
End terminal	14.88	16.81	0.05	1.37	nr	**18.18**
ENERGY METERS						
Ultrasonic						
Energy meter for measuring energy use in chilled water systems; includes ultrasonic flow meter (with sensor and signal converter), energy calculator, pair of temperature sensors with brass pockets, and 3 m of interconnecting cable; includes fixing in position; electrical work elsewhere						
Pipe size (flanged connections to PN16); maximum flow rate						
50 mm, 36 m³/hr	1088.03	1228.81	1.80	42.65	nr	**1271.46**
65 mm, 60 m³/hr	1199.34	1354.53	2.32	54.97	nr	**1409.50**
80 mm, 100 m³/hr	1344.37	1518.32	2.56	60.66	nr	**1578.98**
125 mm, 250 m³/hr	1557.58	1759.12	3.60	85.30	nr	**1844.42**
150 mm, 360 m³/hr	1690.84	1909.62	4.80	113.73	nr	**2023.35**
200 mm, 600 m³/hr	1889.16	2133.61	6.24	147.86	nr	**2281.47**
250 mm, 1000 m³/hr	2179.98	2462.06	9.60	227.47	nr	**2689.53**
300 mm, 1500 m³/hr	2562.52	2894.10	10.80	255.90	nr	**3150.00**
350 mm, 2000 m³/hr	3087.72	3487.26	13.20	312.77	nr	**3800.03**
400 mm, 2500 m³/hr	3534.53	3991.88	15.60	369.64	nr	**4361.52**
500 mm, 3000 m³/hr	4011.13	4530.15	24.00	568.67	nr	**5098.82**
600 mm, 3500 m³/hr	4506.55	5089.68	28.00	663.45	nr	**5753.13**

T: MECHANICAL/COOLING/HEATING SYSTEMS

Item	Net Price £	Material £	Labour hours	Labour £	Unit	Total rate £
T61: CHILLED WATER – cont						
ENERGY METERS – cont						
Electromagnetic						
Energy meter for measuring energy use in chilled water systems; includes electromagnetic flow meter (with sensor and signal converter), energy calculator, pair of temperature sensors with brass pockets, and 3 m of interconnecting cable; includes fixing in position; electrical work elsewhere						
Pipe size (flanged connections to PN40); maximum flow rate						
25 mm, 17.7 m³/hr	780.04	880.97	1.48	35.07	nr	**916.04**
40 mm, 45 m³/hr	787.51	889.41	1.55	36.73	nr	**926.14**
Pipe size (flanged connections to PN16); maximum flow rate						
50 mm, 70 m³/hr	796.32	899.36	1.80	42.65	nr	**942.01**
65 mm, 120 m³/hr	799.72	903.20	2.32	54.97	nr	**958.17**
80 mm, 180 m³/hr	803.79	907.80	2.56	60.66	nr	**968.46**
125 mm, 450 m³/hr	875.01	988.23	3.60	85.30	nr	**1073.53**
150 mm, 625 m³/hr	928.59	1048.74	4.80	113.73	nr	**1162.47**
200 mm, 1100 m³/hr	999.14	1128.43	6.24	147.86	nr	**1276.29**
250 mm, 1750 m³/hr	1121.23	1266.31	9.60	227.47	nr	**1493.78**
300 mm, 2550 m³/hr	1429.18	1614.11	10.80	255.90	nr	**1870.01**
350 mm, 3450 m³/hr	1859.22	2099.79	13.20	312.77	nr	**2412.56**
400 mm, 4500 m³/hr	2121.04	2395.49	15.60	369.64	nr	**2765.13**

T: MECHANICAL/COOLING/HEATING SYSTEMS

Item	Net Price £	Material £	Labour hours	Labour £	Unit	Total rate £
T70: LOCAL COOLING UNITS						
Split system with ceiling void evaporator unit and external condensing unit						
Ceiling mounted 4 way blow cassette heat pump unit with remote fan speed and load control; refrigerant 470C; includes outdoor unit						
Cooling 3.6 kW, heating 4.1 kW	1495.89	1689.45	35.00	829.31	nr	**2518.76**
Cooling 4.9 kW, heating 5.5 kW	1648.46	1861.77	35.00	829.31	nr	**2691.08**
Cooling 7.1 kW, heating 8.2 kW	1995.31	2253.50	35.00	829.31	nr	**3082.81**
Cooling 10 kW, heating 11.2 kW	2360.05	2665.43	35.00	829.31	nr	**3494.74**
Cooling 12.20 kW, heating 14.60 kW	2581.75	2915.81	35.00	829.31	nr	**3745.12**
Ceiling mounted 4 way blow cooling only unit with remote fan speed and load control; refrigerant 470C; includes outdoor unit						
Cooling 3.80 kW	1354.04	1529.25	35.00	829.31	nr	**2358.56**
Cooling 5.20 kW	1512.57	1708.29	35.00	829.31	nr	**2537.60**
Cooling 7.10 kW	1865.39	2106.76	35.00	829.31	nr	**2936.07**
Cooling 10 kW	2166.96	2447.36	35.00	829.31	nr	**3276.67**
Cooling 12.2 kW	2296.88	2594.08	35.00	829.31	nr	**3423.39**
In ceiling, ducted heat pump unit with remote fan speed and load control; refrigerant 407C; includes outdoor unit						
Cooling 3.60 kW, heating 4.10 kW	1096.59	1238.48	35.00	829.31	nr	**2067.79**
Cooling 4.90 kW, heating 5.50 kW	1261.08	1424.26	35.00	829.31	nr	**2253.57**
Cooling 7.10 kW, heating 8.20 kW	1261.08	1424.26	35.00	829.31	nr	**2253.57**
Cooling 10 kW, heating 11.20 kW	1261.08	1424.26	35.00	829.31	nr	**2253.57**
Cooling 12.20 kW, heating 14.50 kW	2542.42	2871.39	35.00	829.31	nr	**3700.70**
In ceiling, ducted cooling only unit with remote fan speed and load control; refrigerant 407C; includes outdoor unit						
Cooling 3.70 kW	989.32	1117.33	35.00	829.31	nr	**1946.64**
Cooling 4.90 kW	1176.45	1328.67	35.00	829.31	nr	**2157.98**
Cooling 7.10 kW	1776.00	2005.81	35.00	829.31	nr	**2835.12**
Cooling 10 kW	2025.11	2287.15	35.00	829.31	nr	**3116.46**
Cooling 12.3 kW	2257.54	2549.65	35.00	829.31	nr	**3378.96**
Room Units						
Ceiling mounted 4 way blow cassette heat pump unit with remote fan speed and load control; refrigerant 407C; excludes outdoor unit						
Cooling 3.6 kW, heating 4.1 kW	944.02	1066.18	17.00	402.81	nr	**1468.99**
Cooling 4.9 kW, heating 5.5 kW	966.66	1091.74	17.00	402.81	nr	**1494.55**
Cooling 7.1 kW, heating 8.2 kW	1046.53	1181.95	17.00	402.81	nr	**1584.76**
Cooling 10 kW, heating 11.2 kW	1133.55	1280.22	17.00	402.81	nr	**1683.03**
Cooling 12.20 kW, heating 14.60 kW	1228.90	1387.91	17.00	402.81	nr	**1790.72**

T: MECHANICAL/COOLING/HEATING SYSTEMS

Item	Net Price £	Material £	Labour hours	Labour £	Unit	Total rate £
T70: LOCAL COOLING UNITS – cont						
Room Units – cont						
Ceiling mounted 4 way blow cooling unit with remote fan speed and load control; refrigerant 407C; excludes outdoor unit						
Cooling 3.80 kW	891.57	1006.93	17.00	402.81	nr	**1409.74**
Cooling 5.20 kW	898.73	1015.02	17.00	402.81	nr	**1417.83**
Cooling 7.10 kW	1046.53	1181.95	17.00	402.81	nr	**1584.76**
Cooling 10 kW	1133.55	1280.22	17.00	402.81	nr	**1683.03**
Cooling 12.2 kW	1228.90	1387.91	17.00	402.81	nr	**1790.72**
In ceiling , ducted heat pump unit with remote fan speed and load control; refrigerant 407C; excludes outdoor unit						
Cooling 3.60 kW, heating 4.10 kW	544.72	615.20	17.00	402.81	nr	**1018.01**
Cooling 4.90 kW, heating 5.50 kW	579.29	654.25	17.00	402.81	nr	**1057.06**
Cooling 7.10 kW, heating 8.20 kW	957.14	1080.99	17.00	402.81	nr	**1483.80**
Cooling 10 kW, heating 11.20 kW	991.69	1120.01	17.00	402.81	nr	**1522.82**
Cooling 12.20 kW, heating 14.50 kW	1189.56	1343.48	17.00	402.81	nr	**1746.29**
In ceiling , ducted cooling unit only with remote fan speed and load control; refrigerant 407C; excludes outdoor unit						
Cooling 3.70 kW	526.84	595.01	17.00	402.81	nr	**997.82**
Cooling 4.90 kW	562.60	635.40	17.00	402.81	nr	**1038.21**
Cooling 7.10 kW	957.14	1080.99	17.00	402.81	nr	**1483.80**
Cooling 10 kW	991.69	1120.01	17.00	402.81	nr	**1522.82**
Cooling 12.3 kW	1189.56	1343.48	17.00	402.81	nr	**1746.29**
External condensing units suitable for connection to multiple indoor units; inverter driven; refrigerant 407C						
Cooling only						
9 kW	2060.88	2327.55	17.00	402.81	nr	**2730.36**
Heat pump						
Cooling 5.20 kW, heating 6.10 kW	1468.47	1658.48	17.00	402.81	nr	**2061.29**
Cooling 6.80 kW, heating 2.50 kW	1872.54	2114.84	17.00	402.81	nr	**2517.65**
Cooling 8 kW, heating 9.60 kW	2168.14	2448.68	17.00	402.81	nr	**2851.49**
Cooling 14.50 kW, heating 16.50 kW	3598.48	4064.11	21.00	497.59	nr	**4561.70**

U: VENTILATION/AIR CONDITIONING SYSTEMS

Item	Net Price £	Material £	Labour hours	Labour £	Unit	Total rate £
U10: DUCTWORK: CIRCULAR						
Y30 – AIR DUCTLINES						
Galvanized sheet metal DW144 class B spirally wound circular section ductwork; including all necessary stiffeners, joints, couplers in the running length and duct supports						
Straight duct						
80 mm dia.	2.94	3.59	0.87	21.38	m	**24.97**
100 mm dia.	3.79	4.64	0.87	21.38	m	**26.02**
160 mm dia.	4.16	5.09	0.87	21.38	m	**26.47**
200 mm dia.	5.30	6.48	0.87	21.38	m	**27.86**
250 mm dia.	6.44	7.87	1.21	29.74	m	**37.61**
315 mm dia.	7.80	9.54	1.21	29.74	m	**39.28**
355 mm dia.	10.66	13.03	1.21	29.74	m	**42.77**
400 mm dia.	11.94	14.60	1.21	29.74	m	**44.34**
450 mm dia.	13.18	16.11	1.21	29.74	m	**45.85**
500 mm dia.	14.34	17.53	1.21	29.74	m	**47.27**
630 mm dia.	27.42	33.52	1.39	34.15	m	**67.67**
710 mm dia.	30.37	37.13	1.39	34.15	m	**71.28**
800 mm dia.	34.89	42.65	1.44	35.38	m	**78.03**
900 mm dia.	43.07	52.65	1.46	35.89	m	**88.54**
1000 mm dia.	53.79	65.76	1.65	40.54	m	**106.30**
1120 mm dia.	63.70	77.88	2.43	59.71	m	**137.59**
1250 mm dia.	69.61	85.10	2.43	59.71	m	**144.81**
1400 mm dia.	78.53	96.01	2.77	68.07	m	**164.08**
1600 mm dia.	101.32	123.87	3.06	75.20	m	**199.07**
Extra over fittings; circular duct class B						
End cap						
80 mm dia.	1.16	1.42	0.15	3.69	nr	**5.11**
100 mm dia.	1.26	1.55	0.15	3.69	nr	**5.24**
160 mm dia.	1.93	2.36	0.15	3.69	nr	**6.05**
200 mm dia.	2.29	2.80	0.20	4.91	nr	**7.71**
250 mm dia.	3.22	3.93	0.29	7.13	nr	**11.06**
315 mm dia.	4.08	4.99	0.29	7.13	nr	**12.12**
355 mm dia.	6.15	7.52	0.44	10.80	nr	**18.32**
400 mm dia.	6.37	7.79	0.44	10.80	nr	**18.59**
450 mm dia.	6.73	8.23	0.44	10.80	nr	**19.03**
500 mm dia.	7.06	8.63	0.44	10.80	nr	**19.43**
630 mm dia.	18.77	22.95	0.58	14.26	nr	**37.21**
710 mm dia.	21.94	26.82	0.69	16.95	nr	**43.77**
800 mm dia.	28.90	35.33	0.81	19.90	nr	**55.23**
900 mm dia.	32.63	39.89	0.92	22.61	nr	**62.50**
1000 mm dia.	44.99	55.00	1.04	25.55	nr	**80.55**
1120 mm dia.	49.76	60.83	1.16	28.51	nr	**89.34**
1250 mm dia.	54.82	67.02	1.16	28.51	nr	**95.53**
1400 mm dia.	69.55	85.03	1.16	28.51	nr	**113.54**
1600 mm dia.	78.61	96.10	1.16	28.51	nr	**124.61**

U: VENTILATION/AIR CONDITIONING SYSTEMS

Item	Net Price £	Material £	Labour hours	Labour £	Unit	Total rate £
U10: DUCTWORK: CIRCULAR – cont						
Extra over fittings – cont						
Reducer						
80 mm dia.	3.75	4.58	0.29	7.13	nr	**11.71**
100 mm dia.	3.91	4.78	0.29	7.13	nr	**11.91**
160 mm dia.	5.13	6.27	0.29	7.13	nr	**13.40**
200 mm dia.	6.02	7.36	0.44	10.80	nr	**18.16**
250 mm dia.	7.16	8.76	0.58	14.26	nr	**23.02**
315 mm dia.	9.10	11.12	0.58	14.26	nr	**25.38**
355 mm dia.	11.93	14.58	0.87	21.38	nr	**35.96**
400 mm dia.	13.42	16.41	0.87	21.38	nr	**37.79**
450 mm dia.	14.42	17.63	0.87	21.38	nr	**39.01**
500 mm dia.	15.11	18.47	0.87	21.38	nr	**39.85**
630 mm dia.	41.94	51.27	0.87	21.38	nr	**72.65**
710 mm dia.	45.66	55.82	0.96	23.60	nr	**79.42**
800 mm dia.	58.49	71.50	1.06	26.05	nr	**97.55**
900 mm dia.	61.93	75.71	1.16	28.51	nr	**104.22**
1000 mm dia.	80.27	98.13	1.25	30.72	nr	**128.85**
1120 mm dia.	88.28	107.92	3.47	85.27	nr	**193.19**
1250 mm dia.	98.12	119.95	3.47	85.27	nr	**205.22**
1400 mm dia.	122.20	149.39	4.05	99.53	nr	**248.92**
1600 mm dia.	115.35	141.02	4.62	113.53	nr	**254.55**
90° segmented radius bend						
80 mm dia.	1.99	2.43	0.29	7.13	nr	**9.56**
100 mm dia.	2.28	2.79	0.29	7.13	nr	**9.92**
160 mm dia.	3.86	4.72	0.29	7.13	nr	**11.85**
200 mm dia.	5.18	6.33	0.44	10.80	nr	**17.13**
250 mm dia.	7.64	9.34	0.58	14.26	nr	**23.60**
315 mm dia.	8.89	10.87	0.58	14.26	nr	**25.13**
355 mm dia.	9.25	11.31	0.87	21.38	nr	**32.69**
400 mm dia.	10.90	13.33	0.87	21.38	nr	**34.71**
450 mm dia.	12.45	15.22	0.87	21.38	nr	**36.60**
500 mm dia.	13.04	15.94	0.87	21.38	nr	**37.32**
630 mm dia.	30.06	36.75	0.87	21.38	nr	**58.13**
710 mm dia.	41.85	51.16	0.96	23.60	nr	**74.76**
800 mm dia.	46.29	56.59	1.06	26.05	nr	**82.64**
900 mm dia.	50.42	61.64	1.16	28.51	nr	**90.15**
1000 mm dia.	93.77	114.64	1.25	30.72	nr	**145.36**
1120 mm dia.	104.21	127.40	3.47	85.27	nr	**212.67**
1250 mm dia.	134.17	164.03	3.47	85.27	nr	**249.30**
1400 mm dia.	239.46	292.75	4.05	99.53	nr	**392.28**
1600 mm dia.	247.81	302.95	4.62	113.53	nr	**416.48**

U: VENTILATION/AIR CONDITIONING SYSTEMS

Item	Net Price £	Material £	Labour hours	Labour £	Unit	Total rate £
45° radius bend						
80 mm dia.	2.08	2.54	0.29	7.13	nr	9.67
100 mm dia.	2.34	2.86	0.29	7.13	nr	9.99
160 mm dia.	3.32	4.06	0.29	7.13	nr	11.19
200 mm dia.	4.31	5.27	0.40	9.83	nr	15.10
250 mm dia.	5.99	7.32	0.58	14.26	nr	21.58
315 mm dia.	8.04	9.83	0.58	14.26	nr	24.09
355 mm dia.	8.60	10.52	0.87	21.38	nr	31.90
400 mm dia.	10.29	12.58	0.87	21.38	nr	33.96
450 mm dia.	10.42	12.74	0.87	21.38	nr	34.12
500 mm dia.	11.16	13.65	0.87	21.38	nr	35.03
630 mm dia.	32.27	39.45	0.87	21.38	nr	60.83
710 mm dia.	40.50	49.51	0.96	23.60	nr	73.11
800 mm dia.	47.83	58.47	1.06	26.05	nr	84.52
900 mm dia.	52.31	63.95	1.16	28.51	nr	92.46
1000 mm dia.	86.66	105.95	1.25	30.72	nr	136.67
1120 mm dia.	94.57	115.62	3.47	85.27	nr	200.89
1250 mm dia.	105.16	128.55	3.47	85.27	nr	213.82
1400 mm dia.	144.55	176.72	4.05	99.53	nr	276.25
1600 mm dia.	141.11	172.50	4.62	113.53	nr	286.03
90° equal twin bend						
80 mm dia.	5.94	7.26	0.58	14.26	nr	21.52
100 mm dia.	6.15	7.52	0.58	14.26	nr	21.78
160 mm dia.	10.89	13.32	0.58	14.26	nr	27.58
200 mm dia.	15.27	18.66	0.87	21.38	nr	40.04
250 mm dia.	23.53	28.77	1.16	28.51	nr	57.28
315 mm dia.	32.65	39.91	1.16	28.51	nr	68.42
355 mm dia.	36.38	44.48	1.73	42.51	nr	86.99
400 mm dia.	40.43	49.43	1.73	42.51	nr	91.94
450 mm dia.	43.76	53.50	1.73	42.51	nr	96.01
500 mm dia.	47.10	57.58	1.73	42.51	nr	100.09
630 mm dia.	90.99	111.24	1.73	42.51	nr	153.75
710 mm dia.	120.79	147.67	1.82	44.72	nr	192.39
800 mm dia.	147.58	180.41	1.93	47.42	nr	227.83
900 mm dia.	171.14	209.22	2.02	49.64	nr	258.86
1000 mm dia.	235.82	288.29	2.11	51.85	nr	340.14
1120 mm dia.	274.63	335.74	4.62	113.53	nr	449.27
1250 mm dia.	328.47	401.56	4.62	113.53	nr	515.09
1400 mm dia.	525.31	642.19	4.62	113.53	nr	755.72
1600 mm dia.	525.28	642.15	4.62	113.53	nr	755.68

U: VENTILATION/AIR CONDITIONING SYSTEMS

Item	Net Price £	Material £	Labour hours	Labour £	Unit	Total rate £
U10: DUCTWORK: CIRCULAR – cont						
Extra over fittings – cont						
Conical branch						
80 mm dia.	7.87	9.62	0.58	14.26	nr	**23.88**
100 mm dia.	8.05	9.84	0.58	14.26	nr	**24.10**
160 mm dia.	8.71	10.65	0.58	14.26	nr	**24.91**
200 mm dia.	9.26	11.32	0.87	21.38	nr	**32.70**
250 mm dia.	12.12	14.82	1.16	28.51	nr	**43.33**
315 mm dia.	13.52	16.53	1.16	28.51	nr	**45.04**
355 mm dia.	14.67	17.93	1.73	42.51	nr	**60.44**
400 mm dia.	15.14	18.51	1.73	42.51	nr	**61.02**
450 mm dia.	18.72	22.89	1.73	42.51	nr	**65.40**
500 mm dia.	19.28	23.57	1.73	42.51	nr	**66.08**
630 mm dia.	38.78	47.41	1.73	42.51	nr	**89.92**
710 mm dia.	43.20	52.81	1.82	44.72	nr	**97.53**
800 mm dia.	55.98	68.43	1.93	47.42	nr	**115.85**
900 mm dia.	59.34	72.54	2.02	49.64	nr	**122.18**
1000 mm dia.	75.26	92.01	2.11	51.85	nr	**143.86**
1120 mm dia.	104.83	128.15	4.62	113.53	nr	**241.68**
1250 mm dia.	104.83	128.15	5.20	127.77	nr	**255.92**
1400 mm dia.	126.29	154.39	5.20	127.77	nr	**282.16**
1600 mm dia.	148.76	181.86	5.20	127.77	nr	**309.63**
45° branch						
80 mm dia.	6.25	7.64	0.58	14.26	nr	**21.90**
100 mm dia.	6.40	7.83	0.58	14.26	nr	**22.09**
160 mm dia.	9.74	11.91	0.58	14.26	nr	**26.17**
200 mm dia.	10.09	12.34	0.87	21.38	nr	**33.72**
250 mm dia.	10.57	12.93	1.16	28.51	nr	**41.44**
315 mm dia.	11.25	13.75	1.16	28.51	nr	**42.26**
355 mm dia.	14.50	17.73	1.73	42.51	nr	**60.24**
400 mm dia.	15.07	18.43	1.73	42.51	nr	**60.94**
450 mm dia.	15.70	19.19	1.73	42.51	nr	**61.70**
500 mm dia.	16.37	20.01	1.73	42.51	nr	**62.52**
630 mm dia.	32.56	39.81	1.73	42.51	nr	**82.32**
710 mm dia.	36.35	44.43	1.82	44.72	nr	**89.15**
800 mm dia.	51.87	63.41	2.13	52.33	nr	**115.74**
900 mm dia.	58.71	71.77	2.31	56.76	nr	**128.53**
1000 mm dia.	72.48	88.61	2.31	56.76	nr	**145.37**
1120 mm dia.	89.81	109.80	4.62	113.53	nr	**223.33**
1250 mm dia.	95.97	117.33	4.62	113.53	nr	**230.86**
1400 mm dia.	110.84	135.51	4.62	113.53	nr	**249.04**
1600 mm dia.	135.50	165.64	4.62	113.53	nr	**279.17**
For galvanized sheet metal DW144 class C rates, refer to galvanized sheet metal DW144 class B						

U: VENTILATION/AIR CONDITIONING SYSTEMS

Item	Net Price £	Material £	Labour hours	Labour £	Unit	Total rate £
U10: DUCTWORK: FLAT OVAL						
Y30 – AIR DUCTLINES						
Galvanized sheet metal DW144 class B spirally wound flat oval section ductwork; including all necessary stiffeners, joints, couplers in the running length and duct supports						
Straight duct						
345 × 102 mm	17.45	21.33	2.71	66.59	m	**87.92**
427 × 102 mm	19.99	24.44	2.99	73.47	m	**97.91**
508 × 102 mm	22.05	26.96	3.14	77.16	m	**104.12**
559 × 152 mm	29.77	36.39	3.43	84.27	m	**120.66**
531 × 203 mm	30.24	36.97	3.43	84.27	m	**121.24**
851 × 203 mm	39.35	48.10	5.72	140.55	m	**188.65**
582 × 254 mm	39.77	48.62	3.62	88.95	m	**137.57**
823 × 254 mm	39.91	48.79	5.80	142.52	m	**191.31**
1303 × 254 mm	94.59	115.64	8.13	199.77	m	**315.41**
632 × 305 mm	36.01	44.02	3.93	96.56	m	**140.58**
1275 × 305 mm	87.27	106.69	8.13	199.77	m	**306.46**
765 × 356 mm	41.74	51.03	5.72	140.55	m	**191.58**
1247 × 356 mm	88.22	107.85	8.13	199.77	m	**307.62**
1727 × 356 mm	118.84	145.28	10.41	255.80	m	**401.08**
737 × 406 mm	42.52	51.98	5.72	140.55	m	**192.53**
818 × 406 mm	44.91	54.90	6.21	152.59	m	**207.49**
978 × 406 mm	49.48	60.49	6.92	170.04	m	**230.53**
1379 × 406 mm	96.10	117.48	8.75	215.01	m	**332.49**
1699 × 406 mm	97.50	119.19	10.41	255.80	m	**374.99**
709 × 457 mm	42.47	51.92	5.72	140.55	m	**192.47**
1189 × 457 mm	81.14	99.20	8.80	216.24	m	**315.44**
1671 × 457 mm	117.13	143.19	10.31	253.34	m	**396.53**
678 × 508 mm	42.05	51.41	5.72	140.55	m	**191.96**
919 × 508 mm	49.48	60.49	7.30	179.37	m	**239.86**
1321 × 508 mm	95.94	117.29	8.75	215.01	m	**332.30**
Extra over fittings; flat oval duct class B						
End cap						
345 × 102 mm	15.31	18.72	0.20	4.91	nr	**23.63**
427 × 102 mm	15.73	19.23	0.20	4.91	nr	**24.14**
508 × 102 mm	16.14	19.73	0.20	4.91	nr	**24.64**
559 × 152 mm	23.37	28.57	0.29	7.13	nr	**35.70**
531 × 203 mm	23.48	28.71	0.29	7.13	nr	**35.84**
851 × 203 mm	26.06	31.86	0.44	10.80	nr	**42.66**
582 × 254 mm	24.11	29.48	0.44	10.80	nr	**40.28**
823 × 254 mm	25.63	31.33	0.44	10.80	nr	**42.13**
1303 × 254 mm	84.68	103.53	0.69	16.95	nr	**120.48**
632 × 305 mm	24.09	29.45	0.69	16.95	nr	**46.40**
1275 × 305 mm	74.70	91.32	0.69	16.95	nr	**108.27**

U: VENTILATION/AIR CONDITIONING SYSTEMS

Item	Net Price £	Material £	Labour hours	Labour £	Unit	Total rate £
U10: DUCTWORK: FLAT OVAL – cont						
Extra over fittings – cont						
End cap – cont						
765 × 356 mm	25.28	30.90	0.69	16.95	nr	**47.85**
1247 × 356 mm	72.55	88.69	0.69	16.95	nr	**105.64**
1727 × 356 mm	107.78	131.76	0.69	16.95	nr	**148.71**
737 × 406 mm	26.01	31.80	1.04	25.55	nr	**57.35**
818 × 406 mm	26.67	32.60	0.69	16.95	nr	**49.55**
978 × 406 mm	27.97	34.20	0.69	16.95	nr	**51.15**
1379 × 406 mm	99.33	121.43	1.04	25.55	nr	**146.98**
1699 × 406 mm	108.42	132.54	1.04	25.55	nr	**158.09**
709 × 457 mm	25.30	30.93	1.04	25.55	nr	**56.48**
1189 × 457 mm	70.83	86.59	1.04	25.55	nr	**112.14**
1671 × 457 mm	105.57	129.06	1.04	25.55	nr	**154.61**
678 × 508 mm	34.14	41.74	1.04	25.55	nr	**67.29**
919 × 508 mm	36.82	45.01	1.04	25.55	nr	**70.56**
1321 × 508 mm	105.54	129.03	1.04	25.55	nr	**154.58**
Reducer						
345 × 102 mm	28.58	34.94	0.95	23.34	nr	**58.28**
427 × 102 mm	29.00	35.45	1.06	26.05	nr	**61.50**
508 × 102 mm	29.55	36.12	1.13	27.76	nr	**63.88**
559 × 152 mm	41.21	50.38	1.26	30.96	nr	**81.34**
531 × 203 mm	41.50	50.74	1.26	30.96	nr	**81.70**
851 × 203 mm	53.45	65.34	1.34	32.92	nr	**98.26**
582 × 254 mm	43.76	53.50	1.34	32.92	nr	**86.42**
823 × 254 mm	54.94	67.17	1.34	32.92	nr	**100.09**
1303 × 254 mm	125.13	152.98	1.34	32.92	nr	**185.90**
632 × 305 mm	48.11	58.81	0.70	17.20	nr	**76.01**
1275 × 305 mm	127.72	156.14	1.16	28.51	nr	**184.65**
765 × 356 mm	51.40	62.84	1.16	28.51	nr	**91.35**
1247 × 356 mm	129.93	158.84	1.16	28.51	nr	**187.35**
1727 × 356 mm	132.15	161.56	1.25	30.72	nr	**192.28**
737 × 406 mm	56.76	69.39	1.16	28.51	nr	**97.90**
818 × 406 mm	57.46	70.25	1.27	31.21	nr	**101.46**
978 × 406 mm	62.36	76.24	1.44	35.38	nr	**111.62**
1379 × 406 mm	131.88	161.23	1.44	35.38	nr	**196.61**
1699 × 406 mm	149.45	182.70	1.44	35.38	nr	**218.08**
709 × 457 mm	57.38	70.14	1.16	28.51	nr	**98.65**
1189 × 457 mm	116.31	142.19	1.34	32.92	nr	**175.11**
1671 × 457 mm	129.31	158.08	1.44	35.38	nr	**193.46**
678 × 508 mm	61.02	74.59	1.16	28.51	nr	**103.10**
919 × 508 mm	65.01	79.47	1.26	30.96	nr	**110.43**
1321 × 508 mm	130.21	159.19	1.44	35.38	nr	**194.57**
90° radius bend						
345 × 102 mm	37.33	45.64	0.29	7.13	nr	**52.77**
427 × 102 mm	35.70	43.64	0.58	14.26	nr	**57.90**
508 × 102 mm	34.41	42.07	0.58	14.26	nr	**56.33**

U: VENTILATION/AIR CONDITIONING SYSTEMS

Item	Net Price £	Material £	Labour hours	Labour £	Unit	Total rate £
559 × 152 mm	47.54	58.12	0.58	14.26	nr	72.38
531 × 203 mm	48.06	58.75	0.87	21.38	nr	80.13
851 × 203 mm	64.94	79.39	0.87	21.38	nr	100.77
582 × 254 mm	45.55	55.68	0.87	21.38	nr	77.06
823 × 254 mm	67.69	82.75	0.87	21.38	nr	104.13
1303 × 254 mm	152.31	186.20	0.96	23.60	nr	209.80
632 × 305 mm	43.00	52.57	0.87	21.38	nr	73.95
1275 × 305 mm	193.01	235.95	0.96	23.60	nr	259.55
765 × 356 mm	50.05	61.18	0.87	21.38	nr	82.56
1247 × 356 mm	194.07	237.25	0.96	23.60	nr	260.85
1727 × 356 mm	236.39	288.99	1.25	30.72	nr	319.71
737 × 406 mm	51.67	63.17	0.96	23.60	nr	86.77
818 × 406 mm	74.49	91.06	0.87	21.38	nr	112.44
978 × 406 mm	66.13	80.84	0.96	23.60	nr	104.44
1379 × 406 mm	173.60	212.23	1.16	28.51	nr	240.74
1699 × 406 mm	296.95	363.02	1.25	30.72	nr	393.74
709 × 457 mm	59.52	72.76	0.87	21.38	nr	94.14
1189 × 457 mm	143.72	175.70	0.96	23.60	nr	199.30
1671 × 457 mm	296.17	362.07	1.25	30.72	nr	392.79
678 × 508 mm	57.51	70.31	0.87	21.38	nr	91.69
919 × 508 mm	64.81	79.23	0.96	23.60	nr	102.83
1321 × 508 mm	179.71	219.70	1.16	28.51	nr	248.21
45° radius bend						
345 × 102 mm	26.42	32.30	0.79	19.42	nr	51.72
427 × 102 mm	25.91	31.67	0.85	20.89	nr	52.56
508 × 102 mm	25.39	31.04	0.95	23.34	nr	54.38
559 × 152 mm	35.25	43.10	0.79	19.42	nr	62.52
531 × 203 mm	35.48	43.37	0.85	20.89	nr	64.26
851 × 203 mm	49.47	60.48	0.98	24.08	nr	84.56
582 × 254 mm	34.58	42.27	0.76	18.68	nr	60.95
823 × 254 mm	51.51	62.97	0.95	23.34	nr	86.31
1303 × 254 mm	149.32	182.55	1.16	28.51	nr	211.06
632 × 305 mm	37.57	45.93	0.58	14.26	nr	60.19
1275 × 305 mm	171.98	210.24	1.16	28.51	nr	238.75
765 × 356 mm	38.56	47.14	0.87	21.38	nr	68.52
1247 × 356 mm	172.58	210.99	1.16	28.51	nr	239.50
1727 × 356 mm	167.85	205.20	1.26	30.96	nr	236.16
737 × 406 mm	40.68	49.73	0.69	16.95	nr	66.68
818 × 406 mm	56.46	69.02	0.78	19.17	nr	88.19
978 × 406 mm	52.91	64.68	0.87	21.38	nr	86.06
1379 × 406 mm	164.50	201.11	1.16	28.51	nr	229.62
1699 × 406 mm	198.40	242.54	1.27	31.21	nr	273.75
709 × 457 mm	45.82	56.01	0.81	19.90	nr	75.91
1189 × 457 mm	135.93	166.18	0.95	23.34	nr	189.52
1671 × 457 mm	195.46	238.95	1.26	30.96	nr	269.91
678 × 508 mm	44.68	54.62	0.92	22.61	nr	77.23
919 × 508 mm	51.95	63.51	1.10	27.03	nr	90.54
1321 × 508 mm	167.17	204.36	1.25	30.72	nr	235.08

U: VENTILATION/AIR CONDITIONING SYSTEMS

Item	Net Price £	Material £	Labour hours	Labour £	Unit	Total rate £
U10: DUCTWORK: FLAT OVAL – cont						
Extra over fittings – cont						
90° hard bend with turning vanes						
345 × 102 mm	38.30	46.82	0.55	13.51	nr	60.33
427 × 102 mm	37.38	45.70	1.16	28.51	nr	74.21
508 × 102 mm	36.46	44.57	1.16	28.51	nr	73.08
559 × 152 mm	38.59	47.17	1.16	28.51	nr	75.68
531 × 203 mm	38.94	47.61	1.73	42.51	nr	90.12
851 × 203 mm	60.26	73.67	1.73	42.51	nr	116.18
582 × 254 mm	44.01	53.81	1.73	42.51	nr	96.32
823 × 254 mm	59.89	73.21	1.73	42.51	nr	115.72
1303 × 254 mm	104.57	127.83	1.82	44.72	nr	172.55
632 × 305 mm	57.93	70.82	1.73	42.51	nr	113.33
1275 × 305 mm	106.48	130.17	1.82	44.72	nr	174.89
765 × 356 mm	56.86	69.51	1.73	42.51	nr	112.02
1247 × 356 mm	107.25	131.12	1.82	44.72	nr	175.84
1727 × 356 mm	180.26	220.37	1.82	44.72	nr	265.09
737 × 406 mm	62.45	76.34	1.73	42.51	nr	118.85
818 × 406 mm	60.29	73.71	1.73	42.51	nr	116.22
978 × 406 mm	65.82	80.46	1.73	42.51	nr	122.97
1379 × 406 mm	141.25	172.68	1.82	44.72	nr	217.40
1699 × 406 mm	221.08	270.27	2.11	51.85	nr	322.12
709 × 457 mm	69.81	85.35	1.73	42.51	nr	127.86
1189 × 457 mm	116.75	142.73	1.82	44.72	nr	187.45
1671 × 457 mm	183.35	224.15	2.11	51.85	nr	276.00
678 × 508 mm	67.49	82.50	1.82	44.72	nr	127.22
919 × 508 mm	74.06	90.54	1.82	44.72	nr	135.26
1321 × 508 mm	147.84	180.73	2.11	51.85	nr	232.58
90° branch						
345 × 102 mm	29.17	35.66	0.58	14.26	nr	49.92
427 × 102 mm	29.73	36.35	0.58	14.26	nr	50.61
508 × 102 mm	30.28	37.02	1.16	28.51	nr	65.53
559 × 152 mm	44.81	54.78	1.16	28.51	nr	83.29
531 × 203 mm	45.58	55.72	1.16	28.51	nr	84.23
851 × 203 mm	59.23	72.41	1.73	42.51	nr	114.92
582 × 254 mm	50.96	62.29	1.73	42.51	nr	104.80
823 × 254 mm	64.38	78.70	1.73	42.51	nr	121.21
1303 × 254 mm	137.04	167.53	1.82	44.72	nr	212.25
632 × 305 mm	56.84	69.48	1.73	42.51	nr	111.99
1275 × 305 mm	137.62	168.24	1.82	44.72	nr	212.96
765 × 356 mm	64.15	78.42	1.73	42.51	nr	120.93
1247 × 356 mm	141.56	173.06	1.82	44.72	nr	217.78
1727 × 356 mm	179.76	219.76	2.11	51.85	nr	271.61
737 × 406 mm	69.19	84.58	1.73	42.51	nr	127.09
818 × 406 mm	70.57	86.27	1.73	42.51	nr	128.78
978 × 406 mm	78.54	96.02	1.82	44.72	nr	140.74
1379 × 406 mm	152.04	185.87	1.93	47.42	nr	233.29
1699 × 406 mm	177.12	216.53	2.11	51.85	nr	268.38

U: VENTILATION/AIR CONDITIONING SYSTEMS

Item	Net Price £	Material £	Labour hours	Labour £	Unit	Total rate £
709 × 457 mm	70.67	86.40	1.73	42.51	nr	**128.91**
1189 × 457 mm	124.83	152.60	1.82	44.72	nr	**197.32**
1671 × 457 mm	177.52	217.02	2.11	51.85	nr	**268.87**
678 × 508 mm	75.45	92.24	1.73	42.51	nr	**134.75**
919 × 508 mm	83.45	102.02	1.82	44.72	nr	**146.74**
1321 × 508 mm	152.67	186.64	2.11	51.85	nr	**238.49**
45° branch						
345 × 102 mm	30.80	37.66	0.58	14.26	nr	**51.92**
427 × 102 mm	32.43	39.64	0.58	14.26	nr	**53.90**
508 × 102 mm	34.23	41.85	1.16	28.51	nr	**70.36**
559 × 152 mm	49.17	60.11	1.73	42.51	nr	**102.62**
531 × 203 mm	48.85	59.72	1.73	42.51	nr	**102.23**
851 × 203 mm	72.32	88.42	1.73	42.51	nr	**130.93**
582 × 254 mm	54.48	66.60	1.73	42.51	nr	**109.11**
823 × 254 mm	73.30	89.61	1.82	44.72	nr	**134.33**
1303 × 254 mm	169.48	207.19	1.92	47.17	nr	**254.36**
632 × 305 mm	60.53	74.00	1.73	42.51	nr	**116.51**
1275 × 305 mm	167.21	204.41	1.82	44.72	nr	**249.13**
765 × 356 mm	69.63	85.12	1.73	42.51	nr	**127.63**
1247 × 356 mm	167.86	205.21	1.82	44.72	nr	**249.93**
1727 × 356 mm	236.12	288.66	1.82	44.72	nr	**333.38**
737 × 406 mm	73.86	90.29	1.73	42.51	nr	**132.80**
818 × 406 mm	77.48	94.72	1.73	42.51	nr	**137.23**
978 × 406 mm	90.90	111.13	1.73	42.51	nr	**153.64**
1379 × 406 mm	182.99	223.71	1.93	47.42	nr	**271.13**
1699 × 406 mm	228.10	278.85	2.19	53.82	nr	**332.67**
709 × 457 mm	73.13	89.40	1.73	42.51	nr	**131.91**
1189 × 457 mm	144.20	176.28	1.82	44.72	nr	**221.00**
1671 × 457 mm	224.43	274.37	2.11	51.85	nr	**326.22**
678 × 508 mm	75.80	92.67	1.73	42.51	nr	**135.18**
919 × 508 mm	90.54	110.68	1.82	44.72	nr	**155.40**
1321 × 508 mm	176.59	215.88	1.93	47.42	nr	**263.30**

For rates for access doors refer to ancillaries in U10: DUCTWORK: RECTANGULAR: CLASS B

U: VENTILATION/AIR CONDITIONING SYSTEMS

Item	Net Price £	Material £	Labour hours	Labour £	Unit	Total rate £
U10: DUCTWORK: FLEXIBLE						
Y30: AIR DUCTLINES						
Aluminium foil flexible ductwork, DW 144 class B; multiply aluminium polyester laminate fabric, with high tensile steel wire helix						
Duct						
102 mm dia.	1.23	1.50	0.33	8.11	m	9.61
152 mm dia.	1.80	2.20	0.33	8.11	m	10.31
203 mm dia.	2.38	2.90	0.33	8.11	m	11.01
254 mm dia.	2.98	3.65	0.33	8.11	m	11.76
304 mm dia.	3.65	4.46	0.33	8.11	m	12.57
355 mm dia.	4.98	6.09	0.33	8.11	m	14.20
406 mm dia.	5.53	6.76	0.33	8.11	m	14.87
Insulated aluminium foil flexible ductwork, DW144 class B; laminate construction of aluminium and polyester multiply inner core with 25 mm insulation; outer layer of multiply aluminium polyester laminate, with high tensile steel wire helix						
Duct						
102 mm dia.	2.76	3.38	0.50	12.29	m	15.67
152 mm dia.	3.60	4.40	0.50	12.29	m	16.69
203 mm dia.	4.37	5.35	0.50	12.29	m	17.64
254 mm dia.	5.38	6.58	0.50	12.29	m	18.87
304 mm dia.	6.92	8.46	0.50	12.29	m	20.75
355 mm dia.	8.58	10.49	0.50	12.29	m	22.78
406 mm dia.	9.50	11.62	0.50	12.29	m	23.91

U: VENTILATION/AIR CONDITIONING SYSTEMS

Item	Net Price £	Material £	Labour hours	Labour £	Unit	Total rate £
U10: DUCTWORK: PLASTIC						
Y30 – AIR DUCTLINES						
Rigid gey PVC DW 154 circular section ductwork; solvent welded or filler rod welded joints; excludes couplers and supports (these are detailed separately); ductwork to conform to curent HSE regulations						
Straight duct (standard length 6 m)						
110 mm	3.41	4.17	1.74	42.76	m	**46.93**
160 mm	6.45	7.89	1.71	42.01	m	**49.90**
200 mm	8.47	10.35	1.79	43.98	m	**54.33**
225 mm	12.07	14.76	1.96	48.16	m	**62.92**
250 mm	11.87	14.51	1.98	48.66	m	**63.17**
315 mm	16.97	20.74	2.23	54.80	m	**75.54**
355 mm	22.73	27.79	2.25	55.29	m	**83.08**
400 mm	28.58	34.94	2.39	58.73	m	**93.67**
450 mm	35.32	43.18	2.95	72.49	m	**115.67**
500 mm	43.49	53.17	2.98	73.23	m	**126.40**
600 mm	67.24	82.20	3.12	76.66	m	**158.86**
Extra for supports (BZP finish)						
Horizontal – Maximum 2.4 m centres						
Vertical – Maximum 4.0 m centres						
Duct Size						
110 mm	21.08	25.77	0.59	14.50	m	**40.27**
160 mm	21.15	25.85	0.59	14.50	m	**40.35**
200 mm	21.23	25.96	0.59	14.50	m	**40.46**
225 mm	22.53	27.54	0.63	15.48	m	**43.02**
250 mm	22.54	27.55	0.63	15.48	m	**43.03**
315 mm	22.70	27.75	0.63	15.48	m	**43.23**
355 mm	21.71	26.54	0.81	19.90	m	**46.44**
400 mm	29.48	36.04	0.78	19.17	m	**55.21**
450 mm	30.30	37.04	0.80	19.66	m	**56.70**
500 mm	30.46	37.23	0.80	19.66	m	**56.89**
600 mm	32.33	39.52	0.85	20.89	m	**60.41**

Note: These are maximum figures and may be reduced subject to local conditions (i.e. a high number of changes of direction).

U: VENTILATION/AIR CONDITIONING SYSTEMS

Item	Net Price £	Material £	Labour hours	Labour £	Unit	Total rate £
U10: DUCTWORK: PLASTIC – cont						
Extra over fittings; Rigid grey PVC						
90° Bend						
110 mm	17.45	21.33	1.11	27.27	m	**48.60**
160 mm	19.25	23.54	1.34	32.92	m	**56.46**
200 mm	16.10	19.68	2.00	49.14	m	**68.82**
225 mm	24.73	30.23	1.79	43.98	m	**74.21**
250 mm	26.32	32.18	2.01	49.39	m	**81.57**
315 mm	45.44	55.55	2.56	62.91	m	**118.46**
355 mm	54.11	66.15	2.71	66.59	m	**132.74**
400 mm	68.73	84.03	3.58	87.97	m	**172.00**
450 mm	207.74	253.97	4.55	111.80	m	**365.77**
500 mm	241.12	294.78	5.01	123.11	m	**417.89**
600 mm	412.73	504.57	5.78	142.03	m	**646.60**
45° Bend						
110 mm	7.18	8.78	0.71	17.45	m	**26.23**
160 mm	8.34	10.20	0.93	22.85	m	**33.05**
200 mm	9.15	11.19	1.14	28.01	m	**39.20**
225 mm	10.33	12.63	1.41	34.65	m	**47.28**
250 mm	11.07	13.53	1.62	39.81	m	**53.34**
315 mm	17.67	21.60	1.93	47.42	m	**69.02**
355 mm	22.99	28.11	2.22	54.55	m	**82.66**
400 mm	26.42	32.30	2.85	70.03	m	**102.33**
450 mm	101.76	124.40	3.77	92.64	m	**217.04**
500 mm	113.80	139.12	4.16	102.23	m	**241.35**
600 mm	185.39	226.64	4.81	118.19	m	**344.83**
Tee						
110 mm	15.23	18.62	1.04	25.55	m	**44.17**
160 mm	18.59	22.72	1.38	33.91	m	**56.63**
200 mm	24.73	30.23	1.79	43.98	m	**74.21**
225 mm	28.49	34.82	2.25	55.29	m	**90.11**
250 mm	34.87	42.63	2.62	64.38	m	**107.01**
315 mm	58.08	71.01	3.17	77.90	m	**148.91**
355 mm	81.23	99.30	3.73	91.65	m	**190.95**
400 mm	79.19	96.81	4.71	115.73	m	**212.54**
450 mm	155.73	190.39	5.72	140.55	m	**330.94**
500 mm	176.99	216.37	6.33	155.54	m	**371.91**
Coupler						
110 mm	4.07	4.97	0.70	17.20	m	**22.17**
160 mm	5.05	6.17	0.91	22.36	m	**28.53**
200 mm	6.26	7.65	1.12	27.52	m	**35.17**
225 mm	8.20	10.02	1.41	34.65	m	**44.67**
250 mm	9.12	11.14	1.60	39.32	m	**50.46**
315 mm	10.87	13.29	1.86	45.71	m	**59.00**
355 mm	11.43	13.98	2.43	59.71	m	**73.69**

U: VENTILATION/AIR CONDITIONING SYSTEMS

Item	Net Price £	Material £	Labour hours	Labour £	Unit	Total rate £
400 mm	12.18	14.89	2.66	65.36	m	**80.25**
450 mm	29.29	35.80	3.20	78.63	m	**114.43**
500 mm	33.05	40.41	3.52	86.49	m	**126.90**
Damper						
110 mm	34.12	41.71	0.96	23.60	m	**65.31**
160 mm	34.64	42.34	1.26	30.96	m	**73.30**
200 mm	35.39	43.26	1.52	37.35	m	**80.61**
225 mm	36.58	44.72	1.88	46.20	m	**90.92**
250 mm	36.44	44.55	2.16	53.09	m	**97.64**
315 mm	41.80	51.10	2.94	72.23	m	**123.33**
355 mm	44.66	54.60	3.42	84.04	m	**138.64**
400 mm	47.94	58.61	3.79	93.13	m	**151.74**
Reducer						
160 × 110 mm	8.92	10.91	0.85	20.89	m	**31.80**
200 × 110 mm	12.98	15.87	0.98	24.08	m	**39.95**
200 × 160 mm	11.24	13.74	1.07	26.30	m	**40.04**
225 × 200 mm	14.75	18.04	1.38	33.91	m	**51.95**
250 × 160 mm	15.60	19.08	1.38	33.91	m	**52.99**
250 × 200 mm	14.82	18.12	1.48	36.37	m	**54.49**
250 × 225 mm	14.54	17.78	1.58	38.83	m	**56.61**
315 × 200 mm	15.87	19.41	1.63	40.05	m	**59.46**
315 × 250 mm	16.32	19.95	1.84	45.21	m	**65.16**
355 × 200 mm	24.18	29.56	1.99	48.90	m	**78.46**
355 × 250 mm	18.04	22.05	1.97	48.41	m	**70.46**
355 × 315 mm	22.48	27.48	2.16	53.09	m	**80.57**
400 × 225 mm	27.30	33.37	2.47	60.70	m	**94.07**
400 × 315 mm	27.58	33.71	2.75	67.57	m	**101.28**
400 × 355 mm	29.04	35.50	2.73	67.08	m	**102.58**
450 × 315 mm	31.34	38.32	2.91	71.50	m	**109.82**
Flange						
110 mm	5.88	7.19	0.70	17.20	m	**24.39**
160 mm	5.95	7.27	0.91	22.36	m	**29.63**
200 mm	6.42	7.85	1.12	27.52	m	**35.37**
225 mm	6.24	7.63	1.35	33.18	m	**40.81**
250 mm	6.67	8.16	1.55	38.09	m	**46.25**
315 mm	9.49	11.60	1.85	45.46	m	**57.06**
355 mm	10.04	12.28	2.06	50.62	m	**62.90**
400 mm	10.06	12.30	2.62	64.38	m	**76.68**

U: VENTILATION/AIR CONDITIONING SYSTEMS

Item	Net Price £	Material £	Labour hours	Labour £	Unit	Total rate £
U10: DUCTWORK: PLASTIC – cont						
Polypropylene (PPS) DW154 circular section ductwork; filler rod welded joints; excludes couplers and supports (these are detailed separately); ductwork to conform to current HSE regulations						
Straight duct (standard length 6 m)						
110 mm	15.96	19.51	0.65	15.97	m	**35.48**
160 mm	17.50	21.39	0.75	18.43	m	**39.82**
200 mm	18.58	22.71	0.85	20.89	m	**43.60**
225 mm	23.82	29.12	1.07	26.30	m	**55.42**
250 mm	24.25	29.64	1.18	28.99	m	**58.63**
315 mm	44.65	54.59	1.41	34.65	m	**89.24**
355 mm	47.42	57.97	1.53	37.59	m	**95.56**
400 mm	57.87	70.75	1.73	42.51	m	**113.26**
Extra for supports (BZP finish)						
Horizontal – Maximum 2.4 m centres						
Vertical – Maximum 4.0 m centres						
Duct Size						
110 mm	29.22	35.72	0.42	10.32	m	**46.04**
160 mm	29.22	35.72	0.42	10.32	m	**46.04**
200 mm	29.22	35.72	0.42	10.32	m	**46.04**
225 mm	29.22	35.72	0.42	10.32	m	**46.04**
250 mm	29.22	35.72	0.42	10.32	m	**46.04**
315 mm	29.22	35.72	0.42	10.32	m	**46.04**
355 mm	29.22	35.72	0.55	13.51	m	**49.23**
400 mm	41.36	50.56	0.55	13.51	m	**64.07**
Note: These are maximum figures and may be reduced subject to local conditions (i.e. a high number of changes of direction)						
Extra over fittings; Polypropylene (DW 154)						
90° Bend						
110 mm	11.97	14.64	1.01	24.81	m	**39.45**
160 mm	14.90	18.21	1.30	31.94	m	**50.15**
200 mm	16.37	20.01	1.58	38.83	m	**58.84**
225 mm	22.82	27.90	2.10	51.60	m	**79.50**
250 mm	23.14	28.28	2.36	58.00	m	**86.28**
315 mm	69.01	84.37	3.14	77.16	m	**161.53**
355 mm	77.51	94.76	3.52	86.49	m	**181.25**
400 mm	75.95	92.85	4.39	107.87	m	**200.72**

U: VENTILATION/AIR CONDITIONING SYSTEMS

Item	Net Price £	Material £	Labour hours	Labour £	Unit	Total rate £
45° Bend						
110 mm	7.47	9.14	0.84	20.63	m	29.77
160 mm	10.88	13.30	1.13	27.76	m	41.06
200 mm	11.58	14.15	1.39	34.15	m	48.30
225 mm	14.20	17.36	1.73	42.51	m	59.87
250 mm	15.29	18.69	1.99	48.90	m	67.59
315 mm	43.32	52.96	2.57	63.16	m	116.12
355 mm	48.29	59.04	2.93	72.00	m	131.04
400 mm	46.34	56.65	3.60	88.46	m	145.11
Tee						
110 mm	33.38	40.81	1.55	38.09	m	78.90
160 mm	41.34	50.54	2.17	53.32	m	103.86
200 mm	47.80	58.43	2.68	65.86	m	124.29
225 mm	56.91	69.58	3.38	83.06	m	152.64
250 mm	66.07	80.77	3.95	97.06	m	177.83
315 mm	93.43	114.22	4.69	115.25	m	229.47
355 mm	116.81	142.80	5.44	133.67	m	276.47
400 mm	136.49	166.86	6.09	149.65	m	316.51
Coupler						
110 mm	7.36	9.00	0.84	20.63	m	29.63
160 mm	7.93	9.69	1.10	27.03	m	36.72
200 mm	8.88	10.86	1.36	33.42	m	44.28
225 mm	9.57	11.70	1.67	41.04	m	52.74
250 mm	9.70	11.86	1.91	46.94	m	58.80
315 mm	13.41	16.40	2.24	55.03	m	71.43
355 mm	17.23	21.06	2.58	63.40	m	84.46
400 mm	17.60	21.52	3.30	81.09	m	102.61
Damper						
110 mm	69.72	85.23	0.79	19.42	m	104.65
160 mm	79.20	96.82	1.13	27.76	m	124.58
200 mm	85.32	104.31	1.44	35.38	m	139.69
225 mm	89.59	109.52	1.81	44.48	m	154.00
250 mm	91.64	112.03	2.08	51.11	m	163.14
315 mm	104.61	127.88	2.40	58.98	m	186.86
355 mm	113.14	138.32	2.75	67.57	m	205.89
400 mm	118.15	144.44	3.57	87.73	m	232.17
Reducer						
160 × 110 mm	25.12	30.70	0.87	21.38	m	52.08
200 × 160 mm	26.00	31.79	1.16	28.51	m	60.30
225 × 200 mm	30.32	37.07	1.46	35.89	m	72.96
250 × 200 mm	35.73	43.68	1.61	39.56	m	83.24
250 × 225 mm	43.79	53.53	1.70	41.78	m	95.31
315 × 200 mm	67.35	82.34	1.66	40.80	m	123.14
315 × 250 mm	60.91	74.46	2.03	49.88	m	124.34

U: VENTILATION/AIR CONDITIONING SYSTEMS

Item	Net Price £	Material £	Labour hours	Labour £	Unit	Total rate £
U10: DUCTWORK: PLASTIC – cont						
Extra over fittings – cont						
Reducer – cont						
355 × 200 mm	72.72	88.90	2.08	51.11	m	**140.01**
355 × 250 mm	77.40	94.63	2.05	50.37	m	**145.00**
355 × 315 mm	88.52	108.21	2.22	54.55	m	**162.76**
400 × 315 mm	105.22	128.64	2.88	70.77	m	**199.41**
400 × 355 mm	112.60	137.65	2.92	71.75	m	**209.40**
Flange						
110 mm	14.13	17.27	0.76	18.68	m	**35.95**
160 mm	15.88	19.42	1.02	25.06	m	**44.48**
200 mm	17.56	21.47	1.27	31.21	m	**52.68**
225 mm	17.93	21.92	1.55	38.09	m	**60.01**
250 mm	19.37	23.68	1.79	43.98	m	**67.66**
315 mm	21.89	26.76	2.08	51.11	m	**77.87**
355 mm	23.95	29.28	2.36	58.00	m	**87.28**
400 mm	25.44	31.10	3.07	75.44	m	**106.54**

U: VENTILATION/AIR CONDITIONING SYSTEMS

Item	Net Price £	Material £	Labour hours	Labour £	Unit	Total rate £
U10: DUCTWORK: RECTANGULAR – CLASS B						
Y30 – AIR DUCTLINES						
Galvanized sheet metal DW144 class B rectangular section ductwork; including all necessary stiffeners, joints, couplers in the running length and duct supports						
Ductwork up to 400 mm longest side						
Sum of two sides 200 mm	10.78	13.17	1.16	28.51	m	**41.68**
Sum of two sides 300 mm	11.53	14.09	1.16	28.51	m	**42.60**
Sum of two sides 400 mm	12.04	14.72	1.19	29.24	m	**43.96**
Sum of two sides 500 mm	12.75	15.58	1.19	29.24	m	**44.82**
Sum of two sides 600 mm	12.16	14.86	1.27	31.21	m	**46.07**
Sum of two sides 700 mm	12.79	15.64	1.27	31.21	m	**46.85**
Sum of two sides 800 mm	13.47	16.47	1.27	31.21	m	**47.68**
Extra over fittings; Rectangular ductwork class B; up to 400 mm longest side						
End Cap						
Sum of two sides 200 mm	7.50	9.17	0.38	9.34	nr	**18.51**
Sum of two sides 300 mm	8.35	10.21	0.38	9.34	nr	**19.55**
Sum of two sides 400 mm	9.19	11.24	0.38	9.34	nr	**20.58**
Sum of two sides 500 mm	10.04	12.28	0.38	9.34	nr	**21.62**
Sum of two sides 600 mm	10.87	13.29	0.38	9.34	nr	**22.63**
Sum of two sides 700 mm	11.71	14.32	0.38	9.34	nr	**23.66**
Sum of two sides 800 mm	12.56	15.36	0.38	9.34	nr	**24.70**
Reducer						
Sum of two sides 200 mm	12.32	15.06	1.40	34.40	nr	**49.46**
Sum of two sides 300 mm	13.91	17.01	1.40	34.40	nr	**51.41**
Sum of two sides 400 mm	22.43	27.42	1.42	34.89	nr	**62.31**
Sum of two sides 500 mm	24.26	29.65	1.42	34.89	nr	**64.54**
Sum of two sides 600 mm	26.10	31.91	1.69	41.53	nr	**73.44**
Sum of two sides 700 mm	27.93	34.14	1.69	41.53	nr	**75.67**
Sum of two sides 800 mm	29.74	36.36	1.92	47.17	nr	**83.53**
Offset						
Sum of two sides 200 mm	18.19	22.24	1.63	40.05	nr	**62.29**
Sum of two sides 300 mm	20.56	25.13	1.63	40.05	nr	**65.18**
Sum of two sides 400 mm	30.10	36.80	1.65	40.54	nr	**77.34**
Sum of two sides 500 mm	32.70	39.97	1.65	40.54	nr	**80.51**
Sum of two sides 600 mm	34.86	42.62	1.92	47.17	nr	**89.79**
Sum of two sides 700 mm	37.24	45.53	1.92	47.17	nr	**92.70**
Sum of two sides 800 mm	39.33	48.08	1.92	47.17	nr	**95.25**

U: VENTILATION/AIR CONDITIONING SYSTEMS

Item	Net Price £	Material £	Labour hours	Labour £	Unit	Total rate £
U10: DUCTWORK: RECTANGULAR – CLASS B – cont						
Extra over fittings; Rectangular ductwork class B; up to 400 mm longest side – cont						
Square to round						
Sum of two sides 200 mm	16.01	19.57	1.63	40.05	nr	59.62
Sum of two sides 300 mm	17.98	21.98	1.63	40.05	nr	62.03
Sum of two sides 400 mm	23.79	29.09	1.65	40.54	nr	69.63
Sum of two sides 500 mm	25.89	31.65	1.65	40.54	nr	72.19
Sum of two sides 600 mm	27.98	34.21	1.92	47.17	nr	81.38
Sum of two sides 700 mm	30.07	36.76	1.92	47.17	nr	83.93
Sum of two sides 800 mm	32.15	39.30	1.92	47.17	nr	86.47
90° radius bend						
Sum of two sides 200 mm	12.04	14.72	1.22	29.97	nr	44.69
Sum of two sides 300 mm	12.98	15.87	1.22	29.97	nr	45.84
Sum of two sides 400 mm	21.55	26.35	1.25	30.72	nr	57.07
Sum of two sides 500 mm	22.92	28.02	1.25	30.72	nr	58.74
Sum of two sides 600 mm	24.74	30.24	1.33	32.68	nr	62.92
Sum of two sides 700 mm	26.33	32.19	1.33	32.68	nr	64.87
Sum of two sides 800 mm	28.11	34.36	1.40	34.40	nr	68.76
45° radius bend						
Sum of two sides 200 mm	13.00	15.89	0.89	21.87	nr	37.76
Sum of two sides 300 mm	14.27	17.45	1.12	27.52	nr	44.97
Sum of two sides 400 mm	22.56	27.58	1.10	27.03	nr	54.61
Sum of two sides 500 mm	24.17	29.55	1.10	27.03	nr	56.58
Sum of two sides 600 mm	26.00	31.79	1.16	28.51	nr	60.30
Sum of two sides 700 mm	27.71	33.88	1.16	28.51	nr	62.39
Sum of two sides 800 mm	29.54	36.11	1.22	29.97	nr	66.08
90° mitre bend						
Sum of two sides 200 mm	20.06	24.52	1.29	31.70	nr	56.22
Sum of two sides 300 mm	21.97	26.86	1.29	31.70	nr	58.56
Sum of two sides 400 mm	31.61	38.65	1.29	31.70	nr	70.35
Sum of two sides 500 mm	34.07	41.65	1.29	31.70	nr	73.35
Sum of two sides 600 mm	37.22	45.51	1.39	34.15	nr	79.66
Sum of two sides 700 mm	40.14	49.07	1.39	34.15	nr	83.22
Sum of two sides 800 mm	43.33	52.97	1.46	35.89	nr	88.86
Branch						
Sum of two sides 200 mm	19.87	24.29	0.92	22.61	nr	46.90
Sum of two sides 300 mm	22.06	26.97	0.92	22.61	nr	49.58
Sum of two sides 400 mm	27.42	33.52	0.95	23.34	nr	56.86
Sum of two sides 500 mm	29.81	36.44	0.95	23.34	nr	59.78
Sum of two sides 600 mm	32.15	39.30	1.03	25.31	nr	64.61
Sum of two sides 700 mm	34.51	42.19	1.03	25.31	nr	67.50
Sum of two sides 800 mm	36.84	45.04	1.03	25.31	nr	70.35

U: VENTILATION/AIR CONDITIONING SYSTEMS

Item	Net Price £	Material £	Labour hours	Labour £	Unit	Total rate £
Grille neck						
Sum of two sides 200 mm	22.67	27.72	1.10	27.03	nr	**54.75**
Sum of two sides 300 mm	25.29	30.92	1.10	27.03	nr	**57.95**
Sum of two sides 400 mm	27.93	34.14	1.16	28.51	nr	**62.65**
Sum of two sides 500 mm	30.56	37.36	1.16	28.51	nr	**65.87**
Sum of two sides 600 mm	33.19	40.57	1.18	28.99	nr	**69.56**
Sum of two sides 700 mm	35.82	43.79	1.18	28.99	nr	**72.78**
Sum of two sides 800 mm	38.46	47.02	1.18	28.99	nr	**76.01**
Ductwork 401 to 600 mm longest side						
Sum of two sides 600 mm	13.70	16.75	1.27	31.21	m	**47.96**
Sum of two sides 700 mm	14.61	17.86	1.27	31.21	m	**49.07**
Sum of two sides 800 mm	15.50	18.95	1.27	31.21	m	**50.16**
Sum of two sides 900 mm	16.34	19.97	1.27	31.21	m	**51.18**
Sum of two sides 1000 mm	17.16	20.98	1.37	33.67	m	**54.65**
Sum of two sides 1100 mm	18.14	22.18	1.37	33.67	m	**55.85**
Sum of two sides 1200 mm	18.96	23.18	1.37	33.67	m	**56.85**
Extra over fittings; Ductwork 401 to 600 mm longest side						
End Cap						
Sum of two sides 600 mm	10.69	13.07	0.38	9.34	nr	**22.41**
Sum of two sides 700 mm	11.53	14.09	0.38	9.34	nr	**23.43**
Sum of two sides 800 mm	12.37	15.12	0.38	9.34	nr	**24.46**
Sum of two sides 900 mm	13.21	16.15	0.58	14.26	nr	**30.41**
Sum of two sides 1000 mm	14.04	17.16	0.58	14.26	nr	**31.42**
Sum of two sides 1100 mm	14.87	18.18	0.58	14.26	nr	**32.44**
Sum of two sides 1200 mm	15.72	19.22	0.58	14.26	nr	**33.48**
Reducer						
Sum of two sides 600 mm	24.65	30.14	1.69	41.53	nr	**71.67**
Sum of two sides 700 mm	26.43	32.31	1.69	41.53	nr	**73.84**
Sum of two sides 800 mm	28.17	34.43	1.92	47.17	nr	**81.60**
Sum of two sides 900 mm	29.95	36.62	1.92	47.17	nr	**83.79**
Sum of two sides 1000 mm	31.71	38.77	2.18	53.57	nr	**92.34**
Sum of two sides 1100 mm	33.61	41.09	2.18	53.57	nr	**94.66**
Sum of two sides 1200 mm	35.38	43.25	2.18	53.57	nr	**96.82**
Offset						
Sum of two sides 600 mm	34.48	42.15	1.92	47.17	nr	**89.32**
Sum of two sides 700 mm	37.01	45.25	1.92	47.17	nr	**92.42**
Sum of two sides 800 mm	39.04	47.73	1.92	47.17	nr	**94.90**
Sum of two sides 900 mm	41.10	50.24	1.92	47.17	nr	**97.41**
Sum of two sides 1000 mm	43.38	53.03	2.18	53.57	nr	**106.60**
Sum of two sides 1100 mm	45.49	55.61	2.18	53.57	nr	**109.18**
Sum of two sides 1200 mm	47.50	58.07	2.18	53.57	nr	**111.64**

U: VENTILATION/AIR CONDITIONING SYSTEMS

Item	Net Price £	Material £	Labour hours	Labour £	Unit	Total rate £
U10: DUCTWORK: RECTANGULAR – CLASS B – cont						
Extra over fittings; Ductwork 401 to 600 mm longest side – cont						
Square to round						
Sum of two sides 600 mm	26.40	32.27	1.33	32.68	nr	**64.95**
Sum of two sides 700 mm	28.43	34.75	1.33	32.68	nr	**67.43**
Sum of two sides 800 mm	30.43	37.20	1.40	34.40	nr	**71.60**
Sum of two sides 900 mm	32.46	39.69	1.40	34.40	nr	**74.09**
Sum of two sides 1000 mm	34.49	42.17	1.82	44.72	nr	**86.89**
Sum of two sides 1100 mm	36.54	44.67	1.82	44.72	nr	**89.39**
Sum of two sides 1200 mm	38.57	47.15	1.82	44.72	nr	**91.87**
90° radius bend						
Sum of two sides 600 mm	23.96	29.29	1.16	28.51	nr	**57.80**
Sum of two sides 700 mm	25.28	30.90	1.16	28.51	nr	**59.41**
Sum of two sides 800 mm	27.19	33.24	1.22	29.97	nr	**63.21**
Sum of two sides 900 mm	29.13	35.61	1.22	29.97	nr	**65.58**
Sum of two sides 1000 mm	30.76	37.61	1.40	34.40	nr	**72.01**
Sum of two sides 1100 mm	32.80	40.10	1.40	34.40	nr	**74.50**
Sum of two sides 1200 mm	34.75	42.48	1.40	34.40	nr	**76.88**
45° bend						
Sum of two sides 600 mm	25.36	31.00	1.16	28.51	nr	**59.51**
Sum of two sides 700 mm	26.92	32.91	1.39	34.15	nr	**67.06**
Sum of two sides 800 mm	28.78	35.18	1.46	35.89	nr	**71.07**
Sum of two sides 900 mm	30.67	37.49	1.46	35.89	nr	**73.38**
Sum of two sides 1000 mm	32.40	39.61	1.88	46.20	nr	**85.81**
Sum of two sides 1100 mm	34.41	42.07	1.88	46.20	nr	**88.27**
Sum of two sides 1200 mm	36.29	44.36	1.88	46.20	nr	**90.56**
90° mitre bend						
Sum of two sides 600 mm	40.36	49.34	1.39	34.15	nr	**83.49**
Sum of two sides 700 mm	42.72	52.22	2.16	53.09	nr	**105.31**
Sum of two sides 800 mm	46.05	56.30	2.26	55.54	nr	**111.84**
Sum of two sides 900 mm	49.43	60.43	2.26	55.54	nr	**115.97**
Sum of two sides 1000 mm	52.47	64.15	3.01	73.96	nr	**138.11**
Sum of two sides 1100 mm	55.99	68.44	3.01	73.96	nr	**142.40**
Sum of two sides 1200 mm	59.44	72.67	3.01	73.96	nr	**146.63**
Branch						
Sum of two sides 600 mm	31.74	38.80	1.03	25.31	nr	**64.11**
Sum of two sides 700 mm	34.07	41.65	1.03	25.31	nr	**66.96**
Sum of two sides 800 mm	36.42	44.53	1.03	25.31	nr	**69.84**
Sum of two sides 900 mm	38.75	47.37	1.03	25.31	nr	**72.68**
Sum of two sides 1000 mm	41.09	50.23	1.29	31.70	nr	**81.93**
Sum of two sides 1100 mm	43.51	53.19	1.29	31.70	nr	**84.89**
Sum of two sides 1200 mm	45.85	56.05	1.29	31.70	nr	**87.75**

U: VENTILATION/AIR CONDITIONING SYSTEMS

Item	Net Price £	Material £	Labour hours	Labour £	Unit	Total rate £
Grille neck						
Sum of two sides 600 mm	32.82	40.12	1.18	28.99	nr	**69.11**
Sum of two sides 700 mm	35.47	43.36	1.18	28.99	nr	**72.35**
Sum of two sides 800 mm	38.11	46.59	1.18	28.99	nr	**75.58**
Sum of two sides 900 mm	40.76	49.83	1.18	28.99	nr	**78.82**
Sum of two sides 1000 mm	43.41	53.07	1.44	35.38	nr	**88.45**
Sum of two sides 1100 mm	46.06	56.31	1.44	35.38	nr	**91.69**
Sum of two sides 1200 mm	48.70	59.53	1.44	35.38	nr	**94.91**
Ductwork 601 to 800 mm longest side						
Sum of two sides 900 mm	18.62	22.76	1.27	31.21	m	**53.97**
Sum of two sides 1000 mm	19.44	23.76	1.37	33.67	m	**57.43**
Sum of two sides 1100 mm	20.27	24.78	1.37	33.67	m	**58.45**
Sum of two sides 1200 mm	21.24	25.97	1.37	33.67	m	**59.64**
Sum of two sides 1300 mm	22.07	26.98	1.40	34.40	m	**61.38**
Sum of two sides 1400 mm	22.89	27.99	1.40	34.40	m	**62.39**
Sum of two sides 1500 mm	23.72	28.99	1.48	36.37	m	**65.36**
Sum of two sides 1600 mm	24.56	30.02	1.55	38.09	m	**68.11**
Extra over fittings: Ductwork 601 to 800 mm longest side						
End Cap						
Sum of two sides 900 mm	13.28	16.23	0.58	14.26	nr	**30.49**
Sum of two sides 1000 mm	14.11	17.25	0.58	14.26	nr	**31.51**
Sum of two sides 1100 mm	14.95	18.27	0.58	14.26	nr	**32.53**
Sum of two sides 1200 mm	15.80	19.31	0.58	14.26	nr	**33.57**
Sum of two sides 1300 mm	16.64	20.34	0.58	14.26	nr	**34.60**
Sum of two sides 1400 mm	17.36	21.22	0.58	14.26	nr	**35.48**
Sum of two sides 1500 mm	20.21	24.71	0.58	14.26	nr	**38.97**
Sum of two sides 1600 mm	23.05	28.18	0.58	14.26	nr	**42.44**
Reducer						
Sum of two sides 900 mm	30.39	37.15	1.92	47.17	nr	**84.32**
Sum of two sides 1000 mm	32.18	39.34	2.18	53.57	nr	**92.91**
Sum of two sides 1100 mm	33.96	41.52	2.18	53.57	nr	**95.09**
Sum of two sides 1200 mm	35.86	43.84	2.18	53.57	nr	**97.41**
Sum of two sides 1300 mm	37.64	46.01	2.30	56.52	nr	**102.53**
Sum of two sides 1400 mm	39.17	47.88	2.30	56.52	nr	**104.40**
Sum of two sides 1500 mm	44.97	54.97	2.47	60.70	nr	**115.67**
Sum of two sides 1600 mm	50.75	62.05	2.47	60.70	nr	**122.75**
Offset						
Sum of two sides 900 mm	42.44	51.88	1.92	47.17	nr	**99.05**
Sum of two sides 1000 mm	44.36	54.23	2.18	53.57	nr	**107.80**
Sum of two sides 1100 mm	46.26	56.56	2.18	53.57	nr	**110.13**
Sum of two sides 1200 mm	48.24	58.98	2.18	53.57	nr	**112.55**
Sum of two sides 1300 mm	50.12	61.27	2.30	56.52	nr	**117.79**
Sum of two sides 1400 mm	51.95	63.51	2.30	56.52	nr	**120.03**
Sum of two sides 1500 mm	59.83	73.14	2.47	60.70	nr	**133.84**
Sum of two sides 1600 mm	67.69	82.75	2.47	60.70	nr	**143.45**

U: VENTILATION/AIR CONDITIONING SYSTEMS

Item	Net Price £	Material £	Labour hours	Labour £	Unit	Total rate £
U10: DUCTWORK: RECTANGULAR – CLASS B – cont						
Extra over fittings: Ductwork 601 to 800 mm longest side – cont						
Square to round						
Sum of two sides 900 mm	32.36	39.56	1.40	34.40	nr	**73.96**
Sum of two sides 1000 mm	34.39	42.04	1.82	44.72	nr	**86.76**
Sum of two sides 1100 mm	36.44	44.55	1.82	44.72	nr	**89.27**
Sum of two sides 1200 mm	38.51	47.08	1.82	44.72	nr	**91.80**
Sum of two sides 1300 mm	40.54	49.56	2.15	52.84	nr	**102.40**
Sum of two sides 1400 mm	42.26	51.66	2.15	52.84	nr	**104.50**
Sum of two sides 1500 mm	49.32	60.30	2.38	58.48	nr	**118.78**
Sum of two sides 1600 mm	56.37	68.92	2.38	58.48	nr	**127.40**
90° radius bend						
Sum of two sides 900 mm	27.56	33.69	1.22	29.97	nr	**63.66**
Sum of two sides 1000 mm	29.54	36.11	1.40	34.40	nr	**70.51**
Sum of two sides 1100 mm	31.52	38.53	1.40	34.40	nr	**72.93**
Sum of two sides 1200 mm	33.56	41.02	1.40	34.40	nr	**75.42**
Sum of two sides 1300 mm	35.54	43.45	1.91	46.94	nr	**90.39**
Sum of two sides 1400 mm	36.88	45.08	1.91	46.94	nr	**92.02**
Sum of two sides 1500 mm	42.88	52.42	2.11	51.85	nr	**104.27**
Sum of two sides 1600 mm	48.87	59.74	2.11	51.85	nr	**111.59**
45° bend						
Sum of two sides 900 mm	30.40	37.16	1.22	29.97	nr	**67.13**
Sum of two sides 1000 mm	32.31	39.50	1.40	34.40	nr	**73.90**
Sum of two sides 1100 mm	34.21	41.82	1.88	46.20	nr	**88.02**
Sum of two sides 1200 mm	36.24	44.30	1.88	46.20	nr	**90.50**
Sum of two sides 1300 mm	38.66	47.27	2.26	55.54	nr	**102.81**
Sum of two sides 1400 mm	39.61	48.42	2.26	55.54	nr	**103.96**
Sum of two sides 1500 mm	45.53	55.66	2.49	61.18	nr	**116.84**
Sum of two sides 1600 mm	51.44	62.88	2.49	61.18	nr	**124.06**
90° mitre bend						
Sum of two sides 900 mm	46.96	57.41	1.22	29.97	nr	**87.38**
Sum of two sides 1000 mm	50.64	61.90	1.40	34.40	nr	**96.30**
Sum of two sides 1100 mm	54.33	66.41	3.01	73.96	nr	**140.37**
Sum of two sides 1200 mm	58.07	70.99	3.01	73.96	nr	**144.95**
Sum of two sides 1300 mm	61.76	75.50	3.67	90.19	nr	**165.69**
Sum of two sides 1400 mm	64.62	79.00	3.67	90.19	nr	**169.19**
Sum of two sides 1500 mm	74.39	90.94	4.07	100.01	nr	**190.95**
Sum of two sides 1600 mm	84.16	102.89	4.07	100.01	nr	**202.90**

U: VENTILATION/AIR CONDITIONING SYSTEMS

Item	Net Price £	Material £	Labour hours	Labour £	Unit	Total rate £
Branch						
Sum of two sides 900 mm	39.96	48.85	1.22	29.97	nr	78.82
Sum of two sides 1000 mm	42.34	51.76	1.40	34.40	nr	86.16
Sum of two sides 1100 mm	44.74	54.69	1.29	31.70	nr	86.39
Sum of two sides 1200 mm	47.23	57.74	1.29	31.70	nr	89.44
Sum of two sides 1300 mm	49.63	60.68	1.39	34.15	nr	94.83
Sum of two sides 1400 mm	51.70	63.20	1.39	34.15	nr	97.35
Sum of two sides 1500 mm	59.44	72.67	1.64	40.29	nr	112.96
Sum of two sides 1600 mm	67.19	82.14	1.64	40.29	nr	122.43
Grille neck						
Sum of two sides 900 mm	40.98	50.10	1.22	29.97	nr	80.07
Sum of two sides 1000 mm	43.64	53.35	1.40	34.40	nr	87.75
Sum of two sides 1100 mm	46.30	56.60	1.44	35.38	nr	91.98
Sum of two sides 1200 mm	48.96	59.85	1.44	35.38	nr	95.23
Sum of two sides 1300 mm	51.62	63.11	1.69	41.53	nr	104.64
Sum of two sides 1400 mm	53.91	65.91	1.69	41.53	nr	107.44
Sum of two sides 1500 mm	62.58	76.51	1.79	43.98	nr	120.49
Sum of two sides 1600 mm	71.26	87.12	1.79	43.98	nr	131.10
Ductwork 801 to 1000 mm longest side						
Sum of two sides 1100 mm	29.55	36.12	1.37	33.67	m	69.79
Sum of two sides 1200 mm	31.19	38.13	1.37	33.67	m	71.80
Sum of two sides 1300 mm	32.82	40.12	1.40	34.40	m	74.52
Sum of two sides 1400 mm	34.59	42.28	1.40	34.40	m	76.68
Sum of two sides 1500 mm	36.21	44.27	1.48	36.37	m	80.64
Sum of two sides 1600 mm	37.85	46.27	1.55	38.09	m	84.36
Sum of two sides 1700 mm	39.48	48.27	1.55	38.09	m	86.36
Sum of two sides 1800 mm	41.25	50.43	1.61	39.56	m	89.99
Sum of two sides 1900 mm	42.88	52.42	1.61	39.56	m	91.98
Sum of two sides 2000 mm	44.51	54.41	1.61	39.56	m	93.97
Extra over fittings; Ductwork 801 to 1000 mm longest side						
End Cap						
Sum of two sides 1100 mm	14.95	18.27	1.44	35.38	nr	53.65
Sum of two sides 1200 mm	15.80	19.31	1.44	35.38	nr	54.69
Sum of two sides 1300 mm	16.64	20.34	1.44	35.38	nr	55.72
Sum of two sides 1400 mm	17.36	21.22	1.44	35.38	nr	56.60
Sum of two sides 1500 mm	20.21	24.71	1.44	35.38	nr	60.09
Sum of two sides 1600 mm	23.05	28.18	1.44	35.38	nr	63.56
Sum of two sides 1700 mm	25.89	31.65	1.44	35.38	nr	67.03
Sum of two sides 1800 mm	28.74	35.13	1.44	35.38	nr	70.51
Sum of two sides 1900 mm	31.59	38.61	1.44	35.38	nr	73.99
Sum of two sides 2000 mm	34.44	42.11	1.44	35.38	nr	77.49

U: VENTILATION/AIR CONDITIONING SYSTEMS

Item	Net Price £	Material £	Labour hours	Labour £	Unit	Total rate £
U10: DUCTWORK: RECTANGULAR – CLASS B – cont						
Extra over fittings; Ductwork 801 to 1000 mm longest side – cont						
Reducer						
Sum of two sides 1100 mm	28.20	34.47	1.44	35.38	nr	**69.85**
Sum of two sides 1200 mm	29.55	36.12	1.44	35.38	nr	**71.50**
Sum of two sides 1300 mm	30.89	37.76	1.69	41.53	nr	**79.29**
Sum of two sides 1400 mm	32.10	39.24	1.69	41.53	nr	**80.77**
Sum of two sides 1500 mm	37.46	45.79	2.47	60.70	nr	**106.49**
Sum of two sides 1600 mm	42.81	52.33	2.47	60.70	nr	**113.03**
Sum of two sides 1700 mm	48.17	58.89	2.47	60.70	nr	**119.59**
Sum of two sides 1800 mm	53.63	65.56	2.59	63.64	nr	**129.20**
Sum of two sides 1900 mm	58.99	72.12	2.71	66.59	nr	**138.71**
Sum of two sides 2000 mm	64.35	78.67	2.71	66.59	nr	**145.26**
Offset						
Sum of two sides 1100 mm	43.58	53.28	1.44	35.38	nr	**88.66**
Sum of two sides 1200 mm	44.45	54.34	1.44	35.38	nr	**89.72**
Sum of two sides 1300 mm	45.19	55.25	1.69	41.53	nr	**96.78**
Sum of two sides 1400 mm	45.53	55.66	1.69	41.53	nr	**97.19**
Sum of two sides 1500 mm	52.03	63.60	2.47	60.70	nr	**124.30**
Sum of two sides 1600 mm	58.42	71.42	2.47	60.70	nr	**132.12**
Sum of two sides 1700 mm	64.69	79.08	2.59	63.64	nr	**142.72**
Sum of two sides 1800 mm	70.87	86.64	2.61	64.13	nr	**150.77**
Sum of two sides 1900 mm	77.76	95.06	2.71	66.59	nr	**161.65**
Sum of two sides 2000 mm	83.68	102.30	2.71	66.59	nr	**168.89**
Square to round						
Sum of two sides 1100 mm	30.58	37.39	1.44	35.38	nr	**72.77**
Sum of two sides 1200 mm	32.19	39.36	1.44	35.38	nr	**74.74**
Sum of two sides 1300 mm	33.79	41.31	1.69	41.53	nr	**82.84**
Sum of two sides 1400 mm	35.11	42.92	1.69	41.53	nr	**84.45**
Sum of two sides 1500 mm	41.71	51.00	2.38	58.48	nr	**109.48**
Sum of two sides 1600 mm	48.34	59.09	2.38	58.48	nr	**117.57**
Sum of two sides 1700 mm	54.95	67.18	2.55	62.67	nr	**129.85**
Sum of two sides 1800 mm	61.57	75.27	2.55	62.67	nr	**137.94**
Sum of two sides 1900 mm	68.19	83.36	2.83	69.54	nr	**152.90**
Sum of two sides 2000 mm	74.81	91.45	2.83	69.54	nr	**160.99**
90° radius bend						
Sum of two sides 1100 mm	22.58	27.60	1.44	35.38	nr	**62.98**
Sum of two sides 1200 mm	24.18	29.56	1.44	35.38	nr	**64.94**
Sum of two sides 1300 mm	25.77	31.51	1.69	41.53	nr	**73.04**
Sum of two sides 1400 mm	27.17	33.22	1.69	41.53	nr	**74.75**
Sum of two sides 1500 mm	32.77	40.06	2.11	51.85	nr	**91.91**
Sum of two sides 1600 mm	38.37	46.91	2.11	51.85	nr	**98.76**
Sum of two sides 1700 mm	43.98	53.77	2.26	55.54	nr	**109.31**

U: VENTILATION/AIR CONDITIONING SYSTEMS

Item	Net Price £	Material £	Labour hours	Labour £	Unit	Total rate £
Sum of two sides 1800 mm	49.63	60.68	2.26	55.54	nr	**116.22**
Sum of two sides 1900 mm	54.26	66.33	2.48	60.95	nr	**127.28**
Sum of two sides 2000 mm	59.86	73.18	2.48	60.95	nr	**134.13**
45° bend						
Sum of two sides 1100 mm	29.43	35.98	1.44	35.38	nr	**71.36**
Sum of two sides 1200 mm	31.14	38.07	1.44	35.38	nr	**73.45**
Sum of two sides 1300 mm	32.83	40.14	1.69	41.53	nr	**81.67**
Sum of two sides 1400 mm	34.39	42.04	1.69	41.53	nr	**83.57**
Sum of two sides 1500 mm	40.10	49.02	2.49	61.18	nr	**110.20**
Sum of two sides 1600 mm	45.80	55.99	2.49	61.18	nr	**117.17**
Sum of two sides 1700 mm	51.51	62.97	2.67	65.61	nr	**128.58**
Sum of two sides 1800 mm	57.33	70.08	2.67	65.61	nr	**135.69**
Sum of two sides 1900 mm	62.52	76.43	3.06	75.20	nr	**151.63**
Sum of two sides 2000 mm	68.23	83.41	3.06	75.20	nr	**158.61**
90° mitre bend						
Sum of two sides 1100 mm	47.11	57.59	1.44	35.38	nr	**92.97**
Sum of two sides 1200 mm	50.31	61.50	1.44	35.38	nr	**96.88**
Sum of two sides 1300 mm	53.49	65.39	1.69	41.53	nr	**106.92**
Sum of two sides 1400 mm	56.34	68.88	1.69	41.53	nr	**110.41**
Sum of two sides 1500 mm	65.54	80.12	2.07	50.87	nr	**130.99**
Sum of two sides 1600 mm	74.75	91.38	2.07	50.87	nr	**142.25**
Sum of two sides 1700 mm	83.94	102.62	2.80	68.80	nr	**171.42**
Sum of two sides 1800 mm	93.17	113.90	2.67	65.61	nr	**179.51**
Sum of two sides 1900 mm	101.29	123.83	2.95	72.49	nr	**196.32**
Sum of two sides 2000 mm	110.53	135.13	2.95	72.49	nr	**207.62**
Branch						
Sum of two sides 1100 mm	44.84	54.82	1.44	35.38	nr	**90.20**
Sum of two sides 1200 mm	47.23	57.74	1.44	35.38	nr	**93.12**
Sum of two sides 1300 mm	49.63	60.68	1.64	40.29	nr	**100.97**
Sum of two sides 1400 mm	51.79	63.31	1.64	40.29	nr	**103.60**
Sum of two sides 1500 mm	59.53	72.78	1.64	40.29	nr	**113.07**
Sum of two sides 1600 mm	67.27	82.24	1.64	40.29	nr	**122.53**
Sum of two sides 1700 mm	75.00	91.69	1.69	41.53	nr	**133.22**
Sum of two sides 1800 mm	82.84	101.27	1.69	41.53	nr	**142.80**
Sum of two sides 1900 mm	90.58	110.74	1.85	45.46	nr	**156.20**
Sum of two sides 2000 mm	98.32	120.20	1.85	45.46	nr	**165.66**
Grille neck						
Sum of two sides 1100 mm	46.30	56.60	1.44	35.38	nr	**91.98**
Sum of two sides 1200 mm	48.96	59.85	1.44	35.38	nr	**95.23**
Sum of two sides 1300 mm	51.62	63.11	1.69	41.53	nr	**104.64**
Sum of two sides 1400 mm	53.91	65.91	1.69	41.53	nr	**107.44**
Sum of two sides 1500 mm	62.58	76.51	1.79	43.98	nr	**120.49**
Sum of two sides 1600 mm	71.26	87.12	1.79	43.98	nr	**131.10**
Sum of two sides 1700 mm	79.93	97.72	1.86	45.71	nr	**143.43**
Sum of two sides 1800 mm	88.60	108.31	2.02	49.64	nr	**157.95**
Sum of two sides 1900 mm	97.27	118.91	2.02	49.64	nr	**168.55**
Sum of two sides 2000 mm	105.95	129.52	2.02	49.64	nr	**179.16**

U: VENTILATION/AIR CONDITIONING SYSTEMS

Item	Net Price £	Material £	Labour hours	Labour £	Unit	Total rate £
U10: DUCTWORK: RECTANGULAR – CLASS B – cont						
Ductwork 1001 to 1250 mm longest side						
Sum of two sides 1300 mm	38.63	47.23	1.40	34.40	m	**81.63**
Sum of two sides 1400 mm	40.62	49.66	1.40	34.40	m	**84.06**
Sum of two sides 1500 mm	42.44	51.88	1.48	36.37	m	**88.25**
Sum of two sides 1600 mm	44.41	54.29	1.55	38.09	m	**92.38**
Sum of two sides 1700 mm	46.24	56.53	1.55	38.09	m	**94.62**
Sum of two sides 1800 mm	48.06	58.75	1.61	39.56	m	**98.31**
Sum of two sides 1900 mm	49.89	60.99	1.61	39.56	m	**100.55**
Sum of two sides 2000 mm	51.86	63.40	1.61	39.56	m	**102.96**
Sum of two sides 2100 mm	53.67	65.61	2.17	53.32	m	**118.93**
Sum of two sides 2200 mm	55.51	67.86	2.19	53.82	m	**121.68**
Sum of two sides 2300 mm	57.63	70.45	2.19	53.82	m	**124.27**
Sum of two sides 2400 mm	59.45	72.68	2.38	58.48	m	**131.16**
Sum of two sides 2500 mm	61.42	75.09	2.38	58.48	m	**133.57**
Extra over fittings; Ductwork 1001 to 1250 mm longest side						
End Cap						
Sum of two sides 1300 mm	16.97	20.74	1.69	41.53	nr	**62.27**
Sum of two sides 1400 mm	17.71	21.65	1.69	41.53	nr	**63.18**
Sum of two sides 1500 mm	20.59	25.17	1.69	41.53	nr	**66.70**
Sum of two sides 1600 mm	23.47	28.70	1.69	41.53	nr	**70.23**
Sum of two sides 1700 mm	26.35	32.21	1.69	41.53	nr	**73.74**
Sum of two sides 1800 mm	29.23	35.73	1.69	41.53	nr	**77.26**
Sum of two sides 1900 mm	32.11	39.25	1.69	41.53	nr	**80.78**
Sum of two sides 2000 mm	34.99	42.78	1.69	41.53	nr	**84.31**
Sum of two sides 2100 mm	37.87	46.30	1.69	41.53	nr	**87.83**
Sum of two sides 2200 mm	40.75	49.82	1.69	41.53	nr	**91.35**
Sum of two sides 2300 mm	43.64	53.35	1.69	41.53	nr	**94.88**
Sum of two sides 2400 mm	46.51	56.86	1.69	41.53	nr	**98.39**
Sum of two sides 2500 mm	49.39	60.38	1.69	41.53	nr	**101.91**
Reducer						
Sum of two sides 1300 mm	30.13	36.83	1.69	41.53	nr	**78.36**
Sum of two sides 1400 mm	31.23	38.18	1.69	41.53	nr	**79.71**
Sum of two sides 1500 mm	36.61	44.75	2.47	60.70	nr	**105.45**
Sum of two sides 1600 mm	42.09	51.46	2.47	60.70	nr	**112.16**
Sum of two sides 1700 mm	47.47	58.03	2.47	60.70	nr	**118.73**
Sum of two sides 1800 mm	52.85	64.61	2.59	63.64	nr	**128.25**
Sum of two sides 1900 mm	58.23	71.18	2.71	66.59	nr	**137.77**
Sum of two sides 2000 mm	63.70	77.88	2.59	63.64	nr	**141.52**
Sum of two sides 2100 mm	69.08	84.45	2.92	71.75	nr	**156.20**
Sum of two sides 2200 mm	74.47	91.04	2.92	71.75	nr	**162.79**
Sum of two sides 2300 mm	79.95	97.74	2.92	71.75	nr	**169.49**
Sum of two sides 2400 mm	85.32	104.31	3.12	76.66	nr	**180.97**
Sum of two sides 2500 mm	90.80	111.00	3.12	76.66	nr	**187.66**

U: VENTILATION/AIR CONDITIONING SYSTEMS

Item	Net Price £	Material £	Labour hours	Labour £	Unit	Total rate £
Offset						
Sum of two sides 1300 mm	59.55	72.80	1.69	41.53	nr	**114.33**
Sum of two sides 1400 mm	61.51	75.20	1.69	41.53	nr	**116.73**
Sum of two sides 1500 mm	68.97	84.32	2.47	60.70	nr	**145.02**
Sum of two sides 1600 mm	76.39	93.39	2.47	60.70	nr	**154.09**
Sum of two sides 1700 mm	83.58	102.18	2.59	63.64	nr	**165.82**
Sum of two sides 1800 mm	90.66	110.83	2.61	64.13	nr	**174.96**
Sum of two sides 1900 mm	97.61	119.33	2.71	66.59	nr	**185.92**
Sum of two sides 2000 mm	104.49	127.74	2.71	66.59	nr	**194.33**
Sum of two sides 2100 mm	111.19	135.93	2.92	71.75	nr	**207.68**
Sum of two sides 2200 mm	118.83	145.27	3.26	80.11	nr	**225.38**
Sum of two sides 2300 mm	127.56	155.94	3.26	80.11	nr	**236.05**
Sum of two sides 2400 mm	136.25	166.57	3.48	85.51	nr	**252.08**
Sum of two sides 2500 mm	144.98	177.24	3.48	85.51	nr	**262.75**
Square to round						
Sum of two sides 1300 mm	32.76	40.05	1.69	41.53	nr	**81.58**
Sum of two sides 1400 mm	34.06	41.64	1.69	41.53	nr	**83.17**
Sum of two sides 1500 mm	40.70	49.76	2.38	58.48	nr	**108.24**
Sum of two sides 1600 mm	47.36	57.90	2.38	58.48	nr	**116.38**
Sum of two sides 1700 mm	54.01	66.02	2.55	62.67	nr	**128.69**
Sum of two sides 1800 mm	60.65	74.15	2.55	62.67	nr	**136.82**
Sum of two sides 1900 mm	67.29	82.27	2.83	69.54	nr	**151.81**
Sum of two sides 2000 mm	73.94	90.39	2.83	69.54	nr	**159.93**
Sum of two sides 2100 mm	80.58	98.51	3.85	94.61	nr	**193.12**
Sum of two sides 2200 mm	87.23	106.64	4.18	102.71	nr	**209.35**
Sum of two sides 2300 mm	93.88	114.77	4.22	103.69	nr	**218.46**
Sum of two sides 2400 mm	100.53	122.90	4.68	115.00	nr	**237.90**
Sum of two sides 2500 mm	107.18	131.03	4.70	115.49	nr	**246.52**
90° radius bend						
Sum of two sides 1300 mm	21.25	25.98	1.69	41.53	nr	**67.51**
Sum of two sides 1400 mm	21.58	26.38	1.69	41.53	nr	**67.91**
Sum of two sides 1500 mm	27.33	33.41	2.11	51.85	nr	**85.26**
Sum of two sides 1600 mm	33.12	40.49	2.11	51.85	nr	**92.34**
Sum of two sides 1700 mm	38.88	47.53	2.19	53.82	nr	**101.35**
Sum of two sides 1800 mm	44.63	54.56	2.19	53.82	nr	**108.38**
Sum of two sides 1900 mm	50.38	61.59	2.48	60.95	nr	**122.54**
Sum of two sides 2000 mm	56.17	68.67	2.26	55.54	nr	**124.21**
Sum of two sides 2100 mm	61.92	75.69	2.48	60.95	nr	**136.64**
Sum of two sides 2200 mm	67.67	82.73	2.48	60.95	nr	**143.68**
Sum of two sides 2300 mm	70.96	86.75	2.48	60.95	nr	**147.70**
Sum of two sides 2400 mm	76.70	93.77	3.90	95.83	nr	**189.60**
Sum of two sides 2500 mm	82.47	100.82	3.90	95.83	nr	**196.65**

U: VENTILATION/AIR CONDITIONING SYSTEMS

Item	Net Price £	Material £	Labour hours	Labour £	Unit	Total rate £
U10: DUCTWORK: RECTANGULAR – CLASS B – cont						
Extra over fittings; Ductwork 1001 to 1250 mm longest side – cont						
45° bend						
Sum of two sides 1300 mm	30.80	37.66	1.69	41.53	nr	79.19
Sum of two sides 1400 mm	31.73	38.79	1.69	41.53	nr	80.32
Sum of two sides 1500 mm	37.54	45.90	2.49	61.18	nr	107.08
Sum of two sides 1600 mm	43.44	53.11	2.49	61.18	nr	114.29
Sum of two sides 1700 mm	49.25	60.20	2.67	65.61	nr	125.81
Sum of two sides 1800 mm	55.06	67.31	2.67	65.61	nr	132.92
Sum of two sides 1900 mm	60.86	74.40	3.06	75.20	nr	149.60
Sum of two sides 2000 mm	66.76	81.62	3.06	75.20	nr	156.82
Sum of two sides 2100 mm	72.57	88.71	4.05	99.53	nr	188.24
Sum of two sides 2200 mm	78.39	95.83	4.05	99.53	nr	195.36
Sum of two sides 2300 mm	82.99	101.45	4.39	107.87	nr	209.32
Sum of two sides 2400 mm	88.79	108.54	4.85	119.17	nr	227.71
Sum of two sides 2500 mm	94.69	115.76	4.85	119.17	nr	234.93
90° mitre bend						
Sum of two sides 1300 mm	64.38	78.70	1.69	41.53	nr	120.23
Sum of two sides 1400 mm	67.03	81.95	1.69	41.53	nr	123.48
Sum of two sides 1500 mm	77.75	95.05	2.80	68.80	nr	163.85
Sum of two sides 1600 mm	88.47	108.16	2.80	68.80	nr	176.96
Sum of two sides 1700 mm	99.19	121.26	2.95	72.49	nr	193.75
Sum of two sides 1800 mm	109.91	134.36	2.95	72.49	nr	206.85
Sum of two sides 1900 mm	120.64	147.49	4.05	99.53	nr	247.02
Sum of two sides 2000 mm	131.36	160.59	4.05	99.53	nr	260.12
Sum of two sides 2100 mm	142.07	173.68	4.07	100.01	nr	273.69
Sum of two sides 2200 mm	152.79	186.79	4.07	100.01	nr	286.80
Sum of two sides 2300 mm	160.68	196.43	4.39	107.87	nr	304.30
Sum of two sides 2400 mm	171.48	209.64	4.85	119.17	nr	328.81
Sum of two sides 2500 mm	182.26	222.81	4.85	119.17	nr	341.98
Branch						
Sum of two sides 1300 mm	50.92	62.25	1.44	35.38	nr	97.63
Sum of two sides 1400 mm	53.05	64.86	1.44	35.38	nr	100.24
Sum of two sides 1500 mm	60.89	74.44	1.64	40.29	nr	114.73
Sum of two sides 1600 mm	68.82	84.13	1.64	40.29	nr	124.42
Sum of two sides 1700 mm	76.66	93.72	1.64	40.29	nr	134.01
Sum of two sides 1800 mm	84.50	103.30	1.64	40.29	nr	143.59
Sum of two sides 1900 mm	92.33	112.88	1.69	41.53	nr	154.41
Sum of two sides 2000 mm	100.26	122.57	1.69	41.53	nr	164.10
Sum of two sides 2100 mm	108.10	132.15	1.85	45.46	nr	177.61
Sum of two sides 2200 mm	115.94	141.74	1.85	45.46	nr	187.20
Sum of two sides 2300 mm	123.87	151.43	2.61	64.13	nr	215.56
Sum of two sides 2400 mm	131.71	161.02	2.61	64.13	nr	225.15
Sum of two sides 2500 mm	139.64	170.71	2.61	64.13	nr	234.84

U: VENTILATION/AIR CONDITIONING SYSTEMS

Item	Net Price £	Material £	Labour hours	Labour £	Unit	Total rate £
Grille neck						
Sum of two sides 1300 mm	53.13	64.95	1.79	43.98	nr	**108.93**
Sum of two sides 1400 mm	55.53	67.89	1.79	43.98	nr	**111.87**
Sum of two sides 1500 mm	64.34	78.66	1.79	43.98	nr	**122.64**
Sum of two sides 1600 mm	73.15	89.42	1.79	43.98	nr	**133.40**
Sum of two sides 1700 mm	81.97	100.21	1.86	45.71	nr	**145.92**
Sum of two sides 1800 mm	90.78	110.98	2.02	49.64	nr	**160.62**
Sum of two sides 1900 mm	99.60	121.77	2.02	49.64	nr	**171.41**
Sum of two sides 2000 mm	108.41	132.53	2.02	49.64	nr	**182.17**
Sum of two sides 2100 mm	117.23	143.31	2.61	64.13	nr	**207.44**
Sum of two sides 2200 mm	126.03	154.08	2.61	64.13	nr	**218.21**
Sum of two sides 2300 mm	134.86	164.87	2.61	64.13	nr	**229.00**
Sum of two sides 2400 mm	143.66	175.63	2.88	70.77	nr	**246.40**
Sum of two sides 2500 mm	152.48	186.41	2.88	70.77	nr	**257.18**
Ductwork 1251 to 1600 mm longest side						
Sum of two sides 1700 mm	60.95	74.51	1.55	38.09	m	**112.60**
Sum of two sides 1800 mm	63.20	77.26	1.61	39.56	m	**116.82**
Sum of two sides 1900 mm	65.47	80.04	1.61	39.56	m	**119.60**
Sum of two sides 2000 mm	67.60	82.64	1.61	39.56	m	**122.20**
Sum of two sides 2100 mm	69.73	85.24	2.17	53.32	m	**138.56**
Sum of two sides 2200 mm	71.87	87.86	2.19	53.82	m	**141.68**
Sum of two sides 2300 mm	74.14	90.64	2.19	53.82	m	**144.46**
Sum of two sides 2400 mm	76.39	93.39	2.38	58.48	m	**151.87**
Sum of two sides 2500 mm	78.52	96.00	2.38	58.48	m	**154.48**
Sum of two sides 2600 mm	80.65	98.59	2.64	64.87	m	**163.46**
Sum of two sides 2700 mm	82.79	101.21	2.66	65.36	m	**166.57**
Sum of two sides 2800 mm	85.03	103.95	2.95	72.49	m	**176.44**
Sum of two sides 2900 mm	95.42	116.65	2.96	72.74	m	**189.39**
Sum of two sides 3000 mm	97.55	119.25	3.15	77.39	m	**196.64**
Sum of two sides 3100 mm	99.80	122.00	3.15	77.39	m	**199.39**
Sum of two sides 3200 mm	101.93	124.61	3.18	78.15	m	**202.76**
Extra over fittings; Ductwork 1251 to 1600 mm longest side						
End Cap						
Sum of two sides 1700 mm	26.07	31.87	0.58	14.26	nr	**46.13**
Sum of two sides 1800 mm	28.92	35.36	0.58	14.26	nr	**49.62**
Sum of two sides 1900 mm	31.78	38.85	0.58	14.26	nr	**53.11**
Sum of two sides 2000 mm	34.62	42.32	0.58	14.26	nr	**56.58**
Sum of two sides 2100 mm	37.47	45.80	0.87	21.38	nr	**67.18**
Sum of two sides 2200 mm	40.32	49.30	0.87	21.38	nr	**70.68**
Sum of two sides 2300 mm	43.18	52.79	0.87	21.38	nr	**74.17**
Sum of two sides 2400 mm	46.02	56.26	0.87	21.38	nr	**77.64**
Sum of two sides 2500 mm	48.87	59.74	0.87	21.38	nr	**81.12**
Sum of two sides 2600 mm	51.72	63.23	0.87	21.38	nr	**84.61**
Sum of two sides 2700 mm	54.58	66.72	0.87	21.38	nr	**88.10**
Sum of two sides 2800 mm	57.42	70.19	1.16	28.51	nr	**98.70**

U: VENTILATION/AIR CONDITIONING SYSTEMS

Item	Net Price £	Material £	Labour hours	Labour £	Unit	Total rate £
U10: DUCTWORK: RECTANGULAR – CLASS B – cont						
Extra over fittings; Ductwork 1251 to 1600 mm longest side – cont						
End Cap – cont						
Sum of two sides 2900 mm	60.27	73.68	1.16	28.51	nr	**102.19**
Sum of two sides 3000 mm	66.64	81.47	1.30	31.94	nr	**113.41**
Sum of two sides 3100 mm	65.38	79.93	1.80	44.23	nr	**124.16**
Sum of two sides 3200 mm	67.42	82.42	1.80	44.23	nr	**126.65**
Reducer						
Sum of two sides 1700 mm	32.08	39.22	2.47	60.70	nr	**99.92**
Sum of two sides 1800 mm	37.05	45.29	2.59	63.64	nr	**108.93**
Sum of two sides 1900 mm	42.09	51.46	2.71	66.59	nr	**118.05**
Sum of two sides 2000 mm	47.04	57.50	2.71	66.59	nr	**124.09**
Sum of two sides 2100 mm	52.00	63.57	2.92	71.75	nr	**135.32**
Sum of two sides 2200 mm	56.95	69.62	2.92	71.75	nr	**141.37**
Sum of two sides 2300 mm	61.98	75.77	2.92	71.75	nr	**147.52**
Sum of two sides 2400 mm	66.96	81.85	3.12	76.66	nr	**158.51**
Sum of two sides 2500 mm	71.91	87.91	3.12	76.66	nr	**164.57**
Sum of two sides 2600 mm	76.87	93.98	3.12	76.66	nr	**170.64**
Sum of two sides 2700 mm	81.81	100.01	3.12	76.66	nr	**176.67**
Sum of two sides 2800 mm	86.80	106.11	3.95	97.06	nr	**203.17**
Sum of two sides 2900 mm	87.03	106.40	3.97	97.55	nr	**203.95**
Sum of two sides 3000 mm	90.97	111.21	4.52	111.06	nr	**222.27**
Sum of two sides 3100 mm	94.65	115.71	4.52	111.06	nr	**226.77**
Sum of two sides 3200 mm	98.00	119.81	4.52	111.06	nr	**230.87**
Offset						
Sum of two sides 1700 mm	81.64	99.81	2.59	63.64	nr	**163.45**
Sum of two sides 1800 mm	90.77	110.96	2.61	64.13	nr	**175.09**
Sum of two sides 1900 mm	96.96	118.53	2.71	66.59	nr	**185.12**
Sum of two sides 2000 mm	102.91	125.80	2.71	66.59	nr	**192.39**
Sum of two sides 2100 mm	108.69	132.87	2.92	71.75	nr	**204.62**
Sum of two sides 2200 mm	114.30	139.73	3.26	80.11	nr	**219.84**
Sum of two sides 2300 mm	119.78	146.44	3.26	80.11	nr	**226.55**
Sum of two sides 2400 mm	128.51	157.11	3.47	85.27	nr	**242.38**
Sum of two sides 2500 mm	133.72	163.47	3.48	85.51	nr	**248.98**
Sum of two sides 2600 mm	142.33	174.00	3.49	85.76	nr	**259.76**
Sum of two sides 2700 mm	150.94	184.52	3.50	86.00	nr	**270.52**
Sum of two sides 2800 mm	159.59	195.10	4.34	106.64	nr	**301.74**
Sum of two sides 2900 mm	159.57	195.07	4.76	116.97	nr	**312.04**
Sum of two sides 3000 mm	168.17	205.59	5.32	130.73	nr	**336.32**
Sum of two sides 3100 mm	174.87	213.78	5.35	131.46	nr	**345.24**
Sum of two sides 3200 mm	181.07	221.36	5.35	131.46	nr	**352.82**
Square to round						
Sum of two sides 1700 mm	38.55	47.12	2.55	62.67	nr	**109.79**
Sum of two sides 1800 mm	44.77	54.73	2.55	62.67	nr	**117.40**
Sum of two sides 1900 mm	50.96	62.29	2.83	69.54	nr	**131.83**

U: VENTILATION/AIR CONDITIONING SYSTEMS

Item	Net Price £	Material £	Labour hours	Labour £	Unit	Total rate £
Sum of two sides 2000 mm	57.17	69.90	2.83	69.54	nr	**139.44**
Sum of two sides 2100 mm	63.37	77.47	3.85	94.61	nr	**172.08**
Sum of two sides 2200 mm	69.57	85.05	4.18	102.71	nr	**187.76**
Sum of two sides 2300 mm	75.76	92.62	4.22	103.69	nr	**196.31**
Sum of two sides 2400 mm	82.00	100.25	4.68	115.00	nr	**215.25**
Sum of two sides 2500 mm	88.21	107.84	4.70	115.49	nr	**223.33**
Sum of two sides 2600 mm	94.41	115.42	4.70	115.49	nr	**230.91**
Sum of two sides 2700 mm	100.62	123.01	4.71	115.73	nr	**238.74**
Sum of two sides 2800 mm	106.84	130.61	8.19	201.25	nr	**331.86**
Sum of two sides 2900 mm	107.23	131.09	8.62	211.81	nr	**342.90**
Sum of two sides 3000 mm	113.43	138.67	8.75	215.01	nr	**353.68**
Sum of two sides 3100 mm	118.07	144.34	8.75	215.01	nr	**359.35**
Sum of two sides 3200 mm	122.26	149.46	8.75	215.01	nr	**364.47**
90° radius bend						
Sum of two sides 1700 mm	79.27	96.91	2.19	53.82	nr	**150.73**
Sum of two sides 1800 mm	86.85	106.17	2.19	53.82	nr	**159.99**
Sum of two sides 1900 mm	98.13	119.96	2.26	55.54	nr	**175.50**
Sum of two sides 2000 mm	109.22	133.52	2.26	55.54	nr	**189.06**
Sum of two sides 2100 mm	120.31	147.08	2.48	60.95	nr	**208.03**
Sum of two sides 2200 mm	131.41	160.65	2.48	60.95	nr	**221.60**
Sum of two sides 2300 mm	142.69	174.44	2.48	60.95	nr	**235.39**
Sum of two sides 2400 mm	149.94	183.30	3.90	95.83	nr	**279.13**
Sum of two sides 2500 mm	161.00	196.82	3.90	95.83	nr	**292.65**
Sum of two sides 2600 mm	172.07	210.36	4.26	104.68	nr	**315.04**
Sum of two sides 2700 mm	183.13	223.88	4.55	111.80	nr	**335.68**
Sum of two sides 2800 mm	190.14	232.45	4.55	111.80	nr	**344.25**
Sum of two sides 2900 mm	190.34	232.69	6.87	168.82	nr	**401.51**
Sum of two sides 3000 mm	201.36	246.16	7.00	172.01	nr	**418.17**
Sum of two sides 3100 mm	209.86	256.55	7.00	172.01	nr	**428.56**
Sum of two sides 3200 mm	217.68	266.11	7.00	172.01	nr	**438.12**
45° bend						
Sum of two sides 1700 mm	38.01	46.46	2.67	65.61	nr	**112.07**
Sum of two sides 1800 mm	41.74	51.03	2.67	65.61	nr	**116.64**
Sum of two sides 1900 mm	47.37	57.91	3.06	75.20	nr	**133.11**
Sum of two sides 2000 mm	52.91	64.68	3.06	75.20	nr	**139.88**
Sum of two sides 2100 mm	58.45	71.45	4.05	99.53	nr	**170.98**
Sum of two sides 2200 mm	63.99	78.23	4.05	99.53	nr	**177.76**
Sum of two sides 2300 mm	69.61	85.10	4.39	107.87	nr	**192.97**
Sum of two sides 2400 mm	73.19	89.48	4.85	119.17	nr	**208.65**
Sum of two sides 2500 mm	78.71	96.22	4.85	119.17	nr	**215.39**
Sum of two sides 2600 mm	84.23	102.97	4.87	119.68	nr	**222.65**
Sum of two sides 2700 mm	89.76	109.74	4.87	119.68	nr	**229.42**
Sum of two sides 2800 mm	93.21	113.95	8.81	216.49	nr	**330.44**
Sum of two sides 2900 mm	101.90	124.58	8.81	216.49	nr	**341.07**
Sum of two sides 3000 mm	98.43	120.33	9.31	228.76	nr	**349.09**
Sum of two sides 3100 mm	102.67	125.52	9.31	228.76	nr	**354.28**
Sum of two sides 3200 mm	106.57	130.28	9.39	230.74	nr	**361.02**

U: VENTILATION/AIR CONDITIONING SYSTEMS

Item	Net Price £	Material £	Labour hours	Labour £	Unit	Total rate £
U10: DUCTWORK: RECTANGULAR – CLASS B – cont						
Extra over fittings; Ductwork 1251 to 1600 mm longest side – cont						
90° mitre bend						
Sum of two sides 1700 mm	86.14	105.31	2.67	65.61	nr	**170.92**
Sum of two sides 1800 mm	92.04	112.52	2.80	68.80	nr	**181.32**
Sum of two sides 1900 mm	102.83	125.71	2.95	72.49	nr	**198.20**
Sum of two sides 2000 mm	113.66	138.95	2.95	72.49	nr	**211.44**
Sum of two sides 2100 mm	124.47	152.16	4.05	99.53	nr	**251.69**
Sum of two sides 2200 mm	135.31	165.42	4.05	99.53	nr	**264.95**
Sum of two sides 2300 mm	146.10	178.61	4.39	107.87	nr	**286.48**
Sum of two sides 2400 mm	152.01	185.83	4.85	119.17	nr	**305.00**
Sum of two sides 2500 mm	162.87	199.11	4.85	119.17	nr	**318.28**
Sum of two sides 2600 mm	173.74	212.40	4.87	119.68	nr	**332.08**
Sum of two sides 2700 mm	184.61	225.68	4.87	119.68	nr	**345.36**
Sum of two sides 2800 mm	190.45	232.83	8.81	216.49	nr	**449.32**
Sum of two sides 2900 mm	191.33	233.90	14.81	363.92	nr	**597.82**
Sum of two sides 3000 mm	202.25	247.25	15.20	373.50	nr	**620.75**
Sum of two sides 3100 mm	211.19	258.18	15.60	383.32	nr	**641.50**
Sum of two sides 3200 mm	219.69	268.57	15.60	383.32	nr	**651.89**
Branch						
Sum of two sides 1700 mm	75.86	92.74	1.69	41.53	nr	**134.27**
Sum of two sides 1800 mm	83.61	102.22	1.69	41.53	nr	**143.75**
Sum of two sides 1900 mm	91.45	111.80	1.85	45.46	nr	**157.26**
Sum of two sides 2000 mm	99.21	121.28	1.85	45.46	nr	**166.74**
Sum of two sides 2100 mm	106.96	130.76	2.61	64.13	nr	**194.89**
Sum of two sides 2200 mm	114.72	140.24	2.61	64.13	nr	**204.37**
Sum of two sides 2300 mm	122.57	149.84	2.61	64.13	nr	**213.97**
Sum of two sides 2400 mm	130.88	160.00	2.88	70.77	nr	**230.77**
Sum of two sides 2500 mm	138.08	168.81	2.88	70.77	nr	**239.58**
Sum of two sides 2600 mm	145.83	178.28	2.88	70.77	nr	**249.05**
Sum of two sides 2700 mm	153.59	187.77	2.88	70.77	nr	**258.54**
Sum of two sides 2800 mm	161.35	197.26	3.94	96.81	nr	**294.07**
Sum of two sides 2900 mm	169.19	206.83	3.94	96.81	nr	**303.64**
Sum of two sides 3000 mm	176.95	216.32	4.83	118.68	nr	**335.00**
Sum of two sides 3100 mm	183.11	223.85	4.83	118.68	nr	**342.53**
Sum of two sides 3200 mm	188.73	230.72	4.83	118.68	nr	**349.40**
Grille neck						
Sum of two sides 1700 mm	81.10	99.15	1.86	45.71	nr	**144.86**
Sum of two sides 1800 mm	89.82	109.81	2.02	49.64	nr	**159.45**
Sum of two sides 1900 mm	98.56	120.49	2.02	49.64	nr	**170.13**
Sum of two sides 2000 mm	107.27	131.14	2.02	49.64	nr	**180.78**
Sum of two sides 2100 mm	116.00	141.81	2.61	64.13	nr	**205.94**
Sum of two sides 2200 mm	124.71	152.46	2.61	64.13	nr	**216.59**
Sum of two sides 2300 mm	133.44	163.13	2.61	64.13	nr	**227.26**

U: VENTILATION/AIR CONDITIONING SYSTEMS

Item	Net Price £	Material £	Labour hours	Labour £	Unit	Total rate £
Sum of two sides 2400 mm	142.15	173.78	2.88	70.77	nr	**244.55**
Sum of two sides 2500 mm	150.88	184.45	2.88	70.77	nr	**255.22**
Sum of two sides 2600 mm	159.60	195.11	2.88	70.77	nr	**265.88**
Sum of two sides 2700 mm	168.32	205.77	2.88	70.77	nr	**276.54**
Sum of two sides 2800 mm	177.04	216.43	3.94	96.81	nr	**313.24**
Sum of two sides 2900 mm	185.76	227.09	4.12	101.24	nr	**328.33**
Sum of two sides 3000 mm	194.48	237.75	5.00	122.86	nr	**360.61**
Sum of two sides 3100 mm	201.42	246.24	5.00	122.86	nr	**369.10**
Sum of two sides 3200 mm	207.73	253.95	5.00	122.86	nr	**376.81**
Ductwork 1601 to 2000 mm longest side						
Sum of two sides 2100 mm	78.64	96.14	2.17	53.32	m	**149.46**
Sum of two sides 2200 mm	81.00	99.02	2.17	53.32	m	**152.34**
Sum of two sides 2300 mm	83.38	101.93	2.19	53.82	m	**155.75**
Sum of two sides 2400 mm	85.70	104.77	2.38	58.48	m	**163.25**
Sum of two sides 2500 mm	88.09	107.69	2.38	58.48	m	**166.17**
Sum of two sides 2600 mm	90.32	110.42	2.64	64.87	m	**175.29**
Sum of two sides 2700 mm	92.56	113.16	2.66	65.36	m	**178.52**
Sum of two sides 2800 mm	94.80	115.90	2.95	72.49	m	**188.39**
Sum of two sides 2900 mm	97.18	118.80	2.96	72.74	m	**191.54**
Sum of two sides 3000 mm	99.41	121.53	2.96	72.74	m	**194.27**
Sum of two sides 3100 mm	107.86	131.86	2.96	72.74	m	**204.60**
Sum of two sides 3200 mm	110.11	134.61	3.15	77.39	m	**212.00**
Sum of two sides 3300 mm	120.75	147.62	3.15	77.39	m	**225.01**
Sum of two sides 3400 mm	122.99	150.36	3.15	77.39	m	**227.75**
Sum of two sides 3500 mm	125.22	153.08	3.15	77.39	m	**230.47**
Sum of two sides 3600 mm	127.46	155.82	3.18	78.15	m	**233.97**
Sum of two sides 3700 mm	129.84	158.73	3.18	78.15	m	**236.88**
Sum of two sides 3800 mm	132.08	161.47	3.18	78.15	m	**239.62**
Sum of two sides 3900 mm	134.32	164.20	3.18	78.15	m	**242.35**
Sum of two sides 4000 mm	136.55	166.93	3.18	78.15	m	**245.08**
Extra over fittings; Ductwork 1601 to 2000 mm longest side						
End cap						
Sum of two sides 2100 mm	37.47	45.80	0.87	21.38	nr	**67.18**
Sum of two sides 2200 mm	40.32	49.30	0.87	21.38	nr	**70.68**
Sum of two sides 2300 mm	43.18	52.79	0.87	21.38	nr	**74.17**
Sum of two sides 2400 mm	46.02	56.26	0.87	21.38	nr	**77.64**
Sum of two sides 2500 mm	48.87	59.74	0.87	21.38	nr	**81.12**
Sum of two sides 2600 mm	51.72	63.23	0.87	21.38	nr	**84.61**
Sum of two sides 2700 mm	54.58	66.72	0.87	21.38	nr	**88.10**
Sum of two sides 2800 mm	57.42	70.19	1.16	28.51	nr	**98.70**
Sum of two sides 2900 mm	60.27	73.68	1.16	28.51	nr	**102.19**
Sum of two sides 3000 mm	63.13	77.18	1.73	42.51	nr	**119.69**
Sum of two sides 3100 mm	65.38	79.93	1.80	44.23	nr	**124.16**
Sum of two sides 3200 mm	67.42	82.42	1.80	44.23	nr	**126.65**
Sum of two sides 3300 mm	69.47	84.92	1.80	44.23	nr	**129.15**
Sum of two sides 3400 mm	71.51	87.43	1.80	44.23	nr	**131.66**
Sum of two sides 3500 mm	73.56	89.93	1.80	44.23	nr	**134.16**

U: VENTILATION/AIR CONDITIONING SYSTEMS

Item	Net Price £	Material £	Labour hours	Labour £	Unit	Total rate £
U10: DUCTWORK: RECTANGULAR – CLASS B – cont						
Extra over fittings; Ductwork 1601 to 2000 mm longest side – cont						
End cap – cont						
Sum of two sides 3600 mm	75.60	92.42	1.80	44.23	nr	**136.65**
Sum of two sides 3700 mm	77.65	94.92	1.80	44.23	nr	**139.15**
Sum of two sides 3800 mm	79.69	97.42	1.80	44.23	nr	**141.65**
Sum of two sides 3900 mm	81.73	99.92	1.80	44.23	nr	**144.15**
Sum of two sides 4000 mm	83.78	102.42	1.80	44.23	nr	**146.65**
Reducer						
Sum of two sides 2100 mm	51.85	63.39	2.61	64.13	nr	**127.52**
Sum of two sides 2200 mm	56.76	69.39	2.61	64.13	nr	**133.52**
Sum of two sides 2300 mm	61.73	75.47	2.61	64.13	nr	**139.60**
Sum of two sides 2400 mm	66.57	81.38	2.88	70.77	nr	**152.15**
Sum of two sides 2500 mm	71.55	87.47	3.12	76.66	nr	**164.13**
Sum of two sides 2600 mm	76.44	93.45	3.12	76.66	nr	**170.11**
Sum of two sides 2700 mm	81.35	99.45	3.12	76.66	nr	**176.11**
Sum of two sides 2800 mm	86.25	105.44	3.95	97.06	nr	**202.50**
Sum of two sides 2900 mm	91.23	111.53	3.97	97.55	nr	**209.08**
Sum of two sides 3000 mm	96.13	117.52	4.52	111.06	nr	**228.58**
Sum of two sides 3100 mm	96.41	117.86	4.52	111.06	nr	**228.92**
Sum of two sides 3200 mm	97.15	118.77	4.52	111.06	nr	**229.83**
Sum of two sides 3300 mm	98.69	120.65	4.52	111.06	nr	**231.71**
Sum of two sides 3400 mm	99.43	121.55	4.52	111.06	nr	**232.61**
Sum of two sides 3500 mm	102.72	125.58	4.52	111.06	nr	**236.64**
Sum of two sides 3600 mm	106.02	129.62	4.52	111.06	nr	**240.68**
Sum of two sides 3700 mm	114.57	140.06	4.52	111.06	nr	**251.12**
Sum of two sides 3800 mm	117.85	144.08	4.52	111.06	nr	**255.14**
Sum of two sides 3900 mm	121.14	148.09	4.52	111.06	nr	**259.15**
Sum of two sides 4000 mm	124.44	152.13	4.52	111.06	nr	**263.19**
Offset						
Sum of two sides 2100 mm	121.44	148.46	2.61	64.13	nr	**212.59**
Sum of two sides 2200 mm	130.63	159.69	2.61	64.13	nr	**223.82**
Sum of two sides 2300 mm	135.97	166.22	2.61	64.13	nr	**230.35**
Sum of two sides 2400 mm	149.01	182.17	2.88	70.77	nr	**252.94**
Sum of two sides 2500 mm	154.18	188.49	3.48	85.51	nr	**274.00**
Sum of two sides 2600 mm	159.08	194.47	3.49	85.76	nr	**280.23**
Sum of two sides 2700 mm	163.80	200.24	3.50	86.00	nr	**286.24**
Sum of two sides 2800 mm	168.35	205.80	4.34	106.64	nr	**312.44**
Sum of two sides 2900 mm	172.76	211.20	4.76	116.97	nr	**328.17**
Sum of two sides 3000 mm	176.93	216.30	5.32	130.73	nr	**347.03**
Sum of two sides 3100 mm	177.18	216.60	5.35	131.46	nr	**348.06**
Sum of two sides 3200 mm	180.86	221.10	5.35	131.46	nr	**352.56**
Sum of two sides 3300 mm	183.42	224.23	5.35	131.46	nr	**355.69**
Sum of two sides 3400 mm	187.10	228.73	5.35	131.46	nr	**360.19**
Sum of two sides 3500 mm	193.34	236.36	5.35	131.46	nr	**367.82**

U: VENTILATION/AIR CONDITIONING SYSTEMS

Item	Net Price £	Material £	Labour hours	Labour £	Unit	Total rate £
Sum of two sides 3600 mm	199.59	244.00	5.35	131.46	nr	**375.46**
Sum of two sides 3700 mm	211.04	257.99	5.35	131.46	nr	**389.45**
Sum of two sides 3800 mm	217.28	265.63	5.35	131.46	nr	**397.09**
Sum of two sides 3900 mm	223.53	273.27	5.35	131.46	nr	**404.73**
Sum of two sides 4000 mm	229.76	280.88	5.35	131.46	nr	**412.34**
Square to round						
Sum of two sides 2100 mm	78.10	95.48	2.61	64.13	nr	**159.61**
Sum of two sides 2200 mm	85.63	104.68	2.61	64.13	nr	**168.81**
Sum of two sides 2300 mm	93.14	113.87	2.61	64.13	nr	**178.00**
Sum of two sides 2400 mm	100.71	123.12	2.88	70.77	nr	**193.89**
Sum of two sides 2500 mm	108.22	132.30	4.70	115.49	nr	**247.79**
Sum of two sides 2600 mm	115.74	141.49	4.70	115.49	nr	**256.98**
Sum of two sides 2700 mm	123.27	150.70	4.71	115.73	nr	**266.43**
Sum of two sides 2800 mm	130.79	159.89	8.19	201.25	nr	**361.14**
Sum of two sides 2900 mm	138.29	169.06	8.19	201.25	nr	**370.31**
Sum of two sides 3000 mm	145.81	178.25	8.19	201.25	nr	**379.50**
Sum of two sides 3100 mm	147.11	179.84	8.19	201.25	nr	**381.09**
Sum of two sides 3200 mm	151.40	185.09	8.19	201.25	nr	**386.34**
Sum of two sides 3300 mm	152.22	186.09	8.19	201.25	nr	**387.34**
Sum of two sides 3400 mm	156.50	191.32	8.62	211.81	nr	**403.13**
Sum of two sides 3500 mm	161.61	197.56	8.62	211.81	nr	**409.37**
Sum of two sides 3600 mm	166.72	203.82	8.62	211.81	nr	**415.63**
Sum of two sides 3700 mm	174.41	213.22	8.62	211.81	nr	**425.03**
Sum of two sides 3800 mm	179.52	219.46	8.75	215.01	nr	**434.47**
Sum of two sides 3900 mm	184.62	225.70	8.75	215.01	nr	**440.71**
Sum of two sides 4000 mm	189.74	231.96	8.75	215.01	nr	**446.97**
90° radius bend						
Sum of two sides 2100 mm	109.81	134.24	2.61	64.13	nr	**198.37**
Sum of two sides 2200 mm	211.91	259.07	2.61	64.13	nr	**323.20**
Sum of two sides 2300 mm	229.26	280.27	2.61	64.13	nr	**344.40**
Sum of two sides 2400 mm	235.45	287.84	2.88	70.77	nr	**358.61**
Sum of two sides 2500 mm	252.73	308.97	3.90	95.83	nr	**404.80**
Sum of two sides 2600 mm	269.73	329.74	4.26	104.68	nr	**434.42**
Sum of two sides 2700 mm	286.71	350.51	4.55	111.80	nr	**462.31**
Sum of two sides 2800 mm	303.70	371.27	4.55	111.80	nr	**483.07**
Sum of two sides 2900 mm	321.00	392.42	6.87	168.82	nr	**561.24**
Sum of two sides 3000 mm	337.99	413.19	6.87	168.82	nr	**582.01**
Sum of two sides 3100 mm	341.25	417.18	6.87	168.82	nr	**586.00**
Sum of two sides 3200 mm	352.35	430.75	6.87	168.82	nr	**599.57**
Sum of two sides 3300 mm	353.41	432.04	6.87	168.82	nr	**600.86**
Sum of two sides 3400 mm	364.51	445.62	7.00	172.01	nr	**617.63**
Sum of two sides 3500 mm	376.67	460.48	7.00	172.01	nr	**632.49**
Sum of two sides 3600 mm	388.83	475.34	7.00	172.01	nr	**647.35**
Sum of two sides 3700 mm	416.84	509.59	7.00	172.01	nr	**681.60**
Sum of two sides 3800 mm	429.00	524.46	7.00	172.01	nr	**696.47**
Sum of two sides 3900 mm	441.17	539.33	7.00	172.01	nr	**711.34**
Sum of two sides 4000 mm	453.33	554.20	7.00	172.01	nr	**726.21**

U: VENTILATION/AIR CONDITIONING SYSTEMS

Item	Net Price £	Material £	Labour hours	Labour £	Unit	Total rate £
U10: DUCTWORK: RECTANGULAR – CLASS B – cont						
Extra over fittings; Ductwork 1601 to 2000 mm longest side – cont						
45° bend						
Sum of two sides 2100 mm	52.78	64.52	2.61	64.13	nr	**128.65**
Sum of two sides 2200 mm	151.70	185.45	2.61	64.13	nr	**249.58**
Sum of two sides 2300 mm	163.38	199.74	2.61	64.13	nr	**263.87**
Sum of two sides 2400 mm	169.14	206.77	2.88	70.77	nr	**277.54**
Sum of two sides 2500 mm	180.78	221.01	4.85	119.17	nr	**340.18**
Sum of two sides 2600 mm	192.19	234.95	4.87	119.68	nr	**354.63**
Sum of two sides 2700 mm	203.60	248.90	4.87	119.68	nr	**368.58**
Sum of two sides 2800 mm	215.01	262.86	8.81	216.49	nr	**479.35**
Sum of two sides 2900 mm	226.65	277.08	8.81	216.49	nr	**493.57**
Sum of two sides 3000 mm	238.05	291.02	9.31	228.76	nr	**519.78**
Sum of two sides 3100 mm	249.92	305.53	9.31	228.76	nr	**534.29**
Sum of two sides 3200 mm	251.21	307.10	9.31	228.76	nr	**535.86**
Sum of two sides 3300 mm	251.88	307.93	9.31	228.76	nr	**536.69**
Sum of two sides 3400 mm	259.39	317.11	9.31	228.76	nr	**545.87**
Sum of two sides 3500 mm	267.59	327.13	9.39	230.74	nr	**557.87**
Sum of two sides 3600 mm	275.78	337.14	9.39	230.74	nr	**567.88**
Sum of two sides 3700 mm	294.56	360.10	9.39	230.74	nr	**590.84**
Sum of two sides 3800 mm	302.75	370.11	9.39	230.74	nr	**600.85**
Sum of two sides 3900 mm	310.95	380.14	9.39	230.74	nr	**610.88**
Sum of two sides 4000 mm	319.13	390.14	9.39	230.74	nr	**620.88**
90° mitre bend						
Sum of two sides 2100 mm	248.54	303.84	2.61	64.13	nr	**367.97**
Sum of two sides 2200 mm	262.71	321.16	2.61	64.13	nr	**385.29**
Sum of two sides 2300 mm	283.56	346.66	2.61	64.13	nr	**410.79**
Sum of two sides 2400 mm	290.40	355.02	2.88	70.77	nr	**425.79**
Sum of two sides 2500 mm	311.29	380.55	4.85	119.17	nr	**499.72**
Sum of two sides 2600 mm	332.29	406.23	4.87	119.68	nr	**525.91**
Sum of two sides 2700 mm	353.28	431.89	4.87	119.68	nr	**551.57**
Sum of two sides 2800 mm	374.27	457.55	8.81	216.49	nr	**674.04**
Sum of two sides 2900 mm	395.17	483.10	14.81	363.92	nr	**847.02**
Sum of two sides 3000 mm	416.16	508.76	15.20	373.50	nr	**882.26**
Sum of two sides 3100 mm	421.36	515.11	15.20	373.50	nr	**888.61**
Sum of two sides 3200 mm	437.53	534.88	15.20	373.50	nr	**908.38**
Sum of two sides 3300 mm	441.18	539.35	15.20	373.50	nr	**912.85**
Sum of two sides 3400 mm	457.35	559.11	15.20	373.50	nr	**932.61**
Sum of two sides 3500 mm	473.51	578.87	15.60	383.32	nr	**962.19**
Sum of two sides 3600 mm	489.68	598.64	15.60	383.32	nr	**981.96**
Sum of two sides 3700 mm	510.92	624.60	15.60	383.32	nr	**1007.92**
Sum of two sides 3800 mm	527.09	644.37	15.60	383.32	nr	**1027.69**
Sum of two sides 3900 mm	543.25	664.12	15.60	383.32	nr	**1047.44**
Sum of two sides 4000 mm	559.43	683.91	15.60	383.32	nr	**1067.23**

U: VENTILATION/AIR CONDITIONING SYSTEMS

Item	Net Price £	Material £	Labour hours	Labour £	Unit	Total rate £
Branch						
Sum of two sides 2100 mm	109.92	134.37	2.61	64.13	nr	**198.50**
Sum of two sides 2200 mm	117.67	143.85	2.61	64.13	nr	**207.98**
Sum of two sides 2300 mm	125.53	153.46	2.61	64.13	nr	**217.59**
Sum of two sides 2400 mm	133.18	162.81	2.88	70.77	nr	**233.58**
Sum of two sides 2500 mm	141.04	172.42	2.88	70.77	nr	**243.19**
Sum of two sides 2600 mm	148.81	181.92	2.88	70.77	nr	**252.69**
Sum of two sides 2700 mm	156.57	191.40	2.88	70.77	nr	**262.17**
Sum of two sides 2800 mm	164.34	200.91	3.94	96.81	nr	**297.72**
Sum of two sides 2900 mm	172.20	210.51	3.94	96.81	nr	**307.32**
Sum of two sides 3000 mm	179.97	220.02	3.94	96.81	nr	**316.83**
Sum of two sides 3100 mm	186.15	227.57	3.94	96.81	nr	**324.38**
Sum of two sides 3200 mm	191.78	234.45	3.94	96.81	nr	**331.26**
Sum of two sides 3300 mm	197.49	241.43	3.94	96.81	nr	**338.24**
Sum of two sides 3400 mm	203.11	248.30	4.83	118.68	nr	**366.98**
Sum of two sides 3500 mm	208.74	255.18	4.83	118.68	nr	**373.86**
Sum of two sides 3600 mm	214.36	262.05	4.83	118.68	nr	**380.73**
Sum of two sides 3700 mm	222.66	272.21	4.83	118.68	nr	**390.89**
Sum of two sides 3800 mm	228.28	279.08	4.83	118.68	nr	**397.76**
Sum of two sides 3900 mm	233.90	285.95	4.83	118.68	nr	**404.63**
Sum of two sides 4000 mm	239.53	292.83	4.83	118.68	nr	**411.51**
Grille neck						
Sum of two sides 2100 mm	116.00	141.81	2.61	64.13	nr	**205.94**
Sum of two sides 2200 mm	124.71	152.46	2.61	64.13	nr	**216.59**
Sum of two sides 2300 mm	133.44	163.13	2.61	64.13	nr	**227.26**
Sum of two sides 2400 mm	142.15	173.78	2.88	70.77	nr	**244.55**
Sum of two sides 2500 mm	150.88	184.45	2.88	70.77	nr	**255.22**
Sum of two sides 2600 mm	159.60	195.11	2.88	70.77	nr	**265.88**
Sum of two sides 2700 mm	168.32	205.77	2.88	70.77	nr	**276.54**
Sum of two sides 2800 mm	177.04	216.43	3.94	96.81	nr	**313.24**
Sum of two sides 2900 mm	185.76	227.09	4.12	101.24	nr	**328.33**
Sum of two sides 3000 mm	194.48	237.75	4.12	101.24	nr	**338.99**
Sum of two sides 3100 mm	201.42	246.24	4.12	101.24	nr	**347.48**
Sum of two sides 3200 mm	207.73	253.95	4.12	101.24	nr	**355.19**
Sum of two sides 3300 mm	214.03	261.65	4.12	101.24	nr	**362.89**
Sum of two sides 3400 mm	220.35	269.38	5.00	122.86	nr	**392.24**
Sum of two sides 3500 mm	226.65	277.08	5.00	122.86	nr	**399.94**
Sum of two sides 3600 mm	232.96	284.79	5.00	122.86	nr	**407.65**
Sum of two sides 3700 mm	239.26	292.50	5.00	122.86	nr	**415.36**
Sum of two sides 3800 mm	245.58	300.22	5.00	122.86	nr	**423.08**
Sum of two sides 3900 mm	251.88	307.93	5.00	122.86	nr	**430.79**
Sum of two sides 4000 mm	258.20	315.65	5.00	122.86	nr	**438.51**

U: VENTILATION/AIR CONDITIONING SYSTEMS

Item	Net Price £	Material £	Labour hours	Labour £	Unit	Total rate £
U10: DUCTWORK: RECTANGULAR – CLASS B – cont						
Ductwork 2001 to 2500 mm longest side						
Sum of two sides 2500 mm	111.49	136.30	2.38	58.48	m	194.78
Sum of two sides 2600 mm	114.27	139.70	2.64	64.87	m	204.57
Sum of two sides 2700 mm	116.55	142.48	2.66	65.36	m	207.84
Sum of two sides 2800 mm	119.59	146.20	2.95	72.49	m	218.69
Sum of two sides 2900 mm	121.88	149.00	2.96	72.74	m	221.74
Sum of two sides 3000 mm	124.03	151.63	3.15	77.39	m	229.02
Sum of two sides 3100 mm	126.28	154.38	3.15	77.39	m	231.77
Sum of two sides 3200 mm	128.43	157.00	3.15	77.39	m	234.39
Sum of two sides 3300 mm	130.71	159.79	3.15	77.39	m	237.18
Sum of two sides 3400 mm	142.21	173.85	2.66	65.36	m	239.21
Sum of two sides 3500 mm	132.87	162.43	3.15	77.39	m	239.82
Sum of two sides 3600 mm	144.36	176.48	3.18	78.15	m	254.63
Sum of two sides 3700 mm	154.69	189.11	3.18	78.15	m	267.26
Sum of two sides 3800 mm	156.85	191.76	3.18	78.15	m	269.91
Sum of two sides 3900 mm	159.00	194.38	3.18	78.15	m	272.53
Sum of two sides 4000 mm	161.94	197.98	3.18	78.15	m	276.13
Extra over fittings; Ductwork 2001 to 2500 mm longest side						
End Cap						
Sum of two sides 2500 mm	49.91	61.02	0.87	21.38	nr	82.40
Sum of two sides 2600 mm	52.83	64.58	0.87	21.38	nr	85.96
Sum of two sides 2700 mm	55.74	68.14	0.87	21.38	nr	89.52
Sum of two sides 2800 mm	58.64	71.69	1.16	28.51	nr	100.20
Sum of two sides 2900 mm	61.55	75.24	1.16	28.51	nr	103.75
Sum of two sides 3000 mm	64.47	78.82	1.73	42.51	nr	121.33
Sum of two sides 3100 mm	66.77	81.63	1.73	42.51	nr	124.14
Sum of two sides 3200 mm	68.85	84.17	1.73	42.51	nr	126.68
Sum of two sides 3300 mm	70.95	86.74	1.73	42.51	nr	129.25
Sum of two sides 3400 mm	73.03	89.28	1.80	44.23	nr	133.51
Sum of two sides 3500 mm	75.13	91.85	1.80	44.23	nr	136.08
Sum of two sides 3600 mm	77.21	94.39	1.80	44.23	nr	138.62
Sum of two sides 3700 mm	79.30	96.94	1.80	44.23	nr	141.17
Sum of two sides 3800 mm	81.39	99.50	1.80	44.23	nr	143.73
Sum of two sides 3900 mm	83.47	102.04	1.80	44.23	nr	146.27
Sum of two sides 4000 mm	85.57	104.61	1.80	44.23	nr	148.84
Reducer						
Sum of two sides 2500 mm	59.53	72.78	3.12	76.66	nr	149.44
Sum of two sides 2600 mm	64.52	78.88	3.12	76.66	nr	155.54
Sum of two sides 2700 mm	69.60	85.09	3.12	76.66	nr	161.75
Sum of two sides 2800 mm	74.48	91.05	3.95	97.06	nr	188.11
Sum of two sides 2900 mm	79.56	97.26	3.97	97.55	nr	194.81
Sum of two sides 3000 mm	84.56	103.37	3.97	97.55	nr	200.92
Sum of two sides 3100 mm	88.25	107.88	3.97	97.55	nr	205.43

U: VENTILATION/AIR CONDITIONING SYSTEMS

Item	Net Price £	Material £	Labour hours	Labour £	Unit	Total rate £
Sum of two sides 3200 mm	91.61	111.99	3.97	97.55	nr	**209.54**
Sum of two sides 3300 mm	95.04	116.18	3.97	97.55	nr	**213.73**
Sum of two sides 3400 mm	96.83	118.38	4.52	111.06	nr	**229.44**
Sum of two sides 3500 mm	97.46	119.15	4.52	111.06	nr	**230.21**
Sum of two sides 3600 mm	98.40	120.29	4.52	111.06	nr	**231.35**
Sum of two sides 3700 mm	100.18	122.47	4.52	111.06	nr	**233.53**
Sum of two sides 3800 mm	100.81	123.24	4.52	111.06	nr	**234.30**
Sum of two sides 3900 mm	104.17	127.35	4.52	111.06	nr	**238.41**
Sum of two sides 4000 mm	107.56	131.49	4.52	111.06	nr	**242.55**
Offset						
Sum of two sides 2500 mm	149.51	182.77	3.48	85.51	nr	**268.28**
Sum of two sides 2600 mm	158.88	194.23	3.49	85.76	nr	**279.99**
Sum of two sides 2700 mm	162.57	198.74	3.50	86.00	nr	**284.74**
Sum of two sides 2800 mm	177.60	217.11	4.34	106.64	nr	**323.75**
Sum of two sides 2900 mm	181.12	221.42	4.76	116.97	nr	**338.39**
Sum of two sides 3000 mm	184.36	225.38	5.32	130.73	nr	**356.11**
Sum of two sides 3100 mm	185.49	226.76	5.35	131.46	nr	**358.22**
Sum of two sides 3200 mm	186.15	227.57	5.32	130.73	nr	**358.30**
Sum of two sides 3300 mm	186.17	227.60	5.32	130.73	nr	**358.33**
Sum of two sides 3400 mm	185.89	227.25	5.35	131.46	nr	**358.71**
Sum of two sides 3500 mm	184.90	226.04	5.32	130.73	nr	**356.77**
Sum of two sides 3600 mm	191.26	233.82	5.35	131.46	nr	**365.28**
Sum of two sides 3700 mm	193.47	236.52	5.35	131.46	nr	**367.98**
Sum of two sides 3800 mm	194.85	238.21	5.35	131.46	nr	**369.67**
Sum of two sides 3900 mm	201.21	245.98	5.35	131.46	nr	**377.44**
Sum of two sides 4000 mm	207.56	253.74	5.35	131.46	nr	**385.20**
Square to round						
Sum of two sides 2500 mm	96.71	118.23	4.70	115.49	nr	**233.72**
Sum of two sides 2600 mm	104.40	127.63	4.70	115.49	nr	**243.12**
Sum of two sides 2700 mm	112.06	136.99	4.71	115.73	nr	**252.72**
Sum of two sides 2800 mm	119.74	146.38	8.19	201.25	nr	**347.63**
Sum of two sides 2900 mm	127.40	155.75	8.19	201.25	nr	**357.00**
Sum of two sides 3000 mm	135.07	165.12	8.19	201.25	nr	**366.37**
Sum of two sides 3100 mm	140.84	172.17	8.19	201.25	nr	**373.42**
Sum of two sides 3200 mm	146.05	178.55	8.62	211.81	nr	**390.36**
Sum of two sides 3300 mm	151.25	184.91	8.62	211.81	nr	**396.72**
Sum of two sides 3400 mm	156.45	191.26	8.62	211.81	nr	**403.07**
Sum of two sides 3500 mm	156.46	191.27	8.62	211.81	nr	**403.08**
Sum of two sides 3600 mm	161.66	197.63	8.75	215.01	nr	**412.64**
Sum of two sides 3700 mm	163.81	200.26	8.75	215.01	nr	**415.27**
Sum of two sides 3800 mm	165.91	202.83	8.75	215.01	nr	**417.84**
Sum of two sides 3900 mm	171.12	209.19	8.75	215.01	nr	**424.20**
Sum of two sides 4000 mm	176.30	215.53	8.75	215.01	nr	**430.54**

U: VENTILATION/AIR CONDITIONING SYSTEMS

Item	Net Price £	Material £	Labour hours	Labour £	Unit	Total rate £
U10: DUCTWORK: RECTANGULAR – CLASS B – cont						
Extra over fittings; Ductwork 2001 to 2500 mm longest side – cont						
90° radius bend						
Sum of two sides 2500 mm	220.58	269.66	3.90	95.83	nr	**365.49**
Sum of two sides 2600 mm	230.70	282.03	4.26	104.68	nr	**386.71**
Sum of two sides 2700 mm	248.28	303.52	4.55	111.80	nr	**415.32**
Sum of two sides 2800 mm	250.40	306.12	4.55	111.80	nr	**417.92**
Sum of two sides 2900 mm	267.91	327.52	6.87	168.82	nr	**496.34**
Sum of two sides 3000 mm	285.14	348.58	6.87	168.82	nr	**517.40**
Sum of two sides 3100 mm	298.48	364.90	6.87	168.82	nr	**533.72**
Sum of two sides 3200 mm	310.78	379.93	6.87	168.82	nr	**548.75**
Sum of two sides 3300 mm	323.37	395.32	6.87	168.82	nr	**564.14**
Sum of two sides 3400 mm	335.66	410.34	6.87	168.82	nr	**579.16**
Sum of two sides 3500 mm	334.94	409.47	7.00	172.01	nr	**581.48**
Sum of two sides 3600 mm	343.33	419.72	7.00	172.01	nr	**591.73**
Sum of two sides 3700 mm	347.23	424.49	7.00	172.01	nr	**596.50**
Sum of two sides 3800 mm	355.62	434.75	7.00	172.01	nr	**606.76**
Sum of two sides 3900 mm	367.92	449.78	7.00	172.01	nr	**621.79**
Sum of two sides 4000 mm	371.30	453.92	7.00	172.01	nr	**625.93**
45° bend						
Sum of two sides 2500 mm	165.19	201.94	4.85	119.17	nr	**321.11**
Sum of two sides 2600 mm	173.15	211.68	4.87	119.68	nr	**331.36**
Sum of two sides 2700 mm	185.00	226.17	4.87	119.68	nr	**345.85**
Sum of two sides 2800 mm	188.79	230.79	8.81	216.49	nr	**447.28**
Sum of two sides 2900 mm	200.60	245.23	8.81	216.49	nr	**461.72**
Sum of two sides 3000 mm	212.19	259.41	9.31	228.76	nr	**488.17**
Sum of two sides 3100 mm	221.22	270.45	9.31	228.76	nr	**499.21**
Sum of two sides 3200 mm	229.53	280.60	9.31	228.76	nr	**509.36**
Sum of two sides 3300 mm	238.06	291.03	9.31	228.76	nr	**519.79**
Sum of two sides 3400 mm	246.36	301.17	9.31	228.76	nr	**529.93**
Sum of two sides 3500 mm	248.09	303.29	9.31	228.76	nr	**532.05**
Sum of two sides 3600 mm	256.29	313.32	9.39	230.74	nr	**544.06**
Sum of two sides 3700 mm	256.39	313.44	9.39	230.74	nr	**544.18**
Sum of two sides 3800 mm	264.59	323.46	9.39	230.74	nr	**554.20**
Sum of two sides 3900 mm	272.90	333.63	9.39	230.74	nr	**564.37**
Sum of two sides 4000 mm	276.74	338.31	9.39	230.74	nr	**569.05**
90° mitre bend						
Sum of two sides 2500 mm	270.28	330.42	4.85	119.17	nr	**449.59**
Sum of two sides 2600 mm	286.90	350.74	4.87	119.68	nr	**470.42**
Sum of two sides 2700 mm	308.31	376.91	4.87	119.68	nr	**496.59**
Sum of two sides 2800 mm	309.67	378.58	8.81	216.49	nr	**595.07**
Sum of two sides 2900 mm	331.12	404.80	14.81	363.92	nr	**768.72**
Sum of two sides 3000 mm	352.70	431.18	14.81	363.92	nr	**795.10**
Sum of two sides 3100 mm	370.36	452.77	15.20	373.50	nr	**826.27**

U: VENTILATION/AIR CONDITIONING SYSTEMS

Item	Net Price £	Material £	Labour hours	Labour £	Unit	Total rate £
Sum of two sides 3200 mm	387.02	473.13	15.20	373.50	nr	**846.63**
Sum of two sides 3300 mm	403.54	493.33	15.20	373.50	nr	**866.83**
Sum of two sides 3400 mm	420.19	513.68	15.20	373.50	nr	**887.18**
Sum of two sides 3500 mm	419.55	512.90	15.20	373.50	nr	**886.40**
Sum of two sides 3600 mm	436.20	533.26	15.60	383.32	nr	**916.58**
Sum of two sides 3700 mm	437.46	534.80	15.60	383.32	nr	**918.12**
Sum of two sides 3800 mm	454.11	555.15	15.60	383.32	nr	**938.47**
Sum of two sides 3900 mm	470.77	575.52	15.60	383.32	nr	**958.84**
Sum of two sides 4000 mm	475.73	581.58	15.60	383.32	nr	**964.90**
Branch						
Sum of two sides 2500 mm	144.31	176.42	2.88	70.77	nr	**247.19**
Sum of two sides 2600 mm	152.23	186.10	2.88	70.77	nr	**256.87**
Sum of two sides 2700 mm	160.26	195.92	2.88	70.77	nr	**266.69**
Sum of two sides 2800 mm	168.08	205.47	3.94	96.81	nr	**302.28**
Sum of two sides 2900 mm	176.10	215.28	3.94	96.81	nr	**312.09**
Sum of two sides 3000 mm	184.03	224.98	3.94	96.81	nr	**321.79**
Sum of two sides 3100 mm	190.35	232.71	3.94	96.81	nr	**329.52**
Sum of two sides 3200 mm	196.09	239.72	3.94	96.81	nr	**336.53**
Sum of two sides 3300 mm	201.92	246.85	3.94	96.81	nr	**343.66**
Sum of two sides 3400 mm	207.66	253.86	3.94	96.81	nr	**350.67**
Sum of two sides 3500 mm	213.69	261.24	4.83	118.68	nr	**379.92**
Sum of two sides 3600 mm	219.43	268.25	4.83	118.68	nr	**386.93**
Sum of two sides 3700 mm	225.27	275.39	4.83	118.68	nr	**394.07**
Sum of two sides 3800 mm	231.01	282.42	4.83	118.68	nr	**401.10**
Sum of two sides 3900 mm	236.75	289.43	4.83	118.68	nr	**408.11**
Sum of two sides 4000 mm	242.57	296.55	4.83	118.68	nr	**415.23**
Grille neck						
Sum of two sides 2500 mm	154.09	188.38	2.88	70.77	nr	**259.15**
Sum of two sides 2600 mm	163.00	199.26	2.88	70.77	nr	**270.03**
Sum of two sides 2700 mm	171.90	210.15	2.88	70.77	nr	**280.92**
Sum of two sides 2800 mm	180.81	221.04	3.94	96.81	nr	**317.85**
Sum of two sides 2900 mm	189.71	231.93	3.94	96.81	nr	**328.74**
Sum of two sides 3000 mm	198.62	242.81	3.94	96.81	nr	**339.62**
Sum of two sides 3100 mm	205.70	251.47	4.12	101.24	nr	**352.71**
Sum of two sides 3200 mm	212.15	259.35	4.12	101.24	nr	**360.59**
Sum of two sides 3300 mm	218.59	267.22	4.12	101.24	nr	**368.46**
Sum of two sides 3400 mm	225.04	275.11	4.12	101.24	nr	**376.35**
Sum of two sides 3500 mm	231.48	282.98	5.00	122.86	nr	**405.84**
Sum of two sides 3600 mm	237.91	290.85	5.00	122.86	nr	**413.71**
Sum of two sides 3700 mm	244.35	298.72	5.00	122.86	nr	**421.58**
Sum of two sides 3800 mm	250.80	306.60	5.00	122.86	nr	**429.46**
Sum of two sides 3900 mm	257.24	314.48	5.00	122.86	nr	**437.34**
Sum of two sides 4000 mm	263.69	322.36	5.00	122.86	nr	**445.22**

U: VENTILATION/AIR CONDITIONING SYSTEMS

Item	Net Price £	Material £	Labour hours	Labour £	Unit	Total rate £
U10: DUCTWORK: RECTANGULAR – CLASS B – cont						
Ductwork 2501 to 4000 mm longest side						
Sum of two sides 3000 mm	183.16	223.91	2.38	58.48	m	**282.39**
Sum of two sides 3100 mm	187.45	229.15	2.38	58.48	m	**287.63**
Sum of two sides 3200 mm	191.48	234.09	2.38	58.48	m	**292.57**
Sum of two sides 3300 mm	195.51	239.01	2.38	58.48	m	**297.49**
Sum of two sides 3400 mm	200.70	245.36	2.64	64.87	m	**310.23**
Sum of two sides 3500 mm	204.73	250.28	2.66	65.36	m	**315.64**
Sum of two sides 3600 mm	208.77	255.22	2.95	72.49	m	**327.71**
Sum of two sides 3700 mm	214.08	261.71	2.96	72.74	m	**334.45**
Sum of two sides 3800 mm	218.11	266.64	3.15	77.39	m	**344.03**
Sum of two sides 3900 mm	230.11	281.31	3.15	77.39	m	**358.70**
Sum of two sides 4000 mm	234.14	286.24	3.15	77.39	m	**363.63**
Sum of two sides 4100 mm	238.31	291.34	3.35	82.31	m	**373.65**
Sum of two sides 4200 mm	242.35	296.28	3.35	82.31	m	**378.59**
Sum of two sides 4300 mm	246.37	301.19	3.60	88.46	m	**389.65**
Sum of two sides 4400 mm	252.26	308.39	3.60	88.46	m	**396.85**
Sum of two sides 4500 mm	256.30	313.33	3.60	88.46	m	**401.79**
Extra over fittings; Ductwork 2501 to 4000 mm longest side						
End Cap						
Sum of two sides 3000 mm	64.73	79.13	1.73	42.51	nr	**121.64**
Sum of two sides 3100 mm	67.04	81.96	1.73	42.51	nr	**124.47**
Sum of two sides 3200 mm	69.13	84.51	1.73	42.51	nr	**127.02**
Sum of two sides 3300 mm	71.23	87.08	1.73	42.51	nr	**129.59**
Sum of two sides 3400 mm	73.32	89.63	1.73	42.51	nr	**132.14**
Sum of two sides 3500 mm	75.43	92.22	1.73	42.51	nr	**134.73**
Sum of two sides 3600 mm	77.52	94.77	1.73	42.51	nr	**137.28**
Sum of two sides 3700 mm	79.62	97.33	1.73	42.51	nr	**139.84**
Sum of two sides 3800 mm	81.72	99.90	1.73	42.51	nr	**142.41**
Sum of two sides 3900 mm	83.82	102.47	1.80	44.23	nr	**146.70**
Sum of two sides 4000 mm	85.92	105.04	1.80	44.23	nr	**149.27**
Sum of two sides 4100 mm	88.01	107.59	1.80	44.23	nr	**151.82**
Sum of two sides 4200 mm	90.11	110.16	1.80	44.23	nr	**154.39**
Sum of two sides 4300 mm	92.21	112.72	1.88	46.20	nr	**158.92**
Sum of two sides 4400 mm	94.30	115.28	1.88	46.20	nr	**161.48**
Sum of two sides 4500 mm	96.41	117.86	1.88	46.20	nr	**164.06**
Reducer						
Sum of two sides 3000 mm	63.00	77.01	3.12	76.66	nr	**153.67**
Sum of two sides 3100 mm	66.03	80.72	3.12	76.66	nr	**157.38**
Sum of two sides 3200 mm	68.64	83.91	3.12	76.66	nr	**160.57**
Sum of two sides 3300 mm	71.25	87.11	3.12	76.66	nr	**163.77**
Sum of two sides 3400 mm	73.84	90.27	3.12	76.66	nr	**166.93**
Sum of two sides 3500 mm	76.44	93.45	3.12	76.66	nr	**170.11**
Sum of two sides 3600 mm	79.05	96.63	3.95	97.06	nr	**193.69**

U: VENTILATION/AIR CONDITIONING SYSTEMS

Item	Net Price £	Material £	Labour hours	Labour £	Unit	Total rate £
Sum of two sides 3700 mm	81.72	99.90	3.97	97.55	nr	197.45
Sum of two sides 3800 mm	82.00	100.25	4.52	111.06	nr	211.31
Sum of two sides 3900 mm	84.33	103.09	4.52	111.06	nr	214.15
Sum of two sides 4000 mm	84.61	103.43	4.52	111.06	nr	214.49
Sum of two sides 4100 mm	87.30	106.73	4.52	111.06	nr	217.79
Sum of two sides 4200 mm	89.91	109.91	4.52	111.06	nr	220.97
Sum of two sides 4300 mm	92.52	113.10	4.92	120.89	nr	233.99
Sum of two sides 4400 mm	95.18	116.36	4.92	120.89	nr	237.25
Sum of two sides 4500 mm	97.78	119.54	5.12	125.80	nr	245.34
Offset						
Sum of two sides 3000 mm	172.17	210.48	3.48	85.51	nr	295.99
Sum of two sides 3100 mm	176.62	215.92	3.48	85.51	nr	301.43
Sum of two sides 3200 mm	180.65	220.84	3.48	85.51	nr	306.35
Sum of two sides 3300 mm	183.58	224.43	3.48	85.51	nr	309.94
Sum of two sides 3400 mm	183.77	224.66	3.50	86.00	nr	310.66
Sum of two sides 3500 mm	188.65	230.63	3.49	85.76	nr	316.39
Sum of two sides 3600 mm	192.94	235.87	4.25	104.42	nr	340.29
Sum of two sides 3700 mm	195.44	238.93	4.76	116.97	nr	355.90
Sum of two sides 3800 mm	189.70	231.90	5.32	130.73	nr	362.63
Sum of two sides 3900 mm	190.06	232.35	5.35	131.46	nr	363.81
Sum of two sides 4000 mm	194.35	237.59	5.35	131.46	nr	369.05
Sum of two sides 4100 mm	199.67	244.10	5.85	143.75	nr	387.85
Sum of two sides 4200 mm	201.25	246.03	5.85	143.75	nr	389.78
Sum of two sides 4300 mm	204.83	250.40	6.15	151.11	nr	401.51
Sum of two sides 4400 mm	205.17	250.83	6.30	154.81	nr	405.64
Sum of two sides 4500 mm	208.41	254.78	6.15	151.11	nr	405.89
Square to round						
Sum of two sides 3000 mm	135.22	165.30	4.70	115.49	nr	280.79
Sum of two sides 3100 mm	141.05	172.43	4.70	115.49	nr	287.92
Sum of two sides 3200 mm	146.23	178.77	4.70	115.49	nr	294.26
Sum of two sides 3300 mm	151.41	185.10	4.70	115.49	nr	300.59
Sum of two sides 3400 mm	156.60	191.45	4.71	115.73	nr	307.18
Sum of two sides 3500 mm	161.78	197.78	4.71	115.73	nr	313.51
Sum of two sides 3600 mm	166.96	204.10	8.19	201.25	nr	405.35
Sum of two sides 3700 mm	172.14	210.44	8.62	211.81	nr	422.25
Sum of two sides 3800 mm	177.32	216.77	8.62	211.81	nr	428.58
Sum of two sides 3900 mm	217.12	265.43	8.62	211.81	nr	477.24
Sum of two sides 4000 mm	223.60	273.35	8.75	215.01	nr	488.36
Sum of two sides 4100 mm	229.46	280.52	11.23	275.95	nr	556.47
Sum of two sides 4200 mm	235.63	288.06	11.23	275.95	nr	564.01
Sum of two sides 4300 mm	241.81	295.61	11.25	276.44	nr	572.05
Sum of two sides 4400 mm	247.99	303.17	11.25	276.44	nr	579.61
Sum of two sides 4500 mm	254.16	310.71	11.26	276.69	nr	587.40

U: VENTILATION/AIR CONDITIONING SYSTEMS

Item	Net Price £	Material £	Labour hours	Labour £	Unit	Total rate £
U10: DUCTWORK: RECTANGULAR – CLASS B – cont						
Extra over fittings; Ductwork 2501 to 4000 mm longest side – cont						
90° radius bend						
Sum of two sides 3000 mm	374.60	457.95	3.90	95.83	nr	**553.78**
Sum of two sides 3100 mm	391.92	479.13	3.90	95.83	nr	**574.96**
Sum of two sides 3200 mm	407.35	497.98	4.26	104.68	nr	**602.66**
Sum of two sides 3300 mm	412.62	504.43	4.26	104.68	nr	**609.11**
Sum of two sides 3400 mm	422.77	516.84	4.26	104.68	nr	**621.52**
Sum of two sides 3500 mm	427.78	522.96	4.55	111.80	nr	**634.76**
Sum of two sides 3600 mm	442.93	541.48	4.55	111.80	nr	**653.28**
Sum of two sides 3700 mm	431.48	527.48	6.87	168.82	nr	**696.30**
Sum of two sides 3800 mm	446.36	545.67	6.87	168.82	nr	**714.49**
Sum of two sides 3900 mm	446.80	546.22	6.87	168.82	nr	**715.04**
Sum of two sides 4000 mm	447.43	546.98	7.00	172.01	nr	**718.99**
Sum of two sides 4100 mm	445.97	545.20	7.20	176.92	nr	**722.12**
Sum of two sides 4200 mm	460.58	563.06	7.20	176.92	nr	**739.98**
Sum of two sides 4300 mm	475.21	580.95	7.41	182.08	nr	**763.03**
Sum of two sides 4400 mm	476.58	582.62	7.41	182.08	nr	**764.70**
Sum of two sides 4500 mm	475.96	581.87	7.55	185.51	nr	**767.38**
45° bend						
Sum of two sides 3000 mm	181.38	221.74	4.85	119.17	nr	**340.91**
Sum of two sides 3100 mm	189.97	232.24	4.85	119.17	nr	**351.41**
Sum of two sides 3200 mm	197.63	241.61	4.85	119.17	nr	**360.78**
Sum of two sides 3300 mm	205.28	250.96	4.87	119.68	nr	**370.64**
Sum of two sides 3400 mm	206.00	251.83	4.87	119.68	nr	**371.51**
Sum of two sides 3500 mm	207.52	253.70	4.87	119.68	nr	**373.38**
Sum of two sides 3600 mm	208.04	254.33	8.81	216.49	nr	**470.82**
Sum of two sides 3700 mm	209.09	255.62	8.81	216.49	nr	**472.11**
Sum of two sides 3800 mm	216.47	264.64	9.31	228.76	nr	**493.40**
Sum of two sides 3900 mm	218.91	267.61	9.31	228.76	nr	**496.37**
Sum of two sides 4000 mm	225.59	275.78	9.31	228.76	nr	**504.54**
Sum of two sides 4100 mm	215.59	263.56	10.01	245.96	nr	**509.52**
Sum of two sides 4200 mm	230.09	281.28	9.31	228.76	nr	**510.04**
Sum of two sides 4300 mm	222.84	272.42	10.01	245.96	nr	**518.38**
Sum of two sides 4400 mm	213.71	261.26	10.52	258.50	nr	**519.76**
Sum of two sides 4500 mm	220.75	269.87	10.52	258.50	nr	**528.37**
90° mitre bend						
Sum of two sides 3000 mm	466.20	569.93	4.85	119.17	nr	**689.10**
Sum of two sides 3100 mm	488.59	597.31	4.85	119.17	nr	**716.48**
Sum of two sides 3200 mm	509.52	622.89	4.87	119.68	nr	**742.57**
Sum of two sides 3300 mm	518.18	633.48	4.87	119.68	nr	**753.16**
Sum of two sides 3400 mm	530.45	648.48	4.87	119.68	nr	**768.16**
Sum of two sides 3500 mm	538.87	658.77	8.81	216.49	nr	**875.26**
Sum of two sides 3600 mm	559.54	684.04	8.81	216.49	nr	**900.53**

U: VENTILATION/AIR CONDITIONING SYSTEMS

Item	Net Price £	Material £	Labour hours	Labour £	Unit	Total rate £
Sum of two sides 3700 mm	549.84	672.19	14.81	363.92	nr	**1036.11**
Sum of two sides 3800 mm	550.30	672.74	14.81	363.92	nr	**1036.66**
Sum of two sides 3900 mm	551.19	673.84	14.81	363.92	nr	**1037.76**
Sum of two sides 4000 mm	561.42	686.34	15.20	373.50	nr	**1059.84**
Sum of two sides 4100 mm	571.41	698.56	16.30	400.53	nr	**1099.09**
Sum of two sides 4200 mm	591.63	723.28	16.50	405.45	nr	**1128.73**
Sum of two sides 4300 mm	592.48	724.31	17.01	417.97	nr	**1142.28**
Sum of two sides 4400 mm	595.16	727.59	17.01	417.97	nr	**1145.56**
Sum of two sides 4500 mm	611.86	748.01	17.01	417.97	nr	**1165.98**
Branch						
Sum of two sides 3000 mm	193.63	236.71	2.88	70.77	nr	**307.48**
Sum of two sides 3100 mm	200.34	244.91	2.88	70.77	nr	**315.68**
Sum of two sides 3200 mm	206.37	252.29	2.88	70.77	nr	**323.06**
Sum of two sides 3300 mm	212.42	259.68	2.88	70.77	nr	**330.45**
Sum of two sides 3400 mm	218.43	267.03	2.88	70.77	nr	**337.80**
Sum of two sides 3500 mm	224.47	274.41	3.94	96.81	nr	**371.22**
Sum of two sides 3600 mm	230.51	281.80	3.94	96.81	nr	**378.61**
Sum of two sides 3700 mm	236.60	289.24	3.94	96.81	nr	**386.05**
Sum of two sides 3800 mm	242.65	296.64	4.83	118.68	nr	**415.32**
Sum of two sides 3900 mm	248.95	304.34	4.83	118.68	nr	**423.02**
Sum of two sides 4000 mm	254.98	311.72	4.83	118.68	nr	**430.40**
Sum of two sides 4100 mm	261.11	319.21	5.44	133.67	nr	**452.88**
Sum of two sides 4200 mm	267.15	326.59	5.44	133.67	nr	**460.26**
Sum of two sides 4300 mm	273.19	333.98	5.85	143.75	nr	**477.73**
Sum of two sides 4400 mm	279.28	341.42	5.85	143.75	nr	**485.17**
Sum of two sides 4500 mm	285.32	348.81	5.85	143.75	nr	**492.56**
Grille neck						
Sum of two sides 3000 mm	200.58	245.21	2.88	70.77	nr	**315.98**
Sum of two sides 3100 mm	207.72	253.94	2.88	70.77	nr	**324.71**
Sum of two sides 3200 mm	214.24	261.91	2.88	70.77	nr	**332.68**
Sum of two sides 3300 mm	220.74	269.86	2.88	70.77	nr	**340.63**
Sum of two sides 3400 mm	227.25	277.81	3.94	96.81	nr	**374.62**
Sum of two sides 3500 mm	233.76	285.77	3.94	96.81	nr	**382.58**
Sum of two sides 3600 mm	240.26	293.71	3.94	96.81	nr	**390.52**
Sum of two sides 3700 mm	246.77	301.68	4.12	101.24	nr	**402.92**
Sum of two sides 3800 mm	253.29	309.65	4.12	101.24	nr	**410.89**
Sum of two sides 3900 mm	259.80	317.61	4.12	101.24	nr	**418.85**
Sum of two sides 4000 mm	266.30	325.55	5.00	122.86	nr	**448.41**
Sum of two sides 4100 mm	272.81	333.51	5.00	122.86	nr	**456.37**
Sum of two sides 4200 mm	279.31	341.46	5.00	122.86	nr	**464.32**
Sum of two sides 4300 mm	285.83	349.43	5.23	128.51	nr	**477.94**
Sum of two sides 4400 mm	292.34	357.39	5.23	128.51	nr	**485.90**
Sum of two sides 4500 mm	298.85	365.35	5.39	132.45	nr	**497.80**

U: VENTILATION/AIR CONDITIONING SYSTEMS

Item	Net Price £	Material £	Labour hours	Labour £	Unit	Total rate £
U10: DUCTWORK: RECTANGULAR – CLASS B – cont						
Y30 – ANCILLARIES						
Access doors, hollow steel construction; 25 mm mineral wool insulation; removeable or hinged; fixed with cams; including sub-frame and integral sealing gaskets						
Rectangular duct						
150 × 150 mm	14.78	18.07	1.25	30.72	nr	**48.79**
200 × 200 mm	16.27	19.89	1.25	30.72	nr	**50.61**
300 × 150 mm	16.72	20.44	1.25	30.72	nr	**51.16**
300 × 300 mm	18.81	23.00	1.25	30.72	nr	**53.72**
400 × 400 mm	21.45	26.22	1.35	33.21	nr	**59.43**
450 × 300 mm	21.45	26.22	1.50	36.89	nr	**63.11**
450 × 450 mm	23.65	28.91	1.50	36.89	nr	**65.80**
Access doors, hollow steel construction; 25 mm mineral wool insulation; removeable or hinged; fixed with cams; including sub-frame and integral sealing gaskets						
Flat oval duct						
235 × 90 mm	30.03	36.71	1.25	30.72	nr	**67.43**
235 × 140 mm	32.01	39.13	1.35	33.21	nr	**72.34**
335 × 235 mm	36.58	44.72	1.50	36.89	nr	**81.61**
535 × 235 mm	41.14	50.29	1.50	36.89	nr	**87.18**

U: VENTILATION/AIR CONDITIONING SYSTEMS

Item	Net Price £	Material £	Labour hours	Labour £	Unit	Total rate £
U10: DUCTWORK: RECTANGULAR – CLASS C						
Y30 – AIR DUCTLINES						
Galvanized sheet metal DW144 class C rectangular section ductwork; including all necessary stiffeners, joints, couplers in the running length and duct supports						
Ductwork up to 400 mm longest side						
Sum of two sides 200 mm	11.57	14.14	1.17	28.76	m	**42.90**
Sum of two sides 300 mm	12.51	15.30	1.17	28.76	m	**44.06**
Sum of two sides 400 mm	12.42	15.18	1.19	29.24	m	**44.42**
Sum of two sides 500 mm	13.30	16.26	1.16	28.51	m	**44.77**
Sum of two sides 600 mm	13.34	16.30	1.19	29.24	m	**45.54**
Sum of two sides 700 mm	14.16	17.31	1.19	29.24	m	**46.55**
Sum of two sides 800 mm	15.04	18.39	1.19	29.24	m	**47.63**
Extra over fittings; Ductwork up to 400 mm longest side						
End Cap						
Sum of two sides 200 mm	7.54	9.22	0.38	9.34	nr	**18.56**
Sum of two sides 300 mm	8.40	10.27	0.38	9.34	nr	**19.61**
Sum of two sides 400 mm	9.25	11.31	0.38	9.34	nr	**20.65**
Sum of two sides 500 mm	10.12	12.37	0.38	9.34	nr	**21.71**
Sum of two sides 600 mm	10.97	13.41	0.38	9.34	nr	**22.75**
Sum of two sides 700 mm	11.84	14.47	0.38	9.34	nr	**23.81**
Sum of two sides 800 mm	12.69	15.51	0.38	9.34	nr	**24.85**
Reducer						
Sum of two sides 200 mm	12.32	15.06	1.40	34.40	nr	**49.46**
Sum of two sides 300 mm	13.91	17.01	1.40	34.40	nr	**51.41**
Sum of two sides 400 mm	22.42	27.41	1.42	34.89	nr	**62.30**
Sum of two sides 500 mm	24.24	29.63	1.42	34.89	nr	**64.52**
Sum of two sides 600 mm	26.08	31.88	1.69	41.53	nr	**73.41**
Sum of two sides 700 mm	27.90	34.10	1.69	41.53	nr	**75.63**
Sum of two sides 800 mm	29.71	36.32	1.92	47.17	nr	**83.49**
Offset						
Sum of two sides 200 mm	18.19	22.24	1.63	40.05	nr	**62.29**
Sum of two sides 300 mm	20.56	25.13	1.63	40.05	nr	**65.18**
Sum of two sides 400 mm	30.07	36.76	1.65	40.54	nr	**77.30**
Sum of two sides 500 mm	32.67	39.94	1.65	40.54	nr	**80.48**
Sum of two sides 600 mm	34.82	42.57	1.92	47.17	nr	**89.74**
Sum of two sides 700 mm	37.18	45.45	1.92	47.17	nr	**92.62**
Sum of two sides 800 mm	39.26	48.00	1.92	47.17	nr	**95.17**

U: VENTILATION/AIR CONDITIONING SYSTEMS

Item	Net Price £	Material £	Labour hours	Labour £	Unit	Total rate £
U10: DUCTWORK: RECTANGULAR – CLASS C – cont						
Extra over fittings; Ductwork up to 400 mm longest side – cont						
Square to round						
Sum of two sides 200 mm	16.01	19.57	1.22	29.97	nr	49.54
Sum of two sides 300 mm	17.98	21.98	1.22	29.97	nr	51.95
Sum of two sides 400 mm	23.78	29.07	1.25	30.72	nr	59.79
Sum of two sides 500 mm	25.87	31.63	1.25	30.72	nr	62.35
Sum of two sides 600 mm	27.95	34.17	1.33	32.68	nr	66.85
Sum of two sides 700 mm	30.05	36.74	1.33	32.68	nr	69.42
Sum of two sides 800 mm	32.12	39.26	1.40	34.40	nr	73.66
90° radius bend						
Sum of two sides 200 mm	12.04	14.72	1.10	27.03	nr	41.75
Sum of two sides 300 mm	12.98	15.87	1.10	27.03	nr	42.90
Sum of two sides 400 mm	21.53	26.32	1.12	27.52	nr	53.84
Sum of two sides 500 mm	22.89	27.99	1.12	27.52	nr	55.51
Sum of two sides 600 mm	24.70	30.20	1.16	28.51	nr	58.71
Sum of two sides 700 mm	26.28	32.13	1.16	28.51	nr	60.64
Sum of two sides 800 mm	28.06	34.30	1.22	29.97	nr	64.27
45° radius bend						
Sum of two sides 200 mm	13.00	15.89	1.29	31.70	nr	47.59
Sum of two sides 300 mm	14.27	17.45	1.29	31.70	nr	49.15
Sum of two sides 400 mm	22.55	27.56	1.29	31.70	nr	59.26
Sum of two sides 500 mm	24.15	29.52	1.29	31.70	nr	61.22
Sum of two sides 600 mm	25.98	31.77	1.39	34.15	nr	65.92
Sum of two sides 700 mm	27.69	33.86	1.39	34.15	nr	68.01
Sum of two sides 800 mm	29.49	36.05	1.46	35.89	nr	71.94
90° mitre bend						
Sum of two sides 200 mm	20.06	24.52	2.04	50.12	nr	74.64
Sum of two sides 300 mm	21.97	26.86	2.04	50.12	nr	76.98
Sum of two sides 400 mm	31.58	38.60	2.09	51.36	nr	89.96
Sum of two sides 500 mm	34.04	41.61	2.09	51.36	nr	92.97
Sum of two sides 600 mm	37.18	45.45	2.15	52.84	nr	98.29
Sum of two sides 700 mm	40.08	49.00	2.15	52.84	nr	101.84
Sum of two sides 800 mm	43.27	52.90	2.26	55.54	nr	108.44
Branch						
Sum of two sides 200 mm	19.93	24.36	0.92	22.61	nr	46.97
Sum of two sides 300 mm	22.14	27.07	0.92	22.61	nr	49.68
Sum of two sides 400 mm	27.55	33.68	0.95	23.34	nr	57.02
Sum of two sides 500 mm	29.97	36.64	0.95	23.34	nr	59.98
Sum of two sides 600 mm	32.36	39.56	1.03	25.31	nr	64.87
Sum of two sides 700 mm	34.74	42.47	1.03	25.31	nr	67.78
Sum of two sides 800 mm	37.11	45.37	1.03	25.31	nr	70.68

U: VENTILATION/AIR CONDITIONING SYSTEMS

Item	Net Price £	Material £	Labour hours	Labour £	Unit	Total rate £
Grille neck						
Sum of two sides 200 mm	22.83	27.91	1.10	27.03	nr	**54.94**
Sum of two sides 300 mm	25.54	31.22	1.10	27.03	nr	**58.25**
Sum of two sides 400 mm	28.26	34.55	1.16	28.51	nr	**63.06**
Sum of two sides 500 mm	30.97	37.86	1.16	28.51	nr	**66.37**
Sum of two sides 600 mm	33.69	41.19	1.18	28.99	nr	**70.18**
Sum of two sides 700 mm	36.41	44.52	1.18	28.99	nr	**73.51**
Sum of two sides 800 mm	39.12	47.82	1.18	28.99	nr	**76.81**
Ductwork 401 to 600 mm longest side						
Sum of two sides 600 mm	18.42	22.52	1.17	28.76	m	**51.28**
Sum of two sides 700 mm	20.13	24.61	1.17	28.76	m	**53.37**
Sum of two sides 800 mm	21.80	26.65	1.17	28.76	m	**55.41**
Sum of two sides 900 mm	23.42	28.63	1.27	31.21	m	**59.84**
Sum of two sides 1000 mm	25.03	30.60	1.48	36.37	m	**66.97**
Sum of two sides 1100 mm	26.79	32.75	1.49	36.62	m	**69.37**
Sum of two sides 1200 mm	28.40	34.72	1.49	36.62	m	**71.34**
Extra over fittings: Ductwork 401 to 600 mm longest side						
End Cap						
Sum of two sides 600 mm	10.69	13.07	0.38	9.34	nr	**22.41**
Sum of two sides 700 mm	11.53	14.09	0.38	9.34	nr	**23.43**
Sum of two sides 800 mm	12.37	15.12	0.38	9.34	nr	**24.46**
Sum of two sides 900 mm	13.21	16.15	0.38	9.34	nr	**25.49**
Sum of two sides 1000 mm	14.04	17.16	0.38	9.34	nr	**26.50**
Sum of two sides 1100 mm	14.87	18.18	0.38	9.34	nr	**27.52**
Sum of two sides 1200 mm	15.72	19.22	0.38	9.34	nr	**28.56**
Reducer						
Sum of two sides 600 mm	22.76	27.82	1.69	41.53	nr	**69.35**
Sum of two sides 700 mm	24.22	29.61	1.69	41.53	nr	**71.14**
Sum of two sides 800 mm	25.65	31.35	1.92	47.17	nr	**78.52**
Sum of two sides 900 mm	27.12	33.16	1.92	47.17	nr	**80.33**
Sum of two sides 1000 mm	28.57	34.93	2.18	53.57	nr	**88.50**
Sum of two sides 1100 mm	30.14	36.84	2.18	53.57	nr	**90.41**
Sum of two sides 1200 mm	31.60	38.64	2.18	53.57	nr	**92.21**
Offset						
Sum of two sides 600 mm	32.27	39.45	1.92	47.17	nr	**86.62**
Sum of two sides 700 mm	34.45	42.12	1.92	47.17	nr	**89.29**
Sum of two sides 800 mm	35.66	43.59	1.92	47.17	nr	**90.76**
Sum of two sides 900 mm	36.80	44.99	1.92	47.17	nr	**92.16**
Sum of two sides 1000 mm	38.32	46.84	2.18	53.57	nr	**100.41**
Sum of two sides 1100 mm	39.33	48.08	2.18	53.57	nr	**101.65**
Sum of two sides 1200 mm	40.12	49.05	2.18	53.57	nr	**102.62**

U: VENTILATION/AIR CONDITIONING SYSTEMS

Item	Net Price £	Material £	Labour hours	Labour £	Unit	Total rate £
U10: DUCTWORK: RECTANGULAR – CLASS C – cont						
Extra over fittings: Ductwork 401 to 600 mm longest side – cont						
Square to round						
Sum of two sides 600 mm	24.51	29.96	1.33	32.68	nr	**62.64**
Sum of two sides 700 mm	26.21	32.04	1.33	32.68	nr	**64.72**
Sum of two sides 800 mm	27.91	34.12	1.40	34.40	nr	**68.52**
Sum of two sides 900 mm	29.61	36.19	1.40	34.40	nr	**70.59**
Sum of two sides 1000 mm	31.33	38.31	1.82	44.72	nr	**83.03**
Sum of two sides 1100 mm	33.08	40.44	1.82	44.72	nr	**85.16**
Sum of two sides 1200 mm	34.78	42.52	1.82	44.72	nr	**87.24**
90° radius bend						
Sum of two sides 600 mm	21.39	26.15	1.16	28.51	nr	**54.66**
Sum of two sides 700 mm	21.90	26.77	1.16	28.51	nr	**55.28**
Sum of two sides 800 mm	23.32	28.51	1.22	29.97	nr	**58.48**
Sum of two sides 900 mm	24.78	30.29	1.22	29.97	nr	**60.26**
Sum of two sides 1000 mm	25.66	31.37	1.40	34.40	nr	**65.77**
Sum of two sides 1100 mm	27.19	33.24	1.40	34.40	nr	**67.64**
Sum of two sides 1200 mm	28.63	35.00	1.40	34.40	nr	**69.40**
45° bend						
Sum of two sides 600 mm	23.87	29.18	1.39	34.15	nr	**63.33**
Sum of two sides 700 mm	24.98	30.54	1.39	34.15	nr	**64.69**
Sum of two sides 800 mm	26.58	32.50	1.46	35.89	nr	**68.39**
Sum of two sides 900 mm	28.18	34.45	1.46	35.89	nr	**70.34**
Sum of two sides 1000 mm	29.49	36.05	1.88	46.20	nr	**82.25**
Sum of two sides 1100 mm	31.22	38.17	1.88	46.20	nr	**84.37**
Sum of two sides 1200 mm	32.81	40.11	1.88	46.20	nr	**86.31**
90° mitre bend						
Sum of two sides 600 mm	37.63	46.00	2.15	52.84	nr	**98.84**
Sum of two sides 700 mm	39.04	47.73	2.15	52.84	nr	**100.57**
Sum of two sides 800 mm	41.86	51.17	2.26	55.54	nr	**106.71**
Sum of two sides 900 mm	44.70	54.64	2.26	55.54	nr	**110.18**
Sum of two sides 1000 mm	46.86	57.29	3.03	74.45	nr	**131.74**
Sum of two sides 1100 mm	49.83	60.91	3.03	74.45	nr	**135.36**
Sum of two sides 1200 mm	52.71	64.44	3.03	74.45	nr	**138.89**
Branch						
Sum of two sides 600 mm	31.74	38.80	1.03	25.31	nr	**64.11**
Sum of two sides 700 mm	34.07	41.65	1.03	25.31	nr	**66.96**
Sum of two sides 800 mm	36.42	44.53	1.03	25.31	nr	**69.84**
Sum of two sides 900 mm	38.75	47.37	1.03	25.31	nr	**72.68**
Sum of two sides 1000 mm	41.09	50.23	1.29	31.70	nr	**81.93**
Sum of two sides 1100 mm	43.51	53.19	1.29	31.70	nr	**84.89**
Sum of two sides 1200 mm	45.85	56.05	1.29	31.70	nr	**87.75**

U: VENTILATION/AIR CONDITIONING SYSTEMS

Item	Net Price £	Material £	Labour hours	Labour £	Unit	Total rate £
Grille neck						
Sum of two sides 600 mm	32.82	40.12	1.18	28.99	nr	**69.11**
Sum of two sides 700 mm	35.47	43.36	1.18	28.99	nr	**72.35**
Sum of two sides 800 mm	38.11	46.59	1.18	28.99	nr	**75.58**
Sum of two sides 900 mm	40.76	49.83	1.18	28.99	nr	**78.82**
Sum of two sides 1000 mm	43.41	53.07	1.44	35.38	nr	**88.45**
Sum of two sides 1100 mm	46.06	56.31	1.44	35.38	nr	**91.69**
Sum of two sides 1200 mm	48.70	59.53	1.44	35.38	nr	**94.91**
Ductwork 601 to 800 mm longest side						
Sum of two sides 900 mm	25.68	31.39	1.27	31.21	m	**62.60**
Sum of two sides 1000 mm	27.30	33.37	1.48	36.37	m	**69.74**
Sum of two sides 1100 mm	28.92	35.36	1.49	36.62	m	**71.98**
Sum of two sides 1200 mm	29.99	36.67	1.49	36.62	m	**73.29**
Sum of two sides 1300 mm	32.28	39.46	1.51	37.10	m	**76.56**
Sum of two sides 1400 mm	33.90	41.45	1.55	38.09	m	**79.54**
Sum of two sides 1500 mm	35.52	43.42	1.61	39.56	m	**82.98**
Sum of two sides 1600 mm	37.14	45.40	1.62	39.81	m	**85.21**
Extra over fittings; Ductwork 601 to 800 mm longest side						
End Cap						
Sum of two sides 900 mm	13.28	16.23	0.38	9.34	nr	**25.57**
Sum of two sides 1000 mm	14.11	17.25	0.38	9.34	nr	**26.59**
Sum of two sides 1100 mm	14.95	18.27	0.38	9.34	nr	**27.61**
Sum of two sides 1200 mm	15.80	19.31	0.38	9.34	nr	**28.65**
Sum of two sides 1300 mm	16.64	20.34	0.38	9.34	nr	**29.68**
Sum of two sides 1400 mm	17.36	21.22	0.38	9.34	nr	**30.56**
Sum of two sides 1500 mm	20.21	24.71	0.38	9.34	nr	**34.05**
Sum of two sides 1600 mm	23.05	28.18	0.38	9.34	nr	**37.52**
Reducer						
Sum of two sides 900 mm	27.55	33.68	1.92	47.17	nr	**80.85**
Sum of two sides 1000 mm	29.01	35.46	2.18	53.57	nr	**89.03**
Sum of two sides 1100 mm	30.48	37.27	2.18	53.57	nr	**90.84**
Sum of two sides 1200 mm	32.06	39.19	2.18	53.57	nr	**92.76**
Sum of two sides 1300 mm	33.52	40.97	2.30	56.52	nr	**97.49**
Sum of two sides 1400 mm	34.74	42.47	2.30	56.52	nr	**98.99**
Sum of two sides 1500 mm	40.21	49.16	2.47	60.70	nr	**109.86**
Sum of two sides 1600 mm	45.68	55.85	2.47	60.70	nr	**116.55**
Offset						
Sum of two sides 900 mm	39.11	47.81	1.92	47.17	nr	**94.98**
Sum of two sides 1000 mm	40.10	49.02	2.18	53.57	nr	**102.59**
Sum of two sides 1100 mm	40.98	50.10	2.18	53.57	nr	**103.67**
Sum of two sides 1200 mm	41.81	51.11	2.18	53.57	nr	**104.68**
Sum of two sides 1300 mm	42.44	51.88	2.47	60.70	nr	**112.58**
Sum of two sides 1400 mm	43.29	52.92	2.47	60.70	nr	**113.62**
Sum of two sides 1500 mm	49.71	60.77	2.47	60.70	nr	**121.47**
Sum of two sides 1600 mm	56.02	68.48	2.47	60.70	nr	**129.18**

U: VENTILATION/AIR CONDITIONING SYSTEMS

Item	Net Price £	Material £	Labour hours	Labour £	Unit	Total rate £
U10: DUCTWORK: RECTANGULAR – CLASS C – cont						
Extra over fittings; Ductwork 601 to 800 mm longest side – cont						
Square to round						
Sum of two sides 900 mm	29.51	36.08	1.40	34.40	nr	70.48
Sum of two sides 1000 mm	31.23	38.18	1.82	44.72	nr	82.90
Sum of two sides 1100 mm	32.96	40.29	1.82	44.72	nr	85.01
Sum of two sides 1200 mm	34.70	42.43	1.82	44.72	nr	87.15
Sum of two sides 1300 mm	36.41	44.52	2.32	57.01	nr	101.53
Sum of two sides 1400 mm	37.83	46.25	2.32	57.01	nr	103.26
Sum of two sides 1500 mm	44.56	54.48	2.56	62.91	nr	117.39
Sum of two sides 1600 mm	51.30	62.72	2.58	63.40	nr	126.12
90° radius bend						
Sum of two sides 900 mm	22.20	27.14	1.22	29.97	nr	57.11
Sum of two sides 1000 mm	23.58	28.83	1.40	34.40	nr	63.23
Sum of two sides 1100 mm	24.96	30.52	1.40	34.40	nr	64.92
Sum of two sides 1200 mm	26.41	32.29	1.40	34.40	nr	66.69
Sum of two sides 1300 mm	27.79	33.97	1.91	46.94	nr	80.91
Sum of two sides 1400 mm	28.16	34.42	1.91	46.94	nr	81.36
Sum of two sides 1500 mm	33.52	40.97	2.11	51.85	nr	92.82
Sum of two sides 1600 mm	38.90	47.56	2.11	51.85	nr	99.41
45° bend						
Sum of two sides 900 mm	27.40	33.50	1.46	35.89	nr	69.39
Sum of two sides 1000 mm	28.97	35.41	1.88	46.20	nr	81.61
Sum of two sides 1100 mm	30.56	37.36	1.88	46.20	nr	83.56
Sum of two sides 1200 mm	32.24	39.42	1.88	46.20	nr	85.62
Sum of two sides 1300 mm	33.81	41.33	2.26	55.54	nr	96.87
Sum of two sides 1400 mm	34.75	42.48	2.44	59.96	nr	102.44
Sum of two sides 1500 mm	40.32	49.30	2.44	59.96	nr	109.26
Sum of two sides 1600 mm	45.90	56.11	2.68	65.86	nr	121.97
90° mitre bend						
Sum of two sides 900 mm	40.94	50.05	2.26	55.54	nr	105.59
Sum of two sides 1000 mm	43.96	53.75	3.03	74.45	nr	128.20
Sum of two sides 1100 mm	46.97	57.42	3.03	74.45	nr	131.87
Sum of two sides 1200 mm	50.04	61.17	3.03	74.45	nr	135.62
Sum of two sides 1300 mm	53.06	64.87	3.85	94.61	nr	159.48
Sum of two sides 1400 mm	54.77	66.96	3.85	94.61	nr	161.57
Sum of two sides 1500 mm	63.83	78.03	4.25	104.42	nr	182.45
Sum of two sides 1600 mm	72.89	89.11	4.26	104.68	nr	193.79
Branch						
Sum of two sides 900 mm	39.96	48.85	1.03	25.31	nr	74.16
Sum of two sides 1000 mm	42.34	51.76	1.29	31.70	nr	83.46
Sum of two sides 1100 mm	44.74	54.69	1.29	31.70	nr	86.39

U: VENTILATION/AIR CONDITIONING SYSTEMS

Item	Net Price £	Material £	Labour hours	Labour £	Unit	Total rate £
Sum of two sides 1200 mm	47.23	57.74	1.29	31.70	nr	89.44
Sum of two sides 1300 mm	49.63	60.68	1.39	34.15	nr	94.83
Sum of two sides 1400 mm	51.70	63.20	1.39	34.15	nr	97.35
Sum of two sides 1500 mm	59.44	72.67	1.64	40.29	nr	112.96
Sum of two sides 1600 mm	67.19	82.14	1.64	40.29	nr	122.43
Grille neck						
Sum of two sides 900 mm	40.98	50.10	1.18	28.99	nr	79.09
Sum of two sides 1000 mm	43.64	53.35	1.44	35.38	nr	88.73
Sum of two sides 1100 mm	46.30	56.60	1.44	35.38	nr	91.98
Sum of two sides 1200 mm	48.96	59.85	1.44	35.38	nr	95.23
Sum of two sides 1300 mm	51.62	63.11	1.69	41.53	nr	104.64
Sum of two sides 1400 mm	53.91	65.91	1.69	41.53	nr	107.44
Sum of two sides 1500 mm	62.58	76.51	1.79	43.98	nr	120.49
Sum of two sides 1600 mm	71.26	87.12	1.79	43.98	nr	131.10
Ductwork 801 to 1000 mm longest side						
Sum of two sides 1100 mm	29.55	36.12	1.49	36.62	m	72.74
Sum of two sides 1200 mm	31.19	38.13	1.49	36.62	m	74.75
Sum of two sides 1300 mm	32.82	40.12	1.51	37.10	m	77.22
Sum of two sides 1400 mm	34.59	42.28	1.55	38.09	m	80.37
Sum of two sides 1500 mm	36.21	44.27	1.61	39.56	m	83.83
Sum of two sides 1600 mm	37.85	46.27	1.62	39.81	m	86.08
Sum of two sides 1700 mm	39.48	48.27	1.74	42.76	m	91.03
Sum of two sides 1800 mm	41.25	50.43	1.76	43.25	m	93.68
Sum of two sides 1900 mm	42.88	52.42	1.81	44.48	m	96.90
Sum of two sides 2000 mm	44.51	54.41	1.82	44.72	m	99.13
Extra over fittings; Ductwork 801 to 1000 mm longest side						
End Cap						
Sum of two sides 1100 mm	14.95	18.27	0.38	9.34	nr	27.61
Sum of two sides 1200 mm	15.80	19.31	0.38	9.34	nr	28.65
Sum of two sides 1300 mm	16.64	20.34	0.38	9.34	nr	29.68
Sum of two sides 1400 mm	17.36	21.22	0.38	9.34	nr	30.56
Sum of two sides 1500 mm	20.21	24.71	0.38	9.34	nr	34.05
Sum of two sides 1600 mm	23.05	28.18	0.38	9.34	nr	37.52
Sum of two sides 1700 mm	25.89	31.65	0.58	14.26	nr	45.91
Sum of two sides 1800 mm	28.74	35.13	0.58	14.26	nr	49.39
Sum of two sides 1900 mm	31.59	38.61	0.58	14.26	nr	52.87
Sum of two sides 2000 mm	34.44	42.11	0.58	14.26	nr	56.37
Reducer						
Sum of two sides 1100 mm	28.20	34.47	2.18	53.57	nr	88.04
Sum of two sides 1200 mm	29.55	36.12	2.18	53.57	nr	89.69
Sum of two sides 1300 mm	30.89	37.76	2.30	56.52	nr	94.28
Sum of two sides 1400 mm	32.10	39.24	2.30	56.52	nr	95.76
Sum of two sides 1500 mm	37.46	45.79	2.47	60.70	nr	106.49
Sum of two sides 1600 mm	42.81	52.33	2.47	60.70	nr	113.03
Sum of two sides 1700 mm	48.17	58.89	2.59	63.64	nr	122.53

U: VENTILATION/AIR CONDITIONING SYSTEMS

Item	Net Price £	Material £	Labour hours	Labour £	Unit	Total rate £
U10: DUCTWORK: RECTANGULAR – CLASS C – cont						
Extra over fittings; Ductwork 801 to 1000 mm longest side – cont						
Reducer – cont						
Sum of two sides 1800 mm	53.63	65.56	2.59	63.64	nr	**129.20**
Sum of two sides 1900 mm	58.99	72.12	2.71	66.59	nr	**138.71**
Sum of two sides 2000 mm	64.35	78.67	2.71	66.59	nr	**145.26**
Offset						
Sum of two sides 1100 mm	43.58	53.28	2.18	53.57	nr	**106.85**
Sum of two sides 1200 mm	44.45	54.34	2.18	53.57	nr	**107.91**
Sum of two sides 1300 mm	45.19	55.25	2.47	60.70	nr	**115.95**
Sum of two sides 1400 mm	45.53	55.66	2.47	60.70	nr	**116.36**
Sum of two sides 1500 mm	52.03	63.60	2.47	60.70	nr	**124.30**
Sum of two sides 1600 mm	58.42	71.42	2.47	60.70	nr	**132.12**
Sum of two sides 1700 mm	64.69	79.08	2.61	64.13	nr	**143.21**
Sum of two sides 1800 mm	70.87	86.64	2.61	64.13	nr	**150.77**
Sum of two sides 1900 mm	77.76	95.06	2.71	66.59	nr	**161.65**
Sum of two sides 2000 mm	83.68	102.30	2.71	66.59	nr	**168.89**
Square to round						
Sum of two sides 1100 mm	30.58	37.39	1.82	44.72	nr	**82.11**
Sum of two sides 1200 mm	32.19	39.36	1.82	44.72	nr	**84.08**
Sum of two sides 1300 mm	33.79	41.31	2.32	57.01	nr	**98.32**
Sum of two sides 1400 mm	35.11	42.92	2.32	57.01	nr	**99.93**
Sum of two sides 1500 mm	41.71	51.00	2.56	62.91	nr	**113.91**
Sum of two sides 1600 mm	48.34	59.09	2.58	63.40	nr	**122.49**
Sum of two sides 1700 mm	54.95	67.18	2.84	69.78	nr	**136.96**
Sum of two sides 1800 mm	61.57	75.27	2.84	69.78	nr	**145.05**
Sum of two sides 1900 mm	68.19	83.36	3.13	76.91	nr	**160.27**
Sum of two sides 2000 mm	74.81	91.45	3.13	76.91	nr	**168.36**
90° radius bend						
Sum of two sides 1100 mm	22.58	27.60	1.40	34.40	nr	**62.00**
Sum of two sides 1200 mm	24.18	29.56	1.40	34.40	nr	**63.96**
Sum of two sides 1300 mm	25.77	31.51	1.91	46.94	nr	**78.45**
Sum of two sides 1400 mm	27.17	33.22	1.91	46.94	nr	**80.16**
Sum of two sides 1500 mm	32.77	40.06	2.11	51.85	nr	**91.91**
Sum of two sides 1600 mm	38.37	46.91	2.11	51.85	nr	**98.76**
Sum of two sides 1700 mm	43.98	53.77	2.55	62.67	nr	**116.44**
Sum of two sides 1800 mm	49.63	60.68	2.55	62.67	nr	**123.35**
Sum of two sides 1900 mm	54.26	66.33	2.80	68.80	nr	**135.13**
Sum of two sides 2000 mm	59.86	73.18	2.80	68.80	nr	**141.98**
45° bend						
Sum of two sides 1100 mm	29.43	35.98	1.88	46.20	nr	**82.18**
Sum of two sides 1200 mm	31.14	38.07	1.88	46.20	nr	**84.27**
Sum of two sides 1300 mm	32.83	40.14	2.26	55.54	nr	**95.68**

U: VENTILATION/AIR CONDITIONING SYSTEMS

Item	Net Price £	Material £	Labour hours	Labour £	Unit	Total rate £
Sum of two sides 1400 mm	34.39	42.04	2.44	59.96	nr	**102.00**
Sum of two sides 1500 mm	40.10	49.02	2.68	65.86	nr	**114.88**
Sum of two sides 1600 mm	45.80	55.99	2.69	66.11	nr	**122.10**
Sum of two sides 1700 mm	51.51	62.97	2.96	72.74	nr	**135.71**
Sum of two sides 1800 mm	57.33	70.08	2.96	72.74	nr	**142.82**
Sum of two sides 1900 mm	62.52	76.43	3.26	80.11	nr	**156.54**
Sum of two sides 2000 mm	68.23	83.41	3.26	80.11	nr	**163.52**
90° mitre bend						
Sum of two sides 1100 mm	47.11	57.59	3.03	74.45	nr	**132.04**
Sum of two sides 1200 mm	50.31	61.50	3.03	74.45	nr	**135.95**
Sum of two sides 1300 mm	53.49	65.39	3.85	94.61	nr	**160.00**
Sum of two sides 1400 mm	56.34	68.88	3.85	94.61	nr	**163.49**
Sum of two sides 1500 mm	65.54	80.12	4.25	104.42	nr	**184.54**
Sum of two sides 1600 mm	74.75	91.38	4.26	104.68	nr	**196.06**
Sum of two sides 1700 mm	83.94	102.62	4.68	115.00	nr	**217.62**
Sum of two sides 1800 mm	93.17	113.90	4.68	115.00	nr	**228.90**
Sum of two sides 1900 mm	101.29	123.83	4.87	119.68	nr	**243.51**
Sum of two sides 2000 mm	110.53	135.13	4.87	119.68	nr	**254.81**
Branch						
Sum of two sides 1100 mm	44.84	54.82	1.29	31.70	nr	**86.52**
Sum of two sides 1200 mm	47.23	57.74	1.29	31.70	nr	**89.44**
Sum of two sides 1300 mm	49.63	60.68	1.39	34.15	nr	**94.83**
Sum of two sides 1400 mm	51.79	63.31	1.39	34.15	nr	**97.46**
Sum of two sides 1500 mm	59.53	72.78	1.64	40.29	nr	**113.07**
Sum of two sides 1600 mm	67.27	82.24	1.64	40.29	nr	**122.53**
Sum of two sides 1700 mm	75.00	91.69	1.69	41.53	nr	**133.22**
Sum of two sides 1800 mm	82.84	101.27	1.69	41.53	nr	**142.80**
Sum of two sides 1900 mm	90.58	110.74	1.85	45.46	nr	**156.20**
Sum of two sides 2000 mm	98.32	120.20	1.85	45.46	nr	**165.66**
Grille neck						
Sum of two sides 1100 mm	46.30	56.60	1.44	35.38	nr	**91.98**
Sum of two sides 1200 mm	48.96	59.85	1.44	35.38	nr	**95.23**
Sum of two sides 1300 mm	51.62	63.11	1.69	41.53	nr	**104.64**
Sum of two sides 1400 mm	53.91	65.91	1.69	41.53	nr	**107.44**
Sum of two sides 1500 mm	62.58	76.51	1.79	43.98	nr	**120.49**
Sum of two sides 1600 mm	71.26	87.12	1.79	43.98	nr	**131.10**
Sum of two sides 1700 mm	79.93	97.72	1.86	45.71	nr	**143.43**
Sum of two sides 1800 mm	88.60	108.31	1.86	45.71	nr	**154.02**
Sum of two sides 1900 mm	97.27	118.91	2.02	49.64	nr	**168.55**
Sum of two sides 2000 mm	105.95	129.52	2.02	49.64	nr	**179.16**
Ductwork 1001 to 1250 mm longest side						
Sum of two sides 1300 mm	49.54	60.56	1.51	37.10	m	**97.66**
Sum of two sides 1400 mm	51.80	63.32	1.55	38.09	m	**101.41**
Sum of two sides 1500 mm	53.92	65.92	1.61	39.56	m	**105.48**

U: VENTILATION/AIR CONDITIONING SYSTEMS

Item	Net Price £	Material £	Labour hours	Labour £	Unit	Total rate £
U10: DUCTWORK: RECTANGULAR – CLASS C – cont						
Ductwork 1001 to 1250 mm longest side – cont						
Sum of two sides 1600 mm	56.18	68.68	1.62	39.81	m	**108.49**
Sum of two sides 1700 mm	58.31	71.29	1.74	42.76	m	**114.05**
Sum of two sides 1800 mm	60.43	73.87	1.76	43.25	m	**117.12**
Sum of two sides 1900 mm	62.55	76.47	1.81	44.48	m	**120.95**
Sum of two sides 2000 mm	64.82	79.24	1.82	44.72	m	**123.96**
Sum of two sides 2100 mm	66.94	81.83	2.53	62.16	m	**143.99**
Sum of two sides 2200 mm	76.98	94.11	2.55	62.67	m	**156.78**
Sum of two sides 2300 mm	79.39	97.06	2.56	62.91	m	**159.97**
Sum of two sides 2400 mm	81.51	99.64	2.76	67.83	m	**167.47**
Sum of two sides 2500 mm	83.78	102.42	2.77	68.07	m	**170.49**
Extra over fittings; Ductwork 1001 to 1250 mm longest side						
End Cap						
Sum of two sides 1300 mm	16.97	20.74	0.38	9.34	nr	**30.08**
Sum of two sides 1400 mm	17.71	21.65	0.38	9.34	nr	**30.99**
Sum of two sides 1500 mm	20.59	25.17	0.38	9.34	nr	**34.51**
Sum of two sides 1600 mm	23.47	28.70	0.38	9.34	nr	**38.04**
Sum of two sides 1700 mm	26.35	32.21	0.58	14.26	nr	**46.47**
Sum of two sides 1800 mm	29.23	35.73	0.58	14.26	nr	**49.99**
Sum of two sides 1900 mm	32.11	39.25	0.58	14.26	nr	**53.51**
Sum of two sides 2000 mm	34.99	42.78	0.58	14.26	nr	**57.04**
Sum of two sides 2100 mm	37.87	46.30	0.87	21.38	nr	**67.68**
Sum of two sides 2200 mm	40.75	49.82	0.87	21.38	nr	**71.20**
Sum of two sides 2300 mm	43.64	53.35	0.87	21.38	nr	**74.73**
Sum of two sides 2400 mm	46.51	56.86	0.87	21.38	nr	**78.24**
Sum of two sides 2500 mm	49.39	60.38	0.87	21.38	nr	**81.76**
Reducer						
Sum of two sides 1300 mm	24.19	29.57	2.30	56.52	nr	**86.09**
Sum of two sides 1400 mm	25.15	30.75	2.30	56.52	nr	**87.27**
Sum of two sides 1500 mm	30.36	37.11	2.47	60.70	nr	**97.81**
Sum of two sides 1600 mm	35.67	43.61	2.47	60.70	nr	**104.31**
Sum of two sides 1700 mm	40.90	50.00	2.59	63.64	nr	**113.64**
Sum of two sides 1800 mm	46.11	56.37	2.59	63.64	nr	**120.01**
Sum of two sides 1900 mm	51.32	62.74	2.71	66.59	nr	**129.33**
Sum of two sides 2000 mm	56.65	69.26	2.71	66.59	nr	**135.85**
Sum of two sides 2100 mm	61.87	75.63	2.92	71.75	nr	**147.38**
Sum of two sides 2200 mm	62.77	76.74	2.92	71.75	nr	**148.49**
Sum of two sides 2300 mm	68.09	83.24	2.92	71.75	nr	**154.99**
Sum of two sides 2400 mm	73.31	89.62	3.12	76.66	nr	**166.28**
Sum of two sides 2500 mm	78.62	96.11	3.12	76.66	nr	**172.77**

U: VENTILATION/AIR CONDITIONING SYSTEMS

Item	Net Price £	Material £	Labour hours	Labour £	Unit	Total rate £
Offset						
Sum of two sides 1300 mm	58.20	71.15	2.47	60.70	nr	**131.85**
Sum of two sides 1400 mm	59.94	73.27	2.47	60.70	nr	**133.97**
Sum of two sides 1500 mm	66.74	81.59	2.47	60.70	nr	**142.29**
Sum of two sides 1600 mm	73.46	89.81	2.47	60.70	nr	**150.51**
Sum of two sides 1700 mm	79.93	97.72	2.61	64.13	nr	**161.85**
Sum of two sides 1800 mm	86.21	105.39	2.61	64.13	nr	**169.52**
Sum of two sides 1900 mm	92.35	112.90	2.71	66.59	nr	**179.49**
Sum of two sides 2000 mm	98.35	120.23	2.71	66.59	nr	**186.82**
Sum of two sides 2100 mm	104.15	127.33	2.92	71.75	nr	**199.08**
Sum of two sides 2200 mm	102.76	125.63	3.26	80.11	nr	**205.74**
Sum of two sides 2300 mm	111.49	136.30	3.26	80.11	nr	**216.41**
Sum of two sides 2400 mm	120.20	146.95	3.47	85.27	nr	**232.22**
Sum of two sides 2500 mm	128.92	157.61	3.48	85.51	nr	**243.12**
Square to round						
Sum of two sides 1300 mm	26.83	32.80	2.32	57.01	nr	**89.81**
Sum of two sides 1400 mm	27.97	34.20	2.32	57.01	nr	**91.21**
Sum of two sides 1500 mm	34.45	42.12	2.56	62.91	nr	**105.03**
Sum of two sides 1600 mm	40.95	50.06	2.58	63.40	nr	**113.46**
Sum of two sides 1700 mm	47.42	57.97	2.84	69.78	nr	**127.75**
Sum of two sides 1800 mm	53.91	65.91	2.84	69.78	nr	**135.69**
Sum of two sides 1900 mm	67.29	82.27	3.13	76.91	nr	**159.18**
Sum of two sides 2000 mm	66.88	81.76	3.13	76.91	nr	**158.67**
Sum of two sides 2100 mm	73.36	89.68	4.26	104.68	nr	**194.36**
Sum of two sides 2200 mm	75.53	92.34	4.27	104.93	nr	**197.27**
Sum of two sides 2300 mm	82.03	100.28	4.30	105.66	nr	**205.94**
Sum of two sides 2400 mm	88.51	108.20	4.77	117.21	nr	**225.41**
Sum of two sides 2500 mm	95.00	116.14	4.79	117.70	nr	**233.84**
90° radius bend						
Sum of two sides 1300 mm	10.15	12.41	1.91	46.94	nr	**59.35**
Sum of two sides 1400 mm	12.80	15.65	1.91	46.94	nr	**62.59**
Sum of two sides 1500 mm	15.24	18.63	2.11	51.85	nr	**70.48**
Sum of two sides 1600 mm	20.70	25.31	2.11	51.85	nr	**77.16**
Sum of two sides 1700 mm	26.15	31.97	2.55	62.67	nr	**94.64**
Sum of two sides 1800 mm	31.60	38.64	2.55	62.67	nr	**101.31**
Sum of two sides 1900 mm	37.04	45.28	2.80	68.80	nr	**114.08**
Sum of two sides 2000 mm	42.51	51.97	2.80	68.80	nr	**120.77**
Sum of two sides 2100 mm	47.96	58.63	2.80	68.80	nr	**127.43**
Sum of two sides 2200 mm	45.06	55.08	2.80	68.80	nr	**123.88**
Sum of two sides 2300 mm	46.48	56.83	4.35	106.89	nr	**163.72**
Sum of two sides 2400 mm	51.90	63.45	4.35	106.89	nr	**170.34**
Sum of two sides 2500 mm	57.32	70.07	4.35	106.89	nr	**176.96**

U: VENTILATION/AIR CONDITIONING SYSTEMS

Item	Net Price £	Material £	Labour hours	Labour £	Unit	Total rate £
U10: DUCTWORK: RECTANGULAR – CLASS C – cont						
Extra over fittings; Ductwork 1001 to 1250 mm longest side – cont						
45° bend						
Sum of two sides 1300 mm	24.76	30.27	2.26	55.54	nr	85.81
Sum of two sides 1400 mm	25.33	30.96	2.44	59.96	nr	90.92
Sum of two sides 1500 mm	30.97	37.86	2.68	65.86	nr	103.72
Sum of two sides 1600 mm	36.70	44.87	2.69	66.11	nr	110.98
Sum of two sides 1700 mm	42.34	51.76	2.96	72.74	nr	124.50
Sum of two sides 1800 mm	47.99	58.67	2.96	72.74	nr	131.41
Sum of two sides 1900 mm	53.62	65.55	3.26	80.11	nr	145.66
Sum of two sides 2000 mm	59.35	72.55	3.26	80.11	nr	152.66
Sum of two sides 2100 mm	64.99	79.45	7.50	184.30	nr	263.75
Sum of two sides 2200 mm	66.09	80.79	7.50	184.30	nr	265.09
Sum of two sides 2300 mm	69.76	85.28	7.55	185.51	nr	270.79
Sum of two sides 2400 mm	75.38	92.15	8.13	199.77	nr	291.92
Sum of two sides 2500 mm	81.09	99.14	8.30	203.95	nr	303.09
90° mitre bend						
Sum of two sides 1300 mm	57.22	69.95	3.85	94.61	nr	164.56
Sum of two sides 1400 mm	59.48	72.72	3.85	94.61	nr	167.33
Sum of two sides 1500 mm	70.27	85.90	4.25	104.42	nr	190.32
Sum of two sides 1600 mm	81.07	99.11	4.26	104.68	nr	203.79
Sum of two sides 1700 mm	99.19	121.26	4.68	115.00	nr	236.26
Sum of two sides 1800 mm	109.91	134.36	4.68	115.00	nr	249.36
Sum of two sides 1900 mm	120.64	147.49	4.87	119.68	nr	267.17
Sum of two sides 2000 mm	124.26	151.90	4.87	119.68	nr	271.58
Sum of two sides 2100 mm	135.06	165.11	7.50	184.30	nr	349.41
Sum of two sides 2200 mm	136.16	166.46	7.50	184.30	nr	350.76
Sum of two sides 2300 mm	142.15	173.78	7.55	185.51	nr	359.29
Sum of two sides 2400 mm	152.99	187.03	8.13	199.77	nr	386.80
Sum of two sides 2500 mm	163.83	200.28	8.30	203.95	nr	404.23
Branch						
Sum of two sides 1300 mm	50.92	62.25	1.39	34.15	nr	96.40
Sum of two sides 1400 mm	53.05	64.86	1.39	34.15	nr	99.01
Sum of two sides 1500 mm	60.89	74.44	1.64	40.29	nr	114.73
Sum of two sides 1600 mm	68.82	84.13	1.64	40.29	nr	124.42
Sum of two sides 1700 mm	76.66	93.72	1.69	41.53	nr	135.25
Sum of two sides 1800 mm	84.50	103.30	1.69	41.53	nr	144.83
Sum of two sides 1900 mm	92.33	112.88	1.85	45.46	nr	158.34
Sum of two sides 2000 mm	100.26	122.57	1.85	45.46	nr	168.03
Sum of two sides 2100 mm	108.10	132.15	2.61	64.13	nr	196.28
Sum of two sides 2200 mm	115.94	141.74	2.61	64.13	nr	205.87
Sum of two sides 2300 mm	123.87	151.43	2.61	64.13	nr	215.56
Sum of two sides 2400 mm	131.71	161.02	2.88	70.77	nr	231.79
Sum of two sides 2500 mm	139.64	170.71	2.88	70.77	nr	241.48

U: VENTILATION/AIR CONDITIONING SYSTEMS

Item	Net Price £	Material £	Labour hours	Labour £	Unit	Total rate £
Grille neck						
Sum of two sides 1300 mm	53.13	64.95	1.69	41.53	nr	**106.48**
Sum of two sides 1400 mm	55.53	67.89	1.69	41.53	nr	**109.42**
Sum of two sides 1500 mm	64.34	78.66	1.79	43.98	nr	**122.64**
Sum of two sides 1600 mm	73.15	89.42	1.79	43.98	nr	**133.40**
Sum of two sides 1700 mm	81.97	100.21	1.86	45.71	nr	**145.92**
Sum of two sides 1800 mm	90.78	110.98	1.86	45.71	nr	**156.69**
Sum of two sides 1900 mm	99.60	121.77	2.02	49.64	nr	**171.41**
Sum of two sides 2000 mm	108.41	132.53	2.02	49.64	nr	**182.17**
Sum of two sides 2100 mm	117.23	143.31	2.61	64.13	nr	**207.44**
Sum of two sides 2200 mm	126.03	154.08	2.80	68.80	nr	**222.88**
Sum of two sides 2300 mm	134.86	164.87	2.80	68.80	nr	**233.67**
Sum of two sides 2400 mm	143.66	175.63	3.06	75.20	nr	**250.83**
Sum of two sides 2500 mm	152.48	186.41	3.06	75.20	nr	**261.61**
Ductwork 1251 to 1600 mm longest side						
Sum of two sides 1700 mm	62.74	76.70	1.74	42.76	m	**119.46**
Sum of two sides 1800 mm	65.09	79.58	1.76	43.25	m	**122.83**
Sum of two sides 1900 mm	67.48	82.49	1.81	44.48	m	**126.97**
Sum of two sides 2000 mm	69.71	85.22	1.82	44.72	m	**129.94**
Sum of two sides 2100 mm	71.95	87.96	2.53	62.16	m	**150.12**
Sum of two sides 2200 mm	74.18	90.68	2.55	62.67	m	**153.35**
Sum of two sides 2300 mm	76.57	93.61	2.56	62.91	m	**156.52**
Sum of two sides 2400 mm	78.92	96.48	2.76	67.83	m	**164.31**
Sum of two sides 2500 mm	81.15	99.21	2.77	68.07	m	**167.28**
Sum of two sides 2600 mm	91.47	111.83	2.97	72.99	m	**184.82**
Sum of two sides 2700 mm	93.72	114.58	2.99	73.47	m	**188.05**
Sum of two sides 2800 mm	96.10	117.48	3.30	81.09	m	**198.57**
Sum of two sides 2900 mm	98.48	120.40	3.31	81.33	m	**201.73**
Sum of two sides 3000 mm	100.72	123.13	3.53	86.74	m	**209.87**
Sum of two sides 3100 mm	103.07	126.00	3.55	87.24	m	**213.24**
Sum of two sides 3200 mm	105.31	128.74	3.56	87.48	m	**216.22**
Extra over fittings; Ductwork 1251 to 1600 mm longest side						
End Cap						
Sum of two sides 1700 mm	26.07	31.87	0.58	14.26	nr	**46.13**
Sum of two sides 1800 mm	28.92	35.36	0.58	14.26	nr	**49.62**
Sum of two sides 1900 mm	31.78	38.85	0.58	14.26	nr	**53.11**
Sum of two sides 2000 mm	34.62	42.32	0.58	14.26	nr	**56.58**
Sum of two sides 2100 mm	37.47	45.80	0.87	21.38	nr	**67.18**
Sum of two sides 2200 mm	40.32	49.30	0.87	21.38	nr	**70.68**
Sum of two sides 2300 mm	43.18	52.79	0.87	21.38	nr	**74.17**
Sum of two sides 2400 mm	46.02	56.26	0.87	21.38	nr	**77.64**
Sum of two sides 2500 mm	48.87	59.74	0.87	21.38	nr	**81.12**
Sum of two sides 2600 mm	51.72	63.23	0.87	21.38	nr	**84.61**
Sum of two sides 2700 mm	54.58	66.72	0.87	21.38	nr	**88.10**
Sum of two sides 2800 mm	57.42	70.19	1.16	28.51	nr	**98.70**

U: VENTILATION/AIR CONDITIONING SYSTEMS

Item	Net Price £	Material £	Labour hours	Labour £	Unit	Total rate £
U10: DUCTWORK: RECTANGULAR – CLASS C – cont						
Extra over fittings; Ductwork 1251 to 1600 mm longest side – cont						
End Cap – cont						
Sum of two sides 2900 mm	60.27	73.68	1.16	28.51	nr	**102.19**
Sum of two sides 3000 mm	63.94	78.17	1.73	42.51	nr	**120.68**
Sum of two sides 3100 mm	65.38	79.93	1.73	42.51	nr	**122.44**
Sum of two sides 3200 mm	67.42	82.42	1.73	42.51	nr	**124.93**
Reducer						
Sum of two sides 1700 mm	35.59	43.51	2.59	63.64	nr	**107.15**
Sum of two sides 1800 mm	40.50	49.51	2.59	63.64	nr	**113.15**
Sum of two sides 1900 mm	45.47	55.59	2.71	66.59	nr	**122.18**
Sum of two sides 2000 mm	54.62	66.77	2.71	66.59	nr	**133.36**
Sum of two sides 2100 mm	55.28	67.58	2.92	71.75	nr	**139.33**
Sum of two sides 2200 mm	60.18	73.57	2.92	71.75	nr	**145.32**
Sum of two sides 2300 mm	65.16	79.66	2.92	71.75	nr	**151.41**
Sum of two sides 2400 mm	70.06	85.64	3.12	76.66	nr	**162.30**
Sum of two sides 2500 mm	74.96	91.64	3.12	76.66	nr	**168.30**
Sum of two sides 2600 mm	75.63	92.46	3.16	77.65	nr	**170.11**
Sum of two sides 2700 mm	78.98	96.55	3.16	77.65	nr	**174.20**
Sum of two sides 2800 mm	83.87	102.54	4.00	98.29	nr	**200.83**
Sum of two sides 2900 mm	93.93	114.83	4.01	98.53	nr	**213.36**
Sum of two sides 3000 mm	98.82	120.81	4.56	112.05	nr	**232.86**
Sum of two sides 3100 mm	102.45	125.25	4.56	112.05	nr	**237.30**
Sum of two sides 3200 mm	105.74	129.26	4.56	112.05	nr	**241.31**
Offset						
Sum of two sides 1700 mm	87.57	107.06	2.61	64.13	nr	**171.19**
Sum of two sides 1800 mm	96.76	118.29	2.61	64.13	nr	**182.42**
Sum of two sides 1900 mm	102.91	125.80	2.71	66.59	nr	**192.39**
Sum of two sides 2000 mm	108.78	132.98	2.71	66.59	nr	**199.57**
Sum of two sides 2100 mm	114.48	139.96	2.92	71.75	nr	**211.71**
Sum of two sides 2200 mm	120.01	146.71	3.26	80.11	nr	**226.82**
Sum of two sides 2300 mm	125.38	153.27	3.26	80.11	nr	**233.38**
Sum of two sides 2400 mm	134.14	163.99	3.47	85.27	nr	**249.26**
Sum of two sides 2500 mm	139.20	170.18	3.48	85.51	nr	**255.69**
Sum of two sides 2600 mm	139.24	170.22	3.49	85.76	nr	**255.98**
Sum of two sides 2700 mm	147.89	180.80	3.50	86.00	nr	**266.80**
Sum of two sides 2800 mm	156.53	191.36	4.33	106.40	nr	**297.76**
Sum of two sides 2900 mm	170.30	208.19	4.74	116.47	nr	**324.66**
Sum of two sides 3000 mm	178.95	218.77	5.31	130.48	nr	**349.25**
Sum of two sides 3100 mm	185.70	227.02	5.34	131.21	nr	**358.23**
Sum of two sides 3200 mm	191.94	234.64	5.35	131.46	nr	**366.10**
Square to round						
Sum of two sides 1700 mm	39.65	48.47	2.84	69.78	nr	**118.25**
Sum of two sides 1800 mm	45.82	56.01	2.84	69.78	nr	**125.79**
Sum of two sides 1900 mm	51.94	63.50	3.13	76.91	nr	**140.41**

U: VENTILATION/AIR CONDITIONING SYSTEMS

Item	Net Price £	Material £	Labour hours	Labour £	Unit	Total rate £
Sum of two sides 2000 mm	58.08	71.01	3.13	76.91	nr	**147.92**
Sum of two sides 2100 mm	64.22	78.51	4.26	104.68	nr	**183.19**
Sum of two sides 2200 mm	70.36	86.02	4.27	104.93	nr	**190.95**
Sum of two sides 2300 mm	76.50	93.52	4.30	105.66	nr	**199.18**
Sum of two sides 2400 mm	82.65	101.04	4.77	117.21	nr	**218.25**
Sum of two sides 2500 mm	88.79	108.54	4.79	117.70	nr	**226.24**
Sum of two sides 2600 mm	89.14	108.97	4.95	121.64	nr	**230.61**
Sum of two sides 2700 mm	95.29	116.49	4.95	121.64	nr	**238.13**
Sum of two sides 2800 mm	101.41	123.97	8.49	208.62	nr	**332.59**
Sum of two sides 2900 mm	110.09	134.59	8.88	218.21	nr	**352.80**
Sum of two sides 3000 mm	116.22	142.08	9.02	221.64	nr	**363.72**
Sum of two sides 3100 mm	120.79	147.67	9.02	221.64	nr	**369.31**
Sum of two sides 3200 mm	124.92	152.72	9.09	223.37	nr	**376.09**
90° radius bend						
Sum of two sides 1700 mm	86.80	106.11	2.55	62.67	nr	**168.78**
Sum of two sides 1800 mm	94.13	115.07	2.55	62.67	nr	**177.74**
Sum of two sides 1900 mm	105.33	128.77	2.80	68.80	nr	**197.57**
Sum of two sides 2000 mm	116.34	142.22	2.80	68.80	nr	**211.02**
Sum of two sides 2100 mm	127.37	155.72	2.61	64.13	nr	**219.85**
Sum of two sides 2200 mm	138.37	169.16	2.62	64.38	nr	**233.54**
Sum of two sides 2300 mm	149.57	182.85	2.63	64.62	nr	**247.47**
Sum of two sides 2400 mm	156.52	191.34	4.34	106.64	nr	**297.98**
Sum of two sides 2500 mm	167.50	204.77	4.35	106.89	nr	**311.66**
Sum of two sides 2600 mm	168.10	205.51	4.53	111.31	nr	**316.82**
Sum of two sides 2700 mm	179.07	218.92	4.53	111.31	nr	**330.23**
Sum of two sides 2800 mm	185.11	226.30	7.13	175.19	nr	**401.49**
Sum of two sides 2900 mm	206.38	252.30	7.17	176.19	nr	**428.49**
Sum of two sides 3000 mm	217.32	265.68	7.26	178.40	nr	**444.08**
Sum of two sides 3100 mm	225.72	275.95	7.26	178.40	nr	**454.35**
Sum of two sides 3200 mm	233.44	285.38	7.31	179.62	nr	**465.00**
45° bend						
Sum of two sides 1700 mm	41.69	50.96	2.96	72.74	nr	**123.70**
Sum of two sides 1800 mm	45.31	55.39	2.96	72.74	nr	**128.13**
Sum of two sides 1900 mm	50.89	62.21	3.26	80.11	nr	**142.32**
Sum of two sides 2000 mm	56.37	68.92	3.26	80.11	nr	**149.03**
Sum of two sides 2100 mm	61.87	75.63	7.50	184.30	nr	**259.93**
Sum of two sides 2200 mm	67.37	82.36	7.50	184.30	nr	**266.66**
Sum of two sides 2300 mm	72.94	89.17	7.55	185.51	nr	**274.68**
Sum of two sides 2400 mm	76.36	93.35	8.13	199.77	nr	**293.12**
Sum of two sides 2500 mm	81.84	100.05	8.30	203.95	nr	**304.00**
Sum of two sides 2600 mm	81.76	99.95	8.56	210.35	nr	**310.30**
Sum of two sides 2700 mm	87.24	106.66	8.62	211.81	nr	**318.47**
Sum of two sides 2800 mm	90.20	110.27	9.09	223.37	nr	**333.64**
Sum of two sides 2900 mm	100.82	123.25	9.09	223.37	nr	**346.62**
Sum of two sides 3000 mm	106.27	129.91	9.62	236.39	nr	**366.30**
Sum of two sides 3100 mm	110.45	135.02	9.62	236.39	nr	**371.41**
Sum of two sides 3200 mm	114.30	139.73	9.62	236.39	nr	**376.12**

U: VENTILATION/AIR CONDITIONING SYSTEMS

Item	Net Price £	Material £	Labour hours	Labour £	Unit	Total rate £
U10: DUCTWORK: RECTANGULAR – CLASS C – cont						
Extra over fittings; Ductwork 1251 to 1600 mm longest side – cont						
90° mitre bend						
Sum of two sides 1700 mm	91.62	112.00	4.68	115.00	nr	**227.00**
Sum of two sides 1800 mm	97.39	119.06	4.68	115.00	nr	**234.06**
Sum of two sides 1900 mm	108.24	132.32	4.87	119.68	nr	**252.00**
Sum of two sides 2000 mm	119.11	145.61	4.87	119.68	nr	**265.29**
Sum of two sides 2100 mm	129.99	158.92	7.50	184.30	nr	**343.22**
Sum of two sides 2200 mm	140.87	172.22	7.50	184.30	nr	**356.52**
Sum of two sides 2300 mm	151.72	185.48	7.55	185.51	nr	**370.99**
Sum of two sides 2400 mm	157.43	192.46	8.13	199.77	nr	**392.23**
Sum of two sides 2500 mm	168.34	205.79	8.30	203.95	nr	**409.74**
Sum of two sides 2600 mm	166.96	204.10	8.56	210.35	nr	**414.45**
Sum of two sides 2700 mm	177.88	217.46	8.62	211.81	nr	**429.27**
Sum of two sides 2800 mm	182.70	223.36	15.20	373.50	nr	**596.86**
Sum of two sides 2900 mm	201.77	246.66	15.20	373.50	nr	**620.16**
Sum of two sides 3000 mm	212.71	260.04	15.60	383.32	nr	**643.36**
Sum of two sides 3100 mm	221.69	271.01	16.04	394.14	nr	**665.15**
Sum of two sides 3200 mm	230.23	281.46	16.04	394.14	nr	**675.60**
Branch						
Sum of two sides 1700 mm	78.25	95.66	1.69	41.53	nr	**137.19**
Sum of two sides 1800 mm	86.01	105.15	1.69	41.53	nr	**146.68**
Sum of two sides 1900 mm	93.87	114.75	1.85	45.46	nr	**160.21**
Sum of two sides 2000 mm	101.63	124.24	1.85	45.46	nr	**169.70**
Sum of two sides 2100 mm	109.40	133.75	2.61	64.13	nr	**197.88**
Sum of two sides 2200 mm	117.16	143.23	2.61	64.13	nr	**207.36**
Sum of two sides 2300 mm	125.02	152.84	2.61	64.13	nr	**216.97**
Sum of two sides 2400 mm	132.78	162.33	2.88	70.77	nr	**233.10**
Sum of two sides 2500 mm	140.55	171.82	2.88	70.77	nr	**242.59**
Sum of two sides 2600 mm	148.32	181.32	2.88	70.77	nr	**252.09**
Sum of two sides 2700 mm	156.09	190.82	2.88	70.77	nr	**261.59**
Sum of two sides 2800 mm	163.83	200.28	3.94	96.81	nr	**297.09**
Sum of two sides 2900 mm	174.24	213.01	3.94	96.81	nr	**309.82**
Sum of two sides 3000 mm	182.01	222.51	4.83	118.68	nr	**341.19**
Sum of two sides 3100 mm	188.19	230.06	4.83	118.68	nr	**348.74**
Sum of two sides 3200 mm	193.81	236.93	4.83	118.68	nr	**355.61**
Grille neck						
Sum of two sides 1700 mm	81.10	99.15	1.86	45.71	nr	**144.86**
Sum of two sides 1800 mm	89.82	109.81	1.86	45.71	nr	**155.52**
Sum of two sides 1900 mm	98.56	120.49	2.02	49.64	nr	**170.13**
Sum of two sides 2000 mm	107.27	131.14	2.02	49.64	nr	**180.78**
Sum of two sides 2100 mm	116.00	141.81	2.80	68.80	nr	**210.61**
Sum of two sides 2200 mm	124.71	152.46	2.80	68.80	nr	**221.26**
Sum of two sides 2300 mm	133.44	163.13	2.80	68.80	nr	**231.93**

U: VENTILATION/AIR CONDITIONING SYSTEMS

Item	Net Price £	Material £	Labour hours	Labour £	Unit	Total rate £
Sum of two sides 2400 mm	142.15	173.78	3.06	75.20	nr	248.98
Sum of two sides 2500 mm	150.88	184.45	3.06	75.20	nr	259.65
Sum of two sides 2600 mm	159.60	195.11	3.08	75.68	nr	270.79
Sum of two sides 2700 mm	168.32	205.77	3.08	75.68	nr	281.45
Sum of two sides 2800 mm	177.04	216.43	4.13	101.48	nr	317.91
Sum of two sides 2900 mm	185.76	227.09	4.13	101.48	nr	328.57
Sum of two sides 3000 mm	194.48	237.75	5.02	123.35	nr	361.10
Sum of two sides 3100 mm	201.42	246.24	5.02	123.35	nr	369.59
Sum of two sides 3200 mm	207.73	253.95	5.02	123.35	nr	377.30
Ductwork 1601 to 2000 mm longest side						
Sum of two sides 2100 mm	83.58	102.18	2.53	62.16	m	164.34
Sum of two sides 2200 mm	86.18	105.36	2.55	62.67	m	168.03
Sum of two sides 2300 mm	88.79	108.54	2.55	62.67	m	171.21
Sum of two sides 2400 mm	91.36	111.69	2.56	62.91	m	174.60
Sum of two sides 2500 mm	93.98	114.89	2.76	67.83	m	182.72
Sum of two sides 2600 mm	96.44	117.89	2.77	68.07	m	185.96
Sum of two sides 2700 mm	98.27	120.14	2.97	72.99	m	193.13
Sum of two sides 2800 mm	101.39	123.95	2.99	73.47	m	197.42
Sum of two sides 2900 mm	104.00	127.14	3.30	81.09	m	208.23
Sum of two sides 3000 mm	106.47	130.16	3.31	81.33	m	211.49
Sum of two sides 3100 mm	123.43	150.90	3.53	86.74	m	237.64
Sum of two sides 3200 mm	125.91	153.92	3.53	86.74	m	240.66
Sum of two sides 3300 mm	128.52	157.12	3.53	86.74	m	243.86
Sum of two sides 3400 mm	130.99	160.13	3.55	87.24	m	247.37
Sum of two sides 3500 mm	133.46	163.15	3.55	87.24	m	250.39
Sum of two sides 3600 mm	135.93	166.18	3.55	87.24	m	253.42
Sum of two sides 3700 mm	138.55	169.37	3.56	87.48	m	256.85
Sum of two sides 3800 mm	141.02	172.40	3.56	87.48	m	259.88
Sum of two sides 3900 mm	143.50	175.43	3.56	87.48	m	262.91
Sum of two sides 4000 mm	145.96	178.44	3.56	87.48	m	265.92
Extra over fittings; Ductwork 1601 to 2000 mm longest side						
End Cap						
Sum of two sides 2100 mm	37.89	46.32	0.87	21.38	nr	67.70
Sum of two sides 2200 mm	40.76	49.83	0.87	21.38	nr	71.21
Sum of two sides 2300 mm	43.62	53.32	0.87	21.38	nr	74.70
Sum of two sides 2400 mm	46.50	56.85	0.87	21.38	nr	78.23
Sum of two sides 2500 mm	49.36	60.35	0.87	21.38	nr	81.73
Sum of two sides 2600 mm	52.23	63.85	0.87	21.38	nr	85.23
Sum of two sides 2700 mm	55.11	67.37	0.87	21.38	nr	88.75
Sum of two sides 2800 mm	57.98	70.88	1.16	28.51	nr	99.39
Sum of two sides 2900 mm	60.84	74.38	1.16	28.51	nr	102.89
Sum of two sides 3000 mm	63.71	77.89	1.73	42.51	nr	120.40
Sum of two sides 3100 mm	65.99	80.67	1.73	42.51	nr	123.18
Sum of two sides 3200 mm	68.05	83.19	1.73	42.51	nr	125.70
Sum of two sides 3300 mm	70.11	85.71	1.73	42.51	nr	128.22
Sum of two sides 3400 mm	72.17	88.23	1.73	42.51	nr	130.74
Sum of two sides 3500 mm	74.24	90.76	1.73	42.51	nr	133.27

U: VENTILATION/AIR CONDITIONING SYSTEMS

Item	Net Price £	Material £	Labour hours	Labour £	Unit	Total rate £
U10: DUCTWORK: RECTANGULAR – CLASS C – cont						
Extra over fittings; Ductwork 1601 to 2000 mm longest side – cont						
End Cap – cont						
Sum of two sides 3600 mm	76.30	93.28	1.73	42.51	nr	**135.79**
Sum of two sides 3700 mm	78.37	95.81	1.73	42.51	nr	**138.32**
Sum of two sides 3800 mm	80.43	98.32	1.73	42.51	nr	**140.83**
Sum of two sides 3900 mm	82.51	100.87	1.73	42.51	nr	**143.38**
Sum of two sides 4000 mm	84.57	103.39	1.73	42.51	nr	**145.90**
Reducer						
Sum of two sides 2100 mm	51.61	63.10	2.92	71.75	nr	**134.85**
Sum of two sides 2200 mm	56.51	69.08	2.92	71.75	nr	**140.83**
Sum of two sides 2300 mm	61.47	75.15	2.92	71.75	nr	**146.90**
Sum of two sides 2400 mm	66.29	81.04	3.12	76.66	nr	**157.70**
Sum of two sides 2500 mm	71.27	87.13	3.12	76.66	nr	**163.79**
Sum of two sides 2600 mm	76.15	93.09	3.16	77.65	nr	**170.74**
Sum of two sides 2700 mm	81.04	99.08	3.16	77.65	nr	**176.73**
Sum of two sides 2800 mm	85.94	105.06	4.00	98.29	nr	**203.35**
Sum of two sides 2900 mm	90.90	111.13	4.01	98.53	nr	**209.66**
Sum of two sides 3000 mm	95.79	117.10	4.01	98.53	nr	**215.63**
Sum of two sides 3100 mm	96.13	117.52	4.01	98.53	nr	**216.05**
Sum of two sides 3200 mm	98.41	120.30	4.56	112.05	nr	**232.35**
Sum of two sides 3300 mm	100.95	123.41	4.56	112.05	nr	**235.46**
Sum of two sides 3400 mm	104.23	127.42	4.56	112.05	nr	**239.47**
Sum of two sides 3500 mm	107.51	131.43	4.56	112.05	nr	**243.48**
Sum of two sides 3600 mm	110.79	135.45	4.56	112.05	nr	**247.50**
Sum of two sides 3700 mm	114.15	139.54	4.56	112.05	nr	**251.59**
Sum of two sides 3800 mm	117.43	143.56	4.56	112.05	nr	**255.61**
Sum of two sides 3900 mm	120.71	147.57	4.56	112.05	nr	**259.62**
Sum of two sides 4000 mm	124.00	151.60	4.56	112.05	nr	**263.65**
Offset						
Sum of two sides 2100 mm	121.22	148.20	2.92	71.75	nr	**219.95**
Sum of two sides 2200 mm	130.40	159.41	3.26	80.11	nr	**239.52**
Sum of two sides 2300 mm	135.71	165.90	3.26	80.11	nr	**246.01**
Sum of two sides 2400 mm	148.77	181.88	3.47	85.27	nr	**267.15**
Sum of two sides 2500 mm	153.90	188.14	3.48	85.51	nr	**273.65**
Sum of two sides 2600 mm	158.77	194.09	3.49	85.76	nr	**279.85**
Sum of two sides 2700 mm	163.44	199.81	3.50	86.00	nr	**285.81**
Sum of two sides 2800 mm	167.94	205.31	4.33	106.40	nr	**311.71**
Sum of two sides 2900 mm	172.30	210.64	4.74	116.47	nr	**327.11**
Sum of two sides 3000 mm	167.83	205.18	5.34	131.21	nr	**336.39**
Sum of two sides 3100 mm	174.06	212.79	5.35	131.46	nr	**344.25**
Sum of two sides 3200 mm	176.43	215.68	5.31	130.48	nr	**346.16**
Sum of two sides 3300 mm	185.49	226.76	5.35	131.46	nr	**358.22**
Sum of two sides 3400 mm	191.71	234.37	5.35	131.46	nr	**365.83**
Sum of two sides 3500 mm	197.94	241.98	5.35	131.46	nr	**373.44**

U: VENTILATION/AIR CONDITIONING SYSTEMS

Item	Net Price £	Material £	Labour hours	Labour £	Unit	Total rate £
Sum of two sides 3600 mm	204.17	249.60	5.35	131.46	nr	**381.06**
Sum of two sides 3700 mm	210.42	257.24	5.35	131.46	nr	**388.70**
Sum of two sides 3800 mm	216.64	264.84	5.35	131.46	nr	**396.30**
Sum of two sides 3900 mm	222.87	272.46	5.35	131.46	nr	**403.92**
Sum of two sides 4000 mm	229.10	280.08	5.35	131.46	nr	**411.54**
Square to round						
Sum of two sides 2100 mm	77.87	95.19	4.26	104.68	nr	**199.87**
Sum of two sides 2200 mm	85.39	104.39	4.27	104.93	nr	**209.32**
Sum of two sides 2300 mm	92.89	113.56	4.30	105.66	nr	**219.22**
Sum of two sides 2400 mm	158.71	194.02	4.77	117.21	nr	**311.23**
Sum of two sides 2500 mm	107.94	131.95	4.79	117.70	nr	**249.65**
Sum of two sides 2600 mm	115.45	141.14	4.95	121.64	nr	**262.78**
Sum of two sides 2700 mm	122.96	150.32	4.95	121.64	nr	**271.96**
Sum of two sides 2800 mm	130.47	159.50	8.49	208.62	nr	**368.12**
Sum of two sides 2900 mm	137.98	168.68	8.88	218.21	nr	**386.89**
Sum of two sides 3000 mm	140.85	172.19	9.02	221.64	nr	**393.83**
Sum of two sides 3100 mm	145.49	177.86	9.02	221.64	nr	**399.50**
Sum of two sides 3200 mm	145.95	178.43	9.09	223.37	nr	**401.80**
Sum of two sides 3300 mm	153.62	187.80	9.09	223.37	nr	**411.17**
Sum of two sides 3400 mm	158.71	194.02	9.09	223.37	nr	**417.39**
Sum of two sides 3500 mm	163.81	200.26	9.09	223.37	nr	**423.63**
Sum of two sides 3600 mm	168.91	206.49	9.09	223.37	nr	**429.86**
Sum of two sides 3700 mm	174.00	212.72	9.09	223.37	nr	**436.09**
Sum of two sides 3800 mm	179.09	218.94	9.09	223.37	nr	**442.31**
Sum of two sides 3900 mm	184.19	225.18	9.09	223.37	nr	**448.55**
Sum of two sides 4000 mm	189.29	231.41	9.09	223.37	nr	**454.78**
90° radius bend						
Sum of two sides 2100 mm	110.66	135.28	2.61	64.13	nr	**199.41**
Sum of two sides 2200 mm	212.77	260.12	2.62	64.38	nr	**324.50**
Sum of two sides 2300 mm	230.44	281.72	2.63	64.62	nr	**346.34**
Sum of two sides 2400 mm	236.32	288.90	4.34	106.64	nr	**395.54**
Sum of two sides 2500 mm	253.95	310.45	4.35	106.89	nr	**417.34**
Sum of two sides 2600 mm	271.29	331.65	4.53	111.31	nr	**442.96**
Sum of two sides 2700 mm	288.62	352.84	4.53	111.31	nr	**464.15**
Sum of two sides 2800 mm	305.96	374.03	7.13	175.19	nr	**549.22**
Sum of two sides 2900 mm	323.60	395.60	7.17	176.19	nr	**571.79**
Sum of two sides 3000 mm	340.93	416.79	7.26	178.40	nr	**595.19**
Sum of two sides 3100 mm	331.02	404.68	7.26	178.40	nr	**583.08**
Sum of two sides 3200 mm	343.52	419.95	7.31	179.62	nr	**599.57**
Sum of two sides 3300 mm	371.87	454.61	7.31	179.62	nr	**634.23**
Sum of two sides 3400 mm	384.38	469.91	7.31	179.62	nr	**649.53**
Sum of two sides 3500 mm	396.88	485.19	7.31	179.62	nr	**664.81**
Sum of two sides 3600 mm	409.40	500.50	7.31	179.62	nr	**680.12**
Sum of two sides 3700 mm	422.20	516.14	7.31	179.62	nr	**695.76**
Sum of two sides 3800 mm	434.71	531.44	7.31	179.62	nr	**711.06**
Sum of two sides 3900 mm	447.22	546.73	7.31	179.62	nr	**726.35**
Sum of two sides 4000 mm	459.72	562.01	7.31	179.62	nr	**741.63**

U: VENTILATION/AIR CONDITIONING SYSTEMS

Item	Net Price £	Material £	Labour hours	Labour £	Unit	Total rate £
U10: DUCTWORK: RECTANGULAR – CLASS C – cont						
Extra over fittings; Ductwork 1601 to 2000 mm longest side – cont						
45° bend						
Sum of two sides 2100 mm	143.25	175.13	7.50	184.30	nr	**359.43**
Sum of two sides 2200 mm	152.17	186.03	7.50	184.30	nr	**370.33**
Sum of two sides 2300 mm	164.02	200.52	7.55	185.51	nr	**386.03**
Sum of two sides 2400 mm	169.62	207.36	8.13	199.77	nr	**407.13**
Sum of two sides 2500 mm	181.45	221.82	8.30	203.95	nr	**425.77**
Sum of two sides 2600 mm	193.04	235.99	8.56	210.35	nr	**446.34**
Sum of two sides 2700 mm	204.63	250.17	8.62	211.81	nr	**461.98**
Sum of two sides 2800 mm	216.23	264.34	9.09	223.37	nr	**487.71**
Sum of two sides 2900 mm	228.06	278.80	9.09	223.37	nr	**502.17**
Sum of two sides 3000 mm	236.38	288.98	9.62	236.39	nr	**525.37**
Sum of two sides 3100 mm	239.66	292.98	9.62	236.39	nr	**529.37**
Sum of two sides 3200 mm	244.76	299.23	9.62	236.39	nr	**535.62**
Sum of two sides 3300 mm	263.73	322.41	9.62	236.39	nr	**558.80**
Sum of two sides 3400 mm	272.10	332.65	9.62	236.39	nr	**569.04**
Sum of two sides 3500 mm	280.48	342.89	9.62	236.39	nr	**579.28**
Sum of two sides 3600 mm	288.87	353.15	9.62	236.39	nr	**589.54**
Sum of two sides 3700 mm	297.47	363.66	9.62	236.39	nr	**600.05**
Sum of two sides 3800 mm	305.85	373.90	9.62	236.39	nr	**610.29**
Sum of two sides 3900 mm	314.22	384.14	9.62	236.39	nr	**620.53**
Sum of two sides 4000 mm	322.60	394.38	9.62	236.39	nr	**630.77**
90° mitre bend						
Sum of two sides 2100 mm	247.97	303.15	7.50	184.30	nr	**487.45**
Sum of two sides 2200 mm	262.10	320.42	7.50	184.30	nr	**504.72**
Sum of two sides 2300 mm	282.92	345.87	7.55	185.51	nr	**531.38**
Sum of two sides 2400 mm	289.67	354.12	8.13	199.77	nr	**553.89**
Sum of two sides 2500 mm	310.52	379.62	8.30	203.95	nr	**583.57**
Sum of two sides 2600 mm	331.48	405.23	8.56	210.35	nr	**615.58**
Sum of two sides 2700 mm	352.45	430.87	8.62	211.81	nr	**642.68**
Sum of two sides 2800 mm	373.41	456.50	15.20	373.50	nr	**830.00**
Sum of two sides 2900 mm	394.27	482.00	15.20	373.50	nr	**855.50**
Sum of two sides 3000 mm	404.12	494.04	16.04	394.14	nr	**888.18**
Sum of two sides 3100 mm	415.23	507.63	15.60	383.32	nr	**890.95**
Sum of two sides 3200 mm	420.26	513.77	16.04	394.14	nr	**907.91**
Sum of two sides 3300 mm	445.34	544.43	16.04	394.14	nr	**938.57**
Sum of two sides 3400 mm	461.47	564.15	16.04	394.14	nr	**958.29**
Sum of two sides 3500 mm	477.61	583.89	16.04	394.14	nr	**978.03**
Sum of two sides 3600 mm	493.74	603.60	16.04	394.14	nr	**997.74**
Sum of two sides 3700 mm	509.78	623.21	16.04	394.14	nr	**1017.35**
Sum of two sides 3800 mm	525.91	642.93	16.04	394.14	nr	**1037.07**
Sum of two sides 3900 mm	542.05	662.66	16.04	394.14	nr	**1056.80**
Sum of two sides 4000 mm	558.19	682.40	16.04	394.14	nr	**1076.54**

U: VENTILATION/AIR CONDITIONING SYSTEMS

Item	Net Price £	Material £	Labour hours	Labour £	Unit	Total rate £
Branch						
Sum of two sides 2100 mm	111.15	135.88	2.61	64.13	nr	**200.01**
Sum of two sides 2200 mm	118.97	145.45	2.61	64.13	nr	**209.58**
Sum of two sides 2300 mm	126.89	155.13	2.61	64.13	nr	**219.26**
Sum of two sides 2400 mm	134.60	164.55	2.88	70.77	nr	**235.32**
Sum of two sides 2500 mm	142.51	174.22	2.88	70.77	nr	**244.99**
Sum of two sides 2600 mm	150.34	183.79	2.88	70.77	nr	**254.56**
Sum of two sides 2700 mm	158.17	193.36	2.88	70.77	nr	**264.13**
Sum of two sides 2800 mm	166.00	202.94	3.94	96.81	nr	**299.75**
Sum of two sides 2900 mm	173.91	212.60	3.94	96.81	nr	**309.41**
Sum of two sides 3000 mm	181.74	222.18	4.83	118.68	nr	**340.86**
Sum of two sides 3100 mm	187.97	229.79	4.83	118.68	nr	**348.47**
Sum of two sides 3200 mm	193.66	236.76	4.83	118.68	nr	**355.44**
Sum of two sides 3300 mm	202.01	246.96	4.83	118.68	nr	**365.64**
Sum of two sides 3400 mm	207.71	253.93	4.83	118.68	nr	**372.61**
Sum of two sides 3500 mm	213.38	260.86	4.83	118.68	nr	**379.54**
Sum of two sides 3600 mm	219.07	267.81	4.83	118.68	nr	**386.49**
Sum of two sides 3700 mm	224.84	274.87	4.83	118.68	nr	**393.55**
Sum of two sides 3800 mm	230.52	281.81	4.83	118.68	nr	**400.49**
Sum of two sides 3900 mm	236.21	288.77	4.83	118.68	nr	**407.45**
Sum of two sides 4000 mm	241.89	295.71	4.83	118.68	nr	**414.39**
Grille neck						
Sum of two sides 2100 mm	118.06	144.33	2.80	68.80	nr	**213.13**
Sum of two sides 2200 mm	126.88	155.11	2.80	68.80	nr	**223.91**
Sum of two sides 2300 mm	135.69	165.88	2.80	68.80	nr	**234.68**
Sum of two sides 2400 mm	144.51	176.67	3.06	75.20	nr	**251.87**
Sum of two sides 2500 mm	153.33	187.45	3.06	75.20	nr	**262.65**
Sum of two sides 2600 mm	162.15	198.23	3.08	75.68	nr	**273.91**
Sum of two sides 2700 mm	158.17	193.36	3.08	75.68	nr	**269.04**
Sum of two sides 2800 mm	166.00	202.94	4.13	101.48	nr	**304.42**
Sum of two sides 2900 mm	173.91	212.60	4.13	101.48	nr	**314.08**
Sum of two sides 3000 mm	197.43	241.36	5.02	123.35	nr	**364.71**
Sum of two sides 3100 mm	204.46	249.95	5.02	123.35	nr	**373.30**
Sum of two sides 3200 mm	210.88	257.80	5.02	123.35	nr	**381.15**
Sum of two sides 3300 mm	217.27	265.62	5.02	123.35	nr	**388.97**
Sum of two sides 3400 mm	223.68	273.45	5.02	123.35	nr	**396.80**
Sum of two sides 3500 mm	230.08	281.27	5.02	123.35	nr	**404.62**
Sum of two sides 3600 mm	236.49	289.11	5.02	123.35	nr	**412.46**
Sum of two sides 3700 mm	242.90	296.95	5.02	123.35	nr	**420.30**
Sum of two sides 3800 mm	249.31	304.79	5.02	123.35	nr	**428.14**
Sum of two sides 3900 mm	255.72	312.62	5.02	123.35	nr	**435.97**
Sum of two sides 4000 mm	262.11	320.43	5.02	123.35	nr	**443.78**

U: VENTILATION/AIR CONDITIONING SYSTEMS

Item	Net Price £	Material £	Labour hours	Labour £	Unit	Total rate £
U10: DUCTWORK: RECTANGULAR – CLASS C – cont						
Ductwork 2001 to 2500 mm longest side						
Sum of two sides 2500 mm	159.32	194.77	2.77	68.07	m	**262.84**
Sum of two sides 2600 mm	163.98	200.47	2.97	72.99	m	**273.46**
Sum of two sides 2700 mm	168.12	205.53	2.99	73.47	m	**279.00**
Sum of two sides 2800 mm	173.02	211.52	3.30	81.09	m	**292.61**
Sum of two sides 2900 mm	177.17	216.59	3.31	81.33	m	**297.92**
Sum of two sides 3000 mm	181.19	221.50	3.53	86.74	m	**308.24**
Sum of two sides 3100 mm	185.32	226.56	3.55	87.24	m	**313.80**
Sum of two sides 3200 mm	189.34	231.47	3.56	87.48	m	**318.95**
Sum of two sides 3300 mm	193.49	236.54	3.56	87.48	m	**324.02**
Sum of two sides 3400 mm	197.50	241.44	3.56	87.48	m	**328.92**
Sum of two sides 3500 mm	216.77	265.00	3.56	87.48	m	**352.48**
Sum of two sides 3600 mm	220.79	269.92	3.56	87.48	m	**357.40**
Sum of two sides 3700 mm	224.93	274.98	3.56	87.48	m	**362.46**
Sum of two sides 3800 mm	228.95	279.89	3.56	87.48	m	**367.37**
Sum of two sides 3900 mm	232.97	284.81	3.56	87.48	m	**372.29**
Sum of two sides 4000 mm	237.78	290.69	3.56	87.48	m	**378.17**
Extra over fittings; Ductwork 2001 to 2500 mm longest side						
End Cap						
Sum of two sides 2500 mm	50.41	61.62	0.87	21.38	nr	**83.00**
Sum of two sides 2600 mm	53.34	65.21	0.87	21.38	nr	**86.59**
Sum of two sides 2700 mm	56.28	68.80	0.87	21.38	nr	**90.18**
Sum of two sides 2800 mm	59.21	72.39	1.16	28.51	nr	**100.90**
Sum of two sides 2900 mm	62.14	75.96	1.16	28.51	nr	**104.47**
Sum of two sides 3000 mm	65.07	79.55	1.73	42.51	nr	**122.06**
Sum of two sides 3100 mm	67.39	82.39	1.73	42.51	nr	**124.90**
Sum of two sides 3200 mm	69.50	84.96	1.73	42.51	nr	**127.47**
Sum of two sides 3300 mm	71.60	87.53	1.73	42.51	nr	**130.04**
Sum of two sides 3400 mm	73.71	90.11	1.73	42.51	nr	**132.62**
Sum of two sides 3500 mm	75.82	92.69	1.73	42.51	nr	**135.20**
Sum of two sides 3600 mm	77.93	95.28	1.73	42.51	nr	**137.79**
Sum of two sides 3700 mm	80.04	97.85	1.73	42.51	nr	**140.36**
Sum of two sides 3800 mm	82.15	100.43	1.73	42.51	nr	**142.94**
Sum of two sides 3900 mm	84.26	103.01	1.73	42.51	nr	**145.52**
Sum of two sides 4000 mm	86.37	105.59	1.73	42.51	nr	**148.10**
Reducer						
Sum of two sides 2500 mm	41.98	51.32	3.12	76.66	nr	**127.98**
Sum of two sides 2600 mm	46.24	56.53	3.16	77.65	nr	**134.18**
Sum of two sides 2700 mm	50.59	61.85	3.16	77.65	nr	**139.50**
Sum of two sides 2800 mm	54.74	66.92	4.00	98.29	nr	**165.21**
Sum of two sides 2900 mm	59.09	72.23	4.01	98.53	nr	**170.76**
Sum of two sides 3000 mm	63.36	77.46	4.56	112.05	nr	**189.51**
Sum of two sides 3100 mm	66.31	81.06	4.56	112.05	nr	**193.11**

U: VENTILATION/AIR CONDITIONING SYSTEMS

Item	Net Price £	Material £	Labour hours	Labour £	Unit	Total rate £
Sum of two sides 3200 mm	68.94	84.27	4.56	112.05	nr	196.32
Sum of two sides 3300 mm	71.64	87.58	4.56	112.05	nr	199.63
Sum of two sides 3400 mm	74.27	90.79	4.56	112.05	nr	202.84
Sum of two sides 3500 mm	75.22	91.96	4.56	112.05	nr	204.01
Sum of two sides 3600 mm	75.49	92.29	4.56	112.05	nr	204.34
Sum of two sides 3700 mm	76.52	93.54	4.56	112.05	nr	205.59
Sum of two sides 3800 mm	79.13	96.74	4.56	112.05	nr	208.79
Sum of two sides 3900 mm	81.75	99.94	4.56	112.05	nr	211.99
Sum of two sides 4000 mm	84.45	103.24	4.56	112.05	nr	215.29
Offset						
Sum of two sides 2500 mm	139.74	170.84	3.48	85.51	nr	256.35
Sum of two sides 2600 mm	148.69	181.77	3.49	85.76	nr	267.53
Sum of two sides 2700 mm	148.51	181.56	3.50	86.00	nr	267.56
Sum of two sides 2800 mm	166.56	203.62	4.33	106.40	nr	310.02
Sum of two sides 2900 mm	159.69	195.23	4.74	116.47	nr	311.70
Sum of two sides 3000 mm	149.13	182.31	5.31	130.48	nr	312.79
Sum of two sides 3100 mm	149.15	182.34	5.34	131.21	nr	313.55
Sum of two sides 3200 mm	149.11	182.29	5.35	131.46	nr	313.75
Sum of two sides 3300 mm	149.62	182.91	5.35	131.46	nr	314.37
Sum of two sides 3400 mm	149.81	183.14	5.35	131.46	nr	314.60
Sum of two sides 3500 mm	150.21	183.63	5.35	131.46	nr	315.09
Sum of two sides 3600 mm	150.60	184.11	5.35	131.46	nr	315.57
Sum of two sides 3700 mm	150.88	184.45	5.35	131.46	nr	315.91
Sum of two sides 3800 mm	156.06	190.79	5.35	131.46	nr	322.25
Sum of two sides 3900 mm	161.25	197.13	5.35	131.46	nr	328.59
Sum of two sides 4000 mm	166.46	203.50	5.35	131.46	nr	334.96
Square to round						
Sum of two sides 2500 mm	71.67	87.62	4.79	117.70	nr	205.32
Sum of two sides 2600 mm	78.34	95.77	4.95	121.64	nr	217.41
Sum of two sides 2700 mm	85.01	103.93	4.95	121.64	nr	225.57
Sum of two sides 2800 mm	91.68	112.07	8.49	208.62	nr	320.69
Sum of two sides 2900 mm	98.36	120.24	8.88	218.21	nr	338.45
Sum of two sides 3000 mm	105.03	128.40	9.02	221.64	nr	350.04
Sum of two sides 3100 mm	109.80	134.23	9.02	221.64	nr	355.87
Sum of two sides 3200 mm	114.00	139.37	9.09	223.37	nr	362.74
Sum of two sides 3300 mm	115.25	140.89	9.09	223.37	nr	364.26
Sum of two sides 3400 mm	118.20	144.50	9.09	223.37	nr	367.87
Sum of two sides 3500 mm	122.42	149.66	9.09	223.37	nr	373.03
Sum of two sides 3600 mm	123.20	150.62	9.09	223.37	nr	373.99
Sum of two sides 3700 mm	126.35	154.46	9.09	223.37	nr	377.83
Sum of two sides 3800 mm	130.55	159.60	9.09	223.37	nr	382.97
Sum of two sides 3900 mm	134.77	164.76	9.09	223.37	nr	388.13
Sum of two sides 4000 mm	138.96	169.88	9.09	223.37	nr	393.25

U: VENTILATION/AIR CONDITIONING SYSTEMS

Item	Net Price £	Material £	Labour hours	Labour £	Unit	Total rate £
U10: DUCTWORK: RECTANGULAR – CLASS C – cont						
Extra over fittings; Ductwork 2001 to 2500 mm longest side – cont						
90° radius bend						
Sum of two sides 2500 mm	189.60	231.79	4.35	106.89	nr	**338.68**
Sum of two sides 2600 mm	195.04	238.43	4.53	111.31	nr	**349.74**
Sum of two sides 2700 mm	211.50	258.56	4.53	111.31	nr	**369.87**
Sum of two sides 2800 mm	204.64	250.18	7.13	175.19	nr	**425.37**
Sum of two sides 2900 mm	220.81	269.94	7.17	176.19	nr	**446.13**
Sum of two sides 3000 mm	236.68	289.35	7.26	178.40	nr	**467.75**
Sum of two sides 3100 mm	248.69	304.03	7.26	178.40	nr	**482.43**
Sum of two sides 3200 mm	259.63	317.39	7.31	179.62	nr	**497.01**
Sum of two sides 3300 mm	263.24	321.81	7.31	179.62	nr	**501.43**
Sum of two sides 3400 mm	270.86	331.12	7.31	179.62	nr	**510.74**
Sum of two sides 3500 mm	274.19	335.20	7.31	179.62	nr	**514.82**
Sum of two sides 3600 mm	281.81	344.51	7.31	179.62	nr	**524.13**
Sum of two sides 3700 mm	301.58	368.69	7.31	179.62	nr	**548.31**
Sum of two sides 3800 mm	312.53	382.07	7.31	179.62	nr	**561.69**
Sum of two sides 3900 mm	320.31	391.59	7.31	179.62	nr	**571.21**
Sum of two sides 4000 mm	323.48	395.46	7.31	179.62	nr	**575.08**
45° bend						
Sum of two sides 2500 mm	155.23	189.77	8.30	203.95	nr	**393.72**
Sum of two sides 2600 mm	161.04	196.87	8.56	210.35	nr	**407.22**
Sum of two sides 2700 mm	172.52	210.90	8.62	211.81	nr	**422.71**
Sum of two sides 2800 mm	172.01	210.28	9.09	223.37	nr	**433.65**
Sum of two sides 2900 mm	183.35	224.15	9.09	223.37	nr	**447.52**
Sum of two sides 3000 mm	194.46	237.72	9.62	236.39	nr	**474.11**
Sum of two sides 3100 mm	203.03	248.21	9.62	236.39	nr	**484.60**
Sum of two sides 3200 mm	210.85	257.77	9.62	236.39	nr	**494.16**
Sum of two sides 3300 mm	218.92	267.64	9.62	236.39	nr	**504.03**
Sum of two sides 3400 mm	219.38	268.19	9.62	236.39	nr	**504.58**
Sum of two sides 3500 mm	226.75	277.20	9.62	236.39	nr	**513.59**
Sum of two sides 3600 mm	227.20	277.75	9.62	236.39	nr	**514.14**
Sum of two sides 3700 mm	246.04	300.78	9.62	236.39	nr	**537.17**
Sum of two sides 3800 mm	253.87	310.36	9.62	236.39	nr	**546.75**
Sum of two sides 3900 mm	261.70	319.93	9.62	236.39	nr	**556.32**
Sum of two sides 4000 mm	262.47	320.88	9.62	236.39	nr	**557.27**
90° mitre bend						
Sum of two sides 2500 mm	202.90	248.04	8.30	203.95	nr	**451.99**
Sum of two sides 2600 mm	212.64	259.95	8.56	210.35	nr	**470.30**
Sum of two sides 2700 mm	231.23	282.68	8.62	211.81	nr	**494.49**
Sum of two sides 2800 mm	220.67	269.77	15.20	373.50	nr	**643.27**
Sum of two sides 2900 mm	238.99	292.17	15.20	373.50	nr	**665.67**
Sum of two sides 3000 mm	257.44	314.73	15.20	373.50	nr	**688.23**
Sum of two sides 3100 mm	271.96	332.47	16.04	394.14	nr	**726.61**

U: VENTILATION/AIR CONDITIONING SYSTEMS

Item	Net Price £	Material £	Labour hours	Labour £	Unit	Total rate £
Sum of two sides 3200 mm	285.48	349.01	16.04	394.14	nr	**743.15**
Sum of two sides 3300 mm	298.87	365.37	16.04	394.14	nr	**759.51**
Sum of two sides 3400 mm	312.39	381.90	16.04	394.14	nr	**776.04**
Sum of two sides 3500 mm	313.73	383.54	16.04	394.14	nr	**777.68**
Sum of two sides 3600 mm	313.86	383.70	16.04	394.14	nr	**777.84**
Sum of two sides 3700 mm	325.64	398.09	16.04	394.14	nr	**792.23**
Sum of two sides 3800 mm	339.16	414.63	16.04	394.14	nr	**808.77**
Sum of two sides 3900 mm	352.68	431.16	16.04	394.14	nr	**825.30**
Sum of two sides 4000 mm	354.08	432.87	16.04	394.14	nr	**827.01**
Branch						
Sum of two sides 2500 mm	153.32	187.44	2.88	70.77	nr	**258.21**
Sum of two sides 2600 mm	161.56	197.51	2.88	70.77	nr	**268.28**
Sum of two sides 2700 mm	169.92	207.73	2.88	70.77	nr	**278.50**
Sum of two sides 2800 mm	178.07	217.69	3.94	96.81	nr	**314.50**
Sum of two sides 2900 mm	186.42	227.90	3.94	96.81	nr	**324.71**
Sum of two sides 3000 mm	194.68	238.00	4.83	118.68	nr	**356.68**
Sum of two sides 3100 mm	201.32	246.12	4.83	118.68	nr	**364.80**
Sum of two sides 3200 mm	207.39	253.53	4.83	118.68	nr	**372.21**
Sum of two sides 3300 mm	213.55	261.06	4.83	118.68	nr	**379.74**
Sum of two sides 3400 mm	219.63	268.50	4.83	118.68	nr	**387.18**
Sum of two sides 3500 mm	225.97	276.25	4.83	118.68	nr	**394.93**
Sum of two sides 3600 mm	232.05	283.68	4.83	118.68	nr	**402.36**
Sum of two sides 3700 mm	240.90	294.50	4.83	118.68	nr	**413.18**
Sum of two sides 3800 mm	246.97	301.92	4.83	118.68	nr	**420.60**
Sum of two sides 3900 mm	253.04	309.34	4.83	118.68	nr	**428.02**
Sum of two sides 4000 mm	259.20	316.87	4.83	118.68	nr	**435.55**
Grille neck						
Sum of two sides 2500 mm	156.59	191.44	3.06	75.20	nr	**266.64**
Sum of two sides 2600 mm	165.60	202.45	3.08	75.68	nr	**278.13**
Sum of two sides 2700 mm	174.61	213.46	3.08	75.68	nr	**289.14**
Sum of two sides 2800 mm	183.62	224.48	4.13	101.48	nr	**325.96**
Sum of two sides 2900 mm	192.63	235.49	4.13	101.48	nr	**336.97**
Sum of two sides 3000 mm	201.63	246.49	5.02	123.35	nr	**369.84**
Sum of two sides 3100 mm	208.81	255.28	5.02	123.35	nr	**378.63**
Sum of two sides 3200 mm	215.36	263.28	5.02	123.35	nr	**386.63**
Sum of two sides 3300 mm	221.90	271.27	5.02	123.35	nr	**394.62**
Sum of two sides 3400 mm	228.44	279.27	5.02	123.35	nr	**402.62**
Sum of two sides 3500 mm	234.98	287.27	5.02	123.35	nr	**410.62**
Sum of two sides 3600 mm	241.52	295.26	5.02	123.35	nr	**418.61**
Sum of two sides 3700 mm	248.06	303.25	5.02	123.35	nr	**426.60**
Sum of two sides 3800 mm	254.62	311.28	5.02	123.35	nr	**434.63**
Sum of two sides 3900 mm	261.16	319.27	5.02	123.35	nr	**442.62**
Sum of two sides 4000 mm	267.69	327.25	5.02	123.35	nr	**450.60**

Y30 – DUCTWORK ANCILLARIES: ACCESS DOORS

Refer to ancillaries in U10: DUCTWORK: RECTANGULAR: CLASS B for details of access doors

U: VENTILATION/AIR CONDITIONING SYSTEMS

Item	Net Price £	Material £	Labour hours	Labour £	Unit	Total rate £
U10: DUCTWORK: VOLUME/FIRE DAMPERS						
Y30 – DUCTWORK ANCILLARIES: VOLUME CONTROL AND FIRE DAMPERS						
Volume control damper; opposed blade; galvanized steel casing; aluminium aerofoil blades; manually operated						
Rectangular						
Sum of two sides 200 mm	23.41	28.62	1.60	39.32	nr	**67.94**
Sum of two sides 300 mm	25.04	30.61	1.60	39.32	nr	**69.93**
Sum of two sides 400 mm	27.38	33.47	1.60	39.32	nr	**72.79**
Sum of two sides 500 mm	29.96	36.63	1.60	39.32	nr	**75.95**
Sum of two sides 600 mm	33.12	40.49	1.70	41.79	nr	**82.28**
Sum of two sides 700 mm	36.40	44.50	2.10	51.62	nr	**96.12**
Sum of two sides 800 mm	39.79	48.65	2.15	52.84	nr	**101.49**
Sum of two sides 900 mm	43.65	53.36	2.30	56.52	nr	**109.88**
Sum of two sides 1000 mm	47.40	57.95	2.40	58.98	nr	**116.93**
Sum of two sides 1100 mm	51.62	63.11	2.60	63.89	nr	**127.00**
Sum of two sides 1200 mm	58.39	71.38	2.80	68.80	nr	**140.18**
Sum of two sides 1300 mm	62.96	76.97	3.10	76.18	nr	**153.15**
Sum of two sides 1400 mm	67.88	82.99	3.25	79.87	nr	**162.86**
Sum of two sides 1500 mm	73.49	89.85	3.40	83.54	nr	**173.39**
Sum of two sides 1600 mm	78.99	96.56	3.45	84.77	nr	**181.33**
Sum of two sides 1700 mm	84.26	103.01	3.60	88.46	nr	**191.47**
Sum of two sides 1800 mm	90.57	110.72	3.90	95.83	nr	**206.55**
Sum of two sides 1900 mm	96.43	117.88	4.20	103.24	nr	**221.12**
Sum of two sides 2000 mm	103.44	126.45	4.33	106.40	nr	**232.85**
Circular						
100 mm dia.	31.25	38.20	0.80	19.66	nr	**57.86**
160 mm dia.	37.21	45.48	0.90	22.11	nr	**67.59**
200 mm dia.	40.37	49.36	1.05	25.81	nr	**75.17**
250 mm dia.	44.94	54.94	1.20	29.50	nr	**84.44**
315 mm dia.	51.95	63.51	1.35	33.21	nr	**96.72**
350 mm dia.	54.65	66.81	1.65	40.55	nr	**107.36**
400 mm dia.	59.67	72.94	1.90	46.71	nr	**119.65**
450 mm dia.	64.48	78.83	2.10	51.62	nr	**130.45**
500 mm dia.	70.22	85.84	2.95	72.48	nr	**158.32**
650 mm dia.	88.70	108.44	4.55	111.80	nr	**220.24**
700 mm dia.	95.84	117.16	5.20	127.77	nr	**244.93**
800 mm dia.	111.30	136.06	5.80	142.52	nr	**278.58**
900 mm dia.	128.03	156.52	6.40	157.26	nr	**313.78**
1000 mm dia.	145.71	178.13	7.00	172.01	nr	**350.14**

U: VENTILATION/AIR CONDITIONING SYSTEMS

Item	Net Price £	Material £	Labour hours	Labour £	Unit	Total rate £
Flat oval						
345 × 102 mm	54.54	66.67	1.20	29.49	nr	96.16
508 × 102 mm	59.45	72.68	1.60	39.32	nr	112.00
559 × 152 mm	67.17	82.11	1.90	46.71	nr	128.82
531 × 203 mm	73.38	89.70	1.90	46.71	nr	136.41
851 × 203 mm	84.38	103.15	4.55	111.80	nr	214.95
582 × 254 mm	81.45	99.57	2.10	51.62	nr	151.19
823 × 254 mm	89.18	109.03	4.10	100.74	nr	209.77
632 × 305 mm	89.76	109.74	2.95	72.48	nr	182.22
765 × 356 mm	100.42	122.77	4.55	111.80	nr	234.57
737 × 406 mm	106.95	130.75	4.55	111.80	nr	242.55
818 × 406 mm	109.19	133.49	5.20	127.77	nr	261.26
978 × 406 mm	118.07	144.34	5.50	135.15	nr	279.49
709 × 457 mm	111.30	136.06	4.50	110.68	nr	246.74
678 × 508 mm	117.73	143.92	4.55	111.80	nr	255.72
919 × 508 mm	127.33	155.66	6.00	147.43	nr	303.09
Fire damper; galvanized steel casing; stainless steel folding shutter; fusible link with manual reset; BS 476 4 hour fire-rated						
Rectangular						
Sum of two sides 200 mm	103.60	126.65	1.60	39.32	nr	165.97
Sum of two sides 300 mm	109.45	133.81	1.60	39.32	nr	173.13
Sum of two sides 400 mm	115.89	141.68	1.60	39.32	nr	181.00
Sum of two sides 500 mm	124.41	152.09	1.60	39.32	nr	191.41
Sum of two sides 600 mm	133.31	162.98	1.70	41.79	nr	204.77
Sum of two sides 700 mm	142.67	174.42	2.10	51.62	nr	226.04
Sum of two sides 800 mm	152.36	186.27	2.15	52.85	nr	239.12
Sum of two sides 900 mm	162.55	198.72	2.30	56.62	nr	255.34
Sum of two sides 1000 mm	173.06	211.56	2.40	59.07	nr	270.63
Sum of two sides 1100 mm	184.06	225.01	2.60	63.98	nr	288.99
Sum of two sides 1200 mm	195.40	238.88	2.80	68.82	nr	307.70
Sum of two sides 1300 mm	207.21	253.32	3.10	76.18	nr	329.50
Sum of two sides 1400 mm	219.37	268.18	3.25	79.87	nr	348.05
Sum of two sides 1500 mm	231.88	283.48	3.40	83.58	nr	367.06
Sum of two sides 1600 mm	244.98	299.49	3.45	84.77	nr	384.26
Sum of two sides 1700 mm	258.30	315.78	3.60	88.46	nr	404.24
Sum of two sides 1800 mm	272.11	332.66	3.90	95.83	nr	428.49
Sum of two sides 1900 mm	286.37	350.09	4.20	103.24	nr	453.33
Sum of two sides 2000 mm	300.99	367.97	4.33	106.40	nr	474.37
Sum of two sides 2100 mm	179.07	218.92	4.43	108.85	nr	327.77
Sum of two sides 2200 mm	189.13	231.21	4.55	111.80	nr	343.01

U: VENTILATION/AIR CONDITIONING SYSTEMS

Item	Net Price £	Material £	Labour hours	Labour £	Unit	Total rate £
U10: DUCTWORK: VOLUME/FIRE DAMPERS – cont						
Fire damper – cont						
Circular						
100 mm dia.	100.51	122.88	0.80	19.66	nr	**142.54**
160 mm dia.	111.18	135.92	0.90	22.11	nr	**158.03**
200 mm dia.	121.50	148.54	1.05	25.81	nr	**174.35**
250 mm dia.	127.64	156.04	1.20	29.50	nr	**185.54**
315 mm dia.	155.07	189.57	1.35	33.21	nr	**222.78**
355 mm dia.	160.95	196.76	1.65	40.55	nr	**237.31**
400 mm dia.	176.03	215.20	1.90	46.71	nr	**261.91**
450 mm dia.	191.64	234.28	2.10	51.62	nr	**285.90**
500 mm dia.	218.74	267.41	2.95	72.48	nr	**339.89**
630 mm dia.	249.95	305.57	4.55	111.80	nr	**417.37**
710 mm dia.	272.33	332.93	5.20	127.77	nr	**460.70**
800 mm dia.	304.39	372.12	5.80	142.52	nr	**514.64**
900 mm dia.	340.00	415.66	6.40	157.26	nr	**572.92**
1000 mm dia.	229.07	280.04	7.00	172.01	nr	**452.05**
Flat oval						
345 × 102 mm	119.98	146.67	1.20	29.50	nr	**176.17**
427 × 102 mm	124.98	152.79	1.35	33.21	nr	**186.00**
508 × 102 mm	127.37	155.72	1.60	39.32	nr	**195.04**
559 × 152 mm	140.87	172.22	1.90	46.71	nr	**218.93**
531 × 203 mm	151.99	185.81	1.90	46.71	nr	**232.52**
851 × 203 mm	167.00	204.16	4.55	111.80	nr	**315.96**
582 × 254 mm	166.38	203.40	2.10	51.62	nr	**255.02**
632 × 305 mm	180.88	221.13	2.95	72.49	nr	**293.62**
765 × 356 mm	198.00	242.06	4.55	111.80	nr	**353.86**
737 × 406 mm	210.13	256.88	4.55	111.80	nr	**368.68**
818 × 406 mm	212.62	259.93	5.20	127.77	nr	**387.70**
978 × 406 mm	223.09	272.73	5.50	135.15	nr	**407.88**
709 × 457 mm	164.92	201.61	4.50	110.68	nr	**312.29**
678 × 508 mm	171.37	209.50	4.55	111.80	nr	**321.30**
Smoke/fire damper; galvanized steel casing; stainless steel folding shutter; fusible link and 24 V DC electro-magnetic shutter release mechanism; spring operated; BS 476 4 hour fire-rating						
Rectangular						
Sum of two sides 200 mm	331.07	404.74	1.60	39.32	nr	**444.06**
Sum of two sides 300 mm	332.00	405.87	1.60	39.32	nr	**445.19**
Sum of two sides 400 mm	332.94	407.03	1.60	39.32	nr	**446.35**
Sum of two sides 500 mm	341.35	417.30	1.60	39.32	nr	**456.62**
Sum of two sides 600 mm	350.13	428.04	1.70	41.79	nr	**469.83**
Sum of two sides 700 mm	359.26	439.20	2.10	51.62	nr	**490.82**
Sum of two sides 800 mm	368.61	450.63	2.15	52.85	nr	**503.48**

U: VENTILATION/AIR CONDITIONING SYSTEMS

Item	Net Price £	Material £	Labour hours	Labour £	Unit	Total rate £
Sum of two sides 900 mm	378.31	462.49	2.30	56.62	nr	**519.11**
Sum of two sides 1000 mm	388.37	474.79	2.40	59.07	nr	**533.86**
Sum of two sides 1100 mm	398.66	487.37	2.60	63.98	nr	**551.35**
Sum of two sides 1200 mm	409.30	500.37	2.80	68.82	nr	**569.19**
Sum of two sides 1300 mm	420.29	513.81	3.10	76.18	nr	**589.99**
Sum of two sides 1400 mm	431.52	527.54	3.25	79.87	nr	**607.41**
Sum of two sides 1500 mm	443.09	541.68	3.40	83.58	nr	**625.26**
Sum of two sides 1600 mm	455.01	556.25	3.45	84.77	nr	**641.02**
Sum of two sides 1700 mm	467.17	571.11	3.60	88.46	nr	**659.57**
Sum of two sides 1800 mm	479.69	586.42	3.90	95.83	nr	**682.25**
Sum of two sides 1900 mm	492.43	602.00	4.20	103.24	nr	**705.24**
Sum of two sides 2000 mm	505.53	618.01	4.33	106.40	nr	**724.41**
Circular						
100 mm dia.	351.20	429.35	0.80	19.66	nr	**449.01**
160 mm dia.	351.20	429.35	0.90	22.11	nr	**451.46**
200 mm dia.	351.20	429.35	1.05	25.81	nr	**455.16**
250 mm dia.	357.70	437.29	1.20	29.50	nr	**466.79**
315 mm dia.	364.44	445.53	1.35	33.21	nr	**478.74**
355 mm dia.	381.95	466.94	1.65	40.55	nr	**507.49**
400 mm dia.	396.41	484.62	1.90	46.71	nr	**531.33**
450 mm dia.	403.49	493.27	2.10	51.62	nr	**544.89**
500 mm dia.	418.31	511.38	2.95	72.48	nr	**583.86**
630 mm dia.	459.83	562.14	4.55	111.80	nr	**673.94**
710 mm dia.	478.06	584.43	5.20	127.77	nr	**712.20**
800 mm dia.	511.29	625.06	5.80	142.52	nr	**767.58**
900 mm dia.	542.09	662.71	6.40	157.26	nr	**819.97**
1000 mm dia.	577.10	705.51	7.00	172.01	nr	**877.52**
Flat oval						
531 × 203 mm	437.51	534.86	1.90	46.71	nr	**581.57**
851 × 203 mm	458.84	560.94	4.55	111.80	nr	**672.74**
582 × 254 mm	452.25	552.88	2.10	51.62	nr	**604.50**
632 × 305 mm	466.99	570.90	2.95	72.49	nr	**643.39**
765 × 356 mm	485.87	593.98	4.55	111.80	nr	**705.78**
737 × 406 mm	497.29	607.94	4.55	111.80	nr	**719.74**
818 × 406 mm	501.18	612.70	5.20	127.77	nr	**740.47**
978 × 406 mm	517.04	632.08	5.50	135.15	nr	**767.23**
709 × 457 mm	504.69	616.99	4.50	110.68	nr	**727.67**
678 × 508 mm	515.87	630.66	4.55	111.80	nr	**742.46**

U: VENTILATION/AIR CONDITIONING SYSTEMS

Item	Net Price £	Material £	Labour hours	Labour £	Unit	Total rate £
U10: GRILLES/DIFFUSERS/LOUVRES						
Y46 – GRILLES/DIFFUSERS/LOUVRES						
Supply grilles; single deflection; extruded aluminium alloy frame and adjustable horizontal vanes; silver grey polyester powder coated; screw fixed						
Rectangular; for duct, ceiling and sidewall applications						
150 × 100 mm	11.26	12.72	0.60	13.62	nr	**26.34**
150 × 150 mm	11.75	13.27	0.60	13.62	nr	**26.89**
200 × 150 mm	12.73	14.38	0.65	14.75	nr	**29.13**
200 × 200 mm	13.22	14.93	0.72	16.34	nr	**31.27**
300 × 100 mm	13.22	14.93	0.72	16.34	nr	**31.27**
300 × 150 mm	13.22	14.93	0.80	18.16	nr	**33.09**
300 × 200 mm	15.67	17.70	0.88	19.98	nr	**37.68**
300 × 300 mm	16.65	18.81	1.04	23.60	nr	**42.41**
400 × 100 mm	14.20	16.03	0.88	19.98	nr	**36.01**
400 × 150 mm	14.69	16.60	0.94	21.34	nr	**37.94**
400 × 200 mm	18.12	20.46	1.04	23.60	nr	**44.06**
400 × 300 mm	21.05	23.78	1.12	25.42	nr	**49.20**
600 × 200 mm	23.01	25.99	1.26	28.61	nr	**54.60**
600 × 300 mm	29.38	33.18	1.40	31.78	nr	**64.96**
600 × 400 mm	36.72	41.47	1.61	36.55	nr	**78.02**
600 × 500 mm	42.11	47.56	1.76	39.95	nr	**87.51**
600 × 600 mm	51.41	58.06	2.17	49.27	nr	**107.33**
800 × 300 mm	35.74	40.36	1.76	39.95	nr	**80.31**
800 × 400 mm	43.57	49.20	2.17	49.27	nr	**98.47**
800 × 600 mm	59.73	67.46	3.00	68.11	nr	**135.57**
1000 × 300 mm	40.15	45.35	2.60	59.02	nr	**104.37**
1000 × 400 mm	49.94	56.41	3.00	68.11	nr	**124.52**
1000 × 600 mm	69.52	78.52	3.80	86.26	nr	**164.78**
1000 × 800 mm	110.30	124.57	3.80	86.26	nr	**210.83**
1200 × 600 mm	111.69	126.14	4.61	104.65	nr	**230.79**
1200 × 800 mm	140.73	158.94	4.61	104.65	nr	**263.59**
1200 × 1000 mm	167.55	189.23	4.61	104.65	nr	**293.88**
Rectangular; for duct, ceiling and sidewall applications; including opposed blade damper volume regulator						
150 × 100 mm	18.12	20.46	0.72	16.35	nr	**36.81**
150 × 150 mm	19.09	21.56	0.72	16.35	nr	**37.91**
200 × 150 mm	21.05	23.78	0.83	18.86	nr	**42.64**
200 × 200 mm	21.54	24.33	0.90	20.43	nr	**44.76**
300 × 100 mm	21.54	24.33	0.90	20.43	nr	**44.76**
300 × 150 mm	23.01	25.99	0.98	22.25	nr	**48.24**
300 × 200 mm	25.46	28.75	1.06	24.07	nr	**52.82**

U: VENTILATION/AIR CONDITIONING SYSTEMS

Item	Net Price £	Material £	Labour hours	Labour £	Unit	Total rate £
300 × 300 mm	27.91	31.52	1.20	27.25	nr	58.77
400 × 100 mm	23.99	27.09	1.06	24.07	nr	51.16
400 × 150 mm	25.95	29.30	1.13	25.65	nr	54.95
400 × 200 mm	27.91	31.52	1.20	27.25	nr	58.77
400 × 300 mm	32.31	36.50	1.34	30.43	nr	66.93
600 × 200 mm	37.70	42.57	1.50	34.09	nr	76.66
600 × 300 mm	42.11	47.56	1.66	37.71	nr	85.27
600 × 400 mm	53.37	60.27	1.80	40.90	nr	101.17
600 × 500 mm	61.20	69.12	2.00	45.40	nr	114.52
600 × 600 mm	66.10	74.65	2.60	59.11	nr	133.76
800 × 300 mm	62.67	70.78	2.00	45.40	nr	116.18
800 × 400 mm	71.48	80.73	2.60	59.11	nr	139.84
800 × 600 mm	92.53	104.50	3.61	81.95	nr	186.45
1000 × 300 mm	72.46	81.83	3.00	68.17	nr	150.00
1000 × 400 mm	87.15	98.43	3.61	81.95	nr	180.38
1000 × 600 mm	115.06	129.95	4.61	104.61	nr	234.56
1000 × 800 mm	196.85	222.32	4.61	104.61	nr	326.93
1200 × 600 mm	144.92	163.67	5.62	127.53	nr	291.20
1200 × 800 mm	232.28	262.33	6.10	138.48	nr	400.81
1200 × 1000 mm	266.33	300.79	6.50	147.55	nr	448.34
Supply grilles; double deflection; extruded aluminium alloy frame and adjustable horizontal and vertical vanes; white polyester powder coated; screw fixed						
Rectangular; for duct, ceiling and sidewall applications						
150 × 100 mm	11.26	12.72	0.88	19.98	nr	32.70
150 × 150 mm	11.26	12.72	0.88	19.98	nr	32.70
200 × 150 mm	12.73	14.38	1.08	24.52	nr	38.90
200 × 200 mm	14.69	16.60	1.25	28.39	nr	44.99
300 × 100 mm	14.69	16.60	1.25	28.39	nr	44.99
300 × 150 mm	14.69	16.60	1.50	34.05	nr	50.65
300 × 200 mm	16.65	18.81	1.75	39.72	nr	58.53
300 × 300 mm	21.05	23.78	2.15	48.82	nr	72.60
400 × 100 mm	18.12	20.46	1.75	39.72	nr	60.18
400 × 150 mm	18.12	20.46	1.95	44.27	nr	64.73
400 × 200 mm	19.58	22.11	2.15	48.82	nr	70.93
400 × 300 mm	24.48	27.65	2.55	57.90	nr	85.55
600 × 200 mm	27.91	31.52	3.01	68.33	nr	99.85
600 × 300 mm	35.25	39.81	3.36	76.28	nr	116.09
600 × 400 mm	43.08	48.65	3.80	86.26	nr	134.91
600 × 500 mm	51.41	58.06	4.20	95.34	nr	153.40
600 × 600 mm	60.22	68.01	4.51	102.38	nr	170.39
800 × 300 mm	48.96	55.29	4.20	95.34	nr	150.63
800 × 400 mm	60.22	68.01	4.51	102.38	nr	170.39
800 × 600 mm	87.15	98.43	5.10	115.77	nr	214.20
1000 × 300 mm	111.69	126.14	4.80	108.97	nr	235.11
1000 × 400 mm	37.16	41.97	5.10	115.77	nr	157.74
1000 × 600 mm	52.93	59.78	5.72	129.85	nr	189.63

U: VENTILATION/AIR CONDITIONING SYSTEMS

Item	Net Price £	Material £	Labour hours	Labour £	Unit	Total rate £
U10: GRILLES/DIFFUSERS/LOUVRES – cont						
Supply grilles; double deflection – cont						
Rectangular – cont						
1000 × 800 mm	62.38	70.45	6.33	143.70	nr	**214.15**
1200 × 600 mm	196.85	222.32	5.72	129.85	nr	**352.17**
1200 × 800 mm	245.61	277.39	6.33	143.70	nr	**421.09**
1200 × 1000 mm	279.34	315.49	6.33	143.70	nr	**459.19**
Rectangular; for duct, ceiling and sidewall applications; including opposed blade damper volume regulator						
150 × 100 mm	19.58	22.11	1.00	22.71	nr	**44.82**
150 × 150 mm	21.05	23.78	1.00	22.71	nr	**46.49**
200 × 150 mm	22.52	25.44	1.26	28.61	nr	**54.05**
200 × 200 mm	25.46	28.75	1.43	32.46	nr	**61.21**
300 × 100 mm	23.99	27.09	1.43	32.46	nr	**59.55**
300 × 150 mm	26.93	30.42	1.68	38.14	nr	**68.56**
300 × 200 mm	28.89	32.63	1.93	43.81	nr	**76.44**
300 × 300 mm	32.31	36.50	2.31	52.44	nr	**88.94**
400 × 100 mm	27.91	31.52	1.93	43.81	nr	**75.33**
400 × 150 mm	30.84	34.83	2.14	48.57	nr	**83.40**
400 × 200 mm	33.78	38.15	2.31	52.44	nr	**90.59**
400 × 300 mm	39.17	44.24	2.77	62.88	nr	**107.12**
600 × 200 mm	43.57	49.20	3.25	73.79	nr	**122.99**
600 × 300 mm	51.41	58.06	3.62	82.17	nr	**140.23**
600 × 400 mm	59.73	67.46	3.99	90.58	nr	**158.04**
600 × 500 mm	64.63	72.99	4.44	100.79	nr	**173.78**
600 × 600 mm	78.83	89.03	4.94	112.14	nr	**201.17**
800 × 300 mm	66.59	75.20	4.44	100.79	nr	**175.99**
800 × 400 mm	80.29	90.68	4.94	112.14	nr	**202.82**
800 × 600 mm	122.89	138.79	5.71	129.62	nr	**268.41**
1000 × 300 mm	85.68	96.77	5.20	118.04	nr	**214.81**
1000 × 400 mm	103.80	117.23	5.71	129.62	nr	**246.85**
1000 × 600 mm	158.14	178.60	6.53	148.24	nr	**326.84**
1000 × 800 mm	229.38	259.06	6.53	148.24	nr	**407.30**
1200 × 600 mm	191.92	216.75	7.34	166.62	nr	**383.37**
1200 × 800 mm	268.64	303.40	8.80	199.77	nr	**503.17**
1200 × 1000 mm	302.51	341.65	8.80	199.77	nr	**541.42**
Floor grille suitable for mounting in raised access floors; heavy duty; extruded alumiinium; standard mill finish; complete with opposed blade volume control damper						
Diffuser						
600 mm × 600 mm	162.55	183.58	0.70	15.89	nr	**199.47**
Extra for nylon coated black finish	23.01	25.99	–	–	nr	**25.99**

U: VENTILATION/AIR CONDITIONING SYSTEMS

Item	Net Price £	Material £	Labour hours	Labour £	Unit	Total rate £
Exhaust grilles; aluminium						
0° fixed blade core						
150 × 150 mm	9.30	10.51	0.60	13.68	nr	24.19
200 × 200 mm	12.05	13.61	0.72	16.35	nr	29.96
250 × 250 mm	13.41	15.15	0.80	18.16	nr	33.31
300 × 300 mm	16.65	18.81	1.00	22.71	nr	41.52
350 × 350 mm	21.37	24.13	1.20	27.24	nr	51.37
0° fixed blade core; including opposed blade damper volume regulator						
150 × 150 mm	14.64	16.53	0.62	14.09	nr	30.62
200 × 200 mm	18.60	21.00	0.72	16.35	nr	37.35
250 × 250 mm	22.28	25.16	0.80	18.16	nr	43.32
300 × 300 mm	26.24	29.63	1.00	22.71	nr	52.34
350 × 350 mm	34.99	39.51	1.20	27.24	nr	66.75
45° fixed blade core						
150 × 150 mm	10.28	11.61	0.62	14.09	nr	25.70
200 × 200 mm	11.75	13.27	0.72	16.35	nr	29.62
250 × 250 mm	13.22	14.93	0.80	18.16	nr	33.09
300 × 300 mm	18.60	21.00	1.00	22.71	nr	43.71
350 × 350 mm	22.03	24.88	1.20	27.24	nr	52.12
45° fixed blade core; including opposed blade damper volume regulator						
150 × 150 mm	17.14	19.36	0.62	14.09	nr	33.45
200 × 200 mm	20.56	23.23	0.72	16.35	nr	39.58
250 × 250 mm	22.03	24.88	0.80	18.16	nr	43.04
300 × 300 mm	28.89	32.63	1.00	22.71	nr	55.34
350 × 350 mm	31.33	35.38	1.20	27.24	nr	62.62
Eggcrate core						
150 × 150 mm	11.75	13.27	0.62	14.09	nr	27.36
200 × 200 mm	13.22	14.93	1.00	22.71	nr	37.64
250 × 250 mm	16.65	18.81	0.80	18.16	nr	36.97
300 × 300 mm	20.56	23.23	1.00	22.71	nr	45.94
350 × 350 mm	23.50	26.54	1.20	27.24	nr	53.78
Eggcrate core; including opposed blade damper volume regulator						
150 × 150 mm	13.22	14.93	0.62	14.09	nr	29.02
200 × 200 mm	19.58	22.11	0.72	16.35	nr	38.46
250 × 250 mm	23.50	26.54	0.80	18.16	nr	44.70
300 × 300 mm	27.42	30.97	1.00	22.71	nr	53.68
350 × 350 mm	33.78	38.15	1.20	27.24	nr	65.39

U: VENTILATION/AIR CONDITIONING SYSTEMS

Item	Net Price £	Material £	Labour hours	Labour £	Unit	Total rate £
U10: GRILLES/DIFFUSERS/LOUVRES – cont						
Exhaust grilles – cont						
Mesh/perforated plate core						
150 × 150 mm	24.97	28.20	0.62	14.08	nr	**42.28**
200 × 200 mm	18.12	20.46	0.72	16.34	nr	**36.80**
250 × 250 mm	21.05	23.78	0.80	18.16	nr	**41.94**
300 × 300 mm	24.97	28.20	1.00	22.71	nr	**50.91**
350 × 350 mm	27.91	31.52	1.20	27.24	nr	**58.76**
Mesh/perforated plate core; including opposed blade damper volume regulator						
150 × 150 mm	14.19	16.02	0.62	14.08	nr	**30.10**
200 × 200 mm	17.93	20.25	0.72	16.34	nr	**36.59**
250 × 250 mm	27.57	31.14	0.80	18.16	nr	**49.30**
300 × 300 mm	33.72	38.09	0.80	18.16	nr	**56.25**
350 × 350 mm	47.15	53.25	1.20	27.24	nr	**80.49**
Plastic air diffusion system						
Eggcrate grilles						
150 × 150 mm	3.83	4.32	0.62	14.70	nr	**19.02**
200 × 200 mm	6.03	6.81	0.72	17.07	nr	**23.88**
250 × 250 mm	6.70	7.57	0.80	18.95	nr	**26.52**
300 × 300 mm	7.42	8.38	1.00	23.69	nr	**32.07**
Single deflection grilles						
150 × 150 mm	6.03	6.81	0.62	14.70	nr	**21.51**
200 × 200 mm	6.03	6.81	0.72	17.07	nr	**23.88**
250 × 250 mm	6.70	7.57	0.80	18.95	nr	**26.52**
300 × 300 mm	7.42	8.38	1.00	23.69	nr	**32.07**
Double deflection grilles						
150 × 150 mm	6.03	6.81	0.62	14.70	nr	**21.51**
200 × 200 mm	9.85	11.13	0.72	17.07	nr	**28.20**
250 × 250 mm	11.13	12.57	0.80	18.95	nr	**31.52**
300 × 300 mm	15.27	17.25	1.00	23.69	nr	**40.94**
Door transfer grilles						
150 × 150 mm	5.61	6.33	0.62	14.70	nr	**21.03**
200 × 200 mm	6.63	7.49	0.72	17.07	nr	**24.56**
250 × 250 mm	10.14	11.45	0.80	18.95	nr	**30.40**
300 × 300 mm	12.26	13.84	1.00	23.69	nr	**37.53**

U: VENTILATION/AIR CONDITIONING SYSTEMS

Item	Net Price £	Material £	Labour hours	Labour £	Unit	Total rate £
Opposed blade dampers						
150 × 150 mm	2.82	3.18	0.62	14.70	nr	**17.88**
200 × 200 mm	3.68	4.15	0.72	17.07	nr	**21.22**
250 × 250 mm	3.68	4.15	0.80	18.95	nr	**23.10**
300 × 300 mm	4.72	5.33	1.00	23.69	nr	**29.02**
Ceiling mounted diffusers; circular aluminium multi-core diffuser						
Circular; for ceiling mounting						
141 mm dia. neck	26.32	29.72	0.80	18.16	nr	**47.88**
197 mm dia. neck	31.82	35.93	1.10	24.97	nr	**60.90**
309 mm dia. neck	48.96	55.29	1.40	31.79	nr	**87.08**
365 mm dia. neck	60.59	68.43	1.50	34.09	nr	**102.52**
477 mm dia. neck	99.76	112.67	2.00	45.40	nr	**158.07**
Circular; for ceiling mounting; including louvre damper volume control						
141 mm dia. neck	28.76	32.48	1.00	22.71	nr	**55.19**
197 mm dia. neck	34.27	38.71	1.20	27.25	nr	**65.96**
309 mm dia. neck	52.63	59.44	1.60	36.32	nr	**95.76**
365 mm dia. neck	67.32	76.03	1.90	43.16	nr	**119.19**
477 mm dia. neck	111.38	125.80	2.40	54.57	nr	**180.37**
Ceiling mounted diffusers; rectangular aluminium multi-cone diffuser; four way flow						
Rectangular; for ceiling mounting						
150 × 150 mm neck	23.50	26.54	1.80	40.90	nr	**67.44**
300 × 150 mm neck	35.65	40.26	2.30	52.21	nr	**92.47**
300 × 300 mm neck	34.27	38.71	2.80	63.57	nr	**102.28**
450 × 150 mm neck	45.06	50.89	2.80	63.57	nr	**114.46**
450 × 300 mm neck	50.62	57.17	3.20	72.64	nr	**129.81**
450 × 450 mm neck	49.45	55.84	3.40	77.21	nr	**133.05**
600 × 150 mm neck	54.46	61.51	3.20	72.64	nr	**134.15**
600 × 300 mm neck	62.22	70.27	3.50	79.46	nr	**149.73**
600 × 600 mm neck	72.46	81.83	4.00	90.80	nr	**172.63**
Rectangular; for ceiling mounting; including opposed blade damper volume regulator						
150 × 150 mm neck	33.78	38.15	1.80	40.90	nr	**79.05**
300 × 150 mm neck	35.65	40.26	2.30	52.21	nr	**92.47**
300 × 300 mm neck	47.49	53.63	2.80	63.59	nr	**117.22**
450 × 150 mm neck	58.04	65.55	2.80	63.57	nr	**129.12**
450 × 300 mm neck	67.21	75.91	3.30	74.91	nr	**150.82**
450 × 450 mm neck	68.54	77.41	3.51	79.65	nr	**157.06**
600 × 150 mm neck	68.59	77.47	3.30	74.91	nr	**152.38**
600 × 300 mm neck	80.77	91.22	4.00	90.80	nr	**182.02**
600 × 600 mm neck	99.88	112.80	5.62	127.53	nr	**240.33**

U: VENTILATION/AIR CONDITIONING SYSTEMS

Item	Net Price £	Material £	Labour hours	Labour £	Unit	Total rate £
U10: GRILLES/DIFFUSERS/LOUVRES – cont						
Slot diffusers; continuous aluminium slot diffuser with flanged frame (1500 mm sections)						
Diffuser						
1 slot	11.59	13.09	3.76	85.33	m	**98.42**
2 slot	20.42	23.06	3.76	85.31	m	**108.37**
3 slot	27.32	30.85	3.76	85.31	m	**116.16**
4 slot	33.94	38.33	4.50	102.26	m	**140.59**
6 slot	48.29	54.54	4.50	102.26	m	**156.80**
Diffuser; including equalizing deflector						
1 slot	23.99	27.09	5.26	119.48	m	**146.57**
2 slot	43.08	48.65	5.26	119.48	m	**168.13**
3 slot	57.28	64.69	5.26	119.48	m	**184.17**
4 slot	70.99	80.17	6.33	143.67	m	**223.84**
6 slot	100.86	113.91	6.33	143.67	m	**257.58**
Extra over for ends						
1 slot	7.34	8.29	1.00	22.71	nr	**31.00**
2 slot	9.64	10.88	1.00	22.71	nr	**33.59**
3 slot	11.02	12.44	1.00	22.71	nr	**35.15**
4 slot	12.39	13.99	1.30	29.51	nr	**43.50**
6 slot	15.15	17.11	1.40	31.78	nr	**48.89**
Plenum boxes; 1.0 m long; circular spigot; including cord operated flap damper						
1 slot	24.48	27.65	2.75	62.54	nr	**90.19**
2 slot	25.15	28.41	2.75	62.54	nr	**90.95**
3 slot	26.11	29.49	2.75	62.54	nr	**92.03**
4 slot	26.52	29.95	3.51	79.65	nr	**109.60**
6 slot	28.15	31.79	3.51	79.65	nr	**111.44**
Plenum boxes; 2.0 m long; circular spigot; including cord operated flap damper						
1 slot	31.42	35.49	3.26	73.94	nr	**109.43**
2 slot	31.42	35.49	3.26	73.94	nr	**109.43**
3 slot	33.05	37.32	3.26	73.94	nr	**111.26**
4 slot	34.27	38.71	3.76	85.33	nr	**124.04**
6 slot	35.09	39.63	3.76	85.33	nr	**124.96**

U: VENTILATION/AIR CONDITIONING SYSTEMS

Item	Net Price £	Material £	Labour hours	Labour £	Unit	Total rate £
Perforated diffusers; rectangular face aluminium perforated diffuser; quick release face plate; for integration with rectangular ceiling tiles						
Circular spigot; rectangular diffuser						
150 mm dia. spigot; 300 × 300 diffuser	64.14	72.44	1.00	22.71	nr	**95.15**
300 mm dia. spigot; 600 × 600 diffuser	117.99	133.25	1.40	31.79	nr	**165.04**
Circular spigot; rectangular diffuser; including louvre damper volume regulator						
150 mm dia. spigot; 300 × 300 diffuser	81.27	91.79	1.00	22.71	nr	**114.50**
300 mm dia. spigot; 600 × 600 diffuser	136.60	154.27	1.60	36.32	nr	**190.59**
Rectangular spigot; rectangular diffuser						
150 × 150 mm dia. spigot; 300 × 300 mm diffuser	35.85	40.48	1.00	23.69	nr	**64.17**
300 × 150 mm dia. spigot; 600 × 300 mm diffuser	47.42	53.56	1.20	28.44	nr	**82.00**
300 × 300 mm dia. spigot; 600 × 600 mm diffuser	51.67	58.35	1.40	33.17	nr	**91.52**
600 × 300 mm dia. spigot; 1200 × 600 mm diffuser	54.90	62.01	1.60	37.91	nr	**99.92**
Rectangular spigot; rectangular diffuser; including opposed blade damper volume regulator						
150 × 150 mm dia. spigot; 300 × 300 mm diffuser	40.90	46.20	1.20	28.44	nr	**74.64**
300 × 150 mm dia. spigot; 600 × 300 mm diffuser	56.60	63.92	1.40	33.18	nr	**97.10**
300 × 300 mm dia. spigot; 600 × 600 mm diffuser	61.79	69.79	1.60	37.91	nr	**107.70**
600 × 300 mm dia. spigot; 1200 × 600 mm diffuser	69.41	78.39	1.80	42.65	nr	**121.04**
Floor swirl diffuser; manual adjustment of air discharge direction; complete with damper and dirt trap						
Plastic Diffuser						
150 dia.	20.76	23.45	0.50	11.35	nr	**34.80**
200 dia.	26.93	30.42	0.50	11.35	nr	**41.77**
Aluminium Diffuser						
150 dia.	25.25	28.52	0.50	11.35	nr	**39.87**
200 dia.	31.42	35.49	0.50	11.35	nr	**46.84**

U: VENTILATION/AIR CONDITIONING SYSTEMS

Item	Net Price £	Material £	Labour hours	Labour £	Unit	Total rate £
U10: GRILLES/DIFFUSERS/LOUVRES – cont						
Plastic air diffusion system						
Cellular diffusers						
300 × 300 mm	6.24	7.05	2.80	66.37	nr	73.42
600 × 600 mm	12.54	14.16	4.00	94.78	nr	108.94
Multi-cone diffusers						
300 × 300 mm	5.07	5.72	2.80	66.37	nr	72.09
450 × 450 mm	9.96	11.25	3.40	80.59	nr	91.84
500 × 500 mm	10.66	12.04	3.80	90.10	nr	102.14
600 × 600 mm	13.53	15.28	4.00	94.78	nr	110.06
625 × 625 mm	23.54	26.58	4.26	100.84	nr	127.42
Opposed blade dampers						
300 × 300 mm	6.10	6.88	1.20	28.45	nr	35.33
450 × 450 mm	8.74	9.87	1.50	35.58	nr	45.45
600 × 600 mm	17.11	19.33	2.60	61.71	nr	81.04
Plenum boxes						
300 mm	3.63	4.10	2.80	66.37	nr	70.47
450 mm	5.22	5.90	3.40	80.59	nr	86.49
600 mm	8.70	9.82	4.00	94.78	nr	104.60
Plenum spigot reducer						
600 mm	4.80	5.42	1.00	23.69	nr	29.11
Blanking kits for cellular diffusers						
300 mm	1.37	1.55	0.88	20.86	nr	22.41
600 mm	4.93	5.57	1.10	26.07	nr	31.64
Blanking kits for multi-cone diffusers						
300 mm	1.37	1.55	0.88	20.86	nr	22.41
450 mm	2.93	3.30	0.90	21.32	nr	24.62
600 mm	4.93	5.57	1.10	26.07	nr	31.64
Acoustic louvres; opening mounted; 300 mm deep steel louvres with blades packed with acoustic infill; 12 mm galvanized mesh birdscreen; screw fixing in opening						
Louvre units; self finished galvanized steel						
900 high × 600 wide	109.40	123.55	3.00	71.08	nr	194.63
900 high × 900 wide	135.74	153.31	3.00	71.08	nr	224.39
900 high × 1200 wide	160.70	181.50	3.34	79.14	nr	260.64

U: VENTILATION/AIR CONDITIONING SYSTEMS

Item	Net Price £	Material £	Labour hours	Labour £	Unit	Total rate £
900 high × 1500 wide	209.73	236.86	3.34	79.14	nr	**316.00**
900 high × 1800 wide	235.15	265.58	3.34	79.14	nr	**344.72**
900 high × 2100 wide	260.57	294.29	3.34	79.14	nr	**373.43**
900 high × 2400 wide	285.53	322.47	3.68	87.20	nr	**409.67**
900 high × 2700 wide	322.76	364.53	3.68	87.20	nr	**451.73**
900 high × 3000 wide	345.45	390.15	3.68	87.20	nr	**477.35**
1200 high × 600 wide	143.45	162.02	3.00	71.08	nr	**233.10**
1200 high × 900 wide	176.59	199.44	3.34	79.14	nr	**278.58**
1200 high × 1200 wide	209.27	236.34	3.34	79.14	nr	**315.48**
1200 high × 1500 wide	277.36	313.25	3.34	79.14	nr	**392.39**
1200 high × 1800 wide	310.05	350.17	3.68	87.20	nr	**437.37**
1200 high × 2100 wide	343.63	388.09	3.68	87.20	nr	**475.29**
1200 high × 2400 wide	376.32	425.01	3.68	87.20	nr	**512.21**
1500 high × 600 wide	177.95	200.98	3.00	71.08	nr	**272.06**
1500 high × 900 wide	218.36	246.62	3.34	79.14	nr	**325.76**
1500 high × 1200 wide	258.75	292.23	3.34	79.14	nr	**371.37**
1500 high × 1500 wide	345.00	389.64	3.68	87.20	nr	**476.84**
1500 high × 1800 wide	385.40	435.26	3.68	87.20	nr	**522.46**
1500 high × 2100 wide	426.26	481.42	4.00	94.78	nr	**576.20**
1800 high × 600 wide	211.53	238.90	3.34	79.14	nr	**318.04**
1800 high × 900 wide	259.66	293.26	3.34	79.14	nr	**372.40**
1800 high × 1200 wide	308.23	348.12	3.68	87.20	nr	**435.32**
1800 high × 1500 wide	413.09	466.54	3.68	87.20	nr	**553.74**
Louvre units; polyester powder coated steel						
900 high × 600 wide	158.89	179.45	3.00	71.08	nr	**250.53**
900 high × 900 wide	209.27	236.34	3.00	71.08	nr	**307.42**
900 high × 1200 wide	258.75	292.23	3.34	79.14	nr	**371.37**
900 high × 1500 wide	332.29	375.29	3.34	79.14	nr	**454.43**
900 high × 1800 wide	382.22	431.67	3.34	79.14	nr	**510.81**
900 high × 2100 wide	432.62	488.60	3.34	79.14	nr	**567.74**
900 high × 2400 wide	482.09	544.47	3.68	87.20	nr	**631.67**
900 high × 2700 wide	543.37	613.68	3.68	87.20	nr	**700.88**
900 high × 3000 wide	590.59	667.01	3.68	87.20	nr	**754.21**
1200 high × 600 wide	208.82	235.84	3.00	71.08	nr	**306.92**
1200 high × 900 wide	274.64	310.18	3.34	79.14	nr	**389.32**
1200 high × 1200 wide	340.46	384.51	3.34	79.14	nr	**463.65**
1200 high × 1500 wide	440.78	497.81	3.34	79.14	nr	**576.95**
1200 high × 1800 wide	506.15	571.64	3.68	87.20	nr	**658.84**
1200 high × 2100 wide	572.43	646.50	3.68	87.20	nr	**733.70**
1200 high × 2400 wide	637.79	720.31	3.68	87.20	nr	**807.51**
1500 high × 600 wide	259.66	293.26	3.00	71.08	nr	**364.34**
1500 high × 900 wide	340.91	385.02	3.34	79.14	nr	**464.16**
1500 high × 1200 wide	422.17	476.80	3.34	79.14	nr	**555.94**
1500 high × 1500 wide	549.28	620.35	3.68	87.20	nr	**707.55**
1500 high × 1800 wide	630.99	712.63	3.68	87.20	nr	**799.83**
1500 high × 2100 wide	712.25	804.41	4.00	94.78	nr	**899.19**
1800 high × 600 wide	309.58	349.64	3.34	79.14	nr	**428.78**
1800 high × 900 wide	406.73	459.36	3.34	79.14	nr	**538.50**
1800 high × 1200 wide	504.34	569.60	3.68	87.20	nr	**656.80**
1800 high × 1500 wide	658.22	743.39	3.68	87.20	nr	**830.59**

U: VENTILATION/AIR CONDITIONING SYSTEMS

Item	Net Price £	Material £	Labour hours	Labour £	Unit	Total rate £
U10: GRILLES/DIFFUSERS/LOUVRES – cont						
Weather louvres; opening mounted; 300 mm deep galvanized steel louvres; screw fixing in position						
Louvre units; including 12 mm galvanized mesh birdscreen						
900 × 600 mm	103.18	116.54	2.25	53.37	nr	**169.91**
900 × 900 mm	149.08	168.37	2.25	53.37	nr	**221.74**
900 × 1200 mm	179.38	202.59	2.50	59.24	nr	**261.83**
900 × 1500 mm	218.30	246.54	2.50	59.24	nr	**305.78**
900 × 1800 mm	257.23	290.52	2.50	59.24	nr	**349.76**
900 × 2100 mm	322.43	364.15	2.50	59.24	nr	**423.39**
900 × 2400 mm	346.37	391.19	2.76	65.46	nr	**456.65**
900 × 2700 mm	391.04	441.64	2.76	65.46	nr	**507.10**
900 × 3000 mm	427.05	482.31	2.76	65.46	nr	**547.77**
1200 × 600 mm	139.74	157.82	2.25	53.37	nr	**211.19**
1200 × 900 mm	195.26	220.52	2.50	59.24	nr	**279.76**
1200 × 1200 mm	251.21	283.71	2.50	59.24	nr	**342.95**
1200 × 1500 mm	293.23	331.17	2.50	59.24	nr	**390.41**
1200 × 1800 mm	383.49	433.11	2.76	65.46	nr	**498.57**
1200 × 2100 mm	436.13	492.56	2.76	65.46	nr	**558.02**
1200 × 2400 mm	455.46	514.40	2.76	65.46	nr	**579.86**
1500 × 600 mm	165.16	186.53	2.25	53.37	nr	**239.90**
1500 × 900 mm	227.59	257.04	2.50	59.24	nr	**316.28**
1500 × 1200 mm	299.79	338.58	2.50	59.24	nr	**397.82**
1500 × 1500 mm	351.48	396.96	2.76	65.46	nr	**462.42**
1500 × 1800 mm	414.44	468.07	2.76	65.46	nr	**533.53**
1500 × 2100 mm	518.34	585.41	3.00	71.16	nr	**656.57**
1800 × 600 mm	184.75	208.65	2.50	59.24	nr	**267.89**
1800 × 900 mm	255.40	288.45	2.50	59.24	nr	**347.69**
1800 × 1200 mm	334.90	378.24	2.76	65.46	nr	**443.70**
1800 × 1500 mm	396.15	447.41	3.00	71.16	nr	**518.57**

U: VENTILATION/AIR CONDITIONING SYSTEMS

Item	Net Price £	Material £	Labour hours	Labour £	Unit	Total rate £
U10: PLANT/EQUIPMENT						
Y41 – FANS						
Axial flow fan; including ancillaries, anti vibration mountings, mounting feet, matching flanges, flexible connectors and clips; 415 V, 3 phase, 50 Hz motor; includes fixing in position; electrical work elsewhere						
Aerofoil blade fan unit; short duct case						
315 mm dia.; 0.47 m³/s duty; 147 Pa	550.80	622.07	4.50	106.63	nr	**728.70**
500 mm dia.; 1.89 m³/s duty; 500 Pa	758.88	857.08	1.00	23.69	nr	**880.77**
560 mm dia.; 2.36 m³/s duty; 147 Pa	737.46	832.88	5.50	130.33	nr	**963.21**
710 mm dia.; 5.67 m³/s duty; 245 Pa	1140.36	1287.91	6.00	142.17	nr	**1430.08**
Aerofoil blade fan unit; long duct case						
315 mm dia.; 0.47 m³/s duty; 147 Pa	633.42	715.38	4.50	106.63	nr	**822.01**
500 mm dia.; 1.89 m³/s duty; 500 Pa	805.80	910.06	5.00	118.47	nr	**1028.53**
560 mm dia.; 2.36 m³/s duty; 147 Pa	867.04	979.23	5.50	130.33	nr	**1109.56**
710 mm dia.; 5.67 m³/s duty; 245 Pa	1206.66	1362.79	6.00	142.17	nr	**1504.96**
Aerofoil blade fan unit; two stage parallel fan arrangement; long duct case						
315 mm; 0.47 m³/s @ 500 Pa	1114.86	1259.12	4.50	106.63	nr	**1365.75**
355 mm; 0.83 m³/s @ 147 Pa	1067.94	1206.13	4.75	112.56	nr	**1318.69**
710 mm; 3.77 m³/s @ 431 Pa	2044.08	2308.58	6.00	142.17	nr	**2450.75**
710 mm; 6.61 m³/s @ 500 Pa	2547.96	2877.65	6.00	142.17	nr	**3019.82**
Axial flow fan; suitable for operation at 300°C for 90 minutes; including ancillaries, anti vibration mountings, mounting feet, matching flanges, flexible connectors and clips; 415 V, 3 phase, 50 Hz motor; includes fixing in position; electrical work elsewhere						
450 mm; 2.0 m³/s @ 300 Pa	642.60	725.75	5.00	118.47	nr	**844.22**
630 mm; 4.6 m³/s @ 200 Pa	1750.32	1976.80	5.50	130.33	nr	**2107.13**
900 mm; 9.0 m³/s @ 300 Pa	2042.04	2306.27	6.50	154.02	nr	**2460.29**
1000 mm; 15.0 m³/s @ 400 Pa	4926.60	5564.08	7.50	177.71	nr	**5741.79**

U: VENTILATION/AIR CONDITIONING SYSTEMS

Item	Net Price £	Material £	Labour hours	Labour £	Unit	Total rate £
U10: PLANT/EQUIPMENT – cont						
Y41 – FANS – cont						
Bifurcated fan; suitable for temperature up to 200°C with motor protection to IP55; including ancillaries, anti vibration mountings, mounting feet, matching flanges, flexible connectors and clips; 415 V, 3 phase, 50 Hz motor; includes fixing in position; electrical work elsewhere						
300 mm; 0.50 m³/s @ 100 Pa	1081.20	1221.10	4.50	106.63	nr	**1327.73**
400 mm; 1.97 m³/s @ 200 Pa	1627.92	1838.56	5.00	118.47	nr	**1957.03**
630 mm; 3.86 m³/s @ 200 Pa	1912.50	2159.97	5.50	130.33	nr	**2290.30**
800 mm; 6.10 m³/s @ 400 Pa	3656.70	4129.86	6.50	154.02	nr	**4283.88**
Duct mounted in line fan with backward curved centrifugal impellor; including ancillaries, matching flanges, flexible connectors and clips; 415 V, 3 phase, 50 Hz motor; includes fixing in position; electrical work elsewhere						
0.5 m³/s @ 200 Pa	1257.24	1419.92	4.50	106.63	nr	**1526.55**
1.0 m³/s @ 300 Pa	1549.93	1750.48	5.00	118.47	nr	**1868.95**
3.0 m³/s @ 500 Pa	2561.06	2892.44	5.50	130.33	nr	**3022.77**
5.0 m³/s @ 750 Pa	3392.56	3831.54	6.50	154.02	nr	**3985.56**
7.0 m³/s @ 1000 Pa	4124.29	4657.95	7.00	165.86	nr	**4823.81**
Twin fan extract unit; belt driven; located internally; complete with anti-vibration mounts and non return shutter; including ancillaries, matching flanges, flexible connectors and clips; 3 phase, 50 Hz motor; includes fixing in position; electrical work elsewhere						
0.25 m³/s @ 150 Pa	1734.86	1959.34	4.50	106.63	nr	**2065.97**
Extra for external unit	232.51	262.60	–	–	nr	**262.60**
0.50 m³/s @ 200 Pa	2018.25	2279.40	5.00	118.47	nr	**2397.87**
Extra for external unit	232.51	262.60	–	–	nr	**262.60**
1.00 m³/s @ 200 Pa	2301.49	2599.29	5.00	118.47	nr	**2717.76**
Extra for external unit	232.51	262.60	–	–	nr	**262.60**
1.50 m³/s @ 250 Pa	7024.36	7933.27	5.50	130.33	nr	**8063.60**
Extra for external unit	232.51	262.60	–	–	nr	**262.60**
2.00 m³/s @ 250 Pa	2984.11	3370.24	6.50	154.02	nr	**3524.26**
Extra for external unit	232.51	262.60	–	–	nr	**262.60**
Extra for auto changeover panel	218.83	247.15	2.50	59.24	nr	**306.39**

U: VENTILATION/AIR CONDITIONING SYSTEMS

Item	Net Price £	Material £	Labour hours	Labour £	Unit	Total rate £
Roof mounted extract fan; including ancillaries, fibreglass cowling, fitted shutters and bird guard; 415 V, 3 phase, 50 Hz motor; includes fixing in position; electrical work elsewhere						
Flat roof installation, fixed to curb						
315 mm; 900 rpm	310.56	350.75	4.50	106.63	nr	457.38
315 mm; 1380 rpm	578.31	653.14	4.50	106.63	nr	759.77
400 mm; 900 rpm	813.62	918.90	5.50	130.33	nr	1049.23
400 mm; 1360 rpm	901.64	1018.31	5.50	130.33	nr	1148.64
800 mm; 530 rpm	2338.80	2641.43	7.00	165.86	nr	2807.29
800 mm; 700 rpm	2197.05	2481.33	7.00	165.86	nr	2647.19
800 mm; 920 rpm	1949.00	2201.19	7.00	165.86	nr	2367.05
1000 mm; 470 rpm	3118.39	3521.90	8.00	189.56	nr	3711.46
1000 mm; 570 rpm	2976.66	3361.83	8.00	189.56	nr	3551.39
1000 mm; 710 rpm	2895.87	3270.58	8.00	189.56	nr	3460.14
Pitched roof installation; including purlin mounting box						
315 mm; 900 rpm	642.90	726.09	4.50	106.63	nr	832.72
315 mm; 1380 rpm	642.90	726.09	4.50	106.63	nr	832.72
400 mm; 900 rpm	891.65	1007.03	5.50	130.33	nr	1137.36
400 mm; 1360 rpm	991.67	1119.99	5.50	130.33	nr	1250.32
800 mm; 530 rpm	2494.25	2817.00	7.00	165.86	nr	2982.86
800 mm; 700 rpm	2352.50	2656.91	7.00	165.86	nr	2822.77
800 mm; 920 rpm	2504.44	2828.50	7.00	165.86	nr	2994.36
1000 mm; 470 rpm	3348.30	3781.56	8.00	189.56	nr	3971.12
1000 mm; 570 rpm	3206.56	3621.47	8.00	189.56	nr	3811.03
1000 mm; 710 rpm	3125.76	3530.22	8.00	189.56	nr	3719.78
Centrifugal fan; single speed for internal domestic kitchens/utility rooms; fitted with standard overload protection; complete with housing; includes placing in position; electrical work elsewhere						
Window mounted						
245 m³/hr	85.98	97.10	0.50	11.85	nr	108.95
500 m³/hr	269.87	609.59	0.50	11.85	nr	621.44
Wall mounted						
245 m³/hr	225.89	365.62	0.83	19.74	nr	385.36
500 m³/hr	444.24	1124.70	0.83	19.74	nr	1144.44

U: VENTILATION/AIR CONDITIONING SYSTEMS

Item	Net Price £	Material £	Labour hours	Labour £	Unit	Total rate £
U10: PLANT/EQUIPMENT – cont						
Y41 – FANS – cont						
Centrifugal fan; various speeds, simultaneous ventilation from separate areas fitted with standard overload protection; complete with housing; includes placing in position; ducting and electrical work elsewhere						
Fan unit						
147-300 m³/hr	141.03	159.28	1.00	23.69	nr	**182.97**
175-411 m³/hr	232.83	262.96	1.00	23.69	nr	**286.65**
Toilet extract units; centrifugal fan; various speeds for internal domestic bathrooms/ W.Cs, with built in filter; complete with housing; includes placing in position; electrical work elsewhere						
Fan unit; fixed to wall; including shutter						
Single speed 85 m³/hr	94.44	106.66	0.75	17.77	nr	**124.43**
Two speed 60-85 m³/hr	117.43	132.62	0.83	19.74	nr	**152.36**
Humidity controlled; autospeed; fixed to wall; including shutter						
30-60-85 m³/hr	190.63	215.30	1.00	23.69	nr	**238.99**
Y42 – AIR FILTRATION						
High efficiency duct mounted filters; 99.997% H13 (EU13); tested to BS 3928						
Standard; 2100 m³/ hr air volume; continuous rating up to 80°C; sealed wood case, aluminium spacers, neoprene gaskets; water repellant filter media.; includes placing in position						
610 × 610 × 292 mm	229.50	259.19	1.00	22.71	nr	**281.90**
Side withdrawl frame	86.29	97.45	2.50	56.75	nr	**154.20**
High capacity; 3400 m³/hr air volume; continuous rating up to 80°C; anti-corrosion coated mild steel frame, polyurethane sealant and neoprene gaskets; water repellant filter media.; includes placing in position						
610 × 610 × 292 mm	245.82	277.62	1.00	22.71	nr	**300.33**
Side withdrawl frame	86.29	97.45	2.50	56.75	nr	**154.20**

U: VENTILATION/AIR CONDITIONING SYSTEMS

Item	Net Price £	Material £	Labour hours	Labour £	Unit	Total rate £
Bag filters; 40/60% F5 (EU5); tested to BSEN 779						
Duct mounted bag filter; continuous rating up to 60°C; rigid filter assembly; sealed into one piece coated mild steel header with sealed pocket separators; includes placing in position						
6 pocket, 592 × 592 × 25 mm header;						
pockets 350 mm long; 1690 m³/hr	17.99	20.32	1.00	22.71	nr	**43.03**
Side withdrawl frame	18.54	20.94	2.00	45.40	nr	**66.34**
6 pocket, 592 × 592 × 25 mm header;						
pockets 500 mm long; 2550 m³/hr	19.64	22.19	1.50	34.05	nr	**56.24**
Side withdrawl frame	18.54	20.94	2.50	56.75	nr	**77.69**
6 pocket, 592 × 592 × 25 mm header;						
pockets 600 mm long; 3380 m³/hr	21.27	24.02	1.50	34.05	nr	**58.07**
Side withdrawl frame	18.54	20.94	3.00	68.11	nr	**89.05**
Bag filters; 80/90% F7, (EU7); tested to BSEN 779						
Duct mounted bag filter; continuous rating up to 60°C; rigid filter assembly; sealed into one piece coated mild steel header with sealed pocket separators; includes placing in position						
6 pocket, 592 × 592 × 25 mm header;						
pockets 500 mm long; 1688 m³/hr	21.27	24.02	1.00	22.71	nr	**46.73**
Side withdrawl frame	18.54	20.94	2.00	45.40	nr	**66.34**
6 pocket, 592 × 592 × 25 mm header;						
pockets 600 mm long; 2047 m³/hr	22.91	25.88	1.50	34.05	nr	**59.93**
Side withdrawl frame	18.54	20.94	2.50	56.75	nr	**77.69**
6 pocket, 592 × 592 × 25 mm header;						
pockets 700 mm long; 2729 m³/hr	24.54	27.71	1.50	34.05	nr	**61.76**
Side withdrawl frame	18.54	20.94	3.00	68.11	nr	**89.05**
Grease filters, washable; minimum 65%						
Double sided extract unit; lightweight stainless steel construction; demountable composite filter media. of woven metal mat and expanded metal mesh supports; for mounting on hood and extract systems (hood not included); includes placing in position						
500 × 686 × 565 mm, 4080 m³/hr	366.49	413.92	2.00	45.40	nr	**459.32**
1000 × 686 × 565 mm, 8160 m³/hr;	558.89	631.21	3.00	68.11	nr	**699.32**
1500 × 686 × 565 mm, 12240 m³/hr;	766.59	865.79	3.50	79.46	nr	**945.25**

U: VENTILATION/AIR CONDITIONING SYSTEMS

Item	Net Price £	Material £	Labour hours	Labour £	Unit	Total rate £
U10: PLANT/EQUIPMENT – cont						
Y42 – AIR FILTRATION – cont						
Panel filters; 82% G3 (EU3); tested to BS EN779						
Modular duct mounted filter panels; continuous rating up to 100°C; graduated density media.; rigid cardboard frame; includes placing in position						
596 × 596 × 47 mm, 2360 m³/hr	5.14	5.80	1.00	22.71	nr	**28.51**
Side withdrawl frame	52.51	59.30	2.50	56.75	nr	**116.05**
596 × 287 × 47 mm, 1140 m³/hr	3.66	4.13	1.00	22.71	nr	**26.84**
Side withdrawl frame	52.51	59.30	2.50	56.75	nr	**116.05**
Panel filters; 90% G4 (EU4); tested to BS EN779						
Modular duct mounted filter panels; continuous rating up to 100°C; pleated media. with wire support; rigid cardboard frame; includes placing in position						
596 × 596 × 47 mm, 2560 m³/hr	10.50	11.86	1.00	22.71	nr	**34.57**
side withdrawl frame	66.23	74.80	3.00	68.11	nr	**142.91**
596 × 287 × 47 mm, 1230 m3/hr	8.10	9.15	1.00	22.71	nr	**31.86**
Side withdrawl frame	52.51	59.30	3.00	68.11	nr	**127.41**
Carbon filters; standard duty disposable carbon filters; steel frame with bonded carbon panels; for fixing to ductwork; including placing in position						
12 panels						
597 × 597 × 298 mm, 1460 m³/hr	312.34	352.76	0.33	7.49	nr	**360.25**
597 × 597 × 451 mm, 2200 m³/hr	352.11	397.67	0.33	7.49	nr	**405.16**
597 × 597 × 597 mm, 2930 m³/hr	392.67	443.48	0.33	7.49	nr	**450.97**
8 panels						
451 × 451 × 298 mm, 740 m³/hr	235.48	265.95	0.29	6.59	nr	**272.54**
451 × 451 × 451 mm, 1105 m³/hr	260.62	294.34	0.29	6.59	nr	**300.93**
451 × 451 × 597 mm, 1460 m³/hr	284.62	321.45	0.29	6.59	nr	**328.04**
6 panels						
298 × 298 × 298 mm, 365 m³/hr	166.85	188.43	0.25	5.68	nr	**194.11**
298 × 298 × 451 mm, 550 m³/hr	178.64	201.76	0.25	5.68	nr	**207.44**
298 × 298 × 597 mm, 780 m³/hr	189.92	214.50	0.25	5.68	nr	**220.18**

U: VENTILATION/AIR CONDITIONING SYSTEMS

Item	Net Price £	Material £	Labour hours	Labour £	Unit	Total rate £
U10: SILENCERS/ACOUSTIC TREATMENT						
Y45 – SILENCERS/ACOUSTIC TREATMENT						
Attenuators; DW144 galvanized construction c/w splitters; self securing; fitted to ductwork						
To suit rectangular ducts; unit length 600 mm						
100 × 100 mm	88.33	99.76	0.75	17.02	nr	**116.78**
150 × 150 mm	92.21	104.14	0.75	17.02	nr	**121.16**
200 × 200 mm	96.07	108.50	0.75	17.02	nr	**125.52**
300 × 300 mm	107.49	121.40	0.75	17.02	nr	**138.42**
400 × 400 mm	120.66	136.27	1.00	22.71	nr	**158.98**
500 × 500 mm	140.82	159.04	1.25	28.39	nr	**187.43**
600 × 300 mm	135.65	153.20	1.25	28.39	nr	**181.59**
600 × 600 mm	189.85	214.41	1.25	28.39	nr	**242.80**
700 × 300 mm	143.82	162.42	1.50	34.05	nr	**196.47**
700 × 700 mm	214.16	241.87	1.50	34.05	nr	**275.92**
800 × 300 mm	148.98	168.26	2.00	45.40	nr	**213.66**
800 × 800 mm	247.07	279.04	2.00	45.40	nr	**324.44**
1000 × 1000 mm	316.27	357.19	3.00	68.11	nr	**425.30**
To suit rectangular ducts; unit length 1200 mm						
200 × 200 mm	105.94	119.65	1.00	22.71	nr	**142.36**
300 × 300 mm	130.98	147.93	1.00	22.71	nr	**170.64**
400 × 400 mm	157.48	177.86	1.33	30.19	nr	**208.05**
500 × 500 mm	190.08	214.68	1.66	37.69	nr	**252.37**
600 × 300 mm	181.24	204.69	1.66	37.69	nr	**242.38**
600 × 600 mm	265.17	299.48	1.66	37.69	nr	**337.17**
700 × 300 mm	194.28	219.42	2.00	45.40	nr	**264.82**
700 × 700 mm	303.45	342.72	2.00	45.40	nr	**388.12**
800 × 300 mm	203.11	229.39	2.66	60.38	nr	**289.77**
800 × 800 mm	350.15	395.45	2.66	60.38	nr	**455.83**
1000 × 1000 mm	473.98	535.31	4.00	90.80	nr	**626.11**
1300 × 1300 mm	813.62	918.90	8.00	181.61	nr	**1100.51**
1500 × 1500 mm	931.28	1051.78	8.00	181.61	nr	**1233.39**
1800 × 1800 mm	1274.21	1439.09	10.66	242.00	nr	**1681.09**
2000 × 2000 mm	1429.82	1614.83	13.33	302.60	nr	**1917.43**
To suit rectangular ducts; unit length 1800 mm						
200 × 200 mm	133.70	151.00	1.00	22.71	nr	**173.71**
300 × 300 mm	177.67	200.66	1.00	22.71	nr	**223.37**
400 × 400 mm	212.37	239.85	1.33	30.19	nr	**270.04**
500 × 500 mm	247.50	279.52	1.66	37.69	nr	**317.21**
600 × 300 mm	258.87	292.36	1.66	37.69	nr	**330.05**
600 × 600 mm	361.71	408.51	1.66	37.69	nr	**446.20**
700 × 300 mm	274.42	309.93	2.00	45.40	nr	**355.33**
700 × 700 mm	432.17	488.09	2.00	45.40	nr	**533.49**
800 × 300 mm	290.19	327.74	2.66	60.38	nr	**388.12**
800 × 800 mm	502.85	567.92	2.66	60.38	nr	**628.30**
1000 × 1000 mm	675.90	763.36	4.00	90.80	nr	**854.16**

U: VENTILATION/AIR CONDITIONING SYSTEMS

Item	Net Price £	Material £	Labour hours	Labour £	Unit	Total rate £
U10: SILENCERS/ACOUSTIC TREATMENT – cont						
Y45 – SILENCERS/ACOUSTIC TREATMENT – cont						
Attenuators – cont						
To suit rectangular ducts; unit length 1800 mm – cont						
1300 × 1300 mm	1129.14	1275.24	8.00	181.61	nr	**1456.85**
1500 × 1500 mm	1279.61	1445.19	8.00	181.61	nr	**1626.80**
1800 × 1800 mm	1776.50	2006.37	13.33	302.60	nr	**2308.97**
2000 × 2000 mm	2011.63	2271.93	10.66	242.00	nr	**2513.93**
2300 × 2300 mm	2514.50	2839.86	16.00	363.21	nr	**3203.07**
2500 × 2500 mm	3507.42	3961.26	18.66	423.61	nr	**4384.87**
To suit rectangular ducts; unit length 2400 mm						
500 × 500 mm	320.07	361.49	2.08	47.29	nr	**408.78**
600 × 300 mm	305.56	345.10	2.08	47.29	nr	**392.39**
600 × 600 mm	455.10	513.99	2.08	47.29	nr	**561.28**
700 × 300 mm	331.21	374.06	2.50	56.75	nr	**430.81**
700 × 700 mm	537.56	607.11	2.50	56.75	nr	**663.86**
800 × 300 mm	350.15	395.45	3.33	75.59	nr	**471.04**
800 × 800 mm	622.11	702.61	3.33	75.59	nr	**778.20**
1000 × 1000 mm	839.34	947.95	5.00	113.51	nr	**1061.46**
1300 × 1300 mm	1359.47	1535.38	10.00	227.01	nr	**1762.39**
1500 × 1500 mm	1551.57	1752.34	10.00	227.01	nr	**1979.35**
1800 × 1800 mm	2213.81	2500.26	13.33	302.55	nr	**2802.81**
2000 × 2000 mm	2484.26	2805.71	16.66	378.20	nr	**3183.91**
2300 × 2300 mm	3073.59	3471.29	20.00	454.01	nr	**3925.30**
2500 × 2500 mm	4315.76	4874.20	23.32	529.47	nr	**5403.67**
To suit circular ducts; unit length 600 mm						
100 mm dia.	76.71	86.64	0.75	17.02	nr	**103.66**
200 mm dia.	90.26	101.94	0.75	17.02	nr	**118.96**
250 mm dia.	102.32	115.56	0.75	17.02	nr	**132.58**
315 mm dia.	119.10	134.51	1.00	22.71	nr	**157.22**
355 mm dia.	137.81	155.64	1.00	22.71	nr	**178.35**
400 mm dia.	148.13	167.30	1.00	22.71	nr	**190.01**
450 mm dia.	178.89	202.04	1.00	22.71	nr	**224.75**
500 mm dia.	196.96	222.44	1.26	28.56	nr	**251.00**
630 mm dia.	232.22	262.27	1.50	34.05	nr	**296.32**
710 mm dia.	252.88	285.60	1.50	34.05	nr	**319.65**
800 mm dia.	281.28	317.68	2.00	45.40	nr	**363.08**
1000 mm dia.	377.57	426.42	3.00	68.11	nr	**494.53**
To suit circular ducts; unit length 1200 mm						
100 mm dia.	112.85	127.45	1.00	22.71	nr	**150.16**
200 mm dia.	123.82	139.84	1.00	22.71	nr	**162.55**
250 mm dia.	139.10	157.10	1.00	22.71	nr	**179.81**

U: VENTILATION/AIR CONDITIONING SYSTEMS

Item	Net Price £	Material £	Labour hours	Labour £	Unit	Total rate £
315 mm dia.	159.10	179.68	1.33	30.19	nr	**209.87**
355 mm dia.	190.73	215.41	1.33	30.19	nr	**245.60**
400 mm dia.	202.34	228.52	1.33	30.19	nr	**258.71**
450 mm dia.	234.39	264.72	1.33	30.19	nr	**294.91**
500 mm dia.	240.20	271.28	1.66	37.69	nr	**308.97**
630 mm dia.	318.06	359.21	2.00	45.40	nr	**404.61**
710 mm dia.	351.60	397.10	2.00	45.40	nr	**442.50**
800 mm dia.	350.96	396.37	2.66	60.38	nr	**456.75**
1000 mm dia.	471.15	532.11	4.00	90.80	nr	**622.91**
1250 mm dia.	887.53	1002.37	8.00	181.61	nr	**1183.98**
1400 mm dia.	964.96	1089.82	8.00	181.61	nr	**1271.43**
1600 mm dia.	1069.79	1208.21	10.66	242.00	nr	**1450.21**
To suit circular ducts; unit length 1800 mm						
200 mm dia.	173.51	195.96	1.25	28.39	nr	**224.35**
250 mm dia.	194.16	219.28	1.25	28.39	nr	**247.67**
315 mm dia.	230.73	260.59	1.66	37.69	nr	**298.28**
355 mm dia.	272.03	307.23	1.66	37.69	nr	**344.92**
400 mm dia.	281.49	317.91	1.66	37.69	nr	**355.60**
450 mm dia.	287.53	324.74	1.66	37.69	nr	**362.43**
500 mm dia.	316.09	356.99	2.08	47.22	nr	**404.21**
630 mm dia.	349.67	394.91	2.50	56.75	nr	**451.66**
710 mm dia.	409.04	461.97	2.50	56.75	nr	**518.72**
800 mm dia.	423.24	478.00	3.33	75.59	nr	**553.59**
1000 mm dia.	1105.00	1247.98	6.66	151.19	nr	**1399.17**
1250 mm dia.	1123.07	1268.39	6.66	151.19	nr	**1419.58**
1400 mm dia.	1240.80	1401.36	5.00	113.51	nr	**1514.87**
1600 mm dia.	1241.54	1402.19	9.90	224.74	nr	**1626.93**

U: VENTILATION/AIR CONDITIONING SYSTEMS

Item	Net Price £	Material £	Labour hours	Labour £	Unit	Total rate £
U10: THERMAL INSULATION						
Y50 – THERMAL INSULATION						
Concealed Ductwork						
Flexible wrap; 20 kg–45 kg Bright Class O aluminium foil faced; Bright Class O foil taped joints; 62 mm metal pins and washers; aluminium bands						
40 mm thick insulation	9.85	11.13	0.40	9.48	m²	**20.61**
Semi-rigid slab; 45 kg Bright Class O aluminium foil faced mineral fibre; Bright Class O foil taped joints; 62 mm metal pins and washers; aluminium bands						
40 mm thick insulation	12.89	14.56	0.65	15.40	m²	**29.96**
Plantroom Ductwork						
Semi-rigid slab; 45 kg Bright Class O aluminium foil faced mineral fibre; Bright Class O foil taped joints; 62 mm metal pins and washers; 22 swg plain/embossed aluminium cladding; pop rivited						
50 mm thick insulation	32.52	36.73	1.50	35.55	m²	**72.28**
External Ductwork						
Semi-rigid slab; 45 kg Bright Class O aluminium foil faced mineral fibre; Bright Class O foil taped joints; 62 mm metal pins and washers; 0.8 mm polyisobutylene sheeting; welded joints						
50 mm thick insulation	23.73	26.81	1.25	29.62	m²	**56.43**

U: VENTILATION/AIR CONDITIONING SYSTEMS

Item	Net Price £	Material £	Labour hours	Labour £	Unit	Total rate £
U14: DUCTWORK: FIRE RATED						
Y30 – DUCTLINES						
The relevant BS requires that the fire-rating of ductwork meets 3 criteria; stability (hours), integrity (hours) and insulation (hours). The least of the 3 periods defines the fire-rating. The BS does however allow stability and integrity to be considered in isolation. Rates are therefore provided for both types of system						
Care should to be taken when using the rates within this section to ensure that the requiremements for stability, integrity and insulation are known and the appropriate rates are used						
High density single layer mineral wool fire-rated ductwork slab, in accordance with BS476, Part 24 (ISO 6944: 1985), ducts 'Type A' and 'Type B'; 165 kg class O foil faced mineral fibre; 100 mm wide bright class O foil taped joints; welded pins; includes protection to all supports						
½ hour stability, integrity and insulation						
25 mm thick, vertical and horizontal						
ductwork	35.39	43.26	1.25	32.06	m²	**75.32**
1 hour stability, integrity and insulation						
30 mm thick, vertical ductwork	41.55	50.80	1.50	38.47	m²	**89.27**
40 mm thick, horizontal ductwork	48.87	59.74	1.50	38.47	m²	**98.21**
1½ hour stability, integrity and insulation						
50 mm thick, vertical ductwork	58.38	71.37	1.75	44.89	m²	**116.26**
70 mm thick, horizontal ductwork	73.03	89.28	1.75	44.89	m²	**134.17**
2 hour stability, integrity and insulation						
70 mm, vertical ductwork	75.25	91.99	2.00	51.29	m²	**143.28**
90 mm horizontal ductwork	89.88	109.88	2.00	51.29	m²	**161.17**
Kitchen extract, 1 hour stability, integrity and insulation						
90 mm, vertical and horizontal	89.88	109.88	2.00	51.29	m²	**161.17**

U: VENTILATION/AIR CONDITIONING SYSTEMS

Item	Net Price £	Material £	Labour hours	Labour £	Unit	Total rate £
U14: DUCTWORK: FIRE RATED – cont						
Galvanized sheet metal rectangular section ductwork to BS476 Part 24 (ISO 6944:1985), ducts 'Type A' and 'Type B'; provides 2 hours stability and 2 hours integrity at 1100°C (no rating for insulation); including all necessary stiffeners, joints and supports in the running length						
Ductwork up to 600 mm longest side						
Sum of two sides 200 mm	60.61	74.10	2.91	71.50	m	**145.60**
Sum of two sides 300 mm	65.80	80.44	2.99	73.47	m	**153.91**
Sum of two sides 400 mm	70.98	86.78	3.17	77.90	m	**164.68**
Sum of two sides 500 mm	76.17	93.12	3.37	82.81	m	**175.93**
Sum of two sides 600 mm	81.35	99.45	3.54	86.98	m	**186.43**
Sum of two sides 700 mm	86.54	105.79	3.72	91.40	m	**197.19**
Sum of two sides 800 mm	91.74	112.16	3.90	95.83	m	**207.99**
Sum of two sides 900 mm	96.91	118.47	5.04	123.84	m	**242.31**
Sum of two sides 1000 mm	102.11	124.83	5.58	137.11	m	**261.94**
Sum of two sides 1100 mm	107.28	131.15	5.84	143.50	m	**274.65**
Sum of two sides 1200 mm	112.48	137.50	6.11	150.13	m	**287.63**
Extra over fittings; Ductwork up to 600 mm longest side						
End Cap						
Sum of two sides 200 mm	17.21	21.04	0.81	19.90	nr	**40.94**
Sum of two sides 300 mm	18.25	22.31	0.84	20.63	nr	**42.94**
Sum of two sides 400 mm	19.30	23.60	0.87	21.38	nr	**44.98**
Sum of two sides 500 mm	20.34	24.86	0.90	22.11	nr	**46.97**
Sum of two sides 600 mm	21.39	26.15	0.93	22.85	nr	**49.00**
Sum of two sides 700 mm	22.44	27.43	0.96	23.60	nr	**51.03**
Sum of two sides 800 mm	23.51	28.74	0.98	24.08	nr	**52.82**
Sum of two sides 900 mm	24.55	30.01	1.17	28.76	nr	**58.77**
Sum of two sides 1000 mm	25.59	31.28	1.22	29.97	nr	**61.25**
Sum of two sides 1100 mm	26.65	32.58	1.25	30.72	nr	**63.30**
Sum of two sides 1200 mm	27.69	33.86	1.28	31.46	nr	**65.32**
Reducer						
Sum of two sides 200 mm	68.25	83.44	2.23	54.80	nr	**138.24**
Sum of two sides 300 mm	71.39	87.27	2.37	58.24	nr	**145.51**
Sum of two sides 400 mm	74.51	91.09	2.51	61.68	nr	**152.77**
Sum of two sides 500 mm	77.67	94.96	2.65	65.13	nr	**160.09**
Sum of two sides 600 mm	80.79	98.77	2.79	68.56	nr	**167.33**
Sum of two sides 700 mm	83.93	102.61	2.87	70.52	nr	**173.13**
Sum of two sides 800 mm	87.06	106.43	3.01	73.96	nr	**180.39**

U: VENTILATION/AIR CONDITIONING SYSTEMS

Item	Net Price £	Material £	Labour hours	Labour £	Unit	Total rate £
Sum of two sides 900 mm	90.20	110.27	3.06	75.20	nr	**185.47**
Sum of two sides 1000 mm	93.32	114.08	3.17	77.90	nr	**191.98**
Sum of two sides 1100 mm	96.45	117.91	3.22	79.12	nr	**197.03**
Sum of two sides 1200 mm	99.60	121.77	3.28	80.60	nr	**202.37**
Offset						
Sum of two sides 200 mm	164.73	201.39	2.95	72.49	nr	**273.88**
Sum of two sides 300 mm	168.81	206.37	3.11	76.42	nr	**282.79**
Sum of two sides 400 mm	172.89	211.36	3.26	80.11	nr	**291.47**
Sum of two sides 500 mm	176.99	216.37	3.42	84.04	nr	**300.41**
Sum of two sides 600 mm	181.07	221.36	3.57	87.73	nr	**309.09**
Sum of two sides 700 mm	189.25	231.36	3.23	79.36	nr	**310.72**
Sum of two sides 800 mm	185.17	226.37	3.73	91.65	nr	**318.02**
Sum of two sides 900 mm	193.35	236.37	3.45	84.77	nr	**321.14**
Sum of two sides 1000 mm	197.43	241.36	3.67	90.19	nr	**331.55**
Sum of two sides 1100 mm	201.51	246.35	3.78	92.89	nr	**339.24**
Sum of two sides 1200 mm	205.62	251.37	3.89	95.58	nr	**346.95**
90° radius bend						
Sum of two sides 200 mm	66.79	81.65	2.06	50.62	nr	**132.27**
Sum of two sides 300 mm	71.86	87.85	2.21	54.30	nr	**142.15**
Sum of two sides 400 mm	76.90	94.01	2.36	58.00	nr	**152.01**
Sum of two sides 500 mm	81.98	100.22	2.51	61.68	nr	**161.90**
Sum of two sides 600 mm	87.04	106.41	2.66	65.36	nr	**171.77**
Sum of two sides 700 mm	92.08	112.57	2.81	69.05	nr	**181.62**
Sum of two sides 800 mm	97.15	118.77	2.97	72.99	nr	**191.76**
Sum of two sides 900 mm	102.19	124.93	2.99	73.47	nr	**198.40**
Sum of two sides 1000 mm	107.27	131.14	3.04	74.70	nr	**205.84**
Sum of two sides 1100 mm	112.31	137.30	3.07	75.44	nr	**212.74**
Sum of two sides 1200 mm	117.37	143.49	3.09	75.93	nr	**219.42**
45° radius bend						
Sum of two sides 200 mm	82.35	100.67	1.52	37.35	nr	**138.02**
Sum of two sides 300 mm	84.41	103.20	1.59	39.07	nr	**142.27**
Sum of two sides 400 mm	86.46	105.70	1.66	40.80	nr	**146.50**
Sum of two sides 500 mm	88.50	108.19	1.73	42.51	nr	**150.70**
Sum of two sides 600 mm	90.54	110.68	1.80	44.23	nr	**154.91**
Sum of two sides 700 mm	92.60	113.21	1.87	45.96	nr	**159.17**
Sum of two sides 800 mm	94.63	115.69	1.93	47.42	nr	**163.11**
Sum of two sides 900 mm	96.68	118.19	1.99	48.90	nr	**167.09**
Sum of two sides 1000 mm	98.72	120.69	2.05	50.37	nr	**171.06**
Sum of two sides 1100 mm	100.76	123.18	2.11	51.85	nr	**175.03**
Sum of two sides 1200 mm	102.80	125.67	2.17	53.32	nr	**178.99**

U: VENTILATION/AIR CONDITIONING SYSTEMS

Item	Net Price £	Material £	Labour hours	Labour £	Unit	Total rate £
U14: DUCTWORK: FIRE RATED – cont						
Extra over fittings; Ductwork up to 600 mm longest side – cont						
90° mitre bend						
Sum of two sides 200 mm	69.24	84.65	2.47	60.70	nr	**145.35**
Sum of two sides 300 mm	80.25	98.11	2.65	65.13	nr	**163.24**
Sum of two sides 400 mm	91.26	111.57	2.84	69.78	nr	**181.35**
Sum of two sides 500 mm	102.28	125.04	3.02	74.20	nr	**199.24**
Sum of two sides 600 mm	113.29	138.49	3.20	78.63	nr	**217.12**
Sum of two sides 700 mm	124.30	151.96	3.38	83.06	nr	**235.02**
Sum of two sides 800 mm	135.31	165.42	3.56	87.48	nr	**252.90**
Sum of two sides 900 mm	146.33	178.89	3.59	88.22	nr	**267.11**
Sum of two sides 1000 mm	157.36	192.37	3.66	89.94	nr	**282.31**
Sum of two sides 1100 mm	168.35	205.80	3.69	90.67	nr	**296.47**
Sum of two sides 1200 mm	179.39	219.31	3.72	91.40	nr	**310.71**
Branch (Side-on Shoe)						
Sum of two sides 200 mm	25.70	31.41	0.98	24.08	nr	**55.49**
Sum of two sides 300 mm	25.85	31.60	1.06	26.05	nr	**57.65**
Sum of two sides 400 mm	25.99	31.78	1.14	28.01	nr	**59.79**
Sum of two sides 500 mm	26.14	31.96	1.22	29.97	nr	**61.93**
Sum of two sides 600 mm	26.26	32.11	1.30	31.94	nr	**64.05**
Sum of two sides 700 mm	26.41	32.29	1.38	33.91	nr	**66.20**
Sum of two sides 800 mm	26.74	32.69	1.37	33.67	nr	**66.36**
Sum of two sides 900 mm	26.54	32.44	1.46	35.89	nr	**68.33**
Sum of two sides 1000 mm	26.81	32.77	1.46	35.89	nr	**68.66**
Sum of two sides 1100 mm	26.96	32.96	1.50	36.86	nr	**69.82**
Sum of two sides 1200 mm	27.10	33.12	1.55	38.09	nr	**71.21**
Ductwork 601 to 800 mm longest side						
Sum of two sides 900 mm	96.90	118.46	5.04	123.84	m	**242.30**
Sum of two sides 1000 mm	102.09	124.81	5.58	137.11	m	**261.92**
Sum of two sides 1100 mm	107.28	131.15	5.84	143.50	m	**274.65**
Sum of two sides 1200 mm	112.48	137.50	6.11	150.13	m	**287.63**
Sum of two sides 1300 mm	117.68	143.86	6.38	156.78	m	**300.64**
Sum of two sides 1400 mm	122.88	150.23	6.65	163.40	m	**313.63**
Sum of two sides 1500 mm	128.08	156.58	6.91	169.80	m	**326.38**
Sum of two sides 1600 mm	133.27	162.93	7.18	176.43	m	**339.36**
Extra over fittings; Ductwork 601 to 800 mm longest side						
End Cap						
Sum of two sides 900 mm	22.81	27.88	1.17	28.76	nr	**56.64**
Sum of two sides 1000 mm	24.44	29.88	1.22	29.97	nr	**59.85**
Sum of two sides 1100 mm	26.08	31.88	1.25	30.72	nr	**62.60**
Sum of two sides 1200 mm	27.70	33.87	1.28	31.46	nr	**65.33**
Sum of two sides 1300 mm	29.33	35.85	1.31	32.19	nr	**68.04**
Sum of two sides 1400 mm	30.96	37.85	1.34	32.92	nr	**70.77**
Sum of two sides 1500 mm	32.59	39.84	1.36	33.42	nr	**73.26**
Sum of two sides 1600 mm	34.22	41.84	1.39	34.15	nr	**75.99**

U: VENTILATION/AIR CONDITIONING SYSTEMS

Item	Net Price £	Material £	Labour hours	Labour £	Unit	Total rate £
Reducer						
Sum of two sides 900 mm	85.36	104.35	3.06	75.20	nr	**179.55**
Sum of two sides 1000 mm	90.12	110.17	3.17	77.90	nr	**188.07**
Sum of two sides 1100 mm	94.86	115.97	3.22	79.12	nr	**195.09**
Sum of two sides 1200 mm	99.61	121.78	3.28	80.60	nr	**202.38**
Sum of two sides 1300 mm	104.37	127.60	3.33	81.82	nr	**209.42**
Sum of two sides 1400 mm	109.12	133.40	3.38	83.06	nr	**216.46**
Sum of two sides 1500 mm	113.86	139.19	3.44	84.52	nr	**223.71**
Sum of two sides 1600 mm	118.61	145.00	3.49	85.76	nr	**230.76**
Offset						
Sum of two sides 900 mm	181.55	221.94	3.45	84.77	nr	**306.71**
Sum of two sides 1000 mm	190.20	232.52	3.67	90.19	nr	**322.71**
Sum of two sides 1100 mm	198.84	243.08	3.78	92.89	nr	**335.97**
Sum of two sides 1200 mm	207.49	253.66	3.89	95.58	nr	**349.24**
Sum of two sides 1300 mm	216.12	264.21	4.00	98.29	nr	**362.50**
Sum of two sides 1400 mm	224.78	274.79	4.11	100.99	nr	**375.78**
Sum of two sides 1500 mm	233.41	285.34	4.23	103.94	nr	**389.28**
Sum of two sides 1600 mm	242.33	296.25	4.34	106.64	nr	**402.89**
90° radius bend						
Sum of two sides 900 mm	90.78	110.98	2.99	73.47	nr	**184.45**
Sum of two sides 1000 mm	99.69	121.87	3.04	74.70	nr	**196.57**
Sum of two sides 1100 mm	108.59	132.76	3.07	75.44	nr	**208.20**
Sum of two sides 1200 mm	117.50	143.64	3.09	75.93	nr	**219.57**
Sum of two sides 1300 mm	126.41	154.54	3.12	76.66	nr	**231.20**
Sum of two sides 1400 mm	135.31	165.42	3.15	77.39	nr	**242.81**
Sum of two sides 1500 mm	144.23	176.33	3.17	77.90	nr	**254.23**
Sum of two sides 1600 mm	153.15	187.22	3.20	78.63	nr	**265.85**
45° bend						
Sum of two sides 900 mm	90.13	110.19	1.74	42.76	nr	**152.95**
Sum of two sides 1000 mm	94.37	115.37	1.85	45.46	nr	**160.83**
Sum of two sides 1100 mm	98.60	120.54	1.91	46.94	nr	**167.48**
Sum of two sides 1200 mm	102.82	125.70	1.96	48.16	nr	**173.86**
Sum of two sides 1300 mm	107.04	130.86	2.02	49.64	nr	**180.50**
Sum of two sides 1400 mm	111.27	136.03	2.07	50.87	nr	**186.90**
Sum of two sides 1500 mm	115.49	141.19	2.13	52.33	nr	**193.52**
Sum of two sides 1600 mm	119.72	146.36	2.18	53.57	nr	**199.93**
90° mitre bend						
Sum of two sides 900 mm	135.90	166.14	3.59	88.22	nr	**254.36**
Sum of two sides 1000 mm	150.40	183.87	3.66	89.94	nr	**273.81**
Sum of two sides 1100 mm	164.88	201.57	3.69	90.67	nr	**292.24**
Sum of two sides 1200 mm	179.39	219.31	3.72	91.40	nr	**310.71**
Sum of two sides 1300 mm	193.85	236.98	3.75	92.15	nr	**329.13**
Sum of two sides 1400 mm	208.35	254.71	3.78	92.89	nr	**347.60**
Sum of two sides 1500 mm	222.83	272.41	3.81	93.62	nr	**366.03**
Sum of two sides 1600 mm	237.33	290.14	3.84	94.35	nr	**384.49**

U: VENTILATION/AIR CONDITIONING SYSTEMS

Item	Net Price £	Material £	Labour hours	Labour £	Unit	Total rate £
U14: DUCTWORK: FIRE RATED – cont						
Extra over fittings; Ductwork 601 to 800 mm longest side – cont						
Branch (Side-on Shoe)						
Sum of two sides 900 mm	25.22	30.83	1.37	33.67	nr	**64.50**
Sum of two sides 1000 mm	25.85	31.60	1.46	35.89	nr	**67.49**
Sum of two sides 1100 mm	26.48	32.37	1.50	36.86	nr	**69.23**
Sum of two sides 1200 mm	27.11	33.15	1.55	38.09	nr	**71.24**
Sum of two sides 1300 mm	27.74	33.91	1.59	39.07	nr	**72.98**
Sum of two sides 1400 mm	27.26	33.32	1.68	41.28	nr	**74.60**
Sum of two sides 1500 mm	28.39	34.71	1.63	40.05	nr	**74.76**
Sum of two sides 1600 mm	29.65	36.25	1.72	42.26	nr	**78.51**
Ductwork 801 to 1000 mm longest side						
Sum of two sides 1100 mm	107.28	131.15	5.84	143.50	m	**274.65**
Sum of two sides 1200 mm	112.48	137.50	6.11	150.13	m	**287.63**
Sum of two sides 1300 mm	117.67	143.85	6.38	156.78	m	**300.63**
Sum of two sides 1400 mm	122.86	150.19	6.65	163.40	m	**313.59**
Sum of two sides 1500 mm	128.05	156.54	6.91	169.80	m	**326.34**
Sum of two sides 1600 mm	133.25	162.89	7.18	176.43	m	**339.32**
Sum of two sides 1700 mm	138.45	169.26	7.45	183.07	m	**352.33**
Sum of two sides 1800 mm	143.64	175.60	7.71	189.45	m	**365.05**
Sum of two sides 1900 mm	148.84	181.96	7.98	196.09	m	**378.05**
Sum of two sides 2000 mm	154.03	188.30	8.25	202.71	m	**391.01**
Extra over fittings; Ductwork 801 to 1000 mm longest side						
End Cap						
Sum of two sides 1100 mm	26.08	31.88	1.25	30.72	nr	**62.60**
Sum of two sides 1200 mm	27.70	33.87	1.28	31.46	nr	**65.33**
Sum of two sides 1300 mm	29.33	35.85	1.31	32.19	nr	**68.04**
Sum of two sides 1400 mm	30.96	37.85	1.34	32.92	nr	**70.77**
Sum of two sides 1500 mm	32.59	39.84	1.36	33.42	nr	**73.26**
Sum of two sides 1600 mm	34.21	41.82	1.39	34.15	nr	**75.97**
Sum of two sides 1700 mm	35.84	43.82	1.42	34.89	nr	**78.71**
Sum of two sides 1800 mm	37.47	45.80	1.45	35.63	nr	**81.43**
Sum of two sides 1900 mm	39.09	47.79	1.48	36.37	nr	**84.16**
Sum of two sides 2000 mm	40.72	49.78	1.51	37.10	nr	**86.88**
Reducer						
Sum of two sides 1100 mm	94.86	115.97	3.22	79.12	nr	**195.09**
Sum of two sides 1200 mm	99.63	121.80	3.28	80.60	nr	**202.40**
Sum of two sides 1300 mm	104.41	127.64	3.33	81.82	nr	**209.46**
Sum of two sides 1400 mm	109.19	133.49	3.38	83.06	nr	**216.55**
Sum of two sides 1500 mm	113.94	139.30	3.44	84.52	nr	**223.82**
Sum of two sides 1600 mm	118.75	145.17	3.49	85.76	nr	**230.93**
Sum of two sides 1700 mm	123.51	150.99	3.54	86.98	nr	**237.97**

U: VENTILATION/AIR CONDITIONING SYSTEMS

Item	Net Price £	Material £	Labour hours	Labour £	Unit	Total rate £
Sum of two sides 1800 mm	128.29	156.84	3.60	88.46	nr	**245.30**
Sum of two sides 1900 mm	133.07	162.68	3.65	89.68	nr	**252.36**
Sum of two sides 2000 mm	140.22	171.42	3.70	90.92	nr	**262.34**
Offset						
Sum of two sides 1100 mm	197.28	241.17	3.78	92.89	nr	**334.06**
Sum of two sides 1200 mm	205.62	251.37	3.89	95.58	nr	**346.95**
Sum of two sides 1300 mm	213.95	261.56	4.00	98.29	nr	**359.85**
Sum of two sides 1400 mm	222.30	271.77	4.11	100.99	nr	**372.76**
Sum of two sides 1500 mm	230.63	281.94	4.23	103.94	nr	**385.88**
Sum of two sides 1600 mm	238.98	292.16	4.34	106.64	nr	**398.80**
Sum of two sides 1700 mm	247.31	302.34	4.45	109.34	nr	**411.68**
Sum of two sides 1800 mm	255.66	312.54	4.56	112.05	nr	**424.59**
Sum of two sides 1900 mm	263.99	322.73	4.67	114.75	nr	**437.48**
Sum of two sides 2000 mm	276.49	338.02	5.53	135.88	nr	**473.90**
90° radius bend						
Sum of two sides 1100 mm	108.42	132.54	3.07	75.44	nr	**207.98**
Sum of two sides 1200 mm	117.35	143.46	3.09	75.93	nr	**219.39**
Sum of two sides 1300 mm	126.29	154.39	3.12	76.66	nr	**231.05**
Sum of two sides 1400 mm	135.22	165.30	3.15	77.39	nr	**242.69**
Sum of two sides 1500 mm	144.15	176.22	3.17	77.90	nr	**254.12**
Sum of two sides 1600 mm	153.07	187.13	3.20	78.63	nr	**265.76**
Sum of two sides 1700 mm	162.00	198.05	3.22	79.12	nr	**277.17**
Sum of two sides 1800 mm	170.93	208.97	3.25	79.87	nr	**288.84**
Sum of two sides 1900 mm	179.87	219.89	3.28	80.60	nr	**300.49**
Sum of two sides 2000 mm	188.53	230.48	3.30	81.09	nr	**311.57**
45° bend						
Sum of two sides 1100 mm	99.24	121.32	1.91	46.94	nr	**168.26**
Sum of two sides 1200 mm	103.39	126.39	1.96	48.16	nr	**174.55**
Sum of two sides 1300 mm	107.55	131.48	2.02	49.64	nr	**181.12**
Sum of two sides 1400 mm	111.70	136.56	2.07	50.87	nr	**187.43**
Sum of two sides 1500 mm	115.85	141.63	2.13	52.33	nr	**193.96**
Sum of two sides 1600 mm	120.00	146.70	2.18	53.57	nr	**200.27**
Sum of two sides 1700 mm	124.15	151.77	2.24	55.03	nr	**206.80**
Sum of two sides 1800 mm	128.30	156.85	2.30	56.52	nr	**213.37**
Sum of two sides 1900 mm	132.44	161.91	2.35	57.74	nr	**219.65**
Sum of two sides 2000 mm	136.60	166.99	2.76	67.83	nr	**234.82**
90° mitre bend						
Sum of two sides 1100 mm	164.88	201.57	3.69	90.67	nr	**292.24**
Sum of two sides 1200 mm	179.39	219.31	3.72	91.40	nr	**310.71**
Sum of two sides 1300 mm	193.85	236.98	3.75	92.15	nr	**329.13**
Sum of two sides 1400 mm	208.35	254.71	3.78	92.89	nr	**347.60**
Sum of two sides 1500 mm	222.86	272.45	3.81	93.62	nr	**366.07**
Sum of two sides 1600 mm	237.33	290.14	3.84	94.35	nr	**384.49**
Sum of two sides 1700 mm	251.82	307.86	3.87	95.10	nr	**402.96**
Sum of two sides 1800 mm	266.31	325.56	3.90	95.83	nr	**421.39**
Sum of two sides 1900 mm	280.80	343.28	3.93	96.56	nr	**439.84**
Sum of two sides 2000 mm	295.29	360.99	3.96	97.31	nr	**458.30**

U: VENTILATION/AIR CONDITIONING SYSTEMS

Item	Net Price £	Material £	Labour hours	Labour £	Unit	Total rate £
U14: DUCTWORK: FIRE RATED – cont						
Extra over fittings; Ductwork 801 to 1000 mm longest side – cont						
Branch (Side-on Shoe)						
Sum of two sides 1100 mm	26.45	32.33	1.50	36.86	nr	69.19
Sum of two sides 1200 mm	27.07	33.09	1.55	38.09	nr	71.18
Sum of two sides 1300 mm	27.73	33.90	1.59	39.07	nr	72.97
Sum of two sides 1400 mm	28.37	34.68	1.63	40.05	nr	74.73
Sum of two sides 1500 mm	29.02	35.47	1.68	41.28	nr	76.75
Sum of two sides 1600 mm	29.65	36.25	1.72	42.26	nr	78.51
Sum of two sides 1700 mm	30.29	37.03	1.77	43.50	nr	80.53
Sum of two sides 1800 mm	30.93	37.81	1.81	44.48	nr	82.29
Sum of two sides 1900 mm	31.58	38.60	1.86	45.71	nr	84.31
Sum of two sides 2000 mm	32.52	39.76	1.90	46.69	nr	86.45
Ductwork 1001 to 1250 mm longest side						
Sum of two sides 1300 mm	133.37	163.05	6.38	156.78	m	319.83
Sum of two sides 1400 mm	139.12	170.07	6.65	163.40	m	333.47
Sum of two sides 1500 mm	144.88	177.12	6.91	169.80	m	346.92
Sum of two sides 1600 mm	149.19	182.38	6.91	169.80	m	352.18
Sum of two sides 1700 mm	153.49	187.65	7.45	183.07	m	370.72
Sum of two sides 1800 mm	159.22	194.65	7.71	189.45	m	384.10
Sum of two sides 1900 mm	164.96	201.66	7.98	196.09	m	397.75
Sum of two sides 2000 mm	170.72	208.71	8.25	202.71	m	411.42
Sum of two sides 2100 mm	176.46	215.72	9.62	236.39	m	452.11
Sum of two sides 2200 mm	180.76	220.98	9.62	236.39	m	457.37
Sum of two sides 2300 mm	185.07	226.25	10.07	247.45	m	473.70
Sum of two sides 2400 mm	190.81	233.26	10.47	257.27	m	490.53
Sum of two sides 2500 mm	196.55	240.29	10.87	267.11	m	507.40
Extra over fittings; Ductwork 1001 to 1250 mm longest side						
End Cap						
Sum of two sides 1300 mm	29.35	35.89	1.31	32.19	nr	68.08
Sum of two sides 1400 mm	30.96	37.85	1.34	32.92	nr	70.77
Sum of two sides 1500 mm	32.56	39.81	1.36	33.42	nr	73.23
Sum of two sides 1600 mm	34.21	41.82	1.39	34.15	nr	75.97
Sum of two sides 1700 mm	35.84	43.82	1.42	34.89	nr	78.71
Sum of two sides 1800 mm	37.44	45.77	1.45	35.63	nr	81.40
Sum of two sides 1900 mm	39.09	47.79	1.48	36.37	nr	84.16
Sum of two sides 2000 mm	40.72	49.78	1.51	37.10	nr	86.88
Sum of two sides 2100 mm	42.35	51.78	2.66	65.36	nr	117.14
Sum of two sides 2200 mm	43.98	53.77	2.80	68.80	nr	122.57
Sum of two sides 2300 mm	45.61	55.75	2.95	72.49	nr	128.24
Sum of two sides 2400 mm	47.22	57.73	3.10	76.18	nr	133.91
Sum of two sides 2500 mm	48.85	59.72	3.24	79.61	nr	139.33

U: VENTILATION/AIR CONDITIONING SYSTEMS

Item	Net Price £	Material £	Labour hours	Labour £	Unit	Total rate £
Reducer						
Sum of two sides 1300 mm	104.37	127.60	3.33	81.82	nr	**209.42**
Sum of two sides 1400 mm	109.07	133.34	3.38	83.06	nr	**216.40**
Sum of two sides 1500 mm	113.76	139.07	3.44	84.52	nr	**223.59**
Sum of two sides 1600 mm	118.49	144.86	3.49	85.76	nr	**230.62**
Sum of two sides 1700 mm	123.22	150.64	3.54	86.98	nr	**237.62**
Sum of two sides 1800 mm	127.94	156.41	3.60	88.46	nr	**244.87**
Sum of two sides 1900 mm	132.66	162.17	3.65	89.68	nr	**251.85**
Sum of two sides 2000 mm	137.38	167.95	3.70	90.92	nr	**258.87**
Sum of two sides 2100 mm	142.08	173.69	3.75	92.15	nr	**265.84**
Sum of two sides 2200 mm	146.80	179.47	3.80	93.38	nr	**272.85**
Sum of two sides 2300 mm	151.52	185.24	3.85	94.61	nr	**279.85**
Sum of two sides 2400 mm	156.25	191.01	3.90	95.83	nr	**286.84**
Sum of two sides 2500 mm	160.97	196.79	3.95	97.06	nr	**293.85**
Offset						
Sum of two sides 1300 mm	213.24	260.68	4.00	98.29	nr	**358.97**
Sum of two sides 1400 mm	221.71	271.04	4.11	100.99	nr	**372.03**
Sum of two sides 1500 mm	230.16	281.38	4.23	103.94	nr	**385.32**
Sum of two sides 1600 mm	238.61	291.71	4.34	106.64	nr	**398.35**
Sum of two sides 1700 mm	247.07	302.05	4.45	109.34	nr	**411.39**
Sum of two sides 1800 mm	255.52	312.38	4.56	112.05	nr	**424.43**
Sum of two sides 1900 mm	263.97	322.71	4.67	114.75	nr	**437.46**
Sum of two sides 2000 mm	272.42	333.04	5.53	135.88	nr	**468.92**
Sum of two sides 2100 mm	280.88	343.38	5.76	141.54	nr	**484.92**
Sum of two sides 2200 mm	289.19	353.54	5.99	147.19	nr	**500.73**
Sum of two sides 2300 mm	297.78	364.04	6.22	152.84	nr	**516.88**
Sum of two sides 2400 mm	306.22	374.35	6.45	158.50	nr	**532.85**
Sum of two sides 2500 mm	314.69	384.72	6.68	164.15	nr	**548.87**
90° radius bend						
Sum of two sides 1300 mm	126.12	154.18	3.12	76.66	nr	**230.84**
Sum of two sides 1400 mm	140.34	171.57	3.15	77.39	nr	**248.96**
Sum of two sides 1500 mm	144.04	176.09	3.17	77.90	nr	**253.99**
Sum of two sides 1600 mm	152.99	187.03	3.20	78.63	nr	**265.66**
Sum of two sides 1700 mm	161.95	197.99	3.22	79.12	nr	**277.11**
Sum of two sides 1800 mm	170.89	208.91	3.25	79.87	nr	**288.78**
Sum of two sides 1900 mm	179.85	219.86	3.28	80.60	nr	**300.46**
Sum of two sides 2000 mm	188.81	230.82	3.30	81.09	nr	**311.91**
Sum of two sides 2100 mm	197.76	241.76	3.32	81.58	nr	**323.34**
Sum of two sides 2200 mm	206.72	252.72	3.34	82.07	nr	**334.79**
Sum of two sides 2300 mm	215.68	263.67	3.36	82.56	nr	**346.23**
Sum of two sides 2400 mm	224.64	274.63	3.38	83.06	nr	**357.69**
Sum of two sides 2500 mm	233.59	285.57	3.40	83.54	nr	**369.11**

U: VENTILATION/AIR CONDITIONING SYSTEMS

Item	Net Price £	Material £	Labour hours	Labour £	Unit	Total rate £
U14: DUCTWORK: FIRE RATED – cont						
Extra over fittings; Ductwork 1001 to 1250 mm longest side – cont						
45° bend						
Sum of two sides 1300 mm	102.16	124.89	2.02	49.64	nr	**174.53**
Sum of two sides 1400 mm	106.75	130.50	2.07	50.87	nr	**181.37**
Sum of two sides 1500 mm	111.35	136.12	2.13	52.33	nr	**188.45**
Sum of two sides 1600 mm	115.96	141.76	2.18	53.57	nr	**195.33**
Sum of two sides 1700 mm	120.55	147.37	2.24	55.03	nr	**202.40**
Sum of two sides 1800 mm	125.15	153.00	2.30	56.52	nr	**209.52**
Sum of two sides 1900 mm	129.74	158.61	2.35	57.74	nr	**216.35**
Sum of two sides 2000 mm	134.36	164.25	2.76	67.83	nr	**232.08**
Sum of two sides 2100 mm	138.96	169.88	2.89	71.02	nr	**240.90**
Sum of two sides 2200 mm	143.56	175.50	3.01	73.96	nr	**249.46**
Sum of two sides 2300 mm	148.15	181.12	3.13	76.91	nr	**258.03**
Sum of two sides 2400 mm	152.75	186.74	3.26	80.11	nr	**266.85**
Sum of two sides 2500 mm	157.36	192.37	3.37	82.81	nr	**275.18**
90° mitre bend						
Sum of two sides 1300 mm	193.42	236.46	3.75	92.15	nr	**328.61**
Sum of two sides 1400 mm	207.77	254.00	3.78	92.89	nr	**346.89**
Sum of two sides 1500 mm	222.69	272.24	3.81	93.62	nr	**365.86**
Sum of two sides 1600 mm	237.31	290.11	3.84	94.35	nr	**384.46**
Sum of two sides 1700 mm	251.93	307.99	3.87	95.10	nr	**403.09**
Sum of two sides 1800 mm	266.53	325.83	3.90	95.83	nr	**421.66**
Sum of two sides 1900 mm	281.17	343.73	3.93	96.56	nr	**440.29**
Sum of two sides 2000 mm	295.80	361.62	3.96	97.31	nr	**458.93**
Sum of two sides 2100 mm	310.41	379.48	3.99	98.05	nr	**477.53**
Sum of two sides 2200 mm	325.05	397.37	4.02	98.78	nr	**496.15**
Sum of two sides 2300 mm	339.66	415.23	4.05	99.53	nr	**514.76**
Sum of two sides 2400 mm	354.27	433.09	4.08	100.26	nr	**533.35**
Sum of two sides 2500 mm	368.89	450.98	4.11	100.99	nr	**551.97**
Branch (Side-on Shoe)						
Sum of two sides 1300 mm	27.78	33.96	1.59	39.07	nr	**73.03**
Sum of two sides 1400 mm	28.41	34.73	1.63	40.05	nr	**74.78**
Sum of two sides 1500 mm	29.04	35.50	1.68	41.28	nr	**76.78**
Sum of two sides 1600 mm	29.67	36.28	1.72	42.26	nr	**78.54**
Sum of two sides 1700 mm	30.32	37.07	1.77	43.50	nr	**80.57**
Sum of two sides 1800 mm	30.96	37.85	1.81	44.48	nr	**82.33**
Sum of two sides 1900 mm	31.59	38.61	1.86	45.71	nr	**84.32**
Sum of two sides 2000 mm	32.24	39.42	1.90	46.69	nr	**86.11**
Sum of two sides 2100 mm	32.87	40.18	2.58	63.40	nr	**103.58**
Sum of two sides 2200 mm	33.50	40.95	2.61	64.13	nr	**105.08**
Sum of two sides 2300 mm	34.14	41.74	2.64	64.87	nr	**106.61**
Sum of two sides 2400 mm	34.78	42.52	2.88	70.77	nr	**113.29**
Sum of two sides 2500 mm	35.41	43.29	2.91	71.50	nr	**114.79**

U: VENTILATION/AIR CONDITIONING SYSTEMS

Item	Net Price £	Material £	Labour hours	Labour £	Unit	Total rate £
Ductwork 1251 to 2000 mm longest side						
Sum of two sides 1700 mm	146.25	178.79	7.45	183.07	m	**361.86**
Sum of two sides 1800 mm	155.63	190.26	7.71	189.45	m	**379.71**
Sum of two sides 1900 mm	165.00	201.72	7.98	196.09	m	**397.81**
Sum of two sides 2000 mm	174.37	213.17	8.25	202.71	m	**415.88**
Sum of two sides 2100 mm	183.74	224.62	9.62	236.39	m	**461.01**
Sum of two sides 2200 mm	193.12	236.09	9.66	237.37	m	**473.46**
Sum of two sides 2300 mm	202.47	247.52	10.07	247.45	m	**494.97**
Sum of two sides 2400 mm	211.84	258.97	10.47	257.27	m	**516.24**
Sum of two sides 2500 mm	221.21	270.43	10.87	267.11	m	**537.54**
Sum of two sides 2600 mm	230.60	281.91	11.27	276.94	m	**558.85**
Sum of two sides 2700 mm	239.96	293.35	11.67	286.76	m	**580.11**
Sum of two sides 2800 mm	249.34	304.82	12.08	296.84	m	**601.66**
Sum of two sides 2900 mm	258.70	316.26	12.48	306.66	m	**622.92**
Sum of two sides 3000 mm	268.07	327.72	12.88	316.50	m	**644.22**
Sum of two sides 3100 mm	277.44	339.17	13.26	325.83	m	**665.00**
Sum of two sides 3200 mm	286.81	350.62	13.69	336.40	m	**687.02**
Sum of two sides 3300 mm	296.19	362.10	14.09	346.22	m	**708.32**
Sum of two sides 3400 mm	305.55	373.54	14.49	356.05	m	**729.59**
Sum of two sides 3500 mm	314.92	384.99	14.89	365.89	m	**750.88**
Sum of two sides 3600 mm	324.28	396.44	15.29	375.71	m	**772.15**
Sum of two sides 3700 mm	333.66	407.90	15.69	385.54	m	**793.44**
Sum of two sides 3800 mm	343.02	419.34	16.09	395.38	m	**814.72**
Sum of two sides 3900 mm	352.41	430.83	16.49	405.20	m	**836.03**
Sum of two sides 4000 mm	361.76	442.25	16.89	415.03	m	**857.28**
Extra over fittings; Ductwork 1251 to 2000 mm longest side						
End Cap						
Sum of two sides 1700 mm	53.24	65.09	1.42	34.89	nr	**99.98**
Sum of two sides 1800 mm	56.66	69.27	1.45	35.63	nr	**104.90**
Sum of two sides 1900 mm	60.09	73.46	1.48	36.37	nr	**109.83**
Sum of two sides 2000 mm	63.51	77.64	1.51	37.10	nr	**114.74**
Sum of two sides 2100 mm	66.94	81.83	2.66	65.36	nr	**147.19**
Sum of two sides 2200 mm	70.36	86.02	2.80	68.80	nr	**154.82**
Sum of two sides 2300 mm	73.77	90.19	2.95	72.49	nr	**162.68**
Sum of two sides 2400 mm	77.20	94.38	3.10	76.18	nr	**170.56**
Sum of two sides 2500 mm	80.62	98.56	3.24	79.61	nr	**178.17**
Sum of two sides 2600 mm	84.05	102.75	3.39	83.31	nr	**186.06**
Sum of two sides 2700 mm	87.46	106.92	3.54	86.98	nr	**193.90**
Sum of two sides 2800 mm	90.89	111.12	3.68	90.42	nr	**201.54**
Sum of two sides 2900 mm	94.34	115.33	3.83	94.11	nr	**209.44**
Sum of two sides 3000 mm	97.75	119.50	3.98	97.80	nr	**217.30**
Sum of two sides 3100 mm	101.17	123.68	4.12	101.24	nr	**224.92**
Sum of two sides 3200 mm	104.60	127.87	4.27	104.93	nr	**232.80**
Sum of two sides 3300 mm	108.01	132.05	4.42	108.60	nr	**240.65**
Sum of two sides 3400 mm	111.44	136.24	4.57	112.30	nr	**248.54**
Sum of two sides 3500 mm	114.26	139.69	4.72	115.98	nr	**255.67**

U: VENTILATION/AIR CONDITIONING SYSTEMS

Item	Net Price £	Material £	Labour hours	Labour £	Unit	Total rate £
U14: DUCTWORK: FIRE RATED – cont						
Extra over fittings; Ductwork 1251 to 2000 mm longest side – cont						
End Cap – cont						
Sum of two sides 3600 mm	116.56	142.50	4.87	119.68	nr	**262.18**
Sum of two sides 3700 mm	119.98	146.67	5.02	123.35	nr	**270.02**
Sum of two sides 3800 mm	115.40	141.08	5.17	127.04	nr	**268.12**
Sum of two sides 3900 mm	128.55	157.16	5.32	130.73	nr	**287.89**
Sum of two sides 4000 mm	131.97	161.34	5.47	134.42	nr	**295.76**
Reducer						
Sum of two sides 1700 mm	114.97	140.55	3.54	86.98	nr	**227.53**
Sum of two sides 1800 mm	128.97	157.66	3.60	88.46	nr	**246.12**
Sum of two sides 1900 mm	142.96	174.77	3.65	89.68	nr	**264.45**
Sum of two sides 2000 mm	170.96	209.00	2.92	71.75	nr	**280.75**
Sum of two sides 2100 mm	156.97	191.90	3.70	90.92	nr	**282.82**
Sum of two sides 2200 mm	184.96	226.12	3.10	76.18	nr	**302.30**
Sum of two sides 2300 mm	198.97	243.24	3.29	80.84	nr	**324.08**
Sum of two sides 2400 mm	212.97	260.35	3.48	85.51	nr	**345.86**
Sum of two sides 2500 mm	226.95	277.45	3.66	89.94	nr	**367.39**
Sum of two sides 2600 mm	240.97	294.59	3.85	94.61	nr	**389.20**
Sum of two sides 2700 mm	254.96	311.69	4.03	99.02	nr	**410.71**
Sum of two sides 2800 mm	268.96	328.81	4.22	103.69	nr	**432.50**
Sum of two sides 2900 mm	282.96	345.93	4.40	108.12	nr	**454.05**
Sum of two sides 3000 mm	303.93	371.55	4.59	112.79	nr	**484.34**
Sum of two sides 3100 mm	303.93	371.55	4.78	117.46	nr	**489.01**
Sum of two sides 3200 mm	317.95	388.69	4.96	121.88	nr	**510.57**
Sum of two sides 3300 mm	338.94	414.36	5.15	126.56	nr	**540.92**
Sum of two sides 3400 mm	352.94	431.47	5.34	131.21	nr	**562.68**
Sum of two sides 3500 mm	366.94	448.59	5.53	135.88	nr	**584.47**
Sum of two sides 3600 mm	380.93	465.69	5.72	140.55	nr	**606.24**
Sum of two sides 3700 mm	394.94	482.81	5.91	145.22	nr	**628.03**
Sum of two sides 3800 mm	408.93	499.92	6.10	149.90	nr	**649.82**
Sum of two sides 3900 mm	422.93	517.04	6.29	154.56	nr	**671.60**
Sum of two sides 4000 mm	436.93	534.15	6.48	159.23	nr	**693.38**
Offset						
Sum of two sides 1700 mm	114.97	140.55	4.45	109.34	nr	**249.89**
Sum of two sides 1800 mm	124.45	152.14	4.56	112.05	nr	**264.19**
Sum of two sides 1900 mm	133.93	163.73	4.67	114.75	nr	**278.48**
Sum of two sides 2000 mm	143.37	175.28	5.53	135.88	nr	**311.16**
Sum of two sides 2100 mm	152.85	186.86	5.76	141.54	nr	**328.40**
Sum of two sides 2200 mm	162.33	198.45	5.99	147.19	nr	**345.64**
Sum of two sides 2300 mm	171.81	210.04	6.22	152.84	nr	**362.88**
Sum of two sides 2400 mm	181.26	221.59	6.45	158.50	nr	**380.09**
Sum of two sides 2500 mm	190.75	233.19	6.68	164.15	nr	**397.34**
Sum of two sides 2600 mm	200.23	244.78	6.91	169.80	nr	**414.58**
Sum of two sides 2700 mm	209.71	256.37	7.14	175.44	nr	**431.81**

U: VENTILATION/AIR CONDITIONING SYSTEMS

Item	Net Price £	Material £	Labour hours	Labour £	Unit	Total rate £
Sum of two sides 2800 mm	219.19	267.96	7.37	181.10	nr	**449.06**
Sum of two sides 2900 mm	228.64	279.51	7.60	186.75	nr	**466.26**
Sum of two sides 3000 mm	238.12	291.10	7.83	192.39	nr	**483.49**
Sum of two sides 3100 mm	247.60	302.70	8.06	198.06	nr	**500.76**
Sum of two sides 3200 mm	257.08	314.28	8.29	203.70	nr	**517.98**
Sum of two sides 3300 mm	266.53	325.83	8.52	209.36	nr	**535.19**
Sum of two sides 3400 mm	276.00	337.41	8.75	215.01	nr	**552.42**
Sum of two sides 3500 mm	285.48	349.01	8.98	220.67	nr	**569.68**
Sum of two sides 3600 mm	294.96	360.59	9.21	226.31	nr	**586.90**
Sum of two sides 3700 mm	304.43	372.17	9.44	231.96	nr	**604.13**
Sum of two sides 3800 mm	313.89	383.74	9.67	237.62	nr	**621.36**
Sum of two sides 3900 mm	323.37	395.32	9.90	243.27	nr	**638.59**
Sum of two sides 4000 mm	332.86	406.92	10.13	248.91	nr	**655.83**
90° radius bend						
Sum of two sides 1700 mm	160.00	195.60	3.22	79.12	nr	**274.72**
Sum of two sides 1800 mm	174.46	213.28	3.25	79.87	nr	**293.15**
Sum of two sides 1900 mm	184.13	225.10	3.32	81.58	nr	**306.68**
Sum of two sides 2000 mm	188.92	230.96	3.28	80.60	nr	**311.56**
Sum of two sides 2100 mm	203.39	248.64	3.30	81.09	nr	**329.73**
Sum of two sides 2200 mm	232.71	284.49	3.32	81.58	nr	**366.07**
Sum of two sides 2300 mm	246.76	301.67	3.36	82.56	nr	**384.23**
Sum of two sides 2400 mm	261.20	319.32	3.38	83.06	nr	**402.38**
Sum of two sides 2500 mm	275.69	337.04	3.40	83.54	nr	**420.58**
Sum of two sides 2600 mm	290.15	354.71	3.42	84.04	nr	**438.75**
Sum of two sides 2700 mm	304.62	372.40	3.44	84.52	nr	**456.92**
Sum of two sides 2800 mm	319.07	390.06	3.46	85.03	nr	**475.09**
Sum of two sides 2900 mm	333.51	407.72	3.48	85.51	nr	**493.23**
Sum of two sides 3000 mm	347.98	425.41	3.50	86.00	nr	**511.41**
Sum of two sides 3100 mm	362.43	443.08	3.52	86.49	nr	**529.57**
Sum of two sides 3200 mm	364.45	445.55	3.54	86.98	nr	**532.53**
Sum of two sides 3300 mm	365.51	446.83	3.56	87.48	nr	**534.31**
Sum of two sides 3400 mm	365.39	446.69	3.58	87.97	nr	**534.66**
Sum of two sides 3500 mm	378.26	462.43	3.60	88.46	nr	**550.89**
Sum of two sides 3600 mm	391.27	478.33	3.62	88.95	nr	**567.28**
Sum of two sides 3700 mm	404.28	494.24	3.64	89.43	nr	**583.67**
Sum of two sides 3800 mm	417.30	510.15	3.66	89.94	nr	**600.09**
Sum of two sides 3900 mm	430.32	526.07	3.68	90.42	nr	**616.49**
Sum of two sides 4000 mm	443.33	541.98	3.70	90.92	nr	**632.90**
45° bend						
Sum of two sides 1700 mm	118.61	145.00	2.24	55.03	nr	**200.03**
Sum of two sides 1800 mm	131.73	161.04	2.30	56.52	nr	**217.56**
Sum of two sides 1900 mm	144.81	177.03	2.35	57.74	nr	**234.77**
Sum of two sides 2000 mm	157.93	193.07	2.76	67.83	nr	**260.90**
Sum of two sides 2100 mm	203.39	248.64	3.01	73.96	nr	**322.60**
Sum of two sides 2200 mm	197.25	241.14	3.13	76.91	nr	**318.05**
Sum of two sides 2300 mm	217.84	266.31	2.89	71.02	nr	**337.33**
Sum of two sides 2400 mm	210.34	257.14	3.26	80.11	nr	**337.25**
Sum of two sides 2500 mm	223.45	273.17	3.37	82.81	nr	**355.98**
Sum of two sides 2600 mm	236.53	289.16	3.49	85.76	nr	**374.92**
Sum of two sides 2700 mm	249.66	305.21	3.62	88.95	nr	**394.16**

U: VENTILATION/AIR CONDITIONING SYSTEMS

Item	Net Price £	Material £	Labour hours	Labour £	Unit	Total rate £
U14: DUCTWORK: FIRE RATED – cont						
Extra over fittings; Ductwork 1251 to 2000 mm longest side – cont						
45° bend – cont						
Sum of two sides 2800 mm	262.75	321.22	3.74	91.90	nr	**413.12**
Sum of two sides 2900 mm	275.85	337.23	3.86	94.85	nr	**432.08**
Sum of two sides 3000 mm	288.95	353.24	3.98	97.80	nr	**451.04**
Sum of two sides 3100 mm	302.07	369.29	4.10	100.74	nr	**470.03**
Sum of two sides 3200 mm	315.15	385.27	4.22	103.69	nr	**488.96**
Sum of two sides 3300 mm	328.27	401.31	4.34	106.64	nr	**507.95**
Sum of two sides 3400 mm	341.36	417.31	4.46	109.60	nr	**526.91**
Sum of two sides 3500 mm	354.47	433.34	4.58	112.55	nr	**545.89**
Sum of two sides 3600 mm	367.58	449.37	4.70	115.49	nr	**564.86**
Sum of two sides 3700 mm	380.67	465.37	4.82	118.44	nr	**583.81**
Sum of two sides 3800 mm	393.78	481.40	4.94	121.39	nr	**602.79**
Sum of two sides 3900 mm	406.86	497.39	5.06	124.34	nr	**621.73**
Sum of two sides 4000 mm	419.99	513.44	5.18	127.29	nr	**640.73**
90° mitre bend						
Sum of two sides 1700 mm	239.64	292.96	3.87	95.10	nr	**388.06**
Sum of two sides 1800 mm	281.06	343.60	3.90	95.83	nr	**439.43**
Sum of two sides 1900 mm	322.48	394.23	3.93	96.56	nr	**490.79**
Sum of two sides 2000 mm	363.87	444.84	3.96	97.31	nr	**542.15**
Sum of two sides 2100 mm	405.30	495.48	3.82	93.86	nr	**589.34**
Sum of two sides 2200 mm	446.72	546.12	3.83	94.11	nr	**640.23**
Sum of two sides 2300 mm	488.13	596.74	3.83	94.11	nr	**690.85**
Sum of two sides 2400 mm	529.55	647.38	3.84	94.35	nr	**741.73**
Sum of two sides 2500 mm	570.96	698.00	3.85	94.61	nr	**792.61**
Sum of two sides 2600 mm	612.35	748.60	3.85	94.61	nr	**843.21**
Sum of two sides 2700 mm	653.78	799.25	3.86	94.85	nr	**894.10**
Sum of two sides 2800 mm	695.19	849.87	3.86	94.85	nr	**944.72**
Sum of two sides 2900 mm	736.61	900.51	3.87	95.10	nr	**995.61**
Sum of two sides 3000 mm	778.02	951.13	3.87	95.10	nr	**1046.23**
Sum of two sides 3100 mm	819.44	1001.77	3.88	95.35	nr	**1097.12**
Sum of two sides 3200 mm	860.84	1052.38	3.88	95.35	nr	**1147.73**
Sum of two sides 3300 mm	902.27	1103.03	3.89	95.58	nr	**1198.61**
Sum of two sides 3400 mm	943.68	1153.65	3.90	95.83	nr	**1249.48**
Sum of two sides 3500 mm	985.10	1204.30	3.91	96.08	nr	**1300.38**
Sum of two sides 3600 mm	1026.51	1254.91	3.92	96.33	nr	**1351.24**
Sum of two sides 3700 mm	1067.93	1305.56	3.93	96.56	nr	**1402.12**
Sum of two sides 3800 mm	1109.33	1356.16	3.94	96.81	nr	**1452.97**
Sum of two sides 3900 mm	1150.76	1406.82	3.95	97.06	nr	**1503.88**
Sum of two sides 4000 mm	1192.17	1457.44	3.96	97.31	nr	**1554.75**

U: VENTILATION/AIR CONDITIONING SYSTEMS

Item	Net Price £	Material £	Labour hours	Labour £	Unit	Total rate £
Branch (Side-on Shoe)						
Sum of two sides 1700 mm	27.17	33.22	1.77	43.50	nr	**76.72**
Sum of two sides 1800 mm	29.69	36.30	1.81	44.48	nr	**80.78**
Sum of two sides 1900 mm	32.22	39.39	1.86	45.71	nr	**85.10**
Sum of two sides 2000 mm	34.76	42.50	1.90	46.69	nr	**89.19**
Sum of two sides 2100 mm	37.28	45.58	2.58	63.40	nr	**108.98**
Sum of two sides 2200 mm	39.80	48.66	2.61	64.13	nr	**112.79**
Sum of two sides 2300 mm	42.33	51.75	2.65	65.13	nr	**116.88**
Sum of two sides 2400 mm	44.85	54.83	2.68	65.86	nr	**120.69**
Sum of two sides 2500 mm	47.39	57.94	2.71	66.59	nr	**124.53**
Sum of two sides 2600 mm	49.92	61.03	2.75	67.57	nr	**128.60**
Sum of two sides 2700 mm	52.43	64.10	2.78	68.31	nr	**132.41**
Sum of two sides 2800 mm	54.96	67.19	2.81	69.05	nr	**136.24**
Sum of two sides 2900 mm	57.50	70.30	2.84	69.78	nr	**140.08**
Sum of two sides 3000 mm	60.02	73.38	2.87	70.52	nr	**143.90**
Sum of two sides 3100 mm	62.54	76.46	2.90	71.26	nr	**147.72**
Sum of two sides 3200 mm	65.08	79.56	2.93	72.00	nr	**151.56**
Sum of two sides 3300 mm	67.60	82.64	2.93	72.00	nr	**154.64**
Sum of two sides 3400 mm	70.13	85.74	3.00	73.72	nr	**159.46**
Sum of two sides 3500 mm	72.66	88.83	3.03	74.45	nr	**163.28**
Sum of two sides 3600 mm	75.17	91.90	3.06	75.20	nr	**167.10**
Sum of two sides 3700 mm	77.71	95.00	3.09	75.93	nr	**170.93**
Sum of two sides 3800 mm	80.24	98.10	3.12	76.66	nr	**174.76**
Sum of two sides 3900 mm	82.78	101.20	3.15	77.39	nr	**178.59**
Sum of two sides 4000 mm	85.28	104.26	3.18	78.15	nr	**182.41**
Rectangular section ductwork to BS476 Part 24 (ISO 6944:1985) (Duraduct LT), ducts 'Type A' and 'Type B'; manufactured from 6 mm thick laminate fire board consisting of steel circular hole punched facings pressed to a fibre cement core; provides upto 4 hours stability, 4 hours integrity and 32 minutes insulation; including all necessary stiffeners, joints and supports in the running length						
Ductwork up to 600 mm longest side						
Sum of two sides 200 mm	202.44	247.49	4.00	98.29	m	**345.78**
Sum of two sides 400 mm	207.38	253.52	4.00	98.29	m	**351.81**
Sum of two sides 600 mm	216.71	264.93	4.00	98.29	m	**363.22**
Sum of two sides 800 mm	285.97	349.60	5.50	135.15	m	**484.75**
Sum of two sides 1000 mm	385.25	470.97	5.50	135.15	m	**606.12**
Sum of two sides 1200 mm	398.21	486.82	6.00	147.43	m	**634.25**

U: VENTILATION/AIR CONDITIONING SYSTEMS

Item	Net Price £	Material £	Labour hours	Labour £	Unit	Total rate £
U14: DUCTWORK: FIRE RATED – cont						
Extra over fittings; Ductwork up to 600 mm longest side						
End Cap						
Sum of two sides 200 mm	48.04	58.73	0.81	19.90	m	**78.63**
Sum of two sides 400 mm	48.04	58.73	0.87	21.38	m	**80.11**
Sum of two sides 600 mm	48.04	58.73	0.93	22.85	m	**81.58**
Sum of two sides 800 mm	77.60	94.86	0.98	24.08	m	**118.94**
Sum of two sides 1000 mm	77.60	94.86	1.22	29.97	m	**124.83**
Sum of two sides 1200 mm	118.24	144.55	1.28	31.46	m	**176.01**
Reducer						
Sum of two sides 200 mm	42.48	51.93	2.23	54.80	m	**106.73**
Sum of two sides 400 mm	42.48	51.93	2.51	61.68	m	**113.61**
Sum of two sides 600 mm	42.48	51.93	2.79	68.56	m	**120.49**
Sum of two sides 800 mm	83.14	101.64	3.01	73.96	m	**175.60**
Sum of two sides 1000 mm	83.14	101.64	3.17	77.90	m	**179.54**
Sum of two sides 1200 mm	105.31	128.74	3.28	80.60	m	**209.34**
Offset						
Sum of two sides 200 mm	42.48	51.93	2.95	72.49	m	**124.42**
Sum of two sides 400 mm	42.48	51.93	3.26	80.11	m	**132.04**
Sum of two sides 600 mm	42.48	51.93	3.57	87.73	m	**139.66**
Sum of two sides 800 mm	83.14	101.64	3.23	79.36	m	**181.00**
Sum of two sides 1000 mm	83.14	101.64	3.67	90.19	m	**191.83**
Sum of two sides 1200 mm	105.31	128.74	3.89	95.58	m	**224.32**
90° radius bend						
Sum of two sides 200 mm	142.26	173.92	2.06	50.62	m	**224.54**
Sum of two sides 400 mm	142.26	173.92	2.36	58.00	m	**231.92**
Sum of two sides 600 mm	142.26	173.92	2.66	65.36	m	**239.28**
Sum of two sides 800 mm	166.28	203.28	2.97	72.99	m	**276.27**
Sum of two sides 1000 mm	166.28	203.28	3.04	74.70	m	**277.98**
Sum of two sides 1200 mm	188.46	230.39	3.09	75.93	m	**306.32**
45° radius bend						
Sum of two sides 200 mm	142.26	173.92	1.52	37.35	m	**211.27**
Sum of two sides 400 mm	142.26	173.92	1.66	40.80	m	**214.72**
Sum of two sides 600 mm	142.26	173.92	1.80	44.23	m	**218.15**
Sum of two sides 800 mm	166.28	203.28	1.93	47.42	m	**250.70**
Sum of two sides 1000 mm	166.28	203.28	2.05	50.37	m	**253.65**
Sum of two sides 1200 mm	188.46	230.39	2.17	53.32	m	**283.71**

U: VENTILATION/AIR CONDITIONING SYSTEMS

Item	Net Price £	Material £	Labour hours	Labour £	Unit	Total rate £
90° mitre bend						
Sum of two sides 200 mm	142.26	173.92	2.47	60.70	m	**234.62**
Sum of two sides 400 mm	142.26	173.92	2.84	69.78	m	**243.70**
Sum of two sides 600 mm	142.26	173.92	3.20	78.63	m	**252.55**
Sum of two sides 800 mm	166.28	203.28	3.56	87.48	m	**290.76**
Sum of two sides 1000 mm	166.28	203.28	3.66	89.94	m	**293.22**
Sum of two sides 1200 mm	188.46	230.39	3.72	91.40	m	**321.79**
Branch						
Sum of two sides 200 mm	51.73	63.24	0.98	24.08	m	**87.32**
Sum of two sides 400 mm	51.73	63.24	1.14	28.01	m	**91.25**
Sum of two sides 600 mm	51.73	63.24	1.30	31.94	m	**95.18**
Sum of two sides 800 mm	64.66	79.04	1.46	35.89	m	**114.93**
Sum of two sides 1000 mm	70.50	86.19	1.46	35.89	m	**122.08**
Sum of two sides 1200 mm	86.82	106.14	1.55	38.09	m	**144.23**
Ductwork 601 to 1000 mm longest side						
Sum of two sides 1000 mm	285.97	349.60	6.00	147.43	m	**497.03**
Sum of two sides 1100 mm	398.21	486.82	6.00	147.43	m	**634.25**
Sum of two sides 1300 mm	404.86	494.95	6.00	147.43	m	**642.38**
Sum of two sides 1500 mm	414.30	506.48	6.00	147.43	m	**653.91**
Sum of two sides 1700 mm	500.71	612.12	6.00	147.43	m	**759.55**
Sum of two sides 1900 mm	573.80	701.47	6.00	147.43	m	**848.90**
Extra over fittings; Ductwork 601 to 1000 mm longest side						
End Cap						
Sum of two sides 1000 mm	77.60	94.86	1.25	30.72	m	**125.58**
Sum of two sides 1100 mm	118.24	144.55	1.25	30.72	m	**175.27**
Sum of two sides 1300 mm	118.24	144.55	1.31	32.19	m	**176.74**
Sum of two sides 1500 mm	118.24	144.55	1.36	33.42	m	**177.97**
Sum of two sides 1700 mm	208.77	255.22	1.42	34.89	m	**290.11**
Sum of two sides 1900 mm	208.77	255.22	1.48	36.37	m	**291.59**
Reducer						
Sum of two sides 1000 mm	83.14	101.64	3.22	79.12	m	**180.76**
Sum of two sides 1100 mm	105.31	128.74	3.22	79.12	m	**207.86**
Sum of two sides 1300 mm	105.31	128.74	3.33	81.82	m	**210.56**
Sum of two sides 1500 mm	105.31	128.74	3.44	84.52	m	**213.26**
Sum of two sides 1700 mm	129.34	158.12	3.54	86.98	m	**245.10**
Sum of two sides 1900 mm	129.34	158.12	3.65	89.68	m	**247.80**
Offset						
Sum of two sides 1000 mm	83.14	101.64	3.78	92.89	m	**194.53**
Sum of two sides 1100 mm	105.31	128.74	3.78	92.89	m	**221.63**
Sum of two sides 1300 mm	105.31	128.74	4.00	98.29	m	**227.03**
Sum of two sides 1500 mm	105.31	128.74	4.23	103.94	m	**232.68**
Sum of two sides 1700 mm	129.34	158.12	4.45	109.34	m	**267.46**
Sum of two sides 1900 mm	129.34	158.12	4.67	114.75	m	**272.87**

U: VENTILATION/AIR CONDITIONING SYSTEMS

Item	Net Price £	Material £	Labour hours	Labour £	Unit	Total rate £
U14: DUCTWORK: FIRE RATED – cont						
Extra over fittings; Ductwork 601 to 1000 mm longest side – cont						
90° radius bend						
Sum of two sides 1000 mm	166.28	203.28	3.78	92.89	m	**296.17**
Sum of two sides 1100 mm	188.46	230.39	3.07	75.44	m	**305.83**
Sum of two sides 1300 mm	188.46	230.39	3.12	76.66	m	**307.05**
Sum of two sides 1500 mm	188.46	230.39	3.17	77.90	m	**308.29**
Sum of two sides 1700 mm	247.57	302.66	3.22	79.12	m	**381.78**
Sum of two sides 1900 mm	247.57	302.66	3.28	80.60	m	**383.26**
45° bend						
Sum of two sides 1000 mm	188.46	230.39	1.91	46.94	m	**277.33**
Sum of two sides 1100 mm	188.46	230.39	2.00	49.14	m	**279.53**
Sum of two sides 1300 mm	188.46	230.39	2.02	49.64	m	**280.03**
Sum of two sides 1500 mm	188.46	230.39	2.13	52.33	m	**282.72**
Sum of two sides 1700 mm	247.57	302.66	2.24	55.03	m	**357.69**
Sum of two sides 1900 mm	247.57	302.66	2.35	57.74	m	**360.40**
90° mitre bend						
Sum of two sides 1000 mm	166.28	203.28	3.78	92.89	m	**296.17**
Sum of two sides 1100 mm	188.46	230.39	3.69	90.67	m	**321.06**
Sum of two sides 1300 mm	188.46	230.39	3.75	92.15	m	**322.54**
Sum of two sides 1500 mm	188.46	230.39	3.81	93.62	m	**324.01**
Sum of two sides 1700 mm	247.57	302.66	3.87	95.10	m	**397.76**
Sum of two sides 1900 mm	247.57	302.66	3.93	96.56	m	**399.22**
Branch						
Sum of two sides 1000 mm	67.23	82.19	2.38	58.48	m	**140.67**
Sum of two sides 1100 mm	86.82	106.14	1.50	36.86	m	**143.00**
Sum of two sides 1300 mm	86.82	106.14	1.59	39.07	m	**145.21**
Sum of two sides 1500 mm	86.82	106.14	1.68	41.28	m	**147.42**
Sum of two sides 1700 mm	110.85	135.52	1.77	43.50	m	**179.02**
Sum of two sides 1900 mm	110.85	135.52	1.86	45.71	m	**181.23**
Ductwork 1001 to 1250 mm longest side						
Sum of two sides 1300 mm	398.21	486.82	6.00	147.43	m	**634.25**
Sum of two sides 1500 mm	423.25	517.43	6.00	147.43	m	**664.86**
Sum of two sides 1700 mm	500.71	612.12	6.00	147.43	m	**759.55**
Sum of two sides 1900 mm	509.05	622.32	6.00	147.43	m	**769.75**
Sum of two sides 2100 mm	667.08	815.51	6.50	159.72	m	**975.23**
Sum of two sides 2300 mm	703.32	859.81	6.50	159.72	m	**1019.53**
Sum of two sides 2500 mm	837.28	1023.58	6.50	159.72	m	**1183.30**

U: VENTILATION/AIR CONDITIONING SYSTEMS

Item	Net Price £	Material £	Labour hours	Labour £	Unit	Total rate £
Extra over fittings; Ductwork 1001 to 1250 mm longest side						
End Cap						
Sum of two sides 1300 mm	118.24	144.55	1.31	32.19	m	**176.74**
Sum of two sides 1500 mm	118.24	144.55	1.36	33.42	m	**177.97**
Sum of two sides 1700 mm	208.77	255.22	1.42	34.89	m	**290.11**
Sum of two sides 1900 mm	208.77	255.22	1.48	36.37	m	**291.59**
Sum of two sides 2100 mm	284.52	347.83	2.66	65.36	m	**413.19**
Sum of two sides 2300 mm	284.52	347.83	2.95	72.49	m	**420.32**
Sum of two sides 2500 mm	284.52	347.83	3.24	79.61	m	**427.44**
Reducer						
Sum of two sides 1300 mm	105.31	128.74	3.33	81.82	m	**210.56**
Sum of two sides 1500 mm	105.31	128.74	3.44	84.52	m	**213.26**
Sum of two sides 1700 mm	129.34	158.12	3.54	86.98	m	**245.10**
Sum of two sides 1900 mm	129.34	158.12	3.65	89.68	m	**247.80**
Sum of two sides 2100 mm	153.34	187.46	3.75	92.15	m	**279.61**
Sum of two sides 2300 mm	153.34	187.46	3.85	94.61	m	**282.07**
Sum of two sides 2500 mm	153.34	187.46	3.95	97.06	m	**284.52**
Offset						
Sum of two sides 1300 mm	105.31	128.74	4.00	98.29	m	**227.03**
Sum of two sides 1500 mm	105.31	128.74	4.23	103.94	m	**232.68**
Sum of two sides 1700 mm	129.34	158.12	4.45	109.34	m	**267.46**
Sum of two sides 1900 mm	129.34	158.12	4.67	114.75	m	**272.87**
Sum of two sides 2100 mm	153.34	187.46	5.76	141.54	m	**329.00**
Sum of two sides 2300 mm	153.34	187.46	6.22	152.84	m	**340.30**
Sum of two sides 2500 mm	153.34	187.46	6.68	164.15	m	**351.61**
90° radius bend						
Sum of two sides 1300 mm	188.46	230.39	3.12	76.66	m	**307.05**
Sum of two sides 1500 mm	188.46	230.39	3.17	77.90	m	**308.29**
Sum of two sides 1700 mm	247.57	302.66	3.22	79.12	m	**381.78**
Sum of two sides 1900 mm	247.57	302.66	3.28	80.60	m	**383.26**
Sum of two sides 2100 mm	253.11	309.43	3.32	81.58	m	**391.01**
Sum of two sides 2300 mm	253.11	309.43	3.36	82.56	m	**391.99**
Sum of two sides 2500 mm	253.11	309.43	3.40	83.54	m	**392.97**
45° bend						
Sum of two sides 1300 mm	188.46	230.39	2.02	49.64	m	**280.03**
Sum of two sides 1500 mm	188.46	230.39	2.13	52.33	m	**282.72**
Sum of two sides 1700 mm	247.57	302.66	2.24	55.03	m	**357.69**
Sum of two sides 1900 mm	247.57	302.66	2.35	57.74	m	**360.40**
Sum of two sides 2100 mm	253.11	309.43	2.89	71.02	m	**380.45**
Sum of two sides 2300 mm	253.11	309.43	3.13	76.91	m	**386.34**
Sum of two sides 2500 mm	253.11	309.43	3.37	82.81	m	**392.24**

U: VENTILATION/AIR CONDITIONING SYSTEMS

Item	Net Price £	Material £	Labour hours	Labour £	Unit	Total rate £
U14: DUCTWORK: FIRE RATED – cont						
Extra over fittings; Ductwork 1001 to 1250 mm longest side – cont						
90° mitre bend						
Sum of two sides 1300 mm	188.46	230.39	3.75	92.15	m	**322.54**
Sum of two sides 1500 mm	188.46	230.39	3.81	93.62	m	**324.01**
Sum of two sides 1700 mm	247.57	302.66	3.87	95.10	m	**397.76**
Sum of two sides 1900 mm	247.57	302.66	3.93	96.56	m	**399.22**
Sum of two sides 2100 mm	253.11	309.43	3.99	98.05	m	**407.48**
Sum of two sides 2300 mm	253.11	309.43	4.05	99.53	m	**408.96**
Sum of two sides 2500 mm	253.11	309.43	4.11	100.99	m	**410.42**
Branch						
Sum of two sides 1300 mm	86.82	106.14	1.59	39.07	m	**145.21**
Sum of two sides 1500 mm	86.82	106.14	1.68	41.28	m	**147.42**
Sum of two sides 1700 mm	110.85	135.52	1.77	43.50	m	**179.02**
Sum of two sides 1900 mm	110.85	135.52	1.86	45.71	m	**181.23**
Sum of two sides 2100 mm	116.38	142.27	2.58	63.40	m	**205.67**
Sum of two sides 2300 mm	116.38	142.27	2.64	64.87	m	**207.14**
Sum of two sides 2500 mm	116.38	142.27	2.91	71.50	m	**213.77**
Ductwork 1251 to 2000 mm longest side						
Sum of two sides 1800 mm	500.71	612.12	6.00	147.43	m	**759.55**
Sum of two sides 2000 mm	584.14	714.12	6.00	147.43	m	**861.55**
Sum of two sides 2200 mm	667.08	815.51	6.50	159.72	m	**975.23**
Sum of two sides 2400 mm	750.17	917.09	6.50	159.72	m	**1076.81**
Sum of two sides 2600 mm	790.32	966.17	6.66	163.66	m	**1129.83**
Sum of two sides 2800 mm	866.90	1059.79	6.66	163.66	m	**1223.45**
Sum of two sides 3000 mm	943.01	1152.84	6.66	163.66	m	**1316.50**
Sum of two sides 3200 mm	950.06	1161.46	9.00	221.15	m	**1382.61**
Sum of two sides 3400 mm	1045.06	1277.59	9.00	221.15	m	**1498.74**
Sum of two sides 3600 mm	1063.93	1300.66	11.70	287.49	m	**1588.15**
Sum of two sides 3800 mm	1158.60	1416.39	11.70	287.49	m	**1703.88**
Sum of two sides 4000 mm	1340.32	1638.55	11.70	287.49	m	**1926.04**
Extra over fittings; Ductwork 1251 to 2000 mm longest sides						
End Cap						
Sum of two sides 1800 mm	208.77	255.22	1.45	35.63	m	**290.85**
Sum of two sides 2000 mm	208.77	255.22	1.51	37.10	m	**292.32**
Sum of two sides 2200 mm	284.52	347.83	2.80	68.80	m	**416.63**
Sum of two sides 2400 mm	284.52	347.83	3.10	76.18	m	**424.01**
Sum of two sides 2600 mm	389.81	476.55	3.39	83.31	m	**559.86**
Sum of two sides 2800 mm	389.81	476.55	3.68	90.42	m	**566.97**
Sum of two sides 3000 mm	389.81	476.55	3.98	97.80	m	**574.35**
Sum of two sides 3200 mm	526.53	643.69	4.27	104.93	m	**748.62**
Sum of two sides 3400 mm	526.53	643.69	4.57	112.30	m	**755.99**
Sum of two sides 3600 mm	648.45	792.74	4.87	119.68	m	**912.42**
Sum of two sides 3800 mm	648.45	792.74	5.17	127.04	m	**919.78**
Sum of two sides 4000 mm	648.45	792.74	5.47	134.42	m	**927.16**

U: VENTILATION/AIR CONDITIONING SYSTEMS

Item	Net Price £	Material £	Labour hours	Labour £	Unit	Total rate £
Reducer						
Sum of two sides 1800 mm	129.34	158.12	3.60	88.46	m	**246.58**
Sum of two sides 2000 mm	129.34	158.12	3.70	90.92	m	**249.04**
Sum of two sides 2200 mm	153.34	187.46	3.10	76.18	m	**263.64**
Sum of two sides 2400 mm	153.34	187.46	3.48	85.51	m	**272.97**
Sum of two sides 2600 mm	212.46	259.74	3.85	94.61	m	**354.35**
Sum of two sides 2800 mm	212.46	259.74	4.22	103.69	m	**363.43**
Sum of two sides 3000 mm	212.46	259.74	4.59	112.79	m	**372.53**
Sum of two sides 3200 mm	236.48	289.10	4.96	121.88	m	**410.98**
Sum of two sides 3400 mm	236.48	289.10	5.34	131.21	m	**420.31**
Sum of two sides 3600 mm	258.65	316.20	5.72	140.55	m	**456.75**
Sum of two sides 3800 mm	258.65	316.20	6.10	149.90	m	**466.10**
Sum of two sides 4000 mm	258.65	316.20	6.48	159.23	m	**475.43**
Offset						
Sum of two sides 1800 mm	129.34	158.12	4.56	112.05	m	**270.17**
Sum of two sides 2000 mm	129.34	158.12	5.53	135.88	m	**294.00**
Sum of two sides 2200 mm	153.34	187.46	5.99	147.19	m	**334.65**
Sum of two sides 2400 mm	153.34	187.46	6.45	158.50	m	**345.96**
Sum of two sides 2600 mm	212.46	259.74	6.91	169.80	m	**429.54**
Sum of two sides 2800 mm	212.46	259.74	7.37	181.10	m	**440.84**
Sum of two sides 3000 mm	212.46	259.74	7.83	192.39	m	**452.13**
Sum of two sides 3200 mm	236.48	289.10	8.29	203.70	m	**492.80**
Sum of two sides 3400 mm	236.48	289.10	8.75	215.01	m	**504.11**
Sum of two sides 3600 mm	258.65	316.20	9.21	226.31	m	**542.51**
Sum of two sides 3800 mm	258.65	316.20	9.67	237.62	m	**553.82**
Sum of two sides 4000 mm	258.65	316.20	10.13	248.91	m	**565.11**
90° radius bend						
Sum of two sides 1800 mm	247.57	302.66	3.25	79.87	m	**382.53**
Sum of two sides 2000 mm	247.57	302.66	3.30	81.09	m	**383.75**
Sum of two sides 2200 mm	253.11	309.43	3.34	82.07	m	**391.50**
Sum of two sides 2400 mm	253.11	309.43	3.38	83.06	m	**392.49**
Sum of two sides 2600 mm	387.98	474.30	3.42	84.04	m	**558.34**
Sum of two sides 2800 mm	387.98	474.30	3.46	85.03	m	**559.33**
Sum of two sides 3000 mm	387.98	474.30	3.50	86.00	m	**560.30**
Sum of two sides 3200 mm	465.57	569.17	3.54	86.98	m	**656.15**
Sum of two sides 3400 mm	465.57	569.17	3.58	87.97	m	**657.14**
Sum of two sides 3600 mm	545.00	666.27	3.62	88.95	m	**755.22**
Sum of two sides 3800 mm	545.00	666.27	3.66	89.94	m	**756.21**
Sum of two sides 4000 mm	545.00	666.27	3.70	90.92	m	**757.19**
45° bend						
Sum of two sides 1800 mm	247.57	302.66	2.30	56.52	m	**359.18**
Sum of two sides 2000 mm	247.57	302.66	2.76	67.83	m	**370.49**
Sum of two sides 2200 mm	253.11	309.43	3.01	73.96	m	**383.39**
Sum of two sides 2400 mm	253.11	309.43	3.26	80.11	m	**389.54**
Sum of two sides 2600 mm	387.98	474.30	3.49	85.76	m	**560.06**
Sum of two sides 2800 mm	387.98	474.30	3.74	91.90	m	**566.20**
Sum of two sides 3000 mm	387.98	474.30	3.98	97.80	m	**572.10**

U: VENTILATION/AIR CONDITIONING SYSTEMS

Item	Net Price £	Material £	Labour hours	Labour £	Unit	Total rate £
U14: DUCTWORK: FIRE RATED – cont						
Extra over fittings; Ductwork 1251 to 2000 mm longest sides – cont						
45° bend – cont						
Sum of two sides 3200 mm	465.57	569.17	4.22	103.69	m	**672.86**
Sum of two sides 3400 mm	465.57	569.17	4.46	109.60	m	**678.77**
Sum of two sides 3600 mm	545.00	666.27	4.70	115.49	m	**781.76**
Sum of two sides 3800 mm	545.00	666.27	4.94	121.39	m	**787.66**
Sum of two sides 4000 mm	545.00	666.27	5.18	127.29	m	**793.56**
90° mitre band						
Sum of two sides 1800 mm	247.57	302.66	3.90	95.83	m	**398.49**
Sum of two sides 2000 mm	247.57	302.66	3.96	97.31	m	**399.97**
Sum of two sides 2200 mm	253.11	309.43	3.83	94.11	m	**403.54**
Sum of two sides 2400 mm	336.55	411.43	3.84	94.35	m	**505.78**
Sum of two sides 2600 mm	387.98	474.30	3.85	94.61	m	**568.91**
Sum of two sides 2800 mm	387.98	474.30	3.86	94.85	m	**569.15**
Sum of two sides 3000 mm	387.98	474.30	3.87	95.10	m	**569.40**
Sum of two sides 3200 mm	465.57	569.17	3.88	95.35	m	**664.52**
Sum of two sides 3400 mm	465.57	569.17	3.90	95.83	m	**665.00**
Sum of two sides 3600 mm	545.00	666.27	3.92	96.33	m	**762.60**
Sum of two sides 3800 mm	545.00	666.27	3.94	96.81	m	**763.08**
Sum of two sides 4000 mm	545.00	666.27	3.96	97.31	m	**763.58**
Branch						
Sum of two sides 1800 mm	110.85	135.52	1.81	44.48	m	**180.00**
Sum of two sides 2000 mm	110.85	135.52	1.90	46.69	m	**182.21**
Sum of two sides 2200 mm	116.38	142.27	2.61	64.13	m	**206.40**
Sum of two sides 2400 mm	116.38	142.27	2.68	65.86	m	**208.13**
Sum of two sides 2600 mm	138.56	169.39	2.75	67.57	m	**236.96**
Sum of two sides 2800 mm	138.56	169.39	2.81	69.05	m	**238.44**
Sum of two sides 3000 mm	138.56	169.39	2.87	70.52	m	**239.91**
Sum of two sides 3200 mm	144.12	176.19	2.93	72.00	m	**248.19**
Sum of two sides 3400 mm	144.12	176.19	3.00	73.72	m	**249.91**
Sum of two sides 3600 mm	184.75	225.86	3.06	75.20	m	**301.06**
Sum of two sides 3800 mm	184.75	225.86	3.12	76.66	m	**302.52**
Sum of two sides 4000 mm	184.75	225.86	3.18	78.15	m	**304.01**

U: VENTILATION/AIR CONDITIONING SYSTEMS

Item	Net Price £	Material £	Labour hours	Labour £	Unit	Total rate £
U30: LOW VELOCITY AIR CONDITIONING						
Y40 – AIR HANDLING UNITS						
Supply air handling unit; inlet with motorized damper, LTHW frost coil (at -5°C to +5°C), panel filter (EU4), bag filter (EU6), cooling coil (at 28°C db/20°C wb to 12°C db/11.5°C wb), LTHW heating coil (at 5°C to 21°C), supply fan, outlet plenum; includes access sections; all units located internally; Includes placing in position and fitting of sections together; electrical work elsewhere						
Volume, external pressure						
2 m³/s at 350 Pa	4928.63	5566.37	40.00	947.78	nr	**6514.15**
2 m³/s at 700 Pa	5265.54	5946.88	40.00	947.78	nr	**6894.66**
5 m³/s at 350 Pa	8310.67	9386.03	65.00	1540.16	nr	**10926.19**
5 m³/s at 700 Pa	8601.39	9714.37	65.00	1540.16	nr	**11254.53**
8 m³/s at 350 Pa	12447.61	14058.27	77.00	1824.49	nr	**15882.76**
8 m/³s at 700 Pa	12641.65	14277.42	77.00	1824.49	nr	**16101.91**
10 m/³ at 350 Pa	13707.22	15480.87	100.00	2369.47	nr	**17850.34**
10 m³/s at 700 Pa	13988.61	15798.67	100.00	2369.47	nr	**18168.14**
13 m³/s at 350 Pa	17129.20	19345.64	108.00	2559.03	nr	**21904.67**
13 m³/s at 700 Pa	17658.24	19943.13	108.00	2559.03	nr	**22502.16**
15 m³/s at 350 Pa	19201.10	21685.63	120.00	2843.36	nr	**24528.99**
15 m³/s at 700 Pa	19550.47	22080.21	120.00	2843.36	nr	**24923.57**
18 m³/s at 350 Pa	21788.32	24607.62	133.00	3151.39	nr	**27759.01**
18 m³/s at 700 Pa	22188.86	25059.99	133.00	3151.39	nr	**28211.38**
20 m³/s at 350 Pa	25248.96	28516.05	142.00	3364.64	nr	**31880.69**
20 m³/s at 700 Pa	25612.07	28926.14	142.00	3364.64	nr	**32290.78**
Extra for inlet and discharge attenuators at 900 mm long						
2 m³/s at 350 Pa	1516.03	1712.19	5.00	118.47	nr	**1830.66**
5 m³/s at 350 Pa	3181.78	3593.49	10.00	236.95	nr	**3830.44**
10 m³/s at 700 Pa	4448.26	5023.85	13.00	308.03	nr	**5331.88**
15 m³/s at 700 Pa	6629.96	7487.84	16.00	379.11	nr	**7866.95**
20 m³/s at 700 Pa	8483.50	9581.23	20.00	473.89	nr	**10055.12**
Extra for locating units externally						
2 m³/s at 350 Pa	1123.30	1268.65	–	–	nr	**1268.65**
5 m³/s at 350 Pa	1425.25	1609.67	–	–	nr	**1609.67**
10 m³/s at 700 Pa	2240.47	2530.38	–	–	nr	**2530.38**
15 m³/s at 700 Pa	3182.46	3594.26	–	–	nr	**3594.26**
20 m³/s at 700 Pa	5469.48	6177.20	–	–	nr	**6177.20**

U: VENTILATION/AIR CONDITIONING SYSTEMS

Item	Net Price £	Material £	Labour hours	Labour £	Unit	Total rate £
U30: LOW VELOCITY AIR CONDITIONING – cont						
Modular air handling unit with supply and extract sections. Supply side; inlet with motorized damper, LTHW frost coil (at -5°C to 5°C), panel filter (EU4), bag filter (EU6), cooling coil at 28°Cdb/20°Cwb to 12°Cdb/ 11.5°C wb), LTHW heating coil (at 5°C to 21°C), supply fan, outlet plenum. Extract side; inlet with motorized damper, extract fan; includes access sections; placing in position and fitting of sections together; electrical work elsewhere						
2 m³/s at 350 Pa	6885.12	7776.02	50.00	1184.74	nr	**8960.76**
2 m³/s at 700 Pa	7218.27	8152.28	50.00	1184.74	nr	**9337.02**
5 m³/s at 350 Pa	11866.16	13401.59	86.00	2037.75	nr	**15439.34**
5 m³/s at 700 Pa	12088.89	13653.13	86.00	2037.75	nr	**15690.88**
8 m³/s at 350 Pa	16477.87	18610.02	105.00	2487.94	nr	**21097.96**
8 m/3s at 700 Pa	16873.41	19056.74	105.00	2487.94	nr	**21544.68**
10 m/3 at 350 Pa	19027.02	21489.02	120.00	2843.36	nr	**24332.38**
10 m/s at 700 Pa	19436.92	21951.96	120.00	2843.36	nr	**24795.32**
13 m³/s at 350 Pa	23848.99	26934.93	130.00	3080.31	nr	**30015.24**
13 m³/s at 700 Pa	24597.02	27779.75	130.00	3080.31	nr	**30860.06**
15 m³/s at 350 Pa	26649.58	30097.91	145.00	3435.74	nr	**33533.65**
15 m³/s at 700 Pa	27319.00	30853.94	145.00	3435.74	nr	**34289.68**
18 m³/s at 350 Pa	29818.25	33676.59	160.00	3791.16	nr	**37467.75**
18 m³/s at 700 Pa	30516.37	34465.03	160.00	3791.16	nr	**38256.19**
20 m³/s at 350 Pa	32590.14	36807.14	175.00	4146.58	nr	**40953.72**
20 m³/s at 700 Pa	33723.12	38086.72	175.00	4146.58	nr	**42233.30**
Extra for inlet and discharge attenuators at 900 mm long						
2 m³/s at 350 Pa	2763.79	3121.41	8.00	189.56	nr	**3310.97**
5 m³/s at 350 Pa	5457.69	6163.89	10.00	236.95	nr	**6400.84**
10 m³/s at 700 Pa	7673.69	8666.62	13.00	308.03	nr	**8974.65**
15 m³/s at 700 Pa	12901.80	14571.23	16.00	379.11	nr	**14950.34**
20 m³/s at 700 Pa	14933.15	16865.42	20.00	473.89	nr	**17339.31**
Extra for locating units externally						
2 m³/s at 350 Pa	2212.29	2498.55	–	–	nr	**2498.55**
5 m³/s at 350 Pa	3141.44	3547.93	–	–	nr	**3547.93**
10 m³/s at 700 Pa	4406.21	4976.36	–	–	nr	**4976.36**
15 m³/s at 700 Pa	8223.21	9287.25	–	–	nr	**9287.25**
20 m³/s at 700 Pa	12534.46	14156.36	–	–	nr	**14156.36**

U: VENTILATION/AIR CONDITIONING SYSTEMS

Item	Net Price £	Material £	Labour hours	Labour £	Unit	Total rate £
Extra for humidifier, self generating type						
2 m³/s at 350 Pa (10 kg/hr)	1815.40	2050.30	5.00	118.47	nr	**2168.77**
5 m³/s at 350 Pa (18 kg/hr)	2246.59	2537.29	5.00	118.47	nr	**2655.76**
10 m³/s at 700 Pa (30 kg/hr)	2556.51	2887.31	6.00	142.17	nr	**3029.48**
15 m³/s at 700 Pa (60 kg/hr)	4734.49	5347.11	8.00	189.56	nr	**5536.67**
20 m³/s at 700 Pa (90 kg/hr)	7118.28	8039.35	10.00	236.95	nr	**8276.30**
Extra for mixing box						
2 m³/s at 350 Pa	909.62	1027.32	4.00	94.78	nr	**1122.10**
5 m³/s at 350 Pa	1260.23	1423.30	4.00	94.78	nr	**1518.08**
10 m³/s at 700 Pa	1750.62	1977.14	5.00	118.47	nr	**2095.61**
15 m³/s at 700 Pa	2336.43	2638.75	6.00	142.17	nr	**2780.92**
20 m³/s at 700 Pa	3502.46	3955.66	6.00	142.17	nr	**4097.83**
Extra for runaround coil, including pump and associated pipework; typical outputs in brackets (based on minimal distance between the supply and extract units)						
2 m³/s @350 Pa (26 kW)	2723.10	3075.45	30.00	710.84	nr	**3786.29**
5 m³/s at 350 Pa (37 kW)	4745.60	5359.66	30.00	710.84	nr	**6070.50**
10 m³/s at 700 Pa (85 kW)	8146.64	9200.78	40.00	947.78	nr	**10148.56**
15 m³/s at 700 Pa (151 kW)	11462.62	12945.83	50.00	1184.74	nr	**14130.57**
20 m³/s at 700 Pa (158 kW)	15364.76	17352.88	60.00	1421.69	nr	**18774.57**
Extra for thermal wheel (typical outputs in brackets)						
2 m³/s at 350 Pa (37 kW)	5888.40	6650.33	12.00	284.34	nr	**6934.67**
5 m³/s at 350 Pa (65 kW)	8457.17	9551.48	12.00	284.34	nr	**9835.82**
10 m³/s at 700 Pa (127 kW)	11278.26	12737.61	15.00	355.42	nr	**13093.03**
15 m³/s at 700 Pa (160 kW)	21130.66	23864.86	17.00	402.81	nr	**24267.67**
20 m³/s at 700 Pa (262 kW)	24683.07	27876.94	19.00	450.20	nr	**28327.14**
Extra for plate heat exchanger, including additional filtration in extract leg (typical outputs in brackets)						
2 m³/s @350 Pa (25 kW)	3284.13	3709.08	12.00	284.34	nr	**3993.42**
5 m³/s at 350 Pa (51 kW)	6420.05	7250.77	12.00	284.34	nr	**7535.11**
10 m³/s at 700 Pa (98 kW)	9419.99	10638.89	15.00	355.42	nr	**10994.31**
15 m³/s at 700 Pa (160 kW)	16267.55	18372.48	17.00	402.81	nr	**18775.29**
20 m³/s at 700 Pa (190 kW)	21008.17	23726.53	19.00	450.20	nr	**24176.73**
Extra for electric heating in lieu of LTHW						
2 m³/s @350 Pa	1237.22	1397.31	–	–	nr	**1397.31**
5 m³/s at 350 Pa	2278.44	2573.26	–	–	nr	**2573.26**
10 m³/s at 700 Pa	3050.17	3444.85	–	–	nr	**3444.85**
15 m³/s at 700 Pa	3198.39	3612.25	–	–	nr	**3612.25**
20 m³/s at 700 Pa	3711.65	4191.91	–	–	nr	**4191.91**

U: VENTILATION/AIR CONDITIONING SYSTEMS

Item	Net Price £	Material £	Labour hours	Labour £	Unit	Total rate £
U31: VAV AIR CONDITIONING						
VAV TERMINAL BOXES						
VAV terminal box; integral acoustic silencer; factory installed and pre wired control components (excluding electronic controller); selected at 200 Pa at entry to unit; includes fixing in position; electrical work elsewhere						
80 l/s – 110 l/s	448.80	506.87	2.00	45.40	nr	**552.27**
Extra for secondary silencer	109.14	123.27	0.50	11.35	nr	**134.62**
Extra for 2 row LTHW heating coil	44.27	50.00	–	–	nr	**50.00**
150 l/s – 190 l/s	472.26	533.37	2.00	45.40	nr	**578.77**
Extra for secondary silencer	119.85	135.35	0.50	11.35	nr	**146.70**
Extra for 2 row LTHW heating coil	53.09	59.96	–	–	nr	**59.96**
250 l/s – 310 l/s	526.83	595.00	2.00	45.40	nr	**640.40**
Extra for secondary silencer	158.61	179.13	0.50	11.35	nr	**190.48**
Extra for 2 row LTHW heating coil	29.66	33.50	–	–	nr	**33.50**
420 l/s – 520 l/s	570.69	644.54	2.00	45.40	nr	**689.94**
Extra for secondary silencer	180.54	203.90	0.50	11.35	nr	**215.25**
Extra for 2 row LTHW heating coil	72.29	81.64	–	–	nr	**81.64**
650 l/s – 790 l/s	691.05	780.47	2.00	45.40	nr	**825.87**
Extra for secondary silencer	231.03	260.93	0.50	11.35	nr	**272.28**
Extra for 2 row LTHW heating coil	94.40	106.61	–	–	nr	**106.61**
1130 l/s – 1370 l/s	802.74	906.61	2.00	45.40	nr	**952.01**
Extra for secondary silencer	322.83	364.60	0.50	11.35	nr	**375.95**
Extra for 2 row LTHW heating coil	120.22	135.77	–	–	nr	**135.77**
Extra for electric heater & thyristor controls, 3 kW/1ph (per box)	350.92	396.33	–	–	nr	**396.33**
Extra for fitting free issue box controller	–	–	2.13	48.35	nr	**48.35**
Fan assisted VAV terminal box; factory installed and pre wired control components (excluding electronic controller); selected at 40 Pa external static pressure; includes fixing in position, electrical work elsewhere						
100 l/s – 175 l/s	993.48	1122.03	3.00	68.11	nr	**1190.14**
Extra for secondary silencer	118.32	133.63	0.50	11.35	nr	**144.98**
Extra for 1 row LTHW heating coil	62.67	70.78	–	–	nr	**70.78**
170 l/s – 360 l/s	1083.75	1223.98	3.00	68.11	nr	**1292.09**
Extra for secondary silencer	149.94	169.34	0.50	11.35	nr	**180.69**
Extra for 1 row LTHW heating coil	71.49	80.74	–	–	nr	**80.74**
300 l/s – 640 l/s	1242.36	1403.11	3.00	68.11	nr	**1471.22**
Extra for secondary silencer	229.50	259.19	0.50	11.35	nr	**270.54**
Extra for 1 row LTHW heating coil	79.63	89.93	–	–	nr	**89.93**
620 l/s – 850 l/s	1242.36	1403.11	3.00	68.11	nr	**1471.22**
Extra for secondary silencer	229.50	259.19	0.50	11.35	nr	**270.54**
Extra for 1 row LTHW heating coil	79.63	89.93	–	–	nr	**89.93**
Extra for electric heater plus thyristor controls' 3 kW/lph (per box)	350.92	396.33	–	–	nr	**396.33**
Extra for fitting free issue controller	–	–	2.13	48.35	nr	**48.35**

U: VENTILATION/AIR CONDITIONING SYSTEMS

Item	Net Price £	Material £	Labour hours	Labour £	Unit	Total rate £
U41: FAN COIL AIR CONDITIONING						
FAN COIL UNITS						
All selections based on summer return air condition of 23°C @ 50% RH, CHW @ 6°/ 12°C, LTHW @ 82°/71°C (where applicable), medium speed, external resistance of 30 Pa						
All selections are based on heating and cooling units. For waterside control units there is no significant reduction in cost between 4 pipe heating and cooling and 2 pipe cooling only units (excluding controls). For airside control units, there is a marginal reduction (less than 5%) between 4 pipe heating and cooling units and 2 pipe cooling only units (excluding controls)						
Ceiling void mounted horizontal waterside control fan coil unit; cooling coil; LTHW heating coil; multi tapped speed transformer; fine wire mesh filter; includes fixing in position; electrical work elsewhere						
Total cooling load, heating load						
2800 W, 1000 W	294.04	332.09	4.00	94.78	nr	**426.87**
4000 W, 1700 W	341.86	386.09	4.00	94.78	nr	**480.87**
4500 W, 1900 W	377.85	426.74	4.00	94.78	nr	**521.52**
6000 W, 2600 W	442.66	499.94	4.00	94.78	nr	**594.72**
Ceiling void mounted horizontal waterside control fan coil unit; cooling coil ; electric heating coil; multi tapped speed transformer; fine wire mesh filter; includes fixing in position; electrical work elsewhere + thyristor and 2 No. HTCO's						
Total cooling load, heating load						
2800 W, 1500 W.	431.74	487.60	4.00	94.78	nr	**582.38**
4000 W, 2000 W	479.56	541.61	4.00	94.78	nr	**636.39**
4500 W, 2000 W	515.55	582.26	4.00	94.78	nr	**677.04**
6000 W, 3000 W	580.36	655.45	4.00	94.78	nr	**750.23**

U: VENTILATION/AIR CONDITIONING SYSTEMS

Item	Net Price £	Material £	Labour hours	Labour £	Unit	Total rate £
U41: FAN COIL AIR CONDITIONING – cont						
FAN COIL UNITS – cont						
Ceiling void mounted horizontal airside control fan coil unit; cooling coil; LHTW heating coil; multi tapped speed transformer; fine wire mesh filter, damper actuator & fixing kit; includes fixing in position; electrical work elsewhere						
Total cooling load, heating load						
2600 W, 2200 W.	600.10	677.75	4.00	94.78	nr	**772.53**
3600 W, 3200 W	643.06	726.27	4.00	94.78	nr	**821.05**
4000 W, 3600 W	708.42	800.09	4.00	94.78	nr	**894.87**
5400 W, 5000 W	775.53	875.88	4.00	94.78	nr	**970.66**
Ceiling void mounted horizontal airside control fan coil unit; cooling coil ; electric heating coil; multi tapped speed transformer; fine wire mesh filter, damper actuator & fixing kit; includes fixing in position; electrical work elsewhere + thyristor and 2 No. HTCO's						
Total cooling load, heating load						
2600 W, 1500 W.	710.26	802.17	4.00	94.78	nr	**896.95**
3600 W, 2000 W	753.22	850.68	4.00	94.78	nr	**945.46**
4000 W, 2000 W	818.58	924.50	4.00	94.78	nr	**1019.28**
5400 W, 3000 W	885.69	1000.29	4.00	94.78	nr	**1095.07**
Ceiling void mounted slimline horizontal waterside control fan coil unit, 170 mm deep; cooilng coil; LTHW heating coil; multi tapped speed transformer; fine wire mesh filter; includes fixing in position; electrical work elsewhere						
Total cooling load, heating load						
1100 W, 1500 W.	478.02	539.88	3.50	82.94	nr	**622.82**
3200 W, 3700 W	585.30	661.03	4.00	94.78	nr	**755.81**
Ceiling void mounted slimline horizontal waterside control fan coil unit, 170 mm deep; cooilng coil; electric heating coil; multi tapped speed transformer; fine wire mesh filter; includes fixing in position; electrical work elsewhere						
Total cooling load, heating load						
1100 W, 1000 W.	586.67	662.58	3.50	82.94	nr	**745.52**
3400 W, 2000 W	693.94	783.74	4.00	94.78	nr	**878.52**

U: VENTILATION/AIR CONDITIONING SYSTEMS

Item	Net Price £	Material £	Labour hours	Labour £	Unit	Total rate £
Ceiling void mounted slimline horizontal airside control fan coil unit, 170 mm deep; cooling coil; LTHW heating coil; multi tapped speed transformer; fine wire mesh filter, damper actuator & fixing kit; includes fixing in position; electrical work elsewhere						
Total cooling load, heating load						
1000 W, 1600 W.	517.27	584.21	3.50	82.94	nr	**667.15**
3000 W, 3300 W	805.94	910.23	4.00	94.78	nr	**1005.01**
4500 W, 4500 W	1033.12	1166.80	4.00	94.78	nr	**1261.58**
Ceiling void mounted slimline horizontal airside control fan coil unit, 170 mm deep; cooilng coil; electric heating coil; multi tapped speed transformer; fine wire mesh filter, damper actuator & fixing kit; includes fixing in position; electrical work elsewhere						
Total cooling load, heating load						
1000 W, 1000 W.	761.99	860.58	3.50	82.94	nr	**943.52**
3000 W, 2000 W	1003.74	1133.62	4.00	94.78	nr	**1228.40**
4500 W, 3000 W	1274.83	1439.79	4.00	94.78	nr	**1534.57**
Low level perimeter waterside control fan coil unit; cooling coil; LTHW heating coil; multi tapped speed transformer; fine wire mesh filter; includes fixing in position; electrical work elsewhere						
Total cooling load, heating load						
1700 W, 1400 W	366.34	413.74	3.50	82.94	nr	**496.68**
Extra over for standard cabinet	194.89	220.11	1.00	23.69	nr	**243.80**
2200 W, 1900 W	424.95	479.94	3.50	82.94	nr	**562.88**
Extra over for standard cabinet	237.39	268.11	1.00	23.69	nr	**291.80**
2600 W, 2200 W	468.91	529.58	3.50	82.94	nr	**612.52**
Extra over for standard cabinet	243.25	274.73	1.00	23.69	nr	**298.42**
3900 W, 3200 W	615.45	695.09	3.50	82.94	nr	**778.03**
Extra over for standard cabinet	265.23	299.55	1.00	23.69	nr	**323.24**
4600 W, 3900 W	703.36	794.38	3.50	82.94	nr	**877.32**
Extra over for standard cabinet	348.76	393.88	1.00	23.69	nr	**417.57**

U: VENTILATION/AIR CONDITIONING SYSTEMS

Item	Net Price £	Material £	Labour hours	Labour £	Unit	Total rate £
U41: FAN COIL AIR CONDITIONING – cont						
FAN COIL UNITS – cont						
Low level perimeter waterside control fan coil unit; cooling coil; electric heating coil; multi tapped speed transformer; fine wire mesh filter; includes fixing in position; electrical work elsewhere						
Total cooling load, heating load						
1700 W, 1500 W	476.25	537.88	3.50	82.94	nr	**620.82**
Extra over for standard cabinet	194.89	220.11	1.00	23.69	nr	**243.80**
2200 W, 2000 W	498.23	562.70	3.50	82.94	nr	**645.64**
Extra over for standard cabinet	237.39	268.11	1.00	23.69	nr	**291.80**
2600 W, 2000 W	622.78	703.36	3.50	82.94	nr	**786.30**
Extra over for standard cabinet	243.25	274.73	1.00	23.69	nr	**298.42**
3800 W, 3000 W	740.01	835.77	3.50	82.94	nr	**918.71**
Extra over for standard cabinet	265.23	299.55	1.00	23.69	nr	**323.24**
4600 W, 4000 W	901.20	1017.81	3.50	82.94	nr	**1100.75**
Extra over for standard cabinet	348.76	393.88	1.00	23.69	nr	**417.57**
Low level perimeter airside control fan coil unit; cooling coil; LTHW heating coil; multi tapped speed transformer; fine wire mesh filter; damper actuator & fixing kit; includes fixing in position; electrical work elsewhere						
Total cooling load, heating load						
1200 W, 1400 W	473.31	534.55	3.50	82.94	nr	**617.49**
Extra over for standard cabinet	194.89	220.11	1.00	23.69	nr	**243.80**
1800 W, 2000 W	529.00	597.45	3.50	82.94	nr	**680.39**
Extra over for standard cabinet	237.39	268.11	1.00	23.69	nr	**291.80**
2200 W, 2400 W	570.03	643.79	3.50	82.94	nr	**726.73**
Extra over for standard cabinet	243.25	274.73	1.00	23.69	nr	**298.42**
3200 W, 3600 W	740.01	835.77	3.50	82.94	nr	**918.71**
Extra over for standard cabinet	265.23	299.55	1.00	23.69	nr	**323.24**

U: VENTILATION/AIR CONDITIONING SYSTEMS

Item	Net Price £	Material £	Labour hours	Labour £	Unit	Total rate £
Low level perimeter airside control fan coil unit; cooling coil; electric heating coil; multi tapped speed transformer; fine wire mesh filter; damper actuator & fixing kit; includes fixing in position; electrical work elsewhere						
Total cooling load, heating load						
1250 W, 1500 W	644.76	728.19	3.50	82.94	nr	**811.13**
Extra over for standard cabinet	194.89	220.11	1.00	23.69	nr	**243.80**
1900 W, 2000 W	696.06	786.12	3.50	82.94	nr	**869.06**
Extra over for standard cabinet	237.39	268.11	1.00	23.69	nr	**291.80**
2300 W, 2000 W	795.69	898.65	3.50	82.94	nr	**981.59**
Extra over for standard cabinet	243.25	274.73	1.00	23.69	nr	**298.42**
3300 W, 3000 W	908.52	1026.08	3.50	82.94	nr	**1109.02**
Extra over for standard cabinet	265.23	299.55	1.00	23.69	nr	**323.24**
Typical DDC control pack for airside control units; controller, return air sensor (damper and damper actuator included in above rates)						
Heating and cooling, per unit	156.06	176.26	2.00	47.39	nr	**223.65**
Cooling only, per unit	183.60	207.36	2.00	47.39	nr	**254.75**
Typical DDC control pack for waterside control units; controller, return air sensor, four port valves and actuators						
Heating and cooling, per unit	137.70	155.52	2.00	47.39	nr	**202.91**
Cooling only, per unit	119.34	134.78	2.00	47.39	nr	**182.17**
Note: Care needs to be taken when using these controls prices, as they can vary significantly, depending on the equipment manufacturer specified and the degree of control required.						

U: VENTILATION/AIR CONDITIONING SYSTEMS

Item	Net Price £	Material £	Labour hours	Labour £	Unit	Total rate £
U70 – AIR CURTAINS						
The selection of air curtains requires consideration of the particular conditions involved; climatic conditions, wind influence, construction and position all influence selection; consultation with a specialist manufacturer is therefore advisable						
Commercial grade air curtains; recessed or exposed units with rigid sheet steel casing; aluminium grilles; high quality motor/centrifugal fan assembly; includes fixing in position; electrical work elsewhere						
Ambient temperature; 240 V single phase supply; mounting height 2.40 m						
1000 × 590 × 270 mm	2363.34	2669.15	12.05	285.48	nr	**2954.63**
1500 × 590 × 270 mm	3025.32	3416.79	12.05	285.48	nr	**3702.27**
2000 × 590 × 270 mm	3660.78	4134.47	12.05	285.52	nr	**4419.99**
2500 × 590 × 270 mm	4068.78	4595.26	13.00	308.03	nr	**4903.29**
Ambient temperature; 240 V single phase supply; mounting height 2.80 m						
1000 × 590 × 270 mm	2731.56	3085.01	16.13	382.17	nr	**3467.18**
1500 × 590 × 270 mm	3486.36	3937.48	16.13	382.17	nr	**4319.65**
2000 × 590 × 270 mm	4346.22	4908.60	16.13	382.17	nr	**5290.77**
2500 × 590 × 270 mm	5026.56	5676.97	17.10	405.18	nr	**6082.15**
Ambient temperature 240 V single phase supply; mounting height 3.30 m						
1000 × 774 × 370 mm	3567.96	4029.63	17.24	408.52	nr	**4438.15**
1500 × 774 × 370 mm	4825.62	5450.03	17.24	408.52	nr	**5858.55**
2000 × 774 × 370 mm	5968.02	6740.25	17.24	408.52	nr	**7148.77**
2500 × 774 × 370 mm	7064.52	7978.63	18.30	433.61	nr	**8412.24**
Ambient temperature; 240 V single phase supply; mounting height 4.00 m						
1000 × 774 × 370 mm	3941.28	4451.26	19.10	452.57	nr	**4903.83**
1500 × 774 × 370 mm	5170.38	5839.40	19.10	452.57	nr	**6291.97**
2000 × 774 × 370 mm	6436.20	7269.01	19.10	452.57	nr	**7721.58**
2500 × 774 × 370 mm	7542.90	8518.92	19.90	471.53	nr	**8990.45**

U: VENTILATION/AIR CONDITIONING SYSTEMS

Item	Net Price £	Material £	Labour hours	Labour £	Unit	Total rate £
Water heated; 240 V single phase supply; mounting height 2.40 m						
1000 × 590 × 270 mm; 2.30–9.40 kW output	2781.54	3141.46	12.05	285.48	nr	3426.94
1500 × 590 × 270 mm; 3.50–14.20 kW output	3559.80	4020.42	12.05	285.48	nr	4305.90
2000 × 590 × 270 mm; 4.70–19.00 kW output	4307.46	4864.83	12.05	285.48	nr	5150.31
2500 × 590 × 270 mm; 5.90–23.70 kW output	4787.88	5407.41	13.00	308.03	nr	5715.44
Water heated; 240 V single phase supply; mounting height 2.80 m						
1000 × 590 × 270 mm; 3.30–11.90 kW output	3214.02	3629.89	16.13	382.17	nr	4012.06
1500 × 590 × 270 mm; 5.00–17.90 kW output	4103.46	4634.43	16.13	382.17	nr	5016.60
2000 × 590 × 270 mm; 6.70–23.90 kW output	5114.28	5776.05	16.13	382.17	nr	6158.22
2500 × 590 × 270 mm; 8.30–29.80 kW output	5913.96	6679.19	17.10	405.18	nr	7084.37
Water heated; 240 V single phase supply; mounting height 3.30 m						
1000 × 774 × 370 mm; 6.10–21.80 kW output	4198.32	4741.56	17.24	408.52	nr	5150.08
1500 × 774 × 370 mm; 9.20–32.80 kW output	5677.32	6411.93	17.24	408.52	nr	6820.45
2000 × 774 × 370 mm; 12.30–43.70 kW output	7021.68	7930.26	17.24	408.52	nr	8338.78
2500 × 774 × 370 mm; 15.30–54.60 kW output	8311.98	9387.51	18.30	433.61	nr	9821.12
Water heated; 240 V single phase supply; mounting height 4.00 m						
1000 × 774 × 370 mm; 7.20–24.20 kW output	4636.92	5236.91	19.10	452.57	nr	5689.48
1500 × 774 × 370 mm; 10.90–36.30 kW output	6083.28	6870.42	19.10	452.57	nr	7322.99
2000 × 774 × 370 mm; 14.50–48.40 kW output	7572.48	8552.32	19.10	452.57	nr	9004.89
2500 × 774 × 370 mm; 18.10–60.60 kW output	8875.02	10023.41	19.90	471.53	nr	10494.94
Electrically heated; 415 V three phase supply; mounting height 2.40 m						
1000 × 590 × 270 mm; 2.30–9.40 kW output	3417.00	3859.14	12.05	285.48	nr	4144.62
1500 × 590 × 270 mm; 3.50–14.20 kW output	4250.34	4800.31	12.05	285.48	nr	5085.79
2000 × 590 × 270 mm; 4.70–19.00 kW output	5066.34	5721.89	12.05	329.16	nr	6051.05
2500 × 590 × 270 mm; 5.90–23.70 kW output	5759.94	6505.24	13.00	355.16	nr	6860.40

U: VENTILATION/AIR CONDITIONING SYSTEMS

Item	Net Price £	Material £	Labour hours	Labour £	Unit	Total rate £
U70 – AIR CURTAINS – cont						
Commercial grade air curtains – cont						
Electrically heated; 415 V three phase supply; mounting height 2.80 m						
1000 × 590 × 270 mm; 3.30–11.90 kW output	4022.88	4543.42	16.13	440.64	nr	4984.06
1500 × 590 × 270 mm; 5.00–17.90 kW output	5047.98	5701.17	16.13	382.17	nr	6083.34
2000 × 590 × 270 mm; 6.70–23.90 kW output	6245.46	7053.60	16.13	382.17	nr	7435.77
2500 × 590 × 270 mm; 8.30–29.80 kW output	6980.88	7884.17	17.10	405.18	nr	8289.35
Electrically heated; 415 V three phase supply; mounting height 3.30 m						
1000 × 774 × 370 mm; 6.10–21.80 kW output	6111.84	6902.68	17.24	471.01	nr	7373.69
1500 × 774 × 370 mm; 9.20–32.80 kW output	8189.58	9249.27	17.24	408.50	nr	9657.77
2000 × 774 × 370 mm; 12.30–43.70 kW output	9883.80	11162.72	17.24	408.50	nr	11571.22
2500 × 774 × 370 mm; 15.30–54.60 kW output	11700.42	13214.40	18.30	433.61	nr	13648.01
Electrically heated; 415 V three phase supply; mounting height 4.00 m						
1000 × 774 × 370 mm; 7.20–24.20 kW output	6745.26	7618.06	19.10	521.81	nr	8139.87
1500 × 774 × 370 mm; 10.90–36.30 kW output	8751.60	9884.01	19.10	452.57	nr	10336.58
2000 × 774 × 370 mm; 14.50–48.40 kW output	10584.54	11954.12	19.10	452.57	nr	12406.69
2500 × 774 × 370 mm; 18.10–60.60 kW output	12537.84	14160.17	19.90	471.53	nr	14631.70
Industrial grade air curtains; recessed or exposed units with rigid sheet steel casing; aluminium grilles; high quality motor/centrifugal fan assembly; includes fixing in position; electrical work elsewhere						
Ambient temperature; 415 V three phase supply; including wiring between multiple units; horizontally or vertically mounted; opening maximum 6.00 m						
1500 × 585 × 853 mm; 1.6 A supply	2463.30	2782.04	17.24	408.52	nr	3190.56
2000 × 585 × 853 mm; 2.1 A supply	3003.90	3392.59	17.24	408.52	nr	3801.11

U: VENTILATION/AIR CONDITIONING SYSTEMS

Item	Net Price £	Material £	Labour hours	Labour £	Unit	Total rate £
Water heated; 415 V three phase supply; including wiring between multiple units; horizontally or vertically mounted; opening maximum 6.00 m						
1500 × 585 × 956 mm; 1.6A supply; 34.80 kW output	2850.90	3219.79	17.24	408.52	nr	**3628.31**
2000 × 585 × 956 mm; 2.1A supply; 50.70 kW output	3442.50	3887.94	17.24	408.52	nr	**4296.46**
Water heated; 415 V three phase supply; including wiring between multiple units; vertically mounted in single bank for openings maximum 6.00 m wide or opposing twin banks for openings maximum 10.00 m wide						
1500 × 585 mm; 1.6A supply; 41.1 kW output	2850.90	3219.79	17.24	408.52	nr	**3628.31**
2000 × 585 mm; 2.1A supply; 57.7 kW output	3442.50	3887.94	17.24	408.52	nr	**4296.46**
Remote mounted electronic controller unit; 415 V three phase supply; excluding wiring to units						
Five speed, 7A	433.50	489.59	5.00	118.47	nr	**608.06**

Managing the Brief for Better Design

2nd Edition

By **Alastair Blyth, John Worthington**

Briefing is not just presenting a set of documents to the design team; it is a process of developing a deep understanding about client needs. This book provides both inspiration to clients and a framework for practitioners. The coverage extends beyond new build, covering briefing for services and fit-outs. Written by an experienced and well-known team of authors, this new edition clearly explains how important the briefing process is to both the construction industry, in delivering well-designed buildings, and to their clients in achieving them. The text is illustrated by excellent examples of effective practice, drawn from DEGW experience, as well as five model briefs and invaluable process charts.

June 2010: 246x189: 272pp
Hb: 978-0-415-46030-9: **£95.00**
Pb: 978-0-415-46031-6: **£34.99**

Life Cycle Assessment in the Built Environment

R. Crawford

Life cycle assessment enables the identification of a broad range of potential environmental impacts occurring across the entire life of a product, from its design through to its eventual disposal or reuse. The need for life cycle assessment to inform environmental design within the built environment is critical, due to the complex range of materials and processes required to construct and manage our buildings and infrastructure systems.

After outlining the framework for life cycle assessment, this book uses a range of case studies to demonstrate the innovative input-output-based hybrid approach for compiling a life cycle inventory. This approach enables a comprehensive analysis of a broad range of resource requirements and environmental outputs so that the potential environmental impacts of a building or infrastructure system can be ascertained. These case studies cover a range of elements that are part of the built environment, including a residential building, a commercial office building and a wind turbine, as well as individual building components such as a residential-scale photovoltaic system.

Comprehensively introducing and demonstrating the uses and benefits of life cycle assessment for built environment projects, this book will show you how to assess the environmental performance of your clients' projects, to compare design options across their entire life and to identify opportunities for improving environmental performance.

March 2011: 272pp
Hb: 978-0-415-55795-5: **£80.00**

Spon's Building Regulations Explained

Eighth Edition

London District Surveyors Association and John Stephenson

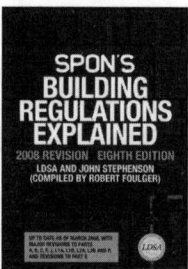

This fully revised, essential reference takes into account all important aspects of building control including new legislation up to the start of 2008, covering major revisions to Parts A, B, C, F, J, L1A, L1B, L2A, L2B and P and revisions to Part E. Each chapter explains in clear terms the appropriate regulation and any other relevant legislation, before explaining the approved document.

Selected Contents:

1. The Development of Building Control
2. Control of Building Work
3. Application of Building Regulations to Inner London
4. Relaxation of Building Regulations
5. Exempt Buildings and Works
6. Notices and Plans
7. Approved Inspectors
8. Work Undertaken by Public Bodies
9. Approved Document to Support Regulation 7

Chapters 10-22. Approved Documents A to N 23. Other Approved Documents

2009: 297x210: 672pp
Hb: 978-0-415-43067-8 **£80.00**

To Order: Tel: +44 (0) 1235 400524 **Fax:** +44 (0) 1235 400525
or Post: Taylor and Francis Customer Services,
Bookpoint Ltd, Unit T1, 200 Milton Park, Abingdon, Oxon, OX14 4TA UK
Email: book.orders@tandf.co.uk

For a complete listing of all our titles visit:
www.tandf.co.uk

Spon's Estimating Costs Guide to Plumbing and Heating

Unit Rates and Project Costs

Fourth Edition

Bryan Spain

Do you work on jobs between £50 and £50,000? - Then this book is for you.

All the cost data you need to keep your estimating accurate, competitive and profitable.

Specially written for contractors and small businesses carrying out small works, *Spon's Estimating Cost Guide to Plumbing and Heating* contains accurate information on thousands of rates each broken down to labour, material overheads and profit.

The first book to include typical project costs for:

- rainwater goods installations
- bathrooms
- external waste systems
- central heating systems
- hot and cold water systems.

July 2008: 216x138: 264pp
Pb: 978-0-415-46905-0: **£31.99**

Spon's Estimating Costs Guide to Electrical Works

Project Costs at a Glance

Fourth Edition

Bryan Spain

Specially written for contractors and small businesses carrying out small works, *Spon's Estimating Costs Guide to Electrical Works* provides accurate information on thousands of rates, each broken down to labour, material overheads and profit for residential, retail and light industrial premises. It is the first book to include typical project costs for new installations, stripping out, rewiring and upgrading for flats and houses.

In addition, vital information and advice is given on setting up and running a business, employing staff, tax, VAT and CIS4's.

For the cost of approximately two hours of your charge-out rate (or less), this book will help you to:

- produce estimates faster
- keep your estimates accurate and competitive
- run your business more effectively
- save time.

August 2008: 216x138: 304pp
Pb: 978-0-415-46904-3: **£31.99**

To Order: Tel: +44 (0) 1235 400524 **Fax:** +44 (0) 1235 400525
or Post: Taylor and Francis Customer Services,
Bookpoint Ltd, Unit T1, 200 Milton Park, Abingdon, Oxon, OX14 4TA UK
Email: book.orders@tandf.co.uk

For a complete listing of all our titles visit:
www.tandf.co.uk

Material Costs/Measured Work Prices

DIRECTIONS

The following explanations are given for each of the column headings and letter codes.

Unit	Prices for each unit are given as singular (i.e. 1 metre, 1 nr) unless stated otherwise.
Net price	Industry tender prices, plus nominal allowance for fixings, waste and applicable trade discounts.
Material cost	Net price plus percentage allowance for overheads, profit and preliminaries.
Labour norms	In man-hours for each operation.
Labour cost	Labour constant multiplied by the appropriate all-in man-hour cost based on gang rate (See also relevant Rates of Wages Section) plus percentage allowance for overheads, profit and preliminaries.
Measured work Price (total rate)	Material cost plus Labour cost.

MATERIAL COSTS

The Material Costs given are based at Second Quarter 2011 but exclude any charges in respect of VAT. The average rate of copper during this quarter is US$9,000 / UK£5,625 per tonne. Users of the book are advised to register on the SPON's website www.pricebooks.co.uk/updates to receive the free quarterly updates – alerts will then be provided by e-mail as changes arise.

MEASURED WORK PRICES

These prices are intended to apply to new work in the London area. The prices are for reasonable quantities of work and the user should make suitable adjustments if the quantities are especially small or especially large. Adjustments may also be required for locality (e.g. outside London – refer to cost indices in approximate estimating section for details of adjustment factors) and for the market conditions (e.g. volume of work secured or being tendered) at the time of use.

ELECTRICAL INSTALLATIONS

The labour rate has been based on average gang rates per man hour effective from 4 January 2010 including allowances for all other emoluments and expenses. The ECA and Unite have not reached agreement to revise the rates at the time of publication. As a result, the rates of pay and other terms and conditions applicable during 2010 will apply. Future changes will be published in the free SPON'S quarterly update by registering on their website. To this rate, has been added 3% and 7.5% to cover site and head office overheads and preliminary items together with a further 2% for profit, resulting in an inclusive rate of £27.48 per man hour. The rate has been calculated on a working year of 2016 hours; a detailed build-up of the rate is given at the end of these directions.

DIRECTIONS

In calculating the 'Measured Work Prices' the following assumptions have been made:

    (a)   That the work is carried out as a subcontract under the Standard Form of Building Contract.

    (b)   That, unless otherwise stated, the work is being carried out in open areas at a height which would not require more than simple scaffolding.

    (c)   That the building in which the work is being carried out is no more than six storey's high.

Where these assumptions are not valid, as for example where work is carried out in ducts and similar confined spaces or in multi-storey structures when additional time is needed to get to and from upper floors, then an appropriate adjustment must be made to the prices. Such adjustment will normally be to the labour element only.

DIRECTIONS

LABOUR RATE – ELECTRICAL

The annual cost of a notional eleven man gang of 11 men

		TECHNICIAN	APPROVED ELECTRICIANS	ELECTRICIANS	LABOURERS	SUB-TOTALS
		1 NR	4 NR	4 NR	2 NR	
Hourly Rate from 4 January 2010		**18.10**	**16.07**	**14.83**	**11.89**	
Working hours per annum per man		1,680.00	1,680.00	1,680.00	1,680.00	
x Hourly rate × nr of men = £ per annum		30,408.00	107,990.40	99,657.60	39,950.40	278,006.40
Overtime Rate		27.15	24.11	22.25	17.84	
Overtime hours per annum per man		336.00	336.00	336.00	336.00	
x Hourly rate × nr of men = £ per annum		9,122.40	32,397.12	29,897.28	11,985.12	83,401.92
Total		39,530.40	140,387.52	129,554.88	51,935.52	361,408.32
Incentive schemes (insert percentage)	0.00%	0.00	0.00	0.00	0.00	0.00
Daily Travel Allowance (15-20 miles each way) effective from 04/01/10		5.03	5.03	5.03	5.03	
Days per annum per man		224.00	224.00	224.00	224.00	
x nr of men = £ per annum		1,126.72	4,506.88	4,506.88	2,253.44	12,393.92
Daily Travel Allowance (15-20 miles each way) effective from 04/01/10		3.45	3.45	3.45	3.45	
Days per annum per man		224.00	224.00	224.00	224.00	
x nr of men = £ per annum		772.80	3,091.20	3,091.20	1,545.60	8,500.80
JIB Pension Scheme @ 2.5%		1,169.90	4,175.29	3,867.22	1,568.41	10,780.82
JIB combined benefits scheme (nr of weeks per man)		52.00	52.00	52.00	52.00	
Benefit Credit effective from 27th September 2010		62.30	56.62	53.01	44.88	
x nr of men = £ per annum		3,239.60	11,776.96	11,026.08	4,667.52	30,710.16
Holiday Top-up Funding		55.76	49.71	46.16	37.31	
x nr of men @ 7.5 hrs per day = £ per annum		2,899.46	10,340.11	9,601.05	3,879.75	26,720.38
National Insurance Contributions:						
Annual gross pay (subject to NI) each		41,429.92	147,985.60	137,152.96	55,734.56	
% of NI Contributions		13.80	13.80	13.80	13.80	
£ Contributions/annum		4,740.24	16,515.62	15,020.71	5,738.17	42,015.24

SUB-TOTAL		492,529.63
TRAINING (INCLUDING ANY TRADE REGISTRATIONS) – SAY	1.00%	4,925.30
SEVERANCE PAY AND SUNDRY COSTS – SAY	1.50%	7,461.82
EMPLOYER'S LIABILITY AND THIRD PARTY INSURANCE – SAY	2.00%	10,098.34
ANNUAL COST OF NOTIONAL GANG		515,015.09

MEN ACTUALLY WORKING = 10.5 THEREFORE ANNUAL COST PER PRODUCTIVE MAN		49,049.06
AVERAGE NR OF HOURS WORKED PER MAN = 2016 THEREFORE ALL IN MAN HOUR		24.33
PRELIMINARY ITEMS – SAY	7.50%	1.82
SITE AND HEAD OFFICE OVERHEADS – SAY	3.00%	0.78
PROFIT – SAY	2.00%	0.54

THEREFORE INCLUSIVE MAN HOUR	**27.48**

Notes:

(1) Hourly wage rates are those effective from 4th January 2010.
(2) The following assumptions have been made in the above calculations: -
  (a) Hourly rates are based on London rate and job reporting own transport.
  (b) The working week of 37.5 hours is made up of 7.5 hours Monday to Friday.
  (c) Five days in the year are lost through sickness or similar reason.
  (d) A working year of 2016 hours.
(3) The incentive scheme addition of 5% is intended to reflect bonus schemes typically in use.
(4) National insurance contributions are those effective from 6th April 2011, paid for 48 weeks. Calculation is based on employer making regular payment into the holiday pay scheme, allowing savings on NI.
(5) Weekly JIB Combined Benefit Credit Scheme are those effective from 27th September 2010.
(6) Paid Holidays with effect from 9th January 2006, for all 30 days + 1 day for Royal Wedding (22 Annual and 9 Public) are to be paid at normal earnings level.
(7) Overtime is paid after 37.5 hours.

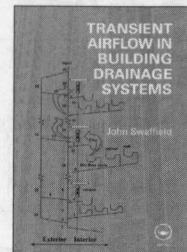

V: ELECTRICAL SUPPLY/POWER/LIGHTING

Item	Net Price £	Material £	Labour hours	Labour £	Unit	Total rate £
V10: ELECTRICAL GENERATION PLANT						
STANDBY GENERATORS						
Standby diesel generating sets; supply and installation; fixing to base; all supports and fixings; all necessary connections to equipment						
Three phase, 400 Volt, four wire 50 Hz packaged standby diesel generating set, complete with radio and television suppressors, daily service fuel tank and associated piping, 4 metres of exhaust pipe and primary exhaust silencer, control panel, mains failure relay, starting battery with charger, all internal wiring, interconnections, earthing and labels. Rated for standby duty; including UK delivery, installation and commissioning						
60 kVA	15544.48	17555.86	100.00	2747.82	nr	**20303.68**
100 kVA	21791.76	24611.51	100.00	2747.82	nr	**27359.33**
150 kVA	24922.41	28147.25	100.00	2747.82	nr	**30895.07**
315 kVA	44933.64	50747.83	120.00	3297.38	nr	**54045.21**
500 kVA	52962.92	59816.05	120.00	3297.38	nr	**63113.43**
750 kVA	90129.39	101791.69	140.00	3846.94	nr	**105638.63**
1000 kVA	119789.68	135289.86	140.00	3846.94	nr	**139136.80**
1500 kVA	140980.87	159223.09	170.00	4671.29	nr	**163894.38**
2000 kVA	217163.24	245263.08	170.00	4671.29	nr	**249934.37**
2500 kVA	297325.84	335798.32	210.00	5770.42	nr	**341568.74**
Extra for residential silencer; peformance 75 dBA at 1 m; including connection to exhaust pipe						
60 kVA	834.58	942.57	10.00	274.78	nr	**1217.35**
100 kVA	997.26	1126.30	10.00	274.78	nr	**1401.08**
150 kVA	1143.64	1291.63	10.00	274.78	nr	**1566.41**
315 kVA	2224.73	2512.60	15.00	412.17	nr	**2924.77**
500 kVA	2845.35	3213.52	15.00	412.17	nr	**3625.69**
750 kVA	4239.23	4787.77	20.00	549.57	nr	**5337.34**
1000 kVA	5709.46	6448.24	20.00	549.57	nr	**6997.81**
1500 kVA	9095.33	10272.22	20.00	549.57	nr	**10821.79**
2000 kVA	9851.10	11125.78	30.00	824.34	nr	**11950.12**
2500 kVA	11263.75	12721.23	30.00	824.34	nr	**13545.57**

V: ELECTRICAL SUPPLY/POWER/LIGHTING

Item	Net Price £	Material £	Labour hours	Labour £	Unit	Total rate £
V10: ELECTRICAL GENERATION PLANT – cont						
STANDBY GENERATORS – cont						
Synchronization panel for paralleling generators – not generators to mains; including interconnecting cables; commissioning and testing; fixing to backgrounds						
2 × 60 kVA	6951.93	7851.47	80.00	2198.25	nr	**10049.72**
2 × 100 kVA	7473.72	8440.79	80.00	2198.25	nr	**10639.04**
2 × 150 kVA	9401.88	10618.43	80.00	2198.25	nr	**12816.68**
2 × 315 kVA	15983.47	18051.65	80.00	2198.25	nr	**20249.90**
2 × 500 kVA	23075.56	26061.42	80.00	2198.25	nr	**28259.67**
2 × 750 kVA	26039.76	29409.17	80.00	2198.25	nr	**31607.42**
2 × 1000 kVA	28064.29	31695.67	120.00	3297.38	nr	**34993.05**
2 × 1500 kVA	33223.20	37522.12	120.00	3297.38	nr	**40819.50**
2 × 2000 kVA	35997.22	40655.08	120.00	3297.38	nr	**43952.46**
2 × 2500 kVA	41966.94	47397.25	120.00	3297.38	nr	**50694.63**
Prefabricated drop-over acoustic housing; performance 85 dBA at 1 m over the range from 60 kVA to 315 kVA, 75 dBA from 500 kVA to 2500 kVA.						
60 kVA	3003.00	3391.57	4.00	109.92	nr	**3501.49**
100 kVA	3106.84	3508.85	7.00	192.35	nr	**3701.20**
150 kVA	5045.04	5697.84	15.00	412.17	nr	**6110.01**
315 kVA	10764.49	12157.36	25.00	686.95	nr	**12844.31**
500 kVA	17779.01	20079.53	40.00	1099.13	nr	**21178.66**
750 kVA	28018.00	31643.39	40.00	1099.13	nr	**32742.52**
1000 kVA	35446.67	40033.30	40.00	1099.13	nr	**41132.43**
1500 kVA	70311.49	79409.45	40.00	1099.13	nr	**80508.58**
2000 kVA	84513.19	95448.77	60.00	1648.69	nr	**97097.46**
2500 kVA	117542.44	132751.85	70.00	1923.48	nr	**134675.33**
COMBINED HEAT AND POWER (CHP) UNITS						
Gas fired engine; acoustic enclosure complete with exhaust fan and attenuators; exhaust gas attenuation; includes 6 m long pipe connections; dry air cooler for secondary water circuit to reject excess heat; controls and panel; commissioning						
Electrical output; Heat output						
82 kW; 132 kW	92344.96	104293.94	–	–	nr	**104293.94**
100 kW, 148 kW	98362.43	111090.04	–	–	nr	**111090.04**
118 kW, 181 kW	103222.69	116579.19	–	–	nr	**116579.19**

V: ELECTRICAL SUPPLY/POWER/LIGHTING

Item	Net Price £	Material £	Labour hours	Labour £	Unit	Total rate £
130 kW, 201 kW	112442.13	126991.58	–	–	nr	**126991.58**
140 kW, 207 kW	119655.00	135137.76	–	–	nr	**135137.76**
150 kW, 208 kW	120349.32	135921.92	–	–	nr	**135921.92**
160 kW, 216 kW	126019.62	142325.92	–	–	nr	**142325.92**
198 kW, 233 kW	148816.56	168072.68	–	–	nr	**168072.68**
210 kW, 319 kW	157438.30	177810.03	–	–	nr	**177810.03**
237 kW, 359 kW	163628.79	184801.54	–	–	nr	**184801.54**
307 kW, 435 kW	221894.62	250606.68	–	–	nr	**250606.68**
380 kW, 500 kW	256899.51	290141.02	–	–	nr	**290141.02**
490 kW, 679 kW	353641.85	399401.34	–	–	nr	**399401.34**
501 kW, 518 kW	326331.81	368557.52	–	–	nr	**368557.52**
600 kW, 873 kW	388532.14	438806.25	–	–	nr	**438806.25**
725 kW, 1019 kW	525371.07	593351.46	–	–	nr	**593351.46**
975 kW, 1293 kW	652374.88	736788.93	–	–	nr	**736788.93**
1160 kW, 1442 kW	738586.65	834156.07	–	–	nr	**834156.07**
1379 kW, 1475 kW	881732.91	995824.74	–	–	nr	**995824.74**
1566 kW, 1647 kW	964530.37	1089335.78	–	–	nr	**1089335.78**
1600 kW, 1625 kW	962215.96	1086721.89	–	–	nr	**1086721.89**
1760 kW, 1821 kW	1056643.89	1193368.33	–	–	nr	**1193368.33**
2430 kW	1372350.00	1549925.23	–	–	nr	**1549925.23**
2931 kW	1659000.00	1873666.31	–	–	nr	**1873666.31**
3500 kW	1785000.00	2015970.07	–	–	nr	**2015970.07**
3900 kW	2100000.00	2371729.50	–	–	nr	**2371729.50**

Note: The costs detailed are based on a specialist subcontract package, as part of the M&E contract works, and include installation.

V: ELECTRICAL SUPPLY/POWER/LIGHTING

Item	Net Price £	Material £	Labour hours	Labour £	Unit	Total rate £
V11: HV SUPPLY						
Y61 – HV CABLES						
Cable; 6350/11000 Volts, 3 core, XLPE; stranded copper conductors; steel wire armoured; LSOH to BS 7835						
Laid in trench/duct including marker tape (cable tiles measured elsewhere)						
95 mm²	37.96	42.87	0.23	6.32	m	**49.19**
120 mm²	41.77	47.18	0.23	6.32	m	**53.50**
150 mm²	49.98	56.45	0.25	6.86	m	**63.31**
185 mm²	54.26	61.28	0.25	6.86	m	**68.14**
240 mm²	63.19	71.37	0.27	7.42	m	**78.79**
300 mm²	72.14	81.48	0.29	7.98	m	**89.46**
Cable tiles; single width; laid in trench above cables on prepared sand bed (cost of excavation excluded); reinforced concrete covers; concave/convex ends						
914 × 152 × 63/38 mm	10.87	12.28	0.11	3.03	m	**15.31**
914 × 229 × 63/38 mm	12.80	14.45	0.11	3.03	m	**17.48**
914 × 305 × 63/38 mm	14.25	16.10	0.11	3.03	m	**19.13**
Clipped direct to backgrounds including cleats						
95 mm²	48.15	54.38	0.47	12.92	m	**67.30**
120 mm²	51.97	58.69	0.50	13.73	m	**72.42**
150 mm²	59.88	67.63	0.53	14.56	m	**82.19**
185 mm²	65.61	74.10	0.55	15.12	m	**89.22**
240 mm²	75.87	85.69	0.60	16.49	m	**102.18**
300 mm²	86.23	97.39	0.68	18.68	m	**116.07**
Terminations for above cables, including heat-shrink kit and glanding off						
95 mm²	466.81	527.22	4.75	130.52	m	**657.74**
120 mm²	464.27	524.34	5.30	145.64	m	**669.98**
150 mm²	499.39	564.01	6.00	164.87	m	**728.88**
185 mm²	500.60	565.38	6.90	189.61	m	**754.99**
240 mm²	523.05	590.73	7.43	204.16	m	**794.89**
300 mm²	564.03	637.01	8.75	240.43	m	**877.44**

V: ELECTRICAL SUPPLY/POWER/LIGHTING

Item	Net Price £	Material £	Labour hours	Labour £	Unit	Total rate £
Cable; 6350/11000 Volts, 3 core, paper insulated; lead sheathed; steel wire armoured; stranded copper conductors; to BS 6480						
Laid in trench/duct including marker tape. (cable tiles measured elsewhere)						
95 mm²	41.34	46.69	0.22	6.04	m	**52.73**
120 mm²	44.62	50.40	0.22	6.04	m	**56.44**
150 mm²	54.59	61.65	0.24	6.60	m	**68.25**
185 mm²	56.97	64.34	0.26	7.15	m	**71.49**
240 mm²	66.10	74.65	0.34	9.34	m	**83.99**
Cable tiles; single width; laid in trench above cables on prepared sand bed (cost of excavation excluded); reinforced concrete covers; concave/convex ends						
914 × 152 × 63/38 mm	10.87	12.28	0.11	3.03	m	**15.31**
914 × 229 × 63/38 mm	12.80	14.45	0.11	3.03	m	**17.48**
914 × 305 × 63/38 mm	14.25	16.10	0.11	3.03	m	**19.13**
Clipped direct to backgrounds including cleats						
95 mm²	52.84	59.68	0.55	15.12	m	**74.80**
120 mm²	56.10	63.36	0.63	17.31	m	**80.67**
150 mm²	65.73	74.24	0.66	18.14	m	**92.38**
185 mm²	69.62	78.63	0.72	19.79	m	**98.42**
240 mm²	80.24	90.63	0.82	22.53	m	**113.16**
Terminations for above cables, including compound joint and glanding off						
95 mm²	510.93	577.04	4.75	130.52	m	**707.56**
120 mm²	520.16	587.47	5.30	145.64	m	**733.11**
150 mm²	543.33	613.63	6.10	167.62	m	**781.25**
185 mm²	544.49	614.95	6.90	189.61	m	**804.56**
240 mm²	572.63	646.72	7.43	204.16	m	**850.88**
300 mm²	607.26	685.84	8.75	240.43	m	**926.27**

V: ELECTRICAL SUPPLY/POWER/LIGHTING

Item	Net Price £	Material £	Labour hours	Labour £	Unit	Total rate £
V11: HV SUPPLY – cont						
Y70 – HV SWITCHGEAR AND TRANSFORMERS						
H.V. Circuit Breakers; installed on prepared foundations including all supports, fixings and inter panel connections where relevant. Excludes main and multi core cabling and heat shrink cable termination kits						
Three phase 11 kV, 630 amp, Air or SF6 insulated, with fixed pattern vacuum or SF6 circuit breaker panels; hand charged spring closing operation; prospective fault level up to 25 kA for 3 seconds. Feeders include ammeter with selector switch, 3 pole IDMT, overcurrent and earth fault relays with necessary current relays with necessary current transformers; incomers include 3 phase VT, voltmeter and phase selector switch; Includes IDMT overcurrent and earth fault relays/CTs						
Single panel with cable chamber	19869.12	22440.08	31.70	871.06	nr	**23311.14**
Three panel with one incomer and two feeders; with cable chambers	59533.83	67237.21	67.83	1863.84	nr	**69101.05**
Five panel with two incoming, two feeders and a bus section; with cable chambers	99205.70	112042.42	99.17	2725.01	nr	**114767.43**
Ring Main Unit (RMU)						
Three phase 11 kV, 630 amp, SF6 insulated RMU with vacuum or SF6 200 amp circuit breaker tee-off, Includes IDMT overcurrent and earth fault relays/CTs for tee-off and cable boxes.						
3 way Ring Main Unit	11979.33	13529.39	67.83	1863.84	nr	**15393.23**
Extra for,						
Remote actuator to ring switches (per switch)	1808.87	2042.93	–	–	nr	**2042.93**
Remote tripping of circuit breaker	904.45	1021.48	–	–	nr	**1021.48**
3 – Phase Neon indicators (per circuit)	106.36	120.13	–	–	nr	**120.13**
Pressure gauge with alarm contacts (for SF6 only)	233.36	263.56	–	–	nr	**263.56**

V: ELECTRICAL SUPPLY/POWER/LIGHTING

Item	Net Price £	Material £	Labour hours	Labour £	Unit	Total rate £
Tripping Batteries						
Battery chargers; switchgear tripping and closing; double wound transformer and earth screen; including fixing to background, commissioning and testing						
Valve regulated lead acid battery						
30 volt; 19 Ah; 3A	2158.10	2437.35	6.50	178.61	nr	**2615.96**
30 volt; 29 Ah; 3A	3457.43	3904.80	8.50	233.57	nr	**4138.37**
110 volt; 19 Ah; 3A	3526.92	3983.28	6.50	178.61	nr	**4161.89**
110 volt; 29 Ah; 3A	3848.29	4346.24	8.50	233.57	nr	**4579.81**
110 volt; 38 Ah; 3A	4112.25	4644.36	10.00	274.78	nr	**4919.14**
Step down transformers; 11/0.415 kV, Dyn 11, 50 Hz. Complete with lifting lugs, mounting skids, provisions for wheels, undrilled gland plates to air-filled cable boxes, off load tapping facility, including UK delivery						
Oil-filled in free breathing ventilated steel tank						
500 kVA	8561.54	9669.37	30.00	824.34	nr	**10493.71**
800 kVA	9750.24	11011.87	30.00	824.34	nr	**11836.21**
1000 kVA	11050.63	12480.53	30.00	824.34	nr	**13304.87**
1250 kVA	13428.00	15165.51	35.00	961.74	nr	**16127.25**
1500 kVA	15912.29	17971.26	35.00	961.74	nr	**18933.00**
2000 kVA	20667.01	23341.22	40.00	1099.13	nr	**24440.35**
MIDEL – filled in gasket-sealed steel tank						
500 kVA	11339.88	12807.20	30.00	824.34	nr	**13631.54**
800 kVA	12924.80	14597.20	30.00	824.34	nr	**15421.54**
1000 kVA	14715.73	16619.87	30.00	824.34	nr	**17444.21**
1250 kVA	17786.47	20087.95	35.00	961.74	nr	**21049.69**
1500 kVA	21162.31	23900.61	35.00	961.74	nr	**24862.35**
2000 kVA	27402.88	30948.68	40.00	1099.13	nr	**32047.81**
Extra for,						
Fluid temperature indicator with 2 N/O contacts	313.59	354.16	2.00	54.96	nr	**409.12**
Winding temperature indicator with 2 N/O contacts	768.30	867.71	2.00	54.96	nr	**922.67**
Dehydrating breather	78.40	88.55	2.00	54.96	nr	**143.51**
Plain rollers	235.19	265.62	2.00	54.96	nr	**320.58**
Pressure relief device with 1 N/O contact	470.38	531.25	2.00	54.96	nr	**586.21**

V: ELECTRICAL SUPPLY/POWER/LIGHTING

Item	Net Price £	Material £	Labour hours	Labour £	Unit	Total rate £
V11: HV SUPPLY – cont						
Y70 – HV SWITCHGEAR AND TRANSFORMERS – cont						
Step down transformers; 11/0.415 kV, Dyn 11, 50Hz. Complete with lifting lugs, mounting skids, provisions for wheels, undrilled gland plates to air-filled cable boxes, off load tapping facility, including delivery						
Cast Resin type in ventilated steel enclosure, AN – Air Natural including winding temperture indicator with 2 N/O contacts						
500 kVA	12132.34	13702.20	40.00	1099.13	nr	**14801.33**
800 kVA	14113.47	15939.68	40.00	1099.13	nr	**17038.81**
1000 kVA	17073.27	19282.47	40.00	1099.13	nr	**20381.60**
1250 kVA	18479.88	20871.09	45.00	1236.52	nr	**22107.61**
1600 kVA	20964.20	23676.86	45.00	1236.52	nr	**24913.38**
2000 kVA	23737.80	26809.35	50.00	1373.91	nr	**28183.26**
Cast Resin type in ventilated steel enclosure with temperature controlled fans to achieve 40% increase to AN/AF rating. Includes winding temperature indicator with 2 N/O contacts						
500/700 kVA	13321.00	15044.67	42.00	1154.08	nr	**16198.75**
800/1120 kVA	15500.26	17505.91	42.00	1154.08	nr	**18659.99**
1000/1400 kVA	19178.02	21659.56	42.00	1154.08	nr	**22813.64**
12501750 kVA	19970.46	22554.54	47.00	1291.47	nr	**23846.01**
1600/2240 kVA	22561.75	25481.13	47.00	1291.47	nr	**26772.60**
2000/2800 kVA	25731.55	29061.09	52.00	1428.87	nr	**30489.96**

V: ELECTRICAL SUPPLY/POWER/LIGHTING

Item	Net Price £	Material £	Labour hours	Labour £	Unit	Total rate £
V20: LV DISTRIBUTION						
Y60 – CONDUIT AND CABLE TRUNKING						
Heavy gauge, screwed drawn steel; surface fixed on saddles to backgrounds, with standard pattern boxes and fittings including all fixings and supports (forming holes, conduit entry, draw wires etc. and components for earth continuity are included)						
Black enamelled						
20 mm dia.	2.06	2.33	0.49	13.46	m	**15.79**
25 mm dia.	2.99	3.38	0.56	15.38	m	**18.76**
32 mm dia.	6.11	6.91	0.64	17.58	m	**24.49**
38 mm dia.	7.27	8.21	0.73	20.05	m	**28.26**
50 mm dia.	11.54	13.04	1.04	28.57	m	**41.61**
Galvanized						
20 mm dia.	1.92	2.17	0.49	13.46	m	**15.63**
25 mm dia.	2.51	2.84	0.56	15.38	m	**18.22**
32 mm dia.	3.91	4.42	0.64	17.58	m	**22.00**
38 mm dia.	6.38	7.20	0.73	20.05	m	**27.25**
50 mm dia.	9.55	10.78	1.04	28.57	m	**39.35**
High impact PVC; surface fixed on saddles to backgrounds; with standard pattern boxes and fittings; including all fixings and supports						
Light gauge						
16 mm dia.	0.63	0.71	0.27	7.42	m	**8.13**
20 mm dia.	0.85	0.96	0.28	7.69	m	**8.65**
25 mm dia.	1.39	1.57	0.33	9.07	m	**10.64**
32 mm dia.	1.83	2.07	0.38	10.44	m	**12.51**
38 mm dia.	2.36	2.66	0.44	12.10	m	**14.76**
50 mm dia.	2.81	3.17	0.48	13.19	m	**16.36**
Heavy gauge						
16 mm dia.	1.00	1.13	0.27	7.42	m	**8.55**
20 mm dia.	1.14	1.29	0.28	7.69	m	**8.98**
25 mm dia.	1.55	1.75	0.33	9.07	m	**10.82**
32 mm dia.	2.45	2.76	0.38	10.44	m	**13.20**
38 mm dia.	3.25	3.67	0.44	12.10	m	**15.77**
50 mm dia.	5.39	6.09	0.48	13.19	m	**19.28**

V: ELECTRICAL SUPPLY/POWER/LIGHTING

Item	Net Price £	Material £	Labour hours	Labour £	Unit	Total rate £
V20: LV DISTRIBUTION – cont						
Y60 – CONDUIT AND CABLE TRUNKING – cont						
Flexible conduits; including adaptors and locknuts (for connections to equipment)						
Metallic, PVC covered conduit; not exceeding 1 m long; including zinc plated mild steel adaptors, lock nuts and earth conductor						
16 mm dia.	8.27	9.34	0.42	11.55	nr	**20.89**
20 mm dia.	7.78	8.78	0.46	12.64	nr	**21.42**
25 mm dia.	12.26	13.84	0.43	11.81	nr	**25.65**
32 mm dia.	18.99	21.45	0.51	14.01	nr	**35.46**
38 mm dia.	23.99	27.09	0.56	15.38	nr	**42.47**
50 mm dia.	83.71	94.54	0.82	22.53	nr	**117.07**
PVC conduit; not exceeding 1 m long; including nylon adaptors, lock nuts						
16 mm dia.	4.62	5.22	0.46	12.64	nr	**17.86**
20 mm dia.	4.62	5.22	0.48	13.19	nr	**18.41**
25 mm dia.	5.79	6.54	0.50	13.73	nr	**20.27**
32 mm.dia.	8.39	9.48	0.58	15.93	nr	**25.41**
PVC adaptable boxes; fixed to backgrounds; including all supports and fixings (cutting and connecting conduit to boxes is included)						
Square pattern						
75 × 75 × 53 mm	3.18	3.59	0.69	18.96	nr	**22.55**
100 × 100 × 75 mm	4.11	4.64	0.71	19.50	nr	**24.14**
150 × 150 × 75 mm	7.36	8.31	0.80	21.98	nr	**30.29**
Terminal strips to be fixed in metal or polythene adaptable boxes)						
20 amp high density polythene						
2 way	1.02	1.15	0.23	6.32	nr	**7.47**
3 way	1.13	1.28	0.23	6.32	nr	**7.60**
4 way	1.22	1.38	0.23	6.32	nr	**7.70**
5 way	1.29	1.46	0.23	6.32	nr	**7.78**
6 way	1.39	1.57	0.25	6.86	nr	**8.43**
7 way	1.49	1.68	0.25	6.86	nr	**8.54**
8 way	1.61	1.82	0.29	7.98	nr	**9.80**
9 way	1.73	1.96	0.30	8.24	nr	**10.20**
10 way	1.99	2.24	0.34	9.34	nr	**11.58**
11 way	2.07	2.34	0.34	9.34	nr	**11.68**
12 way	2.25	2.54	0.34	9.34	nr	**11.88**

V: ELECTRICAL SUPPLY/POWER/LIGHTING

Item	Net Price £	Material £	Labour hours	Labour £	Unit	Total rate £
13 way	2.49	2.82	0.37	10.17	nr	**12.99**
14 way	2.80	3.16	0.37	10.17	nr	**13.33**
15 way	2.89	3.26	0.39	10.72	nr	**13.98**
16 way	2.95	3.34	0.45	12.36	nr	**15.70**
18 way	3.19	3.60	0.45	12.36	nr	**15.96**
TRUNKING						
Galvanized steel trunking; fixed to backgrounds; jointed with standard connectors (including plates for air gap between trunking and background); earth continuity straps included						
Single compartment						
50 × 50 mm	3.28	3.70	0.39	10.72	m	**14.42**
75 × 50 mm	5.68	6.42	0.44	12.10	m	**18.52**
75 × 75 mm	15.84	17.89	0.47	12.92	m	**30.81**
100 × 50 mm	7.33	8.28	0.50	13.73	m	**22.01**
100 × 75 mm	8.97	10.13	0.57	15.67	m	**25.80**
100 × 100 mm	9.61	10.85	0.62	17.03	m	**27.88**
150 × 50 mm	10.73	12.12	0.78	21.44	m	**33.56**
150 × 100 mm	12.28	13.87	0.78	21.44	m	**35.31**
150 × 150 mm	11.71	13.23	0.86	23.62	m	**36.85**
225 × 75 mm	14.90	16.83	0.88	24.18	m	**41.01**
225 × 150 mm	18.62	21.03	0.84	23.08	m	**44.11**
225 × 225 mm	22.15	25.02	0.99	27.20	m	**52.22**
300 × 75 mm	13.50	15.25	0.96	26.39	m	**41.64**
300 × 100 mm	11.58	13.08	0.99	27.20	m	**40.28**
300 × 150 mm	15.74	17.78	0.99	27.20	m	**44.98**
300 × 225 mm	22.54	25.46	1.09	29.95	m	**55.41**
300 × 300 mm	23.58	26.63	1.16	31.88	m	**58.51**
Double compartment						
100 × 50 mm	8.58	9.69	0.54	14.84	m	**24.53**
100 × 75 mm	8.66	9.78	0.62	17.03	m	**26.81**
100 × 100 mm	9.77	11.04	0.66	18.14	m	**29.18**
150 × 50 mm	9.78	11.05	0.70	19.24	m	**30.29**
150 × 100 mm	11.22	12.67	0.83	22.81	m	**35.48**
150 × 150 mm	13.91	15.71	0.92	25.28	m	**40.99**
Triple compartment						
150 × 50 mm	12.92	14.60	0.79	21.71	m	**36.31**
150 × 100 mm	13.27	14.98	0.78	21.44	m	**36.42**
150 × 150 mm	16.25	18.35	1.01	27.75	m	**46.10**

V: ELECTRICAL SUPPLY/POWER/LIGHTING

Item	Net Price £	Material £	Labour hours	Labour £	Unit	Total rate £
V20: LV DISTRIBUTION – cont						
TRUNKING – cont						
Galvanized steel trunking fittings; cutting and jointing trunking to fittings is included						
Stop end						
50 × 50 mm	0.84	0.95	0.19	5.22	nr	**6.17**
75 × 50 mm	1.14	1.29	0.20	5.50	nr	**6.79**
75 × 75 mm	1.14	1.29	0.21	5.77	nr	**7.06**
100 × 50 mm	1.18	1.34	0.31	8.52	nr	**9.86**
100 × 75 mm	1.27	1.44	0.27	7.42	nr	**8.86**
100 × 100 mm	1.24	1.40	0.27	7.42	nr	**8.82**
150 × 50 mm	1.39	1.57	0.28	7.69	nr	**9.26**
150 × 100 mm	1.52	1.71	0.30	8.24	nr	**9.95**
150 × 150 mm	1.56	1.76	0.32	8.80	nr	**10.56**
225 × 75 mm	1.62	1.83	0.35	9.62	nr	**11.45**
225 × 150 mm	2.33	2.63	0.37	10.17	nr	**12.80**
225 × 225 mm	2.77	3.13	0.38	10.44	nr	**13.57**
300 × 75 mm	1.64	1.86	0.42	11.55	nr	**13.41**
300 × 100 mm	1.97	2.22	0.42	11.55	nr	**13.77**
300 × 150 mm	2.06	2.33	0.43	11.81	nr	**14.14**
300 × 225 mm	2.26	2.55	0.45	12.36	nr	**14.91**
300 × 300 mm	2.47	2.78	0.48	13.19	nr	**15.97**
Flanged connector						
50 × 50 mm	1.14	1.29	0.19	5.22	nr	**6.51**
75 × 50 mm	1.74	1.97	0.20	5.50	nr	**7.47**
75 × 75 mm	1.69	1.91	0.21	5.77	nr	**7.68**
100 × 50 mm	1.81	2.04	0.26	7.15	nr	**9.19**
100 × 75 mm	1.87	2.11	0.27	7.42	nr	**9.53**
100 × 100 mm	1.81	2.04	0.27	7.42	nr	**9.46**
150 × 50 mm	1.82	2.06	0.28	7.69	nr	**9.75**
150 × 100 mm	1.96	2.21	0.30	8.24	nr	**10.45**
150 × 150 mm	1.98	2.23	0.32	8.80	nr	**11.03**
225 × 75 mm	1.62	1.83	0.35	9.62	nr	**11.45**
225 × 150 mm	2.33	2.63	0.37	10.17	nr	**12.80**
225 × 225 mm	2.80	3.16	0.38	10.44	nr	**13.60**
300 × 75 mm	1.64	1.86	0.42	11.55	nr	**13.41**
300 × 100 mm	1.97	2.22	0.42	11.55	nr	**13.77**
300 × 150 mm	2.06	2.33	0.43	11.81	nr	**14.14**
300 × 225 mm	2.26	2.55	0.45	12.36	nr	**14.91**
300 × 300 mm	2.47	2.78	0.48	13.19	nr	**15.97**
Bends 90°; single compartment						
50 × 50 mm	5.75	6.50	0.42	11.55	nr	**18.05**
75 × 50 mm	5.96	6.73	0.45	12.36	nr	**19.09**
75 × 75 mm	5.75	6.50	0.48	13.19	nr	**19.69**

V: ELECTRICAL SUPPLY/POWER/LIGHTING

Item	Net Price £	Material £	Labour hours	Labour £	Unit	Total rate £
100 × 50 mm	6.85	7.73	0.53	14.56	nr	**22.29**
100 × 75 mm	6.51	7.35	0.56	15.38	nr	**22.73**
100 × 100 mm	6.24	7.05	0.58	15.93	nr	**22.98**
150 × 50 mm	7.73	8.73	0.64	17.58	nr	**26.31**
150 × 100 mm	10.54	11.90	0.91	25.00	nr	**36.90**
150 × 150 mm	9.23	10.42	0.89	24.45	nr	**34.87**
225 × 75 mm	14.25	16.10	0.76	20.88	nr	**36.98**
225 × 150 mm	18.27	20.63	0.82	22.53	nr	**43.16**
225 × 225 mm	21.04	23.77	0.83	22.81	nr	**46.58**
300 × 75 mm	13.96	15.77	0.85	23.36	nr	**39.13**
300 × 100 mm	14.27	16.12	0.90	24.73	nr	**40.85**
300 × 150 mm	15.07	17.02	0.96	26.39	nr	**43.41**
300 × 225 mm	16.23	18.33	0.98	26.93	nr	**45.26**
300 × 300 mm	17.39	19.65	1.06	29.13	nr	**48.78**
Bends 90°; double compartment						
100 × 50 mm	6.35	7.17	0.53	14.56	nr	**21.73**
100 × 75 mm	8.39	9.48	0.56	15.38	nr	**24.86**
100 × 100 mm	8.49	9.59	0.58	15.93	nr	**25.52**
150 × 50 mm	9.91	11.19	0.65	17.86	nr	**29.05**
150 × 100 mm	9.99	11.28	0.69	18.96	nr	**30.24**
150 × 150 mm	15.77	17.81	0.73	20.05	nr	**37.86**
Bends 90°; triple compartment						
150 × 50 mm	8.99	10.15	0.68	18.68	nr	**28.83**
150 × 100 mm	14.68	16.58	0.73	20.05	nr	**36.63**
150 × 150 mm	21.24	23.99	0.77	21.15	nr	**45.14**
Tees; single compartment						
50 × 50 mm	5.61	6.33	0.56	15.38	nr	**21.71**
75 × 50 mm	6.62	7.48	0.57	15.67	nr	**23.15**
75 × 75 mm	6.46	7.29	0.60	16.49	nr	**23.78**
100 × 50 mm	7.75	8.75	0.65	17.86	nr	**26.61**
100 × 75 mm	7.47	8.44	0.72	19.79	nr	**28.23**
100 × 100 mm	7.86	8.87	0.71	19.50	nr	**28.37**
150 × 50 mm	10.99	12.41	0.82	22.53	nr	**34.94**
150 × 100 mm	11.10	12.54	0.84	23.08	nr	**35.62**
150 × 150 mm	10.67	12.05	0.91	25.00	nr	**37.05**
225 × 75 mm	19.18	21.66	0.94	25.83	nr	**47.49**
225 × 150 mm	25.92	29.27	1.01	27.75	nr	**57.02**
225 × 225 mm	29.66	33.50	1.02	28.03	nr	**61.53**
300 × 75 mm	19.74	22.30	1.07	29.40	nr	**51.70**
300 × 100 mm	20.63	23.30	1.07	29.40	nr	**52.70**
300 × 150 mm	21.46	24.24	1.14	31.33	nr	**55.57**
300 × 225 mm	23.30	26.32	1.19	32.69	nr	**59.01**
300 × 300 mm	25.16	28.42	1.26	34.63	nr	**63.05**

V: ELECTRICAL SUPPLY/POWER/LIGHTING

Item	Net Price £	Material £	Labour hours	Labour £	Unit	Total rate £
V20: LV DISTRIBUTION – cont						
TRUNKING – cont						
Galvanized steel trunking fittings – cont						
Tees; double compartment						
100 × 50 mm	9.64	10.88	0.65	17.86	nr	**28.74**
100 × 75 mm	11.20	12.65	0.71	19.50	nr	**32.15**
100 × 100 mm	10.23	11.56	0.72	19.79	nr	**31.35**
150 × 50 mm	11.64	13.15	0.82	22.53	nr	**35.68**
150 × 100 mm	12.19	13.77	0.85	23.36	nr	**37.13**
150 × 150 mm	13.67	15.44	0.91	25.00	nr	**40.44**
Tees; triple compartment						
150 × 50 mm	12.54	14.16	0.87	23.91	nr	**38.07**
150 × 100 mm	15.64	17.67	0.89	24.45	nr	**42.12**
150 × 150 mm	20.27	22.89	0.96	26.39	nr	**49.28**
Crossovers; single compartment						
50 × 50 mm	7.26	8.20	0.65	17.86	nr	**26.06**
75 × 50 mm	10.54	11.90	0.66	18.14	nr	**30.04**
75 × 75 mm	10.07	11.37	0.69	18.96	nr	**30.33**
100 × 50 mm	12.45	14.07	0.74	20.33	nr	**34.40**
100 × 75 mm	12.54	14.16	0.80	21.98	nr	**36.14**
100 × 100 mm	11.97	13.52	0.81	22.26	nr	**35.78**
150 × 50 mm	14.45	16.32	0.91	25.00	nr	**41.32**
150 × 100 mm	15.08	17.03	0.94	25.83	nr	**42.86**
150 × 150 mm	14.45	16.32	0.99	27.20	nr	**43.52**
225 × 75 mm	26.48	29.91	1.01	27.75	nr	**57.66**
225 × 150 mm	34.42	38.87	1.08	29.68	nr	**68.55**
225 × 225 mm	39.34	44.43	1.09	29.95	nr	**74.38**
300 × 75 mm	26.39	29.80	1.14	31.33	nr	**61.13**
300 × 100 mm	27.17	30.68	1.16	31.88	nr	**62.56**
300 × 150 mm	28.00	31.62	1.19	32.69	nr	**64.31**
300 × 225 mm	30.32	34.24	1.21	33.25	nr	**67.49**
300 × 300 mm	32.70	36.93	1.29	35.46	nr	**72.39**
Crossovers; double compartment						
100 × 50 mm	12.34	13.93	0.74	20.33	nr	**34.26**
100 × 75 mm	12.21	13.79	0.80	21.98	nr	**35.77**
100 × 100 mm	13.37	15.10	0.81	22.26	nr	**37.36**
150 × 50 mm	14.48	16.35	0.86	23.62	nr	**39.97**
150 × 100 mm	14.52	16.40	0.94	25.83	nr	**42.23**
150 × 150 mm	22.74	25.68	1.00	27.48	nr	**53.16**
Crossovers; triple compartment						
150 × 50 mm	13.48	15.23	0.97	26.65	nr	**41.88**
150 × 100 mm	15.33	17.31	0.99	27.20	nr	**44.51**
150 × 150 mm	24.03	27.14	1.06	29.13	nr	**56.27**

V: ELECTRICAL SUPPLY/POWER/LIGHTING

Item	Net Price £	Material £	Labour hours	Labour £	Unit	Total rate £
Galvanized steel flush floor trunking; fixed to backgrounds; supports and fixings; standard coupling joints; earth continuity straps included						
Triple compartment						
350 × 60 mm	41.28	46.62	1.32	36.27	m	**82.89**
Four compartment						
350 × 60 mm	42.98	48.54	1.32	36.27	m	**84.81**
Galvanized steel flush floor trunking; fittings (cutting and jointing trunking to fittings is included)						
Stop end; triple compartment						
350 × 60 mm	4.33	4.89	0.53	14.56	nr	**19.45**
Stop end; four compartment						
350 × 60 mm	10.04	11.34	0.53	14.56	nr	**25.90**
Rising bend; standard; triple compartment						
350 × 60 mm	35.71	40.33	1.30	35.72	nr	**76.05**
Rising bend; standard; four compartment						
350 × 60 mm	37.41	42.25	1.30	35.72	nr	**77.97**
Rising bend; skirting; triple compartment						
350 × 60 mm	70.05	79.11	1.33	36.55	nr	**115.66**
Rising bend; skirting; four compartment						
350 × 60 mm	79.51	89.80	1.33	36.55	nr	**126.35**
Junction box; triple compartment						
350 × 60 mm	46.56	52.58	1.16	31.88	nr	**84.46**
Junction box; four compartment						
350 × 60 mm	48.42	54.68	1.16	31.88	nr	**86.56**
Body coupler (pair)						
3 and 4 Compartment	2.40	2.71	0.16	4.40	nr	**7.11**
Service outlet module comprising flat lid with flanged carpet trim; twin 13 A outlet and drilled plate for mounting 2 telephone outlets; one blank plate; triple compartment						
3 Compartment	57.75	65.22	0.47	12.92	nr	**78.14**

V: ELECTRICAL SUPPLY/POWER/LIGHTING

Item	Net Price £	Material £	Labour hours	Labour £	Unit	Total rate £
V20: LV DISTRIBUTION – cont						
TRUNKING – cont						
Galvanized steel flush floor trunking – cont						
Service outlet module comprising flat lid with flanged carpet trim; twin 13 A outlet and drilled plate for mounting 2 telephone outlets; two blank plates; four compartment						
4 Compartment	64.33	72.65	0.47	12.92	nr	**85.57**
Single compartment PVC trunking; grey finish; clip on lid; fixed to backgrounds; including supports and fixings (standard coupling joints)						
50 × 50 mm	9.94	11.23	0.27	7.42	m	**18.65**
75 × 50 mm	10.78	12.18	0.28	7.69	m	**19.87**
75 × 75 mm	12.21	13.79	0.29	7.98	m	**21.77**
100 × 50 mm	13.83	15.62	0.34	9.34	m	**24.96**
100 × 75 mm	15.16	17.13	0.37	10.17	m	**27.30**
100 × 100 mm	16.02	18.09	0.37	10.17	m	**28.26**
150 × 50 mm	13.90	15.70	0.41	11.27	m	**26.97**
150 × 75 mm	24.77	27.98	0.44	12.10	m	**40.08**
150 × 100 mm	29.82	33.68	0.44	12.10	m	**45.78**
150 × 150 mm	30.51	34.46	0.48	13.19	m	**47.65**
Single compartment PVC trunking; fittings (cutting and jointing trunking to fittings is included)						
Crossover						
50 × 50 mm	16.29	18.40	0.29	7.98	nr	**26.38**
75 × 50 mm	18.06	20.40	0.30	8.24	nr	**28.64**
75 × 75 mm	19.60	22.13	0.31	8.52	nr	**30.65**
100 × 50 mm	26.13	29.51	0.35	9.62	nr	**39.13**
100 × 75 mm	32.95	37.21	0.36	9.89	nr	**47.10**
100 × 100 mm	29.14	32.92	0.40	10.99	nr	**43.91**
150 × 75 mm	37.34	42.17	0.45	12.36	nr	**54.53**
150 × 100 mm	44.80	50.59	0.46	12.64	nr	**63.23**
150 × 150 mm	69.53	78.53	0.47	12.92	nr	**91.45**
Stop end						
50 × 50 mm	0.68	0.77	0.12	3.29	nr	**4.06**
75 × 50 mm	0.97	1.09	0.12	3.29	nr	**4.38**
75 × 75 mm	1.27	1.44	0.13	3.57	nr	**5.01**
100 × 50 mm	1.73	1.96	0.16	4.40	nr	**6.36**
100 × 75 mm	2.71	3.06	0.16	4.40	nr	**7.46**
100 × 100 mm	2.72	3.07	0.18	4.95	nr	**8.02**
150 × 75 mm	7.09	8.01	0.20	5.50	nr	**13.51**

V: ELECTRICAL SUPPLY/POWER/LIGHTING

Item	Net Price £	Material £	Labour hours	Labour £	Unit	Total rate £
150 × 100 mm	8.78	9.91	0.21	5.77	nr	**15.68**
150 × 150 mm	8.95	10.11	0.22	6.04	nr	**16.15**
Flanged coupling						
50 × 50 mm	4.09	4.62	0.32	8.80	nr	**13.42**
75 × 50 mm	4.66	5.26	0.33	9.07	nr	**14.33**
75 × 75 mm	5.60	6.32	0.34	9.34	nr	**15.66**
100 × 50 mm	6.31	7.13	0.44	12.10	nr	**19.23**
100 × 75 mm	7.14	8.07	0.45	12.36	nr	**20.43**
100 × 100 mm	7.61	8.60	0.46	12.64	nr	**21.24**
150 × 75 mm	8.06	9.10	0.57	15.67	nr	**24.77**
150 × 100 mm	8.49	9.59	0.57	15.67	nr	**25.26**
150 × 150 mm	8.90	10.05	0.59	16.21	nr	**26.26**
Internal coupling						
50 × 50 mm	1.44	1.62	0.07	1.92	nr	**3.54**
75 × 50 mm	1.70	1.92	0.07	1.92	nr	**3.84**
75 × 75 mm	1.71	1.93	0.07	1.92	nr	**3.85**
100 × 50 mm	2.29	2.59	0.08	2.20	nr	**4.79**
100 × 75 mm	2.58	2.92	0.08	2.20	nr	**5.12**
100 × 100 mm	2.84	3.20	0.08	2.20	nr	**5.40**
External coupling						
50 × 50 mm	1.58	1.79	0.09	2.47	nr	**4.26**
75 × 50 mm	1.88	2.12	0.09	2.47	nr	**4.59**
75 × 75 mm	1.87	2.11	0.09	2.47	nr	**4.58**
100 × 50 mm	2.51	2.84	0.10	2.74	nr	**5.58**
100 × 75 mm	2.86	3.23	0.10	2.74	nr	**5.97**
100 × 100 mm	3.12	3.52	0.10	2.74	nr	**6.26**
150 × 75 mm	3.57	4.03	0.11	3.03	nr	**7.06**
150 × 100 mm	3.71	4.19	0.11	3.03	nr	**7.22**
150 × 150 mm	3.85	4.35	0.11	3.03	nr	**7.38**
Angle; flat cover						
50 × 50 mm	5.78	6.53	0.18	4.95	nr	**11.48**
75 × 50 mm	7.70	8.70	0.19	5.22	nr	**13.92**
75 × 75 mm	8.87	10.02	0.20	5.50	nr	**15.52**
100 × 50 mm	13.23	14.94	0.23	6.32	nr	**21.26**
100 × 75 mm	20.15	22.76	0.26	7.15	nr	**29.91**
100 × 100 mm	18.38	20.76	0.26	7.15	nr	**27.91**
150 × 75 mm	25.06	28.30	0.30	8.24	nr	**36.54**
150 × 100 mm	29.80	33.66	0.33	9.07	nr	**42.73**
150 × 150 mm	45.16	51.00	0.34	9.34	nr	**60.34**
Angle; internal or external cover						
50 × 50 mm	7.03	7.94	0.18	4.95	nr	**12.89**
75 × 50 mm	9.69	10.94	0.19	5.22	nr	**16.16**
75 × 75 mm	12.45	14.07	0.20	5.50	nr	**19.57**

V: ELECTRICAL SUPPLY/POWER/LIGHTING

Item	Net Price £	Material £	Labour hours	Labour £	Unit	Total rate £
V20: LV DISTRIBUTION – cont						
TRUNKING – cont						
Single compartment PVC trunking – cont						
Angle – cont						
100 × 50 mm	13.41	15.15	0.23	6.32	nr	**21.47**
100 × 75 mm	21.57	24.36	0.26	7.15	nr	**31.51**
100 × 100 mm	21.66	24.46	0.26	7.15	nr	**31.61**
150 × 75 mm	26.48	29.91	0.30	8.24	nr	**38.15**
150 × 100 mm	31.21	35.25	0.33	9.07	nr	**44.32**
150 × 150 mm	43.85	49.52	0.34	9.34	nr	**58.86**
Tee; flat cover						
50 × 50 mm	4.84	5.47	0.24	6.60	nr	**12.07**
75 × 50 mm	7.38	8.33	0.25	6.86	nr	**15.19**
75 × 75 mm	7.59	8.57	0.26	7.15	nr	**15.72**
100 × 50 mm	16.39	18.51	0.32	8.80	nr	**27.31**
100 × 75 mm	17.41	19.67	0.33	9.07	nr	**28.74**
100 × 100 mm	22.72	25.66	0.34	9.34	nr	**35.00**
150 × 75 mm	30.05	33.94	0.41	11.27	nr	**45.21**
150 × 100 mm	38.55	43.53	0.42	11.55	nr	**55.08**
150 × 150 mm	52.43	59.21	0.44	12.10	nr	**71.31**
Tee; internal or external cover						
50 × 50 mm	13.71	15.48	0.24	6.60	nr	**22.08**
75 × 50 mm	15.07	17.02	0.25	6.86	nr	**23.88**
75 × 75 mm	16.75	18.92	0.26	7.15	nr	**26.07**
100 × 50 mm	21.50	24.29	0.32	8.80	nr	**33.09**
100 × 75 mm	24.57	27.75	0.33	9.07	nr	**36.82**
100 × 100 mm	27.57	31.14	0.34	9.34	nr	**40.48**
150 × 75 mm	35.71	40.33	0.41	11.27	nr	**51.60**
150 × 100 mm	43.03	48.59	0.42	11.55	nr	**60.14**
150 × 150 mm	57.45	64.88	0.44	12.10	nr	**76.98**
Division Strip (1.8 m long)						
50 mm	7.44	8.40	0.07	1.92	nr	**10.32**
75 mm	9.52	10.75	0.07	1.92	nr	**12.67**
100 mm	12.11	13.68	0.08	2.20	nr	**15.88**
PVC miniature trunking; white finish; fixed to backgrounds; including supports and fixing; standard coupling joints						
Single compartment						
16 × 16 mm	1.34	1.51	0.20	5.50	m	**7.01**
25 × 16 mm	1.64	1.86	0.21	5.77	m	**7.63**
38 × 16 mm	2.05	2.32	0.24	6.60	m	**8.92**
38 × 25 mm	2.44	2.75	0.25	6.86	m	**9.61**

V: ELECTRICAL SUPPLY/POWER/LIGHTING

Item	Net Price £	Material £	Labour hours	Labour £	Unit	Total rate £
Compartmented						
38 × 16 mm	2.39	2.70	0.24	6.60	m	**9.30**
38 × 25 mm	2.86	3.23	0.25	6.86	m	**10.09**
PVC miniature trunking fittings; single compartment; white finish; cutting and jointing trunking to fittings is included						
Coupling						
16 × 16 mm	0.41	0.46	0.10	2.74	nr	**3.20**
25 × 16 mm	0.41	0.46	0.12	3.16	nr	**3.62**
38 × 16 mm	0.41	0.46	0.12	3.29	nr	**3.75**
38 × 25 mm	0.98	1.11	0.14	3.86	nr	**4.97**
Stop end						
16 × 16 mm	0.41	0.46	0.12	3.29	nr	**3.75**
25 × 16 mm	0.41	0.46	0.13	3.49	nr	**3.95**
38 × 16 mm	0.41	0.46	0.15	4.12	nr	**4.58**
38 × 25 mm	0.98	1.11	0.17	4.67	nr	**5.78**
Bend; flat, internal or external						
16 × 16 mm	0.41	0.46	0.18	4.95	nr	**5.41**
25 × 16 mm	0.41	0.46	0.20	5.45	nr	**5.91**
38 × 16 mm	0.41	0.46	0.21	5.77	nr	**6.23**
38 × 25 mm	0.98	1.11	0.23	6.32	nr	**7.43**
Tee						
16 × 16 mm	0.70	0.80	0.19	5.22	nr	**6.02**
25 × 16 mm	0.70	0.80	0.23	6.32	nr	**7.12**
38 × 16 mm	0.70	0.80	0.26	7.15	nr	**7.95**
38 × 25 mm	0.97	1.09	0.29	7.98	nr	**9.07**
PVC bench trunking; white or grey finish; fixed to backgrounds; including supports and fixings; standard coupling joints						
Trunking						
90 × 90 mm	22.52	25.44	0.33	9.07	m	**34.51**
PVC bench trunking fittings; white or grey finish; cutting and jointing trunking to fittings is included						
Stop end						
90 × 90 mm	4.77	5.39	0.09	2.47	nr	**7.86**
Coupling						
90 × 90 mm	2.95	3.34	0.09	2.47	nr	**5.81**

V: ELECTRICAL SUPPLY/POWER/LIGHTING

Item	Net Price £	Material £	Labour hours	Labour £	Unit	Total rate £
V20: LV DISTRIBUTION – cont						
TRUNKING – cont						
PVC bench trunking fittings – cont						
Internal or external bend						
90 × 90 mm	16.19	18.29	0.28	7.69	nr	**25.98**
Socket plate						
90 × 90 mm – 1 gang	0.91	1.03	0.10	2.74	nr	**3.77**
90 × 90 mm – 2 gang	1.10	1.24	0.10	2.74	nr	**3.98**
PVC underfloor trunking; single compartment; fitted in floor screed; standard coupling joints						
Trunking						
60 × 25 mm	11.05	12.48	0.22	6.04	m	**18.52**
90 × 35 mm	15.91	17.97	0.27	7.42	m	**25.39**
PVC underfloor trunking fittings; single compartment; fitted in floor screed; (cutting and jointing trunking to fittings is included)						
Jointing sleeve						
60 × 25 mm	0.73	0.83	0.08	2.20	nr	**3.03**
90 × 35 mm	1.22	1.38	0.10	2.74	nr	**4.12**
Duct connector						
90 × 35 mm	0.57	0.64	0.17	4.67	nr	**5.31**
Socket reducer						
90 × 35 mm	1.02	1.15	0.12	3.29	nr	**4.44**
Vertical access box; 2 compartment						
Shallow	57.13	64.53	0.37	10.17	nr	**74.70**
Duct bend; vertical						
60 × 25 mm	11.14	12.58	0.27	7.42	nr	**20.00**
90 × 35 mm	12.58	14.21	0.35	9.62	nr	**23.83**
Duct bend; horizontal						
60 × 25 mm	13.14	14.84	0.30	8.24	nr	**23.08**
90 × 35 mm	13.34	15.07	0.37	10.17	nr	**25.24**

V: ELECTRICAL SUPPLY/POWER/LIGHTING

Item	Net Price £	Material £	Labour hours	Labour £	Unit	Total rate £
Zinc coated steel underfloor ducting; fixed to backgrounds; standard coupling joints; earth continuity straps; (Including supports and fixing, packing shims where required)						
Double compartment						
150 × 25 mm	8.99	10.15	0.57	15.67	m	**25.82**
Triple compartment						
225 × 25 mm	15.93	17.99	0.93	25.56	m	**43.55**
Zinc coated steel underfloor ducting fittings (cutting and jointing to fittings is included)						
Stop end; double compartment						
150 × 25 mm	2.83	3.19	0.31	8.52	nr	**11.71**
Stop end; triple compartment						
225 × 25 mm	3.23	3.65	0.37	10.17	nr	**13.82**
Rising bend; double compartment; standard trunking						
150 × 25 mm	18.20	20.55	0.71	19.50	nr	**40.05**
Rising bend; triple compartment; standard trunking						
225 × 25 mm	32.71	36.94	0.85	23.36	nr	**60.30**
Rising bend; double compartment; to skirting						
150 × 25 mm	40.52	45.77	0.90	24.73	nr	**70.50**
Rising bend; triple compartment; to skirting						
225 × 25 mm	47.78	53.96	0.95	26.10	nr	**80.06**
horizontal bend; double compartment						
150 × 25 mm	27.67	31.25	0.64	17.58	nr	**48.83**
Horizontal bend; triple compartment						
225 × 25 mm	31.92	36.05	0.77	21.15	nr	**57.20**
Junction or service outlet boxes; terminal; double compartment						
150 mm	28.63	32.33	0.91	25.00	nr	**57.33**
Junction or service outlet boxes; terminal; triple compartment						
225 mm	32.73	36.96	1.11	30.51	nr	**67.47**

V: ELECTRICAL SUPPLY/POWER/LIGHTING

Item	Net Price £	Material £	Labour hours	Labour £	Unit	Total rate £
V20: LV DISTRIBUTION – cont						
TRUNKING – cont						
Zinc coated steel underfloor ducting fittings – cont						
Junction or service outlet boxes; through or angle; double compartment						
150 mm	38.03	42.95	0.97	26.65	nr	**69.60**
Junction or service outlet boxes; through or angle; triple compartment						
225 mm	42.24	47.71	1.17	32.15	nr	**79.86**
Junction or service outlet boxes; tee; double compartment						
150 mm	38.03	42.95	1.02	28.03	nr	**70.98**
Junction or service outlet boxes; tee; triple compartment						
225 mm	42.24	47.71	1.22	33.52	nr	**81.23**
Junction or service outlet boxes; cross; double compartment						
up to 150 mm	38.03	42.95	1.03	28.30	nr	**71.25**
Junction or service outlet boxes; cross;triple compartment						
225 mm	42.24	47.71	1.23	33.80	nr	**81.51**
Plates for junction/inspection boxes; double and triple compartment						
Blank plate	8.31	9.38	0.92	25.28	nr	**34.66**
Conduit entry plate	10.42	11.77	0.86	23.62	nr	**35.39**
Trunking entry plate	10.42	11.77	0.86	23.62	nr	**35.39**
Service outlet box comprising flat lid with flanged carpet trim; twin 13A outlet and drilled plate for mounting 2 telephone outlets and terminal blocks; terminal outlet box; double compartment						
150 × 25 mm trunking	59.31	66.98	1.68	46.15	nr	**113.13**
Service outlet box comprising flat lid with flanged carpet trim; twin 13 A outlet and drilled plate for mounting 2 telephone outlets and terminal blocks; terminal outlet box; triple compartment						
225 × 25 mm trunking	66.01	74.55	1.93	53.04	nr	**127.59**

V: ELECTRICAL SUPPLY/POWER/LIGHTING

Item	Net Price £	Material £	Labour hours	Labour £	Unit	Total rate £
PVC skirting/dado modular trunking; white (cutting and jointing trunking to fittings and backplates for fixing to walls is included)						
Main carrier/backplate						
50 × 170 mm	17.65	19.93	2.02	55.51	m	**75.44**
62 × 190 mm	19.60	19.60	2.02	49.15	m	**68.75**
Extension carrier/backplate						
50 × 42 mm	10.73	12.12	0.58	15.93	m	**28.05**
Carrier/backplate						
Including cover seal	6.26	7.07	0.53	14.56	m	**21.63**
Chamfered covers for fixing to backplates						
50 × 42 mm	3.63	4.10	0.33	9.07	m	**13.17**
Square covers for fixing to backplates						
50 × 42 mm	7.30	8.24	0.33	9.07	m	**17.31**
Plain covers for fixing to backplates						
85 mm	3.63	4.10	0.34	9.34	m	**13.44**
Retainers-clip to backplates to hold cables						
For chamfered covers	1.01	1.14	0.07	1.92	m	**3.06**
For square-recessed covers	0.87	0.98	0.07	1.92	m	**2.90**
For plain covers	3.36	3.79	0.07	1.92	m	**5.71**
Prepackaged corner assemblies						
Internal ; for 170 × 50 Assy	7.05	7.97	0.51	14.01	nr	**21.98**
Internal ; for 190 × 62 Assy	7.84	7.84	0.51	12.41	nr	**20.25**
Internal ; for 215 × 50 Assy	8.77	9.90	0.53	14.56	nr	**24.46**
Internal ; for 254 × 50 Assy	10.44	11.79	0.53	14.56	nr	**26.35**
External ; for 170 × 50 Assy	7.05	7.97	0.56	15.38	nr	**23.35**
External ; for 190 × 62 Assy	7.84	7.84	0.56	13.62	nr	**21.46**
External ; for 215 × 50 Assy	8.77	9.90	0.58	15.93	nr	**25.83**
External ; for 254 × 50 Assy	10.44	11.79	0.58	15.93	nr	**27.72**
Clip on end caps						
170 × 50 Assy	4.22	4.76	0.11	3.03	nr	**7.79**
215 × 50 Assy	5.00	5.65	0.11	3.03	nr	**8.68**
254 × 50 Assy	5.87	6.63	0.11	3.03	nr	**9.66**
190 × 62 Assy	5.90	6.65	0.11	3.03	nr	**9.68**
Outlet box						
1 Gang; in horizontal trunking; clip in	3.77	4.25	0.34	9.34	nr	**13.59**
2 Gang; in horizontal trunking; clip in	4.70	5.30	0.34	9.34	nr	**14.64**
1 Gang; in vertical trunking; clip in	3.77	4.25	0.34	9.34	nr	**13.59**

V: ELECTRICAL SUPPLY/POWER/LIGHTING

Item	Net Price £	Material £	Labour hours	Labour £	Unit	Total rate £
V20: LV DISTRIBUTION – cont						
TRUNKING – cont						
Sheet steel adaptable boxes; with plain or knockout sides; fixed to backgrounds; including supports and fixings (cutting and connecting conduit to boxes is included)						
Square pattern – black						
75 × 75 × 37 mm	2.02	2.28	0.69	18.96	nr	**21.24**
75 × 75 × 50 mm	1.92	2.17	0.69	18.96	nr	**21.13**
75 × 75 × 75 mm	2.20	2.49	0.69	18.96	nr	**21.45**
100 × 100 × 50 mm	2.02	2.28	0.71	19.50	nr	**21.78**
150 × 150 × 50 mm	3.07	3.47	0.79	21.71	nr	**25.18**
150 × 150 × 75 mm	3.61	4.08	0.80	21.98	nr	**26.06**
150 × 150 × 100 mm	4.81	5.44	0.80	21.98	nr	**27.42**
200 × 200 × 50 mm	6.00	6.78	0.80	19.46	nr	**26.24**
225 × 225 × 50 mm	6.14	6.94	0.93	25.56	nr	**32.50**
225 × 225 × 100 mm	8.15	9.20	0.94	25.83	nr	**35.03**
300 × 300 × 100 mm	8.75	9.88	0.99	27.20	nr	**37.08**
Square pattern – galvanized						
75 × 75 × 37 mm	2.88	3.25	0.69	18.96	nr	**22.21**
75 × 75 × 50 mm	2.60	2.94	0.70	19.24	nr	**22.18**
75 × 75 × 75 mm	3.01	3.40	0.69	18.96	nr	**22.36**
100 × 100 × 50 mm	3.19	3.60	0.71	19.50	nr	**23.10**
150 × 150 × 50 mm	3.70	4.18	0.84	23.08	nr	**27.26**
150 × 150 × 75 mm	4.42	4.99	0.80	21.98	nr	**26.97**
150 × 150 × 100 mm	5.32	6.01	0.80	21.98	nr	**27.99**
225 × 225 × 50 mm	6.83	7.71	0.93	25.56	nr	**33.27**
225 × 225 × 100 mm	8.69	9.81	0.94	25.83	nr	**35.64**
300 × 300 × 100 mm	13.98	15.65	0.96	26.38	nr	**42.03**
Rectangular pattern – black						
100 × 75 × 50 mm	3.23	3.65	0.69	18.96	nr	**22.61**
150 × 75 × 50 mm	3.38	3.81	0.70	19.24	nr	**23.05**
150 × 75 × 75 mm	3.67	4.14	0.71	19.50	nr	**23.64**
150 × 100 × 75 mm	8.17	9.23	0.71	19.50	nr	**28.73**
225 × 75 × 50 mm	7.07	7.99	0.78	21.44	nr	**29.43**
225 × 150 × 75 mm	11.29	12.75	0.81	22.26	nr	**35.01**
225 × 150 × 100 mm	21.19	23.93	0.81	22.26	nr	**46.19**
300 × 150 × 50 mm	21.19	23.93	0.93	25.56	nr	**49.49**
300 × 150 × 75 mm	21.19	23.93	0.94	25.83	nr	**49.76**
300 × 150 × 100 mm	21.19	23.93	0.96	26.39	nr	**50.32**
Rectangular pattern – galvanized						
100 × 75 × 50 mm	4.87	5.50	0.69	18.96	nr	**24.46**
150 × 75 × 50 mm	7.21	8.14	0.70	19.24	nr	**27.38**
150 × 75 × 75 mm	8.85	10.00	0.71	19.50	nr	**29.50**

V: ELECTRICAL SUPPLY/POWER/LIGHTING

Item	Net Price £	Material £	Labour hours	Labour £	Unit	Total rate £
150 × 100 × 75 mm	14.74	16.65	0.71	19.50	nr	**36.15**
225 × 75 × 50 mm	13.29	15.01	0.89	24.45	nr	**39.46**
225 × 150 × 75 mm	18.66	21.07	0.81	22.26	nr	**43.33**
225 × 150 × 100 mm	35.22	39.78	0.81	22.26	nr	**62.04**
300 × 150 × 50 mm	35.22	39.78	0.93	25.56	nr	**65.34**
300 × 150 × 75 mm	35.22	39.78	0.94	25.83	nr	**65.61**
300 × 150 × 100 mm	35.22	39.78	0.96	26.39	nr	**66.17**

Y61 – LV CABLES AND WIRING

ARMOURED CABLE

Cable; XLPE insulated; PVC sheathed; copper stranded conductors to BS 5467; laid in trench/duct including marker tape; (Cable tiles measured elsewhere)

Item	Net Price £	Material £	Labour hours	Labour £	Unit	Total rate £
600/1000 Volt grade; single core (aluminium wire armour)						
25 mm²	1.86	2.10	0.15	4.12	m	**6.22**
35 mm²	2.06	2.33	0.15	4.12	m	**6.45**
50 mm²	3.04	3.44	0.17	4.67	m	**8.11**
70 mm²	4.62	5.22	0.18	4.95	m	**10.17**
95 mm²	6.21	7.02	0.20	5.50	m	**12.52**
120 mm²	7.39	8.34	0.22	6.04	m	**14.38**
150 mm²	9.14	10.32	0.24	6.60	m	**16.92**
185 mm²	10.97	12.39	0.26	7.15	m	**19.54**
240 mm²	14.16	15.99	0.30	8.24	m	**24.23**
300 mm²	17.24	19.47	0.31	8.52	m	**27.99**
400 mm²	22.41	25.31	0.38	10.44	m	**35.75**
500 mm²	28.76	32.48	0.44	12.10	m	**44.58**
630 mm²	36.04	40.71	0.52	14.29	m	**55.00**
800 mm²	46.45	52.46	0.62	17.03	m	**69.49**
1000 mm²	53.79	60.75	0.65	17.86	m	**78.61**
600/1000 Volt grade; two core (galvanized steel wire armour)						
1.5 mm²	0.66	0.74	0.06	1.65	m	**2.39**
2.5 mm²	0.76	0.86	0.06	1.65	m	**2.51**
4 mm²	0.98	1.11	0.08	2.20	m	**3.31**
6 mm²	1.20	1.36	0.08	2.20	m	**3.56**
10 mm²	1.61	1.82	0.10	2.74	m	**4.56**
16 mm²	2.42	2.73	0.10	2.74	m	**5.47**
25 mm²	3.10	3.50	0.15	4.12	nr	**7.62**
35 mm²	4.89	5.52	0.15	4.12	m	**9.64**
50 mm²	6.04	6.82	0.17	4.67	m	**11.49**
70 mm²	8.99	10.15	0.18	4.95	m	**15.10**
95 mm²	12.08	13.65	0.20	5.50	m	**19.15**
120 mm²	14.80	16.72	0.22	6.04	m	**22.76**
150 mm²	18.36	20.74	0.24	6.60	m	**27.34**
185 mm²	23.39	26.42	0.26	7.15	m	**33.57**
240 mm²	29.94	33.81	0.30	8.24	m	**42.05**

V: ELECTRICAL SUPPLY/POWER/LIGHTING

Item	Net Price £	Material £	Labour hours	Labour £	Unit	Total rate £
V20: LV DISTRIBUTION – cont						
ARMOURED CABLE – cont						
Cable – cont						
600/1000 Volt grade – cont						
300 mm²	40.22	45.42	0.31	8.52	m	**53.94**
400 mm²	53.20	60.09	0.35	9.62	m	**69.71**
600/1000 Volt grade; three core (galvanized steel wire armour)						
1.5 mm²	0.72	0.82	0.07	1.92	m	**2.74**
2.5 mm²	0.92	1.04	0.07	1.92	m	**2.96**
4 mm²	1.13	1.28	0.09	2.47	m	**3.75**
6 mm²	1.46	1.65	0.10	2.74	m	**4.39**
10 mm²	2.24	2.53	0.11	3.03	m	**5.56**
16 mm²	3.12	3.52	0.11	3.03	m	**6.55**
25 mm²	4.22	4.76	0.16	4.40	m	**9.16**
35 mm²	5.88	6.64	0.16	4.40	m	**11.04**
50 mm²	8.38	9.47	0.19	5.22	m	**14.69**
70 mm²	12.44	14.05	0.21	5.77	m	**19.82**
95 mm²	17.13	19.35	0.23	6.32	m	**25.67**
120 mm²	21.30	24.05	0.24	6.60	m	**30.65**
150 mm²	26.47	29.90	0.27	7.42	m	**37.32**
185 mm²	32.61	36.83	0.30	8.24	m	**45.07**
240 mm²	41.73	47.13	0.33	9.07	m	**56.20**
300 mm²	52.88	59.72	0.35	9.62	m	**69.34**
400 mm²	62.80	70.93	0.41	11.27	m	**82.20**
600/1000 Volt grade; four core (galvanized steel wire armour)						
1.5 mm²	0.85	0.96	0.08	2.20	m	**3.16**
2.5 mm²	1.07	1.20	0.09	2.47	m	**3.67**
4 mm²	1.32	1.49	0.10	2.74	m	**4.23**
6 mm²	1.94	2.19	0.10	2.74	m	**4.93**
10 mm²	2.72	3.07	0.12	3.29	m	**6.36**
16 mm²	3.93	4.44	0.12	3.29	m	**7.73**
25 mm²	5.67	6.41	0.18	4.95	m	**11.36**
35 mm²	7.39	8.34	0.19	5.22	m	**13.56**
50 mm²	10.70	12.09	0.21	5.77	m	**17.86**
70 mm²	16.51	18.65	0.23	6.32	m	**24.97**
95 mm²	22.12	24.98	0.26	7.15	m	**32.13**
120 mm²	27.78	31.38	0.28	7.69	m	**39.07**
150 mm²	33.63	37.98	0.32	8.80	m	**46.78**
185 mm²	41.27	46.61	0.35	9.62	m	**56.23**
240 mm²	53.35	60.25	0.36	9.89	m	**70.14**
300 mm²	66.93	75.59	0.40	10.99	m	**86.58**
400 mm²	69.58	78.58	0.45	12.36	m	**90.94**

V: ELECTRICAL SUPPLY/POWER/LIGHTING

Item	Net Price £	Material £	Labour hours	Labour £	Unit	Total rate £
600/1000 Volt grade; seven core (galvanized steel wire armour)						
1.5 mm²	1.22	1.38	0.10	2.74	m	**4.12**
2.5 mm²	1.70	1.92	0.10	2.74	m	**4.66**
4 mm²	3.09	3.49	0.11	3.03	m	**6.52**
600/1000 Volt grade; twelve core (galvanized steel wire armour)						
1.5 mm²	1.94	2.19	0.11	3.03	m	**5.22**
2.5 mm²	2.91	3.28	0.11	3.03	m	**6.31**
600/1000 Volt grade; nineteen core (galvanized steel wire armour)						
1.5 mm²	2.93	3.30	0.13	3.57	m	**6.87**
2.5 mm²	4.40	4.97	0.14	3.86	m	**8.83**
600/1000 Volt grade; twenty seven core (galvanized steel wire armour)						
1.5 mm²	4.03	4.55	0.14	3.86	m	**8.41**
2.5 mm²	5.62	6.34	0.16	4.40	m	**10.74**
600/1000 Volt grade; thirty seven core (galvanized steel wire armour)						
1.5 mm²	5.02	5.67	0.15	4.12	m	**9.79**
2.5 mm²	7.11	8.03	0.17	4.67	m	**12.70**
Cable; XLPE insulated; PVC sheathed copper stranded conductors to BS 5467; clipped direct to backgrounds including cleat						
600/1000 Volt grade; single core (aluminium wire armour)						
25 mm²	2.80	3.16	0.35	9.62	m	**12.78**
35 mm²	3.05	3.45	0.36	9.89	m	**13.34**
50 mm²	3.37	3.80	0.37	10.17	m	**13.97**
70 mm²	4.83	5.46	0.39	10.72	m	**16.18**
95 mm²	6.46	7.29	0.42	11.55	m	**18.84**
120 mm²	7.64	8.63	0.47	12.92	m	**21.55**
150 mm²	9.40	10.62	0.51	14.01	m	**24.63**
185 mm²	11.20	12.65	0.59	16.21	m	**28.86**
240 mm²	14.67	16.56	0.68	18.68	m	**35.24**
300 mm²	17.74	20.03	0.74	20.33	m	**40.36**
400 mm²	25.89	29.24	0.88	24.18	m	**53.42**
500 mm²	32.63	36.85	0.88	24.18	m	**61.03**
630 mm²	42.43	47.92	1.05	28.86	m	**76.78**
800 mm²	52.76	59.59	1.33	36.55	m	**96.14**
1000 mm²	59.00	66.64	1.40	38.46	m	**105.10**

V: ELECTRICAL SUPPLY/POWER/LIGHTING

Item	Net Price £	Material £	Labour hours	Labour £	Unit	Total rate £
V20: LV DISTRIBUTION – cont						
ARMOURED CABLE – cont						
Cable – cont						
600/1000 Volt grade; two core (galvanized steel wire armour)						
1.5 mm²	0.83	0.94	0.20	5.50	m	6.44
2.5 mm²	0.95	1.07	0.20	5.50	m	6.57
4.0 mm²	1.20	1.36	0.21	5.77	m	7.13
6.0 mm²	1.40	1.58	0.22	6.04	m	7.62
10.0 mm²	1.81	2.04	0.24	6.60	m	8.64
16.0 mm²	2.62	2.96	0.25	6.86	m	9.82
25 mm²	2.28	2.57	0.35	9.62	m	12.19
35 mm²	5.10	5.76	0.36	9.89	m	15.65
50 mm²	6.28	7.09	0.37	10.17	m	17.26
70 mm²	9.32	10.53	0.39	10.72	m	21.25
95 mm²	12.16	13.73	0.42	11.55	m	25.28
120 mm²	14.94	16.87	0.47	12.92	m	29.79
150 mm²	18.50	20.89	0.51	14.01	m	34.90
185 mm²	23.56	26.61	0.59	16.21	m	42.82
240 mm²	29.31	33.10	0.68	18.68	m	51.78
300 mm²	41.05	46.36	0.74	20.33	m	66.69
400 mm²	51.10	57.71	0.88	24.18	m	81.89
600/1000 Volt grade; three core (galvanized steel wire armour)						
1.5 mm²	0.89	1.01	0.20	5.50	m	6.51
2.5 mm²	1.05	1.18	0.21	5.77	m	6.95
4.0 mm²	1.34	1.51	0.22	6.04	m	7.55
6.0 mm²	1.67	1.89	0.22	6.04	m	7.93
10.0 mm²	2.43	2.74	0.25	6.86	m	9.60
16.0 mm²	3.34	3.77	0.26	7.15	m	10.92
25 mm²	4.68	5.28	0.37	10.17	m	15.45
35 mm²	6.08	6.86	0.39	10.72	m	17.58
50 mm²	8.67	9.79	0.40	10.99	m	20.78
70 mm²	12.57	14.20	0.42	11.55	m	25.75
95 mm²	17.27	19.50	0.45	12.36	m	31.86
120 mm²	21.44	24.21	0.52	14.29	m	38.50
150 mm²	26.69	30.14	0.55	15.12	m	45.26
185 mm²	32.81	37.06	0.63	17.31	m	54.37
240 mm²	43.57	49.20	0.71	19.50	m	68.70
300 mm²	54.85	61.94	0.78	21.44	m	83.38
400 mm²	65.11	73.53	0.87	23.91	m	97.44

V: ELECTRICAL SUPPLY/POWER/LIGHTING

Item	Net Price £	Material £	Labour hours	Labour £	Unit	Total rate £
600/1000 Volt grade; four core (galvanized steel wire armour)						
1.5 mm²	1.02	1.15	0.21	5.77	m	6.92
2.5 mm²	1.26	1.43	0.22	6.04	m	7.47
4.0 mm²	1.54	1.74	0.22	6.04	m	7.78
6.0 mm²	2.15	2.43	0.23	6.32	m	8.75
10.0 mm²	2.94	3.33	0.26	7.15	m	10.48
16.0 mm²	4.14	4.67	0.26	7.15	m	11.82
25 mm²	5.89	6.65	0.39	10.72	m	17.37
35 mm²	7.67	8.66	0.40	10.99	m	19.65
50 mm²	11.00	12.42	0.41	11.27	m	23.69
70 mm²	16.65	18.81	0.45	12.36	m	31.17
95 mm²	22.31	25.19	0.50	13.73	m	38.92
120 mm²	27.96	31.58	0.54	14.84	m	46.42
150 mm²	35.29	39.85	0.60	16.49	m	56.34
185 mm²	43.12	48.69	0.67	18.41	m	67.10
240 mm²	55.83	63.06	0.75	20.61	m	83.67
300 mm²	70.05	79.11	0.83	22.81	m	101.92
400 mm²	72.67	82.07	0.91	25.00	m	107.07
600/1000 Volt grade; seven core (galvanized steel wire armour)						
1.5 mm²	1.43	1.61	0.20	5.50	m	7.11
2.5 mm²	1.95	2.20	0.20	5.50	m	7.70
4.0 mm²	3.52	3.98	0.23	6.32	m	10.30
600/1000 Volt grade; twelve core (galvanized steel wire armour)						
1.5 mm²	2.07	5.89	0.23	6.32	m	12.21
2.5 mm²	3.11	3.51	0.24	6.60	m	10.11
600/1000 Volt grade; nineteen core (galvanized steel wire armour)						
1.5 mm²	3.13	3.54	0.26	7.15	m	10.69
2.5 mm²	4.61	5.20	0.28	7.69	m	12.89
600/1000 Volt grade; twenty seven core (galvanized steel wire armour)						
1.5 mm²	4.23	4.77	0.29	7.98	m	12.75
2.5 mm²	5.89	6.65	0.30	8.24	m	14.89
600/1000 Volt grade; thirty seven core (galvanized steel wire armour)						
1.5 mm²	5.29	5.98	0.32	8.80	m	14.78
2.5 mm²	7.38	8.33	0.33	9.07	m	17.40

V: ELECTRICAL SUPPLY/POWER/LIGHTING

Item	Net Price £	Material £	Labour hours	Labour £	Unit	Total rate £
V20: LV DISTRIBUTION – cont						
ARMOURED CABLE – cont						
Cable termination; brass weatherproof gland with inner and outer seal, shroud, brass locknut and earth ring (including drilling and cutting mild steel gland plate)						
600/1000 Volt grade; single core (aluminium wire armour)						
25 mm²	5.35	6.04	1.70	46.72	nr	**52.76**
35 mm²	5.35	6.04	1.79	49.18	nr	**55.22**
50 mm²	5.86	6.62	2.06	56.61	nr	**63.23**
70 mm²	7.56	8.54	2.12	58.25	nr	**66.79**
95 mm²	7.62	8.61	2.39	65.68	nr	**74.29**
120 mm²	7.64	8.63	2.47	67.88	nr	**76.51**
150 mm²	10.88	12.29	2.73	75.01	nr	**87.30**
185 mm²	11.06	12.50	3.05	83.81	nr	**96.31**
240 mm²	16.91	19.09	3.45	94.80	nr	**113.89**
300 mm²	17.09	19.30	3.84	105.52	nr	**124.82**
400 mm²	24.61	27.80	4.21	115.69	nr	**143.49**
500 mm²	27.32	30.85	5.70	156.62	m	**187.47**
630 mm²	30.78	34.76	6.20	170.37	m	**205.13**
800 mm²	52.50	59.29	7.50	206.08	m	**265.37**
1000 mm²	63.36	71.56	10.00	274.78	m	**346.34**
600/1000 Volt grade; two core (galvanized steel wire armour)						
1.5 mm²	4.85	5.48	0.58	15.93	nr	**21.41**
2.5 mm²	4.85	5.48	0.58	15.97	nr	**21.45**
4 mm²	4.88	5.51	0.58	15.93	nr	**21.44**
6 mm²	4.97	5.61	0.67	18.41	nr	**24.02**
10 mm²	5.30	5.99	1.00	27.48	nr	**33.47**
16 mm²	5.35	6.04	1.11	30.51	nr	**36.55**
25 mm²	6.08	6.86	1.70	46.72	nr	**53.58**
35 mm²	6.00	6.77	1.79	49.18	nr	**55.95**
50 mm²	9.75	11.02	2.06	56.61	nr	**67.63**
70 mm²	10.85	12.25	2.12	58.25	nr	**70.50**
95 mm²	16.13	18.22	2.39	65.68	nr	**83.90**
120 mm²	16.66	18.82	2.47	67.88	nr	**86.70**
150 mm²	23.59	26.64	2.73	75.01	nr	**101.65**
185 mm²	24.02	27.13	3.05	83.81	nr	**110.94**
240 mm²	25.58	28.89	3.45	94.80	nr	**123.69**
300 mm²	34.63	39.11	3.84	105.52	nr	**144.63**
400 mm²	54.86	61.95	4.21	115.69	nr	**177.64**

V: ELECTRICAL SUPPLY/POWER/LIGHTING

Item	Net Price £	Material £	Labour hours	Labour £	Unit	Total rate £
600/1000 Volt grade; three core (galvanized steel wire armour)						
1.5 mm²	5.02	5.67	0.61	16.89	nr	**22.56**
2.5 mm²	5.02	5.67	0.62	17.03	nr	**22.70**
4 mm²	4.88	5.51	0.62	17.03	nr	**22.54**
6 mm²	5.52	6.23	0.71	19.50	nr	**25.73**
10 mm²	6.17	6.97	1.06	29.13	nr	**36.10**
16 mm²	6.34	7.16	1.19	32.69	nr	**39.85**
25 mm²	9.02	10.19	1.81	49.74	nr	**59.93**
35 mm²	9.15	10.33	1.99	54.68	nr	**65.01**
50 mm²	10.11	11.41	2.23	61.28	nr	**72.69**
70 mm²	16.40	18.52	2.40	65.94	nr	**84.46**
95 mm²	16.77	18.94	2.63	72.27	nr	**91.21**
120 mm²	17.57	19.84	2.83	77.75	nr	**97.59**
150 mm²	24.67	27.87	3.22	88.47	nr	**116.34**
185 mm²	25.28	28.55	3.44	94.53	nr	**123.08**
240 mm²	36.26	40.95	3.83	105.23	nr	**146.18**
300 mm²	36.87	41.64	4.28	117.61	nr	**159.25**
400 mm²	49.50	55.91	5.00	137.39	nr	**193.30**
600/1000 Volt grade; four core (galvanized steel wire armour)						
1.5 mm²	5.18	5.85	0.67	18.41	nr	**24.26**
2.5 mm²	5.18	5.85	0.67	18.41	nr	**24.26**
4 mm²	5.23	5.91	0.71	19.50	nr	**25.41**
6 mm²	6.39	7.22	0.76	20.88	nr	**28.10**
10 mm²	6.39	7.22	1.14	31.33	nr	**38.55**
16 mm²	6.51	7.35	1.29	35.46	nr	**42.81**
25 mm²	9.29	10.50	1.99	54.68	nr	**65.18**
35 mm²	9.46	10.68	2.16	59.35	nr	**70.03**
50 mm²	15.10	17.05	2.49	68.42	nr	**85.47**
70 mm²	16.92	19.10	2.65	72.81	nr	**91.91**
95 mm²	24.05	27.16	2.98	81.89	nr	**109.05**
120 mm²	25.10	28.35	3.15	86.56	nr	**114.91**
150 mm²	25.75	29.08	3.50	96.18	nr	**125.26**
185 mm²	35.18	39.73	3.72	102.22	nr	**141.95**
240 mm²	38.32	43.28	4.33	118.98	nr	**162.26**
300 mm²	51.01	57.61	4.86	133.54	nr	**191.15**
400 mm²	54.54	61.60	5.46	150.03	nr	**211.63**
600/1000 Volt grade; seven core (galvanized steel wire armour)						
1.5 mm²	5.65	6.39	0.81	22.26	nr	**28.65**
2.5 mm²	5.65	6.39	0.85	23.36	nr	**29.75**
4 mm²	2.60	2.94	0.93	25.56	nr	**28.50**
600/1000 Volt grade; twelve core (galvanized steel wire armour)						
1.5 mm²	6.78	7.66	1.14	31.33	nr	**38.99**
2.5 mm²	7.46	8.43	1.13	31.05	nr	**39.48**

V: ELECTRICAL SUPPLY/POWER/LIGHTING

Item	Net Price £	Material £	Labour hours	Labour £	Unit	Total rate £
V20: LV DISTRIBUTION – cont						
ARMOURED CABLE – cont						
Cable termination – cont						
600/1000 Volt grade; nineteen core (galvanized steel wire armour)						
1.5 mm²	8.57	9.68	1.54	42.32	nr	**52.00**
2.5 mm²	8.57	9.68	1.54	42.32	nr	**52.00**
600/1000 Volt grade; twenty seven core (galvanized steel wire armour)						
1.5 mm²	9.82	11.09	1.94	53.31	nr	**64.40**
2.5 mm²	12.46	14.08	2.31	63.47	nr	**77.55**
600/1000 Volt grade; thirty seven core (galvanized steel wire armour)						
1.5 mm²	14.09	15.91	2.53	69.51	nr	**85.42**
2.5 mm²	14.09	15.91	2.87	78.87	nr	**94.78**
Cable; XLPE insulated; LSOH sheathed (LSF); copper stranded conductors to BS 6724; laid in trench/duct including marker tape (cable tiles measured elsewhere)						
600/1000 Volt grade; single core (aluminium wire armour)						
50 mm²	3.22	3.64	0.17	4.67	m	**8.31**
70 mm²	4.36	4.93	0.18	4.95	m	**9.88**
95 mm²	5.93	6.70	0.20	5.50	m	**12.20**
120 mm²	7.04	7.96	0.22	6.04	m	**14.00**
150 mm²	8.70	9.82	0.24	6.60	m	**16.42**
185 mm²	10.46	11.81	0.26	7.15	m	**18.96**
240 mm²	13.49	15.24	0.30	8.24	m	**23.48**
300 mm²	16.44	18.56	0.31	8.52	m	**27.08**
400 mm²	21.35	24.11	0.35	9.62	m	**33.73**
500 mm²	27.40	30.95	0.44	12.10	m	**43.05**
630 mm²	34.33	38.77	0.52	14.29	m	**53.06**
800 mm²	45.00	50.83	0.62	17.03	m	**67.86**
1000 mm²	57.43	64.86	0.65	17.86	m	**82.72**
600/1000 Volt grade; two core (galvanized steel wire armour)						
1.5 mm²	0.63	0.71	0.06	1.65	m	**2.36**
2.5 mm²	0.75	0.85	0.06	1.65	m	**2.50**
4 mm²	0.95	1.07	0.08	2.20	m	**3.27**
6 mm²	1.14	1.29	0.08	2.20	m	**3.49**
10 mm²	1.54	1.74	0.10	2.74	m	**4.48**
16 mm²	2.30	2.60	0.10	2.74	m	**5.34**
25 mm²	2.96	3.35	0.15	4.12	m	**7.47**

V: ELECTRICAL SUPPLY/POWER/LIGHTING

Item	Net Price £	Material £	Labour hours	Labour £	Unit	Total rate £
35 mm²	3.32	3.75	0.17	4.67	m	8.42
50 mm²	5.10	5.76	0.15	4.12	m	9.88
70 mm²	8.58	9.69	0.18	4.95	m	14.64
95 mm²	11.51	12.99	0.20	5.50	m	18.49
120 mm²	14.11	15.93	0.22	6.04	m	21.97
150 mm²	17.48	19.74	0.24	6.60	m	26.34
185 mm²	22.27	25.15	0.26	7.15	m	32.30
240 mm²	28.52	32.21	0.30	8.24	m	40.45
300 mm²	38.31	43.27	0.31	8.52	m	51.79
400 mm²	50.66	57.21	0.35	9.62	m	66.83
600/1000 Volt grade; three core (galvanized steel wire armour)						
1.5 mm²	0.70	0.80	0.07	1.92	m	2.72
2.5 mm²	0.87	0.98	0.07	1.92	m	2.90
4 mm²	1.08	1.22	0.09	2.47	m	3.69
6 mm²	1.39	1.57	0.10	2.74	m	4.31
10 mm²	2.13	2.41	0.11	3.03	m	5.44
16 mm²	2.97	3.36	0.11	3.03	m	6.39
25 mm²	4.03	4.55	0.16	4.40	m	8.95
35 mm²	5.59	6.31	0.16	4.40	m	10.71
50 mm²	7.98	9.02	0.19	5.22	m	14.24
70 mm²	11.85	13.38	0.21	5.77	m	19.15
95 mm²	16.31	18.42	0.23	6.32	m	24.74
120 mm²	20.29	22.92	0.24	6.60	m	29.52
150 mm²	25.22	28.48	0.27	7.42	m	35.90
185 mm²	31.08	35.10	0.30	8.24	m	43.34
240 mm²	39.74	44.88	0.33	9.07	m	53.95
300 mm²	50.37	56.89	0.35	9.62	m	66.51
400 mm²	59.81	67.54	0.41	11.27	m	78.81
600/1000 Volt grade; four core (galvanized steel wire armour)						
1.5 mm²	0.82	0.93	0.08	2.20	m	3.13
2.5 mm²	1.02	1.15	0.09	2.47	m	3.62
4 mm²	1.26	1.43	0.10	2.74	m	4.17
6 mm²	1.86	2.10	0.10	2.74	m	4.84
10 mm²	2.59	2.93	0.12	3.29	m	6.22
16 mm²	3.75	4.23	0.12	3.29	m	7.52
25 mm²	5.41	6.11	0.18	4.95	m	11.06
35 mm²	7.04	7.96	0.19	5.22	m	13.18
50 mm²	10.20	11.52	0.21	5.77	m	17.29
70 mm²	15.74	17.78	0.23	6.32	m	24.10
95 mm²	21.07	23.80	0.26	7.15	m	30.95
120 mm²	26.46	29.89	0.28	7.69	m	37.58
150 mm²	32.03	36.18	0.32	8.80	m	44.98
185 mm²	39.31	44.40	0.35	9.62	m	54.02
240 mm²	50.82	57.40	0.36	9.89	m	67.29
300 mm²	63.76	72.01	0.40	10.99	m	83.00
400 mm²	121.60	137.33	0.45	12.36	m	149.69

V: ELECTRICAL SUPPLY/POWER/LIGHTING

Item	Net Price £	Material £	Labour hours	Labour £	Unit	Total rate £
V20: LV DISTRIBUTION – cont						
ARMOURED CABLE – cont						
Cable – cont						
600/1000 Volt grade; seven core (galvanized steel wire armour)						
1.5 mm²	1.17	1.33	0.10	2.74	m	**4.07**
2.5 mm²	1.62	1.83	0.10	2.74	m	**4.57**
4 mm²	2.95	3.34	0.11	3.03	m	**6.37**
600/1000 Volt grade; twelve core (galvanized steel wire armour)						
1.5 mm²	1.86	2.10	0.11	3.03	m	**5.13**
2.5 mm²	2.78	3.14	0.11	3.03	m	**6.17**
600/1000 Volt grade; nineteen core (galvanized steel wire armour)						
1.5 mm²	2.79	3.15	0.13	3.57	m	**6.72**
2.5 mm²	4.19	4.73	0.14	3.86	m	**8.59**
600/1000 Volt grade; twenty seven core (galvanized steel wire armour)						
1.5 mm²	3.83	4.32	0.14	3.86	m	**8.18**
2.5 mm²	5.36	6.05	0.16	4.40	m	**10.45**
600/1000 Volt grade; thirty seven core (galvanized steel wire armour)						
1.5 mm²	4.78	5.40	0.15	4.12	m	**9.52**
2.5 mm²	6.76	7.64	0.17	4.67	m	**12.31**
Cable; XLPE insulated; LSOH sheathed (LSF) copper stranded conductors to BS 6724; clipped direct to backgrounds including cleat						
600/1000 Volt grade; single core (aluminium wire armour)						
50 mm²	3.52	3.98	0.37	10.17	m	**14.15**
70 mm²	4.61	5.20	0.39	10.72	m	**15.92**
95 mm²	6.16	6.96	0.42	11.55	m	**18.51**
120 mm²	7.28	8.22	0.47	12.92	m	**21.14**
150 mm²	8.96	10.12	0.51	14.01	m	**24.13**
185 mm²	10.68	12.07	0.59	16.21	m	**28.28**
240 mm²	13.96	15.77	0.68	18.68	m	**34.45**
300 mm²	16.91	19.09	0.74	20.33	m	**39.42**
400 mm²	24.67	27.87	0.81	22.26	m	**50.13**
500 mm²	31.08	35.10	0.88	24.18	m	**59.28**
630 mm²	40.42	45.66	1.05	28.86	m	**74.52**

V: ELECTRICAL SUPPLY/POWER/LIGHTING

Item	Net Price £	Material £	Labour hours	Labour £	Unit	Total rate £
800 mm²	50.25	56.75	1.33	36.55	m	**93.30**
1000 mm²	56.18	63.45	1.40	38.46	m	**101.91**
600/1000 Volt grade; two core (galvanized steel wire armour)						
1.5 mm²	0.79	0.89	0.20	5.50	m	**6.39**
2.5 mm²	0.90	1.02	0.20	5.50	m	**6.52**
4.0 mm²	1.14	1.29	0.21	5.77	m	**7.06**
6.0 mm²	1.35	1.52	0.22	6.04	m	**7.56**
10.0 mm²	1.74	1.97	0.24	6.60	m	**8.57**
16.0 mm²	2.51	2.84	0.25	6.86	m	**9.70**
25 mm²	3.15	3.56	0.35	9.62	m	**13.18**
35 mm²	4.86	5.49	0.36	9.89	m	**15.38**
50 mm²	5.98	6.75	0.37	10.17	m	**16.92**
70 mm²	8.87	10.02	0.39	10.72	m	**20.74**
95 mm²	11.60	13.10	0.42	11.55	m	**24.65**
120 mm²	14.23	16.08	0.47	12.92	m	**29.00**
150 mm²	17.63	19.91	0.51	14.01	m	**33.92**
185 mm²	22.45	25.36	0.59	16.21	m	**41.57**
240 mm²	27.93	31.55	0.68	18.68	m	**50.23**
300 mm²	39.10	44.16	0.74	20.33	m	**64.49**
400 mm²	48.68	54.98	0.81	22.26	m	**77.24**
600/1000 Volt grade; three core (galvanized steel wire armour)						
1.5 mm²	0.85	0.96	0.20	5.50	m	**6.46**
2.5 mm²	1.00	1.13	0.21	5.77	m	**6.90**
4.0 mm²	1.27	1.44	0.22	6.04	m	**7.48**
6.0 mm²	1.59	1.80	0.22	6.04	m	**7.84**
10.0 mm²	2.33	2.63	0.25	6.86	m	**9.49**
16.0 mm²	3.18	3.59	0.26	7.15	m	**10.74**
25 mm²	4.47	5.05	0.37	10.17	m	**15.22**
35 mm²	5.79	6.54	0.39	10.72	m	**17.26**
50 mm²	8.26	9.33	0.40	10.99	m	**20.32**
70 mm²	11.93	13.47	0.42	11.55	m	**25.02**
95 mm²	16.41	18.53	0.45	12.36	m	**30.89**
120 mm²	20.42	23.06	0.52	14.29	m	**37.35**
150 mm²	25.38	28.66	0.55	15.12	m	**43.78**
185 mm²	31.24	35.28	0.63	17.31	m	**52.59**
240 mm²	41.51	46.88	0.71	19.50	m	**66.38**
300 mm²	52.24	59.00	0.78	21.44	m	**80.44**
400 mm²	62.02	70.04	0.87	23.91	m	**93.95**
600/1000 Volt grade; four core (galvanized steel wire armour)						
1.5 mm²	0.97	1.09	0.21	5.77	m	**6.86**
2.5 mm²	1.21	1.37	0.22	6.04	m	**7.41**
4.0 mm²	1.47	1.66	0.22	6.04	m	**7.70**
6.0 mm²	2.05	2.32	0.23	6.32	m	**8.64**
10.0 mm²	2.80	3.16	0.26	7.15	m	**10.31**
16.0 mm²	3.94	4.45	0.26	7.15	m	**11.60**
25 mm²	5.61	6.33	0.39	10.72	m	**17.05**

V: ELECTRICAL SUPPLY/POWER/LIGHTING

Item	Net Price £	Material £	Labour hours	Labour £	Unit	Total rate £
V20: LV DISTRIBUTION – cont						
ARMOURED CABLE – cont						
Cable – cont						
600/1000 Volt grade – cont						
35 mm²	7.30	8.24	0.40	10.99	m	**19.23**
50 mm²	10.47	11.82	0.41	11.27	m	**23.09**
70 mm²	15.86	17.91	0.45	12.36	m	**30.27**
95 mm²	21.25	24.00	0.50	13.73	m	**37.73**
120 mm²	26.62	30.06	0.54	14.84	m	**44.90**
150 mm²	33.61	37.95	0.60	16.49	m	**54.44**
185 mm²	41.08	46.40	0.67	18.41	m	**64.81**
240 mm²	53.18	60.06	0.75	20.61	m	**80.67**
300 mm²	66.73	75.37	0.83	22.81	m	**98.18**
400 mm²	125.11	141.30	0.91	25.00	m	**166.30**
600/1000 Volt grade; seven core (galvanized steel wire armour)						
1.5 mm²	1.37	1.55	0.20	5.50	m	**7.05**
2.5 mm²	1.87	2.11	0.20	5.50	m	**7.61**
4.0 mm²	3.36	3.79	0.23	6.32	m	**10.11**
600/1000 Volt grade; twelve core (galvanized steel wire armour)						
1.5 mm²	1.98	2.23	0.23	6.32	m	**8.55**
2.5 mm²	2.97	3.36	0.24	6.60	m	**9.96**
600/1000 Volt grade; nineteen core (galvanized steel wire armour)						
1.5 mm²	2.98	3.37	0.26	7.15	m	**10.52**
2.5 mm²	4.38	4.95	0.28	7.69	m	**12.64**
600/1000 Volt grade; twenty seven core (galvanized steel wire armour)						
1.5 mm²	4.04	4.56	0.29	7.98	m	**12.54**
2.5 mm²	5.61	6.33	0.30	8.24	m	**14.57**
600/1000 Volt grade; thirty seven core (galvanized steel wire armour)						
1.5 mm²	5.04	5.69	0.32	8.80	m	**14.49**
2.5 mm²	7.03	7.94	0.33	9.07	m	**17.01**

V: ELECTRICAL SUPPLY/POWER/LIGHTING

Item	Net Price £	Material £	Labour hours	Labour £	Unit	Total rate £
Cable termination; brass weatherproof gland with inner and outer seal, shroud, brass locknut and earth ring (including drilling and cutting mild steel gland plate)						
600/1000 Volt grade; single core (aluminium wire armour)						
25 mm²	4.89	5.52	1.70	46.72	nr	**52.24**
35 mm²	4.89	5.52	1.79	49.18	nr	**54.70**
50 mm²	5.35	6.04	2.06	56.61	nr	**62.65**
70 mm²	6.92	7.81	2.12	58.25	nr	**66.06**
95 mm²	7.03	7.94	2.39	65.68	nr	**73.62**
120 mm²	7.31	8.25	2.47	67.88	nr	**76.13**
150 mm²	10.57	11.93	2.73	75.01	nr	**86.94**
185 mm²	16.41	18.53	3.05	83.81	nr	**102.34**
240 mm²	16.61	18.76	3.45	94.80	nr	**113.56**
300 mm²	23.91	27.00	3.84	105.52	nr	**132.52**
400 mm²	26.53	29.97	4.21	115.69	nr	**145.66**
500 mm²	29.90	33.77	5.70	156.62	m	**190.39**
630 mm²	51.02	57.62	6.20	170.37	m	**227.99**
800 mm²	61.53	69.49	7.50	206.08	m	**275.57**
1000 mm²	80.86	91.32	10.00	274.78	m	**366.10**
600/1000 Volt grade; two core (galvanized steel wire armour)						
1.5 mm²	4.70	5.30	0.52	14.21	nr	**19.51**
2.5 mm²	4.72	5.33	0.58	15.93	nr	**21.26**
4 mm²	4.74	5.36	0.58	15.93	nr	**21.29**
6 mm²	4.83	5.46	0.67	18.41	nr	**23.87**
10 mm²	5.16	5.82	1.00	27.48	nr	**33.30**
16 mm²	5.21	5.89	1.11	30.51	nr	**36.40**
25 mm²	5.91	6.67	1.70	46.72	nr	**53.39**
35 mm²	5.82	6.57	1.79	49.18	nr	**55.75**
50 mm²	9.48	10.71	2.06	56.61	nr	**67.32**
70 mm²	10.55	11.91	2.12	58.25	nr	**70.16**
95 mm²	15.68	17.71	2.39	65.68	nr	**83.39**
120 mm²	16.19	18.29	2.47	67.88	nr	**86.17**
150 mm²	22.93	25.90	2.73	75.01	nr	**100.91**
185 mm²	23.34	26.36	3.05	83.81	nr	**110.17**
240 mm²	24.86	28.08	3.45	94.80	nr	**122.88**
300 mm²	33.64	38.00	3.84	105.52	nr	**143.52**
400 mm²	53.29	60.19	4.21	115.69	nr	**175.88**
600/1000 Volt grade; three core (galvanized steel wire armour)						
1.5 mm²	4.88	5.51	0.62	17.03	nr	**22.54**
2.5 mm²	5.81	6.56	0.62	17.03	nr	**23.59**
4 mm²	6.61	7.47	0.62	17.03	nr	**24.50**
6 mm²	5.36	6.05	0.71	19.50	nr	**25.55**
10 mm²	6.01	6.78	1.06	29.13	nr	**35.91**
16 mm²	6.10	6.88	1.19	32.69	nr	**39.57**
25 mm²	8.78	9.91	1.81	49.74	nr	**59.65**

V: ELECTRICAL SUPPLY/POWER/LIGHTING

Item	Net Price £	Material £	Labour hours	Labour £	Unit	Total rate £
V20: LV DISTRIBUTION – cont						
ARMOURED CABLE – cont						
Cable termination – cont						
600/1000 Volt grade – cont						
35 mm²	8.90	10.05	1.99	54.68	nr	**64.73**
50 mm²	9.82	11.09	2.23	61.28	nr	**72.37**
70 mm²	15.94	18.00	2.40	65.94	nr	**83.94**
95 mm²	16.29	18.40	2.63	72.27	nr	**90.67**
120 mm²	17.06	19.27	2.83	77.75	nr	**97.02**
150 mm²	23.96	27.06	3.22	88.47	nr	**115.53**
185 mm²	24.56	27.73	3.44	94.53	nr	**122.26**
240 mm²	35.23	39.79	3.83	105.23	nr	**145.02**
300 mm²	35.82	40.45	4.28	117.61	nr	**158.06**
400 mm²	48.09	54.32	5.00	137.39	nr	**191.71**
600/1000 Volt grade; four core (galvanized steel wire armour)						
1.5 mm²	5.04	5.69	0.67	18.41	nr	**24.10**
2.5 mm²	5.04	5.69	0.67	18.41	nr	**24.10**
4 mm²	5.08	5.73	0.71	19.50	nr	**25.23**
6 mm²	6.22	7.03	0.76	20.88	nr	**27.91**
10 mm²	6.22	7.03	1.14	31.33	nr	**38.36**
16 mm²	6.33	7.15	1.29	35.46	nr	**42.61**
25 mm²	9.02	10.19	1.99	54.68	nr	**64.87**
35 mm²	9.19	10.38	2.16	59.35	nr	**69.73**
50 mm²	14.68	16.58	2.49	68.42	nr	**85.00**
70 mm²	16.44	18.56	2.65	72.81	nr	**91.37**
95 mm²	23.36	26.39	2.98	81.89	nr	**108.28**
120 mm²	24.40	27.56	3.15	86.56	nr	**114.12**
150 mm²	25.01	28.24	3.50	96.18	nr	**124.42**
185 mm²	34.19	38.62	3.72	102.22	nr	**140.84**
240 mm²	36.95	41.73	4.33	118.98	nr	**160.71**
300 mm²	49.55	55.96	4.86	133.54	nr	**189.50**
400 mm²	52.98	59.83	5.46	150.03	nr	**209.86**
600/1000 Volt grade; seven core (galvanized steel wire armour)						
1.5 mm²	5.48	6.19	0.81	22.26	nr	**28.45**
2.5 mm²	5.48	6.19	0.85	23.36	nr	**29.55**
4 mm²	4.35	4.92	0.93	25.56	nr	**30.48**
600/1000 Volt grade; twelve core (galvanized steel wire armour)						
1.5 mm²	6.59	7.45	1.14	31.33	nr	**38.78**
2.5 mm²	7.24	8.18	1.13	31.05	nr	**39.23**

V: ELECTRICAL SUPPLY/POWER/LIGHTING

Item	Net Price £	Material £	Labour hours	Labour £	Unit	Total rate £
600/1000 Volt grade; nineteen core (galvanized steel wire armour)						
1.5 mm²	8.31	9.38	1.54	42.32	nr	**51.70**
2.5 mm²	10.08	11.38	1.54	42.32	nr	**53.70**
600/1000 Volt grade; twenty seven core (galvanized steel wire armour)						
1.5 mm²	9.55	10.78	1.94	53.31	nr	**64.09**
2.5 mm²	12.17	13.75	2.31	63.47	nr	**77.22**
600/1000 Volt grade; thirty seven core (galvanized steel wire armour)						
1.5 mm²	13.67	15.44	2.53	69.51	nr	**84.95**
2.5 mm²	13.67	15.44	2.87	78.87	nr	**94.31**
UN-ARMOURED CABLE						
Cable: XLPE insulated; PVC sheathed 90c copper to CMA Code 6181e; for internal wiring; clipped to backgrounds; (Supports and fixings included)						
300/500 Volt grade; single core						
6.0 mm²	0.49	0.55	0.09	2.47	m	**3.02**
10 mm²	0.73	0.83	0.10	2.74	m	**3.57**
16 mm²	1.02	1.15	0.12	3.29	m	**4.44**
Cable; LSF insulated to CMA Code 6491B; non-sheathed copper; laid/drawn in trunking/conduit						
450/750 Volt grade; single core						
1.5 mm²	0.09	0.11	0.03	0.83	m	**0.94**
2.5 mm²	0.16	0.19	0.03	0.83	m	**1.02**
4.0 mm²	0.25	0.29	0.03	0.83	m	**1.12**
6.0 mm²	0.35	0.40	0.04	1.09	m	**1.49**
10.0 mm²	0.55	0.62	0.04	1.09	m	**1.71**
16.0 mm²	0.82	0.93	0.05	1.38	m	**2.31**
25.0 mm²	1.92	2.17	0.06	1.65	m	**3.82**
35.0 mm²	1.74	1.97	0.06	1.65	m	**3.62**
50.0 mm²	2.55	2.88	0.07	1.92	m	**4.80**
70.0 mm²	4.06	4.59	0.08	2.20	m	**6.79**
95.0 mm²	5.59	6.31	0.08	2.20	m	**8.51**
120.0 mm²	6.94	7.83	0.10	2.74	m	**10.57**
150.0 mm²	8.63	9.75	0.13	3.57	m	**13.32**
Cable; twin & earth to CMA code 6242Y; clipped to backgrounds						

V: ELECTRICAL SUPPLY/POWER/LIGHTING

Item	Net Price £	Material £	Labour hours	Labour £	Unit	Total rate £
V20: LV DISTRIBUTION – cont						
UN-ARMOURED CABLE – cont						
300/500 Volt grade; PVC/PVC						
1.5 mm² 2C+E	0.40	0.45	0.01	0.28	m	**0.73**
1.5 mm² 3C+E	0.54	0.61	0.02	0.55	m	**1.16**
2.5 mm² 2C+E	0.69	0.78	0.02	0.55	m	**1.33**
4.0 mm² 2C+E	0.77	0.87	0.02	0.55	m	**1.42**
6.0 mm² 2C+E	1.00	1.13	0.02	0.55	m	**1.68**
10.0 mm² 2C+E	2.20	2.49	0.03	0.83	m	**3.32**
16.0 mm² 2C+E	3.16	3.57	0.03	0.83	m	**4.40**
300/500 Volt grade; LSF/LSF						
1.5 mm² 2C+E	0.55	0.62	0.01	0.28	m	**0.90**
1.5 mm² 3C+E	0.95	1.07	0.02	0.55	m	**1.62**
2.5 mm² 2C+E	0.72	0.82	0.02	0.55	m	**1.37**
4.0 mm² 2C+E	1.12	1.26	0.02	0.55	m	**1.81**
6.0 mm² 2C+E	1.48	1.67	0.02	0.55	m	**2.22**
10.0 mm² 2C+E	3.23	3.65	0.03	0.83	m	**4.48**
16.0 mm² 2C+E	4.70	5.30	0.03	0.83	m	**6.13**
EARTH CABLE						
Cable; LSF insulated to CMA Code 6491B; non-sheathed copper; laid/drawn in trunking/conduit						
450/750 Volt grade; single core						
1.5 mm²	0.09	0.11	0.03	0.83	m	**0.94**
2.5 mm²	0.16	0.19	0.03	0.83	m	**1.02**
4.0 mm²	0.25	0.29	0.03	0.83	m	**1.12**
6.0 mm²	0.35	0.40	0.04	1.09	m	**1.49**
10.0 mm²	0.55	0.62	0.04	1.09	m	**1.71**
16.0 mm²	0.82	0.93	0.05	1.38	m	**2.31**
25.0 mm²	1.74	1.97	0.06	1.65	m	**3.62**
35.0 mm²	1.92	2.17	0.06	1.65	m	**3.82**
50.0 mm²	2.55	2.88	0.07	1.92	m	**4.80**
70.0 mm²	4.06	4.59	0.08	2.20	m	**6.79**
95.0 mm²	5.59	6.31	0.08	2.20	m	**8.51**
120.0 mm²	6.94	7.83	0.10	2.74	m	**10.57**
150.0 mm²	8.63	9.75	0.13	3.57	m	**13.32**
185.0 mm²	10.50	11.86	0.16	4.40	m	**16.26**
240.0 mm²	13.70	15.47	0.20	5.50	m	**20.97**

V: ELECTRICAL SUPPLY/POWER/LIGHTING

Item	Net Price £	Material £	Labour hours	Labour £	Unit	Total rate £
FLEXIBLE CABLE						
Flexible cord; PVC insulated; PVC sheathed; copper stranded to CMA Code 218*Y (laid loose)						
300 Volt grade; two core						
0.50 mm²	0.14	0.17	0.07	1.92	m	**2.09**
0.75 mm²	0.18	0.21	0.07	1.92	m	**2.13**
300 Volt grade; three core						
0.50 mm²	0.16	0.19	0.07	1.92	m	**2.11**
0.75 mm²	0.26	0.30	0.07	1.92	m	**2.22**
1.0 mm²	0.27	0.31	0.07	1.92	m	**2.23**
1.5 mm²	0.43	0.49	0.07	1.92	m	**2.41**
2.5 mm²	0.57	0.64	0.08	2.20	m	**2.84**
Flexible cord; PVC insulated; PVC sheathed; copper stranded to CMA Code 318Y (laid loose)						
300/500 Volt grade; two core						
0.75 mm²	0.16	0.19	0.07	1.92	m	**2.11**
1.0 mm²	0.20	0.23	0.07	1.92	m	**2.15**
1.5 mm²	0.27	0.31	0.07	1.92	m	**2.23**
2.5 mm²	0.46	0.52	0.07	1.92	m	**2.44**
300/500 Volt grade; three core						
0.75 mm²	0.20	0.23	0.07	1.92	m	**2.15**
1.0 mm²	0.25	0.29	0.07	1.92	m	**2.21**
1.5 mm²	0.36	0.41	0.07	1.92	m	**2.33**
2.5 mm²	0.59	0.66	0.08	2.20	m	**2.86**
300/500 Volt grade; four core						
0.75 mm²	0.27	0.31	0.08	2.20	m	**2.51**
1.0 mm²	0.37	0.42	0.08	2.20	m	**2.62**
1.5 mm²	0.55	0.62	0.08	2.20	m	**2.82**
2.5 mm²	0.84	0.95	0.09	2.47	m	**3.42**
Flexible cord; PVC insulated; PVC sheathed for use in high temperature zones; copper stranded to CMA Code 309Y (laid loose)						
300/500 Volt grade; two core						
0.50 mm²	0.21	0.24	0.07	1.92	m	**2.16**
0.75 mm²	0.26	0.30	0.07	1.92	m	**2.22**
1.0 mm²	0.30	0.34	0.07	1.92	m	**2.26**
1.5 mm²	0.43	0.49	0.07	1.92	m	**2.41**
2.5 mm²	0.63	0.71	0.07	1.92	m	**2.63**

V: ELECTRICAL SUPPLY/POWER/LIGHTING

Item	Net Price £	Material £	Labour hours	Labour £	Unit	Total rate £
V20: LV DISTRIBUTION – cont						
FLEXIBLE CABLE – cont						
Flexible cord – cont						
300/500 Volt grade; three core						
0.50 mm²	0.27	0.31	0.07	1.92	m	**2.23**
0.75 mm²	0.33	0.38	0.07	1.92	m	**2.30**
1.0 mm²	0.37	0.42	0.07	1.92	m	**2.34**
1.5 mm²	0.48	0.54	0.07	1.92	m	**2.46**
2.5 mm²	0.73	0.83	0.07	1.92	m	**2.75**
Flexible cord; rubber insulated; rubber sheathed; copper stranded to CMA code 318 (laid loose)						
300/500 Volt grade; two core						
0.50 mm²	0.29	0.33	0.07	1.92	m	**2.25**
0.75 mm²	0.32	0.36	0.07	1.92	m	**2.28**
1.0 mm²	0.40	0.45	0.07	1.92	m	**2.37**
1.5 mm²	0.56	0.63	0.07	1.92	m	**2.55**
2.5 mm²	0.72	0.82	0.07	1.92	m	**2.74**
300/500 Volt grade; three core						
0.50 mm²	0.38	0.43	0.07	1.92	m	**2.35**
0.75 mm²	0.42	0.48	0.07	1.92	m	**2.40**
1.0 mm²	0.55	0.62	0.07	1.92	m	**2.54**
1.5 mm²	0.73	0.83	0.07	1.92	m	**2.75**
2.5 mm²	1.13	1.28	0.07	1.92	m	**3.20**
300/500 Volt grade; four core						
0.50 mm²	0.47	0.53	0.08	2.20	m	**2.73**
0.75 mm²	0.54	0.61	0.08	2.20	m	**2.81**
1.0 mm²	0.60	0.67	0.08	2.20	m	**2.87**
1.5 mm²	0.87	0.98	0.08	2.20	m	**3.18**
2.5 mm²	1.14	1.29	0.08	2.20	m	**3.49**
Flexible cord; rubber insulated; rubber sheathed; for 90C operation; copper stranded to CMA Code 318 (laid loose)						
450/750 Volt grade; two core						
0.50 mm²	0.22	0.25	0.07	1.92	m	**2.17**
0.75 mm²	0.34	0.39	0.07	1.92	m	**2.31**
1.0 mm²	0.37	0.42	0.07	1.92	m	**2.34**
1.5 mm²	0.47	0.53	0.07	1.92	m	**2.45**
2.5 mm²	0.72	0.82	0.07	1.92	m	**2.74**

V: ELECTRICAL SUPPLY/POWER/LIGHTING

Item	Net Price £	Material £	Labour hours	Labour £	Unit	Total rate £
450/750 Volt grade; three core						
0.50 mm²	0.36	0.41	0.07	1.92	m	2.33
0.75 mm²	0.41	0.46	0.07	1.92	m	2.38
1.0 mm²	0.45	0.51	0.07	1.92	m	2.43
1.5 mm²	0.48	0.54	0.07	1.92	m	2.46
2.5 mm²	0.65	0.73	0.07	1.92	m	2.65
450/750 Volt grade; four core						
0.75 mm²	0.54	0.61	0.08	2.20	m	2.81
1.0 mm²	0.68	0.77	0.08	2.20	m	2.97
1.5 mm²	0.99	1.12	0.08	2.20	m	3.32
2.5 mm²	1.54	1.74	0.08	2.20	m	3.94
Heavy flexible cable; rubber insulated; rubber sheathed; copper stranded to CMA Code 638P (laid loose)						
450/750 Volt grade; two core						
1.0 mm²	0.38	0.43	0.08	2.20	m	2.63
1.5 mm²	0.45	0.51	0.08	2.20	m	2.71
2.5 mm²	0.62	0.70	0.08	2.20	m	2.90
450/750 Volt grade; three core						
1.0 mm²	0.45	0.51	0.08	2.20	m	2.71
1.5 mm²	0.53	0.60	0.08	2.20	m	2.80
2.5 mm²	0.61	0.69	0.08	2.20	m	2.89
450/750 Volt grade; four core						
1.0 mm²	0.62	0.70	0.08	2.20	m	2.90
1.5 mm²	0.67	0.75	0.08	2.20	m	2.95
2.5 mm²	0.93	1.05	0.08	2.20	m	3.25
FIRE RATED CABLE						
Cable, mineral insulated; copper sheathed with copper conductors; fixed with clips to backgrounds. BASEC approval to BS 6207 Part 1 1995; complies with BS 6387 Category CWZ						
Light duty 500 Volt grade; bare						
2L 1.0	3.56	4.02	0.23	6.32	m	10.34
2L 1.5	4.06	4.59	0.23	6.32	m	10.91
2L 2.5	4.89	5.52	0.25	6.86	m	12.38
2L 4.0	6.90	7.79	0.25	6.86	m	14.65
3L 1.0	4.20	4.74	0.24	6.60	m	11.34
3L 1.5	4.99	5.64	0.25	6.86	m	12.50
3L 2.5	7.13	8.05	0.25	6.86	m	14.91

V: ELECTRICAL SUPPLY/POWER/LIGHTING

Item	Net Price £	Material £	Labour hours	Labour £	Unit	Total rate £
V20: LV DISTRIBUTION – cont						
FIRE RATED CABLE – cont						
Cable, mineral insulated – cont						
Light duty 500 Volt grade; bare – cont						
4L 1.0	4.80	5.42	0.25	6.86	m	**12.28**
4L 1.5	5.38	6.08	0.25	6.86	m	**12.94**
4L 2.5	8.45	9.55	0.26	7.15	m	**16.70**
7L 1.5	8.51	9.61	0.28	7.69	m	**17.30**
7L 2.5	10.81	12.21	0.27	7.42	m	**19.63**
Light duty 500 Volt grade; LSF sheathed						
2L 1.0	3.82	4.31	0.23	6.32	m	**10.63**
2L 1.5	4.25	4.80	0.23	6.32	m	**11.12**
2L 2.5	5.09	5.75	0.25	6.86	m	**12.61**
2L 4.0	7.16	8.09	0.25	6.86	m	**14.95**
3L 1.0	4.54	5.13	0.24	6.60	m	**11.73**
3L 1.5	5.32	6.01	0.25	6.86	m	**12.87**
3L 2.5	7.43	8.39	0.25	6.86	m	**15.25**
4L 1.0	5.16	5.82	0.25	6.86	m	**12.68**
4L 1.5	5.76	6.51	0.25	6.86	m	**13.37**
4L 2.5	8.71	9.83	0.26	7.15	m	**16.98**
7L 1.5	9.36	10.57	0.28	7.69	m	**18.26**
7L 2.5	11.73	13.25	0.27	7.42	m	**20.67**
Heavy duty 750 Volt grade; bare						
1H 10	7.15	8.08	0.25	6.86	m	**14.94**
1H 16	9.68	10.93	0.26	7.15	m	**18.08**
1H 25	13.45	15.19	0.27	7.42	m	**22.61**
1H 35	19.41	21.92	0.32	8.80	m	**30.72**
1H 50	21.25	24.00	0.35	9.62	m	**33.62**
1H 70	28.36	32.03	0.38	10.44	m	**42.47**
1H 95	42.05	47.49	0.41	11.27	m	**58.76**
1H 120	44.24	49.96	0.46	12.64	m	**62.60**
1H 150	54.26	61.28	0.50	13.73	m	**75.01**
1H 185	66.17	74.74	0.56	15.38	m	**90.12**
1H 240	86.09	97.23	0.69	18.96	m	**116.19**
2H 1.5	6.59	7.45	0.25	6.86	m	**14.31**
2H 2.5	7.96	8.99	0.26	7.15	m	**16.14**
2H 4.0	9.75	11.02	0.26	7.15	m	**18.17**
2H 6.0	12.53	14.15	0.29	7.98	m	**22.13**
2H 10.0	16.02	18.09	0.34	9.34	m	**27.43**
2H 16.0	22.91	25.88	0.40	10.99	m	**36.87**
2H 25.0	31.59	35.68	0.44	12.10	m	**47.78**
3H 1.5	7.32	8.27	0.25	6.86	m	**15.13**
3H 2.5	9.08	10.25	0.25	6.86	m	**17.11**
3H 4.0	11.06	12.50	0.27	7.42	m	**19.92**
3H 6.0	13.66	15.43	0.30	8.24	m	**23.67**
3H 10.0	20.25	22.87	0.35	9.62	m	**32.49**

V: ELECTRICAL SUPPLY/POWER/LIGHTING

Item	Net Price £	Material £	Labour hours	Labour £	Unit	Total rate £
3H 16.0	27.28	30.81	0.41	11.27	m	**42.08**
3H 25.0	40.02	45.20	0.47	12.92	m	**58.12**
4H 1.5	8.90	10.05	0.24	6.60	m	**16.65**
4H 2.5	10.79	12.19	0.26	7.15	m	**19.34**
4H 4.0	13.32	15.04	0.29	7.98	m	**23.02**
4H 6.0	16.85	19.03	0.31	8.52	m	**27.55**
4H 10.0	23.94	27.04	0.37	10.17	m	**37.21**
4H 16.0	34.30	38.74	0.44	12.10	m	**50.84**
4H 25.0	48.97	55.30	0.52	14.29	m	**69.59**
7H 1.5	12.00	13.56	0.30	8.24	m	**21.80**
7H 2.5	15.99	18.05	0.32	8.80	m	**26.85**
12H 2.5	27.28	30.81	0.39	10.72	m	**41.53**
19H 1.5	39.99	45.17	0.42	11.55	m	**56.72**
Heavy duty 750 Volt grade; LSF sheathed						
1H 10	7.58	8.56	0.25	6.86	m	**15.42**
1H 16	10.03	11.33	0.26	7.15	m	**18.48**
1H 25	13.80	15.59	0.27	7.42	m	**23.01**
1H 35	20.34	22.97	0.32	8.80	m	**31.77**
1H 50	22.35	25.25	0.35	9.62	m	**34.87**
1H 70	28.59	32.29	0.38	10.44	m	**42.73**
1H 95	42.81	48.35	0.41	11.27	m	**59.62**
1H 120	45.50	51.39	0.46	12.64	m	**64.03**
1H 150	55.46	62.64	0.50	13.73	m	**76.37**
1H 185	67.47	76.20	0.56	15.38	m	**91.58**
1H 240	88.07	99.47	0.68	18.68	m	**118.15**
2H 1.5	6.83	7.71	0.25	6.86	m	**14.57**
2H 2.5	8.18	9.24	0.26	7.15	m	**16.39**
2H 4.0	10.09	11.39	0.26	7.15	m	**18.54**
2H 6.0	12.99	14.67	0.29	7.98	m	**22.65**
2H 10.0	16.68	18.84	0.34	9.34	m	**28.18**
2H 16.0	23.70	26.76	0.40	10.99	m	**37.75**
2H 25.0	32.50	36.71	0.44	12.10	m	**48.81**
3H 1.5	7.43	8.39	0.25	6.86	m	**15.25**
3H 2.5	9.30	10.51	0.25	6.86	m	**17.37**
3H 4.0	11.51	12.99	0.27	7.42	m	**20.41**
3H 6.0	14.46	16.33	0.30	8.24	m	**24.57**
3H 10.0	21.16	23.90	0.35	9.62	m	**33.52**
3H 16.0	28.26	31.92	0.41	11.27	m	**43.19**
3H 25.0	41.04	46.35	0.47	12.92	m	**59.27**
4H 1.5	9.15	10.33	0.24	6.60	m	**16.93**
4H 2.5	11.20	12.65	0.26	7.15	m	**19.80**
4H 4.0	13.90	15.70	0.29	7.98	m	**23.68**
4H 6.0	17.64	19.92	0.31	8.52	m	**28.44**
4H 10.0	25.02	28.25	0.37	10.17	m	**38.42**
4H 16.0	35.32	39.89	0.44	12.10	m	**51.99**
4H 25.0	50.34	56.85	0.52	14.29	m	**71.14**
7H 1.5	12.40	14.00	0.30	8.24	m	**22.24**
7H 2.5	16.48	18.61	0.32	8.80	m	**27.41**
12H 2.5	27.95	31.57	0.39	10.72	m	**42.29**
19H 1.5	40.98	46.29	0.42	11.55	m	**57.84**

V: ELECTRICAL SUPPLY/POWER/LIGHTING

Item	Net Price £	Material £	Labour hours	Labour £	Unit	Total rate £
V20: LV DISTRIBUTION – cont						
FIRE RATED CABLE – cont						
Cable terminations for M.I. Cable; polymeric one piece moulding; containing grey sealing compound; testing; phase marking and connection						
Light duty 500 Volt grade; brass gland; polymeric one moulding containing grey sealing compound; coloured conductor sleeving; Earth tag; plastic gland shroud						
2L 1.5	8.45	9.55	0.27	7.42	m	**16.97**
2L 2.5	8.48	10.82	0.27	7.42	m	**18.24**
3L 1.5	8.78	10.74	0.27	7.42	m	**18.16**
4L 1.5	8.86	10.83	0.27	7.42	m	**18.25**
Cable Terminations; for MI copper sheathed cable. Certified for installation in potentially explosive atmospheres; testing; phase marking and connection; BS 6207 Part 2 1995						
Light duty 500 Volt grade; brass gland; brass pot with earth tail; pot closure; sealing compound; conductor sleving; plastic gland shroud; identification markers						
2L 1.0	9.03	10.20	0.39	10.72	nr	**20.92**
2L 1.5	9.03	10.20	0.41	11.27	nr	**21.47**
2L 2.5	9.03	10.20	0.44	11.98	nr	**22.18**
2L 4.0	9.03	10.20	0.46	12.64	nr	**22.84**
3L 1.0	9.05	10.22	0.43	11.81	nr	**22.03**
3L 1.5	9.05	10.22	0.44	12.02	nr	**22.24**
3L 2.5	9.05	10.22	0.44	12.10	nr	**22.32**
4L 1.0	9.05	10.22	0.47	12.92	nr	**23.14**
4L 1.5	9.05	10.22	0.47	13.04	nr	**23.26**
4L 2.5	9.05	10.22	0.50	13.73	nr	**23.95**
7L 1.0	21.94	24.78	0.69	18.96	nr	**43.74**
7L 1.5	21.94	24.78	0.70	19.24	nr	**44.02**
7L 2.5	21.94	24.78	0.74	20.33	nr	**45.11**

V: ELECTRICAL SUPPLY/POWER/LIGHTING

Item	Net Price £	Material £	Labour hours	Labour £	Unit	Total rate £
Heavy duty 750 Volt grade; brass gland; brass pot with earth tail; pot closure; sealing compound; conductor sleeving; plastic gland shroud; identification markers						
1H 10	9.08	10.25	0.37	10.17	nr	20.42
1H 16	9.08	10.25	0.39	10.72	nr	20.97
1H 25	9.08	10.25	0.56	15.38	nr	25.63
1H 35	9.08	10.25	0.57	15.67	nr	25.92
1H 50	22.01	24.86	0.60	16.49	nr	41.35
1H 70	21.58	24.37	0.67	18.41	nr	42.78
1H 95	21.58	24.37	0.75	20.61	nr	44.98
1H 120	34.96	39.48	0.94	25.83	nr	65.31
1H 150	34.96	39.48	0.99	27.20	nr	66.68
1H 185	34.96	39.48	1.26	34.63	nr	74.11
1H 240	55.15	62.28	1.37	37.64	nr	99.92
2H 1.5	9.62	10.86	0.42	11.55	nr	22.41
2H 2.5	9.62	10.86	0.44	12.18	nr	23.04
2H 4	9.62	10.86	0.47	12.92	nr	23.78
2H 6	9.62	10.86	0.54	14.84	nr	25.70
2H 10	23.32	26.34	0.58	15.93	nr	42.27
2H 16	23.32	26.34	0.69	18.96	nr	45.30
2H 25	37.78	42.67	0.77	21.15	nr	63.82
3H 1.5	9.63	10.87	0.44	12.10	nr	22.97
3H 2.5	9.63	10.87	0.47	12.99	nr	23.86
3H 4	9.63	10.87	0.57	15.67	nr	26.54
3H 6	23.41	26.44	0.61	16.76	nr	43.20
3H 10	23.41	26.44	0.65	17.86	nr	44.30
3H 16	23.41	26.44	0.78	21.44	nr	47.88
3H 25	59.64	67.36	0.85	23.36	nr	90.72
4H 1.5	9.63	10.87	0.52	14.29	nr	25.16
4H 2.5	9.63	10.87	0.53	14.56	nr	25.43
4H 4	23.41	26.44	0.60	16.49	nr	42.93
4H 6	23.41	26.44	0.65	17.86	nr	44.30
4H 10	23.41	26.44	0.69	18.96	nr	45.40
4H 16	37.78	42.67	0.88	24.18	nr	66.85
4H 25	61.39	69.33	0.93	25.56	nr	94.89
7H 1.5	23.32	26.34	0.71	19.50	nr	45.84
7H 2.5	23.32	26.34	0.74	20.33	nr	46.67
12H 1.5	37.78	42.67	0.85	23.36	nr	66.03
12H 2.5	37.78	42.67	1.00	27.48	nr	70.15
19H 2.5	59.64	67.36	1.11	30.51	nr	97.87

V: ELECTRICAL SUPPLY/POWER/LIGHTING

Item	Net Price £	Material £	Labour hours	Labour £	Unit	Total rate £
V20: LV DISTRIBUTION – cont						
FIRE RATED CABLE – cont						
Cable; FP100; LOSH insulated; non sheathed fire-resistant to LPCB Approved to BS 6387 Catergory CWZ; in conduit or trunking including terminations						
450/750 volt grade; single core						
1.0 mm²	0.62	0.70	0.13	3.57	m	**4.27**
1.5 mm²	0.65	0.73	0.13	3.57	m	**4.30**
2.5 mm²	0.79	0.89	0.13	3.57	m	**4.46**
4.0 mm²	0.79	0.89	0.14	3.76	m	**4.65**
6.0 mm²	1.04	1.17	0.13	3.57	m	**4.74**
10 mm²	1.71	1.93	0.16	4.40	m	**6.33**
16 mm²	2.65	2.99	0.16	4.40	m	**7.39**
Cable; FP200; Insudite insulated; LSOH sheathed screened fire-resistant BASEC Approved to BS 7629; fixed with clips to backgrounds						
300/500 volt grade; two core						
1.5 mm²	1.94	2.19	0.20	5.50	m	**7.69**
2.5 mm²	2.36	2.66	0.20	5.50	m	**8.16**
4.0 mm²	3.57	4.03	0.23	6.32	m	**10.35**
300/500 volt grade; three core						
1.5 mm²	2.38	2.69	0.23	6.32	m	**9.01**
2.5 mm²	2.70	3.05	0.23	6.32	m	**9.37**
4.0 mm²	3.75	4.23	0.25	6.86	m	**11.09**
300/500 volt grade; four core						
1.5 mm²	2.66	3.01	0.25	6.86	m	**9.87**
2.5 mm²	3.29	3.71	0.25	6.86	m	**10.57**
4.0 mm	4.49	5.07	0.28	7.69	m	**12.76**
Terminations; including glanding-off, connection to equipment						
Two core						
1.5 mm²	0.73	0.83	0.35	9.62	nr	**10.45**
2.5 mm²	1.06	1.19	0.35	9.62	nr	**10.81**
4.0 mm²	0.94	1.06	0.35	9.62	nr	**10.68**
Three core						
1.5 mm²	0.76	0.86	0.35	9.62	nr	**10.48**
2.5 mm²	0.76	0.86	0.35	9.71	nr	**10.57**
4.0 mm²	0.97	1.09	0.35	9.62	nr	**10.71**

V: ELECTRICAL SUPPLY/POWER/LIGHTING

Item	Net Price £	Material £	Labour hours	Labour £	Unit	Total rate £
Four core						
1.5 mm²	0.99	1.12	0.35	9.62	nr	**10.74**
2.5 mm²	1.88	2.12	0.35	9.62	nr	**11.74**
4.0 mm²	2.85	3.22	0.35	9.62	nr	**12.84**
Cable; FP400; polymeric insulated; LSOH sheathed fire-resistant; armoured; with copper stranded copper conductors; BASEC Approved to BS 7846; fixed with clips to backgrounds						
600/1000 volt grade; two core						
1.5 mm²	2.64	2.98	0.20	5.50	m	**8.48**
2.5 mm²	3.06	3.46	0.20	5.50	m	**8.96**
4.0 mm²	3.39	3.83	0.21	5.77	m	**9.60**
6.0 mm²	4.17	4.71	0.22	6.04	m	**10.75**
10 mm²	4.50	5.08	0.24	6.60	m	**11.68**
16 mm²	6.85	7.73	0.25	6.86	m	**14.59**
25 mm²	9.48	10.71	0.35	9.62	m	**20.33**
600/1000 volt grade; three core						
1.5 mm²	2.93	3.30	0.20	5.50	m	**8.80**
2.5 mm²	3.42	3.87	0.21	5.77	m	**9.64**
4.0 mm²	4.03	4.55	0.22	6.04	m	**10.59**
6.0 mm²	4.27	4.82	0.22	6.04	m	**10.86**
10 mm²	5.19	5.87	0.25	6.86	m	**12.73**
16 mm²	8.49	9.59	0.26	7.15	m	**16.74**
25 mm²	10.59	11.96	0.37	10.17	m	**22.13**
600/1000 volt grade; four core						
1.5 mm²	3.31	3.73	0.21	5.77	m	**9.50**
2.5 mm²	4.00	4.52	0.22	6.04	m	**10.56**
4.0 mm²	4.41	4.98	0.22	6.04	m	**11.02**
6.0 mm²	5.55	6.27	0.23	6.32	m	**12.59**
10 mm²	6.58	7.44	0.26	7.15	m	**14.59**
16 mm²	9.80	11.07	0.26	7.15	m	**18.22**
25 mm²	14.23	16.08	0.39	10.72	m	**26.80**
Terminations; including glanding-off, connection to equipment						
Two core						
1.5 mm²	4.33	4.89	0.58	15.93	nr	**20.82**
2.5 mm²	5.23	5.91	0.58	15.93	nr	**21.84**
4.0 mm²	5.43	6.13	0.61	16.84	nr	**22.97**
6.0 mm²	5.45	6.15	0.67	18.41	nr	**24.56**
10 mm²	5.45	6.15	1.00	27.48	nr	**33.63**
16 mm²	7.67	8.66	1.11	30.51	nr	**39.17**
25 mm²	7.67	8.66	1.70	46.72	nr	**55.38**

V: ELECTRICAL SUPPLY/POWER/LIGHTING

Item	Net Price £	Material £	Labour hours	Labour £	Unit	Total rate £
V20: LV DISTRIBUTION – cont						
FIRE RATED CABLE – cont						
Terminations – cont						
Three core						
1.5 mm²	4.33	4.89	0.62	17.03	nr	**21.92**
2.5 mm²	4.65	5.25	0.62	17.03	nr	**22.28**
4.0 mm²	5.30	5.99	0.66	18.25	nr	**24.24**
6.0 mm²	5.45	6.15	0.71	19.50	nr	**25.65**
10 mm²	5.45	6.15	1.06	29.13	nr	**35.28**
16 mm²	7.67	8.66	1.19	32.69	nr	**41.35**
25 mm²	7.67	8.66	1.81	49.74	nr	**58.40**
Four core						
1.5 mm²	4.33	4.89	0.67	18.41	nr	**23.30**
2.5 mm²	5.15	5.81	0.69	18.83	nr	**24.64**
4.0 mm²	5.45	6.15	0.71	19.50	nr	**25.65**
6.0 mm²	5.45	6.15	0.76	20.88	nr	**27.03**
10 mm²	7.67	8.66	1.14	31.33	nr	**39.99**
16 mm²	8.05	9.09	1.29	35.46	nr	**44.55**
25 mm²	13.28	14.87	1.35	37.09	nr	**51.96**
Cable; Firetuff fire-resistant to BS 6387; fixed with clips to backgrounds						
Two core						
1.5 mm²	1.98	2.23	0.20	5.50	m	**7.73**
2.5 mm²	2.39	2.70	0.20	5.50	m	**8.20**
4.0 mm²	3.10	3.50	0.21	5.77	m	**9.27**
Three core						
1.5 mm²	2.42	2.73	0.20	5.50	m	**8.23**
2.5 mm²	2.73	3.08	0.21	5.77	m	**8.85**
4.0 mm²	3.95	4.46	0.22	6.04	m	**10.50**
Four core						
1.5 mm²	2.70	3.05	0.21	5.77	m	**8.82**
2.5 mm²	3.31	3.73	0.22	6.04	m	**9.77**
4.0 mm²	4.51	5.09	0.22	6.04	m	**11.13**

V: ELECTRICAL SUPPLY/POWER/LIGHTING

Item	Net Price £	Material £	Labour hours	Labour £	Unit	Total rate £
MODULAR WIRING						
Modular wiring systems; including commissioning						
Master distribution box; steel; fixed to backgrounds; 6 Port						
4.0 mm 18 core armoured home run cable	136.08	153.68	0.90	24.73	nr	**178.41**
4.0 mm 24 core armoured cable home run cable	136.08	153.68	0.95	26.10	nr	**179.78**
4.0 mm 18 core armoured home run cable & data cable	144.57	163.28	0.95	26.10	nr	**189.38**
6.0 mm 18 core armoured home run cable	136.08	153.68	1.00	27.48	nr	**181.16**
6.0 mm 24 core armoured home run cable	136.08	153.68	1.10	30.22	nr	**183.90**
6.0 mm 18 core armoured home run cable & data cable	144.57	163.28	1.10	30.22	nr	**193.50**
Master distribution box; steel; fixed to backgrounds; 9 Port						
4.0 mm 27 core armoured home run cable	170.08	192.09	1.30	35.72	nr	**227.81**
4.0 mm 27 core armoured home run cable & data cable	170.08	192.09	1.45	39.84	nr	**231.93**
6.0 mm 27 core armoured home run cable	178.59	201.69	1.45	39.84	nr	**241.53**
6.0 mm 27 core armoured home run cable & data cable	178.59	201.69	1.55	42.58	nr	**244.27**
Metal clad cable; BSEN 60439 Part 2 1993; BASEC approved						
4.0 mm 18 core	11.88	13.41	0.30	8.24	m	**21.65**
4.0 mm 24 core	18.28	20.64	0.32	8.80	m	**29.44**
4.0 mm 27 core	18.42	20.81	0.35	9.62	m	**30.43**
6.0 mm 18 core	15.82	17.87	0.32	8.80	m	**26.67**
6.0 mm 27 core	24.60	27.78	0.35	9.62	m	**37.40**
Metal clad data cable						
Single twisted pair	2.31	2.61	0.18	4.95	m	**7.56**
Twin twisted pair	3.73	4.21	0.18	4.95	m	**9.16**
Distribution cables; armoured; BSEN 60439 Part 2 1993; BASEC approved						
3 wire; 6.1 metre long	35.95	40.61	0.92	25.28	nr	**65.89**
4 wire; 6.1 metre long	43.19	48.78	0.96	26.39	nr	**75.17**

V: ELECTRICAL SUPPLY/POWER/LIGHTING

Item	Net Price £	Material £	Labour hours	Labour £	Unit	Total rate £
V20: LV DISTRIBUTION – cont						
MODULAR WIRING – cont						
Extender cables; armoured; BSEN 60439 Part 2 1993; BASEC approved						
3 Wire						
0.9 metre long	15.72	17.76	0.13	3.57	nr	**21.33**
1.5 metre long	18.43	20.82	0.23	6.32	nr	**27.14**
2.1 metre long	21.14	23.88	0.31	8.52	nr	**32.40**
2.7 metre long	23.82	26.90	0.40	10.99	nr	**37.89**
3.4 metre long	26.53	29.97	0.51	14.01	nr	**43.98**
4.6 metre long	31.98	36.12	0.69	18.96	nr	**55.08**
6.1 metre long	38.72	43.73	0.92	25.28	nr	**69.01**
7.6 metre long	45.51	51.40	1.14	31.33	nr	**82.73**
9.1 metre long	52.29	59.06	1.37	37.64	nr	**96.70**
10.7 metre long	71.28	80.50	1.61	44.24	nr	**124.74**
4 Wire						
0.9 metre long	17.11	19.33	0.14	3.86	nr	**23.19**
1.5 metre long	20.49	23.14	0.24	6.60	nr	**29.74**
2.1 metre long	23.89	26.98	0.32	8.80	nr	**35.78**
2.7 metre long	27.26	30.78	0.43	11.81	nr	**42.59**
3.4 metre long	30.64	34.61	0.51	14.01	nr	**48.62**
4.6 metre long	37.42	42.26	0.67	18.41	nr	**60.67**
6.1 metre long	45.88	51.82	0.92	25.28	nr	**77.10**
7.6 metre long	54.34	61.37	1.22	33.52	nr	**94.89**
9.1 metre long	62.80	70.93	1.46	40.12	nr	**111.05**
10.7 metre long	71.28	80.50	1.71	46.98	nr	**127.48**
3 Wire; including twisted pair						
0.9 metre long	18.71	21.13	0.13	3.57	nr	**24.70**
1.5 metre long	23.01	25.99	0.23	6.32	nr	**32.31**
2.1 metre long	27.30	30.83	0.31	8.52	nr	**39.35**
2.7 metre long	31.59	35.68	0.40	10.99	nr	**46.67**
3.4 metre long	35.87	40.51	0.51	14.01	nr	**54.52**
4.6 metre long	44.46	50.21	0.69	18.96	nr	**69.17**
6.1 metre long	55.19	62.33	0.92	25.28	nr	**87.61**
7.6 metre long	65.90	74.43	1.14	31.33	nr	**105.76**
9.1 metre long	76.63	86.55	1.37	37.64	nr	**124.19**
10.7 metre long	87.36	98.66	1.61	44.24	nr	**142.90**
Extender whip ended cables; armoured; BSEN 60439 Part 2 1993; BASEC approved						
3 wire; 3.0 metre long	21.82	24.64	0.30	8.24	nr	**32.88**
4 wire; 3.0 metre long	25.54	28.85	0.30	8.24	nr	**37.09**

V: ELECTRICAL SUPPLY/POWER/LIGHTING

Item	Net Price £	Material £	Labour hours	Labour £	Unit	Total rate £
T CONNECTORS						
3 Wire						
Snap fix	13.13	14.83	0.10	2.74	nr	**17.57**
0.3 metre flexible cable	14.89	16.82	0.10	2.74	nr	**19.56**
0.3 metre armoured cable	14.88	16.81	0.15	4.12	nr	**20.93**
0.3 metre armoured cable with twisted pair	16.59	18.74	0.15	4.12	nr	**22.86**
4 Wire						
Snap fix	13.70	15.47	0.10	2.74	nr	**18.21**
0.3 metre flexible cable	15.42	17.41	0.10	2.74	nr	**20.15**
0.3 metre armoured cable	15.42	17.41	0.18	4.95	nr	**22.36**
Splitters						
5 wire	16.30	18.41	0.20	5.50	nr	**23.91**
5 wire converter	21.02	23.74	0.20	5.50	nr	**29.24**
Switch modules						
3 wire; 6.1 metre long armoured cable	44.60	50.37	0.75	20.61	nr	**70.98**
4 wire; 6.1 metre long armoured cable	50.19	56.68	0.80	21.98	nr	**78.66**
Distribution cables; unarmoured; IEC 998 DIN/VDE 0628						
3 wire; 6.1 metre long	15.34	17.33	0.70	19.24	nr	**36.57**
4 wire; 6.1 metre long	18.33	20.71	0.75	20.61	nr	**41.32**
Extender cables; unarmoured; IEC 998 DIN/VDE 0628						
3 Wire						
0.9 metre long	10.66	12.04	0.07	1.92	nr	**13.96**
1.5 metre long	11.50	12.98	0.12	3.29	nr	**16.27**
2.1 metre long	12.37	13.97	0.17	4.67	nr	**18.64**
2.7 metre long	14.10	15.92	0.27	7.42	nr	**23.34**
3.4 metre long	15.45	17.45	0.22	6.04	nr	**23.49**
4.6 metre long	15.94	18.00	0.37	10.17	nr	**28.17**
6.1 metre long	18.10	20.44	0.49	13.46	nr	**33.90**
7.6 metre long	20.27	22.89	0.61	16.76	nr	**39.65**
9.1 metre long	22.43	25.34	0.73	20.05	nr	**45.39**
10.7 metre long	24.72	27.92	0.86	23.62	nr	**51.54**
4 Wire						
0.9 metre long	12.18	13.76	0.08	2.20	nr	**15.96**
1.5 metre long	13.27	14.98	0.14	3.86	nr	**18.84**
2.1 metre long	14.34	16.20	0.19	5.22	nr	**21.42**
2.7 metre long	14.10	15.92	0.24	6.60	nr	**22.52**
3.4 metre long	16.51	18.65	0.31	8.52	nr	**27.17**
4.6 metre long	18.85	21.29	0.41	11.27	nr	**32.56**
6.1 metre long	21.57	24.36	0.55	15.12	nr	**39.48**
7.6 metre long	24.29	27.44	0.68	18.68	nr	**46.12**
9.1 metre long	31.79	35.90	0.82	22.53	nr	**58.43**
10.7 metre long	34.67	39.16	0.96	26.39	nr	**65.55**

V: ELECTRICAL SUPPLY/POWER/LIGHTING

Item	Net Price £	Material £	Labour hours	Labour £	Unit	Total rate £
V20: LV DISTRIBUTION – cont						
T CONNECTORS – cont						
Extender cables – cont						
5 Wire						
0.9 metre long	14.36	16.22	0.09	2.47	nr	**18.69**
1.5 metre long	16.12	18.21	0.15	4.12	nr	**22.33**
2.1 metre long	17.86	20.18	0.21	5.77	nr	**25.95**
2.7 metre long	19.59	22.12	0.27	7.42	nr	**29.54**
3.4 metre long	21.35	24.11	0.34	9.34	nr	**33.45**
4.6 metre long	23.70	26.76	0.46	12.64	nr	**39.40**
6.1 metre long	29.48	33.29	0.61	16.76	nr	**50.05**
7.6 metre long	33.82	38.20	0.76	20.88	nr	**59.08**
9.1 metre long	38.18	43.12	0.91	25.00	nr	**68.12**
10.7 metre long	42.85	48.40	1.07	29.40	nr	**77.80**
Extender whip ended cables; armoured; IEC 998 DIN/VDE 0628						
3 wire; 2.5 mm; 3.0 metre long	10.86	12.26	0.30	8.24	nr	**20.50**
4 wire; 2.5 mm; 3.0 metre long	12.76	14.41	0.30	8.24	nr	**22.65**
T Connectors						
3 Wire						
5 pin; direct fix	10.98	12.40	0.10	2.74	nr	**15.14**
5 pin; 1.5 mm flexible cable; 0.3 metre long	14.04	15.86	0.15	4.12	nr	**19.98**
4 Wire						
5 pin; direct fix	12.77	14.42	0.20	5.50	nr	**19.92**
5 pin; 1.5 mm flexible cable; 0.3 metre long	15.91	17.97	0.20	5.50	nr	**23.47**
5 Wire						
5 pin; direct fix	14.55	16.43	0.20	5.50	nr	**21.93**
Splitters						
3 way; 5 pin	8.38	9.47	0.25	6.86	nr	**16.33**
Switch Modules						
3 wire	23.89	26.98	0.20	5.50	nr	**32.48**
4 wire	24.90	28.12	0.22	6.04	nr	**34.16**

V: ELECTRICAL SUPPLY/POWER/LIGHTING

Item	Net Price £	Material £	Labour hours	Labour £	Unit	Total rate £
Y62 – BUSBAR TRUNKING						
MAINS BUSBAR						
Low impedance busbar trunking; fixed to backgrounds including supports, fixings and connections/jointing to equipment						
Straight copper busbar						
1000 amp TP&N	442.49	499.75	3.41	93.71	m	**593.46**
1350 amp TP&N	543.89	614.26	3.58	98.37	m	**712.63**
2000 amp TP&N	682.17	770.44	5.00	137.39	m	**907.83**
2500 amp TP&N	1030.64	1164.00	5.90	162.13	m	**1326.13**
Extra for fittings mains bus bar						
IP54 protection						
1000 amp TP&N	30.18	34.09	2.16	59.35	m	**93.44**
1350 amp TP&N	32.43	36.63	2.61	71.72	m	**108.35**
2000 amp TP&N	44.99	50.82	3.51	96.45	m	**147.27**
2500 amp TP&N	53.41	60.32	3.96	108.81	m	**169.13**
End cover						
1000 amp TP&N	34.70	39.19	0.56	15.38	nr	**54.57**
1350 amp TP&N	36.20	40.88	0.56	15.38	nr	**56.26**
2000 amp TP&N	58.09	65.61	0.66	18.14	nr	**83.75**
2500 amp TP&N	58.82	66.43	0.66	18.14	nr	**84.57**
Edge elbow						
1000 amp TP&N	497.09	561.41	2.01	55.22	nr	**616.63**
1350 amp TP&N	559.43	631.82	2.01	55.22	nr	**687.04**
2000 amp TP&N	867.69	979.96	2.40	65.94	nr	**1045.90**
2500 amp TP&N	1139.64	1287.11	2.40	65.94	nr	**1353.05**
Flat elbow						
1000 amp TP&N	431.27	487.07	2.01	55.22	nr	**542.29**
1350 amp TP&N	467.63	528.14	2.01	55.22	nr	**583.36**
2000 amp TP&N	663.34	749.17	2.40	65.94	nr	**815.11**
2500 amp TP&N	826.14	933.03	2.40	65.94	nr	**998.97**
Offset						
1000 amp TP&N	865.99	978.05	3.00	82.44	nr	**1060.49**
1350 amp TP&N	1065.15	1202.98	3.00	82.44	nr	**1285.42**
2000 amp TP&N	1678.28	1895.45	3.50	96.18	nr	**1991.63**
2500 amp TP&N	1910.34	2157.52	3.50	96.18	nr	**2253.70**
Edge Z unit						
1000 amp TP&N	1297.24	1465.10	3.00	82.44	nr	**1547.54**
1350 amp TP&N	1660.96	1875.88	3.00	82.44	nr	**1958.32**
2000 amp TP&N	2488.84	2810.89	3.50	96.18	nr	**2907.07**
2500 amp TP&N	2842.16	3209.92	3.50	96.18	nr	**3306.10**

V: ELECTRICAL SUPPLY/POWER/LIGHTING

Item	Net Price £	Material £	Labour hours	Labour ·£	Unit	Total rate £
V20: LV DISTRIBUTION – cont						
MAINS BUSBAR – cont						
Extra for fittings mains bus bar – cont						
Flat Z unit						
1000 amp TP&N	1113.67	1257.77	3.00	82.44	nr	**1340.21**
1350 amp TP&N	1402.90	1584.43	3.00	82.44	nr	**1666.87**
2000 amp TP&N	2088.75	2359.03	3.50	96.18	nr	**2455.21**
2500 amp TP&N	2511.35	2836.30	3.50	96.18	nr	**2932.48**
Edge tee						
1000 amp TP&N	1297.24	1465.10	2.20	60.46	nr	**1525.56**
1350 amp TP&N	1660.96	1875.88	2.20	60.46	nr	**1936.34**
2000 amp TP&N	2488.84	2810.89	2.60	71.44	nr	**2882.33**
2500 amp TP&N	2843.88	3211.87	2.60	71.44	nr	**3283.31**
Tap off; TP&N integral contactor/breaker						
18 amp	211.10	238.41	0.82	22.53	nr	**260.94**
Tap off; TP&N fusable with on-load switch; excludes fuses						
32 amp	597.55	674.87	0.82	22.53	nr	**697.40**
63 amp	609.53	688.40	0.88	24.18	nr	**712.58**
100 amp	745.83	842.34	1.18	32.43	nr	**874.77**
160 amp	848.31	958.08	1.41	38.75	nr	**996.83**
250 amp	1095.13	1236.83	1.76	48.36	nr	**1285.19**
315 amp	1292.84	1460.13	2.06	56.61	nr	**1516.74**
Tap off; TP&N MCCB						
63 amp	755.08	852.78	0.88	24.18	nr	**876.96**
125 amp	904.99	1022.09	1.18	32.43	nr	**1054.52**
160 amp	988.14	1116.00	1.41	38.75	nr	**1154.75**
250 amp	1270.87	1435.31	1.76	48.36	nr	**1483.67**
400 amp	1620.06	1829.69	2.06	56.61	nr	**1886.30**
RISING MAINS BUSBAR						
Rising mains busbar; insulated supports, earth continuity bar; including couplers; fixed to backgrounds						
Straight aluminium bar						
200 amp TP&N	157.99	178.43	2.13	58.53	m	**236.96**
315 amp TP&N	176.78	199.65	2.15	59.08	m	**258.73**
400 amp TP&N	204.93	231.45	2.15	59.08	m	**290.53**
630 amp TP&N	253.44	286.23	2.47	67.88	m	**354.11**
800 amp TP&N	380.14	429.33	2.88	79.14	m	**508.47**

V: ELECTRICAL SUPPLY/POWER/LIGHTING

Item	Net Price £	Material £	Labour hours	Labour £	Unit	Total rate £
Extra for fittings rising busbar						
End feed unit						
200 amp TP&N	316.79	357.79	2.57	70.62	nr	428.41
315 amp TP&N	316.79	357.79	2.76	75.84	nr	433.63
400 amp TP&N	354.50	400.37	2.76	75.84	nr	476.21
630 amp TP&N	354.50	400.37	3.64	100.02	nr	500.39
800 amp TP&N	393.84	444.80	4.54	124.76	nr	569.56
Top feeder unit						
200 amp TP&N	316.79	357.79	2.57	70.62	nr	428.41
315 amp TP&N	316.79	357.79	2.76	75.84	nr	433.63
400 amp TP&N	354.50	400.37	2.76	75.84	nr	476.21
630 amp TP&N	354.50	400.37	3.64	100.02	nr	500.39
800 amp TP&N	393.84	444.80	4.54	124.76	nr	569.56
End cap						
200 amp TP&N	27.15	30.66	0.18	4.95	nr	35.61
315 amp TP&N	27.15	30.66	0.27	7.42	nr	38.08
400 amp TP&N	30.18	34.09	0.27	7.42	nr	41.51
630 amp TP&N	30.18	34.09	0.41	11.27	nr	45.36
800 amp TP&N	87.49	98.81	0.41	11.27	nr	110.08
Edge elbow						
200 amp TP&N	37.73	42.62	0.55	15.12	nr	57.74
315 amp TP&N	37.73	42.62	0.94	25.83	nr	68.45
400 amp TP&N	283.60	320.30	0.94	25.83	nr	346.13
630 amp TP&N	283.60	320.30	1.45	39.84	nr	360.14
800 amp TP&N	268.51	303.26	1.45	39.84	nr	343.10
Flat elbow						
200 amp TP&N	122.19	138.00	0.55	15.12	nr	153.12
315 amp TP&N	122.19	138.00	0.94	25.83	nr	163.83
400 amp TP&N	165.93	187.40	0.94	25.83	nr	213.23
630 amp TP&N	165.93	187.40	1.45	39.84	nr	227.24
800 amp TP&N	229.29	258.96	1.45	39.84	nr	298.80
Edge tee						
200 amp TP&N	171.97	194.22	0.61	16.76	nr	210.98
315 amp TP&N	171.97	194.22	1.02	28.03	nr	222.25
400 amp TP&N	241.36	272.60	1.02	28.03	nr	300.63
630 amp TP&N	241.36	272.60	1.57	43.15	nr	315.75
800 amp TP&N	343.93	388.44	1.57	43.15	nr	431.59
Flat tee						
200 amp TP&N	220.24	248.74	0.61	16.76	nr	265.50
315 amp TP&N	171.97	194.22	1.02	28.03	nr	222.25
400 amp TP&N	346.95	391.84	1.02	28.03	nr	419.87
630 amp TP&N	346.95	391.84	1.57	43.15	nr	434.99
800 amp TP&N	484.22	546.87	1.57	43.15	nr	590.02

V: ELECTRICAL SUPPLY/POWER/LIGHTING

Item	Net Price £	Material £	Labour hours	Labour £	Unit	Total rate £
V20: LV DISTRIBUTION – cont						
RISING MAINS BUSBAR – cont						
Tap-off units						
TP&N fusable with on-load switch; excludes fuses						
32 amp	161.41	182.29	0.82	22.53	nr	**204.82**
63 amp	213.34	240.94	0.88	24.18	nr	**265.12**
100 amp	286.33	323.38	1.18	32.43	nr	**355.81**
250 amp	429.49	485.06	1.41	38.75	nr	**523.81**
400 amp	625.98	706.98	2.06	56.61	nr	**763.59**
TP&N MCCB						
32 amp	164.23	185.48	0.82	22.53	nr	**208.01**
63 amp	227.38	256.81	0.88	24.18	nr	**280.99**
100 amp	364.92	412.14	1.18	32.43	nr	**444.57**
250 amp	617.56	697.47	1.41	38.75	nr	**736.22**
400 amp	1083.54	1223.74	2.06	56.61	nr	**1280.35**
LIGHTING BUSBAR						
Prewired busbar, plug-in trunking for lighting; galvanized sheet steel housing (PE); tin-plated copper conductors with tap-off units at 1 m intervals						
Straight lengths – 25 amp						
2 Pole & PE	27.40	30.95	0.16	4.40	m	**35.35**
4 Pole & PE	29.64	33.48	0.16	4.40	m	**37.88**
Straight lengths – 40 amp						
2 Pole & PE	27.09	30.60	0.16	4.40	m	**35.00**
4 Pole & PE	36.39	41.10	0.16	4.40	m	**45.50**
Components for pre-wired busbars, plug-in trunking for lighting						
Plug-in tap off units						
10 amp 4 Pole & PE; 3 m of cable	23.16	26.15	0.10	2.74	nr	**28.89**
16 amp 4 Pole & PE; 3 m of cable	24.98	28.21	0.10	2.74	nr	**30.95**
16 amp with phase selection, 2P & PE; no cable	19.31	21.81	0.10	2.74	nr	**24.55**
Trunking components						
End feed unit & cover; 4P & PE	29.28	33.07	0.23	6.32	nr	**39.39**
Centre feed unit	145.10	163.87	0.29	7.98	nr	**171.85**
Right hand, intermediate terminal box feed unit	30.56	34.52	0.23	6.32	nr	**40.84**

V: ELECTRICAL SUPPLY/POWER/LIGHTING

Item	Net Price £	Material £	Labour hours	Labour £	Unit	Total rate £
End cover (for R/hand feed)	8.46	9.56	0.06	1.65	nr	**11.21**
Flexible elbow unit	69.47	78.46	0.12	3.29	nr	**81.75**
Fixing bracket – universal	5.07	5.72	0.10	2.74	nr	**8.46**
Suspension bracket – flat	4.46	5.04	0.10	2.74	nr	**7.78**
UNDERFLOOR BUSBAR						
Prewired busbar, plug-in trunking for underfloor power distribution; galvanized sheet steel housing (PE); copper conductors with tap-off units at 300 mm intervals						
Straight lengths – 63 amp						
2 pole & PE	17.00	19.20	0.28	7.69	m	**26.89**
3 pole & PE; Clean Earth System	21.50	24.29	0.28	7.69	m	**31.98**
Components for prewired busbars, plug-in trunking for underfloor power distribution						
Plug-in tap-off units						
32 amp 2P & PE; 3 m metal flexible pre-wired conduit	28.16	31.80	0.25	6.86	nr	**38.66**
32 amp 3P & PE; clean earth; 3 m metal flexible prewired conduit	33.91	38.30	0.28	7.69	nr	**45.99**
Trunking components						
End feed unit & cover; 2P & PE	30.34	34.26	0.35	9.62	nr	**43.88**
End feed unit & cover; 3P & PE; clean earth	33.23	37.53	0.38	10.44	nr	**47.97**
End cover; 2P & PE	9.44	10.66	0.11	3.03	nr	**13.69**
End cover; 3P & PE	10.16	11.48	0.11	3.03	nr	**14.51**
Flexible interlink/corner; 2P&PE; 1 m long	52.28	59.05	0.34	9.34	nr	**68.39**
Flexible interlink/corner; 3P&PE; 1 m long	59.03	66.67	0.35	9.62	nr	**76.29**
Flexible interlink/corner; 2P&PE; 2 m long	63.80	72.05	0.37	10.17	nr	**82.22**
Flexible interlink/corner; 3P&PE; 2 m long	69.99	79.05	0.37	10.17	nr	**89.22**
Y63 – CABLE SUPPORTS						
LADDER RACK						
Light duty Galvanized Steel Ladder Rack; fixed to backgrounds; including supports, fixings and brackets; earth continuity straps						
Straight lengths						
150 mm wide ladder	15.11	17.06	0.69	18.96	m	**36.02**
300 mm wide ladder	15.74	17.78	0.88	24.18	m	**41.96**
450 mm wide ladder	16.81	18.98	1.26	34.63	m	**53.61**
600 mm wide ladder	17.90	20.22	1.51	41.49	m	**61.71**

V: ELECTRICAL SUPPLY/POWER/LIGHTING

Item	Net Price £	Material £	Labour hours	Labour £	Unit	Total rate £
V20: LV DISTRIBUTION – cont						
LADDER RACK – cont						
Extra over (cutting and jointing racking to fittings is included)						
Inside riser bend						
150 mm wide ladder	52.34	59.11	0.33	9.07	nr	**68.18**
300 mm wide ladder	54.20	61.21	0.56	15.38	nr	**76.59**
450 mm wide ladder	55.01	62.13	0.85	23.36	nr	**85.49**
600 mm wide ladder	57.01	64.38	0.99	27.20	nr	**91.58**
Outside riser bend						
300 mm wide ladder	52.34	59.11	0.43	11.81	nr	**70.92**
450 mm wide ladder	55.01	62.13	0.73	20.05	nr	**82.18**
600 mm wide ladder	57.01	64.38	0.86	23.62	nr	**88.00**
Equal tee						
300 mm wide ladder	62.82	70.95	0.62	17.03	nr	**87.98**
450 mm wide ladder	64.15	72.45	1.09	29.95	nr	**102.40**
600 mm wide ladder	67.46	76.19	1.12	30.77	nr	**106.96**
Unequal tee						
300 mm wide ladder	65.91	74.44	0.57	15.67	nr	**90.11**
450 mm wide ladder	67.70	76.46	1.17	32.15	nr	**108.61**
600 mm wide ladder	69.54	78.54	1.17	32.15	nr	**110.69**
4 way crossovers						
300 mm wide ladder	102.63	115.91	0.72	19.79	nr	**135.70**
450 mm wide ladder	110.99	125.35	1.13	31.05	nr	**156.40**
600 mm wide ladder	131.60	148.62	1.29	35.46	nr	**184.08**
Heavy duty galvanized steel ladder rack; fixed to backgrounds; including supports, fixings and brackets; earth continuity straps						
Straight lengths						
150 mm wide ladder	21.18	23.92	0.68	18.68	m	**42.60**
300 mm wide ladder	22.32	25.20	0.79	21.71	m	**46.91**
450 mm wide ladder	23.42	26.45	1.07	29.40	m	**55.85**
600 mm wide ladder	25.26	28.53	1.24	34.08	m	**62.61**
750 mm wide ladder	30.47	34.41	1.49	40.94	m	**75.35**
900 mm wide ladder	31.25	35.29	1.67	45.89	m	**81.18**

V: ELECTRICAL SUPPLY/POWER/LIGHTING

Item	Net Price £	Material £	Labour hours	Labour £	Unit	Total rate £
Extra over (cutting and jointing racking to fittings is included)						
Flat bend						
150 mm wide ladder	42.84	48.38	0.34	9.34	nr	**57.72**
300 mm wide ladder	45.40	51.28	0.39	10.72	nr	**62.00**
450 mm wide ladder	50.88	57.47	0.43	11.81	nr	**69.28**
600 mm wide ladder	59.80	67.53	0.61	16.76	nr	**84.29**
750 mm wide ladder	67.28	75.99	0.82	22.53	nr	**98.52**
900 mm wide ladder	71.64	80.91	0.97	26.65	nr	**107.56**
Inside riser bend						
150 mm wide ladder	58.86	66.47	0.27	7.42	nr	**73.89**
300 mm wide ladder	59.72	67.44	0.45	12.36	nr	**79.80**
450 mm wide ladder	64.55	72.90	0.65	17.86	nr	**90.76**
600 mm wide ladder	72.04	81.37	0.81	22.26	nr	**103.63**
750 mm wide ladder	74.39	84.02	0.92	25.28	nr	**109.30**
900 mm wide ladder	80.48	90.89	1.06	29.13	nr	**120.02**
Outside riser bend						
150 mm wide ladder	58.86	66.47	0.27	7.42	nr	**73.89**
300 mm wide ladder	59.72	67.44	0.33	9.07	nr	**76.51**
450 mm wide ladder	64.55	72.90	0.61	16.76	nr	**89.66**
600 mm wide ladder	72.04	81.37	0.76	20.88	nr	**102.25**
750 mm wide ladder	74.39	84.02	0.94	25.83	nr	**109.85**
900 mm wide ladder	80.98	91.46	1.05	28.86	nr	**120.32**
Equal tee						
150 mm wide ladder	65.02	73.43	0.37	10.17	nr	**83.60**
300 mm wide ladder	75.72	85.52	0.57	15.67	nr	**101.19**
450 mm wide ladder	82.05	92.67	0.83	22.81	nr	**115.48**
600 mm wide ladder	95.38	107.72	0.92	25.28	nr	**133.00**
750 mm wide ladder	120.70	136.32	1.13	31.05	nr	**167.37**
900 mm wide ladder	121.29	136.99	1.20	32.98	nr	**169.97**
Unequal tee						
300 mm wide ladder	72.17	81.51	0.57	15.67	nr	**97.18**
450 mm wide ladder	70.58	79.71	1.17	32.15	nr	**111.86**
600 mm wide ladder	79.33	89.60	1.17	32.15	nr	**121.75**
750 mm wide ladder	84.11	94.99	1.25	34.34	Unit	**129.33**
900 mm wide ladder	105.68	119.35	1.33	36.55	nr	**155.90**
4 way crossovers						
150 mm wide ladder	102.41	115.66	0.50	13.73	nr	**129.39**
300 mm wide ladder	107.84	121.80	0.67	18.41	nr	**140.21**
450 mm wide ladder	134.78	152.22	0.92	25.28	nr	**177.50**
600 mm wide ladder	150.67	170.17	1.07	29.40	nr	**199.57**
750 mm wide ladder	158.07	178.52	1.25	34.34	nr	**212.86**
900 mm wide ladder	187.70	211.99	1.36	37.37	nr	**249.36**

V: ELECTRICAL SUPPLY/POWER/LIGHTING

Item	Net Price £	Material £	Labour hours	Labour £	Unit	Total rate £
V20: LV DISTRIBUTION – cont						
LADDER RACK – cont						
Extra heavy duty galvanized steel ladder rack; fixed to backgrounds; including supports, fixings and brackets; earth continuity straps						
Straight lengths						
150 mm wide ladder	25.04	28.28	0.63	17.31	m	**45.59**
300 mm wide ladder	26.28	29.68	0.70	19.24	m	**48.92**
450 mm wide ladder	27.58	31.15	0.83	22.81	m	**53.96**
600 mm wide ladder	27.95	31.57	0.89	24.45	m	**56.02**
750 mm wide ladder	28.03	31.66	1.22	33.52	m	**65.18**
900 mm wide ladder	29.25	33.04	1.44	39.58	m	**72.62**
Extra over (cutting and jointing racking to fittings is included)						
Flat bend						
150 mm wide ladder	54.37	61.40	0.36	9.89	nr	**71.29**
300 mm wide ladder	56.58	63.90	0.39	10.72	nr	**74.62**
450 mm wide ladder	61.09	68.99	0.43	11.81	nr	**80.80**
600 mm wide ladder	69.11	78.05	0.61	16.76	nr	**94.81**
750 mm wide ladder	74.93	84.63	0.82	22.53	nr	**107.16**
900 mm wide ladder	80.76	91.21	0.97	26.65	nr	**117.86**
Inside riser bend						
150 mm wide ladder	64.97	73.38	0.36	9.89	nr	**83.27**
300 mm wide ladder	66.03	74.57	0.39	10.72	nr	**85.29**
450 mm wide ladder	69.79	78.82	0.43	11.81	nr	**90.63**
600 mm wide ladder	76.98	86.94	0.61	16.76	nr	**103.70**
750 mm wide ladder	78.98	89.20	0.82	22.53	nr	**111.73**
900 mm wide ladder	85.16	96.18	0.97	26.65	nr	**122.83**
Outside riser bend						
150 mm wide ladder	65.15	73.58	0.36	9.89	nr	**83.47**
300 mm wide ladder	66.03	74.57	0.39	10.72	nr	**85.29**
450 mm wide ladder	69.79	78.82	0.41	11.27	nr	**90.09**
600 mm wide ladder	76.98	86.94	0.57	15.67	nr	**102.61**
750 mm wide ladder	78.98	89.20	0.82	22.53	nr	**111.73**
900 mm wide ladder	85.16	96.18	0.93	25.56	nr	**121.74**
Equal tee						
150 mm wide ladder	78.53	88.69	0.37	10.17	nr	**98.86**
300 mm wide ladder	86.70	97.92	0.57	15.67	nr	**113.59**
450 mm wide ladder	91.83	103.71	0.83	22.81	nr	**126.52**
600 mm wide ladder	103.49	116.88	0.92	25.28	nr	**142.16**
750 mm wide ladder	114.00	128.75	1.13	31.05	nr	**159.80**
900 mm wide ladder	125.31	141.52	1.20	32.98	nr	**174.50**

V: ELECTRICAL SUPPLY/POWER/LIGHTING

Item	Net Price £	Material £	Labour hours	Labour £	Unit	Total rate £
Unequal tee						
150 mm wide ladder	80.85	91.31	0.37	10.17	nr	**101.48**
300 mm wide ladder	89.04	100.56	0.57	15.67	nr	**116.23**
450 mm wide ladder	94.18	106.37	1.17	32.15	nr	**138.52**
600 mm wide ladder	105.96	119.67	1.17	32.15	nr	**151.82**
750 mm wide ladder	125.10	398.79	1.25	34.34	nr	**433.13**
900 mm wide ladder	130.24	147.09	1.33	36.55	nr	**183.64**
4 way crossovers						
150 mm wide ladder	109.19	123.32	0.50	13.73	nr	**137.05**
300 mm wide ladder	113.48	128.16	0.67	18.41	nr	**146.57**
450 mm wide ladder	132.75	149.93	0.92	25.28	nr	**175.21**
600 mm wide ladder	147.58	166.68	1.07	29.40	nr	**196.08**
750 mm wide ladder	153.41	173.26	1.25	34.34	nr	**207.60**
900 mm wide ladder	160.30	181.04	1.36	37.37	nr	**218.41**
CABLE TRAY						
Galvanized steel cable tray to BS 729; including standard coupling joints, fixings and earth continuity straps (supports and hangers are excluded)						
Light duty tray						
Straight lengths						
50 mm wide	2.13	2.41	0.19	5.22	m	**7.63**
75 mm wide	2.06	2.33	0.23	6.32	m	**8.65**
100 mm wide	2.28	2.57	0.31	8.52	m	**11.09**
150 mm wide	2.90	3.27	0.33	9.07	m	**12.34**
225 mm wide	4.68	5.28	0.39	10.72	m	**16.00**
300 mm wide	9.32	10.53	0.49	13.46	m	**23.99**
450 mm wide	11.41	12.88	0.60	16.49	m	**29.37**
600 mm wide	15.54	17.55	0.79	21.71	m	**39.26**
750 mm wide	19.74	22.30	1.04	28.57	m	**50.87**
900 mm wide	24.59	27.77	1.26	34.63	m	**62.40**
Extra over (cutting and jointing tray to fittings is included)						
Straight reducer						
75 mm wide	6.62	7.48	0.22	6.04	nr	**13.52**
100 mm wide	7.29	8.23	0.25	6.86	nr	**15.09**
150 mm wide	9.68	10.93	0.27	7.42	nr	**18.35**
225 mm wide	12.35	13.94	0.34	9.34	nr	**23.28**
300 mm wide	15.92	17.98	0.39	10.72	nr	**28.70**
450 mm wide	17.63	19.91	0.49	13.46	nr	**33.37**
600 mm wide	21.80	24.62	0.54	14.84	nr	**39.46**
750 mm wide	28.63	32.33	0.61	16.76	nr	**49.09**
900 mm wide	33.26	37.57	0.69	18.96	nr	**56.53**

V: ELECTRICAL SUPPLY/POWER/LIGHTING

Item	Net Price £	Material £	Labour hours	Labour £	Unit	Total rate £
V20: LV DISTRIBUTION – cont						
CABLE TRAY – cont						
Extra over (cutting and jointing tray to fittings is included) – cont						
Flat bend; 90°						
50 mm wide	4.39	4.96	0.19	5.22	nr	**10.18**
75 mm wide	4.49	5.07	0.24	6.60	nr	**11.67**
100 mm wide	5.25	5.93	0.28	7.69	nr	**13.62**
150 mm wide	5.45	6.15	0.30	8.24	nr	**14.39**
225 mm wide	7.54	8.52	0.36	9.89	nr	**18.41**
300 mm wide	10.28	11.61	0.44	12.10	nr	**23.71**
450 mm wide	11.77	13.29	0.57	15.67	nr	**28.96**
600 mm wide	18.38	20.76	0.69	18.96	nr	**39.72**
750 mm wide	26.03	29.40	0.81	22.26	nr	**51.66**
900 mm wide	38.28	43.24	0.94	25.83	nr	**69.07**
Adjustable riser						
50 mm wide	7.93	8.96	0.26	7.15	nr	**16.11**
75 mm wide	8.67	9.79	0.29	7.98	nr	**17.77**
100 mm wide	9.56	10.80	0.32	8.80	nr	**19.60**
150 mm wide	12.18	13.76	0.36	9.89	nr	**23.65**
225 mm wide	14.97	16.91	0.44	12.10	nr	**29.01**
300 mm wide	18.97	21.42	0.52	14.29	nr	**35.71**
450 mm wide	17.21	19.44	0.66	18.14	nr	**37.58**
600 mm wide	21.77	24.58	0.79	21.71	nr	**46.29**
750 mm wide	28.02	31.65	1.03	28.30	nr	**59.95**
900 mm wide	32.97	37.24	1.10	30.22	nr	**67.46**
Inside riser; 90°						
50 mm wide	6.37	7.19	0.28	7.69	nr	**14.88**
75 mm wide	6.71	7.58	0.31	8.52	nr	**16.10**
100 mm wide	7.24	15.76	0.33	9.07	nr	**24.83**
150 mm wide	9.56	10.80	0.37	10.17	nr	**20.97**
225 mm wide	12.08	13.65	0.44	12.10	nr	**25.75**
300 mm wide	16.40	18.52	0.53	14.56	nr	**33.08**
450 mm wide	16.28	18.39	0.67	18.41	nr	**36.80**
600 mm wide	21.72	24.53	0.79	21.71	nr	**46.24**
750 mm wide	26.79	30.25	0.95	26.10	nr	**56.35**
900 mm wide	31.83	35.94	1.11	30.51	nr	**66.45**
Outside riser; 90°						
50 mm wide	6.37	7.19	0.28	7.69	nr	**14.88**
75 mm wide	6.71	7.58	0.31	8.52	nr	**16.10**
100 mm wide	7.24	8.18	0.33	9.07	nr	**17.25**
150 mm wide	9.56	10.80	0.37	10.17	nr	**20.97**
225 mm wide	12.08	13.65	0.44	12.10	nr	**25.75**
300 mm wide	16.40	18.52	0.53	14.56	nr	**33.08**
450 mm wide	16.28	18.39	0.67	18.41	nr	**36.80**

V: ELECTRICAL SUPPLY/POWER/LIGHTING

Item	Net Price £	Material £	Labour hours	Labour £	Unit	Total rate £
600 mm wide	21.72	24.53	0.79	21.71	nr	**46.24**
750 mm wide	26.79	30.25	0.95	26.10	nr	**56.35**
900 mm wide	50.89	57.48	1.11	30.51	nr	**87.99**
Equal tee						
50 mm wide	6.33	7.15	0.30	8.24	nr	**15.39**
75 mm wide	6.56	7.41	0.31	8.52	nr	**15.93**
100 mm wide	7.13	8.05	0.35	9.62	nr	**17.67**
150 mm wide	7.93	8.96	0.36	9.89	nr	**18.85**
225 mm wide	10.98	12.40	0.44	12.10	nr	**24.50**
300 mm wide	13.74	15.51	0.54	14.84	nr	**30.35**
450 mm wide	17.64	19.92	0.71	19.50	nr	**39.42**
600 mm wide	24.25	27.39	0.92	25.28	nr	**52.67**
750 mm wide	36.35	41.06	1.19	32.69	nr	**73.75**
900 mm wide	51.70	58.38	1.44	39.58	nr	**97.96**
Unequal tee						
75 mm wide	6.56	7.41	0.38	10.44	nr	**17.85**
100 mm wide	7.13	8.05	0.39	10.72	nr	**18.77**
150 mm wide	7.93	8.96	0.43	11.81	nr	**20.77**
225 mm wide	10.98	12.40	0.50	13.73	nr	**26.13**
300 mm wide	13.74	15.51	0.63	17.31	nr	**32.82**
450 mm wide	17.64	19.92	0.80	21.98	nr	**41.90**
600 mm wide	24.25	27.39	1.02	28.03	nr	**55.42**
750 mm wide	36.35	41.06	1.12	30.77	nr	**71.83**
900 mm wide	51.70	58.38	1.35	37.10	nr	**95.48**
4 way crossovers						
50 mm wide	8.80	9.93	0.38	10.44	nr	**20.37**
75 mm wide	8.97	10.13	0.40	10.99	nr	**21.12**
100 mm wide	9.84	11.12	0.40	10.99	nr	**22.11**
150 mm wide	11.04	12.46	0.44	12.10	nr	**24.56**
225 mm wide	14.71	16.62	0.53	14.56	nr	**31.18**
300 mm wide	20.91	23.61	0.64	17.58	nr	**41.19**
450 mm wide	24.19	27.32	0.84	23.08	nr	**50.40**
600 mm wide	32.33	36.52	1.03	28.30	nr	**64.82**
750 mm wide	48.67	54.97	1.13	31.05	nr	**86.02**
900 mm wide	70.93	80.11	1.36	37.37	nr	**117.48**
Medium duty tray with return flange						
Straight lengths						
75 mm wide	3.81	4.30	0.33	9.07	m	**13.37**
100 mm wide	4.08	4.61	0.35	9.62	m	**14.23**
150 mm wide	5.08	5.73	0.39	10.72	m	**16.45**
225 mm wide	6.06	6.84	0.45	12.36	m	**19.20**
300 mm wide	8.42	9.51	0.57	15.67	m	**25.18**
450 mm wide	12.90	14.57	0.69	18.96	m	**33.53**
600 mm wide	18.03	20.36	0.91	25.00	m	**45.36**

V: ELECTRICAL SUPPLY/POWER/LIGHTING

Item	Net Price £	Material £	Labour hours	Labour £	Unit	Total rate £
V20: LV DISTRIBUTION – cont						
CABLE TRAY – cont						
Extra over (cutting and jointing tray to fittings is included)						
Straight reducer						
100 mm wide	12.13	13.70	0.25	6.86	nr	**20.56**
150 mm wide	13.25	14.96	0.27	7.42	nr	**22.38**
225 mm wide	15.27	17.25	0.34	9.34	nr	**26.59**
300 mm wide	17.68	19.97	0.39	10.72	nr	**30.69**
450 mm wide	22.84	25.80	0.49	13.46	nr	**39.26**
600 mm wide	28.30	31.97	0.54	14.84	nr	**46.81**
Flat bend; 90°						
75 mm wide	16.17	18.26	0.24	6.60	nr	**24.86**
100 mm wide	18.00	20.33	0.28	7.69	nr	**28.02**
150 mm wide	19.10	21.57	0.30	8.24	nr	**29.81**
225 mm wide	22.05	24.90	0.36	9.89	nr	**34.79**
300 mm wide	27.19	30.71	0.44	12.10	nr	**42.81**
450 mm wide	40.85	46.13	0.57	15.67	nr	**61.80**
600 mm wide	49.39	55.78	0.69	18.96	nr	**74.74**
Adjustable bend						
75 mm wide	18.49	20.88	0.29	7.98	nr	**28.86**
100 mm wide	20.12	22.73	0.32	8.80	nr	**31.53**
150 mm wide	23.01	25.99	0.36	9.89	nr	**35.88**
225 mm wide	25.79	29.13	0.44	12.10	nr	**41.23**
300 mm wide	29.14	32.63	0.52	14.29	nr	**46.92**
Adjustable riser						
75 mm wide	16.49	18.63	0.29	7.98	nr	**26.61**
100 mm wide	16.77	18.94	0.32	8.80	nr	**27.74**
150 mm wide	18.48	20.87	0.36	9.89	nr	**30.76**
225 mm wide	19.60	22.13	0.44	12.10	nr	**34.23**
300 mm wide	20.90	23.60	0.52	14.29	nr	**37.89**
450 mm wide	27.41	30.96	0.66	18.14	nr	**49.10**
600 mm wide	33.62	37.97	0.79	21.71	nr	**59.68**
Inside riser; 90°						
75 mm wide	9.91	11.19	0.31	8.52	nr	**19.71**
100 mm wide	10.00	11.29	0.33	9.07	nr	**20.36**
150 mm wide	11.39	12.86	0.37	10.17	nr	**23.03**
225 mm wide	13.92	15.72	0.44	12.10	nr	**27.82**
300 mm wide	16.91	19.09	0.53	14.56	nr	**33.65**
450 mm wide	24.41	27.57	0.67	18.41	nr	**45.98**
600 mm wide	39.17	44.24	0.79	21.71	nr	**65.95**

V: ELECTRICAL SUPPLY/POWER/LIGHTING

Item	Net Price £	Material £	Labour hours	Labour £	Unit	Total rate £
Outside riser; 90°						
75 mm wide	9.91	11.19	0.31	8.52	nr	**19.71**
100 mm wide	10.00	11.29	0.33	9.07	nr	**20.36**
150 mm wide	11.39	12.86	0.37	10.17	nr	**23.03**
225 mm wide	13.92	15.72	0.44	12.10	nr	**27.82**
300 mm wide	16.91	19.09	0.53	14.56	nr	**33.65**
450 mm wide	24.41	27.57	0.67	18.41	nr	**45.98**
600 mm wide	39.17	44.24	0.79	21.71	nr	**65.95**
Equal tee						
75 mm wide	22.02	24.87	0.31	8.52	nr	**33.39**
100 mm wide	23.48	26.52	0.35	9.62	nr	**36.14**
150 mm wide	25.28	28.55	0.36	9.89	nr	**38.44**
225 mm wide	27.47	31.03	0.74	20.33	nr	**51.36**
300 mm wide	33.34	37.66	0.54	14.84	nr	**52.50**
450 mm wide	43.92	49.60	0.71	19.50	nr	**69.10**
600 mm wide	63.10	71.27	0.92	25.28	nr	**96.55**
Unequal tee						
100 mm wide	23.48	26.52	0.39	10.72	nr	**37.24**
150 mm wide	23.48	26.52	0.43	11.81	nr	**38.33**
225 mm wide	27.47	31.03	0.50	13.73	nr	**44.76**
300 mm wide	33.34	37.66	0.63	17.31	nr	**54.97**
450 mm wide	43.92	49.60	0.80	21.98	nr	**71.58**
600 mm wide	63.10	71.27	1.02	28.03	nr	**99.30**
4 way crossovers						
75 mm wide	31.23	35.27	0.40	10.99	nr	**46.26**
100 mm wide	33.66	38.02	0.40	10.99	nr	**49.01**
150 mm wide	29.25	33.04	0.44	12.10	nr	**45.14**
225 mm wide	42.29	47.77	0.53	14.56	nr	**62.33**
300 mm wide	49.21	55.58	0.64	17.58	nr	**73.16**
450 mm wide	62.63	70.74	0.84	23.08	nr	**93.82**
600 mm wide	91.30	103.11	1.03	28.30	nr	**131.41**
Heavy duty tray with return flange						
Straight lengths						
75 mm	7.79	8.80	0.34	9.34	m	**18.14**
100 mm	8.30	9.37	0.36	9.89	m	**19.26**
150 mm	9.60	10.84	0.40	10.99	m	**21.83**
225 mm	10.76	12.15	0.46	12.64	m	**24.79**
300 mm	13.12	14.82	0.58	15.93	m	**30.75**
450 mm	19.83	22.40	0.70	19.24	m	**41.64**
600 mm	24.04	27.15	0.92	25.28	m	**52.43**
750 mm	30.32	34.24	1.01	27.75	m	**61.99**
900 mm	33.54	37.88	1.14	31.33	m	**69.21**

V: ELECTRICAL SUPPLY/POWER/LIGHTING

Item	Net Price £	Material £	Labour hours	Labour £	Unit	Total rate £
V20: LV DISTRIBUTION – cont						
CABLE TRAY – cont						
Extra over (cutting and jointing tray to fittings is included)						
Straight reducer						
100 mm wide	19.02	21.48	0.25	6.86	nr	**28.34**
150 mm wide	19.75	22.31	0.27	7.42	nr	**29.73**
225 mm wide	22.43	25.34	0.34	9.34	nr	**34.68**
300 mm wide	25.00	28.23	0.39	10.72	nr	**38.95**
450 mm wide	35.76	40.39	0.49	13.46	nr	**53.85**
600 mm wide	39.76	44.90	0.54	14.84	nr	**59.74**
750 mm wide	50.43	56.96	0.60	16.49	nr	**73.45**
900 mm wide	55.31	62.46	0.66	18.14	nr	**80.60**
Flat bend; 90°						
75 mm wide	21.28	24.03	0.24	6.60	nr	**30.63**
100 mm wide	24.19	27.32	0.28	7.69	nr	**35.01**
150 mm wide	25.78	29.11	0.30	8.24	nr	**37.35**
225 mm wide	29.16	32.94	0.36	9.89	nr	**42.83**
300 mm wide	33.22	37.52	0.44	12.10	nr	**49.62**
450 mm wide	48.71	55.01	0.57	15.67	nr	**70.68**
600 mm wide	65.10	73.52	0.69	18.96	nr	**92.48**
750 mm wide	87.24	98.53	0.83	22.81	nr	**121.34**
900 mm wide	98.93	111.73	1.01	27.75	nr	**139.48**
Adjustable bend						
75 mm wide	22.13	24.99	0.29	7.98	nr	**32.97**
100 mm wide	24.47	27.63	0.32	8.80	nr	**36.43**
150 mm wide	25.85	29.19	0.36	9.89	nr	**39.08**
225 mm wide	28.45	32.13	0.44	12.10	nr	**44.23**
300 mm wide	33.75	38.12	0.52	14.29	nr	**52.41**
Adjustable riser						
75 mm wide	19.50	22.02	0.29	7.98	nr	**30.00**
100 mm wide	20.27	22.89	0.32	8.80	nr	**31.69**
150 mm wide	22.45	25.36	0.36	9.89	nr	**35.25**
225 mm wide	24.04	27.15	0.44	12.10	nr	**39.25**
300 mm wide	26.04	29.41	0.52	14.29	nr	**43.70**
450 mm wide	31.69	35.79	0.66	18.14	nr	**53.93**
600 mm wide	37.76	42.65	0.79	21.71	nr	**64.36**
750 mm wide	45.72	51.63	1.03	28.30	nr	**79.93**
900 mm wide	53.09	59.96	1.10	30.22	nr	**90.18**
Inside riser; 90°						
75 mm wide	16.31	18.42	0.31	8.52	nr	**26.94**
100 mm wide	16.52	18.66	0.33	9.07	nr	**27.73**
150 mm wide	17.90	20.22	0.37	10.17	nr	**30.39**

V: ELECTRICAL SUPPLY/POWER/LIGHTING

Item	Net Price £	Material £	Labour hours	Labour £	Unit	Total rate £
225 mm wide	18.71	21.13	0.44	12.10	nr	**33.23**
300 mm wide	19.30	21.80	0.53	14.56	nr	**36.36**
450 mm wide	33.11	37.39	0.67	18.41	nr	**55.80**
600 mm wide	40.62	45.88	0.79	21.71	nr	**67.59**
750 mm wide	50.27	56.77	0.95	26.10	nr	**82.87**
900 mm wide	59.27	66.94	1.11	30.51	nr	**97.45**
Outside riser; 90°						
75 mm wide	16.31	18.42	0.31	8.52	nr	**26.94**
100 mm wide	16.52	18.66	0.33	9.07	nr	**27.73**
150 mm wide	17.90	20.22	0.37	10.17	nr	**30.39**
225 mm wide	18.71	21.13	0.44	12.10	nr	**33.23**
300 mm wide	19.30	21.80	0.53	14.56	nr	**36.36**
450 mm wide	33.11	37.39	0.67	18.41	nr	**55.80**
600 mm wide	40.62	45.88	0.79	21.71	nr	**67.59**
750 mm wide	50.27	56.77	0.95	26.10	nr	**82.87**
900 mm wide	59.27	66.94	1.11	30.51	nr	**97.45**
Equal tee						
75 mm wide	28.41	32.09	0.31	8.52	nr	**40.61**
100 mm wide	31.76	35.87	0.35	9.62	nr	**45.49**
150 mm wide	34.78	39.28	0.36	9.89	nr	**49.17**
225 mm wide	40.02	45.20	0.44	12.10	nr	**57.30**
300 mm wide	43.27	48.87	0.54	14.84	nr	**63.71**
450 mm wide	62.83	70.96	0.71	19.50	nr	**90.46**
600 mm wide	82.96	93.70	0.92	25.28	nr	**118.98**
750 mm wide	109.41	123.56	1.19	32.69	nr	**156.25**
900 mm wide	124.94	141.11	1.45	39.84	nr	**180.95**
Unequal tee						
75 mm wide	28.41	32.09	0.38	10.44	nr	**42.53**
100 mm wide	31.73	35.83	0.39	10.72	nr	**46.55**
150 mm wide	34.78	39.28	0.43	11.81	nr	**51.09**
225 mm wide	40.02	45.20	0.50	13.73	nr	**58.93**
300 mm wide	43.27	48.87	0.63	17.31	nr	**66.18**
450 mm wide	62.83	70.96	0.80	21.98	nr	**92.94**
600 mm wide	82.96	93.70	1.02	28.03	nr	**121.73**
750 mm wide	109.41	123.56	1.12	30.77	nr	**154.33**
900 mm wide	132.34	149.46	1.35	37.10	nr	**186.56**
4 way crossovers						
75 mm wide	28.83	32.56	0.40	10.99	nr	**43.55**
100 mm wide	41.32	46.67	0.40	10.99	nr	**57.66**
150 mm wide	41.32	46.67	0.44	12.10	nr	**58.77**
225 mm wide	57.46	64.89	0.53	14.56	nr	**79.45**
300 mm wide	63.36	71.56	0.64	17.58	nr	**89.14**
450 mm wide	91.13	102.92	0.84	23.08	nr	**126.00**
600 mm wide	123.86	139.88	1.03	28.30	nr	**168.18**
750 mm wide	151.07	170.62	1.13	31.05	nr	**201.67**
900 mm wide	182.23	205.81	1.36	37.37	nr	**243.18**

V: ELECTRICAL SUPPLY/POWER/LIGHTING

Item	Net Price £	Material £	Labour hours	Labour £	Unit	Total rate £
V20: LV DISTRIBUTION – cont						
CABLE TRAY – cont						
GRP cable tray including standard coupling joints and fixings (supports and hangers excluded)						
Tray						
100 mm wide	21.40	24.17	0.34	9.34	m	**33.51**
200 mm wide	27.40	30.95	0.39	10.72	m	**41.67**
400 mm wide	42.97	48.53	0.53	14.56	m	**63.09**
Cover						
100 mm wide	12.06	13.62	0.10	2.74	m	**16.36**
200 mm wide	15.97	18.03	0.11	3.03	m	**21.06**
400 mm wide	27.27	30.79	0.14	3.86	m	**34.65**
Extra for (cutting and jointing to fittings included)						
Reducer						
200 mm wide	56.87	64.23	0.23	6.32	nr	**70.55**
400 mm wide	74.30	83.92	0.30	8.24	nr	**92.16**
Reducer cover						
200 mm wide	35.59	40.20	0.25	6.86	nr	**47.06**
400 mm wide	52.02	58.75	0.28	7.69	nr	**66.44**
Bend						
100 mm wide	47.56	53.71	0.34	9.34	nr	**63.05**
200 mm wide	55.20	62.34	0.40	10.99	nr	**73.33**
400 mm wide	70.94	80.12	0.32	8.80	nr	**88.92**
Bend cover						
100 mm wide	23.63	26.68	0.10	2.74	nr	**29.42**
200 mm wide	31.56	35.64	0.10	2.74	nr	**38.38**
400 mm wide	43.61	49.26	0.13	3.57	nr	**52.83**
Tee						
100 mm wide	60.53	68.36	0.37	10.17	nr	**78.53**
200 mm wide	66.73	75.37	0.43	11.81	nr	**87.18**
400 mm wide	82.78	93.49	0.56	15.38	nr	**108.87**
Tee cover						
100 mm wide	30.49	34.44	0.27	7.42	nr	**41.86**
200 mm wide	36.71	41.46	0.31	8.52	nr	**49.98**
400 mm wide	51.77	58.47	0.37	10.17	nr	**68.64**

V: ELECTRICAL SUPPLY/POWER/LIGHTING

Item	Net Price £	Material £	Labour hours	Labour £	Unit	Total rate £
BASKET TRAY						
Mild steel cable basket; zinc plated including standard coupling joints, fixings and earth continuity straps (supports and hangers are excluded)						
Basket 54 mm deep						
100 mm wide	3.14	3.55	0.22	6.04	m	**9.59**
150 mm wide	3.50	3.96	0.25	6.86	m	**10.82**
200 mm wide	3.83	4.32	0.28	7.69	m	**12.01**
300 mm wide	4.49	5.07	0.34	9.34	m	**14.41**
450 mm wide	5.44	6.14	0.44	12.10	m	**18.24**
600 mm wide	6.66	7.52	0.70	19.24	m	**26.76**
Extra for (cutting and jointing to fittings is included)						
Reducer						
150 mm wide	10.27	11.60	0.25	6.86	nr	**18.46**
200 mm wide	12.05	13.61	0.28	7.69	nr	**21.30**
300 mm wide	12.25	13.83	0.38	10.44	nr	**24.27**
450 mm wide	13.44	15.18	0.48	13.19	nr	**28.37**
600 mm wide	16.75	18.92	0.48	13.19	nr	**32.11**
Bend						
100 mm wide	9.81	11.08	0.23	6.32	nr	**17.40**
150 mm wide	11.49	12.97	0.26	7.15	nr	**20.12**
200 mm wide	11.69	13.20	0.30	8.24	nr	**21.44**
300 mm wide	12.80	14.45	0.35	9.62	nr	**24.07**
450 mm wide	15.94	18.00	0.50	13.73	nr	**31.73**
600 mm wide	18.76	21.19	0.58	15.93	nr	**37.12**
Tee						
100 mm wide	12.33	13.92	0.28	7.69	nr	**21.61**
150 mm wide	13.08	14.77	0.30	8.24	nr	**23.01**
200 mm wide	13.35	15.08	0.33	9.07	nr	**24.15**
300 mm wide	16.79	18.96	0.39	10.72	nr	**29.68**
450 mm wide	21.99	24.84	0.56	15.38	nr	**40.22**
600 mm wide	22.74	25.68	0.65	17.86	nr	**43.54**
Crossover						
100 mm wide	16.82	18.99	0.40	10.99	nr	**29.98**
150 mm wide	17.11	19.33	0.42	11.55	nr	**30.88**
200 mm wide	18.58	20.98	0.46	12.64	nr	**33.62**
300 mm wide	21.11	23.84	0.51	14.01	nr	**37.85**
450 mm wide	24.74	27.94	0.74	20.33	nr	**48.27**
600 mm wide	25.55	28.86	0.82	22.53	nr	**51.39**

V: ELECTRICAL SUPPLY/POWER/LIGHTING

Item	Net Price £	Material £	Labour hours	Labour £	Unit	Total rate £
V20: LV DISTRIBUTION – cont						
BASKET TRAY – cont						
Mild steel cable basket; expoxy coated including standard coupling joints, fixings and earth continuity straps (supports and hangers are excluded)						
Basket 54 mm deep						
100 mm wide	7.03	7.94	0.22	6.04	m	**13.98**
150 mm wide	7.95	8.98	0.25	6.86	m	**15.84**
200 mm wide	8.85	10.00	0.28	7.69	m	**17.69**
300 mm wide	10.05	11.35	0.34	9.34	m	**20.69**
450 mm wide	12.41	14.01	0.44	12.10	m	**26.11**
600 mm wide	14.13	15.96	0.70	19.24	m	**35.20**
Extra for (cutting and jointing to fittings is included)						
Reducer						
150 mm wide	14.70	16.61	0.28	7.69	nr	**24.30**
200 mm wide	16.47	18.60	0.28	7.69	nr	**26.29**
300 mm wide	17.57	19.84	0.38	10.44	nr	**30.28**
450 mm wide	20.54	23.19	0.48	13.19	nr	**36.38**
600 mm wide	25.60	28.92	0.48	13.19	nr	**42.11**
Bend						
100 mm wide	14.25	16.10	0.23	6.32	nr	**22.42**
150 mm wide	15.93	17.99	0.26	7.15	nr	**25.14**
200 mm wide	16.13	18.22	0.30	8.24	nr	**26.46**
300 mm wide	18.14	20.49	0.35	9.62	nr	**30.11**
450 mm wide	23.02	26.00	0.50	13.73	nr	**39.73**
600 mm wide	27.64	31.21	0.58	15.93	nr	**47.14**
Tee						
100 mm wide	16.76	18.93	0.28	7.69	nr	**26.62**
150 mm wide	17.52	19.79	0.30	8.24	nr	**28.03**
200 mm wide	17.78	20.08	0.33	9.07	nr	**29.15**
300 mm wide	22.11	24.97	0.39	10.72	nr	**35.69**
450 mm wide	28.18	31.82	0.56	15.38	nr	**47.20**
600 mm wide	31.60	35.69	0.65	17.86	nr	**53.55**
Crossover						
100 mm wide	21.25	24.00	0.40	10.99	nr	**34.99**
150 mm wide	21.49	24.27	0.42	11.55	nr	**35.82**
200 mm wide	23.02	26.00	0.46	12.64	nr	**38.64**
300 mm wide	26.43	29.85	0.51	14.01	nr	**43.86**
450 mm wide	31.83	35.94	0.74	20.33	nr	**56.27**
600 mm wide	34.39	38.84	0.82	22.53	nr	**61.37**

V: ELECTRICAL SUPPLY/POWER/LIGHTING

Item	Net Price £	Material £	Labour hours	Labour £	Unit	Total rate £
Y71 – LV SWITCHGEAR AND DISTRIBUTION BOARDS						
LV switchboard components, factory-assembled modular construction to IP41; form 4, type 5; 2400 mm high, with front and rear access; top cable entry/exit; includes delivery, offloading, positioning and commissioning (hence separate labour costs are not detailed below); excludes cabling and cable terminations						
Air circuit breakers (ACBs) to BSEN 60947-2, withdrawable type, fitted with adjustable instantaneous and overload protection. Includes enclosure and copper links, assembled into LV switchboard						
ACB-100 kA fault rated						
4 pole, 6300 A (1600 mm wide)	31325.87	35379.28	–	–	nr	**35379.28**
4 pole, 5000 A (1600 mm wide)	23671.14	26734.07	–	–	nr	**26734.07**
4 pole, 4000 A (1600 mm wide)	19858.28	22427.84	–	–	nr	**22427.84**
4 pole, 3200 A (1600 mm wide)	13406.34	15141.05	–	–	nr	**15141.05**
4 pole, 2500 A (1600 mm wide)	10419.44	11767.66	–	–	nr	**11767.66**
4 pole, 2000 A (1600 mm wide)	8122.83	9173.88	–	–	nr	**9173.88**
4 pole, 1600 A (1600 mm wide)	6650.43	7510.96	–	–	nr	**7510.96**
4 pole, 1250 A (1600 mm wide)	5886.10	6647.73	–	–	nr	**6647.73**
4 pole, 1000 A (1600 mm wide)	5814.98	6567.41	–	–	nr	**6567.41**
4 pole, 800 A (1600 mm wide)	5716.62	6456.32	–	–	nr	**6456.32**
3 pole, 6300 A (1600 mm wide)	28384.02	32056.77	–	–	nr	**32056.77**
3 pole, 5000 A (1600 mm wide)	19905.69	22481.39	–	–	nr	**22481.39**
3 pole, 4000 A (1600 mm wide)	16300.20	18409.37	–	–	nr	**18409.37**
3 pole, 3200 A (1600 mm wide)	11193.29	12641.65	–	–	nr	**12641.65**
3 pole, 2500 A (1600 mm wide)	8961.84	10121.46	–	–	nr	**10121.46**
3 pole, 2000 A (1600 mm wide)	6748.20	7621.38	–	–	nr	**7621.38**
3 pole, 1600 A (1600 mm wide)	5486.13	6196.01	–	–	nr	**6196.01**
3 pole, 1250 A (1600 mm wide)	5100.98	5761.02	–	–	nr	**5761.02**
3 pole, 1000 A (1600 mm wide)	4991.38	5637.24	–	–	nr	**5637.24**
3 pole, 800 A (1600 mm wide)	4895.97	5529.48	–	–	nr	**5529.48**

V: ELECTRICAL SUPPLY/POWER/LIGHTING

Item	Net Price £	Material £	Labour hours	Labour £	Unit	Total rate £
V20: LV DISTRIBUTION – cont						
Y71 – LV SWITCHGEAR AND DISTRIBUTION BOARDS – cont						
Air circuit breakers (ACBs) to BSEN 60947-2, withdrawable type, fitted with adjustable instantaneous and overload protection. – cont						
ACB-65 kA fault rated						
4 pole, 4000 A (1600 mm wide)	17426.00	19680.84	–	–	nr	**19680.84**
4 pole, 3200 A (1600 mm wide)	11788.77	13314.18	–	–	nr	**13314.18**
4 pole, 2500 A (1600 mm wide)	9267.00	10466.11	–	–	nr	**10466.11**
4 pole, 2000 A (1600 mm wide)	7370.33	8324.02	–	–	nr	**8324.02**
4 pole, 1600 A (1600 mm wide)	6081.60	6868.53	–	–	nr	**6868.53**
4 pole, 1250 A (1600 mm wide)	5429.84	6132.43	–	–	nr	**6132.43**
4 pole, 1000 A (1600 mm wide)	5326.15	6015.33	–	–	nr	**6015.33**
4 pole, 800 A (1600 mm wide)	5230.76	5907.60	–	–	nr	**5907.60**
3 pole, 4000 A (1600 mm wide)	14762.61	16672.82	–	–	nr	**16672.82**
3 pole, 3200 A (1600 mm wide)	10011.20	11306.60	–	–	nr	**11306.60**
3 pole, 2500 A (1600 mm wide)	7785.70	8793.13	–	–	nr	**8793.13**
3 pole, 2000 A (1600 mm wide)	6188.27	6989.00	–	–	nr	**6989.00**
3 pole, 1600 A (1600 mm wide)	5062.49	5717.55	–	–	nr	**5717.55**
3 pole, 1250 A (1600 mm wide)	4712.90	5322.73	–	–	nr	**5322.73**
3 pole, 1000 A (1600 mm wide)	4606.25	5202.28	–	–	nr	**5202.28**
3 pole, 800 A (1600 mm wide)	4510.85	5094.53	–	–	nr	**5094.53**
Extra for						
Cable box (one per ACB for form 4, types 6 & 7)	292.67	330.54	–	–	nr	**330.54**
Opening Coil	87.31	98.60	–	–	nr	**98.60**
Closing Coil	87.31	98.60	–	–	nr	**98.60**
Undervoltage Release	171.51	193.70	–	–	nr	**193.70**
Motor Operator	530.15	598.75	–	–	nr	**598.75**
Mechnical Interlock (per ACB)	498.97	563.53	–	–	nr	**563.53**
ACB Fortress/Castell Adaptor Kit (one per ACB)	124.74	140.88	–	–	nr	**140.88**
Fortress/Castell ACB Lock (one per ACB)	249.47	281.75	–	–	nr	**281.75**
Fortress/Castell Key	62.37	70.44	–	–	nr	**70.44**
Moulded case circuit breakers (MCCBs) to BS EN 60947-2; plug-in type, fitted with electronic trip unit. Includes metalwork section and copper links, assembled into LV switchboard						
MCCB-150 kA fault rated						
4 Pole, 630 A (800 mm wide, 600 mm high)	2953.58	3335.76	–	–	nr	**3335.76**
4 Pole, 400 A (800 mm wide, 400 mm high)	2113.07	2386.49	–	–	nr	**2386.49**
4 Pole, 250 A (800 mm wide, 400 mm high)	1889.71	2134.23	–	–	nr	**2134.23**

V: ELECTRICAL SUPPLY/POWER/LIGHTING

Item	Net Price £	Material £	Labour hours	Labour £	Unit	Total rate £
4 Pole, 160 A (800 mm wide, 300 mm high)	1284.29	1450.47	–	–	nr	**1450.47**
4 Pole, 100 A (800 mm wide, 200 mm high)	1069.76	1208.18	–	–	nr	**1208.18**
3 Pole, 630 A (800 mm wide, 600 mm high)	2636.20	2977.31	–	–	nr	**2977.31**
3 Pole, 400 A (800 mm wide, 400 mm high)	2013.13	2273.62	–	–	nr	**2273.62**
3 Pole, 250 A (800 mm wide, 400 mm high)	1622.27	1832.19	–	–	nr	**1832.19**
3 Pole, 160 A (800 mm wide, 300 mm high)	1187.32	1340.95	–	–	nr	**1340.95**
3 Pole, 100 A (800 mm wide, 200 mm high)	949.26	1072.09	–	–	nr	**1072.09**
MCCB-70 kA fault rated						
4 Pole, 630 A (800 mm wide, 600 mm high)	2530.39	2857.81	–	–	nr	**2857.81**
4 Pole, 400 A (800 mm wide, 400 mm high)	1739.82	1964.95	–	–	nr	**1964.95**
4 Pole, 250 A (800 mm wide, 400 mm high)	1495.89	1689.45	–	–	nr	**1689.45**
4 Pole, 160 A (800 mm wide, 300 mm high)	1122.65	1267.91	–	–	nr	**1267.91**
4 Pole, 100 A (800 mm wide, 200 mm high)	846.40	955.92	–	–	nr	**955.92**
3 Pole, 630 A (800 mm wide, 600 mm high)	2215.93	2502.66	–	–	nr	**2502.66**
3 Pole, 400 A (800 mm wide, 400 mm high)	1501.78	1696.11	–	–	nr	**1696.11**
3 Pole, 250 A (800 mm wide, 400 mm high)	1325.45	1496.95	–	–	nr	**1496.95**
3 Pole, 160 A (800 mm wide, 300 mm high)	978.64	1105.27	–	–	nr	**1105.27**
3 Pole, 100 A (800 mm wide, 200 mm high)	720.03	813.20	–	–	nr	**813.20**
MCCB-45 kA fault rated						
4 Pole, 630 A (800 mm wide, 600 mm high)	2453.97	2771.50	–	–	nr	**2771.50**
4 Pole, 400 A (800 mm wide, 400 mm high)	1686.94	1905.22	–	–	nr	**1905.22**
3 Pole, 630 A (800 mm wide, 600 mm high)	2121.89	2396.45	–	–	nr	**2396.45**
3 Pole, 400 A (800 mm wide, 400 mm high)	1445.94	1633.04	–	–	nr	**1633.04**
MCCB-36 kA fault rated						
4 Pole, 250 A (800 mm wide, 400 mm high)	1431.24	1616.43	–	–	nr	**1616.43**
4 Pole, 160 A (800 mm wide, 300 mm high)	1084.44	1224.77	–	–	nr	**1224.77**
3 Pole, 250 A (800 mm wide, 400 mm high)	812.94	918.13	–	–	nr	**918.13**
3 Pole, 160 A (800 mm wide, 300 mm high)	715.88	808.51	–	–	nr	**808.51**
Extra for						
Cable box (one per MCCB for form 4, types 6 & 7)	123.74	139.75	–	–	nr	**139.75**
Shunt trip (for ratings 100A to 630A)	43.31	48.91	–	–	nr	**48.91**
Undervoltage release (for ratings 100A to 630A)	61.88	69.89	–	–	nr	**69.89**
Motor operator for 630 A MCCB	587.78	663.84	–	–	nr	**663.84**
Motor operator for 400 A MCCB	587.78	663.84	–	–	nr	**663.84**
Motor operator for 250 A MCCB	488.79	552.03	–	–	nr	**552.03**
Motor operator for 160 A/100A MCCB	312.46	352.89	–	–	nr	**352.89**
Door handle for 630/400 A MCCB	77.35	87.36	–	–	nr	**87.36**
Door handle for 250/160/100 A MCCB	61.88	69.89	–	–	nr	**69.89**
MCCB earth fault protection	464.04	524.09	–	–	nr	**524.09**

V: ELECTRICAL SUPPLY/POWER/LIGHTING

Item	Net Price £	Material £	Labour hours	Labour £	Unit	Total rate £
V20: LV DISTRIBUTION – cont						
LV SWITCHBOARD BUSBAR						
Copper busbar assembled into LV switchboard, ASTA type tested to appropriate fault level. Busbar Length may be estimated by adding the widths of the ACB sections to the width of the MCCB sections. ACB's up to 2000 A rating may be stacked two high; larger ratings are one per section. To determine the number of MCCB sections, add together all the MCCB heights and divide by 1800 mm, rounding up as necessary						
6000 A (6 × 10 mm × 100 mm)	2398.33	2708.66	–	–	nr	**2708.66**
5000 A (4 × 10 mm × 100 mm)	1971.97	2227.13	–	–	nr	**2227.13**
4000 A (4 × 10 mm × 100 mm)	1971.97	2227.13	–	–	nr	**2227.13**
3200 A (3 × 10 mm × 100 mm)	1369.04	1546.19	–	–	nr	**1546.19**
2500 A (2 × 10 mm × 100 mm)	1155.88	1305.45	–	–	nr	**1305.45**
2000 A (2 × 10 mm × 80 mm)	849.41	959.32	–	–	nr	**959.32**
1600 A (2 × 10 mm × 50 mm)	616.23	695.97	–	–	nr	**695.97**
1250 A (2 × 10 mm × 40 mm)	482.99	545.49	–	–	nr	**545.49**
1000 A (2 × 10 mm × 30 mm)	482.99	545.49	–	–	nr	**545.49**
800 A (2 × 10 mm × 20 mm)	376.40	425.11	–	–	nr	**425.11**
630 A (2 × 10 mm × 20 mm)	376.40	425.11	–	–	nr	**425.11**
400 A (2 × 10 mm × 10 mm)	322.85	364.63	–	–	nr	**364.63**
Automatic power factor correction (PFC); floor standing steel enclosure to IP 42, complete with microprocessor based relay and status indication; includes delivery, offloading, positioning and commissioning; excludes cabling and cable terminations						
Standard PFC (no detuning)						
100 kVAr	4702.04	5310.46	–	–	nr	**5310.46**
200 kVAr	6576.29	7427.23	–	–	nr	**7427.23**
400 kVAr	11226.11	12678.71	–	–	nr	**12678.71**
600 kVAr	15377.53	17367.31	–	–	nr	**17367.31**
PFC with detuning reactors						
100 kVAr	7650.70	8640.66	–	–	nr	**8640.66**
200 kVAr	10691.87	12075.34	–	–	nr	**12075.34**
400 kVAr	18426.15	20810.40	–	–	nr	**20810.40**
600 kVAr	27097.55	30603.84	–	–	nr	**30603.84**

V: ELECTRICAL SUPPLY/POWER/LIGHTING

Item	Net Price £	Material £	Labour hours	Labour £	Unit	Total rate £
AUTOMATIC TRANSFER SWITCHES						
Automatic transfer switches; steel enclosure; solenoid operating; programmable controller, keypad and LCD display; fixed to backgrounds; including commissioning and testing						
Panel mounting type 3 pole or 4 pole; overlapping neutral						
100 amp	3149.43	3556.95	2.60	71.44	nr	**3628.39**
250 amp	4207.07	4751.45	3.30	90.68	nr	**4842.13**
400 amp	5560.15	6279.61	4.30	118.16	nr	**6397.77**
630 amp	7144.58	8069.06	5.30	145.64	nr	**8214.70**
800 amp	10303.96	11637.24	5.50	151.12	nr	**11788.36**
1000 amp	16392.66	18513.79	5.83	160.19	nr	**18673.98**
1600 amp	17563.56	19836.20	6.20	170.37	nr	**20006.57**
2000 amp	21578.09	24370.19	6.90	189.61	nr	**24559.80**
Enclosed type 3 pole or 4 pole; over lapping neutral						
63 amp	3043.66	3437.49	2.60	71.44	nr	**3508.93**
100 amp	3149.43	3556.95	2.90	79.69	nr	**3636.64**
125 amp	3227.16	3644.74	2.90	79.69	nr	**3724.43**
160 amp	3380.85	3818.32	2.60	71.44	nr	**3889.76**
250 amp	4207.07	4751.45	3.30	90.68	nr	**4842.13**
400 amp	6615.69	7471.72	4.30	118.16	nr	**7589.88**
630 amp	8905.89	10058.27	4.84	133.00	nr	**10191.27**
800 amp	12293.98	13884.76	5.12	140.69	nr	**14025.45**
1000 amp	18440.96	20827.13	5.50	151.12	nr	**20978.25**
1250 amp	19977.70	22562.72	6.00	164.87	nr	**22727.59**
1600 amp	21514.44	24298.30	6.20	170.37	nr	**24468.67**
2000 amp	25595.97	28907.96	6.90	189.61	nr	**29097.57**
3000 amp	30781.40	34764.36	6.90	189.61	nr	**34953.97**
4000 amp	37678.02	42553.37	6.90	189.61	nr	**42742.98**
Enclosed type 3 pole or 4 pole; over lapping Neutral, with single By-Pass						
63 amp	5496.26	6207.44	2.60	71.44	nr	**6278.88**
100 amp	5655.57	6387.37	2.60	71.44	nr	**6458.81**
125 amp	5814.88	6567.30	2.90	79.69	nr	**6646.99**
160 amp	8363.88	9446.13	2.90	79.69	nr	**9525.82**
250 amp	10865.07	12270.96	3.30	90.68	nr	**12361.64**
400 amp	12187.36	13764.34	4.30	118.16	nr	**13882.50**
630 amp	25330.58	28608.23	4.84	133.00	nr	**28741.23**
800 amp	26764.39	30227.57	5.12	140.69	nr	**30368.26**
1000 amp	50142.60	56630.80	5.50	151.12	nr	**56781.92**
1250 amp	50993.32	57591.60	6.00	164.87	nr	**57756.47**
1600 amp	21514.44	24298.30	6.20	170.37	nr	**24468.67**

V: ELECTRICAL SUPPLY/POWER/LIGHTING

Item	Net Price £	Material £	Labour hours	Labour £	Unit	Total rate £
V20: LV DISTRIBUTION – cont						
AUTOMATIC TRANSFER SWITCHES – cont						
Enclosed type 3 pole or 4 pole; overlapping Neutral, with Dual By-Pass						
63 amp	6466.19	7302.88	2.60	71.44	nr	**7374.32**
100 amp	6653.62	7514.56	2.60	71.44	nr	**7586.00**
125 amp	6841.03	7726.22	2.90	79.69	nr	**7805.91**
160 amp	9839.85	11113.07	2.90	79.69	nr	**11192.76**
250 amp	12782.43	14436.42	3.30	90.68	nr	**14527.10**
400 amp	14338.07	16193.35	4.30	118.16	nr	**16311.51**
630 amp	29800.68	33656.74	4.84	133.00	nr	**33789.74**
800 amp	31487.51	35561.84	5.12	140.69	nr	**35702.53**
1000 amp	58991.30	66624.48	5.50	151.12	nr	**66775.60**
1250 amp	59992.15	67754.84	6.00	164.87	nr	**67919.71**
1600 amp	63389.70	71592.01	6.20	170.37	nr	**71762.38**
BREAKERS/FUSES						
MCCB panelboards; IP4X construction, 50 kA busbars and fully rated neutral; fitted with doorlock, removable glandplate; form 3b Type 2; BSEN 60439-1; including fixing to backgrounds						
Panelboards cubicle with MCCB incomer						
Up to 250 A						
4 Way TPN	887.72	1002.59	1.00	27.48	nr	**1030.07**
Extra over for integral incomer metering	909.20	1026.84	1.50	41.21	nr	**1068.05**
Up to 630 A						
6 Way TPN	1596.89	1803.52	2.00	54.96	nr	**1858.48**
12 Way TPN	1803.53	2036.90	2.50	68.70	nr	**2105.60**
18 Way TPN	2099.43	2371.08	3.00	82.44	nr	**2453.52**
Extra over for integral incomer metering	1074.51	1213.54	1.50	41.21	nr	**1254.75**
Up to 800 A						
6 Way TPN	2732.57	3086.15	2.00	54.96	nr	**3141.11**
12 Way TPN	3278.10	3702.27	2.50	68.70	nr	**3770.97**
18 Way TPN	3489.69	3941.24	3.00	82.44	nr	**4023.68**
Extra over for integral incomer metering	1074.51	1213.54	1.50	41.21	nr	**1254.75**
Up to 1200 A						
20 Way TPN	7311.67	8257.77	3.50	96.18	nr	**8353.95**
Up to 1600A						
28 Way TPN	9468.95	10694.18	3.50	96.18	nr	**10790.36**

V: ELECTRICAL SUPPLY/POWER/LIGHTING

Item	Net Price £	Material £	Labour hours	Labour £	Unit	Total rate £
Up to 2000 A						
28 Way TPN	10307.08	11640.76	4.00	109.92	nr	**11750.68**
Feeder MCCBs						
Single pole						
32 A	74.98	84.68	0.75	20.61	nr	**105.29**
63 A	76.69	86.62	0.75	20.61	nr	**107.23**
100 A	78.39	88.54	0.75	20.61	nr	**109.15**
160 A	83.51	94.32	1.00	27.48	nr	**121.80**
Double pole						
32 A	112.48	127.03	0.75	20.61	nr	**147.64**
63 A	114.17	128.94	0.75	20.61	nr	**149.55**
100 A	167.01	188.62	0.75	20.61	nr	**209.23**
160 A	207.91	234.81	1.00	27.48	nr	**262.29**
Triple pole						
32 A	149.98	169.39	0.75	20.61	nr	**190.00**
63 A	153.37	173.22	0.75	20.61	nr	**193.83**
100 A	199.39	225.19	0.75	20.61	nr	**245.80**
160 A	257.34	290.64	1.00	27.48	nr	**318.12**
250 A	386.86	436.92	1.00	27.48	nr	**464.40**
400 A	523.19	590.89	1.25	34.34	nr	**625.23**
630 A	858.94	970.08	1.50	41.21	nr	**1011.29**
MCB distribution boards; IP3X external protection enclosure; removable earth and neutral bars and DIN rail; 125/250 amp incomers; including fixing to backgrounds						
SP & N						
6 way	72.59	81.99	2.00	54.96	nr	**136.95**
8 way	86.09	97.23	2.50	68.70	nr	**165.93**
12 way	98.80	111.59	3.00	82.44	nr	**194.03**
16 way	117.45	132.65	4.00	109.92	nr	**242.57**
24 way	247.40	279.41	5.00	137.39	nr	**416.80**
TP & N						
4 way	503.03	568.12	3.00	82.44	nr	**650.56**
6 way	520.71	588.09	3.50	96.18	nr	**684.27**
8 way	545.20	615.74	4.00	109.92	nr	**725.66**
12 way	581.33	656.55	4.00	109.92	nr	**766.47**
16 way	665.19	751.26	5.00	137.39	nr	**888.65**
24 way	837.71	946.10	6.40	175.86	nr	**1121.96**

V: ELECTRICAL SUPPLY/POWER/LIGHTING

Item	Net Price £	Material £	Labour hours	Labour £	Unit	Total rate £
V20: LV DISTRIBUTION – cont						
BREAKERS/FUSES – cont						
Miniature circuit breakers for distribution boards; BS EN 60 898; DIN rail mounting; including connecting to circuit						
SP&N; including connecting of wiring						
6 amp	8.67	9.79	0.10	2.74	nr	**12.53**
1–40 amp	9.02	10.19	0.10	2.74	nr	**12.93**
50–63 amp	9.44	10.66	0.14	3.86	nr	**14.52**
TP&N; including connecting of wiring						
6 amp	36.76	41.51	0.30	8.24	nr	**49.75**
10–40 amp	38.21	43.16	0.45	12.36	nr	**55.52**
50–63 amp	40.03	45.21	0.45	12.36	nr	**57.57**
Resdiual current circuit breakers for distribution boards; DIN rail mounting; including connecting to circuit						
SP&N						
10 mA						
6 amp	56.92	64.28	0.21	5.77	nr	**70.05**
10–32 amp	55.85	63.08	0.26	7.15	nr	**70.23**
45 amp	56.76	64.11	0.26	7.15	nr	**71.26**
30 mA						
6 amp	56.92	64.28	0.21	5.77	nr	**70.05**
10–40 amp	55.85	63.08	0.21	5.77	nr	**68.85**
50–63 amp	57.66	65.12	0.26	7.15	nr	**72.27**
100 mA						
6 amp	105.26	118.88	0.21	5.77	nr	**124.65**
10 – 40 amp	105.26	118.88	0.23	6.41	nr	**125.29**
50 -63 amp	105.26	118.88	0.26	7.15	nr	**126.03**
HRC fused distribution boards; IP4X external protection enclosure; including earth and neutral bars; fixing to backgrounds						
SP&N						
20 amp incomer						
4 way	137.13	154.88	1.00	27.48	nr	**182.36**
6 way	165.55	186.98	1.20	32.98	nr	**219.96**
8 way	194.06	219.17	1.40	38.46	nr	**257.63**
12 way	251.14	283.63	1.80	49.46	nr	**333.09**

V: ELECTRICAL SUPPLY/POWER/LIGHTING

Item	Net Price £	Material £	Labour hours	Labour £	Unit	Total rate £
32 amp incomer						
4 way	165.11	186.48	1.00	27.48	nr	**213.96**
6 way	217.11	245.21	1.20	32.98	nr	**278.19**
8 way	255.44	288.50	1.40	38.46	nr	**326.96**
12 way	328.88	371.43	1.80	49.46	nr	**420.89**
TP&N						
20 amp incomer						
4 way	258.72	292.20	1.50	41.21	nr	**333.41**
6 way	327.22	369.56	2.10	57.70	nr	**427.26**
8 way	386.89	436.95	2.70	74.19	nr	**511.14**
12 way	542.79	613.02	3.90	107.17	nr	**720.19**
32 amp incomer						
4 way	309.52	349.57	1.50	41.21	nr	**390.78**
6 way	415.54	469.31	2.10	57.70	nr	**527.01**
8 way	507.61	573.29	2.70	74.19	nr	**647.48**
12 way	704.04	795.14	3.90	107.17	nr	**902.31**
63 amp incomer						
4 way	657.93	743.06	2.17	59.63	nr	**802.69**
6 way	843.79	952.98	2.83	77.75	nr	**1030.73**
8 way	1016.13	1147.61	2.57	70.62	nr	**1218.23**
100 amp incomer						
4 way	1040.52	1175.16	2.40	65.94	nr	**1241.10**
6 way	1360.04	1536.02	2.73	75.01	nr	**1611.03**
8 way	1662.99	1878.18	3.87	106.35	nr	**1984.53**
200 amp incomer						
4 way	2577.25	2910.73	5.36	147.29	nr	**3058.02**
6 way	3406.38	3847.14	6.17	169.54	nr	**4016.68**
HRC fuse; includes fixing to fuse holder						
2–30 amp	2.80	3.16	0.10	2.74	nr	**5.90**
35–63 amp	6.06	6.84	0.12	3.29	nr	**10.13**
80 amp	8.95	10.11	0.15	4.12	nr	**14.23**
100 amp	10.76	12.15	0.15	4.12	nr	**16.27**
125 amp	16.26	18.36	0.15	4.12	nr	**22.48**
160 amp	17.06	19.27	0.15	4.12	nr	**23.39**
200 amp	17.67	19.96	0.15	4.12	nr	**24.08**

V: ELECTRICAL SUPPLY/POWER/LIGHTING

Item	Net Price £	Material £	Labour hours	Labour £	Unit	Total rate £
V20: LV DISTRIBUTION – cont						
BREAKERS/FUSES – cont						
Consumer units; fixed to backgrounds; including supports, fixings, connections/ jointing to equipment						
Switched and insulated; moulded plastic case, 63 amp 230 Volt SP&N; earth and neutral bars; 30 mA RCCB protection; fitted MCB's						
2 way	107.27	121.15	1.67	45.89	nr	**167.04**
4 way	120.37	135.95	1.59	43.69	nr	**179.64**
6 way	130.85	147.78	2.50	68.70	nr	**216.48**
8 way	141.19	159.46	3.00	82.44	nr	**241.90**
12 way	164.02	185.24	4.00	109.92	nr	**295.16**
16 way	198.19	223.84	5.50	151.12	nr	**374.96**
Switched and insulated; moulded plastic case, 100 amp 230 Volt SP&N; earth and neutral bars; 30 mA RCCB protection; fitted MCB's						
2 way	107.27	121.15	1.67	45.89	nr	**167.04**
4 way	120.37	135.95	1.59	43.69	nr	**179.64**
6 way	130.85	147.78	2.50	68.70	nr	**216.48**
8 way	141.19	159.46	3.00	82.44	nr	**241.90**
12 way	164.02	185.24	4.00	109.92	nr	**295.16**
16 way	198.19	223.84	5.50	151.12	nr	**374.96**
Extra for						
Residual current device; double pole; 230 Volt/30 mA tripping current						
16 amp	58.65	66.24	0.22	6.04	nr	**72.28**
30 amp	59.58	67.29	0.22	6.04	nr	**73.33**
40 amp	60.52	68.35	0.22	6.04	nr	**74.39**
63 amp	74.92	84.62	0.22	6.04	nr	**90.66**
80 amp	83.33	94.12	0.22	6.04	nr	**100.16**
100 amp	102.57	115.84	0.25	6.86	nr	**122.70**
Residual current device; double pole; 230 Volt/100 mA tripping current						
63 amp	68.49	77.36	0.22	6.04	nr	**83.40**
80 amp	79.23	89.48	0.22	6.04	nr	**95.52**
100 amp	102.59	115.86	0.25	6.86	nr	**122.72**

V: ELECTRICAL SUPPLY/POWER/LIGHTING

Item	Net Price £	Material £	Labour hours	Labour £	Unit	Total rate £
Heavy duty fuse switches; with HRC fuses BS 5419; short circuit rating 65 kA, 500 volt; including retractable operating switches						
SP&N						
63 amp	264.97	299.26	1.30	35.72	nr	**334.98**
100 amp	387.32	437.44	1.95	53.58	nr	**491.02**
TP&N						
63 amp	333.82	377.01	1.83	50.28	nr	**427.29**
100 amp	469.08	529.78	2.48	68.15	nr	**597.93**
200 amp	723.59	817.22	3.13	86.01	nr	**903.23**
300 amp	1257.08	1419.74	4.45	122.28	nr	**1542.02**
400 amp	1380.04	1558.61	4.45	122.28	nr	**1680.89**
600 amp	2083.09	2352.63	5.72	157.18	nr	**2509.81**
800 amp	3254.02	3675.07	7.88	216.53	nr	**3891.60**
Switch disconnectors to BSEN 60947-3; in sheet steel case; IP41 with door interlock fixed to backgrounds						
Double pole						
20 amp	51.93	58.65	1.02	28.03	nr	**86.68**
32 amp	62.59	70.69	1.02	28.03	nr	**98.72**
63 amp	228.41	257.97	1.21	33.25	nr	**291.22**
100 amp	209.58	236.70	1.86	51.10	nr	**287.80**
TP&N						
20 amp	65.19	73.62	1.29	35.46	nr	**109.08**
32 amp	75.85	85.66	1.83	50.28	nr	**135.94**
63 amp	257.42	290.73	2.48	68.15	nr	**358.88**
100 amp	259.54	293.13	2.48	68.15	nr	**361.28**
125 amp	270.71	305.73	2.48	68.15	nr	**373.88**
160 amp	622.87	703.46	2.48	68.15	nr	**771.61**
Enclosed switch disconnector to BSEN 60947-3; enclosure minimum IP55 rating; complete with earth connection bar; fixed to backgrounds.						
TP						
20 amp	51.93	58.65	1.02	28.03	nr	**86.68**
32 amp	62.59	70.69	1.02	28.03	nr	**98.72**
63 amp	228.41	257.97	1.21	33.25	nr	**291.22**
TP&N						
20 amp	65.19	73.62	1.29	35.46	nr	**109.08**
32 amp	75.85	85.66	1.83	50.28	nr	**135.94**
63 amp	257.42	290.73	2.48	68.15	nr	**358.88**

V: ELECTRICAL SUPPLY/POWER/LIGHTING

Item	Net Price £	Material £	Labour hours	Labour £	Unit	Total rate £
V20: LV DISTRIBUTION – cont						
BREAKERS/FUSES – cont						
Busbar chambers; fixed to background including all supports, fixings, connections/jointing to equipment						
Sheet steel case enclosing 4 pole 550 Volt copper bars, detachable metal end plates						
600 mm long						
200 amp	473.08	534.30	2.62	71.99	nr	**606.29**
300 amp	607.91	686.57	3.03	83.26	nr	**769.83**
500 amp	1041.64	1176.43	4.48	123.10	nr	**1299.53**
900 mm long						
200 amp	681.42	769.59	3.04	83.53	nr	**853.12**
300 amp	803.09	907.00	3.59	98.64	nr	**1005.64**
500 amp	1187.91	1341.62	4.42	121.45	nr	**1463.07**
1350 mm long						
200 amp	930.76	1051.19	3.38	92.88	nr	**1144.07**
300 amp	1095.50	1237.25	3.94	108.26	nr	**1345.51**
500 amp	1753.58	1980.48	4.82	132.45	nr	**2112.93**
Contactor relays; pressed steel enclosure; fixed to backgrounds including supports, fixings, connections/jointing to equipment						
Relays						
6 amp, 415/240 Volt, 4 pole N/O	54.79	61.88	0.52	14.29	nr	**76.17**
6 amp, 415/240 Volt, 8 pole N/O	67.00	75.67	0.85	23.36	nr	**99.03**
Push button stations; heavy gauge pressed steel enclosure; polycarbonate cover; IP65; fixed to backgrounds including supports, fixings,connections/ joining to equipment						
Standard units						
One button (start or stop)	69.33	78.31	0.39	10.72	nr	**89.03**
Two button (start or stop)	73.60	83.12	0.47	12.92	nr	**96.04**
Three button (forward-reverse-stop)	104.30	117.80	0.57	15.67	nr	**133.47**

V: ELECTRICAL SUPPLY/POWER/LIGHTING

Item	Net Price £	Material £	Labour hours	Labour £	Unit	Total rate £
Weatherproof junction boxes; enclosures with rail mounted terminal blocks; side hung door to receive padlock; fixed to backgrounds, including all supports and fixings (suitable for cable up to 2.5 mm²; including glandplates and gaskets)						
Sheet steel with zinc spray finish enclosure						
Overall size 229 × 152; suitable to receive						
3 × 20(A) glands per gland plate	70.22	79.30	1.43	39.29	nr	**118.59**
Overall size 306 × 306; suitable to receive						
14 × 20(A) glands per gland plate	94.21	106.40	2.17	59.63	nr	**166.03**
Overall size 458 × 382; suitable to receive						
18 × 20(A) glands per gland plate	137.17	154.92	3.51	96.45	nr	**251.37**
Overall size 762 x508; suitable to receive						
26 × 20(A) glands per gland plate	145.08	163.85	4.85	133.27	nr	**297.12**
Overall size 914 × 610; suitable to receive						
45 × 20(A) glands per gland plate	160.36	181.11	7.01	192.62	nr	**373.73**
Weatherproof junction boxes; enclosures with rail mounted terminal blocks; screw fixed lid; fixed to backgrounds, including all supports and fixings (suitable for cable up to 2.5 mm²; including glandplates and gaskets)						
Glassfibre reinforced polycarbonate enclosure						
Overall size 190 × 190 × 130	95.91	108.32	1.43	39.29	nr	**147.61**
Overall size 190 × 190 × 180	140.42	158.59	1.53	42.03	nr	**200.62**
Overall size 280 × 190 × 130	158.60	179.12	2.17	59.63	nr	**238.75**
Overall size 280 × 190 × 180	177.68	200.67	2.37	65.12	nr	**265.79**
Overall size 380 × 190 × 130	198.24	223.89	3.33	91.50	nr	**315.39**
Overall size 380 × 190 × 180	212.91	240.45	3.30	90.68	nr	**331.13**
Overall size 380 × 280 × 130	227.60	257.05	4.66	128.05	nr	**385.10**
Overall size 380 × 280 × 180	245.22	276.95	5.36	147.29	nr	**424.24**
Overall size 560 × 280 × 130	295.15	333.34	7.01	192.62	nr	**525.96**
Overall size 560 × 380 × 180	303.97	343.30	7.67	210.75	nr	**554.05**

V: ELECTRICAL SUPPLY/POWER/LIGHTING

Item	Net Price £	Material £	Labour hours	Labour £	Unit	Total rate £
V21: GENERAL LIGHTING						
Y73 – LUMINAIRES (GENERAL)						
LUMINAIRES						
Fluorescent Luminaires; surface fixed to backgrounds						
Batten type; surface mounted						
600 mm Single – 18 W	8.72	9.85	0.58	15.93	nr	**25.78**
600 mm Twin – 18 W	15.45	17.45	0.59	16.21	nr	**33.66**
1200 mm Single – 36 W	11.55	13.05	0.76	20.88	nr	**33.93**
1200 mm Twin – 36 W	13.06	14.75	0.84	23.08	nr	**37.83**
1500 mm Single – 58 W	22.11	24.97	0.77	21.15	nr	**46.12**
1500 mm Twin – 58 W	26.07	29.45	0.85	23.36	nr	**52.81**
1800 mm Single – 70 W	15.75	17.79	1.05	28.86	nr	**46.65**
1800 mm Twin – 70 W	28.82	32.55	1.06	29.13	nr	**61.68**
2400 mm Single – 100 W	21.57	24.36	1.25	34.34	nr	**58.70**
2400 mm Twin – 100 W	37.74	42.63	1.27	34.89	nr	**77.52**
Surface mounted, opal diffuser						
600 mm Twin – 18 W	23.97	27.07	0.62	17.03	nr	**44.10**
1200 mm Single – 36 W	20.47	23.12	0.79	21.71	nr	**44.83**
1200 mm Twin – 36 W	31.88	36.01	0.80	21.98	nr	**57.99**
1500 mm Single – 58 W	23.54	26.58	0.88	24.18	nr	**50.76**
1500 mm Twin – 58 W	37.28	42.11	0.90	24.73	nr	**66.84**
1800 mm Single – 70 W	29.36	33.16	1.09	29.95	nr	**63.11**
1800 mm Twin – 70 W	40.31	45.52	1.10	30.22	nr	**75.74**
2400 mm Single – 100 W	38.57	43.56	1.30	35.72	nr	**79.28**
2400 mm Twin – 100 W	55.69	62.89	1.31	36.00	nr	**98.89**
Surface mounted linear fluorescent; T8 lamp; high frequency control gear; low brightness; 65° cut-off; including wedge style louvre						
1200 mm, 1 × 36 watt	55.22	62.36	1.09	29.95	nr	**92.31**
1200 mm 2 × 36 watt	59.05	66.69	1.09	29.95	nr	**96.64**
Extra for emergency pack	49.81	56.25	0.25	6.86	nr	**63.11**
1500 mm, 1 × 58 watt	64.36	72.69	0.90	24.73	nr	**97.42**
1500 mm 2 × 58 watt	68.79	77.69	0.90	24.73	nr	**102.42**
Extra for emergency pack	49.71	56.14	0.25	6.86	nr	**63.00**
1800 mm, 1 × 70 watt	96.92	109.46	0.90	24.73	nr	**134.19**
1800 mm, 2 × 70 watt	111.26	125.65	0.90	24.73	nr	**150.38**
Extra for emergency pack	82.30	92.95	0.25	6.86	nr	**99.81**

V: ELECTRICAL SUPPLY/POWER/LIGHTING

Item	Net Price £	Material £	Labour hours	Labour £	Unit	Total rate £
Modular recessed linear fluorescent; high frequency control gear; low brightness; 65° cut off; including wedge style louvre; fitted to exposed T grid ceiling						
600 × 600 mm, 3 × 18 watt T8	43.43	49.05	0.84	23.08	nr	**72.13**
600 × 600 mm, 4 × 18 watt T8	44.88	50.68	0.87	23.91	nr	**74.59**
Extra for emergency pack	50.22	56.72	0.25	6.86	nr	**63.58**
300 × 1200 mm, 2 × 36 watt T8	60.61	68.45	0.87	23.91	nr	**92.36**
Extra for emergency pack	63.53	71.75	0.25	6.86	nr	**78.61**
600 × 1200 mm, 3 × 36 watt T8	66.13	74.68	0.89	24.45	nr	**99.13**
600 × 1200 mm, 4 × 36 watt T8	76.07	85.91	0.91	25.00	nr	**110.91**
Extra for emergency pack	58.84	66.45	0.25	6.86	nr	**73.31**
600 × 600, 3 × 14 watt T5	61.69	69.68	0.84	23.08	nr	**92.76**
600 × 600, 4 × 14 watt T5	64.20	72.51	0.87	23.91	nr	**96.42**
Extra for emergency pack	73.85	83.41	0.25	6.86	nr	**90.27**
Modular recessed; T8 lamp; high frequency control gear; cross-blade louvre; fitted to exposed T grid ceiling						
600 × 600 mm, 3 × 18 watt	47.42	53.56	0.84	23.08	nr	**76.64**
600 × 600 mm, 4 × 18 watt	64.17	72.47	0.87	23.91	nr	**96.38**
Extra for emergency pack	57.08	64.46	0.25	6.86	nr	**71.32**
Modular recessed compact fluorescent; TCL lamp; high frequency control gear; low brightness; 65° cut-off; including wedge style louvre; fitted to exposed T grid ceiling						
300 × 300 mm, 2 × 18 watt	96.04	108.47	0.75	20.61	nr	**129.08**
Extra for emergency pack	88.68	100.15	0.25	6.86	nr	**107.01**
500 × 500 mm, 2 × 36 watt	89.81	101.43	0.82	22.53	nr	**123.96**
600 × 600 mm, 2 × 36 watt	93.97	106.13	0.82	22.53	nr	**128.66**
600 × 600 mm, 2 × 40 watt	54.20	61.21	0.82	22.53	nr	**83.74**
Extra for emergency pack	52.62	59.43	0.25	6.86	nr	**66.29**
Ceiling recessed asymetric compact fluorescent downlighter; high frequency control gear; TCD lamp in 200 mm diameter luminaire; for wall-washing application						
1 × 18 watt	152.10	171.78	0.75	20.61	nr	**192.39**
1 × 26 watt	152.10	171.78	0.75	20.61	nr	**192.39**
2 × 18 watt	169.44	191.36	0.75	20.61	nr	**211.97**
2 × 26 watt	169.44	191.36	0.75	20.61	nr	**211.97**
Ceiling recessed asymetric compact fluorescesnt downlights; high frequency control gear; linear 200 mm × 600 mm luminaire with low glare louvre; for wall washing applications						
1 × 55 watt TCL	61.44	69.39	0.75	20.61	nr	**90.00**

V: ELECTRICAL SUPPLY/POWER/LIGHTING

Item	Net Price £	Material £	Labour hours	Labour £	Unit	Total rate £
V21: GENERAL LIGHTING – cont						
LUMINAIRES – cont						
Wall mounted compact fluorescent uplighter; high frequency control gear; TCL lamp in 300 mm × 600 mm luminaire						
2 × 36 watt	238.43	269.28	0.84	23.08	nr	**292.36**
2 × 40 watt	256.42	289.60	0.84	23.08	nr	**312.68**
2 × 55 watt	359.27	405.76	0.84	23.08	nr	**428.84**
Suspended linear fluorescent; T5 lamp; high frequency control gear; low brightness; 65° cut-off; 30% uplight, 70% downlight; including wedge style louvre						
1 × 49 watt	163.50	184.66	0.75	20.61	nr	**205.27**
Extra for emergency pack	90.32	102.01	0.25	6.86	nr	**108.87**
Semi-recessed 'architectural' linear fluorescent; T5 lamp; high frequency control gear; low brightness, delivers direct, ceiling and graduated wall washing illumination						
600 × 600 mm, 2 × 24 watt	133.95	151.29	0.87	23.91	nr	**175.20**
600 × 600 mm, 4 × 14 watt	144.15	162.80	0.87	23.91	nr	**186.71**
500 × 500 mm, 2 × 24 watt	131.71	148.76	0.87	23.91	nr	**172.67**
Extra for emergency pack	64.14	72.44	0.25	6.86	nr	**79.30**
Downlighter, recessed; low voltage; mirror reflector with white/chrome bezel; dimmable transformer; for dichroic lamps						
85 mm dia × 20/50 watt	15.51	17.51	0.66	18.14	nr	**35.65**
118 mm dia × 50 watt	20.50	23.15	0.66	18.14	nr	**41.29**
165 mm dia × 100 watt	93.65	105.76	0.66	18.14	nr	**123.90**
High/Low Bay luminaires						
Compact discharge; aluminium reflector						
150 watt	57.24	64.65	1.50	41.21	nr	**105.86**
250 watt	59.27	66.94	1.50	41.21	nr	**108.15**
400 watt	60.76	68.63	1.50	41.21	nr	**109.84**
Sealed discharge; aluminium reflector						
150 watt	176.87	199.76	1.50	41.21	nr	**240.97**
250 watt	190.90	215.60	1.50	41.21	nr	**256.81**
400 watt	249.60	281.90	1.50	41.21	nr	**323.11**

V: ELECTRICAL SUPPLY/POWER/LIGHTING

Item	Net Price £	Material £	Labour hours	Labour £	Unit	Total rate £
Corrosion resistant GRP body; gasket sealed; acrylic diffuser						
600 mm Single – 18 W	29.59	33.42	0.49	13.46	nr	**46.88**
600 mm Twin – 18 W	38.52	43.50	0.49	13.46	nr	**56.96**
1200 mm Single – 36 W	33.31	37.62	0.64	17.58	nr	**55.20**
1200 mm Twin – 36 W	42.70	48.23	0.64	17.58	nr	**65.81**
1500 mm Single – 58 W	37.12	41.92	0.72	19.79	nr	**61.71**
1500 mm Twin – 58 W	46.15	52.12	0.72	19.79	nr	**71.91**
1800 mm Single – 70 W	54.94	62.05	0.94	25.83	nr	**87.88**
1800 mm Twin – 70 W	68.19	77.01	0.94	25.83	nr	**102.84**
Flameproof to IIA/IIB,I.P. 64; Aluminium Body; BS 229 and 899						
600 mm Single – 18 W	331.75	374.68	1.04	28.57	nr	**403.25**
600 mm Twin – 18 W	413.27	466.74	1.04	28.57	nr	**495.31**
1200 mm Single – 36 W	363.39	410.41	1.31	36.00	nr	**446.41**
1200 mm Twin – 36 W	449.28	507.42	1.18	32.43	nr	**539.85**
1500 mm Single – 58 W	389.04	439.38	1.64	45.06	nr	**484.44**
1500 mm Twin – 58 W	469.77	530.55	1.64	45.06	nr	**575.61**
1800 mm Single – 70 W	426.49	481.67	1.97	54.13	nr	**535.80**
1800 mm Twin – 70 W	492.02	555.69	1.97	54.13	nr	**609.82**
External Lighting						
Ground mounted 50 watt	365.39	412.67	2.25	61.82	nr	**474.49**
Ceiling mounted 50 watt	164.16	185.41	2.25	61.82	nr	**247.23**
Bulkhead; aluminium body and polycarbonate bowl; vandal-resistant; IP65						
60 watt	32.16	36.32	0.75	20.61	nr	**56.93**
Extra for						
Emergency version	86.43	97.61	0.25	6.86	nr	**104.47**
2D 2 pin 16 watt	27.16	30.67	0.66	18.14	nr	**48.81**
2D 2 pin 28 watt	54.24	61.26	0.66	18.14	nr	**79.40**
Extra for						
Emergency version	63.91	72.18	0.25	6.86	nr	**79.04**
Photocell	19.06	21.52	0.75	20.61	nr	**42.13**
1500 mm high circular bollard; polycarbonate visor; vandal-resistant; IP54						
50 watt	201.32	227.37	1.75	48.09	nr	**275.46**
70 watt	204.54	231.01	1.75	48.09	nr	**279.10**
80 watt	244.79	276.46	1.75	48.09	nr	**324.55**
Floodlight; enclosed high performance dischargelight; integeral control gear; reflector; toughened glass; IP65						
70 watt	89.18	100.71	1.25	34.34	nr	**135.05**
100 watt	93.33	105.41	1.25	34.34	nr	**139.75**
150 watt	98.96	111.76	1.25	34.34	nr	**146.10**

V: ELECTRICAL SUPPLY/POWER/LIGHTING

Item	Net Price £	Material £	Labour hours	Labour £	Unit	Total rate £
V21: GENERAL LIGHTING – cont						
LUMINAIRES – cont						
External Lighting – cont						
Floodlight – cont						
250 watt	173.22	195.64	1.25	34.34	nr	**229.98**
400 watt	179.44	202.65	1.25	34.34	nr	**236.99**
Extra for						
Photocell	19.04	21.50	0.75	20.61	nr	**42.11**
Lighting Track						
Single circuit; 25 A 2 P&E steel trunking; low voltage with copper conductors; including couplers and supports; fixed to backgrounds						
Straight track	12.79	14.44	0.50	13.73	m	**28.17**
Live end feed unit complete with end stop	22.03	24.88	0.33	9.07	nr	**33.95**
Flexible couplers 0.5 m	46.80	52.86	0.33	9.07	nr	**61.93**
Tap off complete with 0.8 m of cable	7.36	8.31	0.25	6.86	nr	**15.17**
Three circuit; 25 A 2 P&E steel trunking; low voltage with copper conductors; including couplers and supports incorporating integral twisted pair comms bus bracket; fixed to backgrounds						
Straight track	20.84	23.54	0.75	20.61	m	**44.15**
Live end feed unit complete with end stop	37.53	42.39	0.50	13.73	nr	**56.12**
Flexible couplers 0.5 m	53.46	60.37	0.45	12.36	nr	**72.73**
Tap off complete with 0.8 m of cable	9.66	10.91	0.30	8.24	nr	**19.15**
Y74 – LIGHTING ACCESSORIES						
SWITCHES						
6 amp metal clad surface mounted switch, gridswitch; one way						
1 Gang	5.03	5.68	0.43	11.81	nr	**17.49**
2 Gang	6.90	7.79	0.55	15.12	nr	**22.91**
3 Gang	10.94	12.35	0.77	21.15	nr	**33.50**
4 Gang	12.95	14.63	0.88	24.18	nr	**38.81**
6 Gang	22.25	25.13	1.10	30.22	nr	**55.35**
8 Gang	26.06	29.43	1.28	35.17	nr	**64.60**
10 Gang	39.31	44.40	1.67	45.89	nr	**90.29**

V: ELECTRICAL SUPPLY/POWER/LIGHTING

Item	Net Price £	Material £	Labour hours	Labour £	Unit	Total rate £
Extra for						
10 amp – Two way switch	1.79	2.02	0.03	0.83	nr	**2.85**
20 amp – Two way switch	2.40	2.71	0.04	1.09	nr	**3.80**
20 amp – Intermediate	4.50	5.08	0.08	2.20	nr	**7.28**
20 amp – One way SP switch	1.85	2.09	0.08	2.20	nr	**4.29**
Steel blank plate; 1 Gang	1.24	1.40	0.07	1.92	nr	**3.32**
Steel blank plate; 2 Gang	2.10	2.38	0.08	2.20	nr	**4.58**
6 amp modular type switch; galvanized steel box, bronze or satin chrome coverplate; metalclad switches; flush mounting; one way						
1 Gang	13.64	15.40	0.43	11.81	nr	**27.21**
2 Gang	18.84	21.28	0.55	15.12	nr	**36.40**
3 Gang	27.21	30.73	0.77	21.15	nr	**51.88**
4 Gang	32.39	36.58	0.88	24.18	nr	**60.76**
6 Gang	54.61	61.68	1.18	32.43	nr	**94.11**
8 Gang	65.15	73.58	1.63	44.79	nr	**118.37**
9 Gang	81.72	92.29	1.83	50.28	nr	**142.57**
12 Gang	97.33	109.93	2.29	62.93	nr	**172.86**
6 amp modular type swtich; galvanized steel box; bronze or satin chrome coverplate; flush mounting; two way						
1 Gang	14.13	15.96	0.43	11.81	nr	**27.77**
2 Gang	19.86	22.43	0.55	15.12	nr	**37.55**
3 Gang	28.73	32.45	0.77	21.15	nr	**53.60**
4 Gang	34.44	38.89	0.88	24.18	nr	**63.07**
6 Gang	57.80	65.28	1.18	32.43	nr	**97.71**
8 Gang	69.22	78.17	1.63	44.79	nr	**122.96**
9 Gang	86.29	97.45	1.83	50.28	nr	**147.73**
12 Gang	103.40	116.78	2.22	61.00	nr	**177.78**
Plate switches; 10 amp flush mounted, white plastic fronted; 16 mm metal box; fitted brass earth terminal						
1 Gang 1 Way, Single Pole	2.04	2.31	0.28	7.69	nr	**10.00**
1 Gang 2 Way, Single Pole	2.30	2.60	0.33	9.07	nr	**11.67**
2 Gang 2 Way, Single Pole	3.33	15.88	0.44	12.10	nr	**27.98**
3 Gang 2 Way, Single Pole	6.60	7.46	0.56	15.38	nr	**22.84**
1 Gang Intermediate	7.65	8.64	0.43	11.81	nr	**20.45**
1 Gang 1 Way, Double Pole	6.81	7.69	0.33	9.07	nr	**16.76**
1 Gang Single Pole with bell symbol	5.51	6.22	0.23	6.32	nr	**12.54**
1 Gang Single Pole marked 'PRESS'	4.54	5.13	0.23	6.32	nr	**11.45**
Time delay switch, suppressed	42.56	48.06	0.49	13.46	nr	**61.52**
Plate switches; 6 amp flush mounted white plastic fronted; 25 mm metal box; fitted brass earth terminal						
4 Gang 2 Way, Single Pole	14.96	16.89	0.42	11.55	nr	**28.44**
6 Gang 2 Way, Single Way	23.92	27.02	0.47	12.92	nr	**39.94**

V: ELECTRICAL SUPPLY/POWER/LIGHTING

Item	Net Price £	Material £	Labour hours	Labour £	Unit	Total rate £
V21: GENERAL LIGHTING – cont						
SWITCHES – cont						
Architrave plate switches; 6 amp flush mounted, white plastic fronted; 27 mm metal box; brass earth terminal						
1 Gang 2 Way, Single Pole	2.71	3.06	0.30	8.24	nr	**11.30**
2 Gang 2 Way, Single Pole	5.37	12.13	0.36	9.89	nr	**22.02**
Ceiling switches, white moulded plastic, pull cord; standard unit						
6 amp, 1 Way, Single Pole	3.88	4.39	0.32	8.80	nr	**13.19**
6 amp, 2 Way, Single Pole	4.67	5.27	0.34	9.34	nr	**14.61**
16 amp, 1 Way, Double Pole	6.99	7.89	0.37	10.17	nr	**18.06**
45 amp, 1 Way, Double Pole with neon indicator	10.86	12.26	0.47	12.92	nr	**25.18**
10 amp splash proof moulded switch with plain, threaded or PVC entry						
1 Gang, 2 Way Single Pole	16.67	18.83	0.34	9.34	nr	**28.17**
2 Gang, 1 Way Single Pole	18.84	21.28	0.36	9.89	nr	**31.17**
2 Gang, 2 Way Single Pole	23.87	26.96	0.40	10.99	nr	**37.95**
6 amp watertight switch; metalclad; BS 3676; ingress protected to IP65 surface mounted						
1 Gang, 2 Way; terminal entry	16.09	18.18	0.41	11.27	nr	**29.45**
1 Gang, 2 Way; through entry	16.09	18.18	0.42	11.55	nr	**29.73**
2 Gang, 2 Way; terminal entry	47.43	53.57	0.54	14.84	nr	**68.41**
2 Gang, 2 Way; through entry	47.43	53.57	0.53	14.56	nr	**68.13**
2 Way replacement switch	12.78	14.43	0.10	2.74	nr	**17.17**
15 amp watertight switch; metalclad; BS 3676; ingress protected to IP65; surface mounted						
1 Gang 2 Way, terminal entry	21.54	24.33	0.42	11.55	nr	**35.88**
1 Gang 2 Way, through entry	21.54	24.33	0.43	11.81	nr	**36.14**
2 Gang 2 Way, terminal entry	50.57	57.11	0.55	15.12	nr	**72.23**
2 Gang 2 Way, through entry	50.57	57.11	0.54	14.84	nr	**71.95**
Intermediate interior only	12.78	14.43	0.11	3.03	nr	**17.46**
2 Way interior only	12.78	14.43	0.11	3.03	nr	**17.46**
Double pole interior only	12.78	14.43	0.11	3.03	nr	**17.46**
Electrical accessories; fixed to backgrounds (Including fixings)						
Dimmer switches; rotary action; for individual lights; moulded plastic case; metal backbox; flush mounted						
1 Gang, 1 Way; 250 Watt	13.41	15.15	0.28	7.69	nr	**22.84**
1 Gang, 1 Way; 400 Watt	17.65	19.93	0.28	7.69	nr	**27.62**

V: ELECTRICAL SUPPLY/POWER/LIGHTING

Item	Net Price £	Material £	Labour hours	Labour £	Unit	Total rate £
Dimmer switches; push on/off action; for individual lights; moulded plastic case; metal backbox; flush mounted						
1 Gang, 2 Way; 250 Watt	20.49	23.14	0.34	9.34	nr	**32.48**
3 Gang, 2 Way; 250 Watt	30.19	34.10	0.48	13.19	nr	**47.29**
4 Gang, 2 Way; 250 Watt	39.73	44.87	0.57	15.67	nr	**60.54**
Dimmer switches; rotary action; metal cald; metal backbox; BS 5518 and BS 800; flush mounted						
1 Gang, 1 Way; 400 Watt	35.11	39.66	0.33	9.07	nr	**48.73**
Ceiling Roses						
Ceiling rose: white moulded plastic; flush fixed to conduit box						
Plug in type; ceiling socket with 2 terminals, loop-in and ceiling plug with 3 terminals and cover	6.51	7.35	0.34	9.34	nr	**16.69**
BC lampholder; white moulded plastic; heat resistent PVC insulated and sheathed cable; flush fixed						
2 Core; 0.75 mm²	2.26	2.55	0.33	9.07	nr	**11.62**
Batten holder: white moulded plastic; 3 terminals; BS 5042; fixed to conduit						
Straight pattern; 2 terminals with loop-in and Earth	4.28	4.83	0.29	7.98	nr	**12.81**
Angled pattern; looped in terminal	4.28	4.83	0.29	7.98	nr	**12.81**
LIGHTING CONTROLS						
Lighting control system; including software, commissioning and testing. Typical component parts indicated. System requirements dependent on final lighting design						
Y61 – CABLES						
Cable; Twin twisted bus; LSF sheathed; aluminium conductors	1.19	1.35	0.08	2.20	m	**3.55**
Cable; ELV 4 core 7/0.2; LSF sheathed; alumimium screened; copper conductor	1.56	1.76	0.15	4.12	m	**5.88**
EQUIPMENT						
Central supervisor controller including software	4990.00	5635.68	12.00	329.74	nr	**5965.42**
Area control unit	899.64	1016.05	4.00	109.92	nr	**1125.97**

V: ELECTRICAL SUPPLY/POWER/LIGHTING

Item	Net Price £	Material £	Labour hours	Labour £	Unit	Total rate £
V21: GENERAL LIGHTING – cont						
EQUIPMENT – cont						
Lighting control module; plug in; 9 output, 9 channel switching						
Base and lid assembly	149.94	169.34	2.05	56.33	nr	**225.67**
Lighting control module; plug in; 9 output, 9 channel dimming (DSI)						
Base and lid assembly	169.93	191.91	2.05	56.33	nr	**248.24**
Lighting control module; plug in; 9 output, 9 channel dimming (DALI)						
Base and lid assembly	179.93	203.21	2.05	56.33	nr	**259.54**
Lighting control module; hard wired; 4 circuit switching						
Base and lid assembly	159.94	180.63	1.85	50.84	nr	**231.47**
Compact lighting control module; 3 output 18 ballast drive; dimmable (DALI)						
Base and lid assembly	189.92	214.50	1.85	50.84	nr	**265.34**
Presence detectors						
Flush mounted	51.98	58.70	0.60	16.49	nr	**75.19**
Universal presence detectors with photo cell; flush mounted	57.18	64.58	0.60	16.49	nr	**81.07**
Scene switch plate; anodized aluminium finish						
4 way	62.38	70.45	1.20	32.98	nr	**103.43**

V: ELECTRICAL SUPPLY/POWER/LIGHTING

Item	Net Price £	Material £	Labour hours	Labour £	Unit	Total rate £
V22: GENERAL LV POWER						
Y74 – ACCESSORIES						
OUTLETS						
Socket outlet: unswitched; 13 amp metal clad; BS 1363; galvanized steel box and coverplate with white plastic inserts; fixed surface mounted						
1 Gang	5.95	6.72	0.41	11.27	nr	**17.99**
2 Gang	10.96	12.38	0.41	11.27	nr	**23.65**
Socket outlet: switched; 13 amp metal clad; BS 1363; galvanized steel box and coverplate with white plastic inserts; fixed surface mounted						
1 Gang	6.10	6.88	0.43	11.81	nr	**18.69**
2 Gang	10.92	12.33	0.45	12.36	nr	**24.69**
Socket outlet: switched with neon indicator; 13 amp metal clad; BS 1363; galvanized steel box and coverplate withwhite plastic inserts; fixed surface mounted						
1 Gang	12.94	14.62	0.43	11.81	nr	**26.43**
2 Gang	23.53	26.57	0.45	12.36	nr	**38.93**
Socket outlet: unswitched; 13 amp; BS 1363; white moulded plastic box and coverplate; fixed surface mounted						
1 Gang	4.47	5.05	0.41	11.27	nr	**16.32**
2 Gang	8.83	9.98	0.41	11.27	nr	**21.25**
Socket outlet; switched; 13 amp; BS 1363; white moulded plastic box and coverplate; fixed surface mounted						
1 Gang	5.40	6.10	0.43	11.81	nr	**17.91**
2 Gang	8.67	9.79	0.45	12.36	nr	**22.15**
Socket outlet: switched with neon indicator; 13 amp; BS 1363; white moulded plastic box and coverplate; fixed surface mounted						
1 Gang	11.05	12.48	0.43	11.81	nr	**24.29**
2 Gang	14.89	16.82	0.45	12.36	nr	**29.18**
Socket outlet: switched; 13 amp; BS 1363; galvanized steel box, white moulded coverplate; flush fitted						
1 Gang	5.40	6.10	0.43	11.81	nr	**17.91**
2 Gang	10.45	11.80	0.45	12.36	nr	**24.16**

V: ELECTRICAL SUPPLY/POWER/LIGHTING

Item	Net Price £	Material £	Labour hours	Labour £	Unit	Total rate £
V22: GENERAL LV POWER – cont						
OUTLETS – cont						
Socket outlet: switched with neon indicator; 13 amp; BS 1363; galvanized steel box, white moulded coverplate; flush fixed						
1 Gang	11.05	24.95	0.43	11.81	nr	**36.76**
2 Gang	19.10	43.15	0.45	12.36	nr	**55.51**
Socket outlet: switched; 13 amp; BS 1363; galvanized steel box, satin chrome coverplate; BS 4662; flush fixed						
1 Gang	18.02	20.35	0.43	11.81	nr	**32.16**
2 Gang	25.57	28.88	0.45	12.36	nr	**41.24**
Socket outlet: switched with neon indicator; 13 amp; BS 1363; steel backbox, satin chrome coverplate; BS 4662; flush fixed						
1 Gang	14.17	16.00	0.43	11.81	nr	**27.81**
2 Gang	25.57	28.88	0.45	12.36	nr	**41.24**
RCD protected socket outlets, 13 amp, to BS 1363; galvanized steel box, white moulded cover plate; flush fitted						
2 Gang, 10 mA tripping (active control)	65.87	74.39	0.45	12.36	nr	**86.75**
2 Gang, 30 mA tripping (active control)	58.68	66.27	0.45	12.36	nr	**78.63**
2 Gang, 30 mA tripping (passive control)	58.68	66.27	0.45	12.36	nr	**78.63**
Filtered socket outlets, 13 amp, to BS 1363, with separate 'clean earth' terminal; galvanized steel box, white moulded cover plate; flush fitted						
2 Gang (spike protected)	49.60	56.02	0.50	13.73	nr	**69.75**
2 Gang (spike and RFI protected)	62.78	70.90	0.55	15.12	nr	**86.02**
Replacement filter cassette	17.39	19.65	0.15	4.12	nr	**23.77**
Non-standard socket outlets, 13 amp, to BS 1363, with separate 'clean earth' terminal; for plugs with T-shaped earth pin; galvanized steel box, white moulded cover plate; flush fitted						
1 Gang	9.21	10.40	0.43	11.81	nr	**22.21**
2 Gang	16.56	18.71	0.43	11.81	nr	**30.52**
2 Gang coloured RED	23.01	25.99	0.43	11.81	nr	**37.80**

V: ELECTRICAL SUPPLY/POWER/LIGHTING

Item	Net Price £	Material £	Labour hours	Labour £	Unit	Total rate £
Weatherproof socket outlet: 40 amp; switched; single gang; RCD protected; water and dust protected to I.P.66; surface mounted						
40A 30 mA tripping current protecting 1 socket	74.95	84.65	0.52	14.29	nr	**98.94**
40A 30 mA tripping current protecting 2 sockets	79.94	90.28	0.64	17.58	nr	**107.86**
Plug for weatherproof socket outlet: protected to I.P.66						
13 amp plug	3.28	3.70	0.21	5.77	nr	**9.47**
Floor service outlet box; comprising flat lid with flanged carpet trim; twin 13 A switched socket outlets; punched plate for mounting 2 telephone outlets; one blank plate; triple compartment						
3 Compartment	33.74	38.11	0.88	24.18	nr	**62.29**
Floor service outlet box; comprising flat lid with flanged carpet trim; 2 twin 13A switched socket outlets; punched plate for mounting 1 telephone outlet; one blank plate; triple compartment						
3 Compartment	35.00	39.20	0.88	24.18	nr	63.38
Floor service outlet box; comprising flat lid with flanged carpet trim; twin 13A switched socket outlets; punched plate for mounting 2 telephone outlets; two blank plates; four compartment						
4 Compartment	46.01	51.96	0.88	24.18	nr	**76.14**
Floor service outlet box; comprising flat lid with flanged carpet trim; single 13A unswitched socket outlet; single compartment; circular						
1 Compartment	49.08	55.43	0.79	21.71	nr	**77.14**
Floor service grommet, comprising flat lid with flanged carpet trim; circular						
Floor Grommet	22.20	25.07	0.49	13.46	nr	**38.53**
POWER POSTS/POLES/PILLARS						
Power Post						
Power post; aluminium painted body; PVC-u cover; 5 nr outlets	265.00	299.29	4.00	109.92	nr	**409.21**

V: ELECTRICAL SUPPLY/POWER/LIGHTING

Item	Net Price £	Material £	Labour hours	Labour £	Unit	Total rate £
V22: GENERAL LV POWER – cont						
POWER POSTS/POLES/PILLARS – cont						
Power Pole						
Power pole; 3.6 metres high; aluminium painted body; PVC-u cover; 6 nr outlets	339.80	383.76	4.00	109.92	nr	**493.68**
Extra for						
Power pole extension bar; 900 mm long	37.14	41.94	1.50	41.21	nr	**83.15**
Vertical multi compartment pillar; PVC-u; BS 4678 Part4 EN60529; excludes accessories						
Single						
630 mm long	118.36	133.67	2.00	54.96	nr	**188.63**
3000 mm long	341.08	385.21	2.00	54.96	nr	**440.17**
Double						
630 mm long	118.36	133.67	3.00	82.44	nr	**216.11**
3000 mm long	362.01	408.86	3.00	82.44	nr	**491.30**
CONNECTION UNITS						
Connection units: moulded pattern; BS 5733; moulded plastic box; white coverplate; knockout for flex outlet; surface mounted – standard fused						
DP Switched	6.89	7.78	0.49	13.46	nr	**21.24**
Unswitched	6.32	7.14	0.49	13.46	nr	**20.60**
DP Switched with neon indicator	8.74	9.87	0.49	13.46	nr	**23.33**
Connection units: moulded pattern; BS 5733; galvanized steel box; white coverplate; knockout for flex outlet; surface mounted						
DP Switched	8.68	17.57	0.49	13.46	nr	**31.03**
DP Unswitched	8.12	9.17	0.49	13.46	nr	**22.63**
DP Switched with neon indicator	10.54	11.90	0.49	13.46	nr	**25.36**
Connection units: galvanized pressed steel pattern; galvanized steel box; satin chrome or satin brass finish; white moulded plastic inserts; flush mounted – standard fused						
DP Switched	12.70	14.34	0.49	13.46	nr	**27.80**
Unswitched	11.95	13.49	0.49	13.46	nr	**26.95**
DP Switched with neon indicator	17.14	19.36	0.49	13.46	nr	**32.82**

V: ELECTRICAL SUPPLY/POWER/LIGHTING

Item	Net Price £	Material £	Labour hours	Labour £	Unit	Total rate £
Connection units: galvanized steel box; satin chrome or satin brass finish; white moulded plastic inserts; flex outlet; flush mounted – standard fused						
Switched	12.19	13.77	0.49	13.46	nr	**27.23**
Unswitched	11.57	13.07	0.49	13.46	nr	**26.53**
Switched with neon indicator	15.68	17.71	0.49	13.46	nr	**31.17**
SHAVER SOCKETS						
Shaver unit: self setting overload device; 200/250 voltage supply; white moulded plastic faceplate; unswitched						
Surface type with moulded plastic box	22.91	25.88	0.55	15.12	nr	**41.00**
Flush type with galvanized steel box	23.97	27.07	0.57	15.67	nr	**42.74**
Shaver unit: dual voltage supply unit; white moulded plastic faceplate; unswitched						
Surface type with moulded plastic box	27.62	31.19	0.62	17.03	nr	**48.22**
Flush type with galvanized steel box	28.72	32.44	0.64	17.58	nr	**50.02**
COOKER CONTROL UNITS						
Cooker control unit: BS 4177; 45 amp D.P. main switch; 13 amp switched socket outlet; metal coverplate; plastic inserts; neon indicators						
Surface mounted with mounting box	32.44	36.64	0.61	16.76	nr	**53.40**
Flush mounted with galvanized steel box	31.00	35.01	0.61	16.76	nr	**51.77**
Cooker control unit: BS 4177; 45 amp D.P. main switch; 13 amp switched socket outlet; moulded plastic box and coverplate; surface mounted						
Standard	22.50	25.41	0.61	16.76	nr	**42.17**
With neon indicators	26.41	29.82	0.61	16.76	nr	**46.58**
CONTROL COMPONENTS						
Connector unit: moulded white plastic cover and block; galvanized steel back box; to immersion heaters						
3 Kw up to 915 mm long; fitted to thermostat	24.12	27.24	0.75	20.61	nr	**47.85**
Water heater switch: 20 amp; switched with neon indicator						
DP Switched with neon indicator	11.74	28.09	0.45	12.36	nr	**40.45**

V: ELECTRICAL SUPPLY/POWER/LIGHTING

Item	Net Price £	Material £	Labour hours	Labour £	Unit	Total rate £
V22: GENERAL LV POWER – cont						
SWITCH DISCONNECTORS						
Switch disconnectors; moulded plastic enclosure; fixed to backgrounds						
3 Pole; IP 54; Grey						
16 amp	21.81	24.63	0.80	21.98	nr	**46.61**
25 amp	25.81	29.15	0.80	21.98	nr	**51.13**
40 amp	42.03	47.47	0.80	21.98	nr	**69.45**
63 amp	65.43	73.90	1.00	27.48	nr	**101.38**
80 amp	113.44	128.12	1.25	34.34	nr	**162.46**
6 Pole; IP 54; Grey						
25 amp	36.27	40.96	1.00	27.48	nr	**68.44**
63 amp	61.26	69.19	1.25	34.34	nr	**103.53**
80 amp	116.10	131.12	1.80	49.46	nr	**180.58**
3 Pole; IP 54; Yellow						
16 amp	23.97	27.07	0.80	21.98	nr	**49.05**
25 amp	28.34	32.01	0.80	21.98	nr	**53.99**
40 amp	45.97	51.92	0.80	21.98	nr	**73.90**
63 amp	71.51	80.76	1.00	27.48	nr	**108.24**
6 Pole; IP 54; Yellow						
25 amp	36.27	40.96	1.00	27.48	nr	**68.44**
INDUSTRIAL SOCKETS/PLUGS						
Plugs; Splashproof; 100–130 volts, 50–60 Hz; IP 44 (Yellow)						
2 Pole and earth						
16 amp	2.02	2.28	0.55	15.12	nr	**17.40**
32 amp	7.25	8.19	0.60	16.49	nr	**24.68**
3 Pole and earth						
16 amp	8.03	9.07	0.65	17.86	nr	**26.93**
32 amp	10.80	12.20	0.72	19.79	nr	**31.99**
3 Pole; neutral and earth						
16 amp	8.57	9.68	0.72	19.79	nr	**29.47**
32 amp	12.99	14.67	0.78	21.44	nr	**36.11**

V: ELECTRICAL SUPPLY/POWER/LIGHTING

Item	Net Price £	Material £	Labour hours	Labour £	Unit	Total rate £
Connectors; Splashproof; 100–130 volts, 50–60 Hz; IP 44 (Yellow)						
2 pole and earth						
16 amp	5.20	5.88	0.42	11.55	nr	**17.43**
32 amp	9.00	10.17	0.50	13.73	nr	**23.90**
3 Pole and earth						
16 amp	9.77	11.04	0.48	13.19	nr	**24.23**
32 amp	14.56	16.44	0.58	15.93	nr	**32.37**
3 Pole; neutral and earth						
16 amp	13.67	15.44	0.52	14.29	nr	**29.73**
32 amp	19.18	21.66	0.73	20.05	nr	**41.71**
Angled sockets; surface mounted; Splashproof; 100–130 volts, 50–60 Hz; IP 44 (Yellow)						
2 Pole and earth						
16 amp	4.39	4.96	0.55	15.12	nr	**20.08**
32 amp	10.98	12.40	0.60	16.49	nr	**28.89**
3 Pole and earth						
16 amp	10.47	11.82	0.65	17.86	nr	**29.68**
32 amp	19.66	22.21	0.72	19.79	nr	**42.00**
3 Pole; neutral and earth						
16 amp	14.52	16.40	0.72	19.79	nr	**36.19**
32 amp	19.02	21.48	0.78	21.44	nr	**42.92**
Plugs; Watertight; 100–130 volts, 50–60 Hz; IP 67 (Yellow)						
2 Pole and earth						
16 amp	8.40	9.49	0.55	15.12	nr	**24.61**
32 amp	14.62	16.51	0.60	16.49	nr	**33.00**
63 amp	40.11	45.30	0.75	20.61	nr	**65.91**
Connectors; Watertight; 100–130 volts, 50–60 Hz; IP 67 (Yellow)						
2 Pole and earth						
16 amp	17.14	19.36	0.42	11.55	nr	**30.91**
32 amp	29.06	32.82	0.50	13.73	nr	**46.55**
63 amp	70.33	79.43	0.67	18.41	nr	**97.84**

V: ELECTRICAL SUPPLY/POWER/LIGHTING

Item	Net Price £	Material £	Labour hours	Labour £	Unit	Total rate £
V22: GENERAL LV POWER – cont						
INDUSTRIAL SOCKETS/PLUGS – cont						
Angled sockets; surface mounted; Watertight; 100–130 volts, 50–60 Hz; IP 67 (Yellow)						
2 Pole and earth						
16 amp	14.17	16.00	0.55	15.12	nr	**31.12**
32 amp	27.71	31.29	0.60	16.49	nr	**47.78**
Plugs; Splashproof; 200–250 volts, 50–60 Hz; IP 44 (Blue)						
2 Pole and earth						
16 amp	2.02	2.28	0.55	15.12	nr	**17.40**
32 amp	7.26	8.20	0.60	16.49	nr	**24.69**
63 amp	34.98	39.50	0.75	20.61	nr	**60.11**
3 Pole and earth						
16 amp	8.04	9.08	0.65	17.86	nr	**26.94**
32 amp	10.80	12.20	0.72	19.79	nr	**31.99**
63 amp	35.11	39.66	0.83	22.81	nr	**62.47**
3 Pole; neutral and earth						
16 amp	8.57	9.68	0.72	19.79	nr	**29.47**
32 amp	13.00	14.68	0.78	21.44	nr	**36.12**
Connectors; Splashproof; 200–250 volts, 50–60 Hz; IP 44 (Blue)						
2 Pole and earth						
16 amp	5.00	5.65	0.42	11.55	nr	**17.20**
32 amp	12.41	14.01	0.50	13.73	nr	**27.74**
63 amp	43.83	49.50	0.67	18.41	nr	**67.91**
3 Pole and earth						
16 amp	13.22	14.93	0.48	13.19	nr	**28.12**
32 amp	17.48	19.74	0.58	15.93	nr	**35.67**
63 amp	36.07	40.74	0.75	20.61	nr	**61.35**
3 Pole; neutral and earth						
16 amp	15.37	17.36	0.52	14.29	nr	**31.65**
32 amp	52.66	59.48	0.73	20.05	nr	**79.53**

V: ELECTRICAL SUPPLY/POWER/LIGHTING

Item	Net Price £	Material £	Labour hours	Labour £	Unit	Total rate £
Angled sockets; surface mounted; Splashproof; 200–250 volts, 50–60 Hz; IP 44 (Blue)						
2 Pole and earth						
16 amp	6.44	7.27	0.55	15.12	nr	**22.39**
32 amp	9.48	10.71	0.60	16.49	nr	**27.20**
63 amp	48.99	55.32	0.75	20.61	nr	**75.93**
3 Pole and earth						
16 amp	12.76	14.41	0.65	17.86	nr	**32.27**
32 amp	22.10	24.96	0.72	19.79	nr	**44.75**
63 amp	48.68	54.98	0.83	22.81	nr	**77.79**
3 Pole; neutral and earth						
16 amp	11.12	12.56	0.72	19.79	nr	**32.35**
32 amp	19.45	21.97	0.78	21.44	nr	**43.41**
Plugs; Watertight; 200–250 volts, 50 IP67 (Blue)						
2 Pole and earth						
16 amp	8.40	9.49	0.41	11.27	nr	**20.76**
32 amp	14.61	16.50	0.50	13.73	nr	**30.23**
63 amp	40.11	45.30	0.66	18.14	nr	**63.44**
125 amp	102.83	116.14	0.86	23.62	nr	**139.76**
Connectors; Watertight; 200–250 volts, 50–60 Hz; IP 67 (Blue)						
2 Pole and earth						
16 amp	16.91	19.09	0.42	11.55	nr	**30.64**
32 amp	26.95	30.44	0.50	13.73	nr	**44.17**
63 amp	56.63	63.95	0.67	18.41	nr	**82.36**
125 amp	168.34	190.12	0.87	23.91	nr	**214.03**
Angled sockets; surface mounted; Watertight; 200–250 volts, 50–60 Hz; IP 67 (Blue)						
2 Pole and earth						
16 amp	14.17	16.00	0.55	15.12	nr	**31.12**
32 amp	27.70	31.28	0.60	16.49	nr	**47.77**
125 amp	145.49	164.31	1.00	27.48	nr	**191.79**

V: ELECTRICAL SUPPLY/POWER/LIGHTING

Item	Net Price £	Material £	Labour hours	Labour £	Unit	Total rate £
V32: UNINTERRUPTIBLE POWER SUPPLY						
Uninterruptible power supply; sheet steel enclosure; self contained battery pack; including installation, testing and commissioning						
Single phase input and output; 5 year battery life; standard 13 A socket outlet connection						
1.0 kVA (10 minute supply)	982.28	1109.38	0.30	8.24	nr	**1117.62**
1.0 kVA (30 minute supply)	1719.26	1941.72	0.50	13.73	nr	**1955.45**
2.0 kVA (10 minute supply)	1855.12	2095.16	0.50	13.73	nr	**2108.89**
2.0 kVA (60 minute supply)	3114.48	3517.48	0.50	13.73	nr	**3531.21**
3.0 kVA (10 minute supply)	2423.15	2736.69	0.50	13.73	nr	**2750.42**
3.0 kVA (40 minute supply)	3682.52	4159.02	1.00	27.48	nr	**4186.50**
5.0 kVA (30 minute supply)	5850.73	6607.78	1.00	27.48	nr	**6635.26**
8.0 kVA (10 minute supply)	7468.93	8435.37	2.00	54.96	nr	**8490.33**
8.0 kVA (30 minute supply)	8672.05	9794.17	2.00	54.96	nr	**9849.13**
Uninterruptible power supply; including final connections and testing and commissioning						
Medium size static; single phase input and output; 10 year battery life; in cubicle						
10.0 kVA (10 minutes supply)	7977.35	9009.58	10.00	274.78	nr	**9284.36**
10.0 kVA (30 minutes supply)	9460.25	10684.36	15.00	412.17	nr	**11096.53**
15.0 kVA (10 minutes supply)	10311.37	11645.61	10.00	274.78	nr	**11920.39**
15.0 kVA (30 minutes supply)	12188.80	13765.97	15.00	412.17	nr	**14178.14**
20.0 kVA (10 minutes supply)	13085.44	14778.63	10.00	274.78	nr	**15053.41**
20.0 kVA (30 minutes supply)	14006.92	15819.34	15.00	412.17	nr	**16231.51**
Medium size static; three phase input and output; 10 year battery life; in cubicle						
10.0 kVA (10 minutes supply)	9861.67	11137.72	10.00	274.78	nr	**11412.50**
10.0 kVA (30 minutes supply)	11388.72	12862.36	15.00	412.17	nr	**13274.53**
15.0 kVA (10 minutes supply)	10981.79	12402.78	15.00	412.17	nr	**12814.95**
15.0 kVA (30 minutes supply)	13053.72	14742.80	20.00	549.57	nr	**15292.37**
20.0 kVA (10 minutes supply)	13524.11	15274.06	20.00	549.57	nr	**15823.63**
20.0 kVA (30 minutes supply)	14420.75	16286.73	25.00	686.95	nr	**16973.68**
30.0 kVA (10 minutes supply)	14669.05	16567.16	25.00	686.95	nr	**17254.11**
30.0 kVA (30 minutes supply)	15703.64	17735.62	30.00	824.34	nr	**18559.96**
Large size static; three phase input and output; 10 year battery life; in cubicle						
40 kVA (10 minutes supply)	14531.11	16411.36	30.00	824.34	nr	**17235.70**
40 kVA (30 minutes supply)	19063.97	21530.75	30.00	824.34	nr	**22355.09**
60 kVA (10 minutes supply)	21268.33	24020.35	35.00	961.74	nr	**24982.09**

V: ELECTRICAL SUPPLY/POWER/LIGHTING

Item	Net Price £	Material £	Labour hours	Labour £	Unit	Total rate £
60 kVA (30 minutes supply)	26984.76	30476.46	35.00	961.74	nr	**31438.20**
100 kVA (10 minutes supply)	24483.82	27651.90	40.00	1099.13	nr	**28751.03**
200 kVA (10 minutes supply)	42024.88	47462.69	40.00	1099.13	nr	**48561.82**
300 kVA (10 minutes supply)	62266.90	70323.93	40.00	1099.13	nr	**71423.06**
400 kVA (10 minutes supply)	82289.59	92937.45	50.00	1373.91	nr	**94311.36**
500 kVA (10 minutes supply)	100396.23	113387.00	60.00	1648.69	nr	**115035.69**
600 kVA (10 minutes supply)	115646.02	130610.04	70.00	1923.48	nr	**132533.52**
800 kVA (10 minutes supply)	150853.67	170373.38	80.00	2198.25	nr	**172571.63**
Integral diesel rotary; three phase input and output; no break supply; including ventilation and accoustic attenuation, oil day tank and interconnecting pipework						
100 kVA	127536.62	144039.22	100.00	2747.82	nr	**146787.04**
125 kVA	144795.48	163531.30	100.00	2747.82	nr	**166279.12**
150 kVA	155683.52	175828.19	100.00	2747.82	nr	**178576.01**
180 kVA	166822.78	188408.81	100.00	2747.82	nr	**191156.63**
200 kVA	245786.35	277589.88	100.00	2747.82	nr	**280337.70**
250 kVA	253364.59	286148.70	100.00	2747.82	nr	**288896.52**
300 kVA	260695.37	294428.05	120.00	3297.38	nr	**297725.43**
400 kVA	307419.79	347198.37	120.00	3297.38	nr	**350495.75**
500 kVA	337595.30	381278.45	120.00	3297.38	nr	**384575.83**
630 kVA	408409.40	461255.54	140.00	3846.94	nr	**465102.48**
800 kVA	493102.59	556907.60	140.00	3846.94	nr	**560754.54**
1000 kVA	554131.07	625832.86	140.00	3846.94	nr	**629679.80**
1125 kVA	624873.92	705729.48	160.00	4396.51	nr	**710125.99**
1250 kVA	657755.51	742865.79	160.00	4396.51	nr	**747262.30**
1500 kVA	721915.05	815327.25	170.00	4671.29	nr	**819998.54**
1750 kVA	819132.41	925124.05	170.00	4671.29	nr	**929795.34**

V: ELECTRICAL SUPPLY/POWER/LIGHTING

Item	Net Price £	Material £	Labour hours	Labour £	Unit	Total rate £
V40: EMERGENCY LIGHTING						
Y73 – LUMINAIRES						
24 VOLT/50 VOLT/110 VOLT FLUORESCENT SLAVE LUMINAIRES						
For use with DC central battery systems						
Indoor, 8 Watt	36.70	41.45	0.80	21.98	nr	**63.43**
Indoor, exit sign box	45.72	51.63	0.80	21.98	nr	**73.61**
Outdoor, 8 Watt weatherproof	40.90	46.20	0.80	21.98	nr	**68.18**
Conversion module AC/DC	45.72	51.63	0.25	6.86	nr	**58.49**
Self contained; polycarbonate base and diffuser; LED charging light to European sign directive; 3 hour standby						
Non maintained						
Indoor, 8 Watt	40.52	45.77	1.00	27.48	nr	**73.25**
Outdoor, 8 Watt weatherproof, vandal-resistant IP65	58.13	65.65	1.00	27.48	nr	**93.13**
Maintained						
Indoor, 8 Watt	29.12	32.88	1.00	27.48	nr	**60.36**
Outdoor, 8 Watt weatherproof, vandal-resistant IP65	85.91	97.02	1.00	27.48	nr	**124.50**
Exit signage						
Exit sign; gold effect, pendular including brackets						
Non maintained, 8 Watt	110.29	124.56	1.00	27.48	nr	**152.04**
Maintained, 8 Watt	118.72	134.08	1.00	27.48	nr	**161.56**
Modification kit						
Module and battery for 58 W fluorescent modification from mains fitting to emergency; 3 hour standby	38.02	42.94	0.50	13.73	nr	**56.67**
Extra for remote box (when fitting is too small for modification)	18.88	21.32	0.50	13.73	nr	**35.05**
12 Volt low voltage lighting; non maintained; 3 hour standby						
2 × 20 Watt lamp load	131.39	148.39	1.20	32.98	nr	**181.37**
1 × 50 Watt lamp load	143.43	161.99	1.00	27.48	nr	**189.47**
Maintained 3 hour standby						
2 × 20 Watt lamp load	124.83	140.98	1.20	32.98	nr	**173.96**
1 × 50 Watt lamp load	136.25	153.88	1.00	27.48	nr	**181.36**

V: ELECTRICAL SUPPLY/POWER/LIGHTING

Item	Net Price £	Material £	Labour hours	Labour £	Unit	Total rate £
DC CENTRAL BATTERY SYSTEMS BS5266 COMPLIANT 24/50/110 VOLT						
DC supply to luminaires on mains failure; metal cubicle with battery charger, changeover device and battery as integral unit; ICEL 1001 compliant; 10 year design life valve regulated lead acid battery; 24 hour recharge; LCD display & LED indication; ICEL alarm pack; Includes on-site commissioning on 110 Volt systems only						
24 Volt, wall mounted						
300 W maintained, 1 hour	2088.71	2358.97	4.00	109.92	nr	**2468.89**
635 W maintained, 3 hour	2533.79	2861.65	6.00	164.87	nr	**3026.52**
470 W non maintained, 1 hour	1862.07	2103.02	4.00	109.92	nr	**2212.94**
780 W non maintained, 3 hour	2200.86	2485.64	6.00	164.87	nr	**2650.51**
50 Volt						
935 W maintained, 3 hour	2000.00	2240.00	8.00	219.83	nr	**2459.83**
1965 W maintained, 3 hour	2784.95	3145.31	8.00	219.83	nr	**3365.14**
1311 W non maintained, 3 hour	2726.54	3079.34	8.00	219.83	nr	**3299.17**
2510 W non maintained 3, hour	4163.40	4702.12	8.00	219.83	nr	**4921.95**
110 Volt						
1603 W maintained, 3 hour	3923.93	4431.67	8.00	219.83	nr	**4651.50**
4446 W maintained, 3 hour	4481.15	5060.98	10.00	274.78	nr	**5335.76**
2492 W non maintained, 3 hour	4078.13	4605.82	10.00	274.78	nr	**4880.60**
5429 W non maintained, 3 hour	6794.15	7673.28	12.00	329.74	nr	**8003.02**
DC CENTRAL BATTERY SYSTEMS; BS EN 50171 COMPLIANT; 24/50/110 Volt						
Central power systems						
DC supply to luminaires on mains failure; metal cubicle with battery charger, changeover device and battery as integral unit; 10 year design life valve regulated lead acid battery; 12 hour recharge to 80% of specified duty; low volts discount; LCD display & LED indication; includes on-site commissioning for CPS systems only; battery sized for 'end of life' @ 20°C test pushbutton						
24 Volt, floor standing						
400 W non maintained, 1 hour	1961.39	2215.19	4.00	109.92	nr	**2325.11**
600 W maintained, 3 hour	2522.11	2848.46	6.00	164.87	nr	**3013.33**

V: ELECTRICAL SUPPLY/POWER/LIGHTING

Item	Net Price £	Material £	Labour hours	Labour £	Unit	Total rate £
V40: EMERGENCY LIGHTING – cont						
DC CENTRAL BATTERY SYSTEMS; BS EN 50171 COMPLIANT – cont						
Central power systems – cont						
50 Volt						
2133 W non maintained, 3 hour	4408.73	4979.20	8.00	219.83	nr	**5199.03**
1900 W maintained, 3 hour	4213.63	4758.85	8.00	219.83	nr	**4978.68**
110 Volt						
2200 W non maintained, 3 hour	4078.13	4605.82	8.00	219.83	nr	**4825.65**
4000 W maintained, 3 hour	4551.24	5140.15	12.00	329.74	nr	**5469.89**
Low Power Systems						
DC supply to luminaires on mains failure; metal cubicle with battery charger, changeover device and battery as integral unit; 5 Year design life valve regulated lead acid battery; low volts discount; LED display & LED indication; battery sized for 'end of life' @ 20°C test pushbutton						
24 Volt, floor standing						
300 W non maintained, 1 hour	2151.38	2429.76	4.00	109.92	nr	**2539.68**
600 W maintained, 3 hour	2381.19	2689.30	6.00	164.87	nr	**2854.17**
AC static inverter system; BS5266 compliant; one hour standby						
Central system supplying AC power on mains failure to mains luminaires; ICEL 1001 compliant metal cubicle(s) with changeover device, battery charger, battery & static inverter; 10 year design life valve regulated lead acid battery; 24 hour recharge; LED indication and LCD display; pure sinewave output						
One hour						
750 VA, 600 W single phase I/P & O/P	2543.13	2872.20	6.00	164.87	nr	**3037.07**
3 kVA, 2.55 KW single phase I/P & O/P	4172.75	4712.69	8.00	219.83	nr	**4932.52**
5 kVA, 4.25 KW single phase I/P & O/P	5509.15	6222.01	10.00	274.78	nr	**6496.79**
8 kVA, 6.80 KW single phase I/P & O/P	6490.42	7330.25	12.00	329.74	nr	**7659.99**
10 kVA, 8.5 KW single phase I/P & O/P	8378.21	9462.31	14.00	384.69	nr	**9847.00**
13 kVA, 11.05 KW single phase I/P & O/P	11208.71	12659.06	16.00	439.65	nr	**13098.71**
15 kVA, 12.75 KW single phase I/P & O/P	11616.41	13119.52	30.00	824.34	nr	**13943.86**

V: ELECTRICAL SUPPLY/POWER/LIGHTING

Item	Net Price £	Material £	Labour hours	Labour £	Unit	Total rate £
20 kVA, 17.0 KW 3 phase I/P & single phase O/P	16043.82	18119.81	40.00	1099.13	nr	**19218.94**
30 kVA, 25.5 KW 3 phase I/P & O/P	21681.46	24486.94	60.00	1648.69	nr	**26135.63**
40 kVA, 34.0 KW 3 phase I/P & O/P	27029.41	30526.88	80.00	2198.25	nr	**32725.13**
50 kVA, 42.5 KW 3 phase I/P & O/P	35252.24	39813.70	90.00	2473.04	nr	**42286.74**
65 kVA, 55.25 KW 3 phase I/P & O/P	43244.96	48840.64	100.00	2747.82	nr	**51588.46**
90 kVA, 68.85 KW 3 phase I/P & O/P	57146.32	64540.77	120.00	3297.38	nr	**67838.15**
120 kVA, 102 KW 3 phase I/P & O/P	63789.77	72043.84	150.00	4121.73	nr	**76165.57**
Three hour						
750 VA, 600 W single phase I/P & O/P	2976.53	3361.68	6.00	164.87	nr	**3526.55**
3 kVA, 2.55 KW single phase I/P & O/P	5008.01	5656.02	8.00	219.83	nr	**5875.85**
5 kVA, 4.25 KW single phase I/P & O/P	7938.80	8966.04	10.00	274.78	nr	**9240.82**
8 kVA, 6.80 KW single phase I/P & O/P	9876.99	11155.03	12.00	329.74	nr	**11484.77**
10 kVA, 8.5 KW single phase I/P & O/P	13028.74	14714.59	14.00	384.69	nr	**15099.28**
13 kVA, 11.05 KW single phase I/P & O/P	16064.84	18143.55	16.00	439.65	nr	**18583.20**
15 kVA, 12.75 KW single phase I/P & O/P	16327.69	18440.41	30.00	824.34	nr	**19264.75**
20 kVA, 17.0 KW 3 phase I/P & single phase O/P	23772.52	26848.56	40.00	1099.13	nr	**27947.69**
30 kVA, 25.5 KW 3 phase I/P & O/P	35529.10	40126.39	60.00	1648.69	nr	**41775.08**
40 kVA, 34.0 KW 3 phase I/P & O/P	42478.62	47975.14	80.00	2198.25	nr	**50173.39**
50 kVA, 42.5 KW 3 phase I/P & O/P	57293.51	64707.00	90.00	2473.04	nr	**67180.04**
65 kVA, 55.25 KW 3 phase I/P & O/P	69273.23	78236.84	100.00	2747.82	nr	**80984.66**
90 kVA, 68.85 KW 3 phase I/P & O/P	81984.21	92592.56	120.00	3297.38	nr	**95889.94**
120 kVA, 102 KW 3 phase I/P & O/P	101047.79	114122.87	150.00	4121.73	nr	**118244.60**
AC static inverter system; BS EN 50171 compliant; one hour standby; Low power system (typically wall mounted)						
Central system supplying AC power on mains failure to mains luminaires; metal cubicle(s) with changeover device, battery charger, battery and static inverter; 5 year design life valve regulated lead acid battery; LED indication and LCD display; 12 hour recharge to 80% duty; inverter rated for 120% of load for 100% of duty; battery sized for 'end of life' @ 20°C test pushbutton						
One hour						
300 VA, 240 W single phase I/P & O/P	1144.82	1292.95	3.00	82.44	nr	**1375.39**
600 VA, 480 W single phase I/P & O/P	1378.46	1556.83	4.00	109.92	nr	**1666.75**
750 VA, 600 W single phase I/P & O/P	2543.13	2872.20	6.00	164.87	nr	**3037.07**
Three hour						
150 VA, 120 W single phase I/P & O/P	1261.64	1424.89	3.00	82.44	nr	**1507.33**
450 VA, 360 W single phase I/P & O/P	1495.27	1688.75	4.00	109.92	nr	**1798.67**
750 VA, 600 W single phase I/P & O/P	2976.53	3361.68	6.00	164.87	nr	**3526.55**

V: ELECTRICAL SUPPLY/POWER/LIGHTING

Item	Net Price £	Material £	Labour hours	Labour £	Unit	Total rate £
V40: EMERGENCY LIGHTING – cont						
DC CENTRAL BATTERY SYSTEMS; BS EN 50171 COMPLIANT – cont						
AC static inverter system central power system; CPS BS EN 50171 compliant; one hour standby						
Central system supplying AC power on mains failure to mains luminaires; metal cubicle(s) with changeover device, battery charger, battery & static inverter; LED indication and LCD display; pure sinewave output; 10 year design life valve regulated lead acid battery; 12 hour recharge to 80% duty specified; unverter rated for 120% of load for 100% of duty; battery sized for 'end of life' @ 20°C test push button; includes on-site commissioning						
One hour						
750 VA, 600 W single phase I/P & O/P	2846.86	3215.23	6.00	164.87	nr	**3380.10**
3 kVA, 2.55 KW single phase I/P & O/P	4534.88	5121.67	8.00	219.83	nr	**5341.50**
5 kVA, 4.25 KW single phase I/P & O/P	5871.28	6630.99	10.00	274.78	nr	**6905.77**
8 kVA, 6.80 KW single phase I/P & O/P	6852.57	7739.26	12.00	329.74	nr	**8069.00**
10 kVA, 8.5 KW single phase I/P & O/P	8798.75	9937.27	14.00	384.69	nr	**10321.96**
13 kVA, 11.05 KW single phase I/P & O/P	11629.26	13134.03	16.00	439.65	nr	**13573.68**
15 kVA, 12.75 KW single phase I/P & O/P	12036.95	13594.47	30.00	824.34	nr	**14418.81**
20 kVA, 17.0 KW 3 phase I/P & single phase O/P	16581.18	18726.70	40.00	1099.13	nr	**19825.83**
30 kVA, 25.5 KW 3 phase I/P & O/P	22335.65	25225.77	60.00	1648.69	nr	**26874.46**
40 kVA, 34.0 KW 3 phase I/P & O/P	27683.58	31265.69	80.00	2198.25	nr	**33463.94**
50 kVA, 42.5 KW 3 phase I/P & O/P	36052.45	40717.46	90.00	2473.04	nr	**43190.50**
65 kVA, 55.25 KW 3 phase I/P & O/P	44529.95	50291.91	100.00	2747.82	nr	**53039.73**
90 kVA, 68.85 KW 3 phase I/P & O/P	58431.33	65992.05	120.00	3297.38	nr	**69289.43**
120 kVA, 102 KW 3 phase I/P & O/P	65074.77	73495.12	150.00	4121.73	nr	**77616.85**
Three hour						
750 VA, 600 W single phase I/P & O/P	3280.26	3704.71	6.00	164.87	nr	**3869.58**
3 kVA, 2.55 KW single phase I/P & O/P	5370.13	6065.00	8.00	219.83	nr	**6284.83**
5 kVA, 4.25 KW single phase I/P & O/P	8320.97	9397.66	10.00	274.78	nr	**9672.44**
8 kVA, 6.80 KW single phase I/P & O/P	10239.11	11564.00	12.00	329.74	nr	**11893.74**
10 kVA, 8.5 KW single phase I/P & O/P	13449.28	15189.55	14.00	384.69	nr	**15574.24**
13 kVA, 11.05 KW single phase I/P & O/P	16485.39	18618.52	16.00	439.65	nr	**19058.17**
15 kVA, 12.75 KW single phase I/P & O/P	16748.23	18915.37	30.00	824.34	nr	**19739.71**
20 kVA, 17.0 KW 3 phase I/P & single phase O/P	24309.88	27455.45	40.00	1099.13	nr	**28554.58**
30 kVA, 25.5 KW 3 phase I/P & O/P	36183.28	40865.22	60.00	1648.69	nr	**42513.91**
40 kVA, 34.0 KW 3 phase I/P & O/P	43132.81	48713.98	80.00	2198.25	nr	**50912.23**
50 kVA, 42.5 KW 3 phase I/P & O/P	58093.71	65610.75	90.00	2473.04	nr	**68083.79**
65 kVA, 55.25 KW 3 phase I/P & O/P	70558.23	79688.11	100.00	2747.82	nr	**82435.93**
90 kVA, 68.85 KW 3 phase I/P & O/P	87719.44	99069.90	120.00	3297.38	nr	**102367.28**
120 kVA, 102 KW 3 phase I/P & O/P	102332.79	115574.14	150.00	4121.73	nr	**119695.87**

W: COMMUNICATIONS/SECURITY/CONTROL

Item	Net Price £	Material £	Labour hours	Labour £	Unit	Total rate £
W10: TELECOMMUNICATIONS						
Y61 – CABLES						
Multipair internal telephone cable; BS 6746; loose laid on tray/basket						
0.5 millimetre diameter conductor LSZH insulated and sheathed multipair cables; BT specification CW 1308						
3 pair	0.11	0.13	0.03	0.83	m	**0.96**
4 pair	0.13	0.15	0.03	0.83	m	**0.98**
6 pair	0.20	0.23	0.03	0.83	m	**1.06**
10 pair	0.40	0.45	0.05	1.38	m	**1.83**
15 pair	0.50	0.56	0.06	1.65	m	**2.21**
20 pair + 1 wire	0.67	0.75	0.06	1.65	m	**2.40**
25 pair	0.82	0.93	0.08	2.20	m	**3.13**
40 pair + earth	1.28	1.45	0.08	2.20	m	**3.65**
50 pair + earth	1.67	1.89	0.10	2.74	m	**4.63**
80 pair + earth	2.60	2.94	0.10	2.74	m	**5.68**
100 pair + earth	3.22	3.64	0.12	3.29	m	**6.93**
Multipair internal telephone cable; BS 6746; installed in conduit/trunking						
0.5 millimetre diameter conductor LSZH insulated and sheathed multipair cables; BT specification CW 1308						
3 pair	0.11	0.13	0.05	1.38	m	**1.51**
4 pair	0.13	0.15	0.06	1.65	m	**1.80**
6 pair	0.20	0.23	0.06	1.65	m	**1.88**
10 pair	0.40	0.45	0.07	1.92	m	**2.37**
15 pair	0.50	0.56	0.07	1.92	m	**2.48**
20 pair + 1 wire	0.67	0.75	0.09	2.47	m	**3.22**
25 pair	0.82	0.93	0.10	2.74	m	**3.67**
40 pair + earth	1.28	1.45	0.12	3.29	m	**4.74**
50 pair + earth	1.67	1.89	0.14	3.86	m	**5.75**
80 pair + earth	2.60	2.94	0.14	3.86	m	**6.80**
100 pair + earth	3.22	3.64	0.15	4.12	m	**7.76**
Low speed data; unshielded twisted pair; solid copper conductors; LSOH sheath; nominal impedance 100 Ohm; Category 3 to ISO IS 1801/EIA/TIA 568B and EN50173/ 50174 standards to current revisions						
Installed in riser						
25 pair 24 AWG	1.56	1.76	0.03	0.83	m	**2.59**
50 pair 24 AWG	3.12	3.52	0.06	1.65	m	**5.17**
100 pair 24 AWG	5.25	5.93	0.10	2.74	m	**8.67**

W: COMMUNICATIONS/SECURITY/CONTROL

Item	Net Price £	Material £	Labour hours	Labour £	Unit	Total rate £
W10: TELECOMMUNICATIONS – cont						
Y61 – CABLES – cont						
Low speed data – cont						
Installed below floor						
25 pair 24 AWG	1.56	1.76	0.02	0.55	m	**2.31**
50 pair 24 AWG	3.12	3.52	0.05	1.38	m	**4.90**
100 pair 24 AWG	5.25	5.93	0.08	2.20	m	**8.13**
Y74 – ACCESSORIES						
Telephone outlet: moulded plastic plate with box; fitted and connected; flush or surface mounted						
Single master outlet	6.02	6.80	0.35	9.62	nr	**16.42**
Single secondary outlet	4.44	5.02	0.35	9.62	nr	**14.64**
Telephone outlet: bronze or satin chromeplate; with box; fitted and connected; flush or surface mounted						
Single master outlet	9.80	11.07	0.35	9.62	nr	**20.69**
Single secondary outlet	10.71	12.10	0.35	9.62	nr	**21.72**
Frames and box connections						
Provision and Installation of a Dual Vertical Krone 108A Voice Distribution Frame that can accommodate a total of 138 × Krone 237A Strips	216.67	244.71	1.50	41.21	nr	**285.92**
Label Frame (Traffolyte style)	1.37	1.55	0.27	7.42	nr	**8.97**
Provision and Installation of a Box Connection 301A Voice Termination Unit that can accommodate a total of 10 × Krone 237A Strips	18.79	21.23	0.25	6.86	nr	**28.09**
Label Frame (Traffolyte style)	0.45	0.51	0.08	2.28	nr	**2.79**
Provision and Installation of a Box Connection 201 Voice Termination Unit that can accommodate 20 pairs	10.00	11.29	0.17	4.56	nr	**15.85**
Label Frame (Traffolyte style)	0.45	0.51	0.08	2.28	nr	**2.79**

W: COMMUNICATIONS/SECURITY/CONTROL

Item	Net Price £	Material £	Labour hours	Labour £	Unit	Total rate £
Terminate, Test and Label Voice Multicore System						
Patch Panels						
Voice; 19' wide fully loaded, finished in black including termination and forming of cables (assuming 2 pairs per port)						
25 port – RJ45 UTP – Krone	47.02	53.10	2.60	71.44	nr	**124.54**
50 port – RJ45 UTP – Krone	64.09	72.38	4.65	127.77	nr	**200.15**
900 pair fully loaded Systimax style of frame including forming and termination of 9 × 100 pair cables	529.54	598.06	25.00	686.95	nr	**1285.01**
Installation and termination of Krone Strip (10 pair block – 237 A) including designation label strip	4.21	4.75	0.50	13.73	nr	**18.48**
Patch panel and Outlet labelling per port (Traffolyte style)	0.16	0.19	0.02	0.55	nr	**0.74**
Provision and installation of a voice jumper, for cross termination on Krone Termination Strips	0.05	0.07	0.06	1.65	nr	**1.72**
CW1308/Cat 3 cable circuit test per pair	–	–	0.04	1.09	nr	**1.09**

W: COMMUNICATIONS/SECURITY/CONTROL

Item	Net Price £	Material £	Labour hours	Labour £	Unit	Total rate £
W20: RADIO/TELEVISION						
RADIO						
Y61 – CABLES						
Radio Frequency Cable; BS 2316; PVC sheathed; laid loose						
7/0.41 mm tinned copper inner conductor; solid polyethylene dielectric insulation; bare copper wire braid; PVC sheath; 75 ohm impedance						
Cable	1.16	1.31	0.05	1.38	m	**2.69**
Twin 1/0.58 mm copper covered steel solid core wire conductor; solid polyethylene dielectric insulation; barecopper wire braid; PVC sheath; 75 ohm impedance						
Cable	1.65	1.87	0.05	1.38	m	**3.25**
TELEVISION						
Y61 – CABLES						
Television aerial cable; coaxial; PVC sheathed; fixed to backgrounds						
General purpose TV aerial downlead; copper stranded inner conductor; cellular polythene insulation; copper braid outer conductor; 75 ohm impedance						
7/0.25 mm	0.30	0.34	0.06	1.65	m	**1.99**
Low loss TV aerial downlead; solid copper inner conductor; cellular polythene insulation; copper braid outer; conductor; 75 ohm impedance						
1/1.12 mm	0.49	0.55	0.06	1.65	m	**2.20**
Low loss air spaced; solid copper inner conductor; air spaced polythene insulation; copper braid outer conductor; 75 ohm impedance						
1/1.00 mm	0.29	0.33	0.06	1.65	m	**1.98**

W: COMMUNICATIONS/SECURITY/CONTROL

Item	Net Price £	Material £	Labour hours	Labour £	Unit	Total rate £
Satelite aerial downlead; solid copper inner conductor; air spaced polythene insulation; copper tape and braid outer conductor; 75 ohm impedance						
1/1.00 mm	0.62	0.70	0.06	1.65	m	**2.35**
Satellite TV coaxial; solid copper inner conductor; semi air spaced polyethylene dielectric insulation; plain annealed copper foil and copper braid screen in outer conductor; PVC sheath; 75 ohm impedance						
1/1.25 mm	0.82	0.93	0.08	2.20	m	**3.13**
Satellite TV coaxial; solid copper inner conductor; air spaced polyethylene dielectric insulation; plain annealed copper foil and copper braid screen in outer conductor; PVC sheath; 75 ohm impedance						
1/1.67 mm	1.35	1.52	0.09	2.47	m	**3.99**
Video cable; PVC flame retardant sheath; laid loose						
7/0.1 mm silver coated copper covered annealed steel wire conductor; polyethylene dielectric insulation with tin coated copper wire braid; 75 ohm impedance						
Cable	0.78	0.88	0.05	1.38	m	**2.26**
Y74 – ACCESSORIES						
TV co-axial socket outlet: moulded plastic box; flush or surface mounted						
One way Direct Connection	7.45	8.42	0.35	9.62	nr	**18.04**
Two way Direct Connection	10.39	11.73	0.35	9.62	nr	**21.35**
One way Isolated UHF/VHF	13.12	14.82	0.35	9.62	nr	**24.44**
Two way Isolated UHF/VHF	17.88	20.20	0.35	9.62	nr	**29.82**

W: COMMUNICATIONS/SECURITY/CONTROL

Item	Net Price £	Material £	Labour hours	Labour £	Unit	Total rate £
W23: CLOCKS						
Clock timing systems; master and slave units; fixed to background; excluding supports and fixings						
Quartz master clock with solid state digital readout for parallel loop operation; one minute, half minute and one second pulse; maximum of 160 clocks						
Over two loops only	642.60	725.75	4.40	120.90	nr	**846.65**
Power supplies for above, giving 24 hours power reserve						
2 6 amp hour batteries	187.43	211.68	3.00	82.44	nr	**294.12**
2 15 amp hour batteries	214.20	241.91	5.00	137.39	nr	**379.30**
Radio receiver to accept BBC Rugby Transmitter MSF signal						
To synchronize time of above Quartz master clock	139.23	157.24	7.04	193.51	nr	**350.75**
Wall clocks for slave (impulse) systems; 24 V DC, white dial with black numerals fitted with Axis Polycarbonate disc; BS 467.7 Class O						
305 mm dia. 1 minute impulse	56.27	63.55	1.37	37.64	nr	**101.19**
305 mm dia. 1/2 minute impulse	56.27	63.55	1.37	37.64	nr	**101.19**
227 mm dia. 1 second impulse	80.33	90.73	1.37	37.64	nr	**128.37**
305 mm dia. 1 second impulse	80.33	90.73	1.37	37.64	nr	**128.37**
Quartz battery movement; BS 467.7 Class O; white dial with black numerals and sweep second hand; fitted with Axis Polycarbonate disc; stove enamel case						
305 mm dia.	29.46	33.27	0.77	21.15	nr	**54.42**
Internal wall mounted electric clock; white dial with black numerals; 240v, 50 Hz						
305 mm dia.	37.48	42.33	1.00	27.48	nr	**69.81**
458 mm dia.	160.65	181.44	1.00	27.48	nr	**208.92**
Matching clock; BS 467.7 Class O; 240 V AC, 50/60 Hz mains supply; 12 hour duration; dial with 1–12; IP 66; Axis Polycarbonate disc; spun metal movement cover; semi flush mount on 6 point fixing bezel						
227 mm dia.	214.20	241.91	0.62	17.03	nr	**258.94**

W: COMMUNICATIONS/SECURITY/CONTROL

Item	Net Price £	Material £	Labour hours	Labour £	Unit	Total rate £
Digital clocks; 240 V, 50 Hz supply; with/ without synchronization from masterclock; 12/24 hour display; stand alone operation; 50 mm digits						
Flush – hours/minutes/seconds or minutes/ seconds/10th seconds	214.20	241.91	0.57	15.67	nr	**257.58**
Surface – hours/minutes/seconds or minutes/seconds/10th seconds	187.43	453.59	0.57	15.67	nr	**469.26**
Flush – hours/minutes or minutes/seconds	182.07	641.65	0.57	15.67	nr	**657.32**
Surface – hours/minutes or minutes/ seconds	155.29	405.78	0.57	15.67	nr	**421.45**

W: COMMUNICATIONS/SECURITY/CONTROL

Item	Net Price £	Material £	Labour hours	Labour £	Unit	Total rate £
W30: DATA TRANSMISSION						
Cabinets						
Floor standing; suitable for 19' patch panels with glass lockable doors, metal rear doors, side panels, vertical cable management, 2 × 4 way PDU's, 4 way fan, earth bonding kit; installed on raised floor						
600 wide × 800 deep – 18U	648.52	732.43	3.00	82.44	nr	814.87
600 wide × 800 deep – 24U	675.54	762.95	3.00	82.44	nr	845.39
600 wide × 800 deep – 33U	725.52	819.40	4.00	109.92	nr	929.32
600 wide × 800 deep – 42U	778.23	878.93	4.00	109.92	nr	988.85
600 wide × 800 deep – 47U	821.53	927.83	4.00	109.92	nr	1037.75
800 wide × 800 deep – 42U	880.92	994.91	4.00	109.92	nr	1104.83
800 wide × 800 deep – 47U	918.78	1037.67	4.00	109.92	nr	1147.59
Label cabinet	2.08	2.35	0.25	6.86	nr	9.21
Wall mounted; suitable for 19' patch panels with glass lockable doors, side panels, vertical cable management, 2 × 4 way PDU's, 4 way fan, earth bonding kit; fixed to wall						
19 wide × 500 deep – 9U	432.54	488.51	3.00	82.44	nr	570.95
19 wide × 500 deep – 12U	443.18	500.52	3.00	82.44	nr	582.96
19 wide × 500 deep – 15U	459.27	518.70	3.00	82.44	nr	601.14
19 wide × 500 deep – 18U	540.43	610.36	3.00	82.44	nr	692.80
19 wide × 500 deep – 21U	562.21	634.96	3.00	82.44	nr	717.40
Label cabinet	2.08	2.35	0.25	6.86	nr	9.21
Frames						
Floor standing; suitable for 19' patch panels with supports, vertical cable management, earth bonding kit; installed on raised floor						
19 wide × 500 deep – 25U	486.49	549.44	2.50	68.70	nr	618.14
19 wide × 500 deep – 39U	594.37	671.28	2.50	68.70	nr	739.98
19 wide × 500 deep – 42U	648.57	732.49	2.50	68.70	nr	801.19
19 wide × 500 deep – 47U	702.76	793.69	2.50	68.70	nr	862.39
Label frame	2.08	2.35	0.25	6.86	nr	9.21
Copper Data Cabling						
Unshielded twisted pair; solid copper conductors; PVC insulation; nominal impedance 100 Ohm; Cat 5e to ISO 11801, EIA/TIA 568B and EN 50173/50174 standards to the current revisions						
4 pair 24AWG; nominal outside diameter 5.6 mm; installed above ceiling	0.14	0.17	0.02	0.55	m	0.72
4 pair 24AWG; nominal outside diameter 5.6 mm; installed in riser	0.14	0.17	0.02	0.55	m	0.72

W: COMMUNICATIONS/SECURITY/CONTROL

Item	Net Price £	Material £	Labour hours	Labour £	Unit	Total rate £
4 pair 24AWG; nominal outside diameter 5.6 mm; installed below floor	0.14	0.17	0.01	0.28	m	**0.45**
4 pair 24AWG; nominal outside diameter 5.6 mm; installed in trunking	0.14	0.17	0.02	0.55	m	**0.72**
Unshielded twisted pair; solid copper conductors; LSOH sheathed; nominal impedance 100 Ohm; Cat 5e to ISO 11801, EIA/TIA 568B and EN 50173/50174 standards to the current revisions						
4 pair 24AWG; nominal outside diameter 5.6 mm; installed above ceiling	0.22	0.25	0.02	0.55	m	**0.80**
4 pair 24AWG; nominal outside diameter 5.6 mm; installed in riser	0.22	0.25	0.02	0.55	m	**0.80**
4 pair 24AWG; nominal outside diameter 5.6 mm; installed below floor	0.22	0.25	0.01	0.28	m	**0.53**
4 pair 24AWG; nominal outside diameter 5.6 mm; installed in trunking	0.22	0.25	0.02	0.55	m	**0.80**
Unshielded twisted pair; solid copper conductors; PVC insulation; nominal impedance 100 Ohm; Cat 6 to ISO 11801, EIA/TIA 568B and EN 50173/50174 standards to the current revisions						
4 pair 24AWG; nominal outside diameter 5.6 mm; installed above ceiling	0.23	0.26	0.02	0.41	m	**0.67**
4 pair 24AWG; nominal outside diameter 5.6 mm; installed in riser	0.23	0.26	0.02	0.41	m	**0.67**
4 pair 24AWG; nominal outside diameter 5.6 mm; installed below floor	0.23	0.26	0.01	0.22	m	**0.48**
4 pair 24AWG; nominal outside diameter 5.6 mm; installed in trunking	0.23	0.26	0.02	0.55	m	**0.81**
Unshielded twisted pair; solid copper conductors; LSOH sheathed; nominal impedance 100 Ohm; Cat 6 to ISO 11801, EIA/TIA 568B and EN 50173/50174 standards to the current revisions						
4 pair 24AWG; nominal outside diameter 5.6 mm; installed above ceiling	0.27	0.31	0.02	0.61	m	**0.92**
4 pair 24AWG; nominal outside diameter 5.6 mm; installed in riser	0.27	0.31	0.02	0.61	m	**0.92**
4 pair 24AWG; nominal outside diameter 5.6 mm; installed below floor	0.27	0.31	0.01	0.31	m	**0.62**
4 pair 24AWG; nominal outside diameter 5.6 mm; installed in trunking	0.27	0.31	0.02	0.61	m	**0.92**

W: COMMUNICATIONS/SECURITY/CONTROL

Item	Net Price £	Material £	Labour hours	Labour £	Unit	Total rate £
W30: DATA TRANSMISSION – cont						
Patch Panels						
Category 5e; 19' wide fully loaded, finished in black including termination and forming of cables						
24 port – RJ45 UTP – Krone / 110	68.30	77.14	4.75	130.52	nr	**207.66**
48 port – RJ45 UTP – Krone / 110	128.18	144.77	9.35	256.93	nr	**401.70**
Patch panel labelling per port	0.22	0.25	0.02	0.55	nr	**0.80**
Category 6; 19' wide fully loaded, finished in black including termination and forming of cables						
24 port – RJ45 UTP – Krone / 110	130.65	147.55	5.00	137.39	nr	**284.94**
48 port – RJ45 UTP – Krone / 110	238.05	268.85	9.80	269.28	nr	**538.13**
Patch panel labelling per port	0.22	0.25	0.02	0.55	nr	**0.80**
Work Station						
Category 5e RJ45 data outlet plate and multiway outlet boxes for wall, ceiling and below floor installations including label to ISO 11801 standards						
Wall mounted; fully loaded						
One gang LSOH PVC plate	5.91	6.67	0.15	4.12	nr	**10.79**
Two gang LSOH PVC plate	8.46	9.56	0.20	5.50	nr	**15.06**
Four gang LSOH PVC plate	14.48	16.35	0.40	10.99	nr	**27.34**
One gang satin brass plate	14.79	16.71	0.25	6.86	nr	**23.57**
Two gang satin brass plate	16.67	18.83	0.33	9.07	nr	**27.90**
Ceiling mounted; fully loaded						
One gang metal clad plate	7.85	8.86	0.33	9.07	nr	**17.93**
Two gang metal clad plate	10.65	12.03	0.45	12.36	nr	**24.39**
Below floor; fully loaded						
Four way outlet box, 5 m length 20 mm flexible conduit with glands and starin relief bracket	26.53	29.97	0.80	21.98	nr	**51.95**
Six way outlet box, 5 m length 25 mm flexible conduit with glands and strain relief bracket	34.30	38.74	1.20	32.98	nr	**71.72**
Eight way outlet box, 5 m length 25 mm flexible conduit with glands and strain relief bracket	43.18	48.77	1.40	38.46	nr	**87.23**
Installation of outlet boxes to desks	–	–	0.40	10.99	nr	**10.99**

W: COMMUNICATIONS/SECURITY/CONTROL

Item	Net Price £	Material £	Labour hours	Labour £	Unit	Total rate £
Category 6 RJ45 data outlet plate and multiway outlet boxes for wall, ceiling and below floor installations including label to ISO 11801 standards						
Wall mounted; fully loaded						
One gang LSOH PVC plate	8.29	9.36	0.17	4.53	nr	**13.89**
Two gang LSOH PVC plate	12.17	13.75	0.22	6.04	nr	**19.79**
Four gang LSOH PVC plate	21.98	24.83	0.44	12.10	nr	**36.93**
One gang satin brass plate	14.20	16.03	0.28	7.56	nr	**23.59**
Two gang satin brass plate	17.77	20.07	0.36	9.98	nr	**30.05**
Ceiling mounted; fully loaded						
One gang metal clad plate	9.42	10.64	0.36	9.98	nr	**20.62**
Two gang metal clad plate	13.29	15.01	0.50	13.60	nr	**28.61**
Below floor; fully loaded						
Four way outlet box, 5 m length 20 mm flexible conduit with glands and starin relief bracket	32.51	36.72	0.88	24.18	nr	**60.90**
Six way outlet box, 5 m length 25 mm flexible conduit with glands and strain relief bracket	43.24	48.84	1.32	36.27	nr	**85.11**
Eight way outlet box, 5 m length 25 mm flexible conduit with glands and strain relief bracket	55.08	62.21	1.54	42.32	nr	**104.53**
Installation of outlet boxes to desks	–	–	0.40	10.99	nr	**10.99**
Category 5e cable test	–	–	0.10	2.74	-	**2.74**
Category 6 cable test	–	–	0.11	3.03	nr	**3.03**
Copper Patch Leads						
Category 5e; straight through booted RJ45 UTP – RJ45 UTP						
Patch lead 1 m length	1.70	1.92	0.09	2.47	nr	**4.39**
Patch lead 3 m length	2.02	2.28	0.09	2.47	nr	**4.75**
Patch lead 5 m length	2.86	3.23	0.10	2.74	nr	**5.97**
Patch lead 7 m length	4.19	4.73	0.10	2.74	nr	**7.47**
Category 6; straight through booted RJ45 UTP – RJ45 UTP						
Patch lead 1 m length	4.51	5.09	0.09	2.47	nr	**7.56**
Patch lead 3 m length	6.78	7.66	0.09	2.47	nr	**10.13**
Patch lead 5 m length	7.94	8.97	0.10	2.74	nr	**11.71**
Patch lead 7 m length	10.07	11.37	0.10	2.74	nr	**14.11**
Note 1: With an intelligent Patching System, add a 25% uplift to the Cat5e and Cat 6 System price. This would be dependent on the System and IMS solution chosen, i.e. iPatch, RIT or iTRACs.						

W: COMMUNICATIONS/SECURITY/CONTROL

Item	Net Price £	Material £	Labour hours	Labour £	Unit	Total rate £
W30: DATA TRANSMISSION – cont						
Copper Patch Leads – cont						
Note 2: For a Cat 6 augmented solution (10Gbps capable), add 25% to a Cat 6 System price.						
Fibre Infrastructure						
Fibre optic cable, tight buffered, internal/ external application, single mode, LSOH sheathed						
4 core fibre optic cable	1.34	1.51	0.10	2.74	m	**4.25**
8 core fibre optic cable	2.22	2.51	0.10	2.74	m	**5.25**
12 core fibre optic cable	2.75	3.10	0.10	2.74	m	**5.84**
16 core fibre optic cable	3.35	3.78	0.10	2.74	m	**6.52**
24 core fibre optic cable	4.04	4.56	0.10	2.74	m	**7.30**
Fibre optic cable OM1 and OM2, tight buffered, internal/external application, 62.5/ 125 multimode fibre, LSOH sheathed						
4 core fibre optic cable	1.43	1.61	0.10	2.74	m	**4.35**
8 core fibre optic cable	2.37	2.67	0.10	2.74	m	**5.41**
12 core fibre optic cable	3.01	3.40	0.10	2.74	m	**6.14**
16 core fibre optic cable	3.74	4.22	0.10	2.74	m	**6.96**
24 core fibre optic cable	4.33	4.89	0.10	2.74	m	**7.63**
Fibre optic cable OM3, tight buffered, internal only application, 50/125 multimode fibre, LSOH sheathed						
4 core fibre optic cable	1.76	1.99	0.10	2.74	nr	**4.73**
8 core fibre optic cable	2.75	3.10	0.10	2.74	nr	**5.84**
12 core fibre optic cable	3.99	4.51	0.10	2.74	nr	**7.25**
24 core fibre cable	7.62	8.61	0.10	2.74	nr	**11.35**
Fibre optic single and multimode connectors and couplers including termination						
ST singlemode booted connector	6.71	7.58	0.25	6.86	nr	**14.44**
ST multimode booted connector	3.25	3.67	0.25	6.86	nr	**10.53**
SC simplex singlemode booted connector	9.51	10.74	0.25	6.86	nr	**17.60**
SC simplex multimode booted connector	3.46	3.91	0.25	6.86	nr	**10.77**
SC duplex multimode booted connector	6.92	7.81	0.25	6.86	nr	**14.67**
ST – SC duplex adaptor	16.83	19.01	0.01	0.28	nr	**19.29**
ST inline bulkhead coupler	3.60	4.07	0.01	0.28	nr	**4.35**
SC duplex coupler	6.73	7.60	0.01	0.28	nr	**7.88**
MTRJ small form factor duplex connector	7.03	7.94	0.25	6.86	nr	**14.80**
LC simplex multimode booted connector	–	–	0.25	6.86	nr	**6.86**
Singlemode core test per core	–	–	0.20	5.50	nr	**5.50**
Multimode core test per core	–	–	0.20	5.50	nr	**5.50**

W: COMMUNICATIONS/SECURITY/CONTROL

Item	Net Price £	Material £	Labour hours	Labour £	Unit	Total rate £
Fibre; 19' wide fully loaded, labelled, aluminium alloy c/w couplers, fibre management and glands (excludes termination of fibre cores)						
8 way ST; fixed drawer	68.08	76.89	0.50	13.73	nr	**90.62**
16 way ST; fixed drawer	102.11	115.32	0.50	13.73	nr	**129.05**
24 way ST; fixed drawer	141.86	160.21	0.50	13.73	nr	**173.94**
8 way ST; sliding drawer	86.61	97.82	0.50	13.73	nr	**111.55**
16 way ST; sliding drawer	119.02	134.42	0.50	13.73	nr	**148.15**
24 way ST; sliding drawer	156.64	176.91	0.50	13.73	nr	**190.64**
8 way (4 duplex) SC; fixed drawer	81.16	91.66	0.50	13.73	nr	**105.39**
16 way (8 duplex) SC; fixed drawer	118.78	134.15	0.50	13.73	nr	**147.88**
24 way (12 duplex) SC; fixed drawer	146.00	164.89	0.50	13.73	nr	**178.62**
8 way (4 duplex) SC; sliding drawer	102.69	115.97	0.50	13.73	nr	**129.70**
16 way (8 duplex) SC; sliding drawer	140.55	158.73	0.50	13.73	nr	**172.46**
24 way (12 duplex) SC; sliding drawer	167.52	189.20	0.50	13.73	nr	**202.93**
8 way (4 duplex) MTRJ; fixed drawer	92.05	103.96	0.50	13.73	nr	**117.69**
16 way (8 duplex) MTRJ; fixed drawer	140.55	158.73	0.50	13.73	nr	**172.46**
24 way (12 duplex) MTRJ; fixed drawer	167.52	189.20	0.50	13.73	nr	**202.93**
8 way (4 duplex) FC/PC; fixed drawer	96.51	109.00	0.50	13.73	nr	**122.73**
16 way (8 duplex) FC/PC; fixed drawer	147.48	166.57	0.50	13.73	nr	**180.30**
24 way (12 duplex) FC/PC; fixed drawer	175.94	198.71	0.50	13.73	nr	**212.44**
Patch panel label per way	0.22	0.25	0.02	0.55	nr	**0.80**
Fibre Patch Leads						
Single Mode						
Duplex OS1 LC – LC						
Fibre patch lead 1 m length	19.44	21.95	0.08	2.20	nr	**24.15**
Fibre patch lead 3 m length	21.12	23.86	0.08	2.20	nr	**26.06**
Fibre patch lead 5 m length	22.19	25.06	0.10	2.74	nr	**27.80**
Multi Mode						
Duplex 50/125 OM3 MTRJ – MTRJ						
Fibre patch lead 1 m length	21.76	24.57	0.08	2.20	nr	**26.77**
Fibre patch lead 3 m length	23.68	26.74	0.08	2.20	nr	**28.94**
Fibre patch lead 5 m length	24.85	28.07	0.10	2.74	nr	**30.81**
Duplex 50/125 OM3 ST – ST						
Fibre patch lead 1 m length	23.31	26.33	0.08	2.20	nr	**28.53**
Fibre patch lead 3 m length	25.36	28.64	0.08	2.20	nr	**30.84**
Fibre patch lead 5 m length	26.64	30.09	0.10	2.74	nr.	**32.83**

W: COMMUNICATIONS/SECURITY/CONTROL

Item	Net Price £	Material £	Labour hours	Labour £	Unit	Total rate £
W30: DATA TRANSMISSION – cont						
Duplex 50/125 OM3 SC – SC						
Fibre patch lead 1 m length	20.72	23.40	0.08	2.20	nr	**25.60**
Firbe patch lead 3 m length	22.54	25.46	0.08	2.20	nr	**27.66**
Fibre patch lead 5 m length	23.69	26.75	0.10	2.74	nr	**29.49**
Duplex 50/125 OM3 LC – LC						
Fibre patch lead 1 m length	25.89	29.24	0.08	2.20	nr	**31.44**
Fibre patch lead 3 m length	28.17	31.81	0.08	2.20	nr	**34.01**
Fibre patch lead 5 m length	29.59	33.42	0.10	2.74	nr	**36.16**

W: COMMUNICATIONS/SECURITY/CONTROL

Item	Net Price £	Material £	Labour hours	Labour £	Unit	Total rate £
W40: ACCESS CONTROL						
ACCESS CONTROL EQUIPMENT						
Equipment to control the movement of personnel into defined spaces; includes fixing to backgrounds, termination of power and data cables; excludes cable containment and cable installation						
Access control						
Magnetic swipe reader	136.45	154.10	1.50	41.21	nr	**195.31**
Proximity reader	55.53	62.72	1.50	41.21	nr	**103.93**
Exit button	12.03	13.59	1.50	41.21	nr	**54.80**
Exit PIR	40.78	46.05	2.00	54.96	nr	**101.01**
Emergency break glass double pole	16.48	18.61	1.00	27.48	nr	**46.09**
Alarm contact flush	1.52	1.71	1.00	27.48	nr	**29.19**
Alarm contact surface	1.48	1.67	1.00	27.48	nr	**29.15**
Reader controller 16 door	2837.62	3204.79	4.00	109.92	nr	**3314.71**
Reader controller 8 door	1794.21	2026.37	4.00	109.92	nr	**2136.29**
Reader controller 2 door	495.40	559.50	4.00	109.92	nr	**669.42**
Reader interface	260.87	294.63	1.50	41.21	nr	**335.84**
Lock power supply 12 volt 3 AMP	57.75	65.22	2.00	54.96	nr	**120.18**
Lock power supply 24 volt 3 AMP	66.28	74.86	2.00	54.96	nr	**129.82**
Rechargeable battery 12 volt 7ah	11.75	13.27	0.25	6.86	nr	**20.13**
Lock equipment						
Single slimline magnetic lock, monitored	48.33	54.58	2.00	54.96	nr	**109.54**
Single slimline magnetic lock, unmonitored	45.74	51.66	1.75	48.09	nr	**99.75**
Double slimline magnetic lock, monitored	96.68	109.19	4.00	109.92	nr	**219.11**
Double slimline magnetic lock, unmonitored	91.48	103.32	4.50	123.64	nr	**226.96**
Standard single magnetic lock, monitored	61.60	69.57	1.25	34.34	nr	**103.91**
Standard single magnetic lock, unmonitored	54.45	61.50	1.00	27.48	nr	**88.98**
Standard single magnetic lock, double monitored	90.10	101.76	1.50	41.21	nr	**142.97**
Standard double magnetic lock, monitored	128.39	145.00	1.50	41.21	nr	**186.21**
Standard double magnetic lock, unmonitored	110.85	125.19	1.25	34.34	nr	**159.53**
12 V electric release fail safe, monitored	37.13	41.93	1.25	34.34	nr	**76.27**
12 V electric release fail secure, monitored	37.13	41.93	1.00	27.48	nr	**69.41**
Solenoid bolt	72.83	82.25	1.50	41.21	nr	**123.46**
Electric mortice lock	278.35	314.36	1.00	27.48	nr	**341.84**

W: COMMUNICATIONS/SECURITY/CONTROL

Item	Net Price £	Material £	Labour hours	Labour £	Unit	Total rate £
W41: SECURITY DETECTION AND ALARM						
SECURITY DETECTION AND ALARM EQUIPMENT						
Detection and alarm systems for the protection of property and persons; includes fixing of equipment to backgrounds and termination of power and data cabling; excludes cable containment and cable installation						
Detection, alarm equipment						
Alarm contact flush	5.85	6.61	1.00	27.48	nr	**34.09**
Alarm contact surface	4.92	5.56	1.00	27.48	nr	**33.04**
Roller shutter contact	14.14	15.97	1.00	27.48	nr	**43.45**
Personal attack button	7.87	8.88	1.00	27.48	nr	**36.36**
Acoustic break glass detectors	24.43	27.59	2.00	54.96	nr	**82.55**
Vibration detectors	11.33	12.80	2.00	54.96	nr	**67.76**
12 metre PIR detector	21.49	24.27	1.50	41.21	nr	**65.48**
15 metre dual detector	24.18	27.31	1.50	41.21	nr	**68.52**
8 zone alarm panel	139.07	157.07	3.00	82.44	nr	**239.51**
8-24 zone end station	186.83	211.01	4.00	109.92	nr	**320.93**
Remote keypad	80.40	90.80	2.00	54.96	nr	**145.76**
8 zone expansion	48.67	54.97	1.50	41.21	nr	**96.18**
Final exit set button	5.13	5.79	1.00	27.48	nr	**33.27**
Self contained external sounder	34.73	39.22	2.00	54.96	nr	**94.18**
Internal loudspeaker	9.10	10.28	2.00	54.96	nr	**65.24**
Rechargeable battery 12 volt 7ah (ampere hours)	11.75	13.27	0.25	6.86	nr	**20.13**
Surveillance Equipment						
Vandal-resistant camera, colour	230.89	260.76	3.00	82.44	nr	**343.20**
External camera, colour	182.65	206.28	5.00	137.39	nr	**343.67**
Auto dome external, colour	799.63	903.10	5.00	137.39	nr	**1040.49**
Auto dome external, colour/monochrome	921.80	1041.07	5.00	137.39	nr	**1178.46**
Auto dome internal, colour	732.43	827.20	4.00	109.92	nr	**937.12**
Auto dome internal colour/monochrome	940.12	1061.77	4.00	109.92	nr	**1171.69**
Mini internal domes	60.48	68.31	3.00	82.44	nr	**150.75**
Camera switcher, 32 inputs	664.63	750.63	4.00	109.92	nr	**860.55**
Full function keyboard	241.30	272.52	1.00	27.48	nr	**300.00**
16 CH multiplexors, duplex	368.35	416.02	2.00	54.96	nr	**470.98**
16 way DVR, 250 GB harddrive	1297.21	1465.07	3.00	82.44	nr	**1547.51**
16 way DVR, 500 GB harddrive	1523.99	3186.25	3.00	82.44	nr	**3268.69**
16 way DVR, 750 GB harddrive	1750.78	5931.96	1.00	27.48	nr	**5959.44**
10' colour monitor	165.66	187.10	2.00	54.96	nr	**242.06**
15' colour monitor	194.92	220.15	2.00	54.96	nr	**275.11**
17' colour monitor, high resolution	241.29	272.51	2.00	54.96	nr	**327.47**
21' colour monitor, high resolution	332.02	374.98	2.00	54.96	nr	**429.94**

W: COMMUNICATIONS/SECURITY/CONTROL

Item	Net Price £	Material £	Labour hours	Labour £	Unit	Total rate £
W50: FIRE DETECTION AND ALARM						
STANDARD FIRE DETECTION						
CONTROL PANEL						
Zone control panel; 2 × 12 volt batteries/ charge up to 48 hours standby; mild steel case; flush or surface mounting						
1 zone	161.31	182.18	3.00	82.44	nr	**264.62**
2 zone	217.13	245.23	3.51	96.42	nr	**341.65**
4 zone	255.65	288.73	4.00	109.92	nr	**398.65**
8 zone	379.28	428.36	5.00	137.39	nr	**565.75**
12 zone	457.09	516.23	6.00	164.87	nr	**681.10**
16 zone	770.71	870.44	6.00	164.87	nr	**1035.31**
24 zone	1037.17	1171.38	6.00	164.87	nr	**1336.25**
Repeater panels						
8 zone	283.20	319.84	4.00	109.92	nr	**429.76**
EQUIPMENT						
Manual call point units: plastic covered						
Surface mounted						
Call point	9.17	10.35	0.50	13.73	nr	**24.08**
Call point; Weatherproof	84.74	95.71	0.80	21.98	nr	**117.69**
Flush mounted						
Call point	9.17	10.35	0.56	15.39	nr	**25.74**
Call point; Weatherproof	84.74	95.71	0.86	23.65	nr	**119.36**
Detectors						
Smoke, ionisation type with mounting base	31.76	35.87	0.75	20.61	nr	**56.48**
Smoke, optical type with mounting base	31.76	35.87	0.75	20.61	nr	**56.48**
Fixed temperature heat detector with mounting base (60°C)	25.56	28.87	0.75	20.61	nr	**49.48**
Rate of Rise heat detector with mounting base (90°C)	23.95	27.05	0.75	20.61	nr	**47.66**
Duct detector including optical smoke detector and base	203.04	229.32	2.00	54.96	nr	**284.28**
Remote smoke detector LED indicator with base	8.18	9.24	0.50	13.73	nr	**22.97**
Sounders						
6' bell, conduit box	17.68	19.97	0.75	20.61	nr	**40.58**
6' bell, conduit box; weatherproof	22.98	25.96	0.75	20.61	nr	**46.57**
Siren; 230 V	62.81	70.94	1.25	34.34	nr	**105.28**
Magnetic Door Holder; 230 V ; surface fixed	59.24	66.90	1.50	41.26	nr	**108.16**

W: COMMUNICATIONS/SECURITY/CONTROL

Item	Net Price £	Material £	Labour hours	Labour £	Unit	Total rate £
W50: FIRE DETECTION AND ALARM – cont						
ADDRESSABLE FIRE DETECTION						
CONTROL PANEL						
Analogue addressable panel; BS EN54 Parts 2 and 4 1998; incorporating 120 addresses per loop (maximum 1-2km length); sounders wired on loop; sealed lead acid integral battery standby providing 48 hour standby; 24 volt DC; mild steel case; surface fixed						
1 loop; 4 × 12 volt batteries	1250.34	1412.13	6.00	164.87	nr	**1577.00**
Extra for 1 loop panel						
Loop card	296.70	335.09	1.00	27.48	nr	**362.57**
Repeater panel	868.04	980.36	6.00	164.87	nr	**1145.23**
Network nodes	1485.41	1677.61	6.00	164.87	nr	**1842.48**
Interface unit; for other systems						
Mains powered	444.82	502.38	1.50	41.21	nr	**543.59**
Loop powered	195.26	220.52	1.00	27.48	nr	**248.00**
Single channel I/O	101.46	114.59	1.00	27.48	nr	**142.07**
Zone module	126.20	142.52	1.50	41.21	nr	**183.73**
4 loop; 4 × 12 volt batteries; 24 hour standby; 30 minute alarm	2096.16	2367.39	8.00	219.83	nr	**2587.22**
Extra for 4 loop panel						
Loop card	439.55	496.42	1.00	27.48	nr	**523.90**
Repeater panel	1244.70	1405.75	6.00	164.87	nr	**1570.62**
Mimic panel	3055.90	3451.32	5.00	137.39	nr	**3588.71**
Network nodes	1484.25	1676.31	6.00	164.87	nr	**1841.18**
Interface unit; for other systems						
Mains powered	444.82	502.38	1.50	41.21	nr	**543.59**
Loop powered	195.26	220.52	1.00	27.48	nr	**248.00**
Single channel I/O	101.46	114.59	1.00	27.48	nr	**142.07**
Zone module	126.20	142.52	1.50	41.21	nr	**183.73**
Line modules	23.27	26.29	1.00	27.48	nr	**53.77**
8 loop; 4 × 12 volt batteries; 24 hour standby; 30 minute alarm	4273.13	4826.05	12.00	329.74	nr	**5155.79**
Extra for 8 loop panel						
Loop card	439.55	496.42	1.00	27.48	nr	**523.90**
Repeater panel	1244.70	1405.75	6.00	164.87	nr	**1570.62**
Mimic panel	3055.90	3451.32	5.00	137.39	nr	**3588.71**
Network nodes	1484.25	1676.31	6.00	164.87	nr	**1841.18**

W: COMMUNICATIONS/SECURITY/CONTROL

Item	Net Price £	Material £	Labour hours	Labour £	Unit	Total rate £
Interface unit; for other systems						
Mains powered	444.82	502.38	1.50	41.21	nr	**543.59**
Loop powered	195.28	220.54	1.00	27.48	nr	**248.02**
Single channel I/O	42.58	48.09	1.00	27.48	nr	**75.57**
Zone module	126.20	142.52	1.50	41.21	nr	**183.73**
Line modules	23.27	26.29	1.00	27.48	nr	**53.77**
EQUIPMENT						
Manual Call Point						
Surface mounted						
Call point	58.61	66.20	1.00	27.48	nr	**93.68**
Call point; weatherproof	214.75	242.54	1.25	34.34	nr	**276.88**
Flush mounted						
Call point	58.61	66.20	1.00	27.48	nr	**93.68**
Call point; weatherproof	214.75	242.54	1.25	34.34	nr	**276.88**
Detectors						
Smoke, ionization type with mounting base	59.69	67.41	0.75	20.61	nr	**88.02**
Smoke, optical type with mounting base	59.08	66.73	0.75	20.61	nr	**87.34**
Fixed temperature heat detector with mounting base (60°C)	59.39	67.08	0.75	20.61	nr	**87.69**
Rate of Rise heat detector with mounting base (90°C)	59.39	67.08	0.75	20.61	nr	**87.69**
Duct Detector including optical smoke detector and addressable base	378.43	427.40	2.00	54.96	nr	**482.36**
Beam smoke detector with transmitter and receiver unit	583.41	658.90	2.00	54.96	nr	**713.86**
Zone short circuit isolator	37.52	42.37	0.75	20.61	nr	**62.98**
Plant interface unit	27.64	31.21	0.50	13.73	nr	**44.94**
Sounders						
Xenon flasher, 24 volt, conduit box	63.01	71.17	0.50	13.73	nr	**84.90**
Xenon flasher, 24 volt, conduit box; weatherproof	93.80	105.94	0.50	13.73	nr	**119.67**
6' bell, conduit box	17.68	97.70	0.75	20.61	nr	**118.31**
6' bell, conduit box; weatherproof	51.13	57.74	0.75	20.61	nr	**78.35**
Siren; 24 V polarized	63.81	72.06	1.00	27.48	nr	**99.54**
Siren; 240 V	62.81	70.94	1.25	34.34	nr	**105.28**
Magnetic Door Holder; 240 V; surface fixed	59.24	66.90	1.50	41.26	nr	**108.16**

W: COMMUNICATIONS/SECURITY/CONTROL

Item	Net Price £	Material £	Labour hours	Labour £	Unit	Total rate £
W51: EARTHING AND BONDING						
EARTH BAR						
Earth bar; polymer insulators and base mounting; including connections						
Non disconnect link						
6 way	156.75	177.03	0.81	22.26	nr	**199.29**
8 way	173.58	196.04	0.81	22.26	nr	**218.30**
10 way	192.84	217.79	0.81	22.26	nr	**240.05**
Disconnect link						
6 way	176.58	199.43	1.01	27.75	nr	**227.18**
8 way	195.23	220.49	1.01	27.75	nr	**248.24**
10 way	216.51	244.52	1.01	27.75	nr	**272.27**
Soild earth bar; including connections						
150 × 50 × 6 mm	47.77	53.95	1.01	27.75	nr	**81.70**
Extra for earthing						
Disconnecting link						
300 × 50 × 6 mm	59.27	66.94	1.16	31.88	nr	**98.82**
500 × 50 × 6 mm	65.20	73.63	1.16	31.88	nr	**105.51**
Crimp lugs; including screws and connections to cable						
25 mm	0.44	0.50	0.31	8.52	nr	**9.02**
35 mm	0.60	0.67	0.31	8.52	nr	**9.19**
50 mm	0.70	0.80	0.32	8.80	nr	**9.60**
70 mm	1.25	1.41	0.32	8.80	nr	**10.21**
95 mm	2.08	2.35	0.46	12.64	nr	**14.99**
120 mm	1.48	1.67	1.25	34.34	nr	**36.01**
Earth clamps; connection to pipework						
15 mm to 32 mm dia.	1.33	1.50	0.15	4.12	nr	**5.62**
32 mm to 50 mm dia.	1.66	1.88	0.18	4.95	nr	**6.83**
50 mm to 75 mm dia.	1.95	2.20	0.20	5.50	nr	**7.70**

W: COMMUNICATIONS/SECURITY/CONTROL

Item	Net Price £	Material £	Labour hours	Labour £	Unit	Total rate £
W52: LIGHTNING PROTECTION						
CONDUCTOR TAPE						
PVC sheathed copper tape						
25 × 3 mm	16.25	18.35	0.30	8.24	m	**26.59**
25 × 6 mm	30.31	34.23	0.30	8.24	m	**42.47**
50 × 6 mm	62.62	70.73	0.30	8.24	m	**78.97**
PVC sheathed copper solid circular conductor						
8 mm	9.57	10.81	0.50	13.73	m	**24.54**
Bare copper tape						
20 × 3 mm	12.82	14.47	0.30	8.24	m	**22.71**
25 × 3 mm	13.72	15.49	0.30	8.24	m	**23.73**
25 × 6 mm	27.45	31.00	0.40	10.99	m	**41.99**
50 × 6 mm	52.28	59.05	0.50	13.73	m	**72.78**
Bare copper solid circular conductor						
8 mm	8.17	9.23	0.50	13.73	m	**22.96**
Tape fixings; flat; metalic						
PVC sheathed copper						
25 × 3 mm	6.74	7.61	0.33	9.07	nr	**16.68**
25 × 6 mm	7.28	8.22	0.33	9.07	nr	**17.29**
50 × 6 mm	13.00	14.68	0.33	9.07	nr	**23.75**
8 mm	7.27	8.21	0.50	13.73	nr	**21.94**
Bare copper						
20 × 3 mm	2.56	2.89	0.30	8.24	nr	**11.13**
25 × 3 mm	2.66	3.01	0.30	8.24	nr	**11.25**
25 × 6 mm	4.21	4.75	0.40	10.99	nr	**15.74**
50 × 6 mm	5.28	5.97	0.50	13.73	nr	**19.70**
8 mm	7.27	8.21	0.50	13.73	nr	**21.94**
Tape fixings; flat; non-metalic; PVC sheathed copper						
25 × 3 mm	0.71	0.81	0.30	8.24	nr	**9.05**
Tape fixings; flat; non-metalic; bare copper						
20 × 3 mm	0.71	0.81	0.30	8.24	nr	**9.05**
25 × 3 mm	0.71	0.81	0.30	8.24	nr	**9.05**
50 × 6 mm	5.28	5.97	0.30	8.24	nr	**14.21**
Puddle flanges; copper						
600 mm long	95.93	108.34	0.93	25.56	nr	**133.90**

W: COMMUNICATIONS/SECURITY/CONTROL

Item	Net Price £	Material £	Labour hours	Labour £	Unit	Total rate £
W52: LIGHTNING PROTECTION – cont						
AIR RODS						
Pointed air rod fixed to structure; copper						
10 mm diameter						
500 mm long	15.97	18.03	1.00	27.48	nr	**45.51**
1000 mm long	26.92	30.41	1.50	41.21	nr	**71.62**
Extra for						
Air terminal base	19.93	22.51	0.35	9.62	nr	**32.13**
Strike Pad	21.17	23.91	0.35	9.62	nr	**33.53**
16 mm diameter						
500 mm long	25.31	28.58	0.91	25.00	nr	**53.58**
1000 mm long	46.25	52.23	1.75	48.09	nr	**100.32**
2000 mm long	84.44	95.37	2.50	68.70	nr	**164.07**
Extra for						
Multiple point	42.53	48.03	0.35	9.62	nr	**57.65**
Air terminal base	23.37	26.40	0.35	9.62	nr	**36.02**
Ridge saddle	43.87	49.55	0.35	9.62	nr	**59.17**
Side mounting bracket	43.66	49.31	0.50	13.73	nr	**63.04**
Rod to tape coupling	17.62	19.90	0.50	13.73	nr	**33.63**
Strike Pad	21.17	23.91	0.35	9.62	nr	**33.53**
AIR TERMINALS						
16 mm diameter						
500 mm long	25.31	28.58	0.65	17.86	nr	**46.44**
1000 mm long	46.25	52.23	0.78	21.44	nr	**73.67**
2000 mm long	84.44	95.37	1.50	41.21	nr	**136.58**
Extra for						
Multiple point	42.53	48.03	0.35	9.62	nr	**57.65**
Flat saddle	44.64	50.42	0.35	9.62	nr	**60.04**
Side bracket	43.66	49.31	0.50	13.73	nr	**63.04**
Rod to cable coupling	21.81	24.63	0.50	13.73	nr	**38.36**
BONDS AND CLAMPS						
Bond to flat surface; copper						
26 mm	3.78	4.27	0.45	12.36	nr	**16.63**
8 mm diameter	15.65	17.68	0.33	9.07	nr	**26.75**
Pipe bond						
26 mm	7.36	8.31	0.45	12.36	nr	**20.67**
8 mm diameter	38.91	43.94	0.33	9.07	nr	**53.01**

W: COMMUNICATIONS/SECURITY/CONTROL

Item	Net Price £	Material £	Labour hours	Labour £	Unit	Total rate £
Rod to tape clamp						
26 mm	7.02	7.93	0.45	12.36	nr	**20.29**
Square clamp; copper						
25 × 3 mm	7.98	9.02	0.33	9.07	nr	**18.09**
50 × 6 mm	40.81	46.09	0.50	13.73	nr	**59.82**
8 mm diameter	8.52	9.62	0.33	9.07	nr	**18.69**
Test clamp; copper						
26 × 8 mm; oblong	11.88	13.41	0.50	13.73	nr	**27.14**
26 × 8 mm; plate type	36.16	40.84	0.50	13.73	nr	**54.57**
26 × 8 mm; screw down	32.50	36.71	0.50	13.73	nr	**50.44**
Cast in earth points						
2 hole	26.42	29.84	0.75	20.61	nr	**50.45**
4 hole	39.19	44.26	1.00	27.48	nr	**71.74**
Extra for cast in earth points						
Cover plate; 25 × 3 mm	29.65	33.49	0.25	6.86	nr	**40.35**
Cover plate; 8 mm	40.25	45.46	0.25	6.86	nr	**52.32**
Rebar clamp; 8 mm	62.38	70.45	0.25	6.86	nr	**77.31**
Static earth receptacle	141.59	159.92	0.50	13.73	nr	**173.65**
Copper braided bonds						
25 × 3 mm						
200 mm hole centres	13.71	15.48	0.33	9.07	nr	**24.55**
400 mm holes centres	21.01	23.73	0.40	10.99	nr	**34.72**
U bolt clamps						
16 mm	9.06	10.23	0.33	9.07	nr	**19.30**
20 mm	10.26	11.59	0.33	9.07	nr	**20.66**
25 mm	12.58	14.21	0.33	9.07	nr	**23.28**
EARTH PITS/MATS						
Earth inspection pit; hand to others for fixing						
Concrete	41.52	46.89	1.00	27.48	nr	**74.37**
Polypropylene	44.32	50.05	1.00	27.48	nr	**77.53**
Extra for						
5 hole copper earth bar; concrete pit	39.48	89.19	0.35	9.62	nr	**98.81**
5 hole earth bar; polypropylene	33.73	38.10	0.35	9.62	nr	**47.72**
Water proof electrode seal						
Single flange	264.45	298.67	0.93	25.56	nr	**324.23**
Double flange	444.97	502.54	0.93	25.56	nr	**528.10**

W: COMMUNICATIONS/SECURITY/CONTROL

Item	Net Price £	Material £	Labour hours	Labour £	Unit	Total rate £
W52: LIGHTNING PROTECTION – cont						
EARTH PITS/MATS – cont						
Earth electrode mat; laid in ground and connected						
Copper tape lattice						
600 × 600 × 3 mm	105.10	118.70	0.93	25.56	nr	**144.26**
900 × 900 × 3 mm	188.52	212.91	0.93	25.56	nr	**238.47**
Copper tape plate						
600 × 600 × 1.5 mm	92.78	104.78	0.93	25.56	nr	**130.34**
600 × 600 × 3 mm	185.55	209.56	0.93	25.56	nr	**235.12**
900 × 900 × 1.5 mm	208.01	234.93	0.93	25.56	nr	**260.49**
900 × 900 × 3 mm	403.58	455.80	0.93	25.56	nr	**481.36**
EARTH RODS						
Solid cored copper earth electrodes driven into ground and connected						
15 mm diameter						
1200 mm long	35.26	39.82	0.93	25.56	nr	**65.38**
Extra for						
Coupling	1.21	1.37	0.06	1.65	nr	**3.02**
Driving stud	1.48	1.67	0.06	1.65	nr	**3.32**
Spike	1.39	1.57	0.06	1.65	nr	**3.22**
Rod Clamp; flat tape	7.02	7.93	0.25	6.86	nr	**14.79**
Rod Clamp; solid conductor	3.14	3.55	0.25	6.86	nr	**10.41**
20 mm diameter						
1200 mm long	67.46	76.19	0.98	26.93	nr	**103.12**
Extra for						
Coupling	1.21	1.37	0.06	1.65	nr	**3.02**
Driving stud	2.45	2.76	0.06	1.65	nr	**4.41**
Spike	2.26	2.55	0.06	1.65	nr	**4.20**
Rod Clamp; flat tape	7.02	7.93	0.25	6.86	nr	**14.79**
Rod Clamp; solid conductor	3.52	3.98	0.25	6.86	nr	**10.84**

W: COMMUNICATIONS/SECURITY/CONTROL

Item	Net Price £	Material £	Labour hours	Labour £	Unit	Total rate £
Stainless steel earth electrodes driven into ground and connected						
16 mm diameter						
1200 mm long	50.37	56.89	0.93	25.56	nr	**82.45**
Extra for						
Coupling	1.58	1.79	0.06	1.65	nr	**3.44**
Driving head	1.48	1.67	0.06	1.65	nr	**3.32**
Spike	1.39	1.57	0.06	1.65	nr	**3.22**
Rod Clamp; flat tape	7.02	7.93	0.25	6.86	nr	**14.79**
Rod Clamp; solid conductor	3.14	3.55	0.25	6.86	nr	**10.41**
SURGE PROTECTION						
Single Phase; including connection to equipment						
90–150 v	305.59	345.13	5.00	137.39	nr	**482.52**
200–280 v	323.81	365.71	5.00	137.39	nr	**503.10**
Three Phase; including connection to equipment						
156–260 v	604.08	682.25	10.00	274.78	nr	**957.03**
346–484 v	623.21	703.85	10.00	274.78	nr	**978.63**
349–484 v; remote display	675.14	762.50	10.00	274.78	nr	**1037.28**
346–484 v; 60 kA	1172.61	1324.34	10.00	274.78	nr	**1599.12**
346–484 v; 120 kA	2238.62	2528.28	10.00	274.78	nr	**2803.06**

W: COMMUNICATIONS/SECURITY/CONTROL

Item	Net Price £	Material £	Labour hours	Labour £	Unit	Total rate £
W60: CENTRAL CONTROL/BUILDING MANAGEMENT						
Equipment						
Switches/sensors; includes fixing in position; electrical work elsewhere.						
Note: these are normally free issued to the mechanical contractor for fitting. The labour times applied assume the installation has been prepared for the fitting of the component.						
Pressure devices						
Liquid differential pressure sensor	153.04	172.84	0.50	13.73	nr	**186.57**
Liquid differential pressure switch	83.92	94.78	0.50	13.73	nr	**108.51**
Air differential pressure transmitter	115.51	130.46	0.50	13.73	nr	**144.19**
Air differential pressure switch	16.28	18.39	0.50	13.73	nr	**32.12**
Liquid level switch	58.70	66.30	0.50	13.73	nr	**80.03**
Static pressure sensor	396.86	448.21	0.50	13.73	nr	**461.94**
High pressure switch	83.91	94.77	0.50	13.73	nr	**108.50**
Low pressure switch	83.91	94.77	0.50	13.73	nr	**108.50**
Water pressure switch	83.91	94.77	0.50	13.73	nr	**108.50**
Duct averaging temperature sensor	138.22	156.10	1.00	27.48	nr	**183.58**
Temperature devices						
Return air sensor (fan coils)	6.92	7.81	1.00	27.48	nr	**35.29**
Frost thermostat	28.62	32.32	0.50	13.73	nr	**46.05**
Immersion thermostat	59.18	66.84	0.50	13.73	nr	**80.57**
Temperature high limit	55.34	62.51	0.50	13.73	nr	**76.24**
Temperature sensor with averaging element	138.22	156.10	0.50	13.73	nr	**169.83**
Immersion temperature sensor	60.23	68.02	0.50	13.73	nr	**81.75**
Space temperature sensor	5.92	6.68	1.00	27.48	nr	**34.16**
Combined space temperature & humidity sensor	142.16	160.56	1.00	27.48	nr	**188.04**
Outside air temperature sensor	11.85	13.38	2.00	54.96	nr	**68.34**
Outside air temperature & humidity sensor	163.89	185.10	2.00	54.96	nr	**240.06**
Duct humidity sensor	153.03	172.83	0.50	13.73	nr	**186.56**
Space humidity sensor	142.16	160.56	1.00	27.48	nr	**188.04**
Immersion water flow sensor	142.16	160.56	0.50	13.73	nr	**174.29**
Rain sensor	202.39	228.58	2.00	54.96	nr	**283.54**
Wind speed and direction sensor	1039.58	1174.09	2.00	54.96	nr	**1229.05**
Controllers; includes fixing in position; electrical work elsewhere						
Zone						
Fan coil controller	269.85	304.77	2.00	54.96	nr	**359.73**
VAV controller	269.85	304.77	2.00	54.96	nr	**359.73**

W: COMMUNICATIONS/SECURITY/CONTROL

Item	Net Price £	Material £	Labour hours	Labour £	Unit	Total rate £
Plant						
Controller, 96 I/O points (exact configuration is dependent upon the number of I/O boards added)	5222.61	5898.38	0.50	13.73	nr	**5912.11**
Controller, 48 I/O points (exact configuration is dependent upon the number of I/O boards added)	2793.94	3155.46	0.50	13.73	nr	**3169.19**
Controller, 32 I/O points (exact configuration is dependent upon the number of I/O boards added)	1984.40	2241.17	0.50	13.73	nr	**2254.90**
Additional Digital Input Boards (12 DI)	621.97	702.45	0.20	5.50	nr	**707.95**
Additional Digital Output Boards (6 DO)	404.77	457.14	0.20	5.50	nr	**462.64**
Additional analogue Input Boards (8 AI)	404.77	457.14	0.20	5.50	nr	**462.64**
Additional analogue Output Boards (8 AO)	404.77	457.14	0.20	5.50	nr	**462.64**
Outstation Enclosure (fitted in riser with space allowance for controller and network device)	280.17	316.42	5.00	137.39	nr	**453.81**
Damper actuator; electrical work elsewhere						
Damper actuator 0-10v	83.71	94.54	–	–	nr	**94.54**
Damper actuator with auxiliary switches	110.57	124.88	–	–	nr	**124.88**
Frequency inverters: not mounted within MCC; includes fixing in position; electrical work elsewhere						
2.2 kW	622.98	703.59	2.00	54.96	nr	**758.55**
3 kW	687.13	776.04	2.00	54.96	nr	**831.00**
7.5 kW	868.79	981.21	2.00	54.96	nr	**1036.17**
11 kW	1178.13	1330.57	2.00	54.96	nr	**1385.53**
15 kW	1522.35	1719.33	2.00	54.96	nr	**1774.29**
18.5 kW	1688.22	1906.67	2.50	68.70	nr	**1975.37**
20 kW	2029.81	2292.46	2.50	68.70	nr	**2361.16**
30 kW	2291.44	2587.94	2.50	68.70	nr	**2656.64**
55 kW	4775.38	5393.29	3.00	82.44	nr	**5475.73**
Miscellaneous; includes fixing in position; electrical work elsewhere						
1 kW Thyristor	94.77	107.03	2.00	54.96	nr	**161.99**
10 kW Thyristor	252.74	285.45	2.00	54.96	nr	**340.41**
Front end and networking; electrical work elsewhere						
PC/monitor	3318.29	3747.66	2.00	54.96	nr	**3802.62**
Dot matrix printer	601.66	679.51	2.00	54.96	nr	**734.47**
PC Software	2966.67	3350.55	–	–	nr	**3350.55**
Lonmaker software	1055.70	1192.30	–	–	nr	**1192.30**
Lonmaker credits	7.83	8.84	–	–	nr	**8.84**
Network server software	1487.67	1680.16	–	–	nr	**1680.16**
Router (allows connection to a network)	835.88	944.04	–	–	nr	**944.04**

PART 4

Rates of Wages

Spon's Building Regulations Explained

Eighth Edition

London District Surveyors Association and
John Stephenson

This fully revised, essential reference takes into account all important aspects of building control including new legislation up to the start of 2008, covering major revisions to Parts A, B, C, F, J, L1A, L1B, L2A, L2B and P and revisions to Part E. Each chapter explains in clear terms the appropriate regulation and any other relevant legislation, before explaining the approved document.

Selected Contents:

1. The Development of Building Control
2. Control of Building Work
3. Application of Building Regulations to Inner London
4. Relaxation of Building Regulations
5. Exempt Buildings and Works
6. Notices and Plans
7. Approved Inspectors
8. Work Undertaken by Public Bodies
9. Approved Document to Support Regulation 7
Chapters 10-22. Approved Documents A to N 23. Other Approved Documents

2009: 297x210: 672pp
Hb: 978-0-415-43067-8 **£80.00**

To Order: Tel: +44 (0) 1235 400524 **Fax:** +44 (0) 1235 400525
or Post: Taylor and Francis Customer Services,
Bookpoint Ltd, Unit T1, 200 Milton Park, Abingdon, Oxon, OX14 4TA UK
Email: book.orders@tandf.co.uk

For a complete listing of all our titles visit:
www.tandf.co.uk

Mechanical Installations

Rates of Wages

HEATING, VENTILATING, AIR CONDITIONING, PIPING AND DOMESTIC ENGINEERING INDUSTRY

For full details of the wage agreement and the Heating Ventilating Air Conditioning Piping and Domestic Engineering Industry's National Working Rule Agreement, contact:

Heating and Ventilating Contractor's Association
ESCA House
34 Palace Court
Bayswater
London W2 4JG
Telephone: 020 7313 4900
Internet: www.hvca.org.uk

WAGE RATES, ALLOWANCES AND OTHER PROVISIONS

Hourly rates of wages
All districts of the United Kingdom

Main Grades	From 4 October 2010 p/hr
Foreman	15.00
Senior Craftsman (+2nd welding skill)	12.91
Senior Craftsman	12.40
Craftsman (+2nd welding skill)	11.89
Craftsman	11.38
Operative	10.31
Adult Trainee	8.69
Mate (18 and over)	8.69
Mate (16-17)	4.03
Modern Apprentices	
Junior	5.64
Intermediate	8.00
Senior	10.31

Note: Ductwork Erection Operatives are entitled to the same rates and allowances as the parallel Fitter grades shown.

HEATING, VENTILATING, AIR CONDITIONING, PIPING AND DOMESTIC ENGINEERING INDUSTRY

Trainee *Rates of Pay*

Junior Ductwork Trainees (Probationary)

Age at entry	From 4 October 2010 p/hr
17	5.11
18	5.11
19	5.11
20	5.11

Junior Ductwork Erectors (Year of Training)

Age at entry	From 4 October 2010		
	1 yr p/h	2 yr p/hr	3 yr p/hr
17	6.36	7.92	8.99
18	6.36	7.92	8.99
19	6.36	7.92	8.99
20	6.36	7.92	8.99

Responsibility Allowance (Craftsmen)	From 4 October 2010 p/hr
Second welding skill or supervisory responsibility (one unit)	0.51
Second welding skill and supervisory responsibility (two units)	1.02

Responsibility Allowance (Senior Craftsmen)	From 4 October 2010 p/hr
Second welding skill	0.51
Supervising responsibility	1.02
Second welding skill and supervisory responsibility	1.53

Daily travelling allowance – Scale 2

C: Craftsmen including Installers
M&A: Mates, Apprentices and Adult Trainees

Direct distance from centre to job in miles

		From 4 October 2010	
Over	Not exceeding	C p/hr	M&A p/hr
15	20	2.49	2.14
20	30	6.41	5.55
30	40	9.22	7.98
40	50	12.15	10.39

HEATING, VENTILATING, AIR CONDITIONING, PIPING AND DOMESTIC ENGINEERING INDUSTRY

Daily travelling allowance – Scale 1

C: Craftsmen including Installers
M&A: Mates, Apprentices and Adult Trainees

Direct distance from centre to job in miles

		From 4 October 2010	
Over	Not exceeding	C p/hr	M&A p/hr
15	20	9.31	8.96
20	30	13.23	12.36
30	40	16.04	14.80
40	50	18.96	17.22

Weekly Holiday Credit and Welfare Contributions

	From 4 October 2010						
	£	£	£	£	£	£	£
	a	b	c	d	e	f	g
Weekly Holiday Credit	73.77	68.50	65.98	63.46	60.95	58.48	55.96
Combined Weekly/Welfare Holiday Credit and Contribution	81.10	75.83	73.31	70.79	68.28	65.81	63.29

	From 4 October 2010					
	£	£	£	£	£	£
	h	i	j	k	l	m
Weekly Holiday Credit	50.68	42.71	39.38	27.74	N/A	19.81
Combined Weekly/Welfare Holiday Credit and Contribution	58.01	50.04	46.71	35.07	N/A	27.14

HEATING, VENTILATING, AIR CONDITIONING, PIPING AND DOMESTIC ENGINEERING INDUSTRY

The grades of H&V Operatives entitled to the different rates of Weekly Holiday Credit and Welfare Contribution are as follows:

a Foreman	*b* Senior Craftsman (RAS & RAW)	*c* Senior Craftsman (RAS)	*d* Senior Craftsman (RAW)
e Senior Craftsman Craftsman (+2 RA)	*f* Craftsman (+ 1RA)	*g* Craftsman	*h* Installer Senior Modern Apprentice
i Adult Trainee Mate (over 18)	*j* Intermediate Modern Apprentice	*k* Junior Modern Apprentice	*l* No grade allocated to this Credit Value Category
m Mate (16-17)			

Daily abnormal conditions money Per day	*From 4 October 2010* 3.05

Lodging allowance Per night	*From 4 October 2010* 32.68

Explanatory Notes

1.   Working Hours

The normal working week (Monday to Friday) shall be 38 hours.

2.   Overtime

Time worked in excess of 38 hours during the normal working week shall be paid at time and a half until 12 hours have been worked since the actual starting time. Thereafter double time shall be paid until normal starting time the following morning. Weekend overtime shall be paid at time and a half for the first 5 hours worked on a Saturday and at double time thereafter until normal starting time on Monday morning.

PLUMBING MECHANICAL ENGINEERING SERVICES INDUSTRY

The Joint Industry Board for Plumbing Mechanical Engineering Services has agreed a two year wage agreement for 2011–2012 with effect from 3 January 2011. Current rates of pay for 2012 will be effective from 2 January 2012.

For full details of this wage agreement and the JIB PMES National Working Rules, contact:

The Joint Industry Board for Plumbing Mechanical Engineering Services in England and Wales
Brook House
Brook Street
St Neots
Huntingdon
Cambridge PE19 2HW
Telephone: 01480 476925
E-mail: info@jib-pmes.org.uk

WAGE RATES, ALLOWANCES AND OTHER PROVISIONS

EFFECTIVE FROM **2nd January 2012**

Basic Rates of Hourly Pay

Applicable in England and Wales

	Hourly rate £
Operatives	
Technical plumber and gas service technician	14.99
Advanced plumber and gas service engineer	13.50
Trained plumber and gas service fitter	11.58
Apprentices	
4th year of training with NVQ level 3	11.20
4th year of training with NVQ level 2	10.15
4th year of training	8.94
3rd year of training with NVQ level 2	8.83
3rd year of training	7.26
2nd year of training	6.43
1st year of training	5.61
Adult Trainees	
3rd 6 months of employment	10.09
2nd 6 months of employment	9.69
1st 6 months of employment	9.03

PLUMBING MECHANICAL ENGINEERING SERVICES INDUSTRY

Major Projects Agreement

Where a job is designated as being a Major Project then the following Major Project Performance Payment hourly rate supplement shall be payable:

Employee Category

	National Payment £	London* Payment £
Technical Plumber and Gas Service Technician	2.20	3.57
Advanced Plumber and Gas Service Engineer	2.20	3.57
Trained Plumber and Gas Service Fitter	2.20	3.57
All 4th year apprentices	1.76	2.86
All 3rd year apprentices	1.32	2.68
2nd year apprentice	1.21	1.96
1st year apprentice	0.88	1.43
All adult trainees	1.76	2.86

* The London Payment Supplement applies only to designated Major Projects that are within the M25 London orbital motorway and are effective from 1 February 2007.

* National payment hourly rates are unchanged for 2007 and will continue to be at the rates shown in Promulgation 138A issued 14 October 2003.

Allowances

Daily travel time allowance plus return fares

All daily travel allowances are to be paid at the daily rate as follows:

Over	Not exceeding	All Operatives	3rd & 4th Year Apprentices	1st & 2nd Year Apprentices
20	30	£4.24	£2.74	£1.70
30	40	£9.89	£6.35	£4.07
40	50	£11.30	£6.75	£4.24

PLUMBING MECHANICAL ENGINEERING SERVICES INDUSTRY

Responsibility/Incentive Pay Allowance

As from Monday 2nd January 2012, Employers may, in consultation with the employees concerned, enhance the basic graded rates of pay by the payment of an additional amount, as per the bands shown below, where it is agreed that their work involves extra responsibility, productivity or flexibility.

Band 1 – an additional rate of £ 0.26 per hour
Band 2 – an additional rate of £ 0.46 per hour
Band 3 – an additional rate of £ 0.68 per hour
Band 4 – an additional rate of £ 0.90 per hour

This allowance forms part of an operative's basic rate of pay and shall be used to calculate premium payments.

Mileage allowance	£0.40 per mile
Lodging allowance	£28.35 per night
Subsistence Allowance (London Only)	£4.87 per night

Plumbers welding supplement

Possession of Gas or Arc Certificate	£0.29 per hour
Possession of Gas and Arc Certificate	£0.48 per hour

Weekly Holiday Credit Contributions (38th Issue Stamps Option)

	Public and annual Gross Value, £	Combined Holiday Credit*, £
Technical Plumber and Gas Service Technician	62.50	62.90
Advance Plumber and Gas Service Engineer	56.20	56.60
Trained Plumber and Gas Service Fitter	48.10	48.50
Adult Trainee	45.55	48.50
Apprentice in last year of training	36.30	35.60
Apprentice 3rd year	26.15	25.55
Apprentice 2nd year	23.10	22.60
Apprentice 1st year	20.15	20.30
Working Principal	26.00	26.40
Ancillary Employee	31.90	32.30

* Public and Annual gross value is the Combined Holiday credit value less JIB administration value.

PLUMBING MECHANICAL ENGINEERING SERVICES INDUSTRY

Explanatory Notes

1.   Working Hours

The normal working week (Monday to Friday) shall be 37½ hours, with 45 hours to be worked in the same period before overtime rates become applicable.

2.   Overtime

Overtime shall be paid at time and a half up to 8.00pm (Monday to Friday) and up to 1.00pm (Saturday). Overtime worked after these times shall be paid at double time.

3.   Major Projects Agreement

Under the Major Projects Agreement the normal working week shall be 38 hours (Monday to Friday) with overtime rates payable for all hours worked in excess of 38 hours in accordance with 2 above. However, it should be noted that the hourly rate supplement shall be paid for each hour worked but does <u>not</u> attract premium time enhancement.

4.   Pension

In addition to their hourly rates of pay, plumbing employees are entitled to inclusion within the Industry Pension Scheme (or one providing equivalent benefits). The current levels of industry scheme contributions are 6½% (employers) and 3¼% (employees).

5.   Weekly Holiday Credit Contributions

There has been a fundamental change in the way the JIB-PMES Holiday Pay Schemes shall apply for the 38th issue, in that 60 credits will be paid over a period of 53 weeks.

For full details please refer to the 38th issue of the Holiday Credit Values as published by the JIB for Plumbing Mechanical Engineering Services in England and Wales.

Electrical Installations

Rates of Wages

ELECTRICAL CONTRACTING INDUSTRY

For full details of this wage agreement and the Joint Industry Board for the Electrical Contracting Industry's National Working Rules, contact:

The Joint Industry Board for the Electrical Contracting Industry
Kingswood House
47/51 Sidcup Hill
Sidcup
Kent DA14 6HP
Telephone: 020 8302 0031
Internet: www.jib.org.uk

WAGES (Graded Operatives)

Rates

Since 7 January 2002 two different wage rates have applied to JIB Graded Operatives working on site, depending on whether the Employer transports them to site or whether they provide their own transport. The two categories are:

Job Employed (<u>Transport Provided</u>)

Payable to an Operative who is transported to and from the job by his Employer. The Operative shall also be entitled to payment for Travel Time, when travelling in his own time, as detailed in the appropriate scale.

Job Employed (<u>Own Transport</u>)

Payable to an Operative who travels by his own means to and from the job. The Operative shall be entitled to payment for Travel Allowance and also Travel Time, when travelling in his own time, as detailed in the appropriate scale.

ELECTRICAL CONTRACTING INDUSTRY

The JIB rates of wages are set out below:

From and including 4 January 2010, the JIB hourly rates of wages shall be as set out below:

(i) **National Standard Rate**

Grade	Transport provided	Own transport
Technician (or equivalent specialist grade)	£ 15.38	£ 16.16
Approved Electrician (or equivalent specialist grade)	£ 13.59	£ 14.35
Electrician (or equivalent specialist grade)	£ 12.45	£ 13.23
Senior Graded Electrical Trainee	£ 11.20	£ 11.92
Electrical Improver	£ 11.20	£ 11.92
Labourer	£ 9.89	£ 10.62
Adult Trainee	£ 9.89	£ 10.62

(ii) **London Rate**

Grade	Transport provided	Own transport
Technician (or equivalent specialist grade)	£ 17.23	£ 18.10
Approved Electrician (or equivalent specialist grade)	£ 15.22	£ 16.07
Electrician (or equivalent specialist grade)	£ 13.94	£ 14.83
Senior Graded Electrical Trainee	£ 12.54	£ 13.35
Electrical Improver	£ 12.54	£ 13.35
Labourer	£ 11.08	£ 11.89
Adult Trainee	£ 11.08	£ 11.89

1999 Joint Industry Board Apprentice Training Scheme

From and including 4 January 2010, the JIB hourly rates for Job Employed apprentices shall be:

(i) **National Standard Rates**

	Transport provided	Own transport
Stage 1	£ 4.35	£ 5.09
Stage 2	£ 6.41	£ 7.16
Stage 3	£ 9.28	£ 10.05
Stage 4	£ 9.82	£ 10.59

(ii) **London Rate**

	Transport provided	Own transport
Stage 1	£ 4.87	£ 5.70
Stage 2	£ 7.18	£ 8.02
Stage 3	£ 10.39	£ 11.26
Stage 4	£ 11.00	£ 11.86

ELECTRICAL CONTRACTING INDUSTRY

Travelling Time and Travel Allowances

From and including 4 January 2010

Operatives required to start/finish at the normal starting and finishing time on jobs which are 15 miles and over from the shop – in a straight line – receive payment for Travelling Time and where transport is not provided by the Employer, Travel Allowance, as follows:

Distance	Total Daily Travel Allowance	Total Daily Travelling Time
(a) National Standard Rate		
Up to 15 miles	Nil	Nil
Over 15 & up to 20 miles each way	£ 3.43	£ 4.68
Over 20 & up to 25 miles each way	£ 4.56	£ 5.93
Over 25 & up to 35 miles each way	£ 6.01	£ 7.25
Over 35 & up to 55 miles each way	£ 9.57	£ 9.57
Over 55 & up to 75 miles each way	£ 11.71	£ 11.71

For each additional 10 mile band over 75 miles, additional payment of £2.06 for Daily Travel Allowance and £2.06 for Daily Travel Time will be made.

Note: Special arrangements may apply for work in the Merseyside area.

Distance	Total Daily Travel Allowance	Total Daily Travelling Time
(b) London Rate		
Up to 15 miles	Nil	Nil
Over 15 & up to 20 miles each way	£ 3.45	£ 5.03
Over 20 & up to 25 miles each way	£ 4.59	£ 6.57
Over 25 & up to 35 miles each way	£ 6.05	£ 7.91
Over 35 & up to 55 miles each way	£ 9.65	£ 10.68
Over 55 & up to 75 miles each way	£ 11.79	£ 12.50

For each additional 10 mile band over 75 miles, additional payments of £2.08 for Daily Travel Allowance and £2.08 for Daily Travel Time will be made.

ELECTRICAL CONTRACTING INDUSTRY

Travelling time and travel allowance – Section 8 (Employed permanently at the shop)

Operatives required to start/finish at the normal starting and finishing time on jobs which are 15 miles and over from the shop – in a straight line – receive payment for Travelling Time and where transport is not provided by the Employer, Travel Allowance, as follows:

From and including 4 January 2010

Distance	Total Daily Travel Allowance	Total Daily Travelling Time
(a) National Standard Rate		
Up to 15 miles	Nil	Nil
Over 15 & up to 20 miles each way	£ 3.43	£ 2.75
Over 20 & up to 25 miles each way	£ 4.56	£ 4.13
Over 25 & up to 35 miles each way	£ 6.01	£ 5.51
Over 35 & up to 55 miles each way	£ 9.57	£ 6.89
Over 55 & up to 75 miles each way	£ 11.71	£ 8.27

For each additional 10 mile band over 75 miles, additional payments of £2.06 for Daily Travel Allowance and £1.39 for Daily Travelling Time will be made.

Note: Special arrangements may apply for work in the Merseyside area.

Distance	Total Daily Travel Allowance	Total Daily Travelling Time
(b) London Rate		
Up to 15 miles	Nil	Nil
Over 15 & up to 20 miles each way	£ 3.45	£ 3.06
Over 20 & up to 25 miles each way	£ 4.59	£ 4.59
Over 25 & up to 35 miles each way	£ 6.05	£ 6.11
Over 35 & up to 55 miles each way	£ 9.65	£ 7.62
Over 55 & up to 75 miles each way	£ 11.79	£ 9.17

For each additional 10 mile band over 75 miles, additional payments of £2.08 for Daily Travel Allowance and £1.40 for Daily Travelling Time will be made.

Lodging Allowances
£32.68 from and including 4 January 2011

Lodgings weekend retention fee, maximum reimbursement
£32.68 from and including 4 January 2011

Annual Holiday Lodging Allowance Retention
Maximum £10.00 per night (£70.00 per week) from and including 7 January 2008

ELECTRICAL CONTRACTING INDUSTRY

Responsibility money

From and including 30 March 1998 the minimum payment increased to 10p per hour and the maximum to £1.00 per hour (no change)

From and including 4 January 1992 responsibility payments are enhanced by overtime and shift premiums where appropriate (no change)

Combined JIB Benefits Stamp Value (from week commencing 24 September 2010)

JIB Grade	Weekly JIB combined credit value £	Holiday Value £
Technician	£ 62.30	£46.49
Approved Electrician	£ 56.62	£41.01
Electrician	£ 53.01	£37.63
Senior Graded Electrical Trainee and Electrical Improver	£ 49.04	£33.86
Labourer & Adult Trainee	£ 44.88	£29.89

Explanatory Notes

1. Working Hours

   The normal working week (Monday to Friday) shall be 37½ hours, with 38 hours to be worked in the same period before overtime rates become applicable.

2. Overtime

   Overtime shall be paid at time and a half for all weekday overtime. Saturday overtime shall be paid at time and a half for the first 6 hours, or up to 3.00pm (whichever comes first). Thereafter double time shall be paid until normal starting time on Monday.

Spon's Estimating Costs Guide to Plumbing and Heating

Unit Rates and Project Costs

Fourth Edition

Bryan Spain

Do you work on jobs between £50 and £50,000? - Then this book is for you.

All the cost data you need to keep your estimating accurate, competitive and profitable.

Specially written for contractors and small businesses carrying out small works, *Spon's Estimating Cost Guide to Plumbing and Heating* contains accurate information on thousands of rates each broken down to labour, material overheads and profit.

The first book to include typical project costs for:

- rainwater goods installations
- bathrooms
- external waste systems
- central heating systems
- hot and cold water systems.

July 2008: 216x138: 264pp
Pb: 978-0-415-46905-0: **£31.99**

To Order: Tel: +44 (0) 1235 400524 **Fax:** +44 (0) 1235 400525
or Post: Taylor and Francis Customer Services,
Bookpoint Ltd, Unit T1, 200 Milton Park, Abingdon, Oxon, OX14 4TA UK
Email: book.orders@tandf.co.uk

For a complete listing of all our titles visit:
www.tandf.co.uk

Daywork

When work is carried out in connection with a contract that cannot be valued in any other way, it is usual to assess the value on a cost basis with suitable allowances to cover overheads and profit. The basis of costing is a matter for agreement between the parties concerned but definitions of prime cost for the Heating and Ventilating and Electrical Industries have been published jointly by the Royal Institution of Chartered Surveyors and the appropriate bodies of the industries concerned, for those who wish to use them.

These, together with a schedule of basic plant hire charges are reproduced on the following pages, with the kind permission of the Royal Institution of Chartered Surveyors, who own the copyright.

Spon's Estimating Costs Guide to Electrical Works

Project Costs at a Glance

Fourth Edition

Bryan Spain

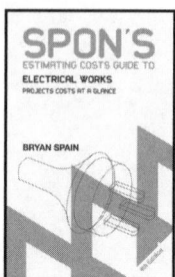

Specially written for contractors and small businesses carrying out small works, *Spon's Estimating Costs Guide to Electrical Work*s provides accurate information on thousands of rates, each broken down to labour, material overheads and profit for residential, retail and light industrial premises. It is the first book to include typical project costs for new installations, stripping out, rewiring and upgrading for flats and houses.

In addition, vital information and advice is given on setting up and running a business, employing staff, tax, VAT and CIS4's.

For the cost of approximately two hours of your charge-out rate (or less), this book will help you to:

- produce estimates faster
- keep your estimates accurate and competitive
- run your business more effectively
- save time.

August 2008: 216x138: 304pp
Pb: 978-0-415-46904-3: **£31.99**

To Order: Tel: +44 (0) 1235 400524 **Fax:** +44 (0) 1235 400525
or Post: Taylor and Francis Customer Services,
Bookpoint Ltd, Unit T1, 200 Milton Park, Abingdon, Oxon, OX14 4TA UK
Email: book.orders@tandf.co.uk

For a complete listing of all our titles visit:
www.tandf.co.uk

HEATING AND VENTILATING INDUSTRY

DEFINITION OF PRIME COST OF DAYWORK CARRIED OUT UNDER A HEATING, VENTILATING, AIR CONDITIONING, REFRIGERATION, PIPEWORK AND/OR DOMESTIC ENGINEERING CONTRACT (JULY 1980 EDITION)

This Definition of Prime Cost is published by the Royal Institution of Chartered Surveyors and the Heating and Ventilating Contractors Association for convenience, and for use by people who choose to use it. Members of the Heating and Ventilating Contractors Association are not in any way debarred from defining Prime Cost and rendering accounts for work carried out on that basis in any way they choose. Building owners are advised to reach agreement with contractors on the Definition of Prime Cost to be used prior to entering into a contract or subcontract.

SECTION 1: APPLICATION

1.1 This Definition provides a basis for the valuation of daywork executed under such heating, ventilating, air conditioning, refrigeration, pipework and or domestic engineering contracts as provide for its use.
1.2 It is not applicable in any other circumstances, such as jobbing or other work carried out as a separate or main contract nor in the case of daywork executed after a date of practical completion.
1.3 The terms 'contract' and 'contractor' herein shall be read as 'subcontract' and 'subcontractor' as applicable.

SECTION 2: COMPOSITION OF TOTAL CHARGES

2.1 The Prime Cost of daywork comprises the sum of the following costs:
    (a)   Labour as defined in Section 3.
    (b)   Materials and goods as defined in Section 4.
    (c)   Plant as defined in Section 5.
2.2 Incidental costs, overheads and profit as defined in Section 6, as provided in the contract and expressed therein as percentage adjustments, are applicable to each of 2.1 (a)–(c).

SECTION 3: LABOUR

3.1 The standard wage rates, emoluments and expenses referred to below and the standard working hours referred to in 3.2 are those laid down for the time being in the rules or decisions or agreements of the Joint Conciliation Committee of the Heating, Ventilating and Domestic Engineering Industry applicable to the works (or those of such other body as may be appropriate) and to the grade of operative concerned at the time when and the area where the daywork is executed.
3.2 Hourly base rates for labour are computed by dividing the annual prime cost of labour, based upon the standard working hours and as defined in 3.4, by the number of standard working hours per annum. See example.
3.3 The hourly rates computed in accordance with 3.2 shall be applied in respect of the time spent by operatives directly engaged on daywork, including those operating mechanical plant and transport and erecting and dismantling other plant (unless otherwise expressly provided in the contract) and handling and distributing the materials and goods used in the daywork.
3.4 The annual prime cost of labour comprises the following:
    (a)   Standard weekly earnings (i.e. the standard working week as determined at the appropriate rate for the operative concerned).
    (b)   Any supplemental payments.
    (c)   Any guaranteed minimum payments (unless included in Section 6.1(a)–(p)).
    (d)   Merit money.
    (e)   Differentials or extra payments in respect of skill, responsibility, discomfort, inconvenience or risk (excluding those in respect of supervisory responsibility – see 3.5)
    (f)   Payments in respect of public holidays.
    (g)   Any amounts which may become payable by the contractor to or in respect of operatives arising from the rules etc. referred to in 3.1 which are not provided for in 3.4 (a)–(f) nor in Section 6.1 (a)–(p).
    (h)   Employers contributions to the WELPLAN, the HVACR Welfare and Holiday Scheme or payments in lieu thereof.
    (i)   Employers National Insurance contributions as applicable to 3.4 (a)–(h).
    (j)   Any contribution, levy or tax imposed by Statute, payable by the contractor in his capacity as an employer.

HEATING AND VENTILATING INDUSTRY

3.5   Differentials or extra payments in respect of supervisory responsibility are excluded from the annual prime cost (see Section 6). The time of principals, staff, foremen, chargehands and the like when working manually is admissible under this Section at the rates for the appropriate grades.

SECTION 4: MATERIALS AND GOODS

4.1   The prime cost of materials and goods obtained specifically for the daywork is the invoice cost after deducting all trade discounts and any portion of cash discounts in excess of 5%.
4.2   The prime cost of all other materials and goods used in the daywork is based upon the current market prices plus any appropriate handling charges.
4.3   The prime cost referred to in 4.1 and 4.2 includes the cost of delivery to site.
4.4   Any Value Added Tax which is treated, or is capable of being treated, as input tax (as defined by the Finance Act 1972, or any re-enactment or amendment thereof or substitution therefore) by the contractor is excluded.

SECTION 5: PLANT

5.1   Unless otherwise stated in the contract, the prime cost of plant comprises the cost of the following:
    (a)   use or hire of mechanically-operated plant and transport for the time employed on and/or provided or retained for the daywork;
    (b)   use of non-mechanical plant (excluding non-mechanical hand tools) for the time employed on and/or provided or retained for the daywork;
    (c)   transport to and from the site and erection and dismantling where applicable.
5.2   The use of non-mechanical hand tools and of erected scaffolding, staging, trestles or the like is excluded (see Section 6), unless specifically retained for the daywork.

SECTION 6: INCIDENTAL COSTS, OVERHEADS AND PROFIT

6.1   The percentage adjustments provided in the contract which are applicable to each of the totals of Sections 3, 4 and 5 comprise the following:
    (a)   Head office charges.
    (b)   Site staff including site supervision.
    (c)   The additional cost of overtime (other than that referred to in 6.2).
    (d)   Time lost due to inclement weather.
    (e)   The additional cost of bonuses and all other incentive payments in excess of any included in 3.4.
    (f)   Apprentices' study time.
    (g)   Fares and travelling allowances.
    (h)   Country, lodging and periodic allowances.
    (i)   Sick pay or insurances in respect thereof, other than as included in 3.4.
    (j)   Third party and employers' liability insurance.
    (k)   Liability in respect of redundancy payments to employees.
    (l)   Employer's National Insurance contributions not included in 3.4.
    (m)   Use and maintenance of non-mechanical hand tools.
    (n)   Use of erected scaffolding, staging, trestles or the like (but see 5.2).
    (o)   Use of tarpaulins, protective clothing, artificial lighting, safety and welfare facilities, storage and the like that may be available on site.
    (p)   Any variation to basic rates required by the contractor in cases where the contract provides for the use of a specified schedule of basic plant charges (to the extent that no other provision is made for such variation – see 5.1).
    (q)   In the case of a subcontract which provides that the subcontractor shall allow a cash discount, such provision as is necessary for the allowance of the prescribed rate of discount.
    (r)   All other liabilities and obligations whatsoever not specifically referred to in this Section nor chargeable under any other Section.
    (s)   Profit.
6.2   The additional cost of overtime where specifically ordered by the Architect/Supervising Officer shall only be chargeable in the terms of a prior written agreement between the parties.

HEATING AND VENTILATING INDUSTRY

MECHANICAL INSTALLATIONS

Calculation of Hourly Base Rate of Labour for Typical Main Grades applicable from 4 October 2010, refer to notes within Section Three – Rates of Wages.

	Foreman	Senior Craftsman (+ 2nd Welding Skill)	Senior Craftsman	Craftsman	Installer	Mate over 18
Hourly Rate from 4 October 2010	15.00	12.91	12.40	11.38	10.31	8.69
Annual standard earnings excluding all holidays, 45.6 weeks × 38 hours	25,992.00	22,370.45	21,486.72	19,719.26	17,865.17	15,058.03
Employers national insurance contributions from 6 April 2011	2,686.03	2,186.26	2,064.30	1,820.39	1,564.53	1,177.14
Weekly holiday credit and welfare contributions (52 weeks) from 6 October 2010	4,217.20	3,681.08	3,550.56	3,291.08	3,016.52	2,602.08
Annual prime cost of labour	32,707.19	28,077.58	26,947.86	24,688.43	22,318.25	18,729.74
Hourly base rate	18.88	16.20	15.55	14.25	12.88	10.81

Notes:

(1)   Annual industry holiday (4.6 weeks × 38 hours) and public holidays (1.8 weeks × 38 hours) are paid through weekly holiday credit and welfare stamp scheme.

(2)   Where applicable, Merit money and other variables (e.g. daily abnormal conditions money), which attract Employer's National Insurance contribution, should be included.

(3)   Contractors in Northern Ireland should add the appropriate amount of CITB Levy to the annual prime cost of labour prior to calculating the hourly base rate.

(4)   Hourly rate based on 1,732.80 hours per annum and calculated as follows

52 Weeks @ 38 hrs/wk	=		1,976.00
Less			
Public Holiday = 9/5 = 1.8 weeks @ 38 hrs/wk	=	68.40	
Annual holidays = 4.6 weeks @ 38 hrs/wk	=	174.80	243.20
Hours	=		1,732.80

(5)   For calculation of Holiday Credits and ENI refer to detailed labour rate evaluation.

(6)   National Insurance contributions are those effective from 6 April 2011.

(7)   Weekly holiday credit/welfare stamp values are those effective from 4 October 2010.

(8)   Hourly rates of wages are those effective from 4 October 2010.

ELECTRICAL INDUSTRY

DEFINITION OF PRIME COST OF DAYWORK CARRIED OUT UNDER AN ELECTRICAL CONTRACT (MARCH 1981 EDITION)

This Definition of Prime Cost is published by The Royal Institution of Chartered Surveyors and The Electrical Contractors' Associations for convenience and for use by people who choose to use it. Members of The Electrical Contractors' Association are not in any way debarred from defining Prime Cost and rendering accounts for work carried out on that basis in any way they choose. Building owners are advised to reach agreement with contractors on the Definition of Prime Cost to be used prior to entering into a contract or subcontract.

SECTION 1: APPLICATION

1.1   This Definition provides a basis for the valuation of daywork executed under such electrical contracts as provide for its use.
1.2   It is not applicable in any other circumstances, such as jobbing, or other work carried out as a separate or main contract, nor in the case of daywork executed after the date of practical completion.
1.3   The terms 'contract' and 'contractor' herein shall be read as 'subcontract' and 'subcontractor' as the context may require.

SECTION 2: COMPOSITION OF TOTAL CHARGES

2.1   The Prime Cost of daywork comprises the sum of the following costs:
      (a)   Labour as defined in Section 3.
      (b)   Materials and goods as defined in Section 4.
      (c)   Plant as defined in Section 5.
2.2   Incidental costs, overheads and profit as defined in Section 6, as provided in the contract and expressed therein as percentage adjustments, are applicable to each of 2.1 (a)–(c).

SECTION 3: LABOUR

3.1   The standard wage rates, emoluments and expenses referred to below and the standard working hours referred to in 3.2 are those laid down for the time being in the rules and determinations or decisions of the Joint Industry Board or the Scottish Joint Industry Board for the Electrical Contracting Industry (or those of such other body as may be appropriate) applicable to the works and relating to the grade of operative con-cerned at the time when and in the area where daywork is executed.
3.2   Hourly base rates for labour are computed by dividing the annual prime cost of labour, based upon the standard working hours and as defined in 3.4 by the number of standard working hours per annum. See examples.
3.3   The hourly rates computed in accordance with 3.2 shall be applied in respect of the time spent by operatives directly engaged on daywork, including those operating mechanical plant and transport and erecting and dismantling other plant (unless otherwise expressly provided in the contract) and handling and distributing the materials and goods used in the daywork.
3.4   The annual prime cost of labour comprises the following:
      (a)   Standard weekly earnings (i.e. the standard working week as determined at the appropriate rate for the operative concerned).
      (b)   Payments in respect of public holidays.
      (c)   Any amounts which may become payable by the Contractor to or in respect of operatives arising from operation of the rules etc. referred to in 3.1 which are not provided for in 3.4(a) and (b) nor in Section 6.
      (d)   Employer's National Insurance Contributions as applicable to 3.4 (a)–(c).
      (e)   Employer's contributions to the Joint Industry Board Combined Benefits Scheme or Scottish Joint Industry Board Holiday and Welfare Stamp Scheme, and holiday payments made to apprentices in compliance with the Joint Industry Board National Working Rules and Industrial Determinations as an employer.
      (f)   Any contribution, levy or tax imposed by Statute, payable by the Contractor in his capacity as an employer.
3.5   Differentials or extra payments in respect of supervisory responsibility are excluded from the annual prime cost (see Section 6). The time of principals and similar categories, when working manually, is admissible under this Section at the rates for the appropriate grades.

ELECTRICAL INDUSTRY

SECTION 4: MATERIALS AND GOODS

4.1 The prime cost of materials and goods obtained specifically for the daywork is the invoice cost after deducting all trade discounts and any portion of cash discounts in excess of 5%.

4.2 The prime cost of all other materials and goods used in the daywork is based upon the current market prices plus any appropriate handling charges.

4.3 The prime cost referred to in 4.1 and 4.2 includes the cost of delivery to site.

4.4 Any Value Added Tax which is treated, or is capable of being treated, as input tax (as defined by the Finance Act 1972, or any re-enactment or amendment thereof or substitution therefore) by the Contractor is excluded.

SECTION 5: PLANT

5.1 Unless otherwise stated in the contract, the prime cost of plant comprises the cost of the following:

(a) Use or hire of mechanically-operated plant and transport for the time employed on and/or provided or retained for the daywork;

(b) Use of non-mechanical plant (excluding non-mechanical hand tools) for the time employed on and/or provided or retained for the daywork;

(c) Transport to and from the site and erection and dismantling where applicable.

5.2 The use of non-mechanical hand tools and of erected scaffolding, staging, trestles or the likes is excluded (see Section 6), unless specifically retained for daywork.

5.3 Note: Where hired or other plant is operated by the Electrical Contractor's operatives, such time is to be included under Section 3 unless otherwise provided in the contract.

SECTION 6: INCIDENTAL COSTS, OVERHEADS AND PROFIT

6.1 The percentage adjustments provided in the contract which are applicable to each of the totals of Sections 3, 4 and 5, compromise the following:

(a) Head Office charges.

(b) Site staff including site supervision.

(c) The additional cost of overtime (other than that referred to in 6.2).

(d) Time lost due to inclement weather.

(e) The additional cost of bonuses and other incentive payments.

(f) Apprentices' study time.

(g) Travelling time and fares.

(h) Country and lodging allowances.

(i) Sick pay or insurance in lieu thereof, in respect of apprentices.

(j) Third party and employers' liability insurance.

(k) Liability in respect of redundancy payments to employees.

(l) Employers' National Insurance Contributions not included in 3.4.

(m) Use and maintenance of non-mechanical hand tools.

(n) Use of erected scaffolding, staging, trestles or the like (but see 5.2.).

(o) Use of tarpaulins, protective clothing, artificial lighting, safety and welfare facilities, storage and the like that may be available on site.

(p) Any variation to basic rates required by the Contractor in cases where the contract provides for the use of a specified schedule of basic plant charges (to the extent that no other provision is made for such variation – see 5.1).

(q) All other liabilities and obligations whatsoever not specifically referred to in this Section nor chargeable under any other Section.

(r) Profit.

(s) In the case of a subcontract which provides that the subcontractor shall allow a cash discount, such provision as is necessary for the allowance of the prescribed rate of discount.

6.2 The additional cost of overtime where specifically ordered by the Architect/Supervising Officer shall only be chargeable in the terms of a prior written agreement between the parties.

ELECTRICAL INDUSTRY

ELECTRICAL INSTALLATIONS

Calculation of Hourly Base Rate of Labour for Typical Main Grades applicable from 4 January 2010.

	Technician	Approved Electrician	Electrician	Labourer
Hourly Rate from 4 January 2010 (London Rates)	18.10	16.07	14.83	11.89
Annual standard earnings excluding all holidays, 45.8 weeks × 37.5 hours	31,086.75	27,600.23	25,470.53	20,421.08
Employers national insurance contributions from 6 April 2011	3,389.11	2,907.97	2,614.07	1,917.24
JIB Combined benefits from 29 September 2009	3,289.00	2,993.64	2,805.92	2,383.16
Holiday top up funding	1,670.65	1,494.04	1,393.46	1,132.67
Annual prime cost of labour	39,386.11	34,946.47	32,234.57	25,804.74
Hourly base rate	22.93	20.35	18.77	15.02

Notes:

(1) Annual industry holiday (4.4 weeks × 37.5 hours) and public holidays (1.8 weeks × 37.5 hours)
(2) It should be noted that all labour costs incurred by the Contractor in his capacity as an Employer, other than those contained in the hourly rate above, must be taken into account under Section 6.
(3) Public Holidays are paid through weekly holiday credit and welfare stamp scheme.
(4) Contractors in Northern Ireland should add the appropriate amount of CITB Levy to the annual prime cost of labour prior to calculating the hourly base rate.
(5) Hourly rate based on 1,717.5 hours per annum and calculated as follows:

52 Weeks @ 37.5 hrs/wk	=		1,950.00
Less			
Public Holiday = 9/5 = 1.8 weeks @ 37.5 hrs/wk	=	67.50	
Annual holidays = 4.4 weeks @ 37.5 hrs/wk	=	165.00	232.50
Hours	=		1,717.50

(6) For calculation of holiday credits and ENI refer to detailed labour rate evaluation.
(7) Hourly wage rates are those effective from 4 January 2010.
(8) National Insurance contributions are those effective from 6 April 2011.
(9) JIB Combined Benefits Values are those effective from 27 September 2010.

BUILDING INDUSTRY PLANT HIRE COSTS

SCHEDULE OF BASIC PLANT CHARGES

This Schedule is published by the Royal Institution of Chartered Surveyors and is for use in connection with Dayworks under a Building Contract.

EXPLANATORY NOTES

1   The rates in the Schedule are intended to apply solely to daywork carried out under and incidental to a Building Contract. They are NOT intended to apply to:
    (i)   jobbing or any other work carried out as a main or separate contract; or
    (ii)   work carried out after the date of commencement of the Defects Liability Period.

2   The rates apply to plant and machinery already on site, whether hired or owned by the Contractor.

3   The rates, unless otherwise stated, include the cost of fuel and power of every description, lubricating oils, grease, maintenance, sharpening of tools, replacement of spare parts, all consumable stores and for licences and insurances applicable to items of plant.

4   The rates, unless otherwise stated, do not include the costs of drivers and attendants (unless otherwise stated).

5   The rates in the Schedule are base costs and may be subject to an overall adjustment for price movement, overheads and profit, quoted by the Contractor prior to the placing of the Contract.

6   The rates should be applied to the time during which the plant is actually engaged in daywork.

7   Whether or not plant is chargeable on daywork depends on the daywork agreement in use and the inclusion of an item of plant in this schedule does not necessarily indicate that item is chargeable.

8   Rates for plant not included in the Schedule or which is not already on site and is specifically provided or hired for daywork shall be settled at prices which are reasonably related to the rates in the Schedule having regard to any overall adjustment quoted by the Contractor in the Conditions of Contract.

NOTE: All rates in the schedule were calculated during the first quarter of 2011.

Daywork

BUILDING INDUSTRY PLANT HIRE COSTS

MECHANICAL PLANT AND TOOLS

Item of plant	Size/Rating	Unit	Rate per hour (£)
PUMPS			
Mobile Pumps			
Including pump hoses, values and strainers, etc.			
Diaphragm	50 mm diameter	Each	1.17
Diaphragm	76 mm diameter	Each	1.89
Diaphragm	102 mm diameter	Each	3.54
Submersible	50 mm diameter	Each	0.76
Submersible	76 mm diameter	Each	0.86
Submersible	102 mm diameter	Each	1.03
Induced Flow	50 mm diameter	Each	0.77
Induced Flow	76 mm diameter	Each	1.67
Centrifugal, self priming	25 mm diameter	Each	1.30
Centrifugal, self priming	50 mm diameter	Each	1.92
Centrifugal, self priming	75 mm diameter	Each	2.74
Centrifugal, self priming	102 mm diameter	Each	3.35
Centrifugal, self priming	152 mm diameter	Each	4.27
SCAFFOLDING, SHORING, FENCING			
Complete Scaffolding			
Mobile working towers, single width	2.0 m x 0.72 m base x 7.45 m high	Each	3.36
Mobile working towers, single width	2.0 m x 0.72 m base x 8.84 m high	Each	3.79
Mobile working towers, double width	2.0 m x 1.35 m x 7.45 m high	Each	3.79
Mobile working towers, double width	2.0 m x 1.35 m x 15.8 m high	Each	7.13
Chimney scaffold, single unit		Each	1.92
Chimney scaffold, twin unit		Each	3.59
Push along access platform	1.63 – 3.1 m	Each	5.00
Push along access platform	1.80 m x 0.70 m	Each	1.79
Trestles			
Trestle, adjustable	Any height	Pair	0.41
Trestle, painters	1.8 m high	Pair	0.31
Trestle, painters	2.4 m high	Pair	0.36
Shoring, Planking and Strutting			
'Acrow' adjustable prop	Sizes up to 4.9 m (open)	Each	0.06
'Strong boy' support attachment		Each	0.22
Adjustable trench strut	Sizes up to 1.67m (open)	Each	0.16
Trench sheet		Metre	0.03
Backhoe trench box	Base unit	Each	1.23
Backhoe trench box	Top unit	Each	0.87
Temporary Fencing			
Including block and coupler			
Site fencing steel grid panel	3.5 m x 2.0 m	Each	0.05
Anti-climb site steel grid fence panel	3.5 m x 2.0 m	Each	0.08
Solid panel Heras	2.0 m x 2.0 m	Each	0.09
Pedestrian gate		Each	0.36
Roadway gate		Each	0.60

BUILDING INDUSTRY PLANT HIRE COSTS

MECHANICAL PLANT AND TOOLS

Item of plant	Size/Rating	Unit	Rate per hour (£)
LIFTING APPLIANCES AND CONVEYORS			
Cranes			
Mobile Cranes			
Rates are inclusive of drivers			
Lorry mounted, telescopic jib			
Two wheel drive	5 tonnes	Each	19.00
Two wheel drive	8 tonnes	Each	42.00
Two wheel drive	10 tonnes	Each	50.00
Two wheel drive	12 tonnes	Each	77.00
Two wheel drive	20 tonnes	Each	89.69
Four wheel drive	18 tonnes	Each	46.51
Four wheel drive	25 tonnes	Each	35.90
Four wheel drive	30 tonnes	Each	38.46
Four wheel drive	45 tonnes	Each	46.15
Four wheel drive	50 tonnes	Each	53.85
Four wheel drive	60 tonnes	Each	61.54
Four wheel drive	70 tonnes	Each	71.79
Static tower crane			
Rates inclusive of driver			

Note: Capacity equals maximum lift in tonnes times maximum radius at which it can be lifted

	Capacity (Metre/tonnes) Up to	Height under hook above ground (m) Up to	Unit	Rate per hour (£)
Tower crane	30	22	Each	22.23
Tower crane	40	22	Each	26.62
Tower crane	40	30	Each	33.33
Tower crane	50	22	Each	29.16
Tower crane	60	22	Each	35.90
Tower crane	60	36	Each	35.90
Tower crane	70	22	Each	41.03
Tower crane	80	22	Each	39.12
Tower crane	90	42	Each	37.18
Tower crane	110	36	Each	47.62
Tower crane	140	36	Each	55.77
Tower crane	170	36	Each	64.11
Tower crane	200	36	Each	71.95
Tower crane	250	36	Each	84.77
Tower crane with luffing jig	30	25	Each	22.23
Tower crane with luffing jig	40	30	Each	26.62
Tower crane with luffing jig	50	30	Each	29.16
Tower crane with luffing jig	60	36	Each	41.03
Tower crane with luffing jig	65	30	Each	33.13
Tower crane with luffing jig	80	22	Each	48.72
Tower crane with luffing jig	100	45	Each	48.72
Tower crane with luffing jig	125	30	Each	53.85
Tower crane with luffing jig	160	50	Each	53.85
Tower crane with luffing jig	200	50	Each	74.36
Tower crane with luffing jig	300	60	Each	100.00

BUILDING INDUSTRY PLANT HIRE COSTS

MECHANICAL PLANT AND TOOLS			
Item of plant	**Size/Rating**	**Unit**	**Rate per hour (£)**
LIFTING APPLIANCES AND CONVEYORS – cont'd			
Crane Equipment			
Muck tipping skip	Up to 200 litres	Each	0.67
Muck tipping skip	500 litres	Each	0.82
Muck tipping skip	750 litres	Each	1.08
Muck tipping skip	1000 litres	Each	1.28
Muck tipping skip	1500 litres	Each	1.41
Muck tipping skip	2000 litres	Each	1.67
Mortar skip	250 litres, plastic	Each	0.41
Mortar skip	350 litres steel	Each	0.77
Boat skip	250 litres	Each	0.92
Boat skip	500 litres	Each	1.08
Boat skip	750 litres	Each	1.23
Boat skip	1000 litres	Each	1.38
Boat skip	1500 litres	Each	1.64
Boat skip	2000 litres	Each	1.90
Boat skip	3000 litres	Each	2.82
Boat skip	4000 litres	Each	3.23
Master flow skip	250 litres	Each	0.77
Master flow skip	500 litres	Each	1.03
Master flow skip	750 litres	Each	1.28
Master flow skip	1000 litres	Each	1.44
Master flow skip	1500 litres	Each	1.69
Master flow skip	2000 litres	Each	1.85
Grand master flow skip	500 litres	Each	1.28
Grand master flow skip	750 litres	Each	1.64
Grand master flow skip	1000 litres	Each	1.69
Grand master flow skip	1500 litres	Each	1.95
Grand master flow skip	2000 litres	Each	2.21
Cone flow skip	500 litres	Each	1.33
Cone flow skip	1000 litres	Each	1.69
Geared rollover skip	500 litres	Each	1.28
Geared rollover skip	750 litres	Each	1.64
Geared rollover skip	1000 litres	Each	1.69
Geared rollover skip	1500 litres	Each	1.95
Geared rollover skip	2000 litres	Each	2.21
Multi skip, rope operated	200mm outlet size, 500 litres	Each	1.49
Multi skip, rope operated	200mm outlet size, 750 litres	Each	1.64
Multi skip, rope operated	200mm outlet size, 1000 litres	Each	1.74
Multi skip, rope operated	200mm outlet size, 1500 litres	Each	2.00
Multi skip, rope operated	200mm outlet size, 2000 litres	Each	2.26
Multi skip, man riding	200mm outlet size, 1000 litres	Each	2.00
Multi skip	4 point lifting frame	Each	0.90
Multi skip	Chain brothers	Set	0.87
Crane Accessories			
Multi-purpose crane forks	1.5 and 2 tonnes S.W.L.	Each	1.13
Self levelling crane forks		Each	1.28
Man cage	1 man, 230kg S.W.L.	Each	1.90
Man cage	2 man, 500kg S.W.L.	Each	1.95
Man cage	4 man, 750kg S.W.L.	Each	2.15

BUILDING INDUSTRY PLANT HIRE COSTS

MECHANICAL PLANT AND TOOLS			
Item of plant	Size/Rating	Unit	Rate per hour (£)
LIFTING APPLIANCES AND CONVEYORS – cont'd			
Crane Equipment – cont'd			
Man cage	8 man, 1000kg S.W.L.	Each	3.33
Stretcher cage	500kg, S.W.L.	Each	2.69
Goods carrying cage	1500kg, S.W.L.	Each	1.33
Goods carrying cage	3000kg, S.W.L.	Each	1.85
Builders' skip lifting cradle	12 tonnes, S.W.L.	Each	2.31
Board/pallet fork	1600kg, S.W.L.	Each	1.90
Gas bottle carrier	500kg, S.W.L.	Each	0.92
Hoists			
Scaffold hoist	200 kg	Each	2.46
Rack and pinion (goods only)	500 kg	Each	4.56
Rack and pinion (goods only)	1100 kg	Each	5.90
Rack and pinion (goods and passenger)	8 person, 80 kg	Each	7.44
Rack and pinion (goods and passenger)	14 person, 1400kg	Each	8.72
Wheelbarrow chain sling		Each	1.67
Conveyors			
Belt conveyors			
Conveyor	8 m long x 450 mm wide	Each	5.90
Miniveyor, control box and loading hopper	3 m unit	Each	4.49
Other Conveying Equipment			
Wheelbarrow		Each	0.62
Hydraulic superlift		Each	4.56
Pavac slab lifter (tile hoist)		Each	4.49
High lift pallet truck		Each	3.08
Lifting Trucks	Payload	Maximum Lift	
Fork lift, two wheel drive	1100 kg	up to 3.0 m	Each	5.64
Fork lift, two wheel drive	2540 kg	up to 3.7 m	Each	5.64
Fork lift, four wheel drive	1524 kg	up to 6.0 m	Each	5.64
Fork lift, four wheel drive	2600 kg	up to 5.4 m	Each	7.44
Fork life, four wheel drive	4000 kg	up to 17 m	Each	10.77
Lifting Platforms			
Hydraulic platform (Cherry picker)	9 m	Each	4.62
Hydraulic platform (Cherry picker)	12 m	Each	7.56
Hydraulic platform (Cherry picker)	15 m	Each	10.13
Hydraulic platform (Cherry picker)	17 m	Each	15.63
Hydraulic platform (Cherry picker)	20 m	Each	18.13
Hydraulic platform (Cherry picker)	25.6 m	Each	32.38
Scissor lift	7.6 m, electric	Each	3.85
Scissor lift	7.8 m, electric	Each	5.13
Scissor lift	9.7 m, electric	Each	4.23
Scissor lift	10 m, diesel	Each	6.41
Telescopic handler	7 m, 2 tonnes	Each	5.13
Telescopic handler	13 m, 3 tonnes	Each	7.18
Lifting and Jacking Gear			
Pipe winch including gantry	1 tonne	Set	1.92
Pipe winch including gantry	3 tonnes	Set	3.21
Chain block	1 tonne	Each	0.35

BUILDING INDUSTRY PLANT HIRE COSTS

MECHANICAL PLANT AND TOOLS			
Item of plant	**Size/Rating**	**Unit**	**Rate per hour (£)**
LIFTING APPLIANCES AND CONVEYORS – cont'd			
Lifting and Jacking Gear – cont'd			
Chain block	2 tonnes	Each	0.58
Chain block	5 tonnes	Each	1.14
Pull lift (Tirfor winch)	1 tonne	Each	0.64
Pull lift (Tirfor winch)	1.6 tonnes	Each	0.90
Pull lift (Tirfor winch)	3.2 tonnes	Each	1.15
Brother or chain slings, two legs	not exceeding 3.1 tonnes	Set	0.21
Brother or chain slings, two legs	not exceeding 4.25 tonnes	Set	0.31
Brother or chain slings, four legs	not exceeding 11.2 tonnes	Set	1.09
CONSTRUCTION VEHICLES			
Lorries			
Plated lorries (Rates are inclusive of driver)			
Platform lorry	7.5 tonnes	Each	16.21
Platform lorry	17 tonnes	Each	22.90
Platform lorry	24 tonnes	Each	30.68
Extra for lorry with crane attachment	up to 2.5 tonnes	Each	3.25
Extra for lorry with crane attachment	up to 5 tonnes	Each	6.00
Extra for lorry with crane attachment	up to 7.5 tonnes	Each	9.10
Tipper Lorries			
(Rates are inclusive of driver)			
Tipper lorry	up to 11 tonnes	Each	15.78
Tipper lorry	up to 17 tonnes	Each	23.95
Tipper lorry	up to 25 tonnes	Each	31.35
Tipper lorry	up to 31 tonnes	Each	37.79
Dumpers			
Site use only (excl. tax, insurance and extra cost of DERV etc. when operating on highway)	Makers Capacity		
Two wheel drive	1 tonnes	Each	1.71
Four wheel drive	2 tonnes	Each	2.43
Four wheel drive	3 tonnes	Each	2.44
Four wheel drive	5 tonnes	Each	3.08
Four wheel drive	6 tonnes	Each	3.85
Four wheel drive	9 tonnes	Each	5.65
Tracked	0.5 tonnes	Each	3.33
Tracked	1.5 tonnes	Each	4.23
Tracked	3.0 tonnes	Each	8.33
Tracked	6.0 tonnes	Each	16.03
Dumper Trucks *(Rates are inclusive of drivers)*			
Dumper truck	up to 15 tonnes	Each	28.56
Dumper truck	up to 17 tonnes	Each	32.82
Dumper truck	up to 23 tonnes	Each	54.64
Dumper truck	up to 30 tonnes	Each	63.50

BUILDING INDUSTRY PLANT HIRE COSTS

MECHANICAL PLANT AND TOOLS

Item of plant	Size/Rating	Unit	Rate per hour (£)
CONSTRUCTION VEHICLES – cont'd			
Dumper Trucks (*Rates are inclusive of drivers*) – cont'd			
Dumper truck	up to 35 tonnes	Each	73.02
Dumper truck	up to 40 tonnes	Each	87.84
Dumper truck	up to 50 tonnes	Each	133.44
Tractors			
Agricultural Type			
Wheeled, rubber-clad tyred	up to 40kW	Each	8.63
Wheeled, rubber-clad tyred	up to 90kW	Each	25.31
Wheeled, rubber-clad tyred	up to 140kW	Each	36.49
Crawler Tractors			
With bull or angle dozer	up to 70kW	Each	29.38
With bull or angle dozer	up to 85kW	Each	38.63
With bull or angle dozer	up to 100kW	Each	52.59
With bull or angle dozer	up to 115kW	Each	55.85
With bull or angle dozer	up to 135kW	Each	60.43
With bull or angle dozer	up to 185kW	Each	76.44
With bull or angle dozer	up to 200kW	Each	96.43
With bull or angle dozer	up to 250kW	Each	117.68
With bull or angle dozer	up to 350kW	Each	160.03
With bull or angle dozer	up to 450kW	Each	219.86
With loading shovel	0.8 m³	Each	26.92
With loading shovel	1.0 m³	Each	32.59
With loading shovel	1.2 m³	Each	37.53
With loading shovel	1.4 m³	Each	42.89
With loading shovel	1.8 m³	Each	52.22
With loading shovel	2.0 m³	Each	57.22
With loading shovel	2.1 m³	Each	60.12
With loading shovel	3.5 m³	Each	87.26
Light Vans			
VW Caddivan or the like		Each	5.26
VW Transport transit or the like	1.0 tonnes	Each	6.03
Luton Box Van or the like	1.8 tonnes	Each	9.87
Water/Fuel Storage			
Mobile water container	110 litres	Each	0.62
Water bowser	1100 litres	Each	0.72
Water bowser	3000 litres	Each	0.87
Mobile fuel container	110 litres	Each	0.62
Fuel bowser	1100 litres	Each	1.23
Fuel bowser	3000 litres	Each	1.87
EXCAVATIONS AND LOADERS			
Excavators			
Wheeled, hydraulic	up to 11 tonnes	Each	25.86
Wheeled, hydraulic	up to 14 tonnes	Each	30.82
Wheeled, hydraulic	up to 16 tonnes	Each	34.50
Wheeled, hydraulic	up to 21 tonnes	Each	39.10
Wheeled, hydraulic	up to 25 tonnes	Each	43.81

BUILDING INDUSTRY PLANT HIRE COSTS

MECHANICAL PLANT AND TOOLS

Item of plant	Size/Rating	Unit	Rate per hour (£)
EXCAVATIONS AND LOADERS – cont'd			
Excavators – cont'd			
Wheeled, hydraulic	up to 30 tonnes	Each	55.30
Crawler, hydraulic	up to 11 tonnes	Each	25.86
Crawler, hydraulic	up to 14 tonnes	Each	30.82
Crawler, hydraulic	up to 17 tonnes	Each	34.50
Crawler, hydraulic	up to 23 tonnes	Each	39.10
Crawler, hydraulic	up to 30 tonnes	Each	43.81
Crawler, hydraulic	up to 35 tonnes	Each	55.30
Crawler, hydraulic	up to 38 tonnes	Each	71.73
Crawler, hydraulic	up to 55 tonnes	Each	95.63
Mini excavator	1000/1500 kg	Each	4.87
Mini excavator	2150/2400 kg	Each	6.67
Mini excavator	2700/3500 kg	Each	7.31
Mini excavator	3500/4500 kg	Each	8.21
Mini excavator	4500/6000 kg	Each	9.23
Mini excavator	7000 kg	Each	14.10
Micro excavator	725mm wide	Each	5.13
Loaders			
Shovel loader	0.4 m³	Each	7.69
Shovel loader	1.57 m³	Each	8.97
Shovel loader, four wheel drive	1.7 m³	Each	4.83
Shovel loader, four wheel drive	2.3 m³	Each	4.38
Shovel loader, four wheel drive	3.3 m³	Each	5.06
Skid steer loader wheeled	300/400 kg payload	Each	7.31
Skid steer loader wheeled	625 kg payload	Each	7.67
Tracked skip loader	650 kg	Each	4.42
Excavator Loaders			
Wheeled tractor type with black-hoe excavator			
Four wheel drive			
Four wheel drive, 2 wheel steer	6 tonnes	Each	6.41
Four wheel drive, 2 wheel steer	8 tonnes	Each	8.59
Attachments			
Breakers for excavator		Each	8.72
Breakers for mini excavator		Each	1.75
Breakers for back-hoe excavator/loader		Each	5.13
COMPACTION EQUIPMENT			
Rollers			
Vibrating roller	368–420 kg	Each	1.43
Single roller	533 kg	Each	1.94
Single roller	750 kg	Each	3.43
Twin roller	up to 650 kg	Each	6.03
Twin roller	up to 950 kg	Each	6.62
Twin roller with seat end steering wheel	up to 1400 kg	Each	7.68
Twin roller with seat end steering wheel	up to 2500 kg	Each	10.61

BUILDING INDUSTRY PLANT HIRE COSTS

MECHANICAL PLANT AND TOOLS

Item of plant	Size/Rating	Unit	Rate per hour (£)
COMPACTION EQUIPMENT – cont'd			
Rollers – cont'd			
Pavement roller	3 – 4 tonnes dead weight	Each	6.00
Pavement roller	4 – 6 tonnes	Each	6.86
Pavement roller	6 – 10 tonnes	Each	7.17
Pavement roller	10 – 13 tonnes	Each	19.86
Rammers			
Tamper rammer 2 stroke–petrol	225 mm – 275 mm	Each	1.52
Soil Compactors			
Plate compactor	75 mm – 400 mm	Each	1.53
Plate compactor rubber pad	375 mm – 1400 mm	Each	1.53
Plate compactor reversible plate–petrol	400 mm	Each	2.44
CONCRETE EQUIPEMENT			
Concrete/Mortar Mixers			
Open drum without hopper	0.09/0.06 m³	Each	0.61
Open drum without hopper	0.12/0.09 m³	Each	1.22
Open drum without hopper	0.15/0.10 m³	Each	0.72
Concrete/Mortar Transport Equipment			
Concrete pump incl. hose, valve and couplers			
Lorry mounted concrete pump	24 m max. distance	Each	50.00
Lorry mounted concrete pump	34 m max. distance	Each	66.00
Lorry mounted concrete pump	42 m max. distance	Each	91.50
Concrete Equipment			
Vibrator, poker, petrol type	up to 75 mm dia.	Each	0.69
Air vibrator (*excluding compressor and hose*)	up to 75 mm dia.	Each	0.64
Extra poker heads	25/36/60 mm diameter	Each	0.76
Vibrating screed unit with beam	5.00 m	Each	2.48
Vibrating screed unit with adjustable beam	3.00 – 5.00 m	Each	3.54
Power float	725 mm – 900 mm	Each	2.56
Power float finishing pan		Each	0.62
Floor grinder	660 x 1016 mm, 110V electric	Each	4.31
Floor plane	450 x 1100 mm	Each	4.31
TESTING EQUIPMENT			
Pipe Testing Equipment			
Pressure testing pump, electric		Set	2.19
Pressure test pump		Set	0.80

BUILDING INDUSTRY PLANT HIRE COSTS

MECHANICAL PLANT AND TOOLS

Item of plant	Size/Rating	Unit	Rate per hour (£)
SITE ACCOMODATION AND TEMPORARY SERVICES			
Heating equipment			
Space heater – propane	80,000 Btu/hr	Each	1.03
Space heater – propane/electric	125,000 Btu/hr	Each	2.09
Space heater – propane/electric	250,000 Btu/hr	Each	2.33
Space heater – propane	125,000 Btu/hr	Each	1.54
Space heater – propane	260,000 Btu/hr	Each	1.88
Cabinet heater		Each	0.82
Cabinet heater, catalytic		Each	0.57
Electric halogen heater		Each	1.27
Ceramic heater	3 kW	Each	0.99
Fan heater	3 kW	Each	0.66
Cooling fan		Each	1.92
Mobile cooling unit, small		Each	3.60
Mobile cooling unit, large		Each	4.98
Air conditioning unit		Each	2.81
Site Lighting and Equipment			
Tripod floodlight	500 W	Each	0.48
Tripod floodlight	1000 W	Each	0.62
Towable floodlight	4 x 100 W	Each	3.85
Hand held floodlight	500 W	Each	0.51
Rechargeable light		Each	0.41
Inspection light		Each	0.37
Plasterer's light		Each	0.65
Lighting mast		Each	2.87
Festoon light string	25 m	Each	0.55
Site Electrical Equipment			
Extension leads	240 V/14 m	Each	0.26
Extension leads	110 V/14 m	Each	0.36
Cable reel	25 m 110 V/240V	Each	0.46
Cable reel	50 m 110 V240V	Each	0.88
4 way junction box	110 V	Each	0.56
Power Generating Units			
Generator – petrol	2 kVA	Each	1.23
Generator – silenced petrol	2 kVA	Each	2.87
Generator – petrol	3 kVA	Each	1.47
Generator – diesel	5 kVA	Each	2.44
Generator – silenced diesel	10 kVA	Each	1.90
Generator – silenced diesel	15 kVA	Each	2.26
Generator – silenced diesel	30 kVA	Each	3.33
Generator – silenced diesel	50 kVA	Each	4.10
Generator – silenced diesel	75 kVA	Each	4.62
Generator – silenced diesel	100 kVA	Each	5.64
Generator – silenced diesel	150 kVA	Each	7.18
Generator – silenced diesel	200 kVA	Each	9.74
Generator – silenced diesel	250 kVA	Each	11.28
Generator – silenced diesel	350 kVA	Each	14.36
Generator – silenced diesel	500 kVA	Each	15.38
Tail adaptor	240 V	Each	0.10

BUILDING INDUSTRY PLANT HIRE COSTS

MECHANICAL PLANT AND TOOLS

Item of plant	Size/Rating	Unit	Rate per hour (£)
SITE ACCOMODATION AND TEMPORARY SERVICES – cont'd			
Transformers			
Transformer	3 kVA	Each	0.32
Transformer	5 kVA	Each	1.23
Transformer	7.5 kVA	Each	0.59
Transformer	10 kVA	Each	2.00
Rubbish Collection and Disposal Equipment			
Rubbish Chutes			
Standard plastic module	1 m section	Each	0.15
Steel liner insert		Each	0.30
Steel top hopper		Each	0.22
Plastic side entry hopper		Each	0.22
Plastic side entry hopper liner		Each	0.22
Dust Extraction Plant			
Dust extraction unit, light duty		Each	2.97
Dust extraction unit, heavy duty		Each	2.97
SITE EQUIPMENT – Welding Equipment			
Arc-(Electric) Complete With Leads			
Welder generator – petrol	200 amp	Each	3.53
Welder generator – diesel	300/350 amp	Each	3.78
Welder generator – diesel	4000 amp	Each	7.92
Extra welding lead sets		Each	0.69
Gas-Oxy Welder			
Welding and cutting set (including oxygen and acetylene, excluding underwater equipment and thermic boring)			
Small		Each	2.24
Large		Each	3.75
Lead burning gun		Each	0.50
Mig welder		Each	1.38
Fume extractor		Each	2.46
Road Works Equipment			
Traffic lights, mains/generator	2-way	Set	10.94
Traffic lights, mains/generator	3-way	Set	11.56
Traffic lights, mains/generator	4-way	Set	12.19
Flashing light		Each	0.10
Road safety cone	450 mm	Each	0.08
Safety cone	750 mm	Each	0.10
Safety barrier plank	1.25 m	Each	0.13
Safety barrier plank	2 m	Each	0.15
Safety barrier plank post		Each	0.13
Safety barrier plank post base		Each	0.10
Safety four gate barrier	1 m each gate	Set	0.77
Guard barrier	2 m	Each	0.19
Road sign	750 mm	Each	0.23
Road sign	900 mm	Each	0.31
Road sign	1200 mm	Each	0.42
Speed ramp/cable protection	500 mm section	Each	0.14
Hose ramp open top	3 m section	Each	0.07
DPC Equipment			
Damp proofing injection machine		Each	2.56

BUILDING INDUSTRY PLANT HIRE COSTS

MECHANICAL PLANT AND TOOLS

Item of plant	Size/Rating	Unit	Rate per hour (£)
SITE ACCOMODATION AND TEMPORARY SERVICES – cont'd			
Cleaning Equipment			
Vacuum cleaner (industrial wet) single motor		Each	1.08
Vacuum cleaner (industrial wet) twin motor	30 litre capacity	Each	1.79
Vacuum cleaner (industrial wet) twin motor	70 litre capacity	Each	2.21
Steam cleaner	Diesel/electric 1 phase	Each	3.33
Steam cleaner	Diesel/electric 3 phase	Each	3.85
Pressure washer, light duty electric	1450 PSI	Each	0.72
Pressure washer, heavy duty, diesel	2500 PSI	Each	1.33
Pressure washer, heavy duty, diesel	4000 PSI	Each	2.18
Cold pressure washer, electric		Each	2.39
Hot pressure washer, petrol		Each	4.19
Hot pressure washer, electric		Each	5.13
Cold pressure washer, petrol		Each	2.92
Sandblast attachment to last washer		Each	1.23
Drain cleaning attachment to last washer		Each	1.03
Surface Preparation Equipment			
Rotavator	5 h.p.	Each	2.46
Rotavator	9 h.p.	Each	5.00
Scabbler, up to three heads		Each	1.53
Scabbler, pole		Each	2.68
Scrabbler, multi-headed floor		Each	3.89
Floor preparation machine		Each	1.05
Compressors and Equipment			
Portable Compressors			
Compressor – electric	4 cfm	Each	1.36
Compressor – electric	8 cfm lightweight	Each	1.31
Compressor – electric	8 cfm	Each	1.36
Compressor – electric	14 cfm	Each	1.56
Compressor – petrol	24 cfm	Each	2.15
Compressor – electric	25 cfm	Each	2.10
Compressor – electric	30 cfm	Each	2.36
Compressor – diesel	100 cfm	Each	2.56
Compressor – diesel	250 cfm	Each	5.54
Compressor – diesel	400 cfm	Each	8.72
Mobile Compressors			
Lorry mounted compressor *(machine plus lorry only)*	up to 3 m³	Each	41.47
(machine plus lorry only)	up to 5 m³	Each	48.94
Tractor mounted compressor *(machine plus rubber tyred tractor)*	Up to 4 m³	Each	21.03
Accessories (Pneumatic Tools) *(with and including up to 15 m of air hose)*			
Demolition pick, medium duty		Each	0.90
Demolition pick, heavy duty		Each	1.03
Breakers (with six steels) light	up to 150 kg	Each	1.19
Breakers (with six steels) medium	295 kg	Each	1.24
Breakers (with six steels) heavy	386 kg	Each	1.44
Rock drill (for use with compressor) hand held		Each	1.18
Additional hoses	15 m	Each	0.09

BUILDING INDUSTRY PLANT HIRE COSTS

MECHANICAL PLANT AND TOOLS			
Item of plant	Size/Rating	Unit	Rate per hour (£)
SITE ACCOMODATION AND TEMPORARY SERVICES – cont'd			
Breakers			
Demolition hammer drill, heavy duty, electric		Each	1.54
Road breaker, electric		Each	2.41
Road breaker, 2 stroke, petrol		Each	4.06
Hydraulic breaker unit, light duty, petrol		Each	3.06
Hydraulic breaker unit, heavy duty, petrol		Each	3.46
Hydraulic breaker unit, heavy duty, diesel		Each	4.62
Quarrying and Tooling Equipment			
Block and stone splitter, hydraulic	600 mm x 600 mm	Each	1.90
Block and stone splitter, manual		Each	1.64
Steel Reinforcement Equipment			
Bar bending machine – manual	up to 13 mm dia. rods	Each	1.03
Bar bending machine – manual	up to 20 mm dia. rods	Each	1.41
Bar shearing machine – electric	up to 38 mm dia. rods	Each	3.08
Bar shearing machine – electric	up to 40 mm dia. rods	Each	4.62
Bar cropper machine – electric	up to 13 mm dia. rods	Each	2.05
Bar cropper machine – electric	up to 20 mm dia. rods	Each	2.56
Bar cropper machine – electric	up to 40 mm dia. rods	Each	4.62
Bar cropper machine – 3 phase	up to 40 mm dia. rods	Each	4.62
Dehumidifiers			
110/240v Water	68 litres extraction per 24 hours	Each	2.46
110/240v Water	90 litres extraction per 24 hours	Each	3.38
SMALL TOOLS			
Saws			
Masonry bench saw	350 mm – 500 mm dia.	Each	1.13
Floor saw	125 mm max. cut	Each	1.15
Floor saw	150 mm max. cut	Each	3.83
Floor saw, reversible	350 mm max. cut	Each	3.32
Wall saw, electric		Each	2.05
Chop/cut off saw, electric	350 mm dia.	Each	1.79
Circular saw, electric	230 mm dia.	Each	0.72
Tyrannosaw		Each	1.74
Reciprocating saw		Each	0.79
Door trimmer		Each	1.17
Stone saw	300 mm	Each	1.44
Chainsaw, petrol	500 mm	Each	3.92
Full chainsaw safety kit		Each	0.41
Worktop jig		Each	1.08
Pipework Equipment			
Pipe bender	15 mm – 22 mm	Each	0.92
Pipe bender, hydraulic	50 mm	Each	1.76
Pipe bender, electric	50 mm – 150 mm dia.	Each	2.19
Pipe cutter, hydraulic		Each	0.46
Tripod pipe vice		Set	0.75
Ratchet threader	12 mm – 32 mm	Each	0.93
Pipe threading machine, electric	12 mm – 75 mm	Each	3.07
Pipe threading machine, electric	12 mm – 100 mm	Each	4.93
Impact wrench, electric		Each	1.33

BUILDING INDUSTRY PLANT HIRE COSTS

MECHANICAL PLANT AND TOOLS

Item of plant	Size/Rating	Unit	Rate per hour (£)
SMALL TOOLS – cont'd			
Hand-held Drills and Equipment			
Impact or hammer drill	up to 25 mm dia.	Each	1.03
Impact or hammer drill	35 mm diameter	Each	1.29
Dry diamond core cutter		Each	0.99
Angle head drill		Each	0.90
Stirrer, mixer drill		Each	1.13
Paint, Insulation Application Equipment			
Airless spray unit		Each	4.13
Portaspray unit		Each	1.16
HPVL turbine spray unit		Each	2.23
Compressor and spray gun		Each	1.91
Other Handtools			
Staple gun		Each	0.96
Air nail gun	110 V	Each	1.01
Cartridge hammer		Each	1.08
Tongue and groove nailer complete with mallet		Each	1.59
Diamond wall chasing machine		Each	2.63
Masonry chain saw	300 mm	Each	5.49
Floor grinder		Each	3.99
Floor plane		Each	1.79
Diamond concrete planer		Each	1.93
Autofeed screwdriver, electric		Each	1.38
Laminate trimmer		Each	0.91
Biscuit jointer		Each	1.49
Random orbital sander		Each	0.97
Floor sander		Each	1.54
Palm, delta, flap or belt sander		Each	0.75
Disc cutter, electric	300 mm	Each	1.49
Disc cutter, 2 stroke petrol	300 mm	Each	1.24
Dust suppressor for petrol disc cutter		Each	0.51
Cutter cart for petrol disc cutter		Each	1.21
Grinder, angle or cutter	up to 225 mm	Each	0.50
Grinder, angle or cutter	300 mm	Each	1.41
Mortar raking tool attachment		Each	0.19
Floor/polisher scrubber	325 mm	Each	1.76
Floor tile stripper		Each	2.44
Wallpaper stripper, electric		Each	0.81
Hot air paint stripper		Each	0.50
Electric diamond tile cutter	all sizes	Each	2.42
Hand tile cutter		Each	0.82
Electric needle gun		Each	1.29
Needle chipping gun		Each	1.85
Pedestrian floor sweeper	250 mm dia.	Each	0.82
Pedestrian floor sweeper	Petrol	Each	2.20
Diamond tile saw		Each	1.84
Blow lamp equipment and glass		Set	0.50

PART 8

Tables and Memoranda

This part of the book contains the following sections:

CONVERSION TABLES

Length	Unit	Conversion factors			
Millimetre	mm	1 in	= 25.4 mm	1 mm	= 0.0394 in
Centimetre	cm	1 in	= 2.54 cm	1 cm	= 0.3937 in
Metre	m	1 ft	= 0.3048 m	1 m	= 3.2808 ft
		1 yd	= 0.9144 m		= 1.0936 yd
Kilometre	km	1 mile	= 1.6093 km	1km	= 0.6214 mile

Note:

1 cm	= 10 mm	1 ft	= 12 in
1 m	= 1 000 mm	1 yd	= 3 ft
1 km	= 1 000 m	1 mile	= 1 760 yd

Area	Unit	Conversion factors			
Square Millimetre	mm^2	$1\ in^2$	$= 645.2\ mm^2$	$1\ mm^2$	$= 0.0016\ in^2$
Square Centimetre	cm^2	$1\ in^2$	$= 6.4516\ cm^2$	$1\ cm^2$	$= 1.1550\ in^2$
Square Metre	m^2	$1\ ft^2$	$= 0.0929\ m^2$	$1\ m^2$	$= 10.764\ ft^2$
		$1\ yd^2$	$= 0.8361\ m^2$	$1\ m^2$	$= 1.1960\ yd^2$
Square Kilometre	km^2	$1\ mile^2$	$= 2.590\ km^2$	$1\ km^2$	$= 0.3861\ mile^2$

Note:

$1\ cm^2$	$= 100\ mm^2$	$1\ ft^2$	$= 144\ in^2$
$1\ m^2$	$= 10\ 000\ cm^2$	$1\ yd^2$	$= 9\ ft^2$
$1\ km^2$	$= 100$ hectares	1 acre	$= 4\ 840\ yd^2$
		$1\ mile^2$	$= 640$ acres

Volume	Unit	Conversion factors			
Cubic Centimetre	cm^3	$1\ cm^3$	$= 0.0610\ in^3$	$1\ in^3$	$= 16.387\ cm^3$
Cubic Decimetre	dm^3	$1\ dm^3$	$= 0.0353\ ft^3$	$1\ ft^3$	$= 28.329\ dm^3$
Cubic Metre	m^3	$1\ m^3$	$= 35.3147\ ft^3$	$1\ ft^3$	$= 0.0283\ m^3$
		$1\ m^3$	$= 1.3080\ yd^3$	$1\ yd^3$	$= 0.7646\ m^3$
Litre	l	1 l	= 1.76 pint	1 pint	= 0.5683 l
			= 2.113 US pt		= 0.4733 US l

Note:

$1\ dm^3$	$= 1\ 000\ cm^3$	$1\ ft^3$	$= 1\ 728\ in^3$	1 pint	= 20 fl oz
$1\ m^3$	$= 1\ 000\ dm^3$	$1\ yd^3$	$= 27\ ft^3$	1 gal	= 8 pints
1 l	$= 1\ dm^3$				

Neither the Centimetre nor Decimetre are SI units, and as such their use, particularly that of the Decimetre, is not widespread outside educational circles.

Mass	Unit	Conversion factors			
Milligram	mg	1 mg	= 0.0154 grain	1 grain	= 64.935 mg
Gram	g	1 g	= 0.0353 oz	1 oz	= 28.35 g
Kilogram	kg	1 kg	= 2.2046 lb	1 lb	= 0.4536 kg
Tonne	t	1 t	= 0.9842 ton	1 ton	= 1.016 t

Note:

1 g	= 1000 mg	1 oz	= 437.5 grains	1 cwt	= 112 lb
1 kg	= 1000 g	1 lb	= 16 oz	1 ton	= 20 cwt
1 t	= 1000 kg	1 stone	= 14 lb		

Force	Unit	Conversion factors			
Newton	N	1 lbf	= 4.448 N	1 kgf	= 9.807 N
Kilonewton	kN	1 lbf	= 0.004448 kN	1 ton f	= 9.964 kN
Meganewton	MN	100 tonf	= 0.9964 MN		

CONVERSION TABLES

Pressure and stress	Unit	Conversion factors
Kilonewton per square metre	kN/m^2	1 lbf/in^2 = 6.895 kN/m^2
		1 bar = 100 kN/m^2
Meganewton per square metre	MN/m^2	1 tonf/ft^2 = 107.3 kN/m^2 = 0.1073 MN/m^2
		1 kgf/cm^2 = 98.07 kN/m^2
		1 lbf/ft^2 = 0.04788 kN/m^2

Coefficient of consolidation (Cv) or swelling	Unit	Conversion factors
Square metre per year	$m^2/year$	1 cm^2/s = 3 154 m^2/year
		1 ft^2/year = 0.0929 m^2/year

Coefficient of permeability	Unit	Conversion factors
Metre per second	m/s	1 cm/s = 0.01 m/s
Metre per year	m/year	1 ft/year = 0.3048 m/year
		= 0.9651 × (10)^{8}m/s

Temperature	Unit	Conversion factors	
Degree Celsius	°C	°C = 5/9 × (°F − 32)	°F = (9 × °C)/ 5 + 32

Power	Unit	Conversion factors
Kilowatt	kW	1 kW = 1.341 HP
Horsepower	HP	1 HP = 0.746 kW

CONVERSION TABLES

SPEED CONVERSION

km/h	m/min	mph	fpm
1	16.7	0.6	54.7
2	33.3	1.2	109.4
3	50.0	1.9	164.0
4	66.7	2.5	218.7
5	83.3	3.1	273.4
6	100.0	3.7	328.1
7	116.7	4.3	382.8
8	133.3	5.0	437.4
9	150.0	5.6	492.1
10	166.7	6.2	546.8
11	183.3	6.8	601.5
12	200.0	7.5	656.2
13	216.7	8.1	710.8
14	233.3	8.7	765.5
15	250.0	9.3	820.2
16	266.7	9.9	874.9
17	283.3	10.6	929.6
18	300.0	11.2	984.3
19	316.7	11.8	1038.9
20	333.3	12.4	1093.6
21	350.0	13.0	1148.3
22	366.7	13.7	1203.0
23	383.3	14.3	1257.7
24	400.0	14.9	1312.3
25	416.7	15.5	1367.0
26	433.3	16.2	1421.7
27	450.0	16.8	1476.4
28	466.7	17.4	1531.1
29	483.3	18.0	1585.7
30	500.0	18.6	1640.4
31	516.7	19.3	1695.1
32	533.3	19.9	1749.8
33	550.0	20.5	1804.5
34	566.7	21.1	1859.1
35	583.3	21.7	1913.8
36	600.0	22.4	1968.5
37	616.7	23.0	2023.2
38	633.3	23.6	2077.9
39	650.0	24.2	2132.5
40	666.7	24.9	2187.2
41	683.3	25.5	2241.9
42	700.0	26.1	2296.6
43	716.7	26.7	2351.3
44	733.3	27.3	2405.9
45	750.0	28.0	2460.6

Tables and Memoranda

CONVERSION TABLES

km/h	m/min	mph	fpm
46	766.7	28.6	2515.3
47	783.3	29.2	2570.0
48	800.0	29.8	2624.7
49	816.7	30.4	2679.4
50	833.3	31.1	2734.0

GEOMETRY

Two dimensional figures

Figure	Diagram of figure	Surface area	Perimeter
Square		a^2	$4a$
Rectangle		ab	$2(a+b)$
Triangle		$\frac{1}{2}ch$	$a+b+c$
Circle		πr^2 $\frac{1}{4}\pi d^2$ where $2r = d$	$2\pi r$ πd
Parallelogram		ah	$2(a+b)$
Trapezium		$\frac{1}{2}h(a+b)$	$a+b+c+d$
Ellipse		Approximately πab	$\pi(a+b)$
Hexagon		$2.6 \times a^2$	

GEOMETRY

Figure	Diagram of figure	Surface area	Perimeter
Octagon		$4.83 \times a^2$	6a
Sector of a circle		$\frac{1}{2}rb$ or $\frac{q}{360}\pi r^2$ note $b =$ angle $\frac{q}{360} \times \pi 2r$	
Segment of a circle		$S - T$ where $S =$ area of sector, $T =$ area of triangle	
Bellmouth		$\frac{3}{14} \times r^2$	

GEOMETRY

Three dimensional figures

Figure	Diagram of figure	Surface area	Volume
Cube		$6a^2$	a^3
Cuboid/ rectangular block		$2(ab + ac + bc)$	abc
Prism/ triangular block		$bd + hc + dc + ad$	$\frac{1}{2}hcd$
Cylinder		$2\pi r^2 + 2\pi h$	$\pi r^2 h$ $\frac{1}{4}\pi d^2 h$
Sphere		$4\pi r^2$	$\frac{4}{3}\pi r^3$
Segment of sphere		$2\pi Rh$	$\frac{1}{6}\pi h(3r^2 + h^2)$ $\frac{1}{3}\pi h^2(3R - H)$
Pyramid		$(a + b)l + ab$	$\frac{1}{3}abh$

GEOMETRY

Figure	Diagram of figure	Surface area	Volume
Frustum of a pyramid		$l(a+b+c+d) + \sqrt{(ab+cd)}$ [rectangular figure only]	$\frac{h}{3}(ab+cd+\sqrt{abcd})$
Cone		πrl (excluding base) $\pi rl + \pi r^2$ (including base)	$\frac{1}{3}\pi r^2 h$ $\frac{1}{12}\pi d^2 h$
Frustum of a cone		$\pi r^2 + \pi R^2 + \pi l(R+r)$	$\frac{1}{3}\pi(R^2+Rr+r^2)$

FORMULAE

FORMULAE

Formula	Description
Pythagoras theorem	$A^2 = B^2 + C^2$ where A is the hypotenuse of a right-angled triangle and B and C are the two adjacent sides
Simpsons Rule	The Area is divided into an even number of strips of equal width, and therefore has an odd number of ordinates at the division points $$\text{area} = \frac{S(A + 2B + 4C)}{3}$$ where S = common interval (strip width) A = sum of first and last ordinates B = sum of remaining odd ordinates C = sum of the even ordinates The Volume can be calculated by the same formula, but by substituting the area of each coordinate rather than its length
Trapezoidal Rule	A given trench is divided into two equal sections, giving three ordinates, the first, the middle and the last $$\text{volume} = \frac{S \times (A + B + 2C)}{2}$$ where S = width of the strips A = area of the first section B = area of the last section C = area of the rest of the sections
Prismoidal Rule	A given trench is divided into two equal sections, giving three ordinates, the first, the middle and the last $$\text{volume} = \frac{L \times (A + 4B + C)}{6}$$ where L = total length of trench A = area of the first section B = area of the middle section C = area of the last section

EARTHWORK

Weights of Typical Materials Handled by Excavators

The weight of the material is that of the state in its natural bed and includes moisture
Adjustments should be made to allow for loose or compacted states

Material	kg/m³	lb/cu yd
Adobe	1914	3230
Ashes	610	1030
Asphalt, rock	2400	4050
Basalt	2933	4950
Bauxite: alum ore	2619	4420
Borax	1730	2920
Caliche	1440	2430
Carnotite	2459	4150
Cement	1600	2700
Chalk (hard)	2406	4060
Cinders	759	1280
Clay: dry	1908	3220
Clay: wet	1985	3350
Coal: bituminous	1351	2280
Coke	510	860
Conglomerate	2204	3720
Dolomite	2886	4870
Earth: dry	1796	3030
Earth: moist	1997	3370
Earth: wet	1742	2940
Feldspar	2613	4410
Felsite	2495	4210
Fluorite	3093	5220
Gabbro	3093	5220
Gneiss	2696	4550
Granite	2690	4540
Gravel, dry	1790	3020
Gypsum	2418	4080
Hardcore (consolidated)	1928	120
Lignite broken	1244	2100
Limestone	2596	4380
Magnesite, magnesium ore	2993	5050
Marble	2679	4520
Marl	2216	3740
Peat	700	1180
Potash	2193	3700
Pumice	640	1080
Quarry waste	1438	90
Quartz	2584	4360
Rhyolite	2400	4050

EARTHWORK

Material	kg/m³	lb/cu yd
Sand: dry	1707	2880
Sand: wet	1831	3090
Sand and gravel – dry	1790	3020
– wet	2092	3530
Sandstone	2412	4070
Schist	2684	4530
Shale	2637	4450
Slag (blast)	2868	4840
Slate	2667	4500
Snow – dry	130	220
– wet	510	860
Taconite	3182	5370
Topsoil	1440	2430
Trachyte	2400	4050
Traprock	2791	4710
Water	1000	62

Transport Capacities

Type of vehicle	Capacity of vehicle	
	Payload	Heaped capacity
Wheelbarrow	150	0.10
1 tonne dumper	1250	1.00
2.5 tonne dumper	4000	2.50
Articulated dump truck (Volvo A20 6 × 4)	18500	11.00
Articulated dump truck (Volvo A35 6 × 6)	32000	19.00
Large capacity rear dumper (Euclid R35)	35000	22.00
Large capacity rear dumper (Euclid R85)	85000	50.00

EARTHWORK

Machine Volumes for Excavating and Filling

Machine type	Cycles per minute	Volume per minute (m³)
1.5 tonne excavator	1	0.04
	2	0.08
	3	0.12
3 tonne excavator	1	0.13
	2	0.26
	3	0.39
5 tonne excavator	1	0.28
	2	0.56
	3	0.84
7 tonne excavator	1	0.28
	2	0.56
	3	0.84
21 tonne excavator	1	1.21
	2	2.42
	3	3.63
Backhoe loader JCB3CX excavator Rear bucket capacity 0.28 m³	1	0.28
	2	0.56
	3	0.84
Backhoe loader JCB3CX loading Front bucket capacity 1.00 m³	1	1.00
	2	2.00

Machine Volumes for Excavating and Filling

Machine type	Loads per hour	Volume per hour (m³)
1 tonne high tip skip loader Volume 0.485 m³	5	2.43
	7	3.40
	10	4.85
3 tonne dumper Max volume 2.40 m³ Available volume 1.9 m³	4	7.60
	5	9.50
	7	13.30
	10	19.00
6 tonne dumper Max volume 3.40 m³ Available volume 3.77 m³	4	15.08
	5	18.85
	7	26.39
	10	37.70

EARTHWORK

Bulkage of Soils (after excavation)

Type of soil	Approximate bulking of 1 m³ after excavation
Vegetable soil and loam	25–30%
Soft clay	30–40%
Stiff clay	10–20%
Gravel	20–25%
Sand	40–50%
Chalk	40–50%
Rock, weathered	30–40%
Rock, unweathered	50–60%

Shrinkage of Materials (on being deposited)

Type of soil	Approximate bulking of 1 m³ after excavation
Clay	10%
Gravel	8%
Gravel and sand	9%
Loam and light sandy soils	12%
Loose vegetable soils	15%

Voids in Material Used as Subbases or Beddings

Material	m³ of voids/m³
Alluvium	0.37
River grit	0.29
Quarry sand	0.24
Shingle	0.37
Gravel	0.39
Broken stone	0.45
Broken bricks	0.42

Angles of Repose

Type of soil		Degrees
Clay	– dry	30
	– damp, well drained	45
	– wet	15–20
Earth	– dry	30
	– damp	45
Gravel	– moist	48
Sand	– dry or moist	35
	– wet	25
Loam		40

EARTHWORK

Slopes and Angles

Ratio of base to height	Angle in degrees
5:1	11
4:1	14
3:1	18
2:1	27
1½:1	34
1:1	45
1:1½	56
1:2	63
1:3	72
1:4	76
1:5	79

Grades (in degrees and percents)

Degrees	Percent	Degrees	Percent
1	1.8	24	44.5
2	3.5	25	46.6
3	5.2	26	48.8
4	7.0	27	51.0
5	8.8	28	53.2
6	10.5	29	55.4
7	12.3	30	57.7
8	14.0	31	60.0
9	15.8	32	62.5
10	17.6	33	64.9
11	19.4	34	67.4
12	21.3	35	70.0
13	23.1	36	72.7
14	24.9	37	75.4
15	26.8	38	78.1
16	28.7	39	81.0
17	30.6	40	83.9
18	32.5	41	86.9
19	34.4	42	90.0
20	36.4	43	93.3
21	38.4	44	96.6
22	40.4	45	100.0

EARTHWORK

Bearing Powers

Ground conditions		Bearing power		
		kg/m²	lb/in²	Metric t/m²
Rock,	broken	483	70	50
	solid	2415	350	240
Clay,	dry or hard	380	55	40
	medium dry	190	27	20
	soft or wet	100	14	10
Gravel,	cemented	760	110	80
Sand,	compacted	380	55	40
	clean dry	190	27	20
Swamp and alluvial soils		48	7	5

Earthwork Support

Maximum depth of excavation in various soils without the use of earthwork support

Ground conditions	Feet (ft)	Metres (m)
Compact soil	12	3.66
Drained loam	6	1.83
Dry sand	1	0.3
Gravelly earth	2	0.61
Ordinary earth	3	0.91
Stiff clay	10	3.05

It is important to note that the above table should only be used as a guide. Each case must be taken on its merits and, as the limited distances given above are approached, careful watch must be kept for the slightest signs of caving in

CONCRETE WORK

Weights of Concrete and Concrete Elements

Type of material		kg/m³	lb/cu ft
Ordinary concrete (dense aggregates)			
Non-reinforced plain or mass concrete			
Nominal weight		2305	144
Aggregate	– limestone	2162 to 2407	135 to 150
	– gravel	2244 to 2407	140 to 150
	– broken brick	2000 (av)	125 (av)
	– other crushed stone	2326 to 2489	145 to 155
Reinforced concrete			
Nominal weight		2407	150
Reinforcement	– 1%	2305 to 2468	144 to 154
	– 2%	2356 to 2519	147 to 157
	– 4%	2448 to 2703	153 to 163
Special concretes			
Heavy concrete			
Aggregates	– barytes, magnetite	3210 (min)	200 (min)
	– steel shot, punchings	5280	330
Lean mixes			
Dry-lean (gravel aggregate)		2244	140
Soil-cement (normal mix)		1601	100

CONCRETE WORK

Type of material		kg/m² per mm thick	lb/sq ft per inch thick
Ordinary concrete (dense aggregates)			
Solid slabs (floors, walls etc.)			
Thickness:	75 mm or 3 in	184	37.5
	100 mm or 4 in	245	50
	150 mm or 6 in	378	75
	250 mm or 10 in	612	125
	300 mm or 12 in	734	150
Ribbed slabs			
Thickness:	125 mm or 5 in	204	42
	150 mm or 6 in	219	45
	225 mm or 9 in	281	57
	300 mm or 12 in	342	70
Special concretes			
Finishes etc.			
	Rendering, screed etc. Granolithic, terrazzo	1928 to 2401	10 to 12.5
	Glass-block (hollow) concrete	1734 (approx)	9 (approx)
Prestressed concrete		Weights as for reinforced concrete (upper limits)	
Air-entrained concrete		Weights as for plain or reinforced concrete	

CONCRETE WORK

Average Weight of Aggregates

Materials	Voids %	Weight kg/m³
Sand	39	1660
Gravel 10–20 mm	45	1440
Gravel 35–75 mm	42	1555
Crushed stone	50	1330
Crushed granite (over 15 mm)	50	1345
(n.e. 15 mm)	47	1440
'All-in' ballast	32	1800–2000

Material	kg/m³	lb/cu yd
Vermiculite (aggregate)	64–80	108–135
All-in aggregate	1999	125

Applications and Mix Design

Site mixed concrete

Recommended mix	Class of work suitable for	Cement (kg)	Sand (kg)	Coarse aggregate (kg)	Nr 25 kg bags cement per m³ of combined aggregate
1:3:6	Roughest type of mass concrete such as footings, road haunching over 300 mm thick	208	905	1509	8.30
1:2.5:5	Mass concrete of better class than 1:3:6 such as bases for machinery, walls below ground etc.	249	881	1474	10.00
1:2:4	Most ordinary uses of concrete, such as mass walls above ground, road slabs etc. and general reinforced concrete work	304	889	1431	12.20
1:1.5:3	Watertight floors, pavements and walls, tanks, pits, steps, paths, surface of 2 course roads, reinforced concrete where extra strength is required	371	801	1336	14.90
1:1:2	Works of thin section such as fence posts and small precast work	511	720	1206	20.40

CONCRETE WORK

Ready mixed concrete

Application	Designated concrete	Standardized prescribed concrete	Recommended consistence (nominal slump class)
Foundations			
Mass concrete fill or blinding	GEN 1	ST2	S3
Strip footings	GEN 1	ST2	S3
Mass concrete foundations			
Single storey buildings	GEN 1	ST2	S3
Double storey buildings	GEN 3	ST4	S3
Trench fill foundations			
Single storey buildings	GEN 1	ST2	S4
Double storey buildings	GEN 3	ST4	S4
General applications			
Kerb bedding and haunching	GEN 0	ST1	S1
Drainage works – immediate support	GEN 1	ST2	S1
Other drainage works	GEN 1	ST2	S3
Oversite below suspended slabs	GEN 1	ST2	S3
Floors			
Garage and house floors with no embedded steel	GEN 3	ST4	S2
Wearing surface: Light foot and trolley traffic	RC30	ST4	S2
Wearing surface: General industrial	RC40	N/A	S2
Wearing surface: Heavy industrial	RC50	N/A	S2
Paving			
House drives, domestic parking and external parking	PAV 1	N/A	S2
Heavy-duty external paving	PAV 2	N/A	S2

CONCRETE WORK

Prescribed Mixes for Ordinary Structural Concrete

Weights of cement and total dry aggregates in kg to produce approximately one cubic metre of fully compacted concrete together with the percentages by weight of fine aggregate in total dry aggregates

Conc. grade	Nominal max size of aggregate (mm)	40		20		14		10	
	Workability	**Med.**	**High**	**Med.**	**High**	**Med.**	**High**	**Med.**	**High**
	Limits to slump that may be expected (mm)	**50–100**	**100–150**	**25–75**	**75–125**	**10–50**	**50–100**	**10–25**	**25–50**
7	Cement (kg)	180	200	210	230	–	–	–	–
	Total aggregate (kg)	1950	1850	1900	1800	–	–	–	–
	Fine aggregate (%)	30–45	30–45	35–50	35–50	–	–	–	–
10	Cement (kg)	210	230	240	260	–	–	–	–
	Total aggregate (kg)	1900	1850	1850	1800	–	–	–	–
	Fine aggregate (%)	30–45	30–45	35–50	35–50	–	–	–	–
15	Cement (kg)	250	270	280	310	–	–	–	–
	Total aggregate (kg)	1850	1800	1800	1750	–	–	–	–
	Fine aggregate (%)	30–45	30–45	35–50	35–50	–	–	–	–
20	Cement (kg)	300	320	320	350	340	380	360	410
	Total aggregate (kg)	1850	1750	1800	1750	1750	1700	1750	1650
	Sand								
	Zone 1 (%)	35	40	40	45	45	50	50	55
	Zone 2 (%)	30	35	35	40	40	45	45	50
	Zone 3 (%)	30	30	30	35	35	40	40	45
25	Cement (kg)	340	360	360	390	380	420	400	450
	Total aggregate (kg)	1800	1750	1750	1700	1700	1650	1700	1600
	Sand								
	Zone 1 (%)	35	40	40	45	45	50	50	55
	Zone 2 (%)	30	35	35	40	40	45	45	50
	Zone 3 (%)	30	30	30	35	35	40	40	45
30	Cement (kg)	370	390	400	430	430	470	460	510
	Total aggregate (kg)	1750	1700	1700	1650	1700	1600	1650	1550
	Sand								
	Zone 1 (%)	35	40	40	45	45	50	50	55
	Zone 2 (%)	30	35	35	40	40	45	45	50
	Zone 3 (%)	30	30	30	35	35	40	40	45

REINFORCEMENT

Weights of Bar Reinforcement

Nominal sizes (mm)	Cross-sectional area (mm²)	Mass (kg/m)	Length of bar (m/tonne)
6	28.27	0.222	4505
8	50.27	0.395	2534
10	78.54	0.617	1622
12	113.10	0.888	1126
16	201.06	1.578	634
20	314.16	2.466	405
25	490.87	3.853	260
32	804.25	6.313	158
40	1265.64	9.865	101
50	1963.50	15.413	65

Weights of Bars (at specific spacings)

Weights of metric bars in kilogrammes per square metre

Size (mm)	Spacing of bars in millimetres									
	75	100	125	150	175	200	225	250	275	300
6	2.96	2.220	1.776	1.480	1.27	1.110	0.99	0.89	0.81	0.74
8	5.26	3.95	3.16	2.63	2.26	1.97	1.75	1.58	1.44	1.32
10	8.22	6.17	4.93	4.11	3.52	3.08	2.74	2.47	2.24	2.06
12	11.84	8.88	7.10	5.92	5.07	4.44	3.95	3.55	3.23	2.96
16	21.04	15.78	12.63	10.52	9.02	7.89	7.02	6.31	5.74	5.26
20	32.88	24.66	19.73	16.44	14.09	12.33	10.96	9.87	8.97	8.22
25	51.38	38.53	30.83	25.69	22.02	19.27	17.13	15.41	14.01	12.84
32	84.18	63.13	50.51	42.09	36.08	31.57	28.06	25.25	22.96	21.04
40	131.53	98.65	78.92	65.76	56.37	49.32	43.84	39.46	35.87	32.88
50	205.51	154.13	123.31	102.76	88.08	77.07	68.50	61.65	56.05	51.38

Basic weight of steelwork taken as 7850 kg/m³
Basic weight of bar reinforcement per metre run = 0.00785 kg/mm²
The value of π has been taken as 3.141592654

REINFORCEMENT

Fabric Reinforcement

Preferred range of designated fabric types and stock sheet sizes

Fabric reference	Longitudinal wires			Cross wires			
	Nominal wire size (mm)	Pitch (mm)	Area (mm²/m)	Nominal wire size (mm)	Pitch (mm)	Area (mm²/m)	Mass (kg/m²)
Square mesh							
A393	10	200	393	10	200	393	6.16
A252	8	200	252	8	200	252	3.95
A193	7	200	193	7	200	193	3.02
A142	6	200	142	6	200	142	2.22
A98	5	200	98	5	200	98	1.54
Structural mesh							
B1131	12	100	1131	8	200	252	10.90
B785	10	100	785	8	200	252	8.14
B503	8	100	503	8	200	252	5.93
B385	7	100	385	7	200	193	4.53
B283	6	100	283	7	200	193	3.73
B196	5	100	196	7	200	193	3.05
Long mesh							
C785	10	100	785	6	400	70.8	6.72
C636	9	100	636	6	400	70.8	5.55
C503	8	100	503	5	400	49.0	4.34
C385	7	100	385	5	400	49.0	3.41
C283	6	100	283	5	400	49.0	2.61
Wrapping mesh							
D98	5	200	98	5	200	98	1.54
D49	2.5	100	49	2.5	100	49	0.77

Stock sheet size 4.8 m × 2.4 m, Area 11.52 m²

Average weight kg/m³ of steelwork reinforcement in concrete for various building elements

Substructure	kg/m³ concrete	Substructure	kg/m³ concrete
Pile caps	110–150	Plate slab	150–220
Tie beams	130–170	Cant slab	145–210
Ground beams	230–330	Ribbed floors	130–200
Bases	125–180	Topping to block floor	30–40
Footings	100–150	Columns	210–310
Retaining walls	150–210	Beams	250–350
Raft	60–70	Stairs	130–170
Slabs – one way	120–200	Walls – normal	40–100
Slabs – two way	110–220	Walls – wind	70–125

Note: For exposed elements add the following %:
Walls 50%, Beams 100%, Columns 15%

FORMWORK

Formwork Stripping Times – Normal Curing Periods

Conditions under which concrete is maturing	Minimum periods of protection for different types of cement					
	Number of days (where the average surface temperature of the concrete exceeds 10°C during the whole period)			Equivalent maturity (degree hours) calculated as the age of the concrete in hours multiplied by the number of degrees Celsius by which the average surface temperature of the concrete exceeds 10°C		
	Other	SRPC	OPC or RHPC	Other	SRPC	OPC or RHPC
1. Hot weather or drying winds	7	4	3	3500	2000	1500
2. Conditions not covered by 1	4	3	2	2000	1500	1000

KEY
OPC – Ordinary Portland Cement
RHPC – Rapid-hardening Portland Cement
SRPC – Sulphate-resisting Portland Cement

Minimum Period before Striking Formwork

	Minimum period before striking		
	Surface temperature of concrete		
	16°C	17°C	t°C (0–25)
Vertical formwork to columns, walls and large beams	12 hours	18 hours	300 hours t+10
Soffit formwork to slabs	4 days	6 days	100 days t+10
Props to slabs	10 days	15 days	250 days t+10
Soffit formwork to beams	9 days	14 days	230 days t+10
Props to beams	14 days	21 days	360 days t+10

MASONRY

Number of Bricks required for Various Types of Work per m² of Walling

Description	Brick size	
	215 × 102.5 × 50 mm	215 × 102.5 × 65 mm
Half brick thick		
Stretcher bond	74	59
English bond	108	86
English garden wall bond	90	72
Flemish bond	96	79
Flemish garden wall bond	83	66
One brick thick and cavity wall of two half brick skins		
Stretcher bond	148	119

Quantities of Bricks and Mortar required per m² of Walling

	Unit	No of bricks required	Mortar required (cubic metres)		
Standard bricks			**No frogs**	**Single frogs**	**Double frogs**
Brick size 215 × 102.5 × 50 mm					
half brick wall (103 mm)	m²	72	0.022	0.027	0.032
2 × half brick cavity wall (270 mm)	m²	144	0.044	0.054	0.064
one brick wall (215 mm)	m²	144	0.052	0.064	0.076
one and a half brick wall (322 mm)	m²	216	0.073	0.091	0.108
Mass brickwork	m³	576	0.347	0.413	0.480
Brick size 215 × 102.5 × 65 mm					
half brick wall (103 mm)	m²	58	0.019	0.022	0.026
2 × half brick cavity wall (270 mm)	m²	116	0.038	0.045	0.055
one brick wall (215 mm)	m²	116	0.046	0.055	0.064
one and a half brick wall (322 mm)	m²	174	0.063	0.074	0.088
Mass brickwork	m³	464	0.307	0.360	0.413
Metric modular bricks			**Perforated**		
Brick size 200 × 100 × 75 mm					
90 mm thick	m²	67	0.016	0.019	
190 mm thick	m²	133	0.042	0.048	
290 mm thick	m²	200	0.068	0.078	
Brick size 200 × 100 × 100 mm					
90 mm thick	m²	50	0.013	0.016	
190 mm thick	m²	100	0.036	0.041	
290 mm thick	m²	150	0.059	0.067	
Brick size 300 × 100 × 75 mm					
90 mm thick	m²	33	–	0.015	
Brick size 300 × 100 × 100 mm					
90 mm thick	m²	44	0.015	0.018	

Note: Assuming 10 mm thick joints

MASONRY

Mortar required per m² Blockwork (9.88 blocks/m²)

Wall thickness	75	90	100	125	140	190	215
Mortar m³/m²	0.005	0.006	0.007	0.008	0.009	0.013	0.014

Mortar Group	Cement: lime: sand	Masonry cement: sand	Cement: sand with plasticizer
1	1:0–0.25:3		
2	1:0.5:4–4.5	1:2.5-3.5	1:3–4
3	1:1:5–6	1:4–5	1:5–6
4	1:2:8–9	1:5.5–6.5	1:7–8
5	1:3:10–12	1:6.5–7	1:8

Group 1: strong inflexible mortar
Group 5: weak but flexible

All mixes within a group are of approximately similar strength
Frost resistance increases with the use of plasticizers
Cement: lime: sand mixes give the strongest bond and greatest resistance to rain penetration
Masonry cement equals ordinary Portland cement plus a fine neutral mineral filler and an air entraining agent

Calcium Silicate Bricks

Type	Strength	Location
Class 2 crushing strength	14.0 N/mm²	not suitable for walls
Class 3	20.5 N/mm²	walls above dpc
Class 4	27.5 N/mm²	cappings and copings
Class 5	34.5 N/mm²	retaining walls
Class 6	41.5 N/mm²	walls below ground
Class 7	48.5 N/mm²	walls below ground

The Class 7 calcium silicate bricks are therefore equal in strength to Class B bricks
Calcium silicate bricks are not suitable for DPCs

Durability of Bricks	
FL	Frost resistant with low salt content
FN	Frost resistant with normal salt content
ML	Moderately frost resistant with low salt content
MN	Moderately frost resistant with normal salt content

MASONRY

Brickwork Dimensions

No. of horizontal bricks	Dimensions (mm)	No. of vertical courses	Height of vertical courses (mm)
½	112.5	1	75
1	225.0	2	150
1½	337.5	3	225
2	450.0	4	300
2½	562.5	5	375
3	675.0	6	450
3½	787.5	7	525
4	900.0	8	600
4½	1012.5	9	675
5	1125.0	10	750
5½	1237.5	11	825
6	1350.0	12	900
6½	1462.5	13	975
7	1575.0	14	1050
7½	1687.5	15	1125
8	1800.0	16	1200
8½	1912.5	17	1275
9	2025.0	18	1350
9½	2137.5	19	1425
10	2250.0	20	1500
20	4500.0	24	1575
40	9000.0	28	2100
50	11250.0	32	2400
60	13500.0	36	2700
75	16875.0	40	3000

TIMBER

Weights of Timber

Material	kg/m³	lb/cu ft
General	806 (avg)	50 (avg)
Douglas fir	479	30
Yellow pine, spruce	479	30
Pitch pine	673	42
Larch, elm	561	35
Oak (English)	724 to 959	45 to 60
Teak	643 to 877	40 to 55
Jarrah	959	60
Greenheart	1040 to 1204	65 to 75
Quebracho	1285	80
Material	kg/m² per mm thickness	lb/sq ft per inch thickness
Wooden boarding and blocks		
Softwood	0.48	2.5
Hardwood	0.76	4
Hardboard	1.06	5.5
Chipboard	0.76	4
Plywood	0.62	3.25
Blockboard	0.48	2.5
Fibreboard	0.29	1.5
Wood-wool	0.58	3
Plasterboard	0.96	5
Weather boarding	0.35	1.8

TIMBER

Conversion Tables (for timber only)

Inches	Millimetres	Feet	Metres
1	25	1	0.300
2	50	2	0.600
3	75	3	0.900
4	100	4	1.200
5	125	5	1.500
6	150	6	1.800
7	175	7	2.100
8	200	8	2.400
9	225	9	2.700
10	250	10	3.000
11	275	11	3.300
12	300	12	3.600
13	325	13	3.900
14	350	14	4.200
15	375	15	4.500
16	400	16	4.800
17	425	17	5.100
18	450	18	5.400
19	475	19	5.700
20	500	20	6.000
21	525	21	6.300
22	550	22	6.600
23	575	23	6.900
24	600	24	7.200

Planed Softwood
The finished end section size of planed timber is usually 3/16" less than the original size from which it is produced. This however varies slightly depending upon availability of material and origin of the species used.

Standards (timber) to cubic metres and cubic metres to standards (timber)

Cubic metres	Cubic metres standards	Standards
4.672	1	0.214
9.344	2	0.428
14.017	3	0.642
18.689	4	0.856
23.361	5	1.070
28.033	6	1.284
32.706	7	1.498
37.378	8	1.712
42.050	9	1.926
46.722	10	2.140
93.445	20	4.281
140.167	30	6.421
186.890	40	8.561
233.612	50	10.702
280.335	60	12.842
327.057	70	14.982
373.779	80	17.122

TIMBER

1 cu metre = 35.3148 cu ft = 0.21403 std

1 cu ft = 0.028317 cu metres

1 std = 4.67227 cu metres

Basic sizes of sawn softwood available (cross-sectional areas)

Thickness (mm)	Width (mm)								
	75	100	125	150	175	200	225	250	300
16	X	X	X	X					
19	X	X	X	X					
22	X	X	X	X					
25	X	X	X	X	X	X	X	X	X
32	X	X	X	X	X	X	X	X	X
36	X	X	X	X					
38	X	X	X	X	X	X	X		
44	X	X	X	X	X	X	X	X	X
47*	X	X	X	X	X	X	X	X	X
50	X	X	X	X	X	X	X	X	X
63	X	X	X	X	X	X	X		
75	X	X	X	X	X	X	X	X	
100		X		X		X		X	X
150				X		X			X
200						X			
250								X	
300									X

* This range of widths for 47 mm thickness will usually be found to be available in construction quality only

Note: The smaller sizes below 100 mm thick and 250 mm width are normally but not exclusively of European origin. Sizes beyond this are usually of North and South American origin

Basic lengths of sawn softwood available (metres)

1.80	2.10	3.00	4.20	5.10	6.00	7.20
	2.40	3.30	4.50	5.40	6.30	
	2.70	3.60	4.80	5.70	6.60	
		3.90			6.90	

Note: Lengths of 6.00 m and over will generally only be available from North American species and may have to be recut from larger sizes

TIMBER

Reductions from basic size to finished size by planning of two opposed faces

Purpose	Reductions from basic sizes for timber			
	15–35 mm	36–100 mm	101–150 mm	over 150 mm
a) Constructional timber	3 mm	3 mm	5 mm	6 mm
b) Matching interlocking boards	4 mm	4 mm	6 mm	6 mm
c) Wood trim not specified in BS 584	5 mm	7 mm	7 mm	9 mm
d) Joinery and cabinet work	7 mm	9 mm	11 mm	13 mm

Note: The reduction of width or depth is overall the extreme size and is exclusive of any reduction of the face by the machining of a tongue or lap joints

Maximum Spans for Various Roof Trusses

Maximum permissible spans for rafters for Fink trussed rafters

Basic size	Actual size	Pitch (degrees)								
(mm)	(mm)	15 (m)	17.5 (m)	20 (m)	22.5 (m)	25 (m)	27.5 (m)	30 (m)	32.5 (m)	35 (m)
38 × 75	35 × 72	6.03	6.16	6.29	6.41	6.51	6.60	6.70	6.80	6.90
38 × 100	35 × 97	7.48	7.67	7.83	7.97	8.10	8.22	8.34	8.47	8.61
38 × 125	35 × 120	8.80	9.00	9.20	9.37	9.54	9.68	9.82	9.98	10.16
44 × 75	41 × 72	6.45	6.59	6.71	6.83	6.93	7.03	7.14	7.24	7.35
44 × 100	41 × 97	8.05	8.23	8.40	8.55	8.68	8.81	8.93	9.09	9.22
44 × 125	41 × 120	9.38	9.60	9.81	9.99	10.15	10.31	10.45	10.64	10.81
50 × 75	47 × 72	6.87	7.01	7.13	7.25	7.35	7.45	7.53	7.67	7.78
50 × 100	47 × 97	8.62	8.80	8.97	9.12	9.25	9.38	9.50	9.66	9.80
50 × 125	47 × 120	10.01	10.24	10.44	10.62	10.77	10.94	11.00	11.00	11.00

TIMBER

Sizes of Internal and External Doorsets

Description	Internal size (mm)	Permissible deviation	External size (mm)	Permissible deviation
Coordinating dimension: height of door leaf height sets	2100		2100	
Coordinating dimension: height of ceiling height set	2300 2350 2400 2700 3000		2300 2350 2400 2700 3000	
Coordinating dimension: width of all door sets S = Single leaf set D = Double leaf set	600 S 700 S 800 S&D 900 S&D 1000 S&D 1200 D 1500 D 1800 D 2100 D		900 S 1000 S 1200 D 1800 D 2100 D	
Work size: height of door leaf height set	2090	± 2.0	2095	± 2.0
Work size: height of ceiling height set	2285 2335 2385 2685 2985	± 2.0	2295 2345 2395 2695 2995	± 2.0
Work size: width of all door sets S = Single leaf set D = Double leaf set	590 S 690 S 790 S&D 890 S&D 990 S&D 1190 D 1490 D 1790 D 2090 D	± 2.0	895 S 995 S 1195 D 1495 D 1795 D 2095 D	± 2.0
Width of door leaf in single leaf sets F = Flush leaf P = Panel leaf	526 F 626 F 726 F&P 826 F&P 926 F&P	± 1.5	806 F&P 906 F&P	± 1.5
Width of door leaf in double leaf sets F = Flush leaf P = Panel leaf	362 F 412 F 426 F 562 F&P 712 F&P 826 F&P 1012 F&P	± 1.5	552 F&P 702 F&P 852 F&P 1002 F&P	± 1.5
Door leaf height for all door sets	2040	± 1.5	1994	± 1.5

ROOFING

Total Roof Loadings for Various Types of Tiles/Slates

	Roof load (slope) kg/m^2		
	Slate/Tile	Roofing underlay and battens^2	Total dead load kg/m
Asbestos cement slate (600 × 300)	21.50	3.14	24.64
Clay tile interlocking	67.00	5.50	72.50
plain	43.50	2.87	46.37
Concrete tile interlocking	47.20	2.69	49.89
plain	78.20	5.50	83.70
Natural slate (18" × 10")	35.40	3.40	38.80
	Roof load (plan) kg/m^2		
Asbestos cement slate (600 × 300)	28.45	76.50	104.95
Clay tile interlocking	53.54	76.50	130.04
plain	83.71	76.50	60.21
Concrete tile interlocking	57.60	76.50	134.10
plain	96.64	76.50	173.14

ROOFING

Tiling Data

Product		Lap (mm)	Gauge of battens	No. slates per m²	Battens (m/m²)	Weight as laid (kg/m²)
CEMENT SLATES						
Eternit slates	600 × 300 mm	100	250	13.4	4.00	19.50
(Duracem)		90	255	13.1	3.92	19.20
		80	260	12.9	3.85	19.00
		70	265	12.7	3.77	18.60
	600 × 350 mm	100	250	11.5	4.00	19.50
		90	255	11.2	3.92	19.20
	500 × 250 mm	100	200	20.0	5.00	20.00
		90	205	19.5	4.88	19.50
		80	210	19.1	4.76	19.00
		70	215	18.6	4.65	18.60
	400 × 200 mm	90	155	32.3	6.45	20.80
		80	160	31.3	6.25	20.20
		70	165	30.3	6.06	19.60
CONCRETE TILES/SLATES						
Redland Roofing						
Stonewold slate	430 × 380 mm	75	355	8.2	2.82	51.20
Double Roman tile	418 × 330 mm	75	355	8.2	2.91	45.50
Grovebury pantile	418 × 332 mm	75	343	9.7	2.91	47.90
Norfolk pantile	381 × 227 mm	75	306	16.3	3.26	44.01
		100	281	17.8	3.56	48.06
Renown interlocking tile	418 × 330 mm	75	343	9.7	2.91	46.40
'49' tile	381 × 227 mm	75	306	16.3	3.26	44.80
		100	281	17.8	3.56	48.95
Plain, vertical tiling	265 × 165 mm	35	115	52.7	8.70	62.20
Marley Roofing						
Bold roll tile	420 × 330 mm	75	344	9.7	2.90	47.00
		100	–	10.5	3.20	51.00
Modern roof tile	420 × 330 mm	75	338	10.2	3.00	54.00
		100	–	11.0	3.20	58.00
Ludlow major	420 × 330 mm	75	338	10.2	3.00	45.00
		100	–	11.0	3.20	49.00
Ludlow plus	387 × 229 mm	75	305	16.1	3.30	47.00
		100	–	17.5	3.60	51.00
Mendip tile	420 × 330 mm	75	338	10.2	3.00	47.00
		100	–	11.0	3.20	51.00
Wessex	413 × 330 mm	75	338	10.2	3.00	54.00
		100	–	11.0	3.20	58.00
Plain tile	267 × 165 mm	65	100	60.0	10.00	76.00
		75	95	64.0	10.50	81.00
		85	90	68.0	11.30	86.00
Plain vertical tiles (feature)	267 × 165 mm	35	110	53.0	8.70	67.00
		34	115	56.0	9.10	71.00

ROOFING

Slate Nails, Quantity per Kilogram

Length	Type			
	Plain wire	Galvanized wire	Copper nail	Zinc nail
28.5 mm	325	305	325	415
34.4 mm	286	256	254	292
50.8 mm	242	224	194	200

Metal Sheet Coverings

Thicknesses and weights of sheet metal coverings								
Lead to BS 1178								
BS Code No	3	4	5	6	7	8		
Colour code	Green	Blue	Red	Black	White	Orange		
Thickness (mm)	1.25	1.80	2.24	2.50	3.15	3.55		
Density (kg/m^2)	14.18	20.41	25.40	30.05	35.72	40.26		
Copper to BS 2870								
Thickness (mm)		0.60	0.70					
Bay width								
Roll (mm)		500	650					
Seam (mm)		525	600					
Standard width to form bay	600	750						
Normal length of sheet	1.80	1.80						
Zinc to BS 849								
Zinc Gauge (Nr)	9	10	11	12	13	14	15	16
Thickness (mm)	0.43	0.48	0.56	0.64	0.71	0.79	0.91	1.04
Density (kg/m^2)	3.1	3.2	3.8	4.3	4.8	5.3	6.2	7.0
Aluminium to BS 4868								
Thickness (mm)	0.5	0.6	0.7	0.8	0.9	1.0	1.2	
Density (kg/m^2)	12.8	15.4	17.9	20.5	23.0	25.6	30.7	

ROOFING

Type of felt	Nominal mass per unit area (kg/10m)	Nominal mass per unit area of fibre base (g/m²)	Nominal length of roll (m)
Class 1			
1B fine granule	14	220	10 or 20
surfaced bitumen	18	330	10 or 20
	25	470	10
1E mineral surfaced bitumen	38	470	10
1F reinforced bitumen	15	160 (fibre) 110 (hessian)	15
1F reinforced bitumen, aluminium faced	13	160 (fibre) 110 (hessian)	15
Class 2			
2B fine granule surfaced bitumen asbestos	18	500	10 or 20
2E mineral surfaced bitumen asbestos	38	600	10
Class 3			
3B fine granule surfaced bitumen glass fibre	18	60	20
3E mineral surfaced bitumen glass fibre	28	60	10
3E venting base layer bitumen glass fibre	32	60*	10
3H venting base layer bitumen glass fibre	17	60*	20

* Excluding effect of perforations

GLAZING

Nominal thickness (mm)	Tolerance on thickness (mm)	Approximate weight (kg/m²)	Normal maximum size (mm)
Float and polished plate glass			
3	+ 0.2	7.50	2140 × 1220
4	+ 0.2	10.00	2760 × 1220
5	+ 0.2	12.50	3180 × 2100
6	+ 0.2	15.00	4600 × 3180
10	+ 0.3	25.00)	6000 × 3300
12	+ 0.3	30.00)	
15	+ 0.5	37.50	3050 × 3000
19	+ 1.0	47.50)	3000 × 2900
25	+ 1.0	63.50)	
Clear sheet glass			
2 *	+ 0.2	5.00	1920 × 1220
3	+ 0.3	7.50	2130 × 1320
4	+ 0.3	10.00	2760 × 1220
5 *	+ 0.3	12.50)	2130 × 2400
6 *	+ 0.3	15.00)	
Cast glass			
3	+ 0.4		
	− 0.2	6.00)	2140 × 1280
4	+ 0.5	7.50)	
5	+ 0.5	9.50	2140 × 1320
6	+ 0.5	11.50)	3700 × 1280
10	+ 0.8	21.50)	
Wired glass			
(Cast wired glass)			
6	+ 0.3	−)	3700 × 1840
	− 0.7	)	
7	+ 0.7	−)	
(Polished wire glass)			
6	+ 1.0	−	330 × 1830

* The 5 mm and 6 mm thickness are known as *thick drawn sheet*. Although 2 mm sheet glass is available it is not recommended for general glazing purposes

METAL

METAL

Weights of Metals

Material	kg/m³	lb/cu ft
Metals, steel construction, etc.		
Iron		
– cast	7207	450
– wrought	7687	480
– ore – general	2407	150
– (crushed) Swedish	3682	230
Steel	7854	490
Copper		
– cast	8731	545
– wrought	8945	558
Brass	8497	530
Bronze	8945	558
Aluminium	2774	173
Lead	11322	707
Zinc (rolled)	7140	446
	g/mm² per metre	**lb/sq ft per foot**
Steel bars	7.85	3.4
Structural steelwork	Net weight of member @ 7854 kg/m³	
riveted	+ 10% for cleats, rivets, bolts, etc.	
welded	+ 1.25% to 2.5% for welds, etc.	
Rolled sections		
beams	+ 2.5%	
stanchions	+ 5% (extra for caps and bases)	
Plate		
web girders	+ 10% for rivets or welds, stiffeners, etc.	
	kg/m	**lb/ft**
Steel stairs: industrial type		
1 m or 3 ft wide	84	56
Steel tubes		
50 mm or 2 in bore	5 to 6	3 to 4
Gas piping		
20 mm or ¾ in	2	1¼

Tables and Memoranda

METAL

Universal Beams BS 4: Part 1: 2005

Designation	Mass (kg/m)	Depth of section (mm)	Width of section (mm)	Thickness		Surface area (m²/m)
				Web (mm)	Flange (mm)	
1016 × 305 × 487	487.0	1036.1	308.5	30.0	54.1	3.20
1016 × 305 × 438	438.0	1025.9	305.4	26.9	49.0	3.17
1016 × 305 × 393	393.0	1016.0	303.0	24.4	43.9	3.15
1016 × 305 × 349	349.0	1008.1	302.0	21.1	40.0	3.13
1016 × 305 × 314	314.0	1000.0	300.0	19.1	35.9	3.11
1016 × 305 × 272	272.0	990.1	300.0	16.5	31.0	3.10
1016 × 305 × 249	249.0	980.2	300.0	16.5	26.0	3.08
1016 × 305 × 222	222.0	970.3	300.0	16.0	21.1	3.06
914 × 419 × 388	388.0	921.0	420.5	21.4	36.6	3.44
914 × 419 × 343	343.3	911.8	418.5	19.4	32.0	3.42
914 × 305 × 289	289.1	926.6	307.7	19.5	32.0	3.01
914 × 305 × 253	253.4	918.4	305.5	17.3	27.9	2.99
914 × 305 × 224	224.2	910.4	304.1	15.9	23.9	2.97
914 × 305 × 201	200.9	903.0	303.3	15.1	20.2	2.96
838 × 292 × 226	226.5	850.9	293.8	16.1	26.8	2.81
838 × 292 × 194	193.8	840.7	292.4	14.7	21.7	2.79
838 × 292 × 176	175.9	834.9	291.7	14.0	18.8	2.78
762 × 267 × 197	196.8	769.8	268.0	15.6	25.4	2.55
762 × 267 × 173	173.0	762.2	266.7	14.3	21.6	2.53
762 × 267 × 147	146.9	754.0	265.2	12.8	17.5	2.51
762 × 267 × 134	133.9	750.0	264.4	12.0	15.5	2.51
686 × 254 × 170	170.2	692.9	255.8	14.5	23.7	2.35
686 × 254 × 152	152.4	687.5	254.5	13.2	21.0	2.34
686 × 254 × 140	140.1	383.5	253.7	12.4	19.0	2.33
686 × 254 × 125	125.2	677.9	253.0	11.7	16.2	2.32
610 × 305 × 238	238.1	635.8	311.4	18.4	31.4	2.45
610 × 305 × 179	179.0	620.2	307.1	14.1	23.6	2.41
610 × 305 × 149	149.1	612.4	304.8	11.8	19.7	2.39
610 × 229 × 140	139.9	617.2	230.2	13.1	22.1	2.11
610 × 229 × 125	125.1	612.2	229.0	11.9	19.6	2.09
610 × 229 × 113	113.0	607.6	228.2	11.1	17.3	2.08
610 × 229 × 101	101.2	602.6	227.6	10.5	14.8	2.07
533 × 210 × 122	122.0	544.5	211.9	12.7	21.3	1.89
533 × 210 × 109	109.0	539.5	210.8	11.6	18.8	1.88
533 × 210 × 101	101.0	536.7	210.0	10.8	17.4	1.87
533 × 210 × 92	92.1	533.1	209.3	10.1	15.6	1.86
533 × 210 × 82	82.2	528.3	208.8	9.6	13.2	1.85
457 × 191 × 98	98.3	467.2	192.8	11.4	19.6	1.67
457 × 191 × 89	89.3	463.4	191.9	10.5	17.7	1.66
457 × 191 × 82	82.0	460.0	191.3	9.9	16.0	1.65
457 × 191 × 74	74.3	457.0	190.4	9.0	14.5	1.64
457 × 191 × 67	67.1	453.4	189.9	8.5	12.7	1.63
457 × 152 × 82	82.1	465.8	155.3	10.5	18.9	1.51
457 × 152 × 74	74.2	462.0	154.4	9.6	17.0	1.50
457 × 152 × 67	67.2	458.0	153.8	9.0	15.0	1.50
457 × 152 × 60	59.8	454.6	152.9	8.1	13.3	1.50
457 × 152 × 52	52.3	449.8	152.4	7.6	10.9	1.48
406 × 178 × 74	74.2	412.8	179.5	9.5	16.0	1.51
406 × 178 × 67	67.1	409.4	178.8	8.8	14.3	1.50
406 × 178 × 60	60.1	406.4	177.9	7.9	12.8	1.49

METAL

Designation	Mass (kg/m)	Depth of section (mm)	Width of section (mm)	Thickness		Surface area (m²/m)
				Web (mm)	Flange (mm)	
406 × 178 × 50	54.1	402.6	177.7	7.7	10.9	1.48
406 × 140 × 46	46.0	403.2	142.2	6.8	11.2	1.34
406 × 140 × 39	39.0	398.0	141.8	6.4	8.6	1.33
356 × 171 × 67	67.1	363.4	173.2	9.1	15.7	1.38
356 × 171 × 57	57.0	358.0	172.2	8.1	13.0	1.37
356 × 171 × 51	51.0	355.0	171.5	7.4	11.5	1.36
356 × 171 × 45	45.0	351.4	171.1	7.0	9.7	1.36
356 × 127 × 39	39.1	353.4	126.0	6.6	10.7	1.18
356 × 127 × 33	33.1	349.0	125.4	6.0	8.5	1.17
305 × 165 × 54	54.0	310.4	166.9	7.9	13.7	1.26
305 × 165 × 46	46.1	306.6	165.7	6.7	11.8	1.25
305 × 165 × 40	40.3	303.4	165.0	6.0	10.2	1.24
305 × 127 × 48	48.1	311.0	125.3	9.0	14.0	1.09
305 × 127 × 42	41.9	307.2	124.3	8.0	12.1	1.08
305 × 127 × 37	37.0	304.4	123.3	7.1	10.7	1.07
305 × 102 × 33	32.8	312.7	102.4	6.6	10.8	1.01
305 × 102 × 28	28.2	308.7	101.8	6.0	8.8	1.00
305 × 102 × 25	24.8	305.1	101.6	5.8	7.0	0.992
254 × 146 × 43	43.0	259.6	147.3	7.2	12.7	1.08
254 × 146 × 37	37.0	256.0	146.4	6.3	10.9	1.07
254 × 146 × 31	31.1	251.4	146.1	6.0	8.6	1.06
254 × 102 × 28	28.3	260.4	102.2	6.3	10.0	0.904
254 × 102 × 25	25.2	257.2	101.9	6.0	8.4	0.897
254 × 102 × 22	22.0	254.0	101.6	5.7	6.8	0.890
203 × 133 × 30	30.0	206.8	133.9	6.4	9.6	0.923
203 × 133 × 25	25.1	203.2	133.2	5.7	7.8	0.915
203 × 102 × 23	23.1	203.2	101.8	5.4	9.3	0.790
178 × 102 × 19	19.0	177.8	101.2	4.8	7.9	0.738
152 × 89 × 16	16.0	152.4	88.7	4.5	7.7	0.638
127 × 76 × 13	13.0	127.0	76.0	4.0	7.6	0.537

METAL

Universal Columns BS 4: Part 1: 2005

Designation	Mass (kg/m)	Depth of section (mm)	Width of section (mm)	Thickness		Surface area (m²/m)
				Web (mm)	Flange (mm)	
356 × 406 × 634	633.9	474.7	424.0	47.6	77.0	2.52
356 × 406 × 551	551.0	455.6	418.5	42.1	67.5	2.47
356 × 406 × 467	467.0	436.6	412.2	35.8	58.0	2.42
356 × 406 × 393	393.0	419.0	407.0	30.6	49.2	2.38
356 × 406 × 340	339.9	406.4	403.0	26.6	42.9	2.35
356 × 406 × 287	287.1	393.6	399.0	22.6	36.5	2.31
356 × 406 × 235	235.1	381.0	384.8	18.4	30.2	2.28
356 × 368 × 202	201.9	374.6	374.7	16.5	27.0	2.19
356 × 368 × 177	177.0	368.2	372.6	14.4	23.8	2.17
356 × 368 × 153	152.9	362.0	370.5	12.3	20.7	2.16
356 × 368 × 129	129.0	355.6	368.6	10.4	17.5	2.14
305 × 305 × 283	282.9	365.3	322.2	26.8	44.1	1.94
305 × 305 × 240	240.0	352.5	318.4	23.0	37.7	1.91
305 × 305 × 198	198.1	339.9	314.5	19.1	31.4	1.87
305 × 305 × 158	158.1	327.1	311.2	15.8	25.0	1.84
305 × 305 × 137	136.9	320.5	309.2	13.8	21.7	1.82
305 × 305 × 118	117.9	314.5	307.4	12.0	18.7	1.81
305 × 305 × 97	96.9	307.9	305.3	9.9	15.4	1.79
254 × 254 × 167	167.1	289.1	265.2	19.2	31.7	1.58
254 × 254 × 132	132.0	276.3	261.3	15.3	25.3	1.55
254 × 254 × 107	107.1	266.7	258.8	12.8	20.5	1.52
254 × 254 × 89	88.9	260.3	256.3	10.3	17.3	1.50
254 × 254 × 73	73.1	254.1	254.6	8.6	14.2	1.49
203 × 203 × 86	86.1	222.2	209.1	12.7	20.5	1.24
203 × 203 × 71	71.0	215.8	206.4	10.0	17.3	1.22
203 × 203 × 60	60.0	209.6	205.8	9.4	14.2	1.21
203 × 203 × 52	52.0	206.2	204.3	7.9	12.5	1.20
203 × 203 × 46	46.1	203.2	203.6	7.2	11.0	1.19
152 × 152 × 37	37.0	161.8	154.4	8.0	11.5	0.912
152 × 152 × 30	30.0	157.6	152.9	6.5	9.4	0.901
152 × 152 × 23	23.0	152.4	152.2	5.8	6.8	0.889

METAL

Joists BS 4: Part 1: 2005 (retained for reference, Corus have ceased manufacture in UK)

Designation	Mass	Depth of section	Width of section	Thickness		Surface area
	(kg/m)	(mm)	(mm)	Web (mm)	Flange (mm)	(m²/m)
254 × 203 × 82	82.0	254.0	203.2	10.2	19.9	1.210
203 × 152 × 52	52.3	203.2	152.4	8.9	16.5	0.932
152 × 127 × 37	37.3	152.4	127.0	10.4	13.2	0.737
127 × 114 × 29	29.3	127.0	114.3	10.2	11.5	0.646
127 × 114 × 27	26.9	127.0	114.3	7.4	11.4	0.650
102 × 102 × 23	23.0	101.6	101.6	9.5	10.3	0.549
102 × 44 × 7	7.5	101.6	44.5	4.3	6.1	0.350
89 × 89 × 19	19.5	88.9	88.9	9.5	9.9	0.476
76 × 76 × 13	12.8	76.2	76.2	5.1	8.4	0.411

Parallel Flange Channels

Designation	Mass	Depth of section	Width of section	Thickness		Surface area
	(kg/m)	(mm)	(mm)	Web (mm)	Flange (mm)	(m²/m)
430 × 100 × 64	64.4	430	100	11.0	19.0	1.23
380 × 100 × 54	54.0	380	100	9.5	17.5	1.13
300 × 100 × 46	45.5	300	100	9.0	16.5	0.969
300 × 90 × 41	41.4	300	90	9.0	15.5	0.932
260 × 90 × 35	34.8	260	90	8.0	14.0	0.854
260 × 75 × 28	27.6	260	75	7.0	12.0	0.79
230 × 90 × 32	32.2	230	90	7.5	14.0	0.795
230 × 75 × 26	25.7	230	75	6.5	12.5	0.737
200 × 90 × 30	29.7	200	90	7.0	14.0	0.736
200 × 75 × 23	23.4	200	75	6.0	12.5	0.678
180 × 90 × 26	26.1	180	90	6.5	12.5	0.697
180 × 75 × 20	20.3	180	75	6.0	10.5	0.638
150 × 90 × 24	23.9	150	90	6.5	12.0	0.637
150 × 75 × 18	17.9	150	75	5.5	10.0	0.579
125 × 65 × 15	14.8	125	65	5.5	9.5	0.489
100 × 50 × 10	10.2	100	50	5.0	8.5	0.382

METAL

Equal Angles BS EN 10056-1

Designation	Mass (kg/m)	Surface area (m²/m)
200 × 200 × 24	71.1	0.790
200 × 200 × 20	59.9	0.790
200 × 200 × 18	54.2	0.790
200 × 200 × 16	48.5	0.790
150 × 150 × 18	40.1	0.59
150 × 150 × 15	33.8	0.59
150 × 150 × 12	27.3	0.59
150 × 150 × 10	23.0	0.59
120 × 120 × 15	26.6	0.47
120 × 120 × 12	21.6	0.47
120 × 120 × 10	18.2	0.47
120 × 120 × 8	14.7	0.47
100 × 100 × 15	21.9	0.39
100 × 100 × 12	17.8	0.39
100 × 100 × 10	15.0	0.39
100 × 100 × 8	12.2	0.39
90 × 90 × 12	15.9	0.35
90 × 90 × 10	13.4	0.35
90 × 90 × 8	10.9	0.35
90 × 90 × 7	9.61	0.35
90 × 90 × 6	8.30	0.35

Unequal Angles BS EN 10056-1

Designation	Mass (kg/m)	Surface area (m²/m)
200 × 150 × 18	47.1	0.69
200 × 150 × 15	39.6	0.69
200 × 150 × 12	32.0	0.69
200 × 100 × 15	33.7	0.59
200 × 100 × 12	27.3	0.59
200 × 100 × 10	23.0	0.59
150 × 90 × 15	26.6	0.47
150 × 90 × 12	21.6	0.47
150 × 90 × 10	18.2	0.47
150 × 75 × 15	24.8	0.44
150 × 75 × 12	20.2	0.44
150 × 75 × 10	17.0	0.44
125 × 75 × 12	17.8	0.40
125 × 75 × 10	15.0	0.40
125 × 75 × 8	12.2	0.40
100 × 75 × 12	15.4	0.34
100 × 75 × 10	13.0	0.34
100 × 75 × 8	10.6	0.34
100 × 65 × 10	12.3	0.32
100 × 65 × 8	9.94	0.32
100 × 65 × 7	8.77	0.32

METAL

Structural Tees Split from Universal Beams BS 4: Part 1: 2005

Designation	Mass (kg/m)	Surface area (m²/m)
305 × 305 × 90	89.5	1.22
305 × 305 × 75	74.6	1.22
254 × 343 × 63	62.6	1.19
229 × 305 × 70	69.9	1.07
229 × 305 × 63	62.5	1.07
229 × 305 × 57	56.5	1.07
229 × 305 × 51	50.6	1.07
210 × 267 × 61	61.0	0.95
210 × 267 × 55	54.5	0.95
210 × 267 × 51	50.5	0.95
210 × 267 × 46	46.1	0.95
210 × 267 × 41	41.1	0.95
191 × 229 × 49	49.2	0.84
191 × 229 × 45	44.6	0.84
191 × 229 × 41	41.0	0.84
191 × 229 × 37	37.1	0.84
191 × 229 × 34	33.6	0.84
152 × 229 × 41	41.0	0.76
152 × 229 × 37	37.1	0.76
152 × 229 × 34	33.6	0.76
152 × 229 × 30	29.9	0.76
152 × 229 × 26	26.2	0.76

Universal Bearing Piles BS 4: Part 1: 2005

Designation	Mass (kg/m)	Depth of Section (mm)	Width of Section (mm)	Thickness Web (mm)	Thickness Flange (mm)
356 × 368 × 174	173.9	361.4	378.5	20.3	20.4
356 × 368 × 152	152.0	356.4	376.0	17.8	17.9
356 × 368 × 133	133.0	352.0	373.8	15.6	15.7
356 × 368 × 109	108.9	346.4	371.0	12.8	12.9
305 × 305 × 223	222.9	337.9	325.7	30.3	30.4
305 × 305 × 186	186.0	328.3	320.9	25.5	25.6
305 × 305 × 149	149.1	318.5	316.0	20.6	20.7
305 × 305 × 126	126.1	312.3	312.9	17.5	17.6
305 × 305 × 110	110.0	307.9	310.7	15.3	15.4
305 × 305 × 95	94.9	303.7	308.7	13.3	13.3
305 × 305 × 88	88.0	301.7	307.8	12.4	12.3
305 × 305 × 79	78.9	299.3	306.4	11.0	11.1
254 × 254 × 85	85.1	254.3	260.4	14.4	14.3
254 × 254 × 71	71.0	249.7	258.0	12.0	12.0
254 × 254 × 63	63.0	247.1	256.6	10.6	10.7
203 × 203 × 54	53.9	204.0	207.7	11.3	11.4
203 × 203 × 45	44.9	200.2	205.9	9.5	9.5

METAL

Hot Formed Square Hollow Sections EN 10210 S275J2H & S355J2H

Size (mm)	Wall thickness (mm)	Mass (kg/m)	Superficial area (m²/m)
40 × 40	2.5	2.89	0.154
	3.0	3.41	0.152
	3.2	3.61	0.152
	3.6	4.01	0.151
	4.0	4.39	0.150
	5.0	5.28	0.147
50 × 50	2.5	3.68	0.194
	3.0	4.35	0.192
	3.2	4.62	0.192
	3.6	5.14	0.191
	4.0	5.64	0.190
	5.0	6.85	0.187
	6.0	7.99	0.185
	6.3	8.31	0.184
60 × 60	3.0	5.29	0.232
	3.2	5.62	0.232
	3.6	6.27	0.231
	4.0	6.90	0.230
	5.0	8.42	0.227
	6.0	9.87	0.225
	6.3	10.30	0.224
	8.0	12.50	0.219
70 × 70	3.0	6.24	0.272
	3.2	6.63	0.272
	3.6	7.40	0.271
	4.0	8.15	0.270
	5.0	9.99	0.267
	6.0	11.80	0.265
	6.3	12.30	0.264
	8.0	15.00	0.259
80 × 80	3.2	7.63	0.312
	3.6	8.53	0.311
	4.0	9.41	0.310
	5.0	11.60	0.307
	6.0	13.60	0.305
	6.3	14.20	0.304
	8.0	17.50	0.299
90 × 90	3.6	9.66	0.351
	4.0	10.70	0.350
	5.0	13.10	0.347
	6.0	15.50	0.345
	6.3	16.20	0.344
	8.0	20.10	0.339
100 × 100	3.6	10.80	0.391
	4.0	11.90	0.390
	5.0	14.70	0.387
	6.0	17.40	0.385
	6.3	18.20	0.384
	8.0	22.60	0.379
	10.0	27.40	0.374
120 × 120	4.0	14.40	0.470

METAL

Size (mm)	Wall thickness (mm)	Mass (kg/m)	Superficial area (m²/m)
	5.0	17.80	0.467
	6.0	21.20	0.465
	6.3	22.20	0.464
	8.0	27.60	0.459
	10.0	33.70	0.454
	12.0	39.50	0.449
	12.5	40.90	0.448
140 × 140	5.0	21.00	0.547
	6.0	24.90	0.545
	6.3	26.10	0.544
	8.0	32.60	0.539
	10.0	40.00	0.534
	12.0	47.00	0.529
	12.5	48.70	0.528
150 × 150	5.0	22.60	0.587
	6.0	26.80	0.585
	6.3	28.10	0.584
	8.0	35.10	0.579
	10.0	43.10	0.574
	12.0	50.80	0.569
	12.5	52.70	0.568
Hot formed from seamless hollow	16.0	65.2	0.559
160 × 160	5.0	24.10	0.627
	6.0	28.70	0.625
	6.3	30.10	0.624
	8.0	37.60	0.619
	10.0	46.30	0.614
	12.0	54.60	0.609
	12.5	56.60	0.608
	16.0	70.20	0.599
180 × 180	5.0	27.30	0.707
	6.0	32.50	0.705
	6.3	34.00	0.704
	8.0	42.70	0.699
	10.0	52.50	0.694
	12.0	62.10	0.689
	12.5	64.40	0.688
	16.0	80.20	0.679
200 × 200	5.0	30.40	0.787
	6.0	36.20	0.785
	6.3	38.00	0.784
	8.0	47.70	0.779
	10.0	58.80	0.774
	12.0	69.60	0.769
	12.5	72.30	0.768
	16.0	90.30	0.759
250 × 250	5.0	38.30	0.987

METAL

Size (mm)	Wall thickness (mm)	Mass (kg/m)	Superficial area (m²/m)
	6.0	45.70	0.985
	6.3	47.90	0.984
	8.0	60.30	0.979
	10.0	74.50	0.974
	12.0	88.50	0.969
	12.5	91.90	0.968
	16.0	115.00	0.959
300 × 300	6.0	55.10	1.18
	6.3	57.80	1.18
	8.0	72.80	1.18
	10.0	90.20	1.17
	12.0	107.00	1.17
	12.5	112.00	1.17
	16.0	141.00	1.16
350 × 350	8.0	85.40	1.38
	10.0	106.00	1.37
	12.0	126.00	1.37
	12.5	131.00	1.37
	16.0	166.00	1.36
400 × 400	8.0	97.90	1.58
	10.0	122.00	1.57
	12.0	145.00	1.57
	12.5	151.00	1.57
	16.0	191.00	1.56
(Grade S355J2H only)	20.00*	235.00	1.55

Note: * SAW process

METAL

Hot Formed Square Hollow Sections JUMBO RHS: JIS G3136

Size (mm)	Wall thickness (mm)	Mass (kg/m)	Superficial area (m²/m)
350 × 350	19.0	190.00	1.33
	22.0	217.00	1.32
	25.0	242.00	1.31
400 × 400	22.0	251.00	1.52
	25.0	282.00	1.51
450 × 450	12.0	162.00	1.76
	16.0	213.00	1.75
	19.0	250.00	1.73
	22.0	286.00	1.72
	25.0	321.00	1.71
	28.0 *	355.00	1.70
	32.0 *	399.00	1.69
500 × 500	12.0	181.00	1.96
	16.0	238.00	1.95
	19.0	280.00	1.93
	22.0	320.00	1.92
	25.0	360.00	1.91
	28.0 *	399.00	1.90
	32.0 *	450.00	1.89
	36.0 *	498.00	1.88
550 × 550	16.0	263.00	2.15
	19.0	309.00	2.13
	22.0	355.00	2.12
	25.0	399.00	2.11
	28.0 *	443.00	2.10
	32.0 *	500.00	2.09
	36.0 *	555.00	2.08
	40.0 *	608.00	2.06
600 × 600	25.0 *	439.00	2.31
	28.0 *	487.00	2.30
	32.0 *	550.00	2.29
	36.0 *	611.00	2.28
	40.0 *	671.00	2.26
700 × 700	25.0 *	517.00	2.71
	28.0 *	575.00	2.70
	32.0 *	651.00	2.69
	36.0 *	724.00	2.68
	40.0 *	797.00	2.68

Note: * SAW process

METAL

Hot Formed Rectangular Hollow Sections: EN10210 S275J2h & S355J2H

Size (mm)	Wall thickness (mm)	Mass (kg/m)	Superficial area (m²/m)
50 × 30	2.5	2.89	0.154
	3.0	3.41	0.152
	3.2	3.61	0.152
	3.6	4.01	0.151
	4.0	4.39	0.150
	5.0	5.28	0.147
60 × 40	2.5	3.68	0.194
	3.0	4.35	0.192
	3.2	4.62	0.192
	3.6	5.14	0.191
	4.0	5.64	0.190
	5.0	6.85	0.187
	6.0	7.99	0.185
	6.3	8.31	0.184
80 × 40	3.0	5.29	0.232
	3.2	5.62	0.232
	3.6	6.27	0.231
	4.0	6.90	0.230
	5.0	8.42	0.227
	6.0	9.87	0.225
	6.3	10.30	0.224
	8.0	12.50	0.219
76.2 × 50.8	3.0	5.62	0.246
	3.2	5.97	0.246
	3.6	6.66	0.245
	4.0	7.34	0.244
	5.0	8.97	0.241
	6.0	10.50	0.239
	6.3	11.00	0.238
	8.0	13.40	0.233
90 × 50	3.0	6.24	0.272
	3.2	6.63	0.272
	3.6	7.40	0.271
	4.0	8.15	0.270
	5.0	9.99	0.267
	6.0	11.80	0.265
	6.3	12.30	0.264
	8.0	15.00	0.259
100 × 50	3.0	6.71	0.292
	3.2	7.13	0.292
	3.6	7.96	0.291
	4.0	8.78	0.290
	5.0	10.80	0.287
	6.0	12.70	0.285
	6.3	13.30	0.284
	8.0	16.30	0.279

METAL

Size (mm)	Wall thickness (mm)	Mass (kg/m)	Superficial area (m²/m)
100 × 60	3.0	7.18	0.312
	3.2	7.63	0.312
	3.6	8.53	0.311
	4.0	9.41	0.310
	5.0	11.60	0.307
	6.0	13.60	0.305
	6.3	14.20	0.304
	8.0	17.50	0.299
120 × 60	3.6	9.70	0.351
	4.0	10.70	0.350
	5.0	13.10	0.347
	6.0	15.50	0.345
	6.3	16.20	0.344
	8.0	20.10	0.339
120 × 80	3.6	10.80	0.391
	4.0	11.90	0.390
	5.0	14.70	0.387
	6.0	17.40	0.385
	6.3	18.20	0.384
	8.0	22.60	0.379
	10.0	27.40	0.374
150 × 100	4.0	15.10	0.490
	5.0	18.60	0.487
	6.0	22.10	0.485
	6.3	23.10	0.484
	8.0	28.90	0.479
	10.0	35.30	0.474
	12.0	41.40	0.469
	12.5	42.80	0.468
160 × 80	4.0	14.40	0.470
	5.0	17.80	0.467
	6.0	21.20	0.465
	6.3	22.20	0.464
	8.0	27.60	0.459
	10.0	33.70	0.454
	12.0	39.50	0.449
	12.5	40.90	0.448
200 × 100	5.0	22.60	0.587
	6.0	26.80	0.585
	6.3	28.10	0.584
	8.0	35.10	0.579
	10.0	43.10	0.574
	12.0	50.80	0.569
	12.5	52.70	0.568
	16.0	65.20	0.559
250 × 150	5.0	30.40	0.787
	6.0	36.20	0.785
	6.3	38.00	0.784
	8.0	47.70	0.779
	10.0	58.80	0.774
	12.0	69.60	0.769
	12.5	72.30	0.768
	16.0	90.30	0.759

METAL

Size (mm)	Wall thickness (mm)	Mass (kg/m)	Superficial area (m²/m)
300 × 200	5.0	38.30	0.987
	6.0	45.70	0.985
	6.3	47.90	0.984
	8.0	60.30	0.979
	10.0	74.50	0.974
	12.0	88.50	0.969
	12.5	91.90	0.968
	16.0	115.00	0.959
400 × 200	6.0	55.10	1.18
	6.3	57.80	1.18
	8.0	72.80	1.18
	10.0	90.20	1.17
	12.0	107.00	1.17
	12.5	112.00	1.17
	16.0	141.00	1.16
450 × 250	8.0	85.40	1.38
	10.0	106.00	1.37
	12.0	126.00	1.37
	12.5	131.00	1.37
	16.0	166.00	1.36
500 × 300	8.0	98.00	1.58
	10.0	122.00	1.57
	12.0	145.00	1.57
	12.5	151.00	1.57
	16.0	191.00	1.56
	20.0	235.00	1.55

METAL

Hot Formed Circular Hollow Sections EN 10210 S275J2H & S355J2H

Outside diameter (mm)	Wall thickness (mm)	Mass (kg/m)	Superficial area (m²/m)
21.3	3.2	1.43	0.067
26.9	3.2	1.87	0.085
33.7	3.0	2.27	0.106
	3.2	2.41	0.106
	3.6	2.67	0.106
	4.0	2.93	0.106
42.4	3.0	2.91	0.133
	3.2	3.09	0.133
	3.6	3.44	0.133
	4.0	3.79	0.133
48.3	2.5	2.82	0.152
	3.0	3.35	0.152
	3.2	3.56	0.152
	3.6	3.97	0.152
	4.0	4.37	0.152
	5.0	5.34	0.152
60.3	2.5	3.56	0.189
	3.0	4.24	0.189
	3.2	4.51	0.189
	3.6	5.03	0.189
	4.0	5.55	0.189
	5.0	6.82	0.189
76.1	2.5	4.54	0.239
	3.0	5.41	0.239
	3.2	5.75	0.239
	3.6	6.44	0.239
	4.0	7.11	0.239
	5.0	8.77	0.239
	6.0	10.40	0.239
	6.3	10.80	0.239
88.9	2.5	5.33	0.279
	3.0	6.36	0.279
	3.2	6.76	0.27
	3.6	7.57	0.279
88.9	4.0	8.38	0.279
	5.0	10.30	0.279
	6.0	12.30	0.279
	6.3	12.80	0.279
114.3	3.0	8.23	0.359
	3.2	8.77	0.359
	3.6	9.83	0.359
	4.0	10.09	0.359
	5.0	13.50	0.359
	6.0	16.00	0.359
	6.3	16.80	0.359

METAL

Outside diameter (mm)	Wall thickness (mm)	Mass (kg/m)	Superficial area (m²/m)
139.7	3.2	10.80	0.439
	3.6	12.10	0.439
	4.0	13.40	0.439
	5.0	16.60	0.439
	6.0	19.80	0.439
	6.3	20.70	0.439
	8.0	26.00	0.439
	10.0	32.00	0.439
168.3	3.2	13.00	0.529
	3.6	14.60	0.529
	4.0	16.20	0.529
	5.0	20.10	0.529
	6.0	24.00	0.529
	6.3	25.20	0.529
	8.0	31.60	0.529
	10.0	39.00	0.529
	12.0	46.30	0.529
	12.5	48.00	0.529
193.7	5.0	23.30	0.609
	6.0	27.80	0.609
	6.3	29.10	0.609
	8.0	36.60	0.609
	10.0	45.30	0.609
193.7	12.0	53.80	0.609
	12.5	55.90	0.609
219.1	5.0	26.40	0.688
	6.0	31.50	0.688
	6.3	33.10	0.688
	8.0	41.60	0.688
	10.0	51.60	0.688
	12.0	61.30	0.688
	12.5	63.70	0.688
	16.0	80.10	0.688
244.5	5.0	29.50	0.768
	6.0	35.30	0.768
	6.3	37.00	0.768
	8.0	46.70	0.768
	10.0	57.80	0.768
	12.0	68.80	0.768
	12.5	71.50	0.768
	16.0	90.20	0.768
273.0	5.0	33.00	0.858
	6.0	39.50	0.858
	6.3	41.40	0.858
	8.0	52.30	0.858
	10.0	64.90	0.858
	12.0	77.20	0.858
	12.5	80.30	0.858
	16.0	101.00	0.858

METAL

Outside diameter (mm)	Wall thickness (mm)	Mass (kg/m)	Superficial area (m²/m)
323.9	5.0	39.30	1.02
	6.0	47.00	1.02
	6.3	49.30	1.02
	8.0	62.30	1.02
	10.0	77.40	1.02
	12.0	92.30	1.02
	12.5	96.00	1.02
	16.0	121.00	1.02
355.6	6.3	54.30	1.12
	8.0	68.60	1.12
	10.0	85.30	1.12
	12.0	102.00	1.12
	12.5	106.00	1.12
	16.0	134.00	1.12
406.4	6.3	62.20	1.28
	8.0	79.60	1.28
	10.0	97.80	1.28
	12.0	117.00	1.28
	12.5	121.00	1.28
	16.0	154.00	1.28
457.0	6.3	70.00	1.44
	8.0	88.60	1.44
	10.0	110.00	1.44
	12.0	132.00	1.44
	12.5	137.00	1.44
	16.0	174.00	1.44
508.0	6.3	77.90	1.60
	8.0	98.60	1.60
	10.0	123.00	1.60
	12.0	147.00	1.60
	12.5	153.00	1.60
	16.0	194.00	1.60

METAL

Spacing of Holes in Angles

Nominal leg length (mm)	Spacing of holes						Maximum diameter of bolt or rivet		
	A	B	C	D	E	F	A	B and C	D, E and F
200		75	75	55	55	55		30	20
150		55	55					20	
125		45	60					20	
120									
100	55						24		
90	50						24		
80	45						20		
75	45						20		
70	40						20		
65	35						20		
60	35						16		
50	28						12		
45	25								
40	23								
30	20								
25	15								

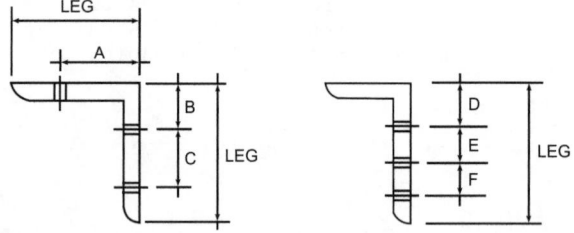

KERBS, PAVING, ETC.

KERBS/EDGINGS/CHANNELS

Precast Concrete Kerbs to BS 7263

Straight kerb units: length from 450 to 915 mm

150 mm high × 125 mm thick		
bullnosed	type BN	
half battered	type HB3	
255 mm high × 125 mm thick		
45° splayed	type SP	
half battered	type HB2	
305 mm high × 150 mm thick		
half battered	type HB1	
Quadrant kerb units		
150 mm high × 305 and 455 mm radius to match	type BN	type QBN
150 mm high × 305 and 455 mm radius to match	type HB2, HB3	type QHB
150 mm high × 305 and 455 mm radius to match	type SP	type QSP
255 mm high × 305 and 455 mm radius to match	type BN	type QBN
255 mm high × 305 and 455 mm radius to match	type HB2, HB3	type QHB
225 mm high × 305 and 455 mm radius to match	type SP	type QSP
Angle kerb units		
305 × 305 × 225 mm high × 125 mm thick		
bullnosed external angle	type XA	
splayed external angle to match type SP	type XA	
bullnosed internal angle	type IA	
splayed internal angle to match type SP	type IA	
Channels		
255 mm wide × 125 mm high flat	type CS1	
150 mm wide × 125 mm high flat type	CS2	
255 mm wide × 125 mm high dished	type CD	

KERBS, PAVING, ETC.

Transition kerb units			
from kerb type SP to HB	left handed	type TL	
	right handed	type TR	
from kerb type BN to HB	left handed	type DL1	
	right handed	type DR1	
from kerb type BN to SP	left handed	type DL2	
	right handed	type DR2	

Number of kerbs required per quarter circle (780mm kerb lengths)

Radius (m)	Number in quarter circle
12	24
10	20
8	16
6	12
5	10
4	8
3	6
2	4
1	2

Precast Concrete Edgings

Round top type ER	Flat top type EF	Bullnosed top type EBN
150 × 50 mm	150 × 50 mm	150 × 50 mm
200 × 50 mm	200 × 50 mm	200 × 50 mm
250 × 50 mm	250 × 50 mm	250 × 50 mm

BASES

Cement Bound Material for Bases and Subbases

CBM1:	very carefully graded aggregate from 37.5–75 ym, with a 7-day strength of 4.5 N/mm^2
CBM2:	same range of aggregate as CBM1 but with more tolerance in each size of aggregate with a 7-day strength of 7.0 N/mm^2
CBM3:	crushed natural aggregate or blast furnace slag, graded from 37.5 mm – 150 ym for 40 mm aggregate, and from 20–75 ym for 20 mm aggregate, with a 7-day strength of 10 N/mm^2
CBM4:	crushed natural aggregate or blast furnace slag, graded from 37.5 mm – 150 ym for 40 mm aggregate, and from 20–75 ym for 20 mm aggregate, with a 7-day strength of 15 N/mm^2

INTERLOCKING BRICK/BLOCK ROADS/PAVINGS

Sizes of Precast Concrete Paving Blocks

Type R blocks	Type S
200 × 100 × 60 mm	Any shape within a 295 mm space
200 × 100 × 65 mm	
200 × 100 × 80 mm	
200 × 100 × 100 mm	

Sizes of clay brick pavers
200 × 100 × 50 mm
200 × 100 × 65 mm
210 × 105 × 50 mm
210 × 105 × 65 mm
215 × 102.5 × 50 mm
215 × 102.5 × 65 mm

Type PA: 3 kN
Footpaths and pedestrian areas, private driveways, car parks, light vehicle traffic and over-run

Type PB: 7 kN
Residential roads, lorry parks, factory yards, docks, petrol station forecourts, hardstandings, bus stations

KERBS, PAVING, ETC.

PAVING AND SURFACING

Weights and Sizes of Paving and Surfacing

Description of item	Size	Quantity per tonne
Paving 50 mm thick	900 × 600 mm	15
Paving 50 mm thick	750 × 600 mm	18
Paving 50 mm thick	600 × 600 mm	23
Paving 50 mm thick	450 × 600 mm	30
Paving 38 mm thick	600 × 600 mm	30
Path edging	914 × 50 × 150 mm	60
Kerb (including radius and tapers)	125 × 254 × 914 mm	15
Kerb (including radius and tapers)	125 × 150 × 914 mm	25
Square channel	125 × 254 × 914 mm	15
Dished channel	125 × 254 × 914 mm	15
Quadrants	300 × 300 × 254 mm	19
Quadrants	450 × 450 × 254 mm	12
Quadrants	300 × 300 × 150 mm	30
Internal angles	300 × 300 × 254 mm	30
Fluted pavement channel	255 × 75 × 914 mm	25
Corner stones	300 × 300 mm	80
Corner stones	360 × 360 mm	60
Cable covers	914 × 175 mm	55
Gulley kerbs	220 × 220 × 150 mm	60
Gulley kerbs	220 × 200 × 75 mm	120

KERBS, PAVING, ETC.

Weights and Sizes of Paving and Surfacing

Material	kg/m^3	lb/cu yd
Tarmacadam	2306	3891
Macadam (waterbound)	2563	4325
Vermiculite (aggregate)	64–80	108–135
Terracotta	2114	3568
Cork – compressed	388	24
	kg/m^2	lb/sq ft
Clay floor tiles, 12.7 mm	27.3	5.6
Pavement lights	122	25
Damp-proof course	5	1
	kg/m^2 per mm thickness	lb/sq ft per inch thickness
Paving Slabs (stone)	2.3	12
Granite setts	2.88	15
Asphalt	2.30	12
Rubber flooring	1.68	9
Polyvinyl chloride	1.94 (avg)	10 (avg)

Coverage (m^2) Per Cubic Metre of Materials Used as Subbases or Capping Layers

Consolidated thickness laid in (mm)	Square metre coverage		
	Gravel	Sand	Hardcore
50	15.80	16.50	–
75	10.50	11.00	–
100	7.92	8.20	7.42
125	6.34	6.60	5.90
150	5.28	5.50	4.95
175	–	–	4.23
200	–	–	3.71
225	–	–	3.30
300	–	–	2.47

KERBS, PAVING, ETC.

Approximate Rate of Spreads

Average thickness of course (mm)	Description	Approximate rate of spread			
		Open Textured		Dense, Medium & Fine Textured	
		(kg/m²)	(m²/t)	(kg/m²)	(m²/t)
35	14 mm open textured or dense wearing course	60–75	13–17	70–85	12–14
40	20 mm open textured or dense base course	70–85	12–14	80–100	10–12
45	20 mm open textured or dense base course	80–100	10–12	95–100	9–10
50	20 mm open textured or dense, or 28 mm dense base course	85–110	9–12	110–120	8–9
60	28 mm dense base course, 40 mm open textured of dense base course or 40 mm single course as base course		8–10	130–150	7–8
65	28 mm dense base course, 40 mm open textured or dense base course or 40 mm single course	100–135	7–10	140–160	6–7
75	40 mm single course, 40 mm open textured or dense base course, 40 mm dense roadbase	120–150	7–8	165–185	5–6
100	40 mm dense base course or roadbase	–	–	220–240	4–4.5

KERBS, PAVING, ETC.

Surface Dressing Roads: Coverage (m^2) per Tonne of Material

Size in mm	Sand	Granite chips	Gravel	Limestone chips
Sand	168	–	–	–
3	–	148	152	165
6	–	130	133	144
9	–	111	114	123
13	–	85	87	95
19	–	68	71	78

Sizes of Flags

Reference	Nominal size (mm)	Thickness (mm)
A	600 × 450	50 and 63
B	600 × 600	50 and 63
C	600 × 750	50 and 63
D	600 × 900	50 and 63
E	450 × 450	50 and 70 chamfered top surface
F	400 × 400	50 and 65 chamfered top surface
G	300 × 300	50 and 60 chamfered top surface

Sizes of Natural Stone Setts

Width (mm)		Length (mm)		Depth (mm)
100	×	100	×	100
75	×	150 to 250	×	125
75	×	150 to 250	×	150
100	×	150 to 250	×	100
100	×	150 to 250	×	150

SEEDING/TURFING AND PLANTING

Topsoil Quality

Topsoil grade	Properties
Premium	Natural topsoil, high fertility, loamy texture, good soil structure, suitable for intensive cultivation.
General purpose	Natural or manufactured topsoil of lesser quality than Premium, suitable for agriculture or amenity landscape, may need fertilizer or soil structure improvement.
Economy	Selected subsoil, natural mineral deposit such as river silt or greensand. The grade comprises two subgrades; 'Low clay' and 'High clay' which is more liable to compaction in handling. This grade is suitable for low-production agricultural land and amenity woodland or conservation planting areas.

Forms of Trees

Standards:	Shall be clear with substantially straight stems. Grafted and budded trees shall have no more than a slight bend at the union. Standards shall be designated as Half, Extra light, Light, Standard, Selected standard, Heavy, and Extra heavy.
Sizes of Standards	
Heavy standard	12–14 cm girth × 3.50 to 5.00 m high
Extra Heavy standard	14–16 cm girth × 4.25 to 5.00 m high
Extra Heavy standard	16–18 cm girth × 4.25 to 6.00 m high
Extra Heavy standard	18–20 cm girth × 5.00 to 6.00 m high
Semi-mature trees:	Between 6.0 m and 12.0 m tall with a girth of 20 to 75 cm at 1.0 m above ground.
Feathered trees:	Shall have a defined upright central leader, with stem furnished with evenly spread and balanced lateral shoots down to or near the ground.
Whips:	Shall be without significant feather growth as determined by visual inspection.
Multi-stemmed trees:	Shall have two or more main stems at, near, above or below ground.

Seedlings grown from seed and not transplanted shall be specified when ordered for sale as:

1+0	one year old seedling
2+0	two year old seedling
1+1	one year seed bed, one year transplanted = two year old seedling
1+2	one year seed bed, two years transplanted = three year old seedling
2+1	two year seed bed, one year transplanted = three year old seedling
1u1	two years seed bed, undercut after 1 year = two year old seedling
2u2	four years seed bed, undercut after 2 years = four year old seedling

SEEDING/TURFING AND PLANTING

Cuttings

The age of cuttings (plants grown from shoots, stems, or roots of the mother plant) shall be specified when ordered for sale. The height of transplants and undercut seedlings/cuttings (which have been transplanted or undercut at least once) shall be stated in centimetres. The number of growing seasons before and after transplanting or undercutting shall be stated.

0 + 1	one year cutting
0 + 2	two year cutting
0 + 1 + 1	one year cutting bed, one year transplanted = two year old seedling
0 + 1 + 2	one year cutting bed, two years transplanted = three year old seedling

Grass Cutting Capacities in m² per hour

Speed	Width of cut in metres												
mph	0.5	0.7	1.0	1.2	1.5	1.7	2.0	2.0	2.1	2.5	2.8	3.0	3.4
1.0	724	1127	1529	1931	2334	2736	3138	3219	3380	4023	4506	4828	5472
1.5	1086	1690	2293	2897	3500	4104	4707	4828	5069	6035	6759	7242	8208
2.0	1448	2253	3058	3862	4667	5472	6276	6437	6759	8047	9012	9656	10944
2.5	1811	2816	3822	4828	5834	6840	7846	8047	8449	10058	11265	12070	13679
3.0	2173	3380	4587	5794	7001	8208	9415	9656	10139	12070	13518	14484	16415
3.5	2535	3943	5351	6759	8167	9576	10984	11265	11829	14082	15772	16898	19151
4.0	2897	4506	6115	7725	9334	10944	12553	12875	13518	16093	18025	19312	21887
4.5	3259	5069	6880	8690	10501	12311	14122	14484	15208	18105	20278	21726	24623
5.0	3621	5633	7644	9656	11668	13679	15691	16093	16898	20117	22531	24140	27359
5.5	3983	6196	8409	10622	12834	15047	17260	17703	18588	22128	24784	26554	30095
6.0	4345	6759	9173	11587	14001	16415	18829	19312	20278	24140	27037	28968	32831
6.5	4707	7322	9938	12553	15168	17783	20398	20921	21967	26152	29290	31382	35566
7.0	5069	7886	10702	13518	16335	19151	21967	22531	23657	28163	31543	33796	38302

Number of Plants per m²: For Plants Planted on an Evenly Spaced Grid

Planting distances

mm	0.10	0.15	0.20	0.25	0.35	0.40	0.45	0.50	0.60	0.75	0.90	1.00	1.20	1.50
0.10	100.00	66.67	50.00	40.00	28.57	25.00	22.22	20.00	16.67	13.33	11.11	10.00	8.33	6.67
0.15	66.67	44.44	33.33	26.67	19.05	16.67	14.81	13.33	11.11	8.89	7.41	6.67	5.56	4.44
0.20	50.00	33.33	25.00	20.00	14.29	12.50	11.11	10.00	8.33	6.67	5.56	5.00	4.17	3.33
0.25	40.00	26.67	20.00	16.00	11.43	10.00	8.89	8.00	6.67	5.33	4.44	4.00	3.33	2.67
0.35	28.57	19.05	14.29	11.43	8.16	7.14	6.35	5.71	4.76	3.81	3.17	2.86	2.38	1.90
0.40	25.00	16.67	12.50	10.00	7.14	6.25	5.56	5.00	4.17	3.33	2.78	2.50	2.08	1.67
0.45	22.22	14.81	11.11	8.89	6.35	5.56	4.94	4.44	3.70	2.96	2.47	2.22	1.85	1.48
0.50	20.00	13.33	10.00	8.00	5.71	5.00	4.44	4.00	3.33	2.67	2.22	2.00	1.67	1.33
0.60	16.67	11.11	8.33	6.67	4.76	4.17	3.70	3.33	2.78	2.22	1.85	1.67	1.39	1.11
0.75	13.33	8.89	6.67	5.33	3.81	3.33	2.96	2.67	2.22	1.78	1.48	1.33	1.11	0.89
0.90	11.11	7.41	5.56	4.44	3.17	2.78	2.47	2.22	1.85	1.48	1.23	1.11	0.93	0.74
1.00	10.00	6.67	5.00	4.00	2.86	2.50	2.22	2.00	1.67	1.33	1.11	1.00	0.83	0.67
1.20	8.33	5.56	4.17	3.33	2.38	2.08	1.85	1.67	1.39	1.11	0.93	0.83	0.69	0.56
1.50	6.67	4.44	3.33	2.67	1.90	1.67	1.48	1.33	1.11	0.89	0.74	0.67	0.56	0.44

SEEDING/TURFING AND PLANTING

Grass Clippings Wet: Based on 3.5 m³/tonne

Annual kg/100 m²	Average 20 cuts kg/100m²	m²/tonne	m²/m³
32.0	1.6	61162.1	214067.3

Nr of cuts	22	20	18	16	12	4
kg/cut	1.45	1.60	1.78	2.00	2.67	8.00
Area capacity of 3 tonne vehicle per load						
m²	206250	187500	168750	150000	112500	37500
Load m³	100 m² units/m³ of vehicle space					
1	196.4	178.6	160.7	142.9	107.1	35.7
2	392.9	357.1	321.4	285.7	214.3	71.4
3	589.3	535.7	482.1	428.6	321.4	107.1
4	785.7	714.3	642.9	571.4	428.6	142.9
5	982.1	892.9	803.6	714.3	535.7	178.6

Transportation of Trees

To unload large trees a machine with the necessary lifting strength is required. The weight of the trees must therefore be known in advance. The following table gives a rough overview. The additional columns with root ball dimensions and the number of plants per trailer provide additional information for example about preparing planting holes and calculating unloading times.

Girth in cm	Rootball diameter in cm	Ball height in cm	Weight in kg	Numbers of trees per trailer
16–18	50–60	40	150	100–120
18–20	60–70	40–50	200	80–100
20–25	60–70	40–50	270	50–70
25–30	80	50–60	350	50
30–35	90–100	60–70	500	12–18
35–40	100–110	60–70	650	10–15
40–45	110–120	60–70	850	8–12
45–50	110–120	60–70	1100	5–7
50–60	130–140	60–70	1600	1–3
60–70	150–160	60–70	2500	1
70–80	180–200	70	4000	1
80–90	200–220	70–80	5500	1
90–100	230–250	80–90	7500	1
100–120	250–270	80–90	9500	1

Data supplied by Lorenz von Ehren GmbH
The information in the table is approximate; deviations depend on soil type, genus and weather

FENCING AND GATES

Types of Preservative

Creosote (tar oil) can be 'factory' applied	by pressure to BS 144: pts 1&2 by immersion to BS 144: pt 1 by hot and cold open tank to BS 144: pts 1&2
Copper/chromium/arsenic (CCA)	by full cell process to BS 4072 pts 1&2
Organic solvent (OS)	by double vacuum (vacvac) to BS 5707 pts 1&3 by immersion to BS 5057 pts 1&3
Pentachlorophenol (PCP)	by heavy oil double vacuum to BS 5705 pts 2&3
Boron diffusion process (treated with disodium octaborate to BWPA Manual 1986)	

Note: Boron is used on green timber at source and the timber is supplied dry

Cleft Chestnut Pale Fences

Pales	Pale spacing	Wire lines	
900 mm	75 mm	2	temporary protection
1050 mm	75 or 100 mm	2	light protective fences
1200 mm	75 mm	3	perimeter fences
1350 mm	75 mm	3	perimeter fences
1500 mm	50 mm	3	narrow perimeter fences
1800 mm	50 mm	3	light security fences

Close-Boarded Fences

Close-boarded fences 1.05 to 1.8 m high
Type BCR (recessed) or BCM (morticed) with concrete posts 140 × 115 mm tapered and Type BW with timber posts

Palisade Fences

Wooden palisade fences
Type WPC with concrete posts 140 × 115 mm tapered and Type WPW with timber posts

For both types of fence:
Height of fence 1050 mm: two rails
Height of fence 1200 mm: two rails
Height of fence 1500 mm: three rails
Height of fence 1650 mm: three rails
Height of fence 1800 mm: three rails

FENCING AND GATES

Post and Rail Fences

Wooden post and rail fences
Type MPR 11/3 morticed rails and Type SPR 11/3 nailed rails
Height to top of rail 1100 mm
Rails: three rails 87 mm, 38 mm

Type MPR 11/4 morticed rails and Type SPR 11/4 nailed rails
Height to top of rail 1100 mm
Rails: four rails 87 mm, 38 mm

Type MPR 13/4 morticed rails and Type SPR 13/4 nailed rails
Height to top of rail 1300 mm
Rail spacing 250 mm, 250 mm, and 225 mm from top
Rails: four rails 87 mm, 38 mm

Steel Posts

Rolled steel angle iron posts for chain link fencing

Posts	Fence height	Strut	Straining post
1500 × 40 × 40 × 5 mm	900 mm	1500 × 40 × 40 × 5 mm	1500 × 50 × 50 × 6 mm
1800 × 40 × 40 × 5 mm	1200 mm	1800 × 40 × 40 × 5 mm	1800 × 50 × 50 × 6 mm
2000 × 45 × 45 × 5 mm	1400 mm	2000 × 45 × 45 × 5 mm	2000 × 60 × 60 × 6 mm
2600 × 45 × 45 × 5 mm	1800 mm	2600 × 45 × 45 × 5 mm	2600 × 60 × 60 × 6 mm
3000 × 50 × 50 × 6 mm	1800 mm	2600 × 45 × 45 × 5 mm	3000 × 60 × 60 × 6 mm
with arms			

Concrete Posts

Concrete posts for chain link fencing

Posts and straining posts	Fence height	Strut
1570 mm 100 × 100 mm	900 mm	1500 mm × 75 × 75 mm
1870 mm 125 × 125 mm	1200 mm	1830 mm × 100 × 75 mm
2070 mm 125 × 125 mm	1400 mm	1980 mm × 100 × 75 mm
2620 mm 125 × 125 mm	1800 mm	2590 mm × 100 × 85 mm
3040 mm 125 × 125 mm	1800 mm	2590 mm × 100 × 85 mm (with arms)

FENCING AND GATES

Rolled Steel Angle Posts

Rolled steel angle posts for rectangular wire mesh (field) fencing

Posts	Fence height	Strut	Straining post
1200 × 40 × 40 × 5 mm	600 mm	1200 × 75 × 75 mm	1350 × 100 × 100 mm
1400 × 40 × 40 × 5 mm	800 mm	1400 × 75 × 75 mm	1550 × 100 × 100 mm
1500 × 40 × 40 × 5 mm	900 mm	1500 × 75 × 75 mm	1650 × 100 × 100 mm
1600 × 40 × 40 × 5 mm	1000 mm	1600 × 75 × 75 mm	1750 × 100 × 100 mm
1750 × 40 × 40 × 5 mm	1150 mm	1750 × 75 × 100 mm	1900 × 125 × 125 mm

Concrete Posts

Concrete posts for rectangular wire mesh (field) fencing

Posts	Fence height	Strut	Straining post
1270 × 100 × 100 mm	600 mm	1200 × 75 × 75 mm	1420 × 100 × 100 mm
1470 × 100 × 100 mm	800 mm	1350 × 75 × 75 mm	1620 × 100 × 100 mm
1570 × 100 × 100 mm	900 mm	1500 × 75 × 75 mm	1720 × 100 × 100 mm
1670 × 100 × 100 mm	600 mm	1650 × 75 × 75 mm	1820 × 100 × 100 mm
1820 × 125 × 125 mm	1150 mm	1830 × 75 × 100 mm	1970 × 125 × 125 mm

Cleft Chestnut Pale Fences

Timber Posts

Timber posts for wire mesh and hexagonal wire netting fences

Round timber for general fences

Posts	Fence height	Strut	Straining post
1300 × 65 mm dia.	600 mm	1200 × 80 mm dia.	1450 × 100 mm dia.
1500 × 65 mm dia.	800 mm	1400 × 80 mm dia.	1650 × 100 mm dia.
1600 × 65 mm dia.	900 mm	1500 × 80 mm dia.	1750 × 100 mm dia.
1700 × 65 mm dia.	1050 mm	1600 × 80 mm dia.	1850 × 100 mm dia.
1800 × 65 mm dia.	1150 mm	1750 × 80 mm dia.	2000 × 120 mm dia.

Squared timber for general fences

Posts	Fence height	Strut	Straining post
1300 × 75 × 75 mm	600 mm	1200 × 75 × 75 mm	1450 × 100 × 100 mm
1500 × 75 × 75 mm	800 mm	1400 × 75 × 75 mm	1650 × 100 × 100 mm
1600 × 75 × 75 mm	900 mm	1500 × 75 × 75 mm	1750 × 100 × 100 mm
1700 × 75 × 75 mm	1050 mm	1600 × 75 × 75 mm	1850 × 100 × 100 mm
1800 × 75 × 75 mm	1150 mm	1750 × 75 × 75 mm	2000 × 125 × 100 mm

FENCING AND GATES

Steel Fences to BS 1722: Part 9: 1992

	Fence height	Top/bottom rails and flat posts	Vertical bars
Light	1000 mm	40 × 10 mm 450 mm in ground	12 mm dia. at 115 mm cs
	1200 mm	40 × 10 mm 550 mm in ground	12 mm dia. at 115 mm cs
	1400 mm	40 × 10 mm 550 mm in ground	12 mm dia. at 115 mm cs
Light	1000 mm	40 × 10 mm 450 mm in ground	16 mm dia. at 120 mm cs
	1200 mm	40 × 10 mm 550 mm in ground	16 mm dia. at 120 mm cs
	1400 mm	40 × 10 mm 550 mm in ground	16 mm dia. at 120 mm cs
Medium	1200 mm	50 × 10 mm 550 mm in ground	20 mm dia. at 125 mm cs
	1400 mm	50 × 10 mm 550 mm in ground	20 mm dia. at 125 mm cs
	1600 mm	50 × 10 mm 600 mm in ground	22 mm dia. at 145 mm cs
	1800 mm	50 × 10 mm 600 mm in ground	22 mm dia. at 145 mm cs
Heavy	1600 mm	50 × 10 mm 600 mm in ground	22 mm dia. at 145 mm cs
	1800 mm	50 × 10 mm 600 mm in ground	22 mm dia. at 145 mm cs
	2000 mm	50 × 10 mm 600 mm in ground	22 mm dia. at 145 mm cs
	2200 mm	50 × 10 mm 600 mm in ground	22 mm dia. at 145 mm cs

Notes: Mild steel fences: round or square verticals; flat standards and horizontals. Tops of vertical bars may be bow-top, blunt, or pointed. Round or square bar railings

Timber Field Gates to BS 3470: 1975

Gates made to this standard are designed to open one way only
All timber gates are 1100 mm high
Width over stiles 2400, 2700, 3000, 3300, 3600, and 4200 mm
Gates over 4200 mm should be made in two leaves

Steel Field Gates to BS 3470: 1975

All steel gates are 1100 mm high
Heavy duty: width over stiles 2400, 3000, 3600 and 4500 mm
Light duty: width over stiles 2400, 3000, and 3600 mm

FENCING AND GATES

Domestic Front Entrance Gates to BS 4092: Part 1: 1966

Metal gates:	Single gates are 900 mm high minimum, 900 mm, 1000 mm and 1100 mm wide

Domestic Front Entrance Gates to BS 4092: Part 2: 1966

Wooden gates:	All rails shall be tenoned into the stiles
	Single gates are 840 mm high minimum, 801 mm and 1020 mm wide
	Double gates are 840 mm high minimum, 2130, 2340 and 2640 mm wide

Timber Bridle Gates to BS 5709:1979 (Horse or Hunting Gates)

Gates open one way only	
Minimum width between posts	1525 mm
Minimum height	1100 mm

Timber Kissing Gates to BS 5709:1979

Minimum width	700 mm
Minimum height	1000 mm
Minimum distance between shutting posts	600 mm
Minimum clearance at mid-point	600 mm

Metal Kissing Gates to BS 5709:1979

Sizes are the same as those for timber kissing gates	
Maximum gaps between rails 120 mm	

Categories of Pedestrian Guard Rail to BS 3049:1976

Class A for normal use
Class B where vandalism is expected
Class C where crowd pressure is likely

DRAINAGE

Width required for Trenches for Various Diameters of Pipes

Pipe diameter (mm)	Trench n.e. 1.50 m deep	Trench over 1.50 m deep
n.e. 100 mm	450 mm	600 mm
100–150 mm	500 mm	650 mm
150–225 mm	600 mm	750 mm
225–300 mm	650 mm	800 mm
300–400 mm	750 mm	900 mm
400–450 mm	900 mm	1050 mm
450–600 mm	1100 mm	1300 mm

Weights and Dimensions – Vitrified Clay Pipes

Product	Nominal diameter (mm)	Effective length (mm)	BS 65 limits of tolerance min (mm)	max (mm)	Crushing strength (kN/m)	Weight (kg/pipe)	(kg/m)
Supersleve	100	1600	96	105	35.00	14.71	9.19
	150	1750	146	158	35.00	29.24	16.71
Hepsleve	225	1850	221	236	28.00	84.03	45.42
	300	2500	295	313	34.00	193.05	77.22
	150	1500	146	158	22.00	37.04	24.69
Hepseal	225	1750	221	236	28.00	85.47	48.84
	300	2500	295	313	34.00	204.08	81.63
	400	2500	394	414	44.00	357.14	142.86
	450	2500	444	464	44.00	454.55	181.63
	500	2500	494	514	48.00	555.56	222.22
	600	2500	591	615	57.00	796.23	307.69
	700	3000	689	719	67.00	1111.11	370.45
	800	3000	788	822	72.00	1351.35	450.45
Hepline	100	1600	95	107	22.00	14.71	9.19
	150	1750	145	160	22.00	29.24	16.71
	225	1850	219	239	28.00	84.03	45.42
	300	1850	292	317	34.00	142.86	77.22
Hepduct (conduit)	90	1500	–	–	28.00	12.05	8.03
	100	1600	–	–	28.00	14.71	9.19
	125	1750	–	–	28.00	20.73	11.84
	150	1750	–	–	28.00	29.24	16.71
	225	1850	–	–	28.00	84.03	45.42
	300	1850	–	–	34.00	142.86	77.22

DRAINAGE

Weights and Dimensions – Vitrified Clay Pipes

Nominal internal diameter (mm)	Nominal wall thickness (mm)	Approximate weight (kg/m)
150	25	45
225	29	71
300	32	122
375	35	162
450	38	191
600	48	317
750	54	454
900	60	616
1200	76	912
1500	89	1458
1800	102	1884
2100	127	2619

Wall thickness, weights and pipe lengths vary, depending on type of pipe required
The particulars shown above represent a selection of available diameters and are applicable to strength class 1 pipes with flexible rubber ring joints
Tubes with Ogee joints are also available

DRAINAGE

Weights and Dimensions – PVC-u Pipes

	Nominal size	Mean outside diameter (mm)		Wall thickness	Weight
		min	max	(mm)	(kg/m)
Standard pipes	82.4	82.4	82.7	3.2	1.2
	110.0	110.0	110.4	3.2	1.6
	160.0	160.0	160.6	4.1	3.0
	200.0	200.0	200.6	4.9	4.6
	250.0	250.0	250.7	6.1	7.2
Perforated pipes heavy grade	As above	As above	As above	As above	As above
thin wall	82.4	82.4	82.7	1.7	–
	110.0	110.0	110.4	2.2	–
	160.0	160.0	160.6	3.2	–

Width of Trenches Required for Various Diameters of Pipes

Pipe diameter (mm)	Trench n.e. 1.5 m deep (mm)	Trench over 1.5 m deep (mm)
n.e. 100	450	600
100–150	500	650
150–225	600	750
225–300	650	800
300–400	750	900
400–450	900	1050
450–600	1100	1300

DRAINAGE

DRAINAGE BELOW GROUND AND LAND DRAINAGE

Flow of Water Which Can Be Carried by Various Sizes of Pipe

Clay or concrete pipes

	Gradient of pipeline							
	1:10	1:20	1:30	1:40	1:50	1:60	1:80	1:100
Pipe size	Flow in litres per second							
DN 100 15.0	8.5	6.8	5.8	5.2	4.7	4.0	3.5	
DN 150 28.0	19.0	16.0	14.0	12.0	11.0	9.1	8.0	
DN 225 140.0	95.0	76.0	66.0	58.0	53.0	46.0	40.0	

Plastic pipes

	Gradient of pipeline							
	1:10	1:20	1:30	1:40	1:50	1:60	1:80	1:100
Pipe size	Flow in litres per second							
82.4 mm i/dia.	12.0	8.5	6.8	5.8	5.2	4.7	4.0	3.5
110 mm i/dia.	28.0	19.0	16.0	14.0	12.0	11.0	9.1	8.0
160 mm i/dia.	76.0	53.0	43.0	37.0	33.0	29.0	25.0	22.0
200 mm i/dia.	140.0	95.0	76.0	66.0	58.0	53.0	46.0	40.0

Vitrified (Perforated) Clay Pipes and Fittings to BS En 295-5 1994

Length not specified		
75 mm bore	250 mm bore	600 mm bore
100	300	700
125	350	800
150	400	1000
200	450	1200
225	500	

Precast Concrete Pipes: Prestressed Non-pressure Pipes and Fittings: Flexible Joints to BS 5911: Pt. 103: 1994

Rationalized metric nominal sizes: 450, 500	
Length:	500–1000 by 100 increments
	1000–2200 by 200 increments
	2200–2800 by 300 increments
Angles: length:	450–600 angles 45, 22.5,11.25°
	600 or more angles 22.5, 11.25°

DRAINAGE

Precast Concrete Pipes: Un-reinforced and Circular Manholes and Soakaways to BS 5911: Pt. 200: 1994

Nominal sizes:	
Shafts:	675, 900 mm
Chambers:	900, 1050, 1200, 1350, 1500, 1800, 2100, 2400, 2700, 3000 mm
Large chambers:	To have either tapered reducing rings or a flat reducing slab in order to accept the standard cover
Ring depths:	1. 300–1200 mm by 300 mm increments except for bottom slab and rings below cover slab, these are by 150 mm increments
	2. 250–1000 mm by 250 mm increments except for bottom slab and rings below cover slab, these are by 125 mm increments
Access hole:	750 × 750 mm for DN 1050 chamber
	1200 × 675 mm for DN 1350 chamber

Calculation of Soakaway Depth

The following formula determines the depth of concrete ring soakaway that would be required for draining given amounts of water.

$$h = \frac{4ar}{3\pi D^2}$$

h = depth of the chamber below the invert pipe
a = The area to be drained
r = The hourly rate of rainfall (50 mm per hour)
π = pi
D = internal diameter of the soakaway

This table shows the depth of chambers in each ring size which would be required to contain the volume of water specified. These allow a recommended storage capacity of ⅓ (one third of the hourly rainfall figure).

Table Showing Required Depth of Concrete Ring Chambers in Metres

Area m²	50	100	150	200	300	400	500
Ring size							
0.9	1.31	2.62	3.93	5.24	7.86	10.48	13.10
1.1	0.96	1.92	2.89	3.85	5.77	7.70	9.62
1.2	0.74	1.47	2.21	2.95	4.42	5.89	7.37
1.4	0.58	1.16	1.75	2.33	3.49	4.66	5.82
1.5	0.47	0.94	1.41	1.89	2.83	3.77	4.72
1.8	0.33	0.65	0.98	1.31	1.96	2.62	3.27
2.1	0.24	0.48	0.72	0.96	1.44	1.92	2.41
2.4	0.18	0.37	0.55	0.74	1.11	1.47	1.84
2.7	0.15	0.29	0.44	0.58	0.87	1.16	1.46
3.0	0.12	0.24	0.35	0.47	0.71	0.94	1.18

DRAINAGE

Precast Concrete Inspection Chambers and Gullies to BS 5911: Part 230: 1994

Nominal sizes:	375 diameter, 750, 900 mm deep
	450 diameter, 750, 900, 1050, 1200 mm deep
Depths:	from the top for trapped or un-trapped units:
	centre of outlet 300 mm
	invert (bottom) of the outlet pipe 400 mm
Depth of water seal for trapped gullies:	
	85 mm, rodding eye int. dia. 100 mm
Cover slab:	65 mm min

Bedding Flexible Pipes: PVC-u Or Ductile Iron

Type 1 =	100 mm fill below pipe, 300 mm above pipe: single size material
Type 2 =	100 mm fill below pipe, 300 mm above pipe: single size or graded material
Type 3 =	100 mm fill below pipe, 75 mm above pipe with concrete protective slab over
Type 4 =	100 mm fill below pipe, fill laid level with top of pipe
Type 5 =	200 mm fill below pipe, fill laid level with top of pipe
Concrete =	25 mm sand blinding to bottom of trench, pipe supported on chocks,100 mm concrete under the pipe, 150 mm concrete over the pipe

DRAINAGE

Bedding Rigid Pipes: Clay or Concrete
(for vitrified clay pipes the manufacturer should be consulted)

Class D:	Pipe laid on natural ground with cut-outs for joints, soil screened to remove stones over 40 mm and returned over pipe to 150 m min depth. Suitable for firm ground with trenches trimmed by hand.
Class N:	Pipe laid on 50 mm granular material of graded aggregate to Table 4 of BS 882, or 10 mm aggregate to Table 6 of BS 882, or as dug light soil (not clay) screened to remove stones over 10 mm. Suitable for machine dug trenches.
Class B:	As Class N, but with granular bedding extending half way up the pipe diameter.
Class F:	Pipe laid on 100 mm granular fill to BS 882 below pipe, minimum 150 mm granular fill above pipe: single size material. Suitable for machine dug trenches.
Class A:	Concrete 100 mm thick under the pipe extending half way up the pipe, backfilled with the appropriate class of fill. Used where there is only a very shallow fall to the drain. Class A bedding allows the pipes to be laid to an exact gradient.
Concrete surround:	25 mm sand blinding to bottom of trench, pipe supported on chocks, 100 mm concrete under the pipe, 150 mm concrete over the pipe. It is preferable to bed pipes under slabs or wall in granular material.

PIPED SUPPLY SYSTEMS

Identification of Service Tubes From Utility to Dwellings

Utility	Colour	Size	Depth
British Telecom	grey	54 mm od	450 mm
Electricity	black	38 mm od	450 mm
Gas	yellow	42 mm od rigid 60 mm od convoluted	450 mm
Water	may be blue	(normally untubed)	750 mm

ELECTRICAL SUPPLY/POWER/LIGHTING SYSTEMS

Electrical Insulation Class En 60.598 BS 4533

Class 1:	luminaires comply with class 1 (I) earthed electrical requirements
Class 2:	luminaires comply with class 2 (II) double insulated electrical requirements
Class 3:	luminaires comply with class 3 (III) electrical requirements

Protection to Light Fittings

BS EN 60529:1992 Classification for degrees of protection provided by enclosures.
(IP Code – International or ingress Protection)

1st characteristic: against ingress of solid foreign objects		
The figure	2	indicates that fingers cannot enter
	3	that a 2.5 mm diameter probe cannot enter
	4	that a 1.0 mm diameter probe cannot enter
	5	the fitting is dust proof (no dust around live parts)
	6	the fitting is dust tight (no dust entry)
2nd characteristic: ingress of water with harmful effects		
The figure	0	indicates unprotected
	1	vertically dripping water cannot enter
	2	water dripping 15° (tilt) cannot enter
	3	spraying water cannot enter
	4	splashing water cannot enter
	5	jetting water cannot enter
	6	powerful jetting water cannot enter
	7	proof against temporary immersion
	8	proof against continuous immersion
Optional additional codes:		A–D protects against access to hazardous parts
	H	High voltage apparatus
	M	fitting was in motion during water test
	S	fitting was static during water test
	W	protects against weather
Marking code arrangement:		(example) IPX5S = IP (International or Ingress Protection)
		X (denotes omission of first characteristic)
		5 = jetting
		S = static during water test

RAIL TRACKS

	kg/m of track	lb/ft of track
Standard gauge		
Bull-head rails, chairs, transverse timber (softwood) sleepers etc.	245	165
Main lines		
Flat-bottom rails, transverse prestressed concrete sleepers, etc.	418	280
Add for electric third rail	51	35
Add for crushed stone ballast	2600	1750
	kg/m²	**lb/sq ft**
Overall average weight – rails connections, sleepers, ballast, etc.	733	150
	kg/m of track	**lb/ft of track**
Bridge rails, longitudinal timber sleepers, etc.	112	75

RAIL TRACKS

Heavy Rails

British Standard Section No.	Rail height (mm)	Foot width (mm)	Head width (mm)	Min web thickness (mm)	Section weight (kg/m)
Flat Bottom Rails					
60A	114.30	109.54	57.15	11.11	30.62
70A	123.82	111.12	60.32	12.30	34.81
75A	128.59	114.30	61.91	12.70	37.45
80A	133.35	117.47	63.50	13.10	39.76
90A	142.88	127.00	66.67	13.89	45.10
95A	147.64	130.17	69.85	14.68	47.31
100A	152.40	133.35	69.85	15.08	50.18
110A	158.75	139.70	69.85	15.87	54.52
113A	158.75	139.70	69.85	20.00	56.22
50 'O'	100.01	100.01	52.39	10.32	24.82
80 'O'	127.00	127.00	63.50	13.89	39.74
60R	114.30	109.54	57.15	11.11	29.85
75R	128.59	122.24	61.91	13.10	37.09
80R	133.35	127.00	63.50	13.49	39.72
90R	142.88	136.53	66.67	13.89	44.58
95R	147.64	141.29	68.26	14.29	47.21
100R	152.40	146.05	69.85	14.29	49.60
95N	147.64	139.70	69.85	13.89	47.27
Bull Head Rails					
95R BH	145.26	69.85	69.85	19.05	47.07

Light Rails

British Standard Section No.	Rail height (mm)	Foot width (mm)	Head width (mm)	Min web thickness (mm)	Section weight (kg/m)
Flat Bottom Rails					
20M	65.09	55.56	30.96	6.75	9.88
30M	75.41	69.85	38.10	9.13	14.79
35M	80.96	76.20	42.86	9.13	17.39
35R	85.73	82.55	44.45	8.33	17.40
40	88.11	80.57	45.64	12.3	19.89
Bridge Rails					
13	48.00	92	36.00	18.0	13.31
16	54.00	108	44.50	16.0	16.06
20	55.50	127	50.00	20.5	19.86
28	67.00	152	50.00	31.0	28.62
35	76.00	160	58.00	34.5	35.38
50	76.00	165	58.50	–	50.18
Crane Rails					
A65	75.00	175.00	65.00	38.0	43.10
A75	85.00	200.00	75.00	45.0	56.20
A100	95.00	200.00	100.00	60.0	74.30
A120	105.00	220.00	120.00	72.0	100.00
175CR	152.40	152.40	107.95	38.1	86.92

RAIL TRACKS

Fish Plates

British Standard Section No.	Overall plate length		Hole diameter	Finished weight per pair	
	4 Hole (mm)	6 Hole (mm)	(mm)	4 Hole (kg/pair)	6 Hole (kg/pair)
For British Standard Heavy Rails: Flat Bottom Rails					
60A	406.40	609.60	20.64	9.87	14.76
70A	406.40	609.60	22.22	11.15	16.65
75A	406.40	–	23.81	11.82	17.73
80A	406.40	609.60	23.81	13.15	19.72
90A	457.20	685.80	25.40	17.49	26.23
100A	508.00	–	pear	25.02	–
110A (shallow)	507.00	–	27.00	30.11	54.64
113A (heavy)	507.00	–	27.00	30.11	54.64
50 'O' (shallow)	406.40	–	–	6.68	10.14
80 'O' (shallow)	495.30	–	23.81	14.72	22.69
60R (shallow)	406.40	609.60	20.64	8.76	13.13
60R (angled)	406.40	609.60	20.64	11.27	16.90
75R (shallow)	406.40	–	23.81	10.94	16.42
75R (angled)	406.40	–	23.81	13.67	–
80R (shallow)	406.40	609.60	23.81	11.93	17.89
80R (angled)	406.40	609.60	23.81	14.90	22.33
For British Standard Heavy Rails: Bull head rails					
95R BH (shallow)	–	457.20	27.00	14.59	14.61
For British Standard Light Rails: Flat Bottom Rails					
30M	355.6	–	–	–	2.72
35M	355.6	–	–	–	2.83
40	355.6	–	–	3.76	–

FRACTIONS, DECIMALS AND MILLIMETRE EQUIVALENTS

Fractions	Decimals	(mm)	Fractions	Decimals	(mm)
1/64	0.015625	0.396875	33/64	0.515625	13.096875
1/32	0.03125	0.79375	17/32	0.53125	13.49375
3/64	0.046875	1.190625	35/64	0.546875	13.890625
1/16	0.0625	1.5875	9/16	0.5625	14.2875
5/64	0.078125	1.984375	37/64	0.578125	14.684375
3/32	0.09375	2.38125	19/32	0.59375	15.08125
7/64	0.109375	2.778125	39/64	0.609375	15.478125
1/8	0.125	3.175	5/8	0.625	15.875
9/64	0.140625	3.571875	41/64	0.640625	16.271875
5/32	0.15625	3.96875	21/32	0.65625	16.66875
11/64	0.171875	4.365625	43/64	0.671875	17.065625
3/16	0.1875	4.7625	11/16	0.6875	17.4625
13/64	0.203125	5.159375	45/64	0.703125	17.859375
7/32	0.21875	5.55625	23/32	0.71875	18.25625
15/64	0.234375	5.953125	47/64	0.734375	18.653125
1/4	0.25	6.35	3/4	0.75	19.05
17/64	0.265625	6.746875	49/64	0.765625	19.446875
9/32	0.28125	7.14375	25/32	0.78125	19.84375
19/64	0.296875	7.540625	51/64	0.796875	20.240625
5/16	0.3125	7.9375	13/16	0.8125	20.6375
21/64	0.328125	8.334375	53/64	0.828125	21.034375
11/32	0.34375	8.73125	27/32	0.84375	21.43125
23/64	0.359375	9.128125	55/64	0.859375	21.828125
3/8	0.375	9.525	7/8	0.875	22.225
25/64	0.390625	9.921875	57/64	0.890625	22.621875
13/32	0.40625	10.31875	29/32	0.90625	23.01875
27/64	0.421875	10.71563	59/64	0.921875	23.415625
7/16	0.4375	11.1125	15/16	0.9375	23.8125
29/64	0.453125	11.50938	61/64	0.953125	24.209375
15/32	0.46875	11.90625	31/32	0.96875	24.60625
31/64	0.484375	12.30313	63/64	0.984375	25.003125
1/2	0.5	12.7	1.0	1	25.4

IMPERIAL STANDARD WIRE GAUGE (SWG)

IMPERIAL STANDARD WIRE GAUGE (SWG)

SWG No.	Diameter (inches)	Diameter (mm)	SWG No.	Diameter (inches)	Diameter (mm)
7/0	0.5	12.7	23	0.024	0.61
6/0	0.464	11.79	24	0.022	0.559
5/0	0.432	10.97	25	0.02	0.508
4/0	0.4	10.16	26	0.018	0.457
3/0	0.372	9.45	27	0.0164	0.417
2/0	0.348	8.84	28	0.0148	0.376
1/0	0.324	8.23	29	0.0136	0.345
1	0.3	7.62	30	0.0124	0.315
2	0.276	7.01	31	0.0116	0.295
3	0.252	6.4	32	0.0108	0.274
4	0.232	5.89	33	0.01	0.254
5	0.212	5.38	34	0.009	0.234
6	0.192	4.88	35	0.008	0.213
7	0.176	4.47	36	0.008	0.193
8	0.16	4.06	37	0.007	0.173
9	0.144	3.66	38	0.006	0.152
10	0.128	3.25	39	0.005	0.132
11	0.116	2.95	40	0.005	0.122
12	0.104	2.64	41	0.004	0.112
13	0.092	2.34	42	0.004	0.102
14	0.08	2.03	43	0.004	0.091
15	0.072	1.83	44	0.003	0.081
16	0.064	1.63	45	0.003	0.071
17	0.056	1.42	46	0.002	0.061
18	0.048	1.22	47	0.002	0.051
19	0.04	1.016	48	0.002	0.041
20	0.036	0.914	49	0.001	0.031
21	0.032	0.813	50	0.001	0.025
22	0.028	0.711			

PIPES, WATER, STORAGE, INSULATION

WATER PRESSURE DUE TO HEIGHT

Imperial

Head (Feet)	Pressure (lb/in²)		Head (Feet)	Pressure (lb/in²)
1	0.43		70	30.35
5	2.17		75	32.51
10	4.34		80	34.68
15	6.5		85	36.85
20	8.67		90	39.02
25	10.84		95	41.18
30	13.01		100	43.35
35	15.17		105	45.52
40	17.34		110	47.69
45	19.51		120	52.02
50	21.68		130	56.36
55	23.84		140	60.69
60	26.01		150	65.03
65	28.18			

Metric

Head (m)	Pressure (bar)		Head (m)	Pressure (bar)
0.5	0.049		18.0	1.766
1.0	0.098		19.0	1.864
1.5	0.147		20.0	1.962
2.0	0.196		21.0	2.06
3.0	0.294		22.0	2.158
4.0	0.392		23.0	2.256
5.0	0.491		24.0	2.354
6.0	0.589		25.0	2.453
7.0	0.687		26.0	2.551
8.0	0.785		27.0	2.649
9.0	0.883		28.0	2.747
10.0	0.981		29.0	2.845
11.0	1.079		30.0	2.943
12.0	1.177		32.5	3.188
13.0	1.275		35.0	3.434
14.0	1.373		37.5	3.679
15.0	1.472		40.0	3.924
16.0	1.57		42.5	4.169
17.0	1.668		45.0	4.415

1 bar	=	14.5038 lbf/in²
1 lbf/in²	=	0.06895 bar
1 metre	=	3.2808 ft or 39.3701 in
1 foot	=	0.3048 metres
1 in wg	=	2.5 mbar (249.1 N/m²)

PIPES, WATER, STORAGE, INSULATION

Dimensions and Weights of Copper Pipes to BSEN 1057, BSEN 12499, BSEN 14251

Outside Diameter (mm)	Internal Diameter (mm)	Weight per Metre (kg)	Internal Diameter (mm)	Weight per Metre (kg)	Internal Diameter (mm)	Weight per Metre (kg)
	Formerly Table X		Formerly Table Y		Formerly Table Z	
6	4.80	0.0911	4.40	0.1170	5.00	0.0774
8	6.80	0.1246	6.40	0.1617	7.00	0.1054
10	8.80	0.1580	8.40	0.2064	9.00	0.1334
12	10.80	0.1914	10.40	0.2511	11.00	0.1612
15	13.60	0.2796	13.00	0.3923	14.00	0.2031
18	16.40	0.3852	16.00	0.4760	16.80	0.2918
22	20.22	0.5308	19.62	0.6974	20.82	0.3589
28	26.22	0.6814	25.62	0.8985	26.82	0.4594
35	32.63	1.1334	32.03	1.4085	33.63	0.6701
42	39.63	1.3675	39.03	1.6996	40.43	0.9216
54	51.63	1.7691	50.03	2.9052	52.23	1.3343
76.1	73.22	3.1287	72.22	4.1437	73.82	2.5131
108	105.12	4.4666	103.12	7.3745	105.72	3.5834
133	130.38	5.5151	–	–	130.38	5.5151
159	155.38	8.7795	–	–	156.38	6.6056

Dimensions of Stainless Steel Pipes to BS 4127

Outside Diameter (mm)	Maximum Outside Diameter (mm)	Minimum Outside Diameter (mm)	Wall Thickness (mm)	Working Pressure (bar)
6	6.045	5.940	0.6	330
8	8.045	7.940	0.6	260
10	10.045	9.940	0.6	210
12	12.045	11.940	0.6	170
15	15.045	14.940	0.6	140
18	18.045	17.940	0.7	135
22	22.055	21.950	0.7	110
28	28.055	27.950	0.8	121
35	35.070	34.965	1.0	100
42	42.070	41.965	1.1	91
54	54.090	53.940	1.2	77

PIPES, WATER, STORAGE, INSULATION

Dimensions of Steel Pipes to BS 1387

Nominal Size	Approx. Outside Diameter	Outside Diameter				Thickness		
		Light		Medium & Heavy		Light	Medium	Heavy
		Max	Min	Max	Min			
(mm)	(mm)	(mm)	(mm)	(mm)	(mm)	(mm)	(mm)	(mm)
6	10.20	10.10	9.70	10.40	9.80	1.80	2.00	2.65
8	13.50	13.60	13.20	13.90	13.30	1.80	2.35	2.90
10	17.20	17.10	16.70	17.40	16.80	1.80	2.35	2.90
15	21.30	21.40	21.00	21.70	21.10	2.00	2.65	3.25
20	26.90	26.90	26.40	27.20	26.60	2.35	2.65	3.25
25	33.70	33.80	33.20	34.20	33.40	2.65	3.25	4.05
32	42.40	42.50	41.90	42.90	42.10	2.65	3.25	4.05
40	48.30	48.40	47.80	48.80	48.00	2.90	3.25	4.05
50	60.30	60.20	59.60	60.80	59.80	2.90	3.65	4.50
65	76.10	76.00	75.20	76.60	75.40	3.25	3.65	4.50
80	88.90	88.70	87.90	89.50	88.10	3.25	4.05	4.85
100	114.30	113.90	113.00	114.90	113.30	3.65	4.50	5.40
125	139.70	–	–	140.60	138.70	–	4.85	5.40
150	165.1*	–	–	166.10	164.10	–	4.85	5.40

* 165.1 mm (6.5in) outside diameter is not generally recommended except where screwing to BS 21 is necessary
All dimensions are in accordance with ISO R65 except approximate outside diameters which are in accordance with ISO R64
Light quality is equivalent to ISO R65 Light Series II

Approximate Metres Per Tonne of Tubes to BS 1387

Nom. Size	BLACK						GALVANIZED					
	Plain/screwed ends			Screwed & socketed			Plain/screwed ends			Screwed & socketed		
	L	M	H	L	M	H	L	M	H	L	M	H
(mm)	(m)	(m)	(m)	(m)	(m)	(m)	(m)	(m)	(m)	(m)	(m)	(m)
6	2765	2461	2030	2743	2443	2018	2604	2333	1948	2584	2317	1937
8	1936	1538	1300	1920	1527	1292	1826	1467	1254	1811	1458	1247
10	1483	1173	979	1471	1165	974	1400	1120	944	1386	1113	939
15	1050	817	688	1040	811	684	996	785	665	987	779	661
20	712	634	529	704	628	525	679	609	512	673	603	508
25	498	410	336	494	407	334	478	396	327	474	394	325
32	388	319	260	384	316	259	373	308	254	369	305	252
40	307	277	226	303	273	223	296	268	220	292	264	217
50	244	196	162	239	194	160	235	191	158	231	188	157
65	172	153	127	169	151	125	167	149	124	163	146	122
80	147	118	99	143	116	98	142	115	97	139	113	96
100	101	82	69	98	81	68	98	81	68	95	79	67
125	–	62	56	–	60	55	–	60	55	–	59	54
150	–	52	47	–	50	46	–	51	46	–	49	45

The figures for 'plain or screwed ends' apply also to tubes to BS 1775 of equivalent size and thickness
Key:
L – Light
M – Medium
H – Heavy

PIPES, WATER, STORAGE, INSULATION

Flange Dimension Chart to BS 4504 & BS 10

Normal Pressure Rating (PN 6) 6 Bar

Nom. Size	Flange Outside Dia.	Table 6/2 Forged Welding Neck	Table 6/3 Plate Slip on	Table 6/4 Forged Bossed Screwed	Table 6/5 Forged Bossed Slip on	Table 6/8 Plate Blank	Raised Face Dia.	Raised Face T'ness	Nr. Bolt Hole	Size of Bolt
15	80	12	12	12	12	12	40	2	4	M10 × 40
20	90	14	14	14	14	14	50	2	4	M10 × 45
25	100	14	14	14	14	14	60	2	4	M10 × 45
32	120	14	16	14	14	14	70	2	4	M12 × 45
40	130	14	16	14	14	14	80	3	4	M12 × 45
50	140	14	16	14	14	14	90	3	4	M12 × 45
65	160	14	16	14	14	14	110	3	4	M12 × 45
80	190	16	18	16	16	16	128	3	4	M16 × 55
100	210	16	18	16	16	16	148	3	4	M16 × 55
125	240	18	20	18	18	18	178	3	8	M16 × 60
150	265	18	20	18	18	18	202	3	8	M16 × 60
200	320	20	22	–	20	20	258	3	8	M16 × 60
250	375	22	24	–	22	22	312	3	12	M16 × 65
300	440	22	24	–	22	22	365	4	12	M20 × 70

Normal Pressure Rating (PN 16) 16 Bar

Nom. Size	Flange Outside Dia.	Table 6/2 Forged Welding Neck	Table 6/3 Plate Slip on	Table 6/4 Forged Bossed Screwed	Table 6/5 Forged Bossed Slip on	Table 6/8 Plate Blank	Raised Face Dia.	Raised Face T'ness	Nr. Bolt Hole	Size of Bolt
15	95	14	14	14	14	14	45	2	4	M12 × 45
20	105	16	16	16	16	16	58	2	4	M12 × 50
25	115	16	16	16	16	16	68	2	4	M12 × 50
32	140	16	16	16	16	16	78	2	4	M16 × 55
40	150	16	16	16	16	16	88	3	4	M16 × 55
50	165	18	18	18	18	18	102	3	4	M16 × 60
65	185	18	18	18	18	18	122	3	4	M16 × 60
80	200	20	20	20	20	20	138	3	8	M16 × 60
100	220	20	20	20	20	20	158	3	8	M16 × 65
125	250	22	22	22	22	22	188	3	8	M16 × 70
150	285	22	22	22	22	22	212	3	8	M20 × 70
200	340	24	24	–	24	24	268	3	12	M20 × 75
250	405	26	26	–	26	26	320	3	12	M24 × 90
300	460	28	28	–	28	28	378	4	12	M24 × 90

PIPES, WATER, STORAGE, INSULATION

Minimum Distances Between Supports/Fixings

Material	BS Nominal Pipe Size		Pipes – Vertical	Pipes – Horizontal on to low gradients
	(inch)	(mm)	Support distance in metres	Support distance in metres
Copper	0.50	15.00	1.90	1.30
	0.75	22.00	2.50	1.90
	1.00	28.00	2.50	1.90
	1.25	35.00	2.80	2.50
	1.50	42.00	2.80	2.50
	2.00	54.00	3.90	2.50
	2.50	67.00	3.90	2.80
	3.00	76.10	3.90	2.80
	4.00	108.00	3.90	2.80
	5.00	133.00	3.90	2.80
	6.00	159.00	3.90	2.80
muPVC	1.25	32.00	1.20	0.50
	1.50	40.00	1.20	0.50
	2.00	50.00	1.20	0.60
Polypropylene	1.25	32.00	1.20	0.50
	1.50	40.00	1.20	0.50
uPVC	–	82.40	1.20	0.50
	–	110.00	1.80	0.90
	–	160.00	1.80	1.20
Steel	0.50	15.00	2.40	1.80
	0.75	20.00	3.00	2.40
	1.00	25.00	3.00	2.40
	1.25	32.00	3.00	2.40
	1.50	40.00	3.70	2.40
	2.00	50.00	3.70	2.40
	2.50	65.00	4.60	3.00
	3.00	80.40	4.60	3.00
	4.00	100.00	4.60	3.00
	5.00	125.00	5.50	3.70
	6.00	150.00	5.50	4.50
	8.00	200.00	8.50	6.00
	10.00	250.00	9.00	6.50
	12.00	300.00	10.00	7.00
	16.00	400.00	10.00	8.25

PIPES, WATER, STORAGE, INSULATION

Litres of Water Storage Required Per Person Per Building Type

Type of Building	Storage (litres)
Houses and flats (up to 4 bedrooms	120/bedroom
Houses and flats (more than 4 bedrooms)	100/bedroom
Hostels	90/bed
Hotels	200/bed
Nurses homes and medical quarters	120/bed
Offices with canteen	45/person
Offices without canteen	40/person
Restaurants	7/meal
Boarding schools	90/person
Day schools – Primary	15/person
Day schools – Secondary	20/person

Recommended Air Conditioning Design Loads

Building Type	Design Loading
Computer rooms	500 W/m² of floor area
Restaurants	150 W/m² of floor area
Banks (main area)	100 W/m² of floor area
Supermarkets	25 W/m² of floor area
Large office block (exterior zone)	100 W/m² of floor area
Large office block (interior zone)	80 W/m² of floor area
Small office block (interior zone)	80 W/m² of floor area

PIPES, WATER, STORAGE, INSULATION

Capacity and Dimensions of Galvanized Mild Steel Cisterns – BS 417

Capacity (litres)	BS type (SCM)	Length (mm)	Dimensions Width (mm)	Depth (mm)
18	45	457	305	305
36	70	610	305	371
54	90	610	406	371
68	110	610	432	432
86	135	610	457	482
114	180	686	508	508
159	230	736	559	559
191	270	762	584	610
227	320	914	610	584
264	360	914	660	610
327	450/1	1220	610	610
336	450/2	965	686	686
423	570	965	762	787
491	680	1090	864	736
709	910	1070	889	889

Capacity of Cold Water Polypropylene Storage Cisterns – BS 4213

Capacity (litres)	BS type (PC)	Maximum height (mm)
18	4	310
36	8	380
68	15	430
91	20	510
114	25	530
182	40	610
227	50	660
273	60	660
318	70	660
455	100	760

PIPES, WATER, STORAGE, INSULATION

Minimum Insulation Thickness to Protect Against Freezing for Domestic Cold Water Systems (8 Hour Evaluation Period)

Pipe size (mm)	Insulation thickness (mm)					
	Condition 1			Condition 2		
	λ = 0.020	λ = 0.030	λ = 0.040	λ = 0.020	λ = 0.030	λ = 0.040
Copper pipes						
15	11	20	34	12	23	41
22	6	9	13	6	10	15
28	4	6	9	4	7	10
35	3	5	7	4	5	7
42	3	4	5	8	4	6
54	2	3	4	2	3	4
76	2	2	3	2	2	3
Steel pipes						
15	9	15	24	10	18	29
20	6	9	13	6	10	15
25	4	7	9	5	7	10
32	3	5	6	3	5	7
40	3	4	5	3	4	6
50	2	3	4	2	3	4
65	2	2	3	2	3	3

Condition 1: water temperature 7°C; ambient temperature –6°C; evaluation period 8 h; permitted ice formation 50%; normal installation, i.e. inside the building and inside the envelope of the structural insulation
Condition 2: water temperature 2°C; ambient temperature –6°C; evaluation period 8 h; permitted ice formation 50%; extreme installation, i.e. inside the building but outside the envelope of the structural insulation
λ = thermal conductivity [W/(mK)]

Insulation Thickness for Chilled And Cold Water Supplies to Prevent Condensation

On a Low Emissivity Outer Surface (0.05, i.e. Bright Reinforced Aluminium Foil) with an Ambient Temperature of +25°C and a Relative Humidity of 80%

Steel pipe size (mm)	t = +10			t = +5			t = 0		
	Insulation thickness (mm)			Insulation thickness (mm)			Insulation thickness (mm)		
	λ = 0.030	λ = 0.040	λ = 0.050	λ = 0.030	λ = 0.040	λ = 0.050	λ = 0.030	λ = 0.040	λ = 0.050
15	16	20	25	22	28	34	28	36	43
25	18	24	29	25	32	39	32	41	50
50	22	28	34	30	39	47	38	49	60
100	26	34	41	36	47	57	46	60	73
150	29	38	46	40	52	64	51	67	82
250	33	43	53	46	60	74	59	77	94
Flat surfaces	39	52	65	56	75	93	73	97	122

t = temperature of contents (°C)
λ = thermal conductivity at mean temperature of insulation [W/(mK)]

PIPES, WATER, STORAGE, INSULATION

Insulation Thickness for Non-domestic Heating Installations to Control Heat Loss

Steel pipe size (mm)	t = 75			t = 100			t = 150		
	Insulation thickness (mm)			Insulation thickness (mm)			Insulation thickness (mm)		
	λ = 0.030	λ = 0.040	λ = 0.050	λ = 0.030	λ = 0.040	λ = 0.050	λ = 0.030	λ = 0.040	λ = 0.050
10	18	32	55	20	36	62	23	44	77
15	19	34	56	21	38	64	26	47	80
20	21	36	57	23	40	65	28	50	83
25	23	38	58	26	43	68	31	53	85
32	24	39	59	28	45	69	33	55	87
40	25	40	60	29	47	70	35	57	88
50	27	42	61	31	49	72	37	59	90
65	29	43	62	33	51	74	40	63	92
80	30	44	62	35	52	75	42	65	94
100	31	46	63	37	54	76	45	68	96
150	33	48	64	40	57	77	50	73	100
200	35	49	65	42	59	79	53	76	103
250	36	50	66	43	61	80	55	78	105

t = hot face temperature (°C)
λ = thermal conductivity at mean temperature of insulation [W/(mK)]

Index

Davis Langdon,
An AECOM Company

Program, Cost, Consultancy
www.davislangdon.com
www.aecom.com

Davis Langdon
An AECOM Company

A leading light in
our industry

Spon's Asia-Pacific Construction Costs Handbook

4th Edition

Edited by **Davis Langdon & Seah**

Spon's Asia Pacific Construction Costs Handbook includes construction cost data for 20 countries. This new edition has been extended to include Pakistan and Cambodia. Australia, UK and America are also included, to facilitate comparison with construction costs elsewhere. Information is presented for each country in the same way, as follows:

- key data on the main economic and construction indicators.

- an outline of the national construction industry, covering structure, tendering and contract procedures, materials cost data, regulations and standards

- labour and materials cost data

- Measured rates for a range of standard construction work items

- Approximate estimating costs per unit area for a range of building types

- price index data and exchange rate movements against £ sterling, $US and Japanese Yen

The book also includes a Comparative Data section to facilitate country-to-country comparisons. Figures from the national sections are grouped in tables according to national indicators, construction output, input costs and costs per square metre for factories, offices, warehouses, hospitals, schools, theatres, sports halls, hotels and housing.

This unique handbook will be an essential reference for all construction professionals involved in work outside their own country and for all developers or multinational companies assessing comparative development costs.

April 2010: 234x156: 480pp
Hb: 978-0-415-46565-6: **£120.00**

To Order: Tel: +44 (0) 1235 400524 **Fax:** +44 (0) 1235 400525
or Post: Taylor and Francis Customer Services,
Bookpoint Ltd, Unit T1, 200 Milton Park, Abingdon, Oxon, OX14 4TA UK
Email: book.orders@tandf.co.uk

For a complete listing of all our titles visit:
www.tandf.co.uk

Spon's First Stage Estimating Handbook

3rd Edition

By **Bryan Spain**

Have you ever had to provide accurate costs for a new supermarket or a pub "just an idea…a ballpark figure…" ?

The earlier a pricing decision has to be made, the more difficult it is to estimate the cost and the more likely the design and the specs are to change. And yet a rough-and-ready estimate is more likely to get set in stone.

Spon's First Stage Estimating Handbook is the only comprehensive and reliable source of first stage estimating costs. Covering the whole spectrum of building costs and a wide range of related M&E work and landscaping work, vital cost data is presented as:

• Costs per square metre   • Elemental cost analyses

• Principal rates   • Composite rates

Compact and clear, *Spon's First Stage Estimating Handbook* is ideal for those key early meetings with clients. And with additional sections on whole life costing and general information, this is an essential reference for all construction professionals and clients making early judgements on the viability of new projects.

January 2010: 216x138: 244pp
Pb: 978-0-415-54715-4: **£45.00**

Software and eBook Single-User Licence Agreement

We welcome you as a user of this Spon Price Book Software and eBook and hope that you find it a useful and valuable tool. Please read this document carefully. **This is a legal agreement** between you (hereinafter referred to as the "Licensee") and Taylor and Francis Books Ltd. (the "Publisher"), which defines the terms under which you may use the Product. **By breaking the seal and opening the document inside the back cover of the book containing the access code you agree to these terms and conditions outlined herein. If you do not agree to these terms you must return the Product to your supplier intact, with the seal on the document unbroken.**

1. Definition of the Product

The product which is the subject of this Agreement, *Spon's External Works and Landscape Price Book 2010* Software and eBook (the "Product") consists of:

1.1 Underlying data comprised in the product (the "Data")

1.2 A compilation of the Data (the "Database")

1.3 Software (the "Software") for accessing and using the Database

1.4 An electronic book containing the data in the price book (the "eBook")

2. Commencement and licence

2.1 This Agreement commences upon the breaking open of the document containing the access code by the Licensee (the "Commencement Date").

2.2 This is a licence agreement (the "Agreement") for the use of the Product by the Licensee, and not an agreement for sale.

2.3 The Publisher licenses the Licensee on a non-exclusive and non-transferable basis to use the Product on condition that the Licensee complies with this Agreement. The Licensee acknowledges that it is only permitted to use the Product in accordance with this Agreement.

3. Multiple use

For more than one user or for a wide area network or consortium, use is only permissible with the purchase from the Publisher of a multiple-user licence and adherence to the terms and conditions of that licence.

4. Installation and Use

4.1 The Licensee may provide access to the Product for individual study in the following manner: The Licensee may install the Product on a secure local area network on a single site for use by one user.

4.2 The Licensee shall be responsible for installing the Product and for the effectiveness of such installation.

4.3 Text from the Product may be incorporated in a coursepack. Such use is only permissible with the express permission of the Publisher in writing and requires the payment of the appropriate fee as specified by the Publisher and signature of a separate licence agreement.

4.4 The Product is a free addition to the book and no technical support will be provided.

5. Permitted Activities

5.1 The Licensee shall be entitled:

5.1.1 to use the Product for its own internal purposes;

5.1.2 to download onto electronic, magnetic, optical or similar storage medium reasonable portions of the Database provided that the purpose of the Licensee is to undertake internal research or study and provided that such storage is temporary;

5.2 The Licensee acknowledges that its rights to use the Product are strictly set out in this Agreement, and all other uses (whether expressly mentioned in Clause 6 below or not) are prohibited.

6. Prohibited Activities

The following are prohibited without the express permission of the Publisher:

6.1 The commercial exploitation of any part of the Product.

6.2 The rental, loan, (free or for money or money's worth) or hire purchase of this product, save with the express consent of the Publisher.

6.3 Any activity which raises the reasonable prospect of impeding the Publisher's ability or opportunities to market the Product.

6.4 Any networking, physical or electronic distribution or dissemination of the product save as expressly permitted by this Agreement.

6.5 Any reverse engineering, decompilation, disassembly or other alteration of the Product save in accordance with applicable national laws.

6.6 The right to create any derivative product or service from the Product save as expressly provided for in this Agreement.

6.7 Any alteration, amendment, modification or deletion from the Product, whether for the purposes of error correction or otherwise.

7. General Responsibilities of the License

7.1 The Licensee will take all reasonable steps to ensure that the Product is used in accordance with the terms and conditions of this Agreement.

7.2 The Licensee acknowledges that damages may not be a sufficient remedy for the Publisher in the event of breach of this Agreement by the Licensee, and that an injunction may be appropriate.

7.3 The Licensee undertakes to keep the Product safe and to use its best endeavours to ensure that the product does not fall into the hands of third parties, whether as a result of theft or otherwise.

7.4 Where information of a confidential nature relating to the product of the business affairs of the Publisher comes into the possession of the Licensee pursuant to this Agreement (or otherwise), the Licensee agrees to use such information solely for the purposes of this Agreement, and under no circumstances to disclose any element of the information to any third party save strictly as permitted under this Agreement. For the avoidance of doubt, the Licensee's obligations under this sub-clause 7.4 shall survive the termination of this Agreement.

8. Warrant and Liability

8.1 The Publisher warrants that it has the authority to enter into this agreement and that it has secured all rights and permissions necessary to enable the Licensee to use the Product in accordance with this Agreement.

8.2 The Publisher warrants that the Product as supplied on the Commencement Date shall be free of defects in materials and workmanship, and undertakes to replace any defective Product within 28 days of notice of such defect being received provided such notice is received within 30 days of such supply. As an alternative to replacement, the Publisher agrees fully to refund the Licensee in such circumstances, if the Licensee so requests, provided that the Licensee returns this copy of *Spon's External Works and Landscape Price Book 2010* to the Publisher. The provisions of this sub-clause 8.2 do not apply where the defect results from an accident or from misuse of the product by the Licensee.

8.3 Sub-clause 8.2 sets out the sole and exclusive remedy of the Licensee in relation to defects in the Product.

8.4 The Publisher and the Licensee acknowledge that the Publisher supplies the Product on an "as is" basis. The Publisher gives no warranties:

8.4.1 that the Product satisfies the individual requirements of the Licensee; or

8.4.2 that the Product is otherwise fit for the Licensee's purpose; or

8.4.3 that the Data are accurate or complete or free of errors or omissions; or

8.4.4 that the Product is compatible with the Licensee's hardware equipment and software operating environment.

8.5 The Publisher hereby disclaims all warranties and conditions, express or implied, which are not stated above.

8.6 Nothing in this Clause 8 limits the Publisher's liability to the Licensee in the event of death or personal injury resulting from the Publisher's negligence.

8.7 The Publisher hereby excludes liability for loss of revenue, reputation, business, profits, or for indirect or consequential losses, irrespective of whether the Publisher was advised by the Licensee of the potential of such losses.

8.8 The Licensee acknowledges the merit of independently verifying Data prior to taking any decisions of material significance (commercial or otherwise) based on such data. It is agreed that the Publisher shall not be liable for any losses which result from the Licensee placing reliance on the Data or on the Database, under any circumstances.

8.9 Subject to sub-clause 8.6 above, the Publisher's liability under this Agreement shall be limited to the purchase price.

9. Intellectual Property Rights

9.1 Nothing in this Agreement affects the ownership of copyright or other intellectual property rights in the Data, the Database of the Software.

9.2 The Licensee agrees to display the Publishers' copyright notice in the manner described in the Product.

9.3 The Licensee hereby agrees to abide by copyright and similar notice requirements required by the Publisher, details of which are as follows:

"© 2010 Taylor & Francis. All rights reserved. All materials in *Spon's External Works and Landscape Price Book 2010* are copyright protected. All rights reserved. No such materials may be used, displayed, modified, adapted, distributed, transmitted, transferred, published or otherwise reproduced in any form or by any means now or hereafter developed other than strictly in accordance with the terms of the licence agreement enclosed with *Spon's External Works and Landscape Price Book 2010*. However, text and images may be printed and copied for research and private study within the preset program limitations. Please note the copyright notice above, and that any text or images printed or copied must credit the source."

9.4 This Product contains material proprietary to and copyedited by the Publisher and others. Except for the licence granted herein, all rights, title and interest in the Product, in all languages, formats and media throughout the world, including copyrights therein, are and remain the property of the Publisher or other copyright holders identified in the Product.

10. Non-assignment

This Agreement and the licence contained within it may not be assigned to any other person or entity without the written consent of the Publisher.

11. Termination and Consequences of Termination.

11.1 The Publisher shall have the right to terminate this Agreement if:

11.1.1 the Licensee is in material breach of this Agreement and fails to remedy such breach (where capable of remedy) within 14 days of a written notice from the Publisher requiring it to do so; or

11.1.2 the Licensee becomes insolvent, becomes subject to receivership, liquidation or similar external administration; or

11.1.3 the Licensee ceases to operate in business.

11.2 The Licensee shall have the right to terminate this Agreement for any reason upon two month's written notice. The Licensee shall not be entitled to any refund for payments made under this Agreement prior to termination under this sub-clause 11.2.

11.3 Termination by either of the parties is without prejudice to any other rights or remedies under the general law to which they may be entitled, or which survive such termination (including rights of the Publisher under sub-clause 7.4 above).

11.4 Upon termination of this Agreement, or expiry of its terms, the Licensee must destroy all copies and any back up copies of the product or part thereof.

12. General

12.1 **Compliance with export provisions**

The Publisher hereby agrees to comply fully with all relevant export laws and regulations of the United Kingdom to ensure that the Product is not exported, directly or indirectly, in violation of English law.

12.2 **Force majeure**

The parties accept no responsibility for breaches of this Agreement occurring as a result of circumstances beyond their control.

12.3 **No waiver**

Any failure or delay by either party to exercise or enforce any right conferred by this Agreement shall not be deemed to be a waiver of such right.

12.4 **Entire agreement**

This Agreement represents the entire agreement between the Publisher and the Licensee concerning the Product. The terms of this Agreement supersede all prior purchase orders, written terms and conditions, written or verbal representations, advertising or statements relating in any way to the Product.

12.5 **Severability**

If any provision of this Agreement is found to be invalid or unenforceable by a court of law of competent jurisdiction, such a finding shall not affect the other provisions of this Agreement and all provisions of this Agreement unaffected by such a finding shall remain in full force and effect.

12.6 **Variations**

This agreement may only be varied in writing by means of variation signed in writing by both parties.

12.7 **Notices**

All notices to be delivered to: Spon's Price Books, Taylor & Francis Books Ltd., 2 Park Square, Milton Park, Abingdon, Oxfordshire, OX14 4RN, UK.

12.8 **Governing law**

This Agreement is governed by English law and the parties hereby agree that any dispute arising under this Agreement shall be subject to the jurisdiction of the English courts.

If you have any queries about the terms of this licence, please contact:

Spon's Price Books
Taylor & Francis Books Ltd.
2 Park Square, Milton Park, Abingdon, Oxfordshire, OX14 4RN
Tel: +44 (0) 20 7017 6672
Fax: +44 (0) 20 7017 6702
www.tandfbuiltenvironment.com

Multiple-user use of the Spon Press Software and eBook

To buy a licence to install your Spon Press Price Book Software and eBook on a secure local area network or a wide area network, and for the supply of network key files, for an agreed number of users please contact:

Spon's Price Books
Taylor & Francis Books Ltd.
2 Park Square, Milton Park, Abingdon, Oxfordshire, OX14 4RN
Tel: +44 (0) 207 017 6672
Fax: +44 (0) 207 017 6072
www.pricebooks.co.uk

Number of users	Licence cost
2–5	£450
6–10	£915
11–20	£1400
21–30	£2150
31–50	£4200
51–75	£5900
76–100	£7100
Over 100	Please contact Spon for details

Software Installation and Use Instructions

System requirements

Minimum

- Pentium processor
- 512 MB of RAM
- 50 MB available hard disk space
- Microsoft Windows XP/Vista/Win 7
- SVGA screen
- Internet connection

Recommended

- 1GB of RAM (2GB for Vista/Win 7)
- 100 MB available hard disk space
- XVGA screen or better
- Broadband Internet connection

Microsoft® is a registered trademark and Windows™ is a trademark of the Microsoft Corporation.

Installation

Spon's Mechanical and Electrical Services Price Book 2012 Electronic Version is supplied solely by internet download. No CD-ROM is supplied.

In your internet browser type in www.pricebooks.co.uk/downloads and follow the instructions on screen. Then type in the unique access code which is sealed inside the back cover of this book.

When the access code is successfully validated, click on the download links to download and then save the files.

Please note: you will only be allowed one download of these files and onto one computer.

A folder called *PriceBook* will be added to your desktop, which will need to be unzipped. Then click on the application called *install*.

Use

- The installation process will create a folder containing the price book program links as well as a program icon on your desktop.
- Double click the icon (from the folder or desktop) installed by the Setup program.
- Follow the instructions on screen.

Technical Support

Further guidance on the installation is provided on www.pricebooks.co.uk/downloads.

The *Electronic Version* is a free addition to the book. For help with the running of the software please visit www. pricebooks.co.uk

All materials in *Spon's Mechanical and Electrical Services Price Book 2012 Electronic Version* are copyright protected. No such materials may be used, displayed, modified, adapted, distributed, transmitted, transferred, published or otherwise reproduced in any form or by any means now or hereafter developed other than strictly in accordance with the terms of the above licence agreement.

The software used in *Spon's Mechanical and Electrical Services Price Book 2012 Electronic Version* is furnished under a single user licence agreement. The software may be used only in accordance with the terms of the licence agreement, unless the additional multi-user licence agreement is purchased from Spon Press, 2 Park Square, Milton Park, Abingdon, Oxon OX14 4RN Tel: +44 (0) 20 7017 6000

Free Updates

with three easy steps…

1. Register today on www.pricebooks.co.uk/updates

2. We'll alert you by email when new updates are posted on our website

3. Then go to www.pricebooks.co.uk/updates
    and download the update.

All four Spon Price Books – *Architects' and Builders'*, *Civil Engineering and Highway Works*, *External Works and Landscape* and *Mechanical and Electrical Services* – are supported by an updating service. Three updates are loaded on our website during the year, in November, February and May. Each gives details of changes in prices of materials, wage rates and other significant items, with regional price level adjustments for Northern Ireland, Scotland and Wales and regions of England. The updates terminate with the publication of the next annual edition.

As a purchaser of a Spon Price Book you are entitled to this updating service for the 2012 edition – free of charge. Simply register via the website www.pricebooks.co.uk/updates and we will send you an email when each update becomes available.

If you haven't got internet access we can supply the updates by an alternative means. Please write to us for details: Spon Price Book Updates, Spon Press Marketing Department, 2 Park Square, Milton Park, Abingdon, Oxfordshire, OX14 4RN.

Find out more about Spon books
Visit www.sponpress.com for more details.